MECHANICAL BEHAVIOUR OF MATERIALS

Proceedings of the Third International Conference
held in Cambridge, England, 20-24 August 1979

(in three volumes)

Volume 3

INTERNATIONAL SERIES ON THE
STRENGTH AND FRACTURE OF MATERIALS AND STRUCTURES

Other Titles in the Series

EASTERLING	Mechanisms of Deformation and Fracture
HAASEN AND GEROLD	Strength of Metals and Alloys (3 volumes)
PIGGOT	Load Bearing Fibre Composites
SMITH	Fracture Mechanics, Current status - future prospects
TAPLIN	Advances in Research on the Strength and Fracture of Materials (6 volumes)

Related Pergamon Journals

Engineering Fracture Mechanics

Fatigue of Engineering Materials and Structures

Materials Research Bulletin

MECHANICAL BEHAVIOUR OF MATERIALS

Proceedings of the Third International Conference
held in Cambridge, England, 20-24 August 1979

Editors

K. J. MILLER
Sheffield University, England

R. F. SMITH
The Welding Institute, Cambridge, England

Volume 3

ICM
3

PERGAMON PRESS

OXFORD · NEW YORK · TORONTO · SYDNEY · PARIS · FRANKFURT

U.K.	Pergamon Press Ltd., Headington Hill Hall, Oxford OX3 0BW, England
U.S.A.	Pergamon Press Inc., Maxwell House, Fairview Park, Elmsford, New York 10523, U.S.A.
CANADA	Pergamon of Canada, Suite 104, 150 Consumers Road, Willowdale, Ontario M2J 1P9, Canada
AUSTRALIA	Pergamon Press (Aust.) Pty. Ltd., P.O. Box 544, Potts Point, N.S.W. 2011, Australia
FRANCE	Pergamon Press SARL, 24 rue des Ecoles, 75240 Paris, Cedex 05, France
FEDERAL REPUBLIC OF GERMANY	Pergamon Press GmbH, 6242 Kronberg-Taunus, Hammerweg 6, Federal Republic of Germany

British Library Cataloguing in Publication Data

International Conference on Mechanical Behaviour of Materials, *3rd, Cambridge, 1979*
Mechanical behaviour of materials. -
(International series on the strength and fracture of materials and structures).
1. Materials - Congresses
I. Title II. Miller, Keith John
III. Smith, R F IV. Series
620.1'1292 TA401.3 80-41052

ISBN 0-08-024739-3

In order to make this volume available as economically and as rapidly as possible the authors' typescripts have been reproduced in their original forms. This method unfortunately has its typographical limitations but it is hoped that they in no way distract the reader.

Printed in Great Britain by A. Wheaton & Co. Ltd, Exeter

These proceedings are dedicated to the memory of

Shuji Taira

the initiator of I.C.M. and organiser of the first
I.C.M. Conference in Kyoto, Japan 1971

CONTENTS OF VOLUME 3

Workshop 2: Crack Growth, Tearing, and R-Curves

CONTENTS OF VOLUME 1

CONTENTS OF VOLUME 2

THEME III: FORMING AND MACHINING OF MATERIALS

Workshop 1: Forming Processes

(a) Bulk Deformation

(b) Sheet Forming

Workshop 2: Cutting and Material Properties

(a) Cutting

(b) Properties

PREFACE

The Third International Conference on Mechanical Behaviour of Materials (ICM3) dealt with six main themes:

I High temperature deformation and failure;
II Environmental and time dependent effects;
III Forming and machining of materials;
IV Designing with brittle solids;
V Designing with new materials; and
VI Fast fracture in ductile materials.

The conference, third in the series after Kyoto, Japan 1971 and Boston, U.S.A. 1976, was held in Cambridge, England during August 1979. The intention of organising ICM3 only three years after Boston was to assist our sister conference, ICF, by having alternative conferences every two years.

The technical programme included the presentation of 15 invited plenary papers and a series of linked workshop sessions in which 127 research papers were discussed. Some 340 delegates from 25 countries attended the five day conference aimed at defining, examining and offering solutions to engineering problems. Plenary sessions were generally divided into three parts (a) the examination of the engineering problem area followed by (b) design considerations and finally, (c) a paper on the contribution made by materials engineers to assist solutions to the problem being discussed. This volume collates all the plenary papers of the conference whilst volumes II and III record the workshop papers. A timetable of the conference is given on page *xxx* of Volume I.

The first theme "High temperature deformation and failure" attracted more submissions of papers than any other and hence it was necessary to have four rather than the initially intended three workshop sessions. These sessions were (1) thermomechanical loading effects on deformation and failure; (2) fatigue life assessment; (3) effects of service loading and history in the creep range; and (4) creep life assessment. Each workshop had approximately eight papers.

The second theme probably suffered more than any other in the selection process because submissions did not always relate to the two principal topics that emerged in the majority of papers, corrosion fatigue and hydrogen embrittlement of materials. In consequence, after refereeing, only two workshops were deemed necessary, namely; (1) cyclic loading effects; and (2) degradation of materials, including hydrogen embrittlement; the latter workshop being somewhat oversubscribed.

The third theme, however, fell into a nearly perfect format with two workshops; (1) forming processes and (2) cutting and material properties, with both workshops being subdivided into two groups, i.e. (1a) bulk deformation, (1b) sheet forming; and (2a) cutting, (2b) properties. The workshops of these first three themes are recorded in Volume II of proceedings whilst the workshops of the fourth, fifth and sixth themes are recorded in Volume III.

It had been intended originally that themes IV and V be one joint theme, but the interest generated was such that it was necessary not only to give each topic a principal place in the conference proceedings but also that each theme be subdivided into well defined topics.

Thus the fourth theme had eight papers for the first workshop on ceramics, eight papers for the second workshop on cement and concrete and finally three papers concerned with "other" materials.

Similarly, the fifth theme was divided into three with the first workshop concerned with composites (ten papers), the second workshop with wood (five papers) and the final workshop with other materials (seven papers).

The last theme of 18 papers was subdivided into two workshops of nine papers each. The first workshop discussed testing, metallurgical factors and combined stress effects, whilst the second workshop is about crack growth, tearing and R-curves.

In selecting papers and formulating workshops, the basic policy of the conference was adhered to as strictly as possible. Briefly, this policy aimed:

(a) To bring together all scientific skills to discuss an engineering problem irrespective of the discipline of the authors and delegates, thereby achieving an interdisciplinary debate and a possible re-evaluation of personal interpretations of physical phenomena; and

(b) To avoid multi-parallel sessions as far as possible so as to permit experts to have input into areas not necessarily their major concern, as well as to learn from experts in other fields of activity.

During the course of the meeting the Charter of the International Conference was formally agreed and the international committee structure formed with many countries represented on the full committee. The Charter and committee structure are reported overleaf.

It is hoped that delegates enjoyed participating in all the sessions of the conference, and we look forward to meeting again at ICM 4 to be held in Sweden in 1983.

Finally, I personally would wish to record my gratitude to all those who made this conference possible. Most of the day to day work fell on the shoulders of Ken J. Pascoe and our hard working secretary Sue Venn. They were aided by the skills of all members of the local committee who worked exceedingly hard for long periods of time. To these and many others far too numerous to list I give my warmest and most sincere thanks. Finally, I record my most sincere thanks to my co-editor of Proceedings, Dr Richard Smith, who was largely responsible for assembling all the papers, and tables of contents, indexes etc.

K.J. Miller
Chairman of ICM 3
and President of ICM

Sheffield 1980.

ACKNOWLEDGEMENTS

Acknowledgements are due to the many who have made possible the conference and the publication of these proceedings. This includes particularly the theme chairmen and referees and the officers of the organising committees, and above all Sue Venn, the secretarial and editorial assistant to the conference, whose unstinting efforts and patience have done much to assist the overall success of ICM 3 and of this publication.

The sponsorship of the following organisations is gratefully acknowledged:

British Gas
British Rail
Central Electricity Generating Board
Delta Metals
Institute of Physics
Institution of Mechanical Engineers
Institution of Metallurgists
Institution of Production Engineers
Instron Ltd
International Congress on Fracture
Imperial Chemical Industries
General Electric Company
ICM-2
Metal Fracture Consultants
Metals Society
MTS Systems Corporation
Rockwell International
Society of Environmental Engineers
The Welding Institute
The Plastics and Rubber Institute
Atomic Energy Authority
US Army European Research Office
US Air Force European Office of Aerospace (R&D)

COMMITTEES

CONFERENCE EXECUTIVE COMMITTEE

J.D. Alexander
J.M. Alexander
M.F. Ashby
A.G. Atkins
P.W.R. Beaumont *(Secretary)*
P.P. Benham
L.M. Brown
H.C. Cotton
R.W. Davidge
B. Edmondson
B.L. Eyre
G. Fearnehough
N.E. Frost
P. Gaskell
B. Harris
R.W.K. Honeycombe
I. Jenkins
W. Johnson
D.R.H. Jones
A. Kelly
J.F. Knott
J.C. Levy
M.J. May
D. Mclean
K.J. Miller *(Chairman)*
R.W. Nichols
R.N. Parkins
K.J. Pascoe *(Treasurer)*
A.R.S. Ponter
R.A. Smith
R.F. Smith
D. Tabor
B. Tomkins
C.H.A. Townley
A.A. Wells
J.G. Williams

LOCAL ORGANISING COMMITTEE

M.F. Ashby
P.W.R. Beaumont
L.M. Brown
P. Gaskell
D.R.H. Jones
K.J. Miller
K.J. Pascoe
R.A. Smith
R.F. Smith
D. Tabor

EDITORIAL BOARD

Theme I - B. Tomkins
Theme II - R.N. Parkins
Theme III - J.M. Alexander
Theme IV - R.W. Davidge
Theme V - I.M. Ward
Theme VI - B.A. Bilby

EDITORS

K.J. Miller
R.F. Smith

NOMENCLATURE

A	Area of cross section of a specimen
A_o	Area of cross section of a specimen at the start of testing
A_f	Area of cross section of a specimen at fracture
a	Crack length - one half the total length of an internal crack or depth of a surface crack
a_o	Original crack length - one half of total length of an internal crack at the start of a fracture toughness test, or depth of a surface crack at the start of a fracture toughness test
a_p	Measured crack length - one half the effective total length of an internal crack or effective depth of a surface crack as measured by physical methods
a_e	Effective crack length - one half the effective total length of an internal crack or effective depth of a surface crack (adjusted for the influences of a crack tip plastic zone)
Δa, Δa_p, . .	Crack growth increment
da/dN	Rate of fatigue crack propagation
B	Testpiece thickness
b	Atomic interval (Burgers vector magnitude)
d	Average grain diameter
D_L	Lattice diffusion rate
D_B	Grain boundary diffusion rate
D_S	Surface diffusion rate
E	Young's modulus of elasticity
exp	Exponential base of natural logarithms
G	Strain energy release rate with crack extension per unit length of crack border, or crack extension force
G_I G_{II} G_{III}	Crack extension forces for various modes of crack opening
h	Planck's constant
I	Moment of inertia

J	Path-independent integral characterising elastic/plastic deformation field intensity at crack tip; also, energy release rate for non-linear elastic material
K	Stress intensity factor - a measure of the stress field intensity near the tip of a perfect crack in a linear elastic solid
K_c	Fracture toughness - the largest value of the stress intensity factor that exists prior to the onset of rapid fracture
K_{max}	Maximum stress intensity factor
K_{min}	Minimum stress intensity factor
K_{th}	Threshold stress intensity factor below which fatigue crack growth will not occur
K_I	Opening mode stress intensity factor
K_{IC}	Plane strain fracture toughness
K_{Ii}	Elastic stress intensity factor at the start of a sustained load flaw growth test
K_{ISCC}	Plane strain K_I threshold above which sustained-load flaw-growth occurs
K_{II}	Edge sliding mode stress intensity factor
K_{III}	Tearing mode stress intensity factor
$\dot{K}$	Rate of change of stress intensity factor with time
ΔK	Stress intensity range
k	Boltzmann's constant
k_y	Parameter that determines grain size dependence of yield strength
l_o	Gauge length
ln	Natural logarithm
log	Common logarithm
m	Strain rate sensitivity exponent
N_f	Number of cycles to failure
n	Strain hardening exponent
P	Force
P_{max}	Maximum force
Q	Activation energy
Q_a	Activation energy for crack growth

Q_c	Activation energy for creep
Q_d	Activation energy for self-diffusion
T	Temperature
T_M	Absolute melting temperature
T_D	Ductile-brittle transformation temperature
t	Time
t_o	Time at the onset of a test
t_f	Fracture time
U	Potential energy
γ_s	True surface energy
γ_B	Grain boundary surface energy
γ_p	Effective surface energy of plastic layer
δ	Value of crack opening displacement
δ_c	Critical crack opening displacement, being one of the following: (1) Crack opening displacement at fracture (2) Crack opening displacement at first instability or discontinuity (3) Crack opening displacement at which an amount of crack growth commences
δ_m	Crack opening displacement at first attainment of maximum force
δ_z	Thickness of grain boundary layer
ε	Normal strain
ε_e	Normal strain, elastic
ε_p	Normal strain, plastic
ε_T	Normal strain, total
ε_{max}	Normal strain at maximum tensile load
ε_E	Engineering normal strain
ε_f	Normal strain, critical value at failure
ε_i	Principal strains (i = 1, 2, 3)
ε_{pi}	Principal strains, plastic
ε_x ε_y ε_z	Cartesian strain components

ε_{ij}	Strain tensor
$\dot{\varepsilon}$	Strain rate
$\dot{\varepsilon}_e$	Strain rate, elastic
$\dot{\varepsilon}_p$	Strain rate, plastic
$\dot{\varepsilon}_o$	Strain rate, initial value
$\Delta\varepsilon$	Strain range
$\Delta\varepsilon_p$	Plastic strain range
ν	Poisson's ratio
σ	Normal stress
σ_y	Yield stress under uniaxial tension
$\sigma_1\ \sigma_2\ \sigma_3$	Principal normal stresses
σ_e	Fatigue strength, endurance limit
σ_f	Fracture stress
σ_{max}	Maximum stress
$\sigma_x\ \sigma_y\ \sigma_z$	Cartesian components of normal stress
$\dot{\sigma}$	Stress rate
$\Delta\sigma$	Stress range
τ	Shear stress
τ_o	Critical shear stress
$\tau_1\ \tau_2\ \tau_3$	Principal shear stresses
τ_{max}	Shear stress, maximum value
Ω	Atomic volume

CONVERSION UNITS

To convert from	to	multiply by
inch	metre (m)	2.54×10^{-2}
pound force	newton (N)	4.448
kilogram force	newton (N)	9.807
kilogram force/metre^2	pascal (Pa)	9.807
pound mass	kilogram mass (kg)	4.536×10^{-1}
ksi	pascal (Pa)	6.895×10^{6}
ksi $\sqrt{in}$	MN $m^{-3/2}$	1.099
torr	pascal (Pa)	1.333×10^{2}
bar	pascal (Pa)	1×10^{5}
angstrom	metre (m)	1×10^{-10}
calorie	joule (J)	4.184
foot-pound	joule (J)	1.356
degree Celsius	kelvin (K)	$T_K = T_C + 273.15$

Multiplication factor	Prefix	Symbol
10^{-12}	pico	p
10^{-9}	nano	n
10^{-6}	micro	μ
10^{-3}	milli	m
10^{3}	kilo	k
10^{6}	mega	M
10^{9}	giga	G

ERRATA

Page	Line	
57	23	Delete - 'but they are usually conservative'
62	18	Sentence reads - 'The two edge conditions of the glass plates tested do not ...'
62	20	Delete sentence - 'Thus the ... not warranted.'
65	15	Colvin and Lightfoot reference is - 'Private communication'
68	24	Sentence should read - 'The chain survival expression may be written, taking V_i as the volume of the ith link as:

$$P_{survive} = \exp-[\int_{vol} (V_i) f(\sigma) dV] = \exp-[\int g(\sigma) dV] \quad (2)$$

Page	Line	
69	8	Equation (5) should read:

$$\sigma_n = (\sigma_1 \cos^2\phi + \sigma_2 \sin^2\phi) \cos^2\psi + \sigma_3 \sin^2\psi$$

Page	Line	
74	4	'width' should be 'thickness'
78	6	'process' should be 'deformation'
79,80		Transpose pages
81		In Fig. 2 label on uppermost contour should be '-250μ'
301		Caption to Fig. 1 should read - 'Stresses: solid, 10; dashed, 20; dotted, 30 MNm^{-2}'
319		Equation (2) should read:

$$\sigma_f^{n+1} = 2(n+1)\, E\, \dot{\varepsilon}\sigma_{IC}^{(n-2)} / AY^2 K_{IC}^{(n-2)} \quad (n-2)$$

Page	Line	
321		Equation (4) should read:

$$G_I = \left(\frac{A_{11}A_{22}}{2}\right) \left[\left(\frac{A_{22}}{A_{11}}\right)^{\frac{1}{2}} + \frac{2A_{12}+A_{66}}{2A_{11}} \right]^{\frac{1}{2}} K_I^2$$

Page	Line	
358		Table 3 elongation values should read: 0.6, 0.6, 0.11, 0.6, 0.6, 0.5, ≤ 1, ≤ 1
550		Unnumbered equation should read:

$$\frac{dR}{da} = \frac{a + R}{a}\{ (\frac{\hat{R}}{\Delta} - \frac{1}{2} \log [\frac{2eR(2a + R)}{a\Delta}]\} + \frac{R}{a}$$

Page	Line	
552		Equation (6) should read: $$\frac{dR}{da} = \frac{R}{a}\{\frac{\hat{R}}{\Delta} - \frac{1}{2}\log(\frac{2R^2}{ea\Delta})\}$$
553		Equations (9c), (9d) and (9e) should read: $$\{\frac{2R(2a + R)}{a\Delta}\} \quad R_f = a_f\,[\cos Q_f - 1] \quad = \exp[(\frac{2\hat{R}}{\Delta}) - 1] \qquad (9c)$$ $$\frac{2R_f^2}{a_f\Delta} = \exp\,[(\frac{2\hat{R}}{\Delta}) - 1] \qquad (9d)$$ $$a_f = a_o\,[2n + 1 + \log(a_o/\Delta) - 2\log(R_i/\Delta)] \qquad (9e)$$
	26	4.11×10^3 should be 550
	27	536% should be 475%
	26	Equations (10) and (11) should read: $$\frac{dR}{da} = \frac{R}{a}\{\frac{\hat{R}}{\Delta} - \frac{1}{2}\log(\frac{2R^2}{ea\Delta})\} \qquad (10)$$ $$\frac{dJ_R}{da} = \frac{\hat{J}}{\Delta} - \frac{4\sigma^2{}_Y}{\pi E}\propto\log(\frac{2a}{e\Delta}) \qquad (11)$$
554		Equations applicable to LSY range should read: $$\lambda_i = n - (1/2)\log(R_i{}^2/a_o\Delta)$$ $$a_f = a_o\,(1 + 2\lambda_i)$$ $$R_f = R_i\,(1 + 2\lambda_i)^{1/2}\exp(\lambda_i)$$ $$Q_i = \cos^{-1}[1 + (R_i/a_o)]^{-1}$$ $$Q_f = \cos^{-1}[1 + (R_i/a_o)(1 + 2\lambda_i)^{-1/2}\exp(\lambda_i)]^{-1}$$
555		Equations (12) should read: $$R = R_i\,(\frac{a}{a_o})^{1/2}\exp\{\frac{1}{2}(1 + 2\lambda_i)(1 - \frac{a_o}{a})\}$$ $$J_R = (a/a_o)\,J_i + \frac{4\sigma_Y^2}{\pi a}\alpha a\,\{(1 + 2\lambda_i)(1 - \frac{a_o}{a}) - \log(\frac{a_o}{a})\}$$ Equations (13) and (14) should read: $$\lambda_i = \frac{\hat{R}}{\Delta} - 1 - \frac{1}{2}\log(\frac{2R_i^2}{ea_o\Delta}) \qquad (13)$$ $$\lambda_i = n - \frac{1}{2}\log(\frac{R_i^2}{a_o\Delta}) \qquad (14)$$

Page	Line	
557		Equations (20a) and (20c) should read:

$$\mathcal{G}^{\Delta} = \frac{8\sigma_Y^2}{\pi E} \begin{cases} \left\{\frac{a\Delta}{a+R}\left(\frac{dR}{da} - \frac{R}{a}\right) + \frac{\Delta}{2}\log\left[\frac{2eR(2a+R)}{a\Delta}\right]\right. & (20a) \\ \left\{\frac{a\Delta}{R}\frac{dR}{da} + \frac{\Delta}{2}\log\left(\frac{2R^2}{ea\Delta}\right)\right. & (20c) \end{cases}$$

559		Equations (30) and (31) should read:

$$R_f = R_i\,(1 + 2\lambda_i)^{1/2}\exp(\lambda_i) \qquad (30)$$

$$\left(\frac{\hat{w}}{\hat{w}_R}\right)_f = \frac{2}{\pi}(1 + 2\lambda_i)^{-1/2}\exp(-\lambda_i) \qquad (31)$$

560		Equation (33) should read:

$$\left(\frac{\hat{w}_\Delta}{\hat{w}_R}\right)_f = \left\{\frac{\pi}{2}\exp(\lambda_i) + \frac{a_o}{R_i}(1 + 2\lambda_i)^{1/2}\left[\log\left(\frac{R_i}{a_o}\right) - \frac{1}{2}\log(1 + 2\lambda_i) + \lambda_i\right]\right\}^{-1} \qquad (33)$$

THEME IV

DESIGNING WITH BRITTLE SOLIDS

Volume 3, ICM 3, Cambridge, England, August 1979

FRACTURE TOUGHNESS OF ALUMINA AT ROOM TEMPERATURE TO 1600°C

A. S. Kobayashi, A. F. Emery, A. E. Gorum* and T. Basu*

Department of Mechanical Engineering, University of Washington, Seattle, Washington, U.S.A.
***Department of Mining, Metallurgical and Ceramic Engineering, University of Washington, Seattle, Washington, U.S.A.**

INTRODUCTION

The recent use of fracture mechanics for assessing the strength of structural ceramics has often been hampered by the lack of accurate measurements of the fracture mechanics parameters. Since modern fracture mechanics relates to metal fracture, despite its historical origin in glass (Griffith, 1921), there is a tendency to transfer the extensive know-how in the fracture mechanics testing techniques of metals to ceramic materials. Evans (1974), in an excellent survey article, discussed some of these standard fracture testing techniques which are also suitable for testing ceramics. Non-standard fracture specimens include the notched ring fracture specimen (Rowcliffe, Jones and Gran, 1978) and the short rod specimen (Barker, 1978), both of which can be cut off from as-fabricated ceramic components with little additional machining. Unlike metallic specimens, reduction in additional machining is a particularly important consideration in fracture toughness testing of brittle ceramic material where the surface finish can have a considerable influence on its apparent fracture toughness.

Another important consideration in fracture mechanics studies of ceramic material is the requirement for high temperature testing, e.g. 1000 $\sim$1400°C, where most metals fail as a structural material. Fracture testing at such temperature requires simple testing procedures where the test fixture itself is fabricated from the same or similar high temperature ceramics. The most commonly used fracture specimen for such high temperature testing of ceramics is the double-torsion (DT) specimen (Outwater and others, 1974) which was first analyzed by Williams and Evans (1973) who assumed that the double torsion specimen can be treated as two independent torsion bars with a straight, through crack. Despite the well known curved crack front of a double torsion specimen (Evans, 1973), the measured fracture toughness, which was obtained by using the above simplified formula, and that obtained by the thoroughly analyzed double cantilever beam (DCB) specimen were in good agreement with each other (Williams and Evans, 1973) thus prompting others to adopt the double torsion specimen for not only high temperature but also room temperature fracture testing of ceramic materials. The recent approximate three-dimensional finite element analysis of a double torsion specimen by Trantina (1977) showed that the stress intensity factor was constant along a naturally observed curved crack front but the more complex 3-D finite element analysis by Tseng and Berry (1978) showed substantial variations in the stress intensity factors along assumed quarter elliptical crack fronts. Both results, however, confirmed the validity of Williams and Evans (1973) strength-of-material analysis for a hypothetical straight, through crack in a DT specimen. The uncertainty in the variation or the lack of such variation along the curved crack front of a DT specimen thus requires further studies before this specimen can be used to generate basic data on fracture toughness.

Recognizing the need for a simpler fracture specimen for high temperature testing of ceramic materials, Kobayashi and others (1978) developed a tapered DCB specimen with a stress intensity factor which is relatively insensitive to the crack length and which is loaded in simple compression through a wedging pin. The salient features of this specimen are the need to monitor only the applied compression load, similar to the popular DT specimen, and the well defined two-dimensional state of stress characterized by an opening mode stress intensity factor, K_I. In this paper, the friction effect, which dominated the previous specimen (Kobayashi and other, 1978) is shown to be eliminated by the steep notch face which allows the loading pin to slide during the continuous opening of the starter section notch with increasing applied load. This acute notch angle also increases the pin-load efficiency. The redesigned fracture specimen is then used to generate preliminary fracture toughness and subcritical crack growth data of commercial grade alumina at room temperature to 1600°C.

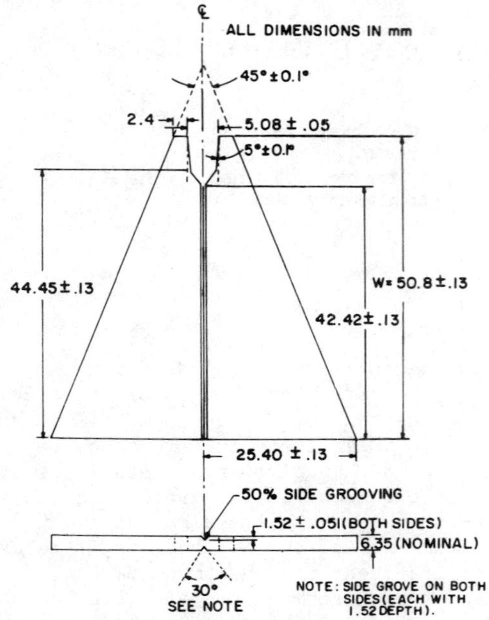

Fig. 1. Wedge loaded tapered DCB specimen.

WEDGE-LOADED TAPERED DCB SPECIMEN

The redesign of the wedge loaded tapered DCB specimen (Kobayashi and others, 1978) by altering the specimen geometry involved a series of finite element analyses of various notch geometries in an attempt to

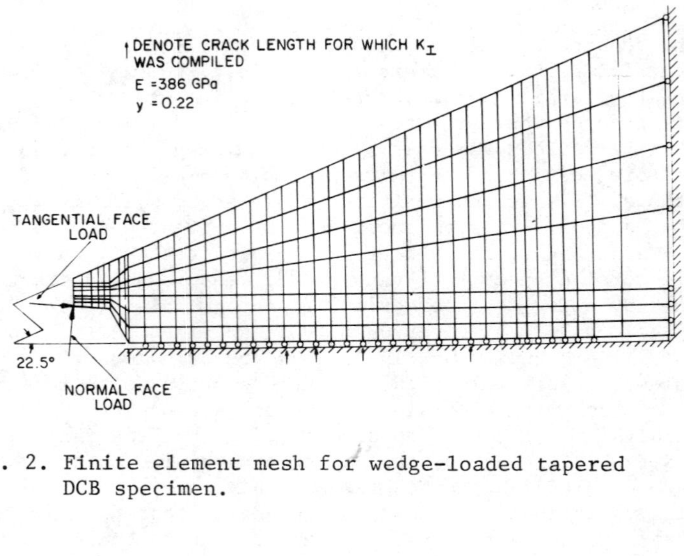

Fig. 2. Finite element mesh for wedge-loaded tapered DCB specimen.

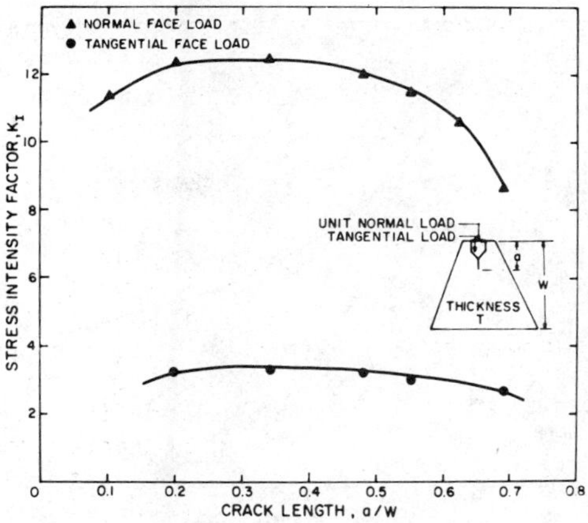

Fig. 3. Stress intensity factors due to unit normal and tangential loads at loading pin-notch face.

reduce the variations in stress intensity factors at the crack tip due to the wedging friction force. The final design involves an acute, included notch angle of $\theta = 10°$, as shown in Fig. 1, as well as a larger pin diameter of 6.4 mm in comparison to $\theta = 60°$ and pin diameter of 3.2 mm used previously (Kobayashi and others, 1978). Figure 2 shows the finite element nodal breakdown which models one half of the specimen, including the starter section notch, and is a more accurate modelling than that used previously. Both stress intensity factors for unit forces normal and tangential to the notch face at the theoretical contact point, as marked in Fig. 2, were computed. A nominal modulus of elasticity of E = 386GPa and a nominal Poisson's ratio of $\nu = 0.22$ were used in the numerical analysis. Although most of the stress intensity factors were computed directly by the procedure of crack opening displacement, the final K_I values were adjusted upward by the more accurate K_I values obtained by the strain energy release rate procedure for the shortest and longest crack lengths.

Figure 3 shows the variations in stress intensity factors, K_I, with crack length, a, for unit normal and tangential face loads applied at the theoretical contact point. The stress intensity factor for a unit tangential face load has been reduced subtantially in comparison to the corresponding values in the previous design. For

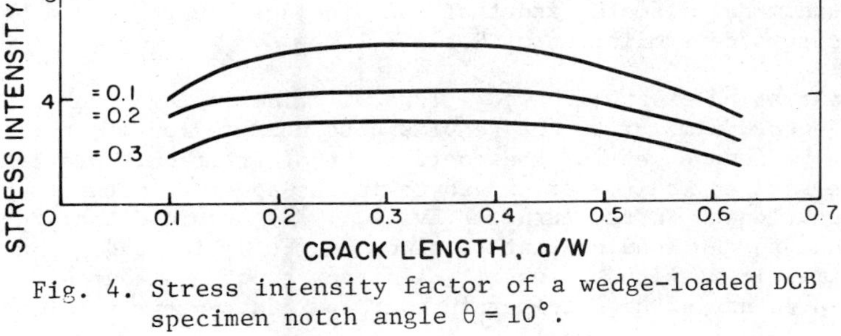

Fig. 4. Stress intensity factor of a wedge-loaded DCB specimen notch angle $\theta = 10°$.

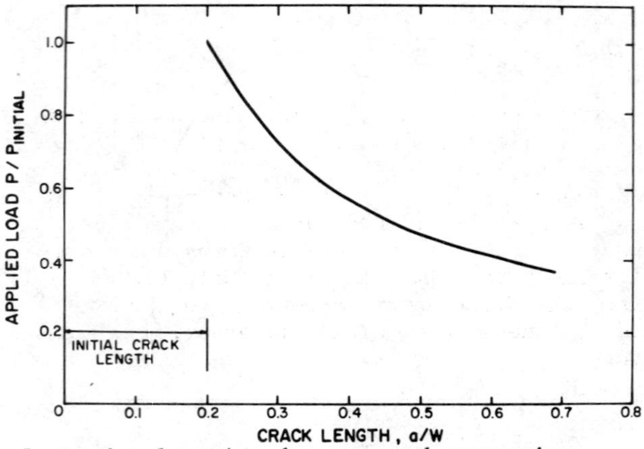

Fig. 5. Load relaxation due to crack extension.

a nominal coefficient of static friction of $\mu = 0.1$, the static friction force will thus increase the theoretical stress intensity factor by only 2.5 percent. The static friction coefficient, however, will substantially reduce the applied normal face load F_n according to

$$F_n = \frac{P}{2(\sin\frac{\theta}{2} + \mu \cos\frac{\theta}{2})}$$

where P is the total pin load. For a notch angle of $\theta = 10°$ and hypothetical static friction coefficients of $\mu = 0$, 0.1, 0.2, and 0.3, the applied normal face loads become $F_n = 5.737P$, 2.677P, 1.745P, and 1.295P, respectively. The resultant stress intensity factor varies substantially with the coefficient of friction as shown in Fig. 4. Fortunately, the larger nominal face load and the smaller beam width in the starter section will force the notch angle to open with increasing applied load and thus eliminate the friction effect. In a later section, a comparison of fracture toughness data generated by this specimen and the previous specimen will show that indeed this frictionless mechanism prevails in the redesigned specimen.

Since the wedging pin will continue to move downwards with increasing pin load, the sensitivity of the resultant stress intensity factor to variations in the contact point of the loading pin and notch face either due to variations in pin diameter or notch face dimensions or both was studied. This induced error was assessed by comparing the stress intensity factor for the shortest crack length, a/W = 0.1, with that obtained by shifting the normal face load, one nodal distance, i.e. 1.8 mm, below the nominal point of load application, Fig. 2. Although the resultant stress intensity increased to that of a/W = 0.2, subsequent increase in stress intensity factor for longer cracks, $a/W \geq 0.2$, was not noticeable. A one nodal point shift in load application corresponded to a 4 percent reduction in loading pin diameter and thus it was concluded that the loading pin diameter should be accurately machined within ±5% of its nominal diameter and that the precrack length should be at least $a/W \geq 0.2$ for accurate determination of K_I.

Figure 5 shows relaxation of applied wedge load with increasing crack length for a fixed pin displacement. Similar to the results obtained previously, this relaxation load was found to be independent of the coefficient of friction. This load relaxation curve can be used to monitor crack growth in the specimen without observing the instantaneous crack length in the furnace (Evans, 1973) provided that the amount of load relaxation due to the loading fixture alone can be delineated. Since the latter could be substantial in magnitude, subcritical crack growth data generated by this procedure in the presence of high temperature creep requires further scrutiny at this time.

FRACTURE TESTING

Upon completing the specimen design, over twenty alumina specimens were fabricated from commercial grade 99%, sintered alumina disks, COOR AD-998, with nominal dimensions of 152 mm diameter x 6.35 mm thickness. Preliminary fracture testing showed that a minimum of 50 percent sidegrooving, as shown in Fig. 1, was necessary to contain the crack in a straight line. The change in stress intensity factor along the crack front through the thickness of the specimens due to the sidegrooving is estimated to be small since three-dimensional photoelastic analysis by Kirkwood and Prado (1976) showed that the sidegrooves tend to raise the boundary region stresses to the same as the mid-thickness value. This sidegrooving increases the apparent stress intensity factor by

$$K_I = \sqrt{\frac{t_{net}}{t}}\, K_{I\ apparent}$$

where t and t_{net} are the original specimen thickness and the net section thickness at the sidegrooving section respectively, and $K_{I\ apparent}$ is the stress intensity factor of the specimen without sidegrooving.

The loading pin was a silicon nitride pin of 6.4 mm diameter, the base supporting fixture was machined from silicon carbide and the compression rod was fabricated from tungsten. The coefficient of static friction between the silicon nitride pin in contact with the alumina notch face was measured using a fixture developed by Staley (1978). Although a low coefficient of friction of about 0.04 was obtained at room temperature testing following the procedure of Kobayashi and others (1978), this coefficient of friction was not used in the actual computation of K_I for the reason mentioned previously. No measurement of the coefficient of friction at elevated temperature was made.

The entire specimen and loading fixture were enclosed in a furnace which was evacuated to a 10^{-4} torr vacuum. This soft vacuum was necessary to prevent the tungsten heating elements from oxidizing prematurely. The specimen was loaded through a 10 Mg Instron testing machine.

Precracking was accomplished by loading the specimen with a slow cross head speed where the machine rigidity was used to arrest the crack which popped upon reaching criticality. This in situ precracking procedure was considered more reliable with less specimen spoilage than the precracking procedure used previously (Kobayashi and others, 1978).

FRACTURE TOUGHNESS

Fracture toughness of fourteen specimens were obtained at room temperature through 1600°C. Table 1 shows a summary of the test result where the test temperature was gradually increased above 1000°C in a futile search for a temperature at which a noticeable increase, such as that in silicon nitride (Evans and Wiederhorn, 1974), in fracture toughness could be observed. Unfortunately, such increase in fracture toughness was not observed up to 1600°C.

TABLE 1 Fracture Toughness of Commercial 99% Alumina

Test Temperature	Rm	Rm	Rm	Rm	Rm	Rm	Rm
Fracture Toughness	8.35 $MNm^{-3/2}$	4.52	5.2	5.16	4.70	4.27	5.27
Remark	Blunt Crack	-	-	-	-	-	-

Test Temperature	1000°C	1200°C	1200°C	1370°C	1500°C	1500°C	1600°C	1600°C
Fracture Toughness	4.7	5.27	5.27	5.22	4.72	4.72	4.61	4.61
Remark	-	-	-	-	-	-	-	-

SUBCRITICAL CRACK GROWTH

Subcritical crack growth at 1600°C was also investigated in one specimen where a fixed loading head displacement was maintained for 3 hours. During this period a 10 percent reduction in applied load was observed indicating a 5 percent increase in crack length from Fig. 5. Visual observation of the crack length after removal of load and at room temperature, however, did not confirm this crack growth thus verifying our suspicion that the measured load relaxation was probably caused by high temperature creep of the various components in the loading fixtures and not by crack extension as originally intended

DISCUSSION

The average fracture toughness of $K_{IC} = 4.91$ $MNm^{-3/2}$ at room temperature is 7 percent higher than the corresponding value obtained previously (Kobayashi and others, 1978). This K_{IC} value falls within the low $K_{IC} = 4$ $MNm^{-3/2}$ (Mai, 1976) and high $K_{IC} = 5.3$ $MNm^{-3/2}$ (Davidge and Tappin, 1970) reported for polycrystalline alumina. Should any sizeable friction force existed between the loading pin and notch face, Fig. 5 shows that the computed stress intensity factor would have been 30∿50 percent lower which is clearly beyond the normal scatter in data. This comparison of previous data thus shows that the postulated slipping mechanism leading to an apparent vanishing coefficient of friction is a reasonable one.

The lack of any noticeable subcritical crack growth was disappointing since we were not able to test this fracture specimen for crack growth response. The quoted reduction in flexure strength from 330 MPa at 25°C to 193 MPa at 1000°C (Coors/Ceramic, 1970) provided hope that subcritical crack growth would have occurred. Similar attempts to measure subcritical crack growth at room temperature and 50% relative humidity as well as sprayed-on tap water failed to produce any slow crack growth over a 24 hours period of sustained loading (Staley, 1978). Since subcritical crack growth measurements made on commercial 95% alumina at room temperature and 50% relative humidity showed measurable crack growth at an applied stress intensity factor as low as $K_I = 0.22$ $MNm^{-3/2}$ (Evans, 1972), the differences could be attributed to the purity of alumina used in the two investigations.

CONCLUSIONS

1. A fracture specimen, convenient for measuring fracture toughness and subcritical crack growth at elevated temperature is presented.
2. Fracture toughness of commercial 99% alumina was found to vary from 4.9 to 4.6 $MNm^{-3/2}$ at room temperature to 1600°C.

3. No subcritical crack growth in commercial 99% alumina at 1600°C under a sustained $K_I = 0.95\ K_{IC}$ was observed in 3 hours.

ACKNOWLEDGEMENT

This investigation was supported by NASA grant NGL 48-002-004. The authors wish to thank Professor J.I. Mueller, University of Washington, Division of Ceramic Engineering, for his encouragement and many helpful discussions throughout the course of this investigation.

REFERENCES

Barker, L.M. (1978). Short rod K_{IC} measurements of Al_2O_3. In R.C. Bradt, D.P.H. Hasselman and F.F. Lange (Ed.), Fracture Mechanics of Ceramics, 3, Plenum Press, New York-London, pp. 483-494.

Coors/Ceramic (1970). Mullite tubes, high-alumina tubes and labware. Coors/Ceramic Catalog 201.

Davidge, R.W. and Tappin, G. (1970). The effects of temperature and environment on the strength of two polycrystalline aluminas. Proc. of the Brit. Ceramic Soc., 15, p. 47.

Evans, A.G. (1972). A method for evaluating the time-dependent failure characteristics of brittle materials — and its application to polycrystalline alumina. J. of Materials Science, I, 1137-1146.

Evans, A.G. (1973). A simple method for evaluating slow crack growth in brittle materials. Int. J. of Fracture, 9, No. 3, 267-275.

Evans, A.G. (1974). Fracture mechanics determination. In R.C. Bradt, D.P.H. Hasselman, and F.F. Lange (Ed.), Fracture Mechanics of Ceramics, 2, Plenum Press, pp. 17-48.

Evans, A.G. and Wiederhorn, S.M. (1974). Crack propagation and failure prediction in silicon nitride at elevated temperature. J. of Materials Science, 9, 270-278.

Griffith, A.A. (1921). The phenomena of rupture and flow in solids. Phil. Trans. of Royal Soc. of London, 221, 163-198.

Kirkwood, H.F. and Prado, M.E. (1976). Three-dimensional stress distribution of a double-cantilever beam with a side notch. Cracks and Fracture ASTM STP 601, pp. 267-273.

Kobayashi, A.S., Staley, L.I., Emery, A.F., and Love, W.J. (1978). A fracture specimen for high temperature testing. In R.C. Bradt, D.P.H. Hasselman and F.F. Lange (Ed.), Fracture Mechanics of Ceramics, 3, Plenum Press, New York-London, pp. 451-461.

Mai, Y.W. (1976). Thermal Stress resistance and fracture toughness of two tool ceramics. J. of Material Science, II, 1430-1438.

Outwater, J.O., Murphy, M.C., Kumble, R.G., and Berry, J.T. (1974). Double torsion technique as a universal fracture toughness test method. Fracture Toughness and Slow-Stable Cracking, ASTM STP 449, pp. 127-138.

Rowcliffe, D.J., Jones, R.L. and Gran, J.K. (1978). A notched ring fracture toughness test for ceramics. In R.C. Bradt, D.P.H. Hasselman and F.F. Lange (Ed.), Fracture Mechanics of Ceramics, 3, Plenum Press, New York-Lond, pp. 473-482.

Staley, L.I. (1978). Design and preliminary developmental testing of a fracture specimen suitable for elevated temperature testing. M.S. thesis submitted to the University of Washington.

Trantina, G.G. (1977). Stress analysis of the double-torsion specimen. J. of American Ceramic Society, 60, 338-341.

Tseng, A.A. and Berry, J.T. (1978). A three-dimensional finite element analysis of the double torsion test. A paper presented at the ASME-CSEM Joint Pressure Vessel and Piping Conference, Montreal, Canada, June 25-28, 1978.

Williams, D.P. and Evans, A.G. (1973). A simple method for studying slow crack growth. J. of Testing and Evaluation, 1, No. 4, 264-270.

Volume 3, ICM 3, Cambridge, England, August 1979

SIZE EFFECTS IN BRITTLE FRACTURE

K. E. Puttick

Physics Department, University of Surrey, Guildford, Surrey GU2 5XH, U.K.

ABSTRACT

Size effects, in which the apparent critical stress for fracture depends on some dimension imposed by the method of testing, are inherent in the energy theory of fracture, since the *surface* energy density requirement of the crack is supplied from the *volume* strain energy density stored in the matrix. In many cases of importance the crack or flaw length is geometrically similar to the linear scale of an inhomogeneous stress field, which accordingly determines the apparent fracture stress; this was first recognized by Roesler (1956) as the true explanation of Auerbach's law of Hertzian fracture. This energy scaling principle may be applied to inhomogeneous stress fields of all kinds, including for instance those associated with plastic-elastic deformation at notch roots or indentations, and fracture transitions in such circumstances are shown to occur at a critical value of a length $\alpha E\Gamma/\sigma_y^2 = \alpha K_c^2/2\sigma_y^2$, where α is a function of the geometry of the test. This is the basis of a new fracture diagram proposed to interpret size effects in any test. The use of energy scaling in the elucidation of size effects in abrasion and crushing is reviewed.

KEYWORDS

Fracture size effects, energy scaling, fracture transitions.

INTRODUCTION

The apparent critical stress for crack initiation or propagation in brittle fracture generally varies with some dimension of geometrically similar specimens. This effect, which is of great practical importance, is inherent in the energy theory of fracture, and indeed the situation investigated by Griffith (1920) of a crack in a field of prior homogeneous tensile stress is a special case. We may deduce the relation between critical stress and size by scaling the stress intensity factor (Irwin 1958; Sih 1973) or the stresses (Gurney and Hunt, 1967). I prefer, however, to start by considering the scaling of the energy balance associated with crack growth in the way proposed by Roesler (1956), since it emphasizes the relation of fracture mechanics to thermodynamics, simplifies the dimensional analysis, and incidentally gives due credit to a pioneer development of Griffith's hypothesis.

HERTZIAN FRACTURE

The classic instance of a size effect is due to Auerbach (1891), who noticed that the critical stress for the development of a cone crack beneath a spherical indenter on a plane glass surface decreased with increasing indenter radius. This effect was subsequently rediscovered by several independent workers in variants of the original static experiment, and was usually interpreted in terms of a statistical distribution of flaws on the surface. However, Roesler (1956) pointed out that all the results could be brought into the same form of a constant strain energy in the field just before fracture, and showed that for cracks geometrically similar to the stress field propagating near equilibrium this criterion is a logical consequence of the energy balance theory of brittle fracture: the strain energy increases as the cube of the linear dimension of the field, whereas the surface energy requirement of a propagating crack increases as the square of the same dimension. The quantity we now call the strain energy release rate is thus proportional to the linear scale of the field. In the particular case of Hertzian indentation this scale is the radius of the circle of contact between indenter and specimen surface, which dimension is in turn a function of indenter radius.

Seen in this light the nucleation of a crack at a surface flaw is analogous to the formation of a liquid droplet from supersaturated vapour, where the free energy supply associated with the phase change increases as the cube of a length, the droplet radius, and the free energy requirement of the surface increases as the square of the same length. There is therefore a characteristic radius at which droplets are stable.

In a later study of Hertzian fracture Frank and Lawn (1967) showed that Roesler's condition of reversibility of crack propagation is not necessary: a Griffith flaw in these circumstances will grow under increasing load to a characteristic size related to that of the stress field, after which it may propagate unstably into an observable cone crack.

FRACTURE MECHANICS

Let us now develop these ideas in current terms. The strain energy release rate of a propagating crack is dimensionally proportional to a length ℓ times the prior strain energy density W per unit volume:

$$G = \beta \ell W \tag{1}$$

In most formulations ℓ is identified with the crack length, but in certain situations such as the double cantilever test (Gurney and Hunt, 1967) G is independent of crack length and directly proportional to a specimen dimension.

Consider now the case in linear elasticity in which the prior stress is inhomogeneous, of the form $\sigma = \sigma_{max} f(x_0/x)$. The strain energy density is then proportional to $(\sigma_{max})^2/2E$. We now have two characteristic lengths, of the crack and of the stress (or strain energy) field. We may then write

$$G = \frac{(\sigma_{max})^2}{2E} \, x_0 \, F\left(\frac{x_0}{\ell}\right) \tag{2}$$

The critical condition for crack propagation is then

$$(\sigma_{max})_c = \left\{\frac{4E\Gamma}{x_0 F\left(\frac{x_\alpha}{\ell}\right)}\right\}^{\frac{1}{2}} \tag{3}$$

where Γ is the effective surface energy of crack growth ($= G_c/2$) or

$$x_{oc} = \frac{4E\Gamma}{(\sigma_{max})_c^2 \, F\left(\frac{x_{oc}}{\ell}\right)} \qquad (4)$$

where x_x is the critical size of stress field.

Thus the apparent critical stress for fracture is a function of the scale of the stress field. The variation of the non-dimensional function F depends on the way that the crack or flaw size ℓ varies with x_0; if it is geometrically similar, so that ℓ/x_0 = constant, then $F(x_0/\ell)$ = constant and $(\sigma_{max})_c \propto x_0^{-\frac{1}{2}}$. In the special case of a crack perturbing a uniform tension and preserving geometrical similarity with itself as it grows, we have

$$(\sigma_{max})_c = \left\{\frac{4E\Gamma}{\beta\ell}\right\}^{\frac{1}{2}} \qquad (5)$$

in agreement with Griffith's relation.

We may in principle extend these ideas to other characteristic lengths, for instance the specimen width ω. We might then write

$$G = \frac{(\sigma_{max})^2}{2E} \, \omega\phi\left(\frac{x_0}{\ell}, \frac{\ell}{\omega}\right) \qquad (6)$$

and observe two ranges of behaviour in which the crack is comparable in size respectively to x_0 and ω. In real specimens there will in fact be a number of characteristic lengths such as surface flaw size, inclusion size, grain size and so on, each with its own critical length for fracture under given conditions. In a surprising number of cases, however, the approximation of a prior flaw length scaling geometrically with one other characteristic length is useful. In the fracture of single crystals, for instance, the stress concentrating flaws may be slip bands or growth steps (Marsh, 1963) which are proportional to the specimen diameter; crazes in amorphous polymers are proportional in length to that of the tensile stress field, and therefore to the size of the plastic zone in plastic-elastic deformation, and it may be that ductile fracture in tensile specimens of metals deformed to geometrically similar neck shapes, or in the plastic zone at a notch root, is also geometrically similar to the size of specimen or zone length.

FRACTURE TRANSITIONS

Various changes in fracture character with size are implicit in energy scaling. As the characteristic length decreases, for instance, the critical stress level eventually equals the yield stress of the material:

$$x_{oc} = \frac{4E\Gamma}{(\sigma_y)^2 F\left(\frac{x_{oc}}{\ell}\right)} \qquad (7)$$

Fracture is then initiated in a plastic-elastic, rather than purely elastic, strain field. For example, Gurney and Hunt (1967) showed that in the double cantilever beam test the relevant critical dimension is the half depth d_c of each arm, so that this type of transition occurs when

$$d_c = \frac{3E\Gamma}{\sigma_y^{\,2}} \tag{8}$$

A similar transition takes place in Hertzian indentation when the pressure on the contact area equals the yield stress. I call these *lower transitions*, and of course for constant specimen size may occur if σ_y is a function of test conditions such as temperature or strain rate.

Suppose now that the critical dimension in (4) is further increased by decreasing σ_y or increasing $E\Gamma$. Fracture will then initiate only after yielding has taken place in a part of the specimen, and under these conditions the scale of the stress field is that of the plastic zone, x_p. This is also the critical length which determines fracture

$$x_{pc} = \frac{4E\Gamma}{\sigma_y g\left(\dfrac{x_{pc}}{\sigma_y^{\,2}}\right)} \tag{9}$$

For example, the critical size of the plastic zone at the tip of a crack propagating under plane strain conditions is (Irwin, 1958)

$$x_{pc} = r_Y \simeq \frac{K_{1c}^2}{\sigma_y^{\,2}} = \frac{2}{\pi(1-\nu^2)}\,\frac{E\Gamma}{\sigma_y^{\,2}} \tag{10}$$

In plastic-elastic indentation by a sphere, in which cracks radial to the impression are formed, the plastic zone in the surface is concentric with the impression, and its critical radius is (Puttick, 1978)

$$x_{pc} \simeq \frac{50E\Gamma}{\sigma_y^{\,2}} \tag{11}$$

In the notched-bar test, on the other hand, the critical zone size is very much smaller because of the high value of σ_{max}, about 2.5 σ_y, compared with $\sigma_{max} \simeq \sigma_y/2$ for indentation. We find (Puttick, 1979; Griffiths and Oates, 1969)

$$x_{pc} \simeq \frac{4E\Gamma}{50\sigma_y^{\,2}} \tag{12}$$

Now as the critical plastic zone size increases a second limit is encountered when it becomes equal to some maximum dimension imposed by the specimen or the test. The most familiar instance is the change from plane strain to plane stress (mode I to mode III) fracture when the size of plastic zone at a crack tip equals the specimen width (Irwin, 1958). In the notched-bar test, the maximum size of plastic zone at the notch root occurs just before the specimen is at the point of general yield, which for a Charpy bend specimen occurs when the zone is about 0.1 of the width of the minimum section.

During plastic-elastic indentation, the maximum size of plastic zone corresponds to penetration of a spherical indenter to a depth of one radius. The associated size effect is illustrated by some recent experiments by Razzak Yousif (private communication), in which indentations were made in blocks 25.4 mm thick of 'Perspex' grade PMMA by spherical indenters of radii 1.6 mm and 0.5 mm to depths of one radius. Fig. 1 shows the percentage of uncracked indentations in each case as a function of temperature: the ductile-brittle transition (i.e. the initiation of radial cracks from crazes in the plastic zone surrounding the indentation) takes place at about + 45°C for the 1.6 mm ball (corresponding to a plastic zone radius of 2.64 mm) but is depressed to − 30°C for the 0.5 mm indenter (plastic zone radius = 0.82 mm).

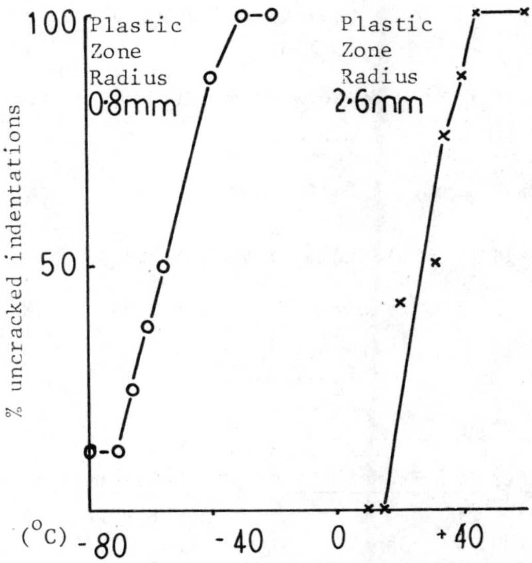

Fig. 1. Fracture Transitions for indentation of PMMA by spheres of radii 1.6 mm (x) and 0.5 mm (O).

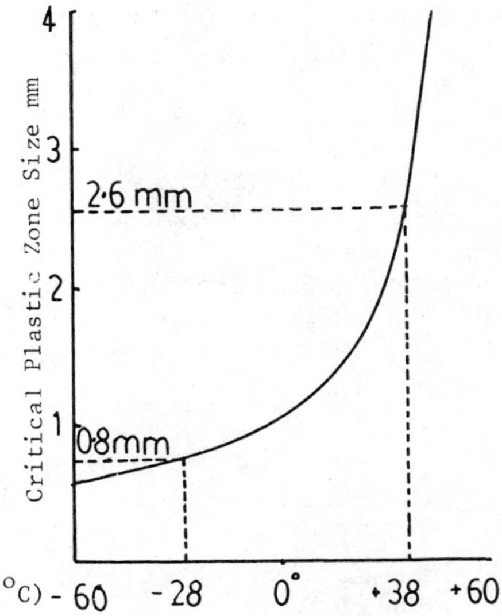

Fig. 2. Transition Diagram for indentation of PMMA.

These changes corresponding to maximum critical zone size may be called 'upper' transitions, and are of particular technological importance. Note that they may occur in circumstances, such as decreasing temperature in PMMA testing, in which the fracture toughness is *increasing*. In general, we see that fracture transitions are characterized by a critical length x which is equal to a multiple α of $E\Gamma/\sigma_y^2 = 0.7\ K_c^2/\sigma_y^2$, and we may take this parameter as the basis of a new fracture diagram. This is illustrated by Fig. 2 for the case of ball indentation of PMMA: the full line represents the value of the critical plastic zone size, $50E\Gamma/\sigma_y^2$, calculated from the published data of Berry (1963) and Bauwens-Crowet (1973), and the upper transition from uncracked to cracked indentation corresponds to the intersection of this curve with the actual zone size for a given indenter. The shift in transition temperature predicted by the diagram is consistent with the experimental value.

Similar diagrams may be constructed in principle for any material and test: Table 1 gives values of $E\Gamma/\sigma_y^2$ for a wide range of materials, and Table II values of the multiplier α for a number of tests.

TABLE 1 Values of $E\Gamma/\sigma_y^2$

Material	T°C	$E\Gamma/\sigma_y^2$m	References
Glass	20	2.4×10^{-8}	Swain & Hagan (1976) Lawn & Wilshaw (1975)
Silicon	20	7.5×10^{-8}	Puttick, Shahid & Hosseini (1979)
PMMA	+40	6.2×10^{-5}	Berry (1963
	-40	1.5×10^{-5}	Bauwens-Crowet (1973)
Low carbon steel (ABS	0	1.7×10^{-2}	Irwin, Kies
ship plate)	-40	1.2×10^{-3}	& Smith (1958)
Water quenched high carbon steel	20	6.5×10^{-6}	Studman & Field (1976) Lawn & Wilshaw (1975)
Alloy steel	20	5×10^{-4}	Irwin, Kies & Smith (1958)

TABLE 2 Values of α

Test	α	Nature of Transition	References
Notched bar (Charpy bend)	$\sim$0.08	Upper (Below to above gross yield)	Griffiths & Oates (1969) Puttick (1979)
Plane strain fracture	$\sim$0.7	Upper (Mode I to Mode III)	Irwin (1958)
Double cantilever	3	Lower (elastic to elastic-plastic stress field)	Gurney & Hunt (1967)
Indentation	$\sim$50	Upper (Radial fracture to no fracture)	Puttick (1978) Swain & Hagan (1976)

Thus an understanding of size effects enables us to rationalize fracture testing by relating different tests via their characteristic lengths; it has recently been pointed out, for example, how the rapid and convenient indentation test may be used to characterize fracture transitions associated with irradiation and molecular orientation in polymers (Puttick, 1978a, 1978b). In addition, it casts new light on old problems. For instance, why is it that in the abrasion of brittle materials like glass or silicon, large abrasive particles remove material by chipping, while small ones of the order of a micron in diameter merely smooth the surface by polishing? Inspection of Tables I and II show that plastic-elastic indentations in these materials have a critical plastic zone size for fracture of this order; particles smaller than this would not be expected to initiate the radial fracture which leads to chipping (Puttick, Shahid and Hosseini, 1979).

Again, why does the stress required to crush particles of such materials increase as the particle size decreases and finally attains the yield stress? Kendall (1978) has suggested that here also energy balance arguments provide the answer. Using an interesting variant of Gurney and Hunt's double cantilever fracture test, in which the specimen is compressed parallel to the crack direction, he deduced a value for α of about 10; the corresponding fracture transition is observed in specimens of the predicted size. No doubt other size effects, perhaps even of geophysical significance, remain to be discovered.

REFERENCES

Auerbach, F. (1891). Ann. Phys. Chem., 43, 61.
Bauwens-Crowet, C. (1973). J. Mat. Sci., 8, 968.
Berry, J.P. (1963). J. Polymer Sci., (A)1, 993.
Frank, F.C., and B.R. Lawn (1967). Proc. Roy. Soc., A299, 291.
Griffith, A.A. (1920). Phil. Trans. Roy. Soc., A221, 163.
Griffiths, J.R., and G. Oates (1969). Proc. 2nd Int. Conference on Fracture, paper 19.
Gurney, C., and J. Hunt (1967). Proc. Roy. Soc., A299, 508.
Irwin, G.R. (1958). Handb. d. Physik, 6, 551.
Irwin, G.R., J.A. Kies, and H.L. Smith (1958). Proc. A.S.T.M., 58, 640.
Kendall, K. (1978). Nature, 272, 510.
Lawn, B.R., and T.R. Wilshaw (1975). J. Mat. Sci., 10, 1049.
Marsh, D.M. (1963). in Fracture of Solids (eds. D.C. Drucker, and J.J. Gilman) John Wiley, New York, p.143.
Puttick, K.E. (1978a). J. Phys. D: App. Phys., 11, 595.
Puttick, K.E. (1978b). J. Phys. D: App. Phys., 11, L69.
Puttick, K.E., M.A. Shahid, and M.M. Hosseini (1979). J. Phys. D: App. Phys., 12, 195.
Puttick, K.E. (1979). J. Phys. D: App. Phys., (in press).
Roesler, F.C. (1956). Proc. Phys. Soc., B69, 981.
St. John, C. (1975). Phil. Mag., 32, 1193.
Sih, G.C. (1973). Handbook of Stress Intensity Factors, Lehigh University, Pa.
Studman, C.J., and J.E. Field (1976). J. Phys. D: App. Phys. 9, 857.
Swain, M.V., and J.T. Hagan (1976). J. Phys. D: App. Phys. 9, 2201.

Volume 3, ICM 3, Cambridge, England, August 1979

FRACTURE TOUGHNESS OF NODULAR CAST IRON

W. L. Bradley and H. E. Mead*

Professor of Mechanical Engineering, Texas A&M University, College Station, TX 77843, U.S.A.
***Graduate Student, Texas A&M University, College Station, TX 77843, U.S.A.**

INTRODUCTION

The advantages of nodular cast iron over other cast irons and cast steels have led to rapid expansion of its application throughout the manufacturing industry. This expanded application has produced a need for more precise characterization of its mechanical properties so that designers can better insure fail-safe behavior in service. Such information would also allow ductile iron to be considered for new applications where its use is presently precluded by code specification and/or lack of adequate characterization of its mechanical properties.

The tensile properties of nodular cast iron are well known. However, only recently have the fracture toughness studies begun to be reported in the literature (Nanstad, Worzala and Loper, Jr., 1974; Holdsworth and Jolley, 1974; Little and Heine, 1974). These studies have usually measured K_{1c} at low temperatures (173K) and K_Q at more ambient temperature levels. Recent work by the author (Bradley; 1978) has shown that K_Q values sometimes give a poor estimate of the relative toughness of various nodular cast irons, and furthermore, fail to provide the K_{1c} value the designer needs. Thus, a research program to determine the K_{1c} values at ambient temperatures (222K to 297K) for nodular cast iron for various microstructures and compositions has been initiated.

EXPERIMENTAL PROGRAM

The experimental program consisted of three parts: first, the measurement of J_{1c} for nodular cast iron with various compositions and microstructures; second, fractographic analysis of the specimens using SEM to relate the macroscopic fracture toughness to microscopic deformation and fracture events; and third, the testing of notched tensile bars to see if an empirical correlation of fracture toughness with behavior in a simple notched tensile test coupon could be found. The cost and complexity of fracture toughness measurements using compact tension coupons necessitates that a simpler means be found to monitor quality control of this property at production installations. The J_{1c} determinations have been made as per ASTM-E24 recommendations, March, 1977. Crack extensions were measured using compliance measurements, with approximately 30-40 compliance measurements made per test. Several early tests were made using both single specimen and multi-specimen J-integral test procedures to verify that resolution in the compliance measurements was sufficient to give an accurate indication of crack extension. The agreement was quite good.

The three microstructural variables studied were nodule count, silicon content and percentage of matrix that is pearlite. The various combinations tested are summarized in Table 1, with the microstructures for various percentages of pearlite shown in Fig. 1. All the specimens shown in Table 1 were tested at 297K. Specimens 1A and 5A (1.5% and 2.99% silicon) were also tested at reduced temperatures down to 222K.

TABLE 1 Chemical Compositions and Microstructures of Nodular Cast Iron Studied*

Plate	C(%)	Si(%)	Ni(%)	Mn(%)	S(%)	P(%)	Pearlite in Matrix	Nodule Count (#/mm^2)
1A	3.60	1.55	1.50	0.12	0.01	0.02	<2%	60
2A	3.60	1.85	0.38	0.35	0.01	0.02	<2%	60
3A	3.60	2.50	0.38	0.35	0.01	0.02	<2%	60
5A	3.60	2.99	0.38	0.35	0.01	0.02	<2%	60
3A	3.60	2.50	0.38	0.38	0.01	0.02	<2%	60
3C	3.60	2.50	0.38	0.35	0.01	0.02	10%	60
4A	3.60	2.55	0.38	0.35	0.01	0.02	25%	60
4C	3.60	2.55	0.38	0.35	0.01	0.02	>75%	60
6A	3.60	2.50	0.38	0.35	0.01	0.02	<2%	40
3A	3.60	2.50	0.38	0.35	0.01	0.02	<2%	60
3AA	3.60	2.50	0.30	0.35	0.01	0.02	<2%	85

* The first group of specimen vary the silicon content, the second group vary the percentage of the matrix that is pearlite and the third group vary the nodule count.

Standard unnotched tensile tests were made on all compositions and at all temperatures where compact tension specimens were tested. Notched tensile bars were tested selectively on specimens with substantial differences in (K_{1c}/σ_y) to see if a quantitative correlation could be found between toughness and notched tensile behavior.

Examination of the fractured surfaces of all compact tension specimens as well as the notched and unnotched tensile bars was made using a stero-microscope to 75X. Further fractographic examination was made on all compact tension specimens using a scanning electron microscope.

EXPERIMENTAL RESULTS AND DISCUSSION

The results of the fracture toughness tests and the unnotched tensile tests are summarized in Table 2. All results are for a temperature of 297K except where otherwise noted. K_{1c} values were calculated from J_{1c} results in the usual way.

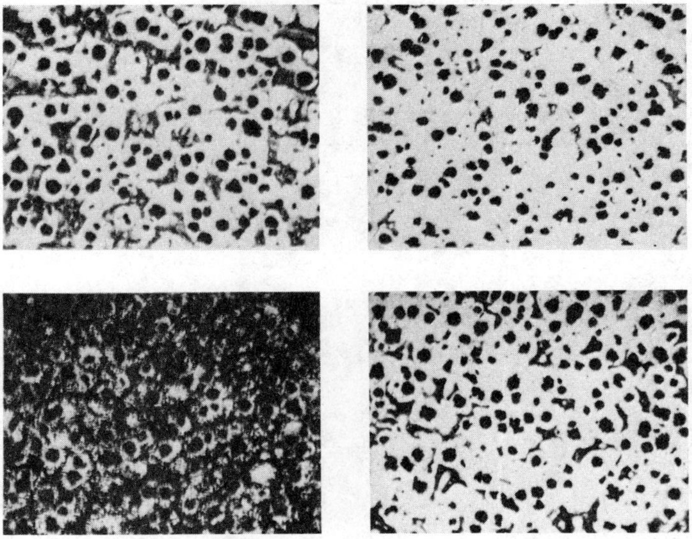

Fig. 1. Microstructure of nodular cast iron containing various amounts of pearlite (see Table 1).

(Landes and Begley, 1974). The 25.4mm (1.0") thick compact tension specimens used in this investigation easily satisfy the minimum thickness requirements for valid J_{1c} (Landes and Begley, 1974); namely,

$$b, B > 50 \frac{J}{\sigma_y} \qquad (1)$$

For our material b, B must be greater than 2.9 mm (0.125") maximum for our results. To satisfy the minimum thickness requirements for ASTM-E399-74, specimens as thick as 63 mm would be required. Since earlier work has been done principally on 25.4 mm or thinner compact tension specimens, the results have only been K_Q values.

A careful examination of the data in Table 2 clearly shows the inadequacy of K_Q for characterizing toughness. Specimens 4C at 297K temperature and 5A at 222K both failed in a quite brittle fashion; yet, comparison of their K_Q values with other tests in their group gave no indication of substandard behavior.

Furthermore, where quite ductile behavior is observed, as is the case with most of our tests, K_Q greatly underestimates actual crack extension loads and K_{1c}. The initial nonlinearity in the load displacement curve is associated with plastic deformation initiating at the crack tip, quite apart from any crack extension. Finally, some authors in earlier studies of the fracture toughness of nodular cast iron have calculated K_{max} values from P_{max} results and associated these numbers with crack extension (for example, Nanstad, Worzala and Loper, Jr.; 1974). This approach has two problems. First, for any significant crack tip plasticity the relationships used to convert P_{max} to a critical stress intensity for crack extension is no longer valid. Furthermore, in all of our tests crack extension was

TABLE 2 Toughness and Tensile Properties of Various Nodular Cast Irons

Plate	K_Q (MPa$\sqrt{m}$)	K_{1c} (MPa$\sqrt{m}$)	σ_y (MPa)	σ_{UTS} (MPa)	El (%)	$\frac{P_{max}}{P_Q}$	$(K_{1c}/\sigma_y)^2$ (10^{-2}m)
1A (1.55% Si)*	24.9	42.9	269	421	22	2.2	2.54
2A (1.85% Si)	23.4	40.1	283	434	18	2.2	2.00
3A (2.50% Si)	26.5	40.7	331	448	19	2.3	1.51
5A (2.99% Si)	27.2	49.5	352	469	24	2.4	1.98
3A (<2% pearlite)	26.5	40.7	331	448	19	2.3	1.51
3C (10% pearlite)	27.4	41.8	345	483	21	2.4	1.46
4A (25% pearlite)	29.0	47.3	386	524	12	1.9	1.50
4C (>75% pearlite)	28.2	29.7	483	745	7	2.0	0.38
6A (40 nodules/mm^2)	26.3	41.3	317	448	19	2.3	1.70
3A (60 nodules/mm^2)	26.5	40.7	331	448	19	2.3	1.51
3AA (85 nodules/mm^2)	26.2	39.0	324	448	26	2.4	1.45
5A (2.99% Si)							
222K	40.4	40.4	409	531	25	1.1	0.98
260K	34.1	61.0	379	503	25	2.0	2.59
297K	27.2	49.5	352	469	24	2.4	1.98
1A (1.55% Si)							
222K	28.4	48.4	310	469	26	2.0	2.43
297K	24.9	42.9	269	421	22	2.2	2.54

* All results are for a temperature of 297K unless otherwise noted.

initiated at loads well below the load maximum; for example, specimen 1A had a P_Q of 1.0×10^4 N and a P_{max} of 2.18×10^4 N, with crack extension as determined by compliance measurements beginning at 1.69×10^4 N. If one calculates K_{max} values from the data in Table 2 by multiplying $P_{max}/P_Q \times K_Q$, the K_{max} values are always greater than the actual K_{1c} except where the fracture was quite brittle, in which case K_{max} was approximately equal to K_{1c}. In summary, K_Q and K_{max} may be considered upper and lower bounds for actual K_{1c} values in nodular cast iron. Because of the considerable ductility of nodular cast iron, K_{max} and K_Q may be quite different, giving considerable ambiguity concerning the actual value of K_{1c}.

Microstructural Effects on Fracture Toughness

The effect of nodule count on the fracture toughness as measured by K_{1c} is seen to be small; however, the range of nodule counts studied was relatively small. It seems safe to conclude that nodule count does not greatly affect fracture toughness as long as the volume fraction remains constant. Higher volume fractions of nodules would be expected to reduce the toughness.

The absence of a strong effect of silicon content on fracture toughness at room temperature was somewhat surprising. Earlier studies by the author (Bradley, 1978) had indicated some effect of silicon on room temperature fracture toughness. In this work the effect of silicon seemed to be to increase the yield strength and the critical stress intensity (K_{1c}) in a similar manner so that the plastic zone size at the crack tip was not much effected. The higher flow stress imparted to the iron by the solid solution strengthening of the silicon will cause the transition from void coalescence to cleavage to occur at a higher temperature. However, apart from a shift in this transition temperature, the silicon does not appear to adversely affect the toughness. The results for the high silicon nodular cast iron show a sharp break between 260K and 222K (see $[K_{1c}/\sigma_y]^2$ in Table 2) in the toughness. The fractured surface changes from 90% dimpled to about 90% cleavage over this same temperature range. By contrast the low silicon nodular cast iron retained its toughness at 222K, and also had a dimpled fractured surface.

The expected deletrious role of pearlite was observed, though smaller amounts of pearlite (25% or less) did not seem to create any problem. Much of the fractured surface for the 75% pearlite showed cleavage, even at room temperature, confirming the macroscopic measurement of poor toughness. It is interesting to note that the 297K value of $(K_{1c}/\sigma_y)^2$ for nodular cast iron with 75% pearlite was much lower than the 2.99% silicon, ferritic nodular cast iron at 222K. Thus, good room temperature toughness seems to require a low amount of pearlite while good low temperature toughness seems to also require low silicon.

Notched Tensile Bar Results

Notched tensile bars were tested for specimens 3A and 4C at room temperature and 1A and 5A at 222K. In general the notched tensile bars showed a greater percentage of cleavage in fracture than did the compact tension specimens of the same material. Specimen 3A ($<$2% pearlite) had 30% cleavage while specimen 4C had 100% cleavage. The low temperature tests of 1A (1.5% Si) and 5A (2.99% Si) gave 75% and 100% cleavage respectively. The ratios of notched to unnotched ultimate tensile strengths for all specimens except 4C were in the range of 1.2-1.3. 4C had a ratio of 0.92. No pattern of correlation between ratios of tensile strengths, notched to unnotched, or percentage of cleavage on the fractured surface emerged. In fact the old "rule of thumb" that a ratio of notched to unnotched tensile strength of greater than one indicates ductile behavior only identified one of the two materials in a condition where brittle fracture did indeed occur in the compact tension tests. Thus, the approach does not appear to be promising.

Fractographic Results

The macroscopic indications of poor toughness in some specimen were substantiated by cleavage facets visable to the naked eye and easily seen in the SEM. Figs. 2 and 3 show fractographs of compact tension specimens 3A and 4C, the high percentage pearlite matrix specimen showing considerable cleavage. Where nodule spacing was non-uniform, cleavage was more likely to be observed, as seen in Fig. 4. The presence of nodules either locally prevents development of restraint sufficient to give plane strain or else provide prior effective "voids" which facilitate void coalescence, and may inhibit cleavage to some degree. The crack growth path during fatigue precracking seems to ignore the graphite nodules. By contrast, the much larger crack opening displacement that occurs during crack extension in a K_{1c} determination gives ductile rupture on a large scale, where the graphite nodules are all void nucleation sites, as seen in Fig. 5.

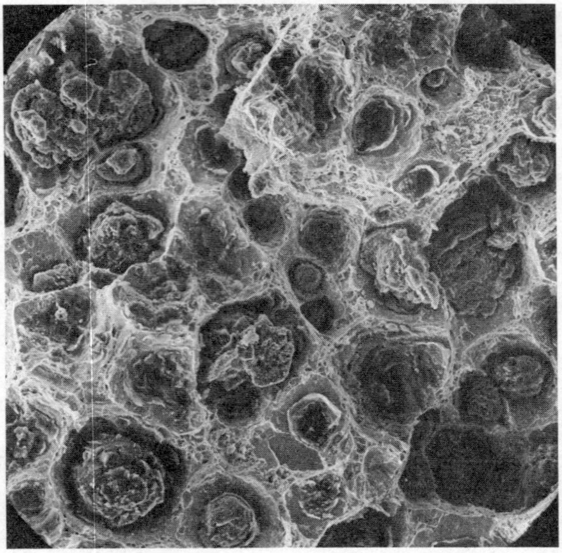

Fig. 2. Fracture surface of a compact tension specimen containing less than 2% pearlite in the matrix, tested at 297K.

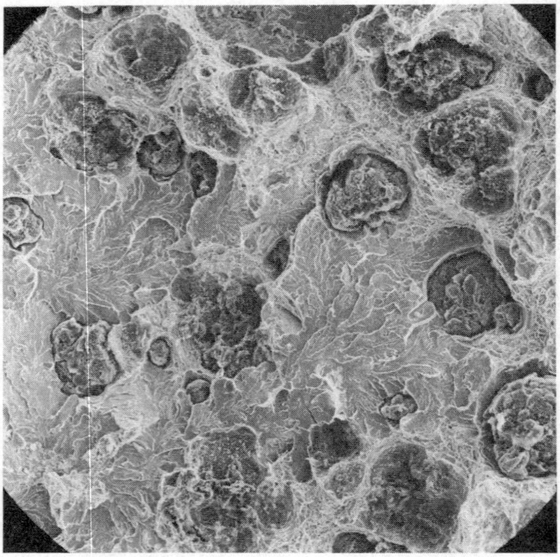

Fig. 3. Fracture surface of a compact tension specimen containing greater than 75% pearlite in the matrix, tested at 297K.

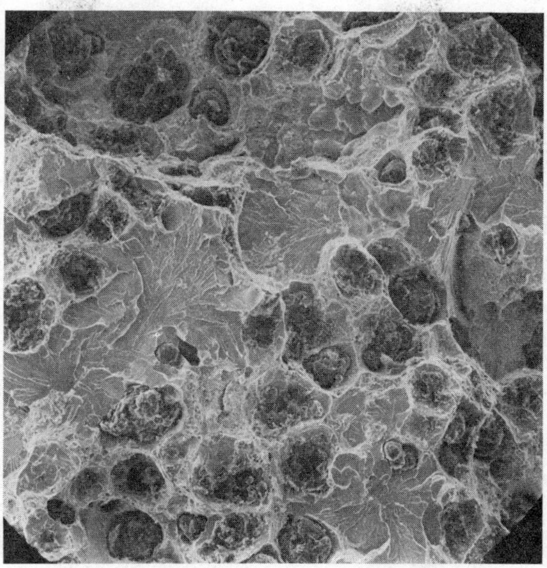

Fig. 4. Fracture surface of a compact tension specimen containing 2.5% Si and 25% pearlite, tested at 297K (specimen 4A).

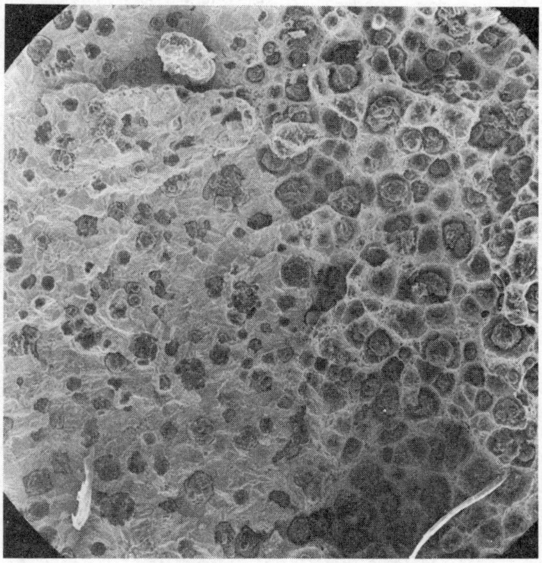

Fig. 5. Fracture surface of compact tension specimen 5A, tested at 297K and viewed at the boundary between the fatigue precracked zone and the crack extension during subsequent testing.

CONCLUSIONS

1. Ambient temperature fracture toughness is best described using K_{1c} values rather than the often used values of K_Q or K_{max}.
2. Graphite nodule count for a fixed volume fraction of nodules has little influence on K_{1c}.
3. Silicon content appears to have a greater effect on the temperature at which the fracture mode changes from void coalescence to cleavage than on the K_{1c}/σ_y ratio for void coalescence type fracture.
4. Pearlite in large amounts is clearly deleterious to fracture toughness.
5. Fatigue crack growth ignores the graphite nodules while ductile fracture occurs almost exclusively around them.
6. Regions without a uniform nodule distribution appear to be more susceptible to cleavage.

ACKNOWLEDGEMENTS

The authors are pleased to acknowlegde the support of this work by the Iron Casting Society. The H.C. Macaulay Foundry Company is also gratefully acknowledged for supplying the castings from which the test coupons were taken for this study.

REFERENCES

Bradley, W.L. (1978). Fracture toughness of nodular cast iron. Oral research report presented to the Iron Casting Society.

Holdsworth, S.R., and Jolley, G. (1974). The influence of graphite nodule number and volume fraction on the fracture toughness of ferritic nodular cast iron. Proceedings of the Second International Symposium on Metallurgy of Cast Iron. Geneva, Switzerland.

Landes, J.D., and Begley, J.A. (1974). Fracture toughness measurements using the J-integral approach. ASTM, STP 560, pp. 170-185.

Little, A., and Heine, H.J. (1974). The fracture toughness of malleable and ductile iron. Proceedings of the Second International Symposium on Metallurgy of Cast Iron. Geneva, Switzerland, pp. 767-788.

Hanstad, R.K., Worzala, F.J., and Loper, C.R., Jr. (1974). The fracture characteristics of nodular cast iron. Proceedings of the Second International Symposium on Metallurgy of Cast Iron. Geneva, Switzerland, pp. 789-807.

Volume 3, ICM 3, Cambridge, England, August 1979

THE INFLUENCE OF THE SURFACE ON THE ULTIMATE STRENGTH OF CERAMIC MATERIALS

K. Heckel and H. Heigl

University of the Federal Armed Forces Munich, 39 Werner Heisenberg Weg, 8014 Neubiberg, F.R.G.

ABSTRACT

The experiments reported below were performed with the aim to show the great influence of the microgeometry of surface on the ultimate strength of unnotched round bar specimens. The specimens used for the tests were made from reaction-bonded silicon nitride by grinding. Four series of specimens were tested to examine the theoretical considerations. The series differ in the length of the specimen (size-effect), in the grinding direction and the grain size of the grinding wheel (machining effect).

INTRODUCTION

It is a well-known fact that strength values of brittle ceramic materials determined by test specimens cannot be used for dimensioning of structural parts. This is caused by a very clear size effect shown by these materials. The following paper deals with that part of the size effect, which is caused by the flaws of the material. The different effect of a flaw being inside or on the surface of a part is considered and treated with statistical and fracture mechanics methods. The theoretical results are compared with experiments performed with specimens made from reaction-bonded silicon nitride.

THEORETICAL CONSIDERATION

To simplify the problem, a round bar specimen loaded in tension with a uniform stress σ_n is considered. In brittle materials the strength is determined by the largest flaw in the volume. In a great number of materials the distribution of the largest flaws in a volume element V_0 can be described by

$$F_{V_0}(a) = \exp\left[-\left(\frac{a}{a_1}\right)^{-\frac{m}{2}}\right] \tag{1}$$

where a_1 and m are material constants /1/. The function $F(a=2a_c)$ for an internal flaw, or $F(a=a_c)$ for a surface flaw, can be regarded as the probability that the largest flaw in a volume element is smaller than a critical one, which causes fracture. The quantity "a" is now identified as a crack length with the same efficiency as the flaw. Therefore, fracture will occur when the crack length becomes $2a_c$ or a_c respectively. This crack length is determined by

$$K_{IC} = \sigma_n \sqrt{a_c}\, Y \tag{2}$$

where K_{IC} is the fracture toughness of the material and Y is the function, which considers the geometry and position of the crack. Inserting a_c from eq (1) finally indicates the probability (for an internal crack) that fracture will occur in a volume element V_0 which is loaded with the stress σ_n

$$P_{f,V_0} = 1 - \exp\left[-\left(\frac{\sigma_n Y \sqrt{a_1/2}}{K_{IC}}\right)^m\right] = 1 - \exp\left[-\left(\frac{\sigma_n}{\sigma_0}\right)^m\right] \tag{3}$$

where $\sigma_0 = K_{IC}/(Y\sqrt{a_1/2})$

For a volume V from eq (3) follows:

$$P_f = 1 - \exp\left[-\frac{1}{V_0}\int_V \left(\frac{\sigma(x,y,z)}{\sigma_0}\right)^m dV\right] \tag{4}$$

where σ is a function of the coordinates.

To apply eq (4) to specimens, one has to consider that the distribution of the crack length on the surface is different from the distribution of internal cracks (Fig. 1).

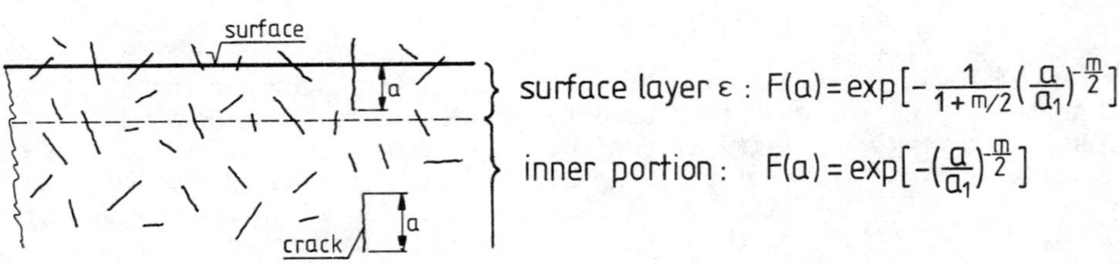

Fig. 1. Subdivision of the distribution of the largest flaws

In order to consider these facts, the cylindrical rod can be divided into a surface layer (suffix s) of the thickness ε and the inner portion (suffix i). On the whole specimen a_1, K_{IC} and m are regarded to be constant. From eq (4) follows

$$P_f = 1 - \exp\left[-\frac{1}{V_0}\left(\int_{V_i}\left(\frac{\sigma}{K_{IC}} Y_i \sqrt{a_1/2}\right)^m dV + \int_{V_s}\left(\frac{\sigma}{K_{IC}} Y_s f(m)\sqrt{a_1}\right)^m dV\right)\right] \tag{5}$$

where $f(m) = (1+m/2)^{-1/m}$ considers the different distribution of the surface cracks. The calculation of the fracture probability of the whole specimen loaded in tension finally results in the following equation:

$$P_f = 1 - \exp[-\frac{V}{V_0}(\frac{\sigma_n}{\sigma_{0,i}})^m[(1-2\varepsilon/R) + \lambda^{-m} 2\varepsilon/R]] \tag{6}$$
$$= 1 - \exp[-(B_i + B_s)]$$

where V = volume of the specimen
R = radius of the specimen
m = Weibull exponent
$\sigma_{0,i} = K_{IC}/(Y_i\sqrt{a_1/2})$
$\lambda = \sigma_{0,s}/\sigma_{0,i} = Y_i(1+m/2)^{1/m}/(Y_s \cdot \sqrt{2})$
$B_{i,s}$ = risk of fracture
σ_n = applied stress

To estimate Y which is defined by eq (2), the following equation, given by Irwin /2/ (1964), was used:

$$K_{IC} = \sigma_n\sqrt{a_c}\sqrt{\frac{z\pi}{\Phi^2 - 0,212(\sigma/\sigma_s)^2}} = \sigma_n\sqrt{a_c}\sqrt{\frac{z\pi}{Q}} \tag{7}$$

where z = 1 for an elliptical crack inside an infinite volume
z = 1.21 for a semi-elliptical surface crack
Q = flaw shape parameter.

Comparison of eq (2) and (7) results in

$$Y_i = \sqrt{\frac{z\pi}{Q}} \tag{8}$$

The cracks inside the inner portion of the specimen can be regarded as penny-shaped. This is justified by the fact that the pores of the material are nearly spherical. The flaw shape parameter therefore becomes Q = 2.7 (ref. 3) and

$$Y_i = 1.08 \tag{9}$$

The most severe surface cracks caused by the pores of the material are semicircular cracks ("initial cracks" Fig. 2). This shape is not influenced by grinding if the grinding direction is parallel to the axis of the specimen and the depth of the ghost line caused by grinding is less than "a" (Fig. 2). According to Q = 2.7 and z = 1.21 one obtains for a polished or a parallel ground surface

$$Y_{s,p} = 1.19 \tag{10}$$

If the grinding direction is normal to the axis of the specimen, the length 2a of the surface crack is enlarged to 2c. This crack can now be treated as a semi-elliptical crack with a depth of "a" and a length of 2c (Fig. 2). The flaw shape parameter becomes Q = 0.9, z = 1.21, and therefore

$$Y_{s,n} = 2.06 \tag{11}$$

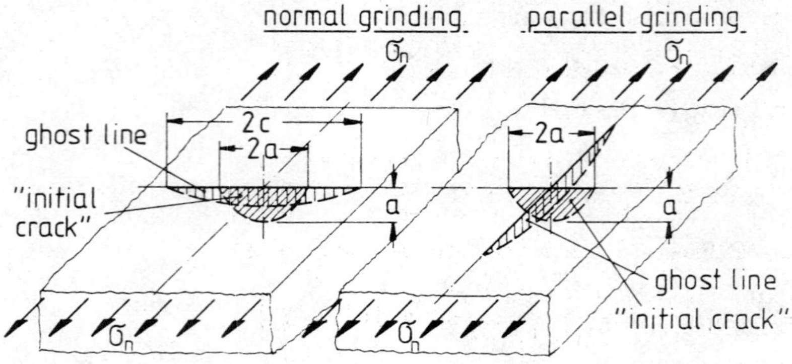

Fig. 2. Graph of the superposition of a flaw with a ghost line (schematic).

According to eq (6) one obtains for an ideal polished or parallel ground surface

$$\lambda_{parallel} \approx 0.75 \tag{12}$$

and for a normal ground surface

$$\lambda_{normal} \approx 0.43 \tag{13}$$

respectively

$$\frac{\lambda_{normal}}{\lambda_{parallel}} = \frac{\sigma_{0,s,n}}{\sigma_{0,i}} : \frac{\sigma_{0,s,p}}{\sigma_{0,i}} = 0,58 \tag{14}$$

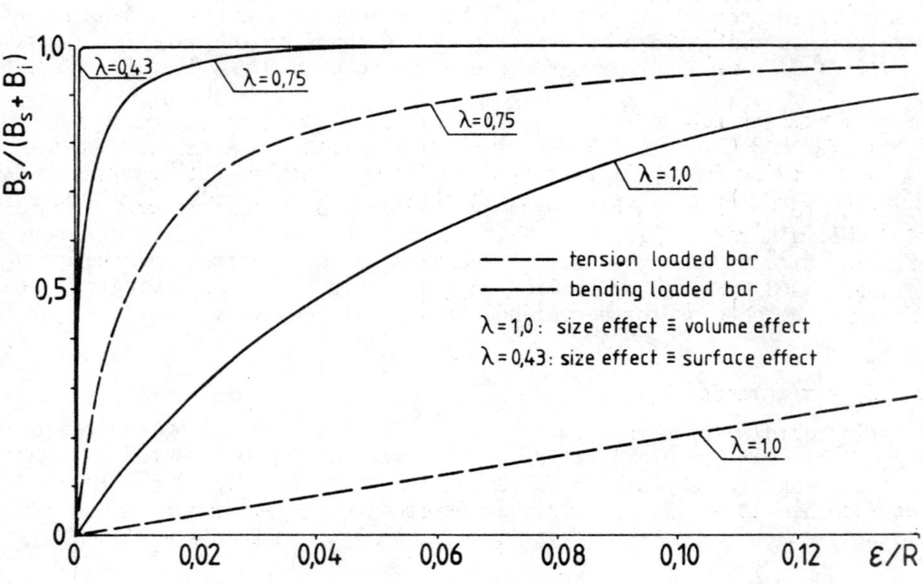

Fig. 3. Influence of the surface layer on the risk of fracture (for m = 14).

It must be mentioned that the estimation of the above shown values is based on the assumption that the cracks are in an extremely unfavorable position. The influence of λ on the risk of fracture $B_s/(B_s+B_i)$ in dependency of ε/R is shown in Fig. 3, where R is the radius of the round bar specimens.

EXPERIMENTS

The material used for the tests was reaction-bonded silicon nitride. A great number of round bar specimens were fabricated by grinding. The specimens were examined under fixed conditions by 4-point-bending tests.

The probability of fracture can be calculated from eq (5).

$$P_f = 1 - \exp\left[-\frac{R^2 L}{V_0}\,\frac{1+mL_0/L}{(m+1)(m+2)}\int_0^{\pi}\sin^m\varphi\, d\varphi\,(\sigma_b/\sigma_{0,i})^m\cdot\right. \qquad (15)$$

$$\left.\cdot\left[(1-\varepsilon/R)^{m+2} + \lambda^{-m}(1-(1-\varepsilon/R)^{m+2})\right]\right]$$

where: R = Radius of the round bar specimens
L, L_o = bearing distances (see Fig. 4.)
σ_b = max. bending stress.

From eq (15) it can be seen, that the surface layer has much more influence on the probability of fracture by bending than by tension (eq (6)). The influence of the surface layer on the total risk of fracture is shown in Fig. 3.

To estimate the thickness ε of the surface layer, it is assumed that ε is slightly greater than a_1. From eq (3) therefore follows for the test material used:

$$\varepsilon > a_1 = 2\,(K_{IC}/(Y\sigma_0))^2 = 0.09 \text{ mm}$$

where: K_{IC} = 79 $Nmm^{-3/2}$ (ref. 4.).
Y = 1.20 (assumption)
σ_0 = 310 Nmm^{-2} (Table 1)

From Fig. 3. it can now be seen that for parallel ground specimens (λ = 0.75) with R = 2.5mm (ε/R = 0.036) the relative risk of fracture is about 0.98 for specimens tested by a bending load. It therefore suffices to extend the integration according to eq (4) only to the surface layer. It therefore follows with $dV = \varepsilon\, dS$ and $\sigma(x,y,z) = \sigma_b \cdot g(x,y,z)$.

$$P_f = 1 - \exp\left[-\frac{1}{S_0}\left(\frac{\sigma_b}{\sigma_{0S}}\right)^m \int_S g(x,y,z)^m\, dS\right] \qquad (16)$$

where: S_o = reference area
σ_{0S} = strength parameter

From eq (16) the strength ratio for specimens of different sizes can be determined. At constant probability of fracture P_f = constant it follows (ref.5):

$$\frac{\sigma_{b,1}}{\sigma_{b,2}} = \left[\frac{\int_{S,2} g_2(x,y,z)^m dS}{\int_{S,1} g_1(x,y,z)^m dS} \right]^{\frac{1}{m}} \tag{17}$$

This relationship was used to calculate the bending strength of specimens with a length of 120 mm taking as a basis the specimens with a length of 60 mm (see Fig. 4.). The bending strengths determined by the shorter specimens (32 tests) ground parallel to the axis are denoted in Fig. 4. by o . "m" was determined by curve fitting to these test results. The graph "A" calculated by eq(17)shows very good agreement with the test results determined by the longer specimens (10 tests).

To examine the influence of the depth of the ghost line, 20 specimens were machined by a grinding wheel with relatively fine grain. The results are denoted in Fig. 4. by ● . It can be seen that this procedure has no evident influence on the test results.

To improve eq (14), 24 specimens of the shorter length were machined by grinding normal to the axis. The test results are denoted in Fig. 4. by □ . Table 1 contains the material constants which were calculated from the test results by curve fitting. From these values follows

$$\frac{\sigma_{0S,n}}{\sigma_{0S,p}} = \frac{213.9}{311.0} = 0.69$$

in satisfactory agreement with the theoretical value of 0.58.

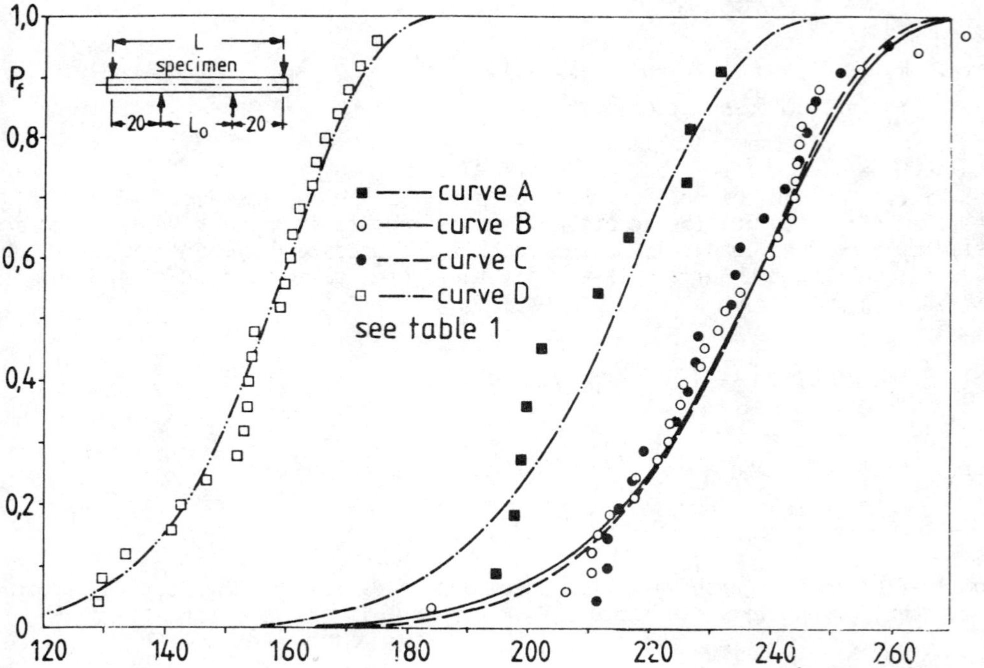

Fig. 4. Test results

Table 1 Test results

Curve	Grinding* direction	Grain size (Fa.E.Winter&S.)	Specimen dimensions mm	m	σ_{OS} N/mm^2	S_{eff} mm^2
A ■	parallel	D 91	Ø 5 x 120	↑	↑	136.0
B ○	parallel	D 91	Ø 5 x 60	14.0	311.0	37.3
C ●	parallel	D 15 C	Ø 5 x 60	14.9	305.6	36.1
D □	normal	D 91	Ø 5 x 60	12.9	213.9	39.1

*Relative to the axis of the specimen

CONCLUSION

In the bending tests described above, the risk of fracture is mainly determined by the surface of the test specimen. Great influence on the bending strength is caused by the grinding direction. In common cases (small stress gradient) the risk of fracture caused by flaws inside the volume of a material must also be taken into account.

The following equation gives the possibility to consider volume as well as surface effects caused by statistically distributed flaws in brittle materials:

$$P_f = 1 - \exp\left[-\frac{1}{V_o}\sigma_{0,i}^{-m}\left(\int_V \sigma(x,y,z)^m dV + \varepsilon^* \int_S \sigma(x,y,z)^m dS\right]\right. \tag{18}$$

where $\varepsilon^* = \varepsilon\,(\lambda^{-m} - 1)$.

REFERENCES

/1/ Frechet, M. (1923). Ann. Soc. Polon. Mat. Cracow, 6, 93
/2/ Irwin, G. R. (1962). Journ.Appl.Mech. 84E, 4, p 651
Mat.Res.& Stds. 4 (1964), p 107
/3/ Tiffany, C.F., I. N. Masters (1965). ASTM-STP 381, p. 249
/4/ Maier, H. R. Technical information by Fa. Rosenthal, Selb
/5/ Ziebart, W., K. Heckel. VDI-Z 120 (1978), No. 14 - July (II)

Volume 3, ICM 3, Cambridge, England, August 1979

THE EFFECT OF COMPRESSIVE STRENGTH ON THE MECHANICAL PERFORMANCE OF STRONG CERAMICS

J. Lankford and D. L. Davidson

Department of Materials Sciences, Southwest Research Institute, San Antonio, Texas, U.S.A.

INTRODUCTION

Most studies of ceramic strength have emphasized tensile modes of loading, since these materials are especially susceptible to failure under tension. However, certain aspects of mechanical behavior, including even tensile failure, can be controlled or at least influenced significantly by compressive strength or compression-induced damage. In this paper, recent experiments aimed at determining the compressive strength and damage behavior of two strong ceramics are summarized. The results are then considered in terms of current knowledge of lifetime prediction under tensile loading, for material which has suffered prior compressive loads (such as might occur in preloading of ceramic turbine components) , and also in terms of current models of indentation/particle impact cracking, as a function of temperature and strain rate (i.e., loading rate).

EXPERIMENTAL PROCEDURES

The compression test setup and procedures have been described in detail elsewhere (Lankford, 1977), and are described here only briefly. Cylindrical specimens of Lucalox[1] Al_2O_3 and sintered α-SiC[2] were compressed at strain rates ranging from 7×10^{-6} to $2\times10^{3}sec^{-1}$, the latter achieved by means of the Hopkinson pressure bar; temperatures varied from -200°C to 1600°C. Acoustic emission (AE) was monitored within the frequency domain 100 kHz to 1 MHz, using a transducer resonant at 160 kHz. Scanning electron microscopy was used to characterize damage induced by loading specimens above the damage threshold established through AE (σ_{AE}).

RESULTS

Compressive strength of Al_2O_3 as a function of temperature is shown in Fig. 1 for a strain rate of $7 \times 10^{-5}sec^{-1}$; also shown is σ_{AE}. The strength generally decreases with temperature, a trend modified by a peak at 350°C, which has previously

---

[1] Type GW, General Electric Lamp Glass Division, Cleveland, Ohio

[2] Carborundum Company, Research and Development Division, Niagara Falls, NY.

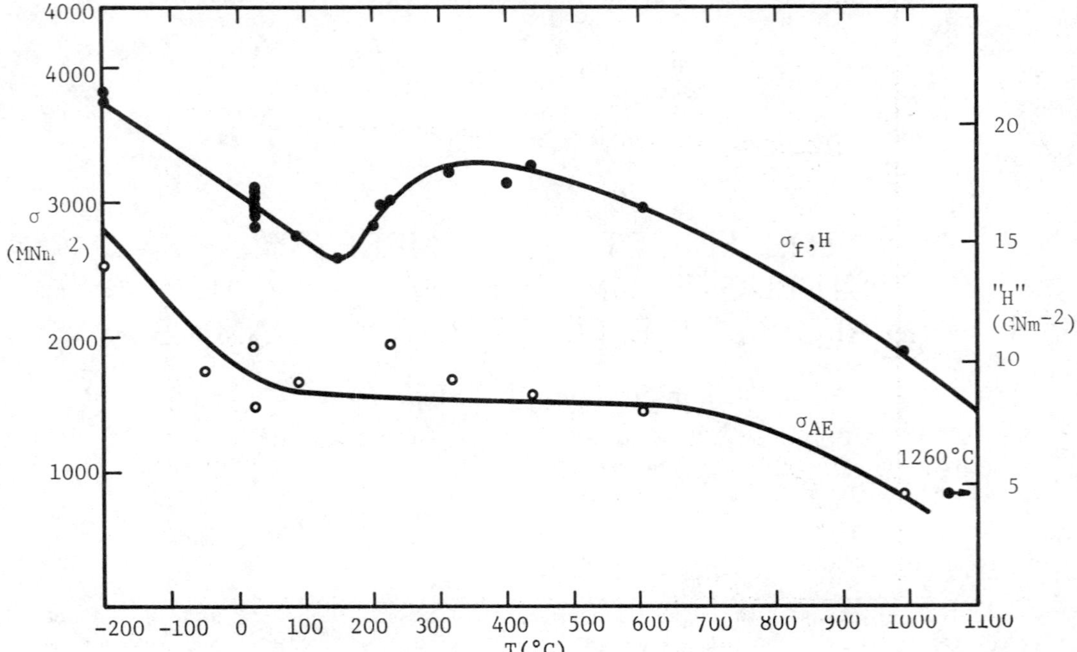

Fig. 1. Compressive strength, acoustic emission stress level, and hardness, versus temperature for Al_2O_3.

(Lankford, 1978) been identified with twinning effects. Damage commences at stress levels on the order of $0.5\sigma_f$, consisting of twinning and more importantly, twin-nucleated cracking, as shown in Fig. 2a. The size of the crack shown, approximately 55 μm, is typical of the early stages of damage, i.e., for stresses just above σ_{AE}.

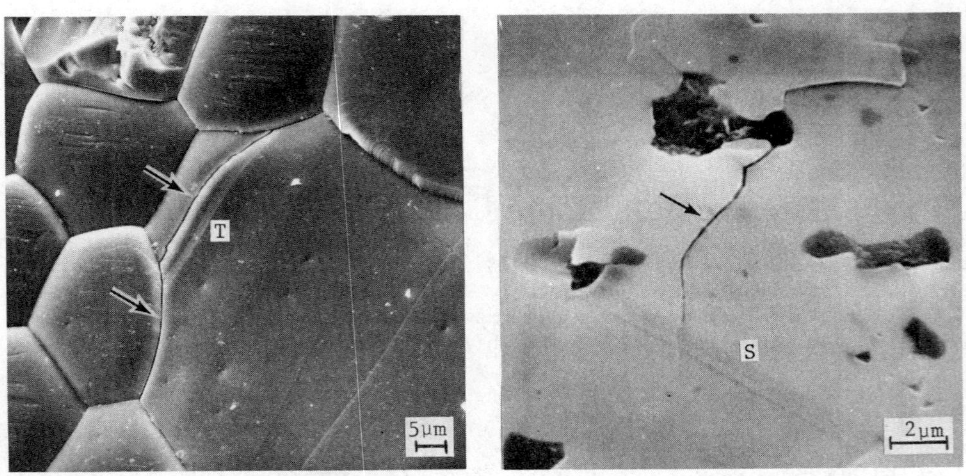

(a) Al_2O_3, T = 1260°C, twin T, combined transgranular-intergranular crack (arrows).

(b) α-SiC, T = 500°C, slip band S, pore-nucleated crack (arrow).

Fig. 2. Compressive damage and microfracture above σ_{AE}.

For α-SiC, on the other hand, strength at the same $\dot{\varepsilon}$ increases monotonically with temperature as shown in Fig. 3, while σ_{AE} peaks at room temperature. For low

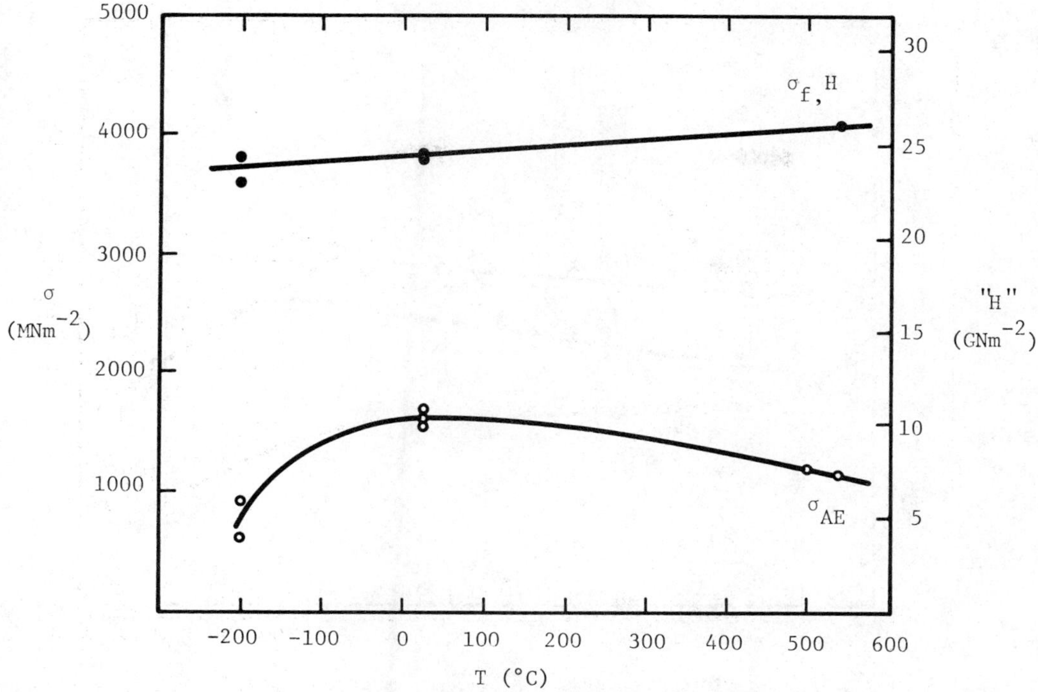

Fig. 3. Compressive strength acoustic emission stress level, and hardness versus temperature for α-SiC.

temperatures (hence high strain rates), σ_{AE} is very low; in fact, it barely exceeds the tensile strength (Kraft and Dooher, 1976). Damage typical of this level is shown in Fig. 2b, evidenced by a ~20μm crack.

As a function of strain rate, the room temperature compressive strengths for both Al_2O_3 and SiC increase monotonically (Fig. 4). However, near the upper end of the strain rate range, the strength for SiC undergoes a dramatic increase, nearly doubling within less than two decades in $\dot{\varepsilon}$.

Plotted in Figs 1,3,4 as co-functional with σ_f is an empirically established hardness "H", which is scaled based on room temperature, quasi-static hardness measurements. The assumed proportionality between "H" and σ_f over a range in thermally activated states (T, $\dot{\varepsilon}$) is based on empirical evidence, as summarized by Rice (1970), that the yield stress (microhardness/3) is the upper limit of both ambient and elevated temperature compressive strengths. It should be noted that at 23°C, the hardnesses of the Al_2O_3 and SiC are 17 and 24 GNm^{-2}, respectively, in proportional agreement with their corresponding slow strain rate compressive strengths of 3000 and 3800 MNm^{-2} ($\simeq$ H/6). Hardness tests over the T-$\dot{\varepsilon}$ range of interest would be extremely difficult to perform, with the adopted procedure, i.e., assuming "$H(T,\dot{\varepsilon})$" $\propto \sigma_f(T,\dot{\varepsilon})$ representing a current best approximation to the actual $H(T,\dot{\varepsilon})$ (note that in Fig. 4, "H" was normalized to σ_f using the Al_2O_3 values).

DISCUSSION

Tensile Lifetimes Following Compressive Loading

It is clear that compressive loads well below the compressive failure level may

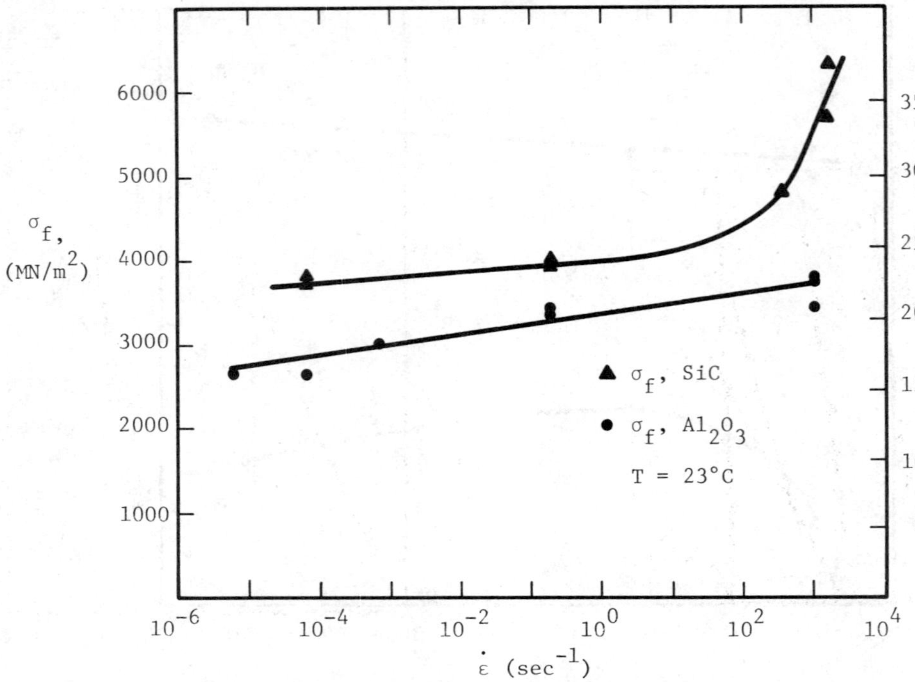

Fig. 4. Compressive strength and hardness versus strain rate for Al_2O_3 and SiC at T = 23°C.

nevertheless introduce damage. The significance of the damage during subsequent static tensile loading depends on the size of the flaws, their orientation relative to the tensile field, temperature, loading rate, and environment. It is instructive to consider a few cases based on the preceding results.

Based on a typical compressive crack, in Lucalox, of length 55 μm, one can calculate the stress intensity in a subsequent tensile field according to

$$K_I = k \sqrt{a}\, \sigma \tag{1}$$

where a conservative value for k is about 1.4. Assuming an operating tensile stress level of, say, 200 MNm^{-2}, then $K_I = 1.6 MNm^{-3/2}$. If one then considers slow crack velocity V versus K_I data for Lucalox (Fig. 5), it is apparent that at lower temperatures, the specified operating conditions are safe for the given set of compression induced flaws. At 1300°C, however, V would be nearly a micron per second. Since K_{IC} at this temperature is only ~2.4$MNm^{-3/2}$, meaning a critical flaw size a_c of 75μm, the specimen would have only seconds to live. It is clear that at high temperatures, cracks of the size introduced during compression would act to reduce permissible operating stress levels in Lucalox; 200 MNm^{-2} is well below the nominal failure stress at 1300°C.

Similar considerations can be seen to hold for SiC, if it can be assumed that V-K_I for hot-pressed SiC (Fig. 6) is equivalent to that for the sintered material. For an initial compressive crack of length 20μm, suppose a tensile operating stress of 350 MNm^{-2} (due to the higher strength in general of SiC). This yields K_I=1.55$MNm^{-3/2}$ which, on the basis of Fig. 6, with T=1400°C, corresponds to V=0.1μm/sec. This would yield at least a finite (but still very short) lifetime, since for α-SiC, K_{IC} at T=1400°C is approximately 7$MNm^{-3/2}$, or a crack length of ~250μm. Again it

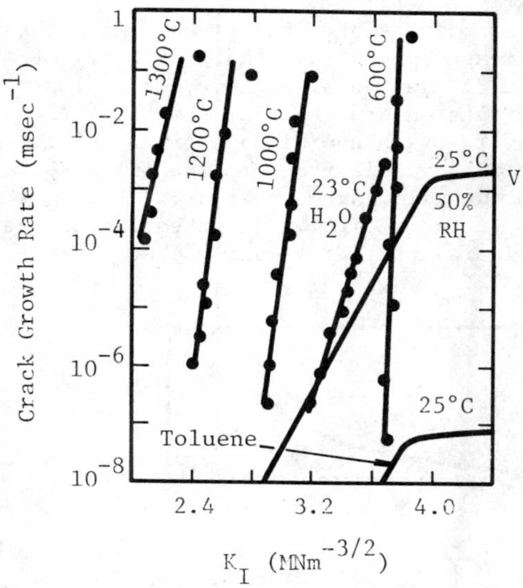

Fig. 5. Slow crack velocity versus stress intensity factor for Lucalox Al_2O_3, various temperatures and environments. Data from Evans (1973), Evans, Linzer, and Russell (1974) and Williams and Evans (1973).

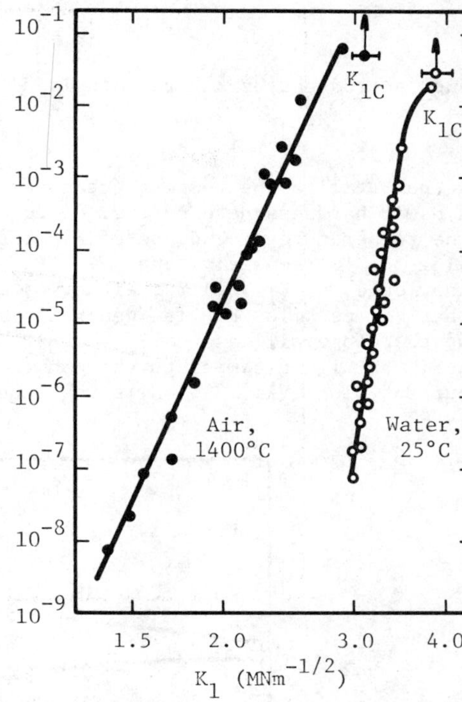

Fig. 6. Slow crack velocity versus stress intensity factor for hot-pressed SiC (Evans and Lange, 1975).

would seem that the tensile operating stress level would be compromised, since 350 MNm^{-2} is only about 60% of the nominal tensile strength at 1400°C.

Effect of Compressive Strength (Hardness) on Indentation Fracture

Particle impacts can cause two principal kinds of damage: (1) erosion, i.e., volume loss, through lateral cracking, and (2) strength degradation, through radial or median crack nucleation. With regard to the former, Evans, Gulden, and Rosenblatt (1978) have shown that the amount of material removed per particle impact $\bar{V}_i$ is given by

$$\bar{V}_i \propto V_o^{2.5} R_p^4 \rho_p^{0.3} H^{-0.3} K_c^{1.5} \qquad (2)$$

where V_o is the particle velocity, R_p the particle radius, and ρ_p the particle density. In our treatment, we shall hold these factors constant, and consider only the effect of H and K_c.

Similarly, Lawn and Evans (1977) have shown that during indentation, the threshold crack size

$$C^* \propto \left(\frac{K_c}{H}\right)^2 \qquad (3)$$

and is produced by a threshold load

$$P^* \propto \left(\frac{K_c}{H}\right)^3 K_c \qquad (4)$$

We now utilize the temperature dependent hardness results of Figs. 1 and 3, and the dynamic hardness data of Fig. 4, together with measurements (by others, Fig. 7) of the temperature dependence of K_{1C} for sintered α-SiC and Lucalox Al_2O_3, in evaluating the T,$\dot{\varepsilon}$ sensitivity of $\bar{V}_i$, C*, and P*. No data exist, to the authors' knowledge, of K_{1C} ($\dot{\varepsilon}$) for either material; however, Evans and Wilshaw (1977) state that for ceramics K_{1C} is "generally similar" to the dynamic fracture toughness K_D. We therefore will assume $K_{1C} = K_D$. In addition, we will assume that $\sigma_f(T)$ for α-SiC at higher temperatures than those already tested is simply an extension of the current plot (Fig. 3), as is the case for tensile failure (Kraft and Dooher, 1976).

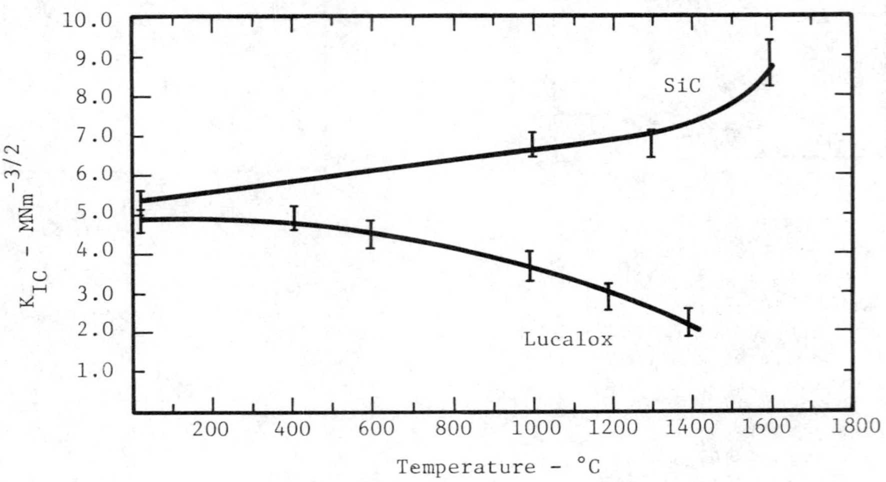

Fig. 7. Fracture toughness versus temperature for α-SiC (Kraft and Dooher, 1976) and Lucalox Al_2O_3 (Evans, Linzer, and Russell, 1974).

These data are used to calculate, for α-SiC and Lucalox, the quantitites $\bar{V}_i/\bar{V}_{io}$, C^*/C^*_o, and P^*/P^*_o as functions of T and $\dot{\varepsilon}$, where $\bar{V}_{io}$, C^*_o, and P^*_o are simply the values of Equations (2), (3), and (4) for T=23°C, $\dot{\varepsilon}=7\times10^{-5}sec^{-1}$. The results are summarized in Figs. 8 to 12.

Temperature has a large effect on C*, P*, and $\bar{V}_i$. In Fig. 8, the large increase in C* is caused by the rapid decrease in H; this is counteracted by the increase, Fig. 9, in P* due to the fact that K_{1C} does not decrease as rapidly as H, and in addition, enters through the fourth power rather than the third, as for H. Although the increases in C* and P* are not as spectacular as those for Al_2O_3, they still are sizeable; P* at 1600°C is four times that at room temperature. It is interesting to note that for Al_2O_3, $\bar{V}_i$ increases by a factor of nearly eight from 23° to 1400°C, (Fig. 10), while over the same range, the removal rate for SiC decreases by nearly 50%.

Increases in strain rate (impact velocity) cause the same effect, a steady decrease, upon C*, P*, and $\bar{V}_i$ in both materials, see Fig. 11 and 12. The threshold crack size decreases with increasing $\dot{\varepsilon}$ most rapidly in Al_2O_3, while the net decrease is greatest for the SiC. Unfortunately, these beneficial effects are counteracted

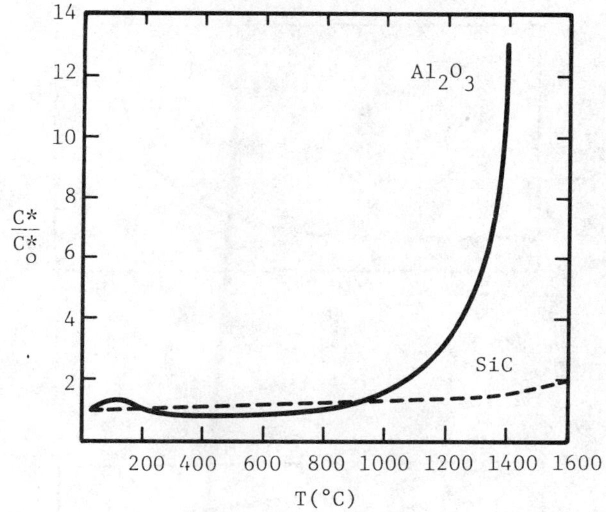

Fig. 8. Threshold crack size versus temperature during quasistatic indentation.

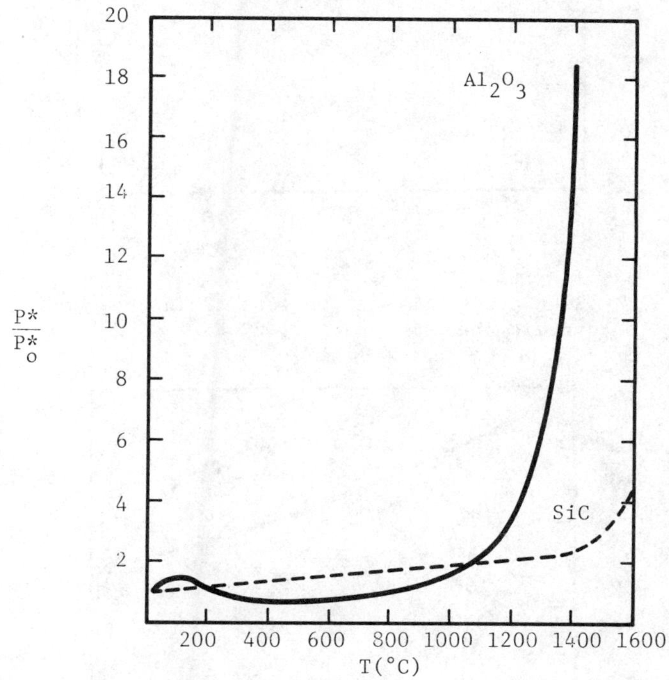

Fig. 9. Threshold load versus temperature during quasistatic indentation.

by corresponding decreases in the minimum load required to nucleate the cracks. No plot is shown for $\bar{V}_i$ versus $\dot{\varepsilon}$, since the dependence was not strong; however, recent unpublished results indicate that $\bar{V}_i$ may have a significant dependence upon strain rate at higher temperature.

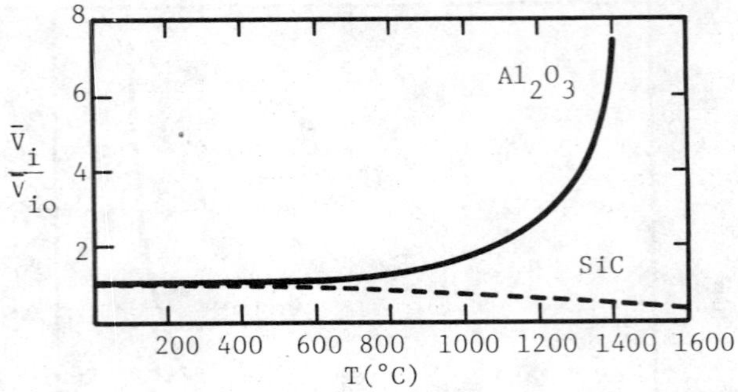

Fig. 10. Material volume removed versus temperature during quasistatic impact.

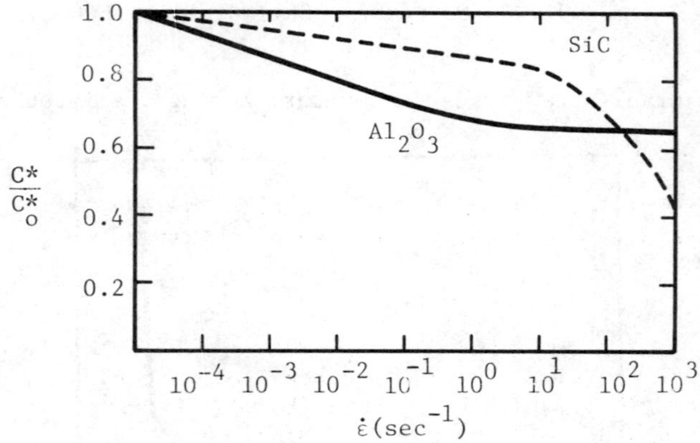

Fig. 11. Threshold crack size versus strain rate during dynamic indentation at 23°C.

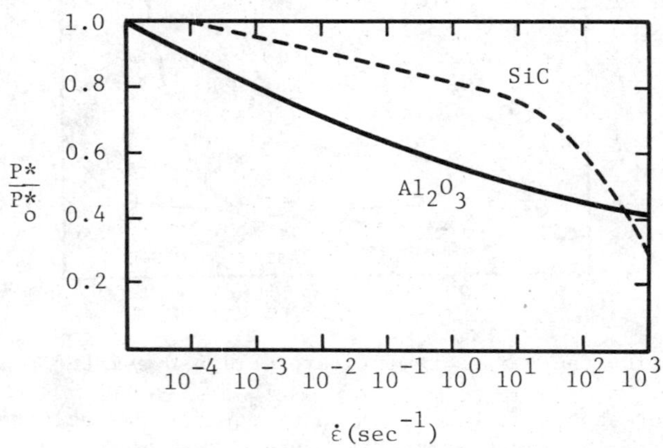

Fig. 12. Threshold load versus strain rate during dynamic indentation at 23°C.

CONCLUSIONS

It is clear from the foregoing results that if the relationship between "H" and σ_f is as assumed, profound alterations may be affected by $\dot{\varepsilon}$ and T in the tensile and indentation/impact characteristics of strong ceramics. Because of the fashion in which they are incorporated into the indentation relationships in particular, T or $\dot{\varepsilon}$-induced hardness changes may exert significantly out-of-proportion influence.

At present, the predicted results must be treated with caution, because of the assumptions involved. Critical experiments remain to be carried out: (1) actual hardness as a function of T and $\dot{\varepsilon}$, especially where both T and $\dot{\varepsilon}$ are high, simulating high velocity impact of a hot target; (2) actual measurement of threshold crack size and load as a function of T and $\dot{\varepsilon}$; (3) K_c as a function of T within grains. Application of the "H" and K_c data used in the present work may suffer from the physical circumstances under which it was obtained, i.e., bulk compression tests, and standard fracture mechanics tests, in which grain boundary effects are important at high temperature. More relevant to the indentation/impact process would be microhardness tests within the interiors of grains, to determine both H and K_c at high temperature.

ACKNOWLEDGEMENTS

The authors are grateful for the support of the U.S. Office of Naval Research, Contract No. N00014-75-C-0668, during the course of this work.

REFERENCES

Evans, A. G. (1973). High-temperature slow crack growth in ceramic materials. In J. J. Burke, A. E. Gorum, and R. N. Katz (Eds.), Ceramics for High-Performance Applications, Brook Hill, Chestnut Hill, Mass. pp. 373-396.
Evans, A. G., Gulden, M. E., and Rosenblatt, M. (1978). Impact damage in brittle materials in the elastic-plastic response regime. Proc. Roy. Soc. London, 361, 343-365.
Evans, A. G., and Lange, F. F. (1975). Crack propagation and fracture in silicon carbide. Jour. Mat. Sci., 10, 1659-1664.
Evans, A. G., Linzer, M., and Russell, L. R. (1974). Acoustic emission and crack propagation in polycrystalline alumina. Mat. Sci. Eng., 15, 253-261.
Evans, A. G., and Wilshaw, T. R. (1977). Dynamic solid particle damage in brittle materials: an appraisal. Jour. Mat. Sci., 12, 97-116.
Kraft, E. H., and Dooher, G. I. (1976). Mechanical response of high performance silicon carbides. Presented at Second International Conference on Mechanical Behavior of Materials, 16-20 August, 1976, Boston, Mass. (unpublished).
Lankford, J. (1977). Compressive strength and microplasticity in polycrystalline alumina. Jour. Mat. Sci., 12, 791-796.
Lankford, J. (1978). Tensile failure of unflawed polycrystalline Al_2O_3. Jour. Mat. Sci., 13, 351-357.
Lawn, B. R., and Evans, A. G. (1977). A model for crack initiation in elastic/plastic indentation fields. Jour. Mat. Sci., 12, 2195-2199.
Rice, R. W. (1970). The compressive strength of ceramics. In W. W. Kriegel and H. Palmour (Eds.), Ceramics in Severe Environments, Vol. 5, Plenum Press, New York, 195-229.
Williams, D. P., and Evans, A. G. (1973). A simple method for studying slow crack growth. Jour. Test. Eval., 1, 264-270.

Volume 3, ICM 3, Cambridge, England, August 1979

PHASE RETENTION AND FRACTURE TOUGHNESS OF MATERIALS CONTAINING TETRAGONAL ZrO_2

F. F. Lange

Rockwell International Science Center, Thousand Oaks, California 91360, U.S.A.

INTRODUCTION

Work on ZrO_2 based materials has shown that a stress-induced, martensitic transformation can increase the fracture toughness of brittle materials (Garvie, Hennick and Pascoe, 1975; Garvie and Pascoe, 1978; Porter and Heuer, 1977; Gupta and co-workers, 1977; Lange, 1978, 1979). Metastable, tetragonal ZrO_2 is the toughening agent. Transformation to its stable monoclinic structure in the vicinity of the crack front is responsible for the increased fracture toughness.

Two questions arise from these observations: 1) How can the tetragonal structure be retained upon cooling from the fabrication temperature when it usually undergoes a diffusionless transformation? 2) How does the stress-induced transformation contribute to the fracture toughness? These two questions have directed inquiries at the Science Center for the past two years. Theoretical and experimental results of this inquiry will be presented herein.

RETENTION OF THE TETRAGONAL PHASE

Retention of the tetragonal structure is the key factor for fabricating a tougher material based on the transformation-toughening concept. In brittle materials, the transformation usually occurs as the polycrystalline body is cooled through its transformation temperature; this produces a highly microcracked body. Stresses that arise from the shape change associated with the transformation cause these microcracks. Retention of the tetragonal phase is viewed here in terms of constraint to this shape change. Constraint can arise from two sources. First, if the polycrystalline body were single phase, neighboring grains, each with a different crystallographic orientation, could constrain the anisotropic shape change of one another. Second, a particle of the material were surrounded by a matrix of another material, the matrix material could constrain the particle's shape change. If the transformation were to take place, stresses that arise from the constraint would alter the transformation thermodynamics from the familiar, unconstrained case. These stresses can also cause microcrack development which must be avoided to maintain the constraint.

The following sections present the thermodynamics of the constraint transformation for predicting the conditions required for phase retention. The energetics of microcracking are also presented to predict the microstructure required to maintain constraint.

Thermodynamics of the Constrained Transformations

The $ZrO_2(t) \rightarrow ZrO_2(m)$ transformation is athermal, and diffusionless, and involves both a shear strain (~8%) and a volume increase (3-5%, depending on temperature),(Sabbarao, Maiti and Srivastava, 1974; Heuer and Nord, 1976). Based on analogous transformations observed in metallic systems, this transformation is termed martensitic. Extensive twinning of the monoclinic phase accommodates much of the shear strain (Bailey, 1964; Bansal and Heuer, 1974: Buljan, McKinistry and Stubican, 1976), whereas intergranular cracking usually results from the volume change.

The thermodynamics of the transformation under the application of an externally applied pressure has been a subject of extensive theoretical and experimental study. Whitney (1962) and Berezhnoi (1962) have calculated dP/dT of pure ZrO_2 to be -3.3×10^{-2}Kbar/°C, and -2.8×10^{-2}kbar/°C, respectively. Experimental work appears to confirm their results. This information (Fig. 1) shows that the tetragonal state can be retained at room temperature by the application of a pressure ⩾3700 MPa.
Oxides that substitute into the ZrO_2 structure lower the ambient pressure (1 bar), transformation temperature. The effect of Y_2O_3 has received the most study (Strivastaba and co-workers, 1974; Scott, 1975; Stubican, Hink and Ray, 1978); the transformation temperature can be lowered to ~560°C for the eutectoid composition containing ~3.0 mol% (m/o) Y_2O_3. If the same dP/dT value is assumed for the ZrO_2/Y_2O_3 materials up to the eutectoid composition, then a pressure of only ~1700 MPa is required for $ZrO_2(t)$ to be retained at room temperature. Thus, unless solid-solution alloying significantly increases dP/dT, the applied pressure required for retaining $ZrO_2(t)$ is lowered.

The temperature-pressure-compositional dependence on the thermodynamics of the transformation can be represented in T-P-X space, as shown in Fig. 2. The boundary surface in this space separates the tetragonal state from the monoclinic state. It represents the condition where the free energy of both states are equal to one another, viz. $G_t(P,T,X) = G_m(P,T,X)$ or

$$\Delta G_{t\rightarrow m}(P,T,X) = \Delta G_{t\rightarrow m}(1,T,X) + \int_1^P \left.\frac{\delta\Delta G(P,T,X)}{\delta P}\right|_T dP = 0 \qquad (1)$$

where $\Delta G_{t\rightarrow m}(1,T,X)$ is the chemical free energy change associated with the transformation at ambient pressure (P = 1 bar). The second term on the RHS of Eq. (1) is the work performed by the external pressure that causes the spontaneous shape change. When it is assumed that the transformation only involves a volume change ($\Delta V/V$), then $\delta\Delta G(P,T,X)/\delta P = \Delta V(P,T,X)/V$. With this assumption, the conditions (T,P) required for the transformation to be promoted can be expressed as,

$$-\Delta G_{t\rightarrow m}(1,T,X) \geqslant \int_1^P \left.\frac{\Delta V(P,T,X)}{V}\right|_T dP \qquad (2)$$

This was the approach Whitney (1962) and Berezhnoi (1962) used in calculating the P-T diagram (X=0) shown in Fig. 2.

The thermodynamics of the constrained transformation differ from the case of an external applied pressure in two respects (Lange, 1979). First, in general, the stress arising within the constrained, transformed particle is a tensor quantity and not a scalar pressure. Second, the strain energy associated with the transformed particle and the surrounding matrix material must be included in the general thermodynamics. Thus, the free energy change of the constrained transformation can be written as:

$$\Delta G^c_{t\rightarrow m} = \Delta G_{t\rightarrow m}(1,T,X) + \int_1^{\tilde{\sigma}_t} \left.\frac{\delta\Delta G(\sigma_t,T,X)}{V}\right|_T d\tilde{\sigma} + \Delta U_{SE} \qquad (3)$$

where $\tilde{\sigma}_t$ is the stress tensor describing the stress state within the particle due to the constraining matrix. ΔU_{SE} is the change in the strain energy of the particle and matrix; it is a function of $\tilde{\sigma}_t$ and the elastic properties of the two materials.

Upon cooling, the t→m reaction will proceed when $\Delta G_{t\to m} \leqslant 0$, or when

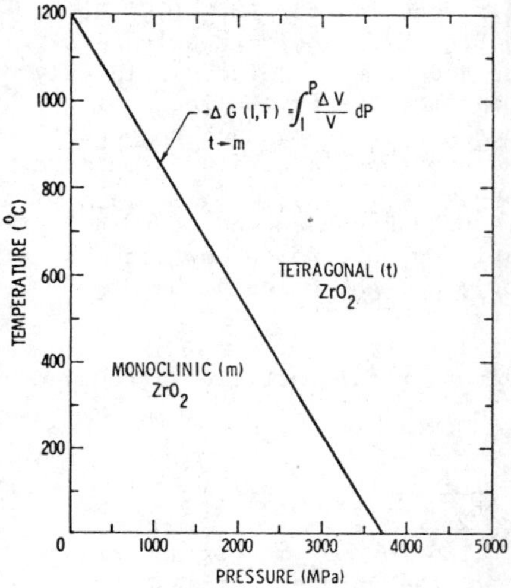

Fig. 1 The temperature-pressure phase equilibria for pure ZrO_2 after Whitney (1962).

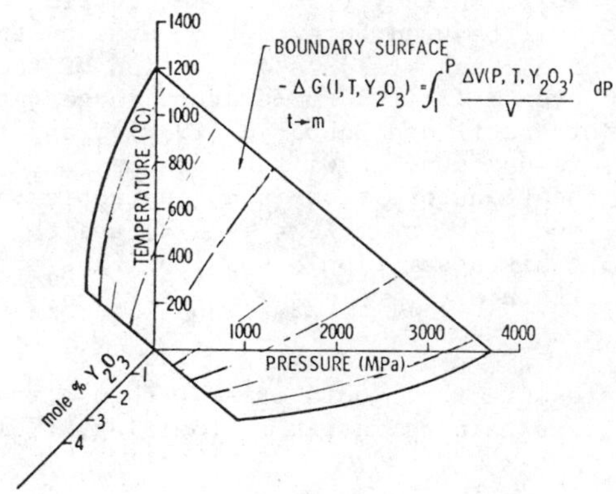

Fig. 2 The apparent temperature-pressure Y_2O_3 phase diagram for ZrO_2-Y_2O_3 compositions up to the eutectoid composition.

$$-\Delta G_{t\to m}(1,T,X) \geqslant \int_1^{\tilde{\sigma}_t} \left.\frac{\delta \Delta G(\tilde{\sigma}_t,T,X)}{\delta \sigma}\right|_T d\tilde{\sigma} + \Delta U_{SE} \quad (4)$$

Inspection of Eq. (4) shows that in the absence of constraint, the conditions for transformation reverts to ambient pressure conditions, viz $\Delta G_{t\to m}(1,T,X)<0$; for pure ZrO_2 (X=0) and the ZrO_2-Y_2O_3 eutectoid composition (X≃3 m/o Y_2O_3), this conditions occurs at ~1200°C and ~560°C, respectively. Below these temperatures, $\Delta G_{t\to m}(1,T,X)$ has a negative value. Thus, by inspecting Eq. (4) again, it can be seen that the constrained transformation temperature is lower than the unconstrained transformation temperature.*

The stress state, $\tilde{\sigma}_t$, within the constrained, transformed particle depends on the transformation strain tensor, $\tilde{\varepsilon}_t$, the elastic properties of the transformed material and the constraining matrix, the shape of the particle, and the orientation of the particle's shape with respect to its crystal coordinate system.

---

* Since $\int_1^{\tilde{\sigma}_t} \frac{\delta \Delta G}{\delta \sigma} d\tilde{\sigma}$ is work associated with the spontaneous shape change and ΔU_{SE} is the strain energy change, both are positive. It should be noted here that Eshelby (1961) neglects the work term in his analysis of the free energy change associated with a constrained transformation.

For ellipsoidal shapes, $\tilde{\sigma}_t$ is uniform from point to point within the particle (Eshelby, 1961). The strain energy, ΔU_{SE}, can be computed with knowledge of $\tilde{\varepsilon}_t$ (or $\tilde{\sigma}_t$) and the elastic properties of the transformed material and matrix material.

Present attempts to calculate the constrained transformation temperature (from Eq. (4)) are limited to available thermodynamic and crystallographic data for pure ZrO_2. Whitney (1962) has reported an explicit function for $\Delta G_{t\to m}(1,T,0)$ which will be used here. Both terms on the RHS of Eq. (4) require knowledge of the shape change ($\tilde{\varepsilon}_t$) as a function of temperature and stress. Since $\tilde{\varepsilon}_t$ is only known over a limited temperature range (as computed from the crystallographic data of Pratil and Subbarao (1969)), and the effect of stress on each component of $\tilde{\varepsilon}_t$ is unknown, it will be assumed here that the spontaneous strain associated with the transformation is an isotropic volume change ($\varepsilon_{11} = \varepsilon_{22} = \varepsilon_{33}$, $\varepsilon_{12} = 0$, etc., and $\varepsilon_{11} + \varepsilon_{22} + \varepsilon_{33} = \Delta V/V$) and that $\Delta V/V$ is only dependent on temperature. This assumption also simplifies the calulation of $\tilde{\sigma}_t$, which now reduces to a pressure ($\sigma_{11} = \sigma_{22} = \sigma_{33} = P_t$, $\sigma_{12} = 0$, etc.), and the calculation of ΔU_{SE}.

The pressure that would arise within a transformed spherical particle surrounded by a constraining matrix is (Selsing, 1961),

$$P_t = \frac{k}{3}\frac{\Delta V}{V}, \quad \text{where} \quad k = \frac{2E_1E_2}{(1+\nu_1)E_2 + 2(1-2\nu_2)E_1}, \tag{5}$$

and $E_{1,2}$, $\nu_{1,2}$ are Young's modulus and Poisson's ratio for the matrix and transformed particle. Assuming that the strain energy in the untransformed state is zero, the strain energy per unit volume within the transformed sphere and surrounding matrix is (Davidge and Green, 1968),

$$\Delta U_{SE} = \frac{3}{2k}P_t^2 = \frac{k}{6}\left(\frac{\Delta V}{V}\right)^2 . \tag{6}$$

With the assumption outlined above, $\delta\Delta G(\tilde{\sigma}_t,T,X)/\delta\sigma = \Delta V(T,X)/V$ and the first term on the RHS of Eq. (4) reduces to,

$$\int_1^{\tilde{\sigma}_t} \frac{\delta\Delta G(\tilde{\sigma},T,X)}{\delta s}\, d\tilde{\sigma} = \int_1^{P_t} \frac{\Delta V(T)}{V}\, dP = \frac{k}{3}\left(\frac{\Delta V(T)}{V}\right)^2 . \tag{7}$$

By combining Eqs. (4), (6) and (7), the constrained transformation temperature can be estimated from

$$-\Delta G_{t\to m}(1,T,X) \geqslant \frac{k}{2}\left(\frac{\Delta V(T)}{V}\right)^2 . \tag{8}$$

Using Whitney's (1962) derivation of $\Delta G_{t\to m}(1,T)$ and Patil and Subbarao's (1969) crystallographic data* to compute $\Delta V(T)/V$ for pure ZrO_2 ($X=0$), the constrained transformation temperature can be estimated for different constraints. For the case where ZrO_2 is the constraining matrix (case of self constraint), k = 186 GPa ($E_1 = E_2 = 206$ GPa, $\nu = 0.25$), the transformation temperature is 650°C, which is in agreement with experimental lower bound values of ~700°C generally observed for polycrystalline ZrO_2 (Sabbarao, Maiti and Srivastava, 1974). When the constraining matrix is Al_2O_3, k = 251 GPa (E_1 = 390 GPa, E_2 = 206 GPa, $\nu_1 = \nu_2 = 0.25$), the transformation temperature is now estimated to be ~120°C.

In the above paragraphs, the thermodynamics of the constrained transformation were examined to determine the equilibrium transformation temperature as a function of the constraining parameter, k. It was shown that constrained transformation temperature can be considerably below the ambient pressure transformation temperature. As it will be shown in the proceeding section, microcracking must be avoided in order to take full advantage of the lowered transformation temperature.

Microstructural Considerations for Phase Retention

Tensile stresses would arise within the matrix material if the $ZrO_2(t) \rightarrow ZrO_2(m)$ transformation were to proceed. When conditions are suitable, the tensile stresses will cause the extension of a small pre-existing crack. Crack extension will release some (or all) of the constraint. Thus, if constraint is to be maintained, microcracking must be avoided. The conditions for microcracking will be examined and shown to be related to microstructural features.

Let us re-examine the free energy change of the constrained transformation to include crack extension. For illustrative purposes, it will simply be assumed that crack extension decreases the effect of the constraint by lowering the magnitude of the transformed stress from what it would be without microcracking. For the case of the isotropic volume change, the pressure within the spherical, transformed particle will be reduced from P_t, the pressure for the uncracked system. The pressure for the cracked system will be denoted by P(c), indicating its dependence on crack length, c; thus $0 < P(c) \leqslant P_t$.

The dependence of P(c) on temperature and composition and other features of the transformation can be determined by rewriting Eq. (4) with the assumed isotropic volume change, viz.,

$$\Delta G_{t\rightarrow m}(P(c),T,X) = \Delta G(1,T,X) + \frac{9}{2k} P(c)^2 + \frac{A\gamma}{V} \quad , \tag{9}$$

where the last term is the surface energy of the crack per unit volume of transformed material. Transformation and crack extension will take place when

$$-\Delta G(1,T,X) \geqslant \frac{9}{2k} P(c)^2 + \frac{A\gamma}{V} \quad . \tag{10}$$

Rearranging Eq. (10), and substituting $A\gamma/V = 2\pi D^2 g(c)\gamma/(\pi/6D^3) = 12g(c)\gamma/D$, (where g(c) is a dimensionaless number relating the crack surface area (A) to the surface of the sphere, and D is the sphere diameter),

$$P(c) = \left[-k \frac{8}{3D} g(c)\gamma + \frac{2}{9} \Delta G(1,T,X))\right]^{1/2} . \tag{11}$$

It can be seen that P(c) only has real meaning below the ambient pressure transformation temperature when,

$$-\Delta G(I,T,X) \geqslant \frac{12g(c)\gamma}{D} \quad . \tag{12}$$

Equation (12) shows that as the temperature is lowered below the ambient pressure transformation temperature, large particles will be the first to transform and microcrack. Further, it can be shown that a critical particle size (D_c) exists, below which microcracking can be avoided over the complete temperature range where it is thermodynamically possible to retain the high temperature phase by constraint (Lange, 1978).

---

* Scott's (1975) crystallographic data for both structures at room temperature are in agreement.

Experimental Observations Concerning Phase Retention

Phase retention has been investigated for materials fabricated in the ZrO_2-Y_2O_3 (up to 7.5 m/o Y_2O_3) system (Lange, 1979) and the ZrO_2-Al_2O_3 system. The phase diagram for the ZrO_2 rich portion of the ZrO_2-Y_2O_3 system is shown in Fig. 3. This diagram shows that different ZrO_2 (tetragonal)/ZrO_2 (cubic) composites can be fabricated by varying the amount of Y_2O_3 within the range of 3 to 6.5 m/o Y_2O_3. Phase studies in the ZrO_2-Al_2O_3 indicate no binary phase formation (Levin and McMurdle, 1975). Ultra-submicron size powders were used in these studies. Yttria was added as yttrium nitride dissolved in methanol, which was subsequently decomposed to Y_2O_3 by heat treatment at 400°C. The ZrO_2/Al_2O_3 composite powders were milled together with Al_2O_3 milling media in plastic containers. Densification was accomplished by either sintering or hot-pressing at temperatures between 900°C and 1650°C. X-ray diffraction analysis was used at room temperature for phase identification. Grain size measurements were made by observing fracture surfaces with a scanning electron microscope.

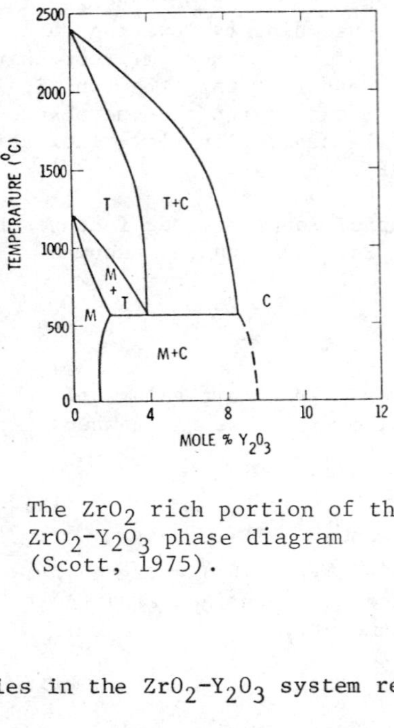

Fig. 3 The ZrO_2 rich portion of the ZrO_2-Y_2O_3 phase diagram (Scott, 1975).

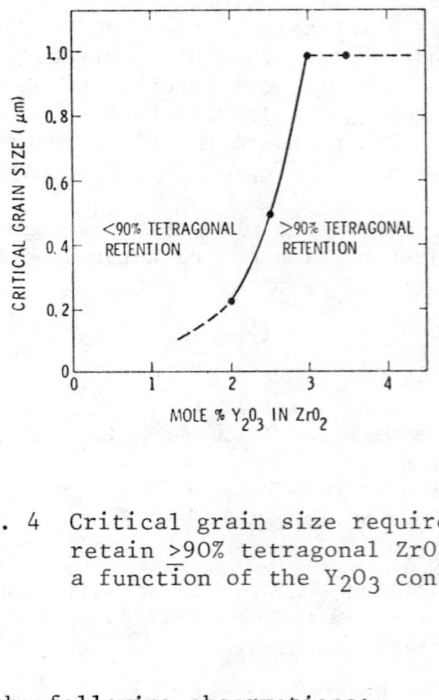

Fig. 4 Critical grain size required to retain ≥90% tetragonal ZrO_2 as a function of the Y_2O_3 content.

Studies in the ZrO_2-Y_2O_3 system resulted in the following observations:

1. Densities ⩾85% of theoretical were required if a high proportion (⩾90%) of the tetragonal phase was to be retained. This observation is consistent with the argument that constraint (in this case, self constraint) is required for phase retention.
2. The critical grain size (D_c) for which ⩾90% of the tetragonal phase could be retained increased with increasing Y_2O_3 contents up to the eutectoid composition (~3% Y_2O_3) as shown in Fig. 4. This observation is consistent with Eq. (12), which indicates the grain size/ compositional dependence for microcracking. Namely, the temperature where $\Delta G(1,T,X) \leqslant 0$ will decrease with increasing $X(Y_2O_3)$ up to the eutectoid composition, thus increasing the grain size required to produce a microcrack.
3. Pure ZrO_2 was completely monoclinic under all sintering conditions; this is consistent with estimated transformation temperature of ~650°C for self constraint.

4. Materials containing a large fraction (>30%) of the monoclinic phase were highly microcracked as observed by their firable nature. This observation is consistent with the obvious hypothesis that microcrack relieves constraint.

Studies in the ZrO_2-Y_2O_3 system resulted in the following observations:

1. Composites containing up to 10 v/o of pure ZrO_2 retained a large fraction (>80%) of the tetragonal phase. This observation is consistent with the theoretical result that higher elastic modulus of the Al_2O_3 constraining matrix will significantly lower the transformation temperature relative to the condition of self-constraint. (The calculated transformation temperature is 120°C.)
2. Under the same fabrication conditions, monoclinic ZrO_2 became the dominant phase for composites containing >10 v/o pure ZrO_2. This is consistent with the fact that the elastic modulus of the constraining matrix decreases with increasing volume content of the ZrO_2, thus increasing the transformation temperature toward that of self-constraint.
3. Composite Al_2O_3-ZrO_2 materials in which 2 m/o Y_2O_3 was added were totally tetragonal up to ZrO_2 volume contents of 60%, illustrating (as also observed for the ZrO_2-Y_2O_3 system) that alloy additions that lower the ambient pressure transformation temperature also lower the constraint transformation temperature.

The above observations are consistent with the theoretical arguments concerning the thermodynamics of the constrained phase transformation and the microstructural control if microcracking is to be prevented. In addition, the assumption that the shape change associated with the transformation can be treated as an isotropic volume expansion appears to give reasonable first order estimate of the constrained transformation temperature.

FRACTURE TOUGHNESS

Investigations into the fracture of ceramics containing tetragonta ZrO_2 have shown that transformation to the monoclinic structure occurs in the volume of material adjacent to the fracture surface. The depth of the transformed layer is uncertain; Porter and Heuer's (1977) TEM studies for one material indicates a depth of ≤2μm. The transformation presumably occured during crack propagation and was induced by the stress field of the crack.

In view of the foregoing arguments that the prevention of the martensitic reaction is due to constraint, it is consistent to argue that the stress field of the crack front can release this constraint and allow the reaction to proceed. As will be shown, work is performed to release this constraint which contributes to the fracture toughness of the material.

Contribution of Stress-Induced Phase Transformation

Without defining the mechanics of the phenomena responsible for the energy absorbed by the stress-induced transformation in the vicinity of the crack front, an energetic approach has been presented that accounts for energy absorbed by the transformation contained within a process zone surrounding a penny-shaped crack (Lange, 1979). The analysis resulted in a general expression for the critical stress intensity factor of the composite (or single phase) material containing a given volume fraction (V_p) of the metastable phase:

$$K_c = \left[K_o^2 + \frac{2U_o E_c V_o R}{(1-\nu_c^2)} \right]^{1/2} \qquad (13)$$

K_o is the critical stress intensity factor for the composite when the contribution of the transformation is neglected, U_o is the energy per unit volume absorbed during the transformation, R is the radius of the process zone, E_c, ν_c are the elastic properties of the composite material, and V_p is the volume fraction of the material undergoing transformation.

To use Eq. (13), a mechanism of energy dissipation associated with the stress-induced transformation must be chosen. Since retention of the metastable phase is due to matrix (or self) constraint, Lange, (1979) proposed that the stress field of the crack can release this constraint and allow the spontaneous shape change of every metastable particle within a volume of material adjacent to the crack front.

In a preceding section, it was shown that if the transformation were to take place in a constraining matrix, an internal pressure, P_t, would arise (for the case of an isotropic volume expansion) within the transformed particle that would reverse the transformation. As a tensile load is applied to the cracked body, the stress at the crack front will increase until it counters the effect of the matrix and allows the particle to transform as if it were in an unconstrained state. It can be seen that this point is reached when the magnitude of the crack's tensile stress is equal and opposite to the internal pressure that would arise in the constrained state, viz. $\sigma_t = -P_t$. If the crack's stress were now reduced by decreasing the applied load, the transformation would reverse when $\sigma<\sigma_t$. Keeping in mind that the transformation involves a spontaneous volume change, the stress-strain behavior of material within the process zone for the cyclic loading and unloading of the crack is shown by the curve abcd in Fig. 5. It can be seen that as particles within the process zone are cycled through their transformation by increasing and decreasing the applied stress (i.e., abcd → dcba), energy is neither lost nor gained. If, on the other hand, the particles were left in their transformed state after the fracture process, the work done by the crack's stress field to unconstrain the spontaneous shape change must be considered as energy absorbed during fracture. This situation can be accomplished if the crack were to extend after its stress field unconstrained the transformation (e.g., crack extension after the material in the process zone was brought to state d, Fig. 5). Crack extension, either through or adjacent to the transformed particles, will prevent the matrix from reconstraining and reversing the transformation. Since the transformation is not reversed, the material adjacent to the newly formed fracture surfaces will follow the stress-strain curve de in Fig. 5 as the stress within the fractured, process zone relaxes to zero. The area abce in Fig. 5 represents the work done by the crack's stress field (or the applied stress) to unconstrain the transformation. Since σ_t was the stress that caused the spontaneous volume change associated with the transformation, the energy absorbed per unit volume of transformed material is*

$$U_o = -\sigma_t \frac{\Delta V}{V} = P_t \frac{\Delta V}{V} \quad . \tag{14}$$

The critical stress intensity factor can be restated by combining Eqs. (5), (13), and (14)

$$K_c = \left[K_o^2 + \frac{2k(\frac{\Delta V}{V})^2 E_c V_p R}{3(1-\nu_c^2)} \right]^{1/2} \tag{15}$$

---

* The work done by the crack's tensile stress is equilibrated to the energy absorbed, accounting for the negative sign in Eq. (14).

It can be seen that the contribution of the stress-induced fracture toughness will depend on the elastic constraining factor, k, the shape change associated with the transformation, $\Delta V/V$, the elastic constants of the constraining material, E_c, ν_c, the volume fraction of the metastable phase, V_p, and the radius of the process zone, R.

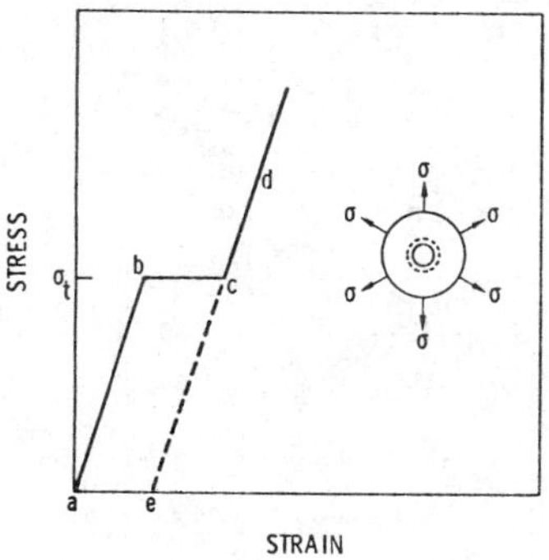

Fig. 5 The stress strain behavior of the composite material (see insert) ahead of the crack front.

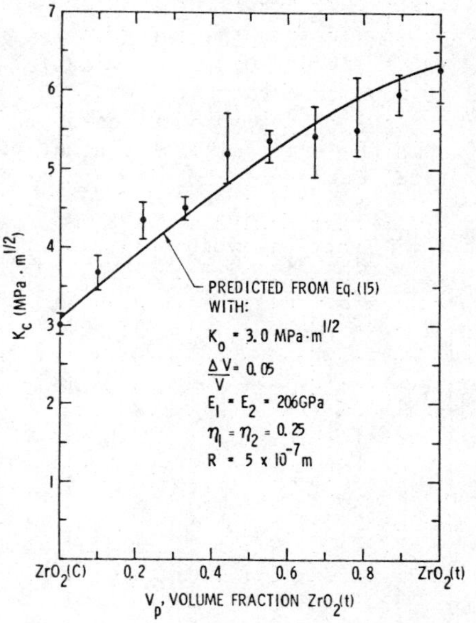

Fig. 6 The critical stress intensity factor of $ZrO_2(t)/ZrO_2(c)$ materials as a function of the volume fraction of $ZrO_2(t)$

The radius of the process zone defines the distance from the fracture surface in which the particles cannot be reconstrained by the matrix once they have been transformed by the stress field of the crack. An upper limit for R can be calculated, assuming that all particles transformed within the stress field of the crack are also those that cannot be reconstrained as the crack extends. This can be accomplished by estimating the zone surrounding the crack front where the tensile stresses are $\geqslant \sigma_t$:

$$R = \frac{1}{2\pi}\left(\frac{K_c}{\sigma_t}\right)^2 = \frac{9}{2\pi}\left(\frac{K_c}{k\frac{\Delta V}{V}}\right)^2 . \qquad (16)$$

Substituting values into Eq. (16), R ≃ 0.7μm for the case of self constraint (K_c = 6.5 MPa $m^{1/2}$, k = 186 GPa, $\Delta V/V$ = 0.05) and R ≃ 0.4μm for constraint by Al_2O_3 (K_c = 6.5 MPa $m^{1/2}$, k = 250 GPa, $\Delta V/V$ = 0.05). Thus, R is estimated to be <1μm.

Experiments Concerning Fracture Toughness and Strength

Two composite systems have been fabricated: 1) ZrO_2 (tetragonal)/ ZrO_2 (cubic) composites were sintered by using different mole fractions of Y_2O_3 (between 2.5 and 7.5 m/o% Y_2O_3) to move across the tetragonal plus cubic phase field in the ZrO_2/Y_2O_3 binary system (see Fig. 3); the densities of these composite materials ranged between 88 and 92% of theoretical. 2) ZrO_2 (tetragonal)/Al_2O_3 composites were hot-pressed from one end-member to the other; additions of 2 m/o Y_2O_3 were used so that the tetragonal ZrO_2 could be retained over the full range of compositions. Densities of the latter materials were >98% of theoretical.

K_c was measured on polished specimens using the indentation technique described previously (Evans and Charles, 1976). Flexural strengths of certain $ZrO_2(t)/Al_2O_3$ composites were measured in 4 point loading on bars (approx. 0.3 x 0.3 x 3.0 cm) that were diamond cut and ground from the hot-pressed billets.

Figure 6 illustrates the K_c data for the $ZrO_2(t)/ZrO_2(c)$ composite materials. As illustrated, nearly perfect agreement is obtained between theory (Eq. (15)) and measurements. Values of $K_o = 3.0$ MPa.m$^{1/2}$ (the fracture toughness of the cubic ZrO_2), $\Delta V/V = 0.05$ (the measured volume change of the respective unit cells at room temperature, Scott, 1973), $E_1 = E_2 = 206$ GPa, $\nu_1 = \nu_2 = 0.25$ (assuming the elastic properties of the two forms of ZrO_2 to be the same), and $R = 0.5\mu m$ (the grain size of the material) were used for obtaining the theoretical curve.

Figure 7 illustrates the K_c data for the $ZrO_2(t)/Al_2O_3$ composite materials. For this case, agreement between theory and experiments is not as good. Predicted

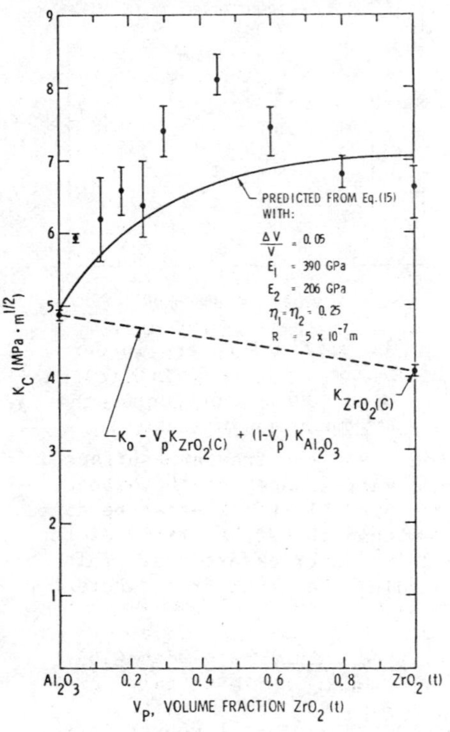

Fig. 7 The critical stress intensity factor of $ZrO_2(t)/Al_2O_3$ materials as a function of the volume fraction of $ZrO_2(t)$.

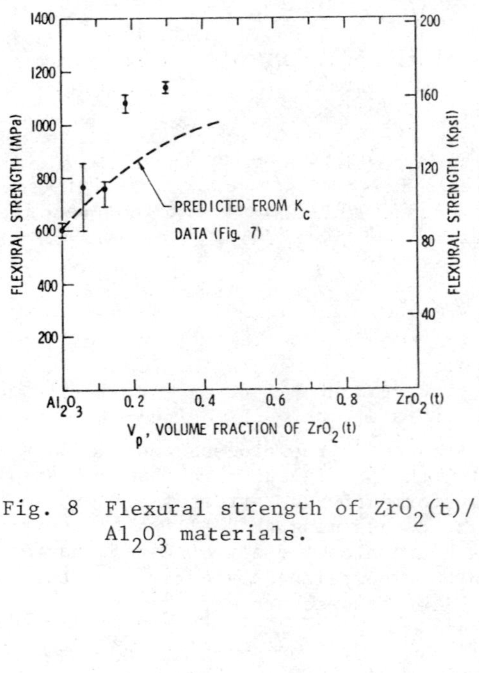

Fig. 8 Flexural strength of $ZrO_2(t)/Al_2O_3$ materials.

values were generally lower than measured values and the apparent maximum at $V_p \simeq .45$ was not predicted. Despite these differences, the larger increase in K_c for small volume fractions of $ZrO_2(t)$ in the Al_2O_3 matrix relative to the smaller increase in the ZrO_2 (cubic) matrix is predicted by Eq. (15). Flexural strength data for $ZrO_2(t)/Al_2O_3$ composites are shown in Fig. 8. Several composites had flexural strengths >1100 MPa, which is extraordinary for any ceramic material. The broken line indicates predicted values based on the average strength of the Al_2O_3 single phase material ($\sigma_{Al_2O_3}$) with the assumption that the crack size distribution was the same for all materials:

$$\sigma_{predicted} = \sigma_{Al_2O_3} \left(\frac{K_c}{K_{Al_2O_3}}\right) \quad . \qquad (17)$$

Since the higher strength values are not in agreement with those predicted, it must be concluded that either the crack size distribution is not constant from material to material or some other phenomenon is also increasing the strength. Garvie and Pascoe (1978) have shown that surface grinding increases the flexural strength of a ZrO_2(t) based material. Grinding causes the surface to transform. Since the volume of the transform surface layer is larger, surface compressive stresses are likely to develop; this can explain the strengthening caused by surface grinding. It is reasonable to suspect that this same phenomenon might add to the strengthening of the Al_2O_3/ZrO_2(t) composite specimens; this would explain the difference between theoretical and experimental strength values reported in Fig. 8.

SUMMARY

The elastic constraint of a spontaneous shape change accompanying a diffusionless phase transformation alters the thermodynamics of the transformation relative to the unconstrained case by two terms: 1) the work performed by the stress field that arises during the spontaneous shape change, and 2) the increased strain energy imparted to the system by the same stress field. The effect of these two terms is to lower the transformation temperature relative to the unconstrained case. The greater the elastic modulus of the constraining matrix, the lower the transformation temperature.

The unconstrained transformation temperature for the ZrO_2(t) → ZrO(m) reaction is ~1200°C. By assuming that the shear strains associated with this transformation could be neglected, the transformation temperature for the self-constraint (ZrO_2 constraining ZrO_2) of pure ZrO_2 was estimated to be ~650°C. Constraint with Al_2O_3, which has about twice the modulus of ZrO_2, further lowers the estimated transformation temperature to ~120°C. Alloying additions (e.g., Y_2O_3, CaO, MgO, etc.) that lower the unconstrained transformation temperature, will further lower the constrained transformation temperature. It is therefore reasonable to expect that if constraint can be maintained, tetragonal ZrO_2 can be retained in a metastable form below room temperature if the proper constraining material and/or alloy additions are used in fabrication.

Microcracking, which can accompany the transformation during cooling, must be avoided if constraint is to be maintained. Microcracking can be avoided by fabricating a microstructure in which the grain size (the case of self-constraint) or the particle size (the case of constraint by another matrix material) is less than a critical size. This requires fabricating a material, either with a very fine particle size powder and avoiding extensive grain growth, or by carefully precipitating small particles of the metastable phase within a suitable matrix material. Proper control over the size of the metastable phase is a key factor in phase retention.

The contribution of the stress-induced phase transformation to fracture toughness is viewed in terms of the work done by the stress field of the crack to overcome the effect of the matrix constraint. Transformation is induced ahead of the crack when the crack's stress field counters the stress field that would arise for the constrained transformation. When this situation is satisfied, particles ahead of the crack will transform. When this occurs, the chemical free energy change will be equal to the work done by the external load (or the crack's stress field) which promotes the spontaneous shape change accompanying the transformation. That is, when the stress field of the crack is sufficient to unconstrain the transformation, the free energy change of the system is zero. After the crack's stress field is sufficient to allow the transformation, crack extension will prohibit the reconstraint of particles either on or close to the fracture surface. The work done to unconstrain the particles that remain in their transformed state must be included in the energy required to cause crack extension. When this idea is formalized, the contribution of the stress induced phase transformation to fracture toughness was found to depend on the magnitude of the transformation shape change, the elastic properties of the transformed and constraining materials, the volume fraction of the metastable phase and the radius of the zone ahead of the crack which remains in the transformed state.

Observations concerning phase retention and fracture toughness of materials fabricated in two composite systems, each containing tetragonal ZrO_2, support the theoretical concepts and conclusions.

REFERENCES

Bailey, J.E. (1964). The Monoclinic-Tetragonal Transformation and Associated Twinning in Thin Films of ZrO_2. Prog. Roy. Soc. 279A, 395.

Bansal, G.K., and A.H. Heuer, (1974). On a Martensitic Phase Transformation in ZrO_2-II. Crystallographic Aspects. Acta Met. 22, 409.

Berezhnoi, A.S. (1962). Dop. Akad. Nauk. Ukr. SSR 1, 65.

Buljan, S.T., H.A. McKinstry and V.S. Stubican (1976). Optical and X-ray Single Crystal Studies of the Monoclinic/Tetragonal Transition in ZrO_2. J. Am. Ceram. Soc. 59, 351.

Davidge, R.W. and T.J. Green (1968). Strength of Two-Phase Ceramic/Glass Materials. J. Mat. Sci. 3, 629.

Evans, A.G. and E.A. Charles (1976). Fracture Toughness Determinations by Indentation. J. Am. Ceram. Soc. 59, 371.

Eshelby, J.D. (1961). In, I.N. Sneddon and R. Hill (Ed.), Progress in Solid Mechanics, Vol. 2, North-Holland, Chap. 3, pp. 88-140.

Garvie, R.C., R.H. Hennick and R.T. Pascoe (1975). Ceramic Steel. Nature 258, 703.

Garvie, R.C. and R.T. Pascoe (1978). In, H. Palmour III, R.F. Davis, and T.M. Hare (Eds). Processing of Crystalline Ceramics, Plenum Press, p. 263.

Gupta, T.K., J.H. Bechtold, R.C. Kuznichi, L.H. Cadoff and B.R. Rossing (1977). Stabilization of Tetragonal Phase in Polycrystalline ZrO_2. J. Mat. Sci. 12, 2421.

Gupta, T.K., F.F. Lange and J.H. Bechtold (1978). Effect of Stress-Induced Phase Transformation on Properties of Polycrystalline ZrO_2 containing Metastable Tetragonal Phase, J. Mat. Sci. 13, 1461.

Heuer, A.H., and G.L. Nord Jr (1976). In, H.R. Weuk Electron Microscopy in Minerology, Springer-Verlag, p. 274.

Lange, F.F. (1978). In, R.C. Bradt, D.P.H. Hasselman and F.F. Lange, Fracture Mechanics of Ceramics, Vol. 4, Plenum Press, p. 1464.

Lange, F.F. (1979). Stress-Induced Martensitic Reactions: I: Theory of Metastable Phase Retention and Contribution to Fracture Toughness: II: Experiments in the ZrO_2-Y_2O_3 System. J. Mat. Sci. (in press).

Levin, E.M. and H.F. McMurdle (1975). Phase Diagrams for Ceramists, 1975 Supplement, The American Ceramic Society, p. 135.

Patil, R.N. and E.C. Subbarao (1969). Axial Thermal Expansion of ZrO_2 and HfO_2 in the Range Room Temperature to 1400°C. J. Appl. Cryst. 2, 281.

Porter, D.L. and A.H. Heuer (1977). Mechanism of Toughening Partially Stabilized ZrO_2. J. Am. Ceram. Soc. 60, 183.

Sabbarao, E.E., H.S. Maiti and K.K. Srivastava (1974). Martensitic Transformation in ZrO_2. Phys. Stat. Sol. 21, 9.

Scott, H.G. (1975). Phase Relationships in the Zirconia-Yttria System. J. Mat. Sci. 10, 1527.

Selsing, J. (1961). Internal Stresses in Ceramics. J. Am. Ceram. Soc. 44, 419.

Srivastaba, K.K., R.N. Patil, C.B. Chandary, K.V.G.K. Gokhale and E.C. Subbarao (1974). Revised Phase Diagram of the System ZrO_2-$YO_{1.5}$. Trans. Brit. Ceram. Soc. 73, 85.

Stubican, V.S., R.C. Hink and S.P. Ray (1978). Phase Equilibria and Ordering in the System ZrO_2-Y_2O_3. J. Am. Ceram. Soc. 61, 17.

Whitney, E.D. (1962). Effect of Pressure on Monoclinic-Tetragonal Transition of Zirconia. J. Am. Ceram. Soc. 45, 612.

Volume 3, ICM 3, Cambridge, England, August 1979

THERMAL FRACTURE OF BUILDING GLASS SUBJECTED TO SOLAR RADIATION

Y. W. Mai and L. J. S. Jacob

Department of Mechanical Engineering, University of Sydney, Sydney, N.S.W. 2006, Australia

INTRODUCTION

In modern architectural design there is an increasing tendency to feature "all glass walls" as the outer periphery of modern buildings. Solar control glasses (either heat absorbing or heat reflective types) are often used for this purpose to reduce the heat load on air-conditioning systems. However, glasses as typical brittle materials are susceptible to thermal stress fracture. In particular the breakage potential is much higher for solar control glasses than for clear glass (Blight, 1974). Thus the use of large areas of solar control glasses in buildings has greatly increased the incidence of thermal fracture and hence the cost for glass replacement. Design against thermal fracture of building glass due to solar radiation absorption is consequently an engineering problem of practical importance and for which a solution is immediately required.

The development of thermal stresses in glass fixed in buildings is well understood. During sunshine hours the central exposed areas become hot due to solar radiation but the edges remain cool as they are shielded by the edge cover. As a consequence of this temperature differential tensile stresses are set up at the edges and if at any point the glass strength is exceeded fracture ensues. A similar differential heating effect may also result from partial shading of the glass by building overhangs, protruding columns and windows with deep mullions. A common solution to this brittle fracture problem is to use toughened glass. However, in many cases, this alternative is uneconomical and in large commercial buildings impossible due to the size limitations of commercially available toughened glass. Pilkington Brothers Limited (1964, 1972) has published some useful guildelines for designing against thermal fracture but they are usually conservative. Recently, Blight (1974) has conducted an extensive investigation on a number of factors that affect the cool edge strains (e.g. edge strip width, glass sheet size and proportion of total glass area subjected to solar radiation) and has shown how effectively control of these factors may be used to minimise thermal fracture. However the application of these methods in in situ full size building glasses has not been studied. Although it is well-known that cut glasses contain many edge flaws and that their sizes largely determine the actual fracture stress/strain none of these previous studies has seriously considered using fracture mechanics to analyse thermal fracture. The present paper reports the results of an experimental investigation on a new proposed method to minimise thermal fracture in glass. In addition to the two-dimensional stress analysis solutions, the application of fracture mechanics principles to determine fracture stresses is also considered.

THERMAL FRACTURE - THEORETICAL CONSIDERATIONS

Fracture stress at the cool edge of the glass pane may be determined numerically from two dimensional stress solutions and from fracture mechanics analysis. Details of these methods have been described by Jacob (1978) and for completeness of this paper they are given briefly as follows.

Two Dimensional Stress Solutions

Roberts' technique. The thermal stress problem for flat glass plates used in windows is one of thin plates with no significant thickness temperature variation. The governing partial differential equation, after coordinate transformations, is given by (see Fig. 1):

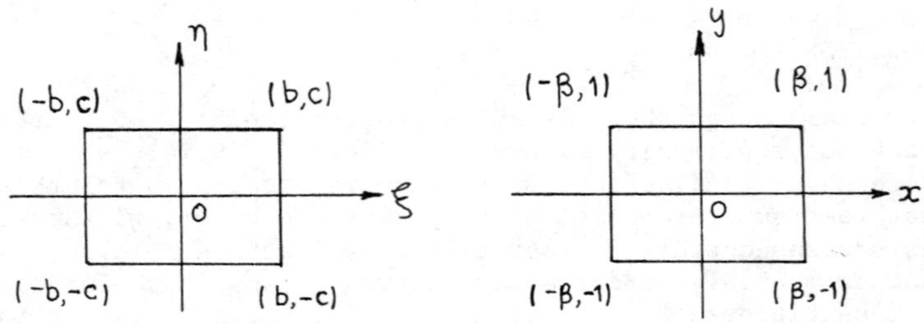

Fig. 1. Coordinate systems used to calculate thermal stresses.

$$\nabla^4\Omega(x,y) = -\nabla^2\Lambda(x,y) \tag{1}$$

where $\Lambda = E\alpha T/E_o\alpha_o T_o$. α is the coefficient of linear thermal expansion and Ω is the Airy stress function. Note that the subscript 'o' is used to denote the reference state and the (ξ,η) and (x,y) relations are given by: $x = \xi/c$, $y = \eta/c$ and $\beta = b/c$. Roberts (1962) has used a boundary collocation method to obtain solutions for equation (1) so that the elastic stresses developed in a rectangular plate of uniform thickness subjected to a thermal gradient in two directions can be computed. Thus

$$\sigma_x = E_o\alpha_o T_o \frac{\partial^2}{\partial y^2}\Omega(x,y) = E_o\alpha_o T_o \sum_{i=1}^{3} G_i\theta_i , \tag{2}$$

$$\sigma_y = E_o\alpha_o T_o \frac{\partial^2}{\partial x^2}\Omega(x,y) = E_o\alpha_o T_o \sum_{i=1}^{3} G_i\phi_i . \tag{3}$$

G_i clearly depends on the temperature distribution for a given problem. For temperature variation in either x or y direction only, say $T = T(y)$, it is assumed that $T(y)$ can be separated into an even and an odd function so that $T(y) = T_{even}(y) + T_{odd}(y)$. Thus,

$$\sum G_i\theta_i = \sum[G_i\theta_i]_{even} + \sum[G_i\theta_i]_{odd} \quad (4)$$

where
$$G_{even\,1} = \left.\frac{\nabla^2 E\alpha T_{even}(y)}{E_o\alpha_o T_o}\right]_{y=\frac{1}{6}}$$

$$G_{odd\,1} = \left.\frac{\nabla^2 E\alpha T_{odd}(y)}{E_o\alpha_o T_o}\right]_{y=\frac{1}{6}} \quad (5)$$

$G_2\theta_2$ and $G_3\theta_3$ are computed at y = 3/6 and 5/6 respectively. The stress parameters θ_i and ϕ_i have also been computed as a function of (x,y) for even and odd temperature distributions and for various β values (Roberts, 1962). As will be shown later, for the present experiments with a square glass plate $\beta = 1$, and the tensile stress required at the cool edge (σ_x) is that at x = 0 and y = 1. Note that σ_y at this position is zero.

<u>Pilkington-Duhamel method</u>. Colvin and Lightfoot (1974) have applied successfully the Duhamel's method (1838) to predict the maximum thermal stress along the edge of a plate by calculating and summing the "bridling" and "flexing" stresses. For a square plate with side (c) and edge cover of width (d) the stress at the center of the edge is given by:

$$\sigma_x = 0.61\ [E\alpha(T_c - T_E)(c - 2d)/c + 0.4\ E\alpha(T_c - T_E)(c - 2d)/c] \quad (6)$$

where T_c and T_E are the temperatures at the center and edge of the glass plate. The first and second terms inside the square bracket are respectively the bridling and flexing stresses. 0.61 is an efficiency factor for a square plate when compared with a rectangular plate. Equation (4) may be simplified to:

$$\sigma_x = 0.854\ E\alpha(T_c - T_E)(c - 2d)/c\ . \quad (7)$$

When T_c and T_E are known at the moment of fracture σ_f can be computed.

<u>Fracture Mechanics Techniques</u>

As shown in Fig. 2 the fracture surface is typified by a corner flaw with three distinct regions of mirror, mist and hackle. Liu (1972) has shown that for such a flaw geometry the stress intensity factor (K) is given by

$$K = 2\sigma\sqrt{a/\pi}\ \delta_{fo}\delta_{f90}\ . \quad (8)$$

Substitution of $\delta_{fo} = 1.03$ and $\delta_{f90} = 1.22$ (Smith, Emery and Kobayashi, 1967) into equation (8) gives the fracture stress as:

$$\sigma_f = K_{IC}/\sqrt{2a_f}\ . \quad (9)$$

K_{IC} may be determined easily from separate experiments but the critical flaw size (a_f) is difficult to measure. However, it has been shown (Mecholsky, Rice and Freiman, 1974) that $a_f \simeq 0.1$ r, where r is the mirror radius which can be conveniently measured from fracture surfaces. This therefore allows σ_f to be calculated from equation (9).

On a more empirical approach, studies by various researchers, e.g. Bansal (1977), have shown that

$$\sigma_f = Br^{-\frac{1}{2}} \quad (10)$$

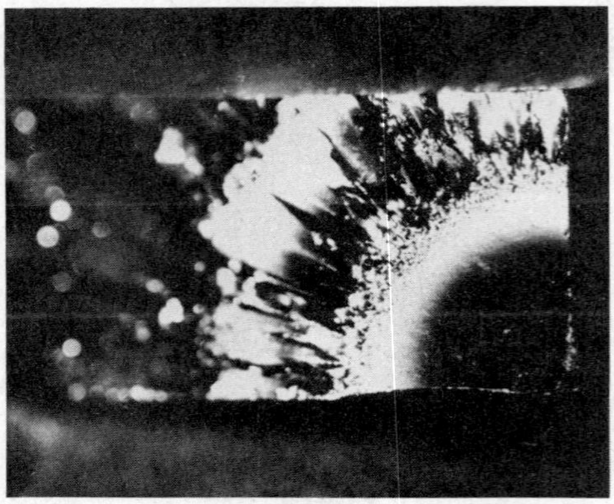

Fig. 2. Typical fracture surface.

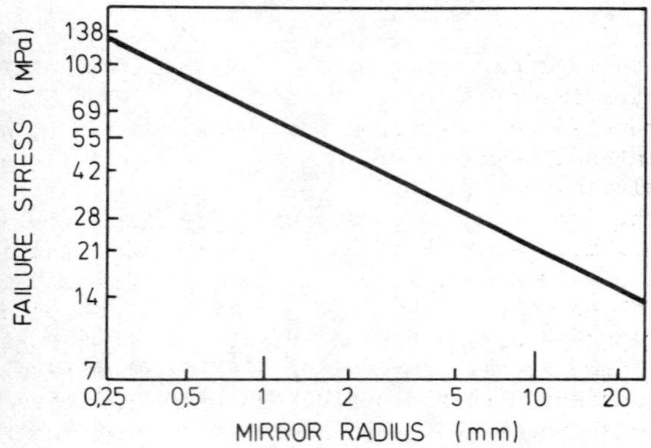

Fig. 3. Relation between mirror radius and fracture stress.

where B is a "mirror" constant. This equation is in fact similar to equation (9) if a_f is replaced by 0.1 r. Pittsburg Plate Glass Company (1974) has published a design curve, Fig. 3, based on equation (10) so that if r is known σ_f can be determined.

THE NEW GLAZING TECHNIQUE

To minimise the incidence of thermal fracture without the use of toughened glass it is necessary to increase the cool edge temperature. This may be achieved by utilizing a new glazing technique as shown in Fig. 4 where a specially developed thermal conducting sealant, made from mainly aluminium oxide, mineral oils and wetting agents, is used in the rebate (together with the primary conventional sealant) and the external face of the aluminium frame is painted in black. It is possible to estimate the temperature rise at the cool edge if the thermal conductivity of the new sealant is known. However due to environmental effects, surface conductance, external wind speeds and framing details this computation can become very complex. Consequently an experimental evaluation of performance is adopted in the present work.

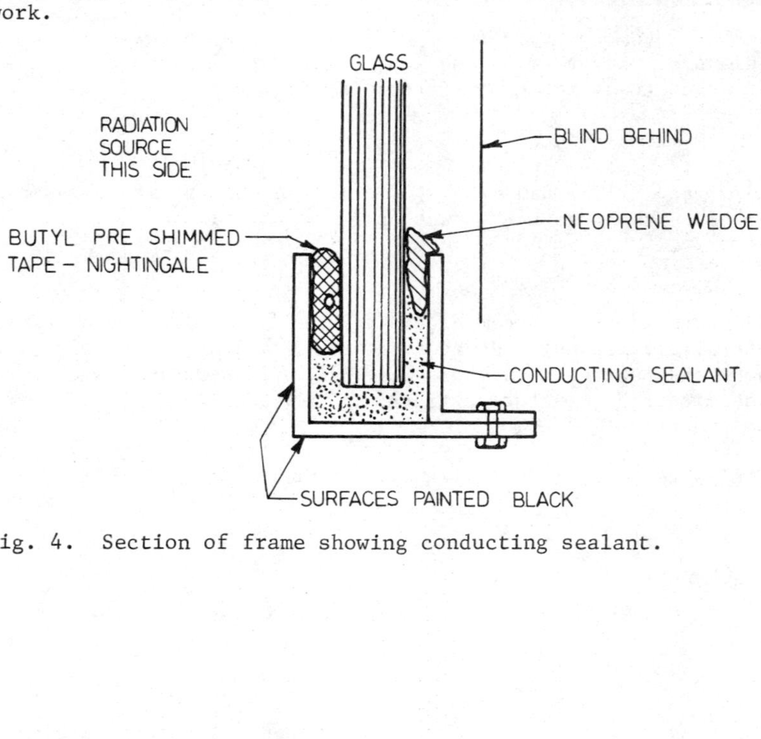

Fig. 4. Section of frame showing conducting sealant.

EXPERIMENTAL WORK

Fig. 5 shows the general arrangement of equipment used for the thermal fracture experiments. The solar source was simulated by a bank of 19 infra-red lamps mounted on a vertical board in a geometric arrangement to give a uniform radiation intensity of ∿ 5000 Watt/m^2 on a glass plate (455 x 455 x 6 mm^3) glazed in an aluminium frame. The frame was supported by a wooden stand placed 600 mm away from the lamps. An edge cover (d) of 20 mm all around the glass plate was permitted. A white venetian blind was hung behind the frame to ensure that the transmitted energy was reflected back on to the sample. Two special high temperature strain gauges were placed at the center of two adjacent edges and their outputs were recorded by a 2-pen recorder. (Fracture usually occurred at or near the location of the strain gauges.) A 10-channel digital thermometer equipped with copper-constantan thermocouples was used to record the temperature rise at various locations on the sample at half minute intervals until fracture occurred. These locations were so chosen to give the thermal gradients across two axes of the sample and two adjacent edges.

Fracture experiments were conducted on samples with clean cut and ground edges using the conventional butyl tape sealant and the new thermal conducting sealant. The solar control glass samples used in this work were the heat absorbing "Antisun" type with 51% absorption, 44% reflection and 5% transmission. K_{IC} = 0.85 MPa$\sqrt{m}$ (Jacob, 1978), E = 69 GPa and α = 8.5 x 10^{-6}/°C.

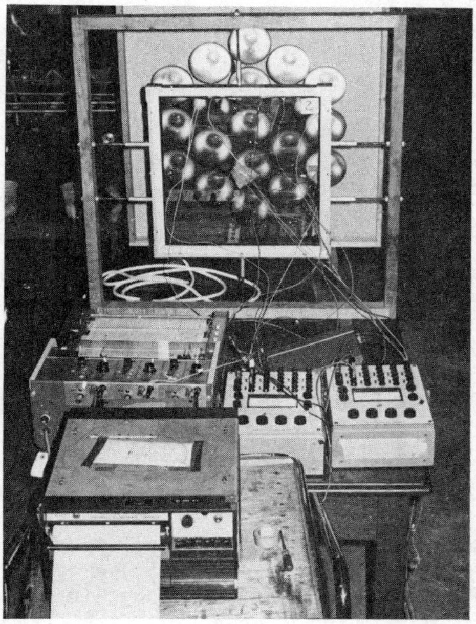

Fig. 5. General arrangement of equipment for fracture experiment.

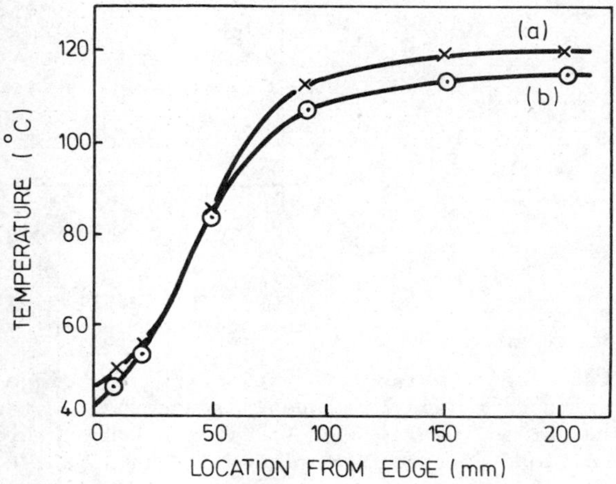

Fig. 6. Temperature profile for (a) ground edge and (b) clean out edge samples in conventional sealant.

RESULTS AND DISCUSSION

Fracture Stress Analysis

Fracture stress calculations using the Pilkington method and fracture mechanics are simple since T_C, T_E and r at fracture are readily available from experimental measurements. In order to use Roberts' stress analysis it is necessary to obtain the thermal gradients at fracture along the $0\text{-}\eta$ and $0\text{-}\xi$ axes. Because of symmetry these gradients are approximately equal. Figs. 6 and 7 show these typical thermal gradients along the $0\text{-}\eta$ axis for glass panes with clean cut and ground edges glazed respectively with the conventional and new conducting sealant. A third order polynomial, $T = T(\eta)$, and after coordinate transformation, $T = T(y)$, was used to give a good fit to these experimental data. This allows G_i and θ_i to be obtained and σ_f to be calculated at the center of the edge using equations (2), (4) and (5). The temperature distribution $T = T(\xi)$ does not affect σ_f. Tables 1 and 2 give the calculated fracture stresses for each of the samples tested according to these various analyses. It may be noted that there is good agreement between the computed σ_f results using the Pilkington method and fracture mechanics techniques with those calculated from experimental fracture strains measured by the gauges. The

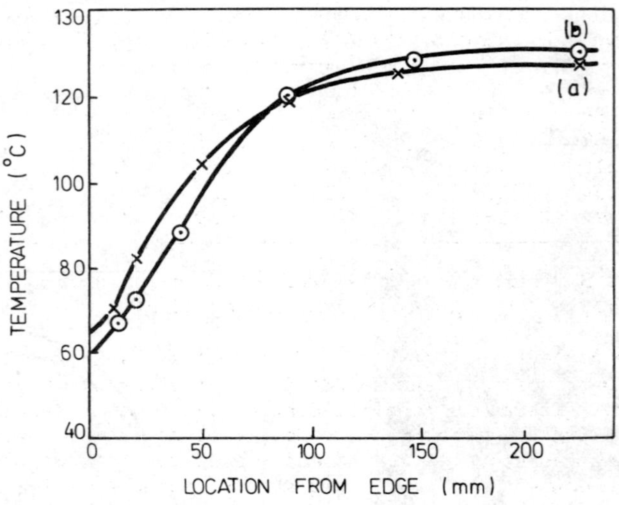

Fig. 7. Temperature profile for (a) ground edge and (b) clean cut edge samples in conducting sealant.

stresses obtained by the Roberts' technique are usually higher than the other methods. This is probably caused by the fact that G_i is very sensitive to the 3rd order polynomial used to fit the temperature distribution curve. The edge conditions of the glass plates do not appear to have any significant effects in controlling the fracture stress (see Tables 1 and 2). Thus the much publicized requirement of glass manufacturers for clean cut edges for solar control glasses seems not warranted.

Comparison Between the New and the Conventional Glazing Technique

The primary aim in using the conducting sealant was to raise the edge temperature of the glass pane and consequently reduce the thermal gradient. This in turn should decrease the thermal stress built up at the edge. By comparing Figs. 6 and 7 it is clear that the edge temperature in the conducting sealant is indeed substantially

TABLE 1 Comparison of Fracture Stresses for Glass Panes Glazed with Conventional Sealant

Sample No.*	T_c-T_E (°C)	r (mm)	Fracture strain (μ strain)	Fracture stress (MPa)	Fracture stress calculations (MPa) Eq. (7)	Eq. (2)	Eq. (10)	Eq. (9)
1	75	3.69	492	33.9	34.3	40.5	33.7	31.3
2	74	4.85	414	28.6	33.8	52.0	29.4	27.3
3	62	6.15	374	25.8	28.3	32.9	26.1	24.3
1G	77	4.46	482	33.3	35.2	29.8	30.7	28.5
2G	83	4.15	446	30.5	37.9	33.3	31.8	29.5
3G	74	4.39	444	30.6	33.8	37.7	30.9	28.7

TABLE 2 Comparison of Fracture Stresses for Glass Panes Glazed with Conducting Sealant

Sample No.*	T_c-T_E (°C)	r (mm)	Fracture strain (μ strain)	Fracture stress (MPa)	Fracture stress calculations (MPa) Eq. (7)	Eq. (2)	Eq. (10)	Eq. (9)
4	56	7.62	300	22.8	25.6	37.1	23.5	21.8
5	49	6.96	310	21.4	22.4	31.8	24.6	22.8
6	63	5.00	376	25.9	28.8	34.8	26.5	26.9
4G	70	4.15	414	28.6	31.9	33.9	31.8	29.5
5G	69	5.54	366	25.3	31.5	39.7	27.6	25.6
6G	63	4.92	368	25.4	28.8	30.4	29.2	27.1

* 1, 2, 3, 4, 5, 6 - samples with clean cut edges; 1G, 2G, 3G, 4G, 5G, 6G - samples with ground edges.

higher than in the conventional sealant. This increase is normally in the range 15°C to 30°C and is of a sufficient magnitude for the engineers to recommend the use of solar control glasses in many buildings which otherwise would have required the much more expensive toughened glass. Note that at fracture T_c is also higher with the new glazing technique. Figs. 8 and 9 show the rate of thermal strain development for the conventional and new glazing techniques. It is apparent that at any given time $d\varepsilon/dt$ was much smaller in the conducting sealant and that the time to fracture was also longer. All these positive results confirm the advantages and indicate the effectiveness of the new glazing technique. It should however be pointed out that in the conducting sealant all the samples failed at lower stress and strain levels (see Tables 1 and 2). This undesirable effect was not completely understood but it was thought to be caused by some form of slow crack growth (activated by the sealant) giving rise to a slightly larger mirror radius at fracture.

While the new proposed glazing technique has been shown in these experiments to be better than conventional glazing with regard to minimising incidence of thermal fracture, numerous problems have yet to be solved before the method can be fully accepted and used in critical commercial projects. Some of these are mentioned as follows: (a) A suitable thermal conducting sealant has to be further developed so that it possesses physical characteristics of good adhesion, non-ageing, being impervious to water vapour penetration and good resistance to U-V degradation. (b) Since the tests reported here were carried out under laboratory conditions and the frame was not installed into a wall as would be in practice the performances exhibited by the conducting sealant might not be realistic. To counteract the heat losses from the frame to the building structure it would be necessary to insulate

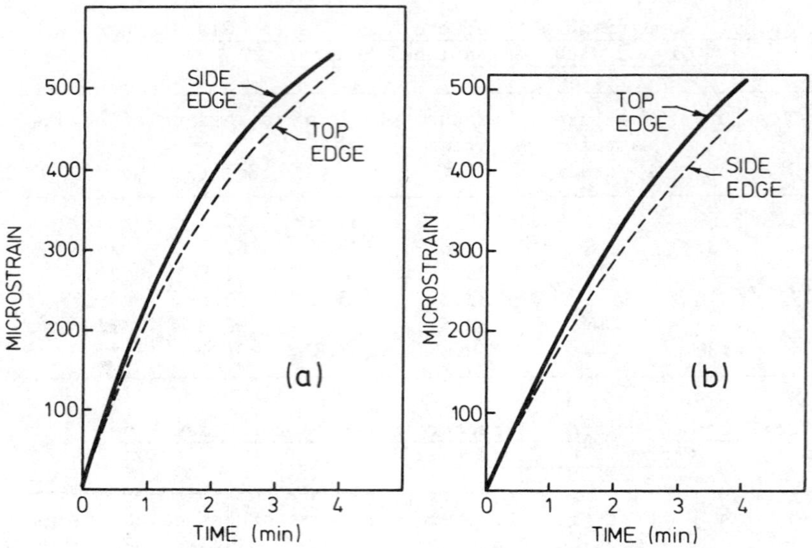

Fig. 8. Measured strain for (a) clean cut edge and (b) ground edge samples in conventional sealant.

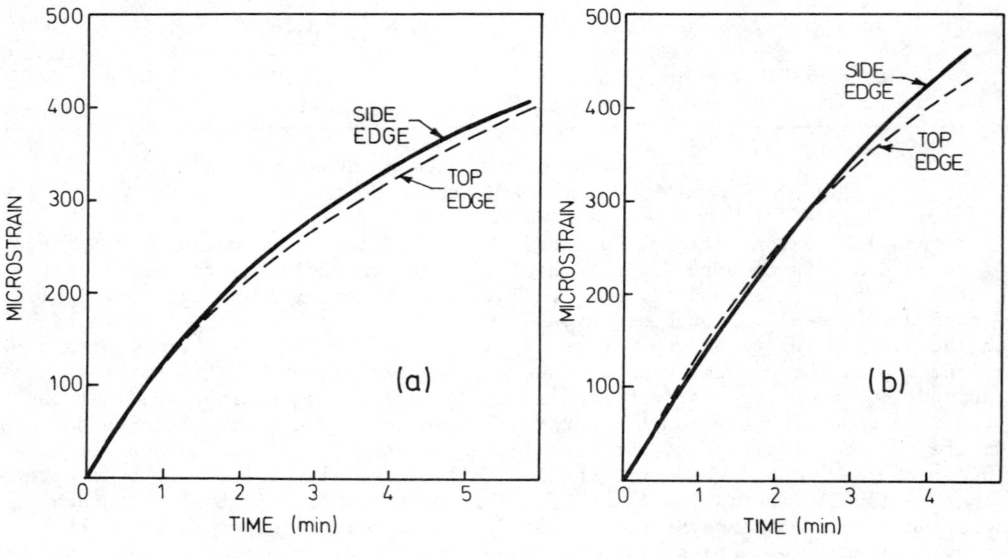

Fig. 9. Measured strain for (a) clean cut edge and (b) ground edge samples in conducting sealant.

the perimeter of the frame. In addition effects of external shades and edge cover widths must be carefully evaluated in future experiments. (c) The whole glazing system hinges on speed and economy of installations so a very detailed cost benefit study has to be carried out prior to the use of this new glazing technique.

CONCLUSIONS

Fracture stress of glass panes fixed in window frames due to solar radiation absorption can be accurately determined from the mirror radius and K_{IC} using fracture mechanics analysis. The methods used by commercial glass manufacturers such as Pilkington Brothers Ltd. and Pittsburg Plate Glass Co. give good agreement with experimental σ_f values. The concept of using a thermal conducting sealant in the rebate of a black metal frame to minimise thermal fracture has been successfully proven to be valid in this study. With some further research and development work it is hoped that this new glazing technique can be made commercially available to benefit the building industry.

REFERENCES

Bansal, C.K. (1977). On fracture mirror formation in glass and polycrystalline ceramics. Philosophic Magazine, 35, 935-944.

Blight, G.E. (1974). Thermal strain and fracture of building glass. First Australian Conference on Engineering Materials. N.S.W. University, N.S.W. pp. 685-700.

Colvin, J.B., and D.G. Lightfoot (1974). Thermal stress in non-rectangular glazing. Pilkington, Technical Sales Laboratory, Report No. 382.

Duhamel, J.M. (1838). Memoire sur le calcul des actions moleculaires developpees par les changements de temperature dans les corps solides. Mem. pres. Acad. Sci., Paris, 5, 440.

Jacob, L.J.S. (1978). Aspects of thermally induced fracture of solar control glasses used in buildings. M.Eng.Sc. Thesis, University of Sydney.

Liu, A.F. (1972). Stress intensity factor for a corner flaw. Engr. Fract. Mech., 4, 175-179.

Mecholsky, J.J., R.W. Rice, and S.W. Freiman (1974). Prediction of fracture energy and flaw size in glass from measurements of mirror size. J. Am. Ceram. Soc., 57, 440-443.

Pilkington Brothers Limited (1964). The fracture of glass by solar radiation. Glass and Windows Bulletin, 4, pp. 1-3.

Pilkington Borthers Limited (1972). The application of solar control glasses. Glass and Windows Bulletin, 10, pp. 6-29.

Pittsburg Plate Glass Company (1974). Installation recommendations in tinted glass. Technical Service Report No. 104C, Glass Division, P.P.G. Technical Services.

Roberts, E. (1962). Elastic design charts for thin plates with spanwise and chordwise variations in temperature. NASA, Technical Note D-1182.

Smith, F.W., A.F. Emery, and A.S. Kobayashi (1967). Stress intensity factors for semi-circular cracks - II. Semi-infinite Solid. Trans ASME, J. Appl. Mech., 34, 952-959.

Volume 3, ICM 3, Cambridge, England, August 1979

FUNDAMENTAL APPROACHES TO THE STATISTICAL FRACTURE ANALYSIS OF BRITTLE SOLIDS

D. H. Pankow and I. Finnie*

Space Science Laboratory, University of California, Berkeley, California 94720, U.S.A.
***Department of Mechanical Engineering, University of California, Berkeley, California 94720, U.S.A.**

Although brittle materials have many very attractive properties, their use in mechanical designs is often limited by low tensile strengths and a lack of ductility. Theoretical predictions based on their crystalline structure suggest tensile strengths roughly one hundred times greater than the normally observed values. Griffith (1920) was the first to suggest that this discrepancy was the result of small flaws dispersed throughout these brittle materials. Weibull (1939) astutely observed strength variations which he logically attributed to a distribution of flaws in the material. Yet to this day, these "inherent flaws" often cannot be systematically detected by non-destructive methods. We can only infer the presence of a distribution of flaws of varying severity from the variation in strength of nominally identical specimens and the "size effect." That is, for geometrically similar specimens, the average stress at fracture increases as size decreases. Perhaps because these "inherent flaws" are generally undetectable, relatively little effort has been given to predicting mechanical behavior from the microstructural features.

As a low tensile strength is often the single greatest shortcoming from a list of desirable material properties, it is increasingly urgent that we understand, and are able to predict this behavior. The familiar procedures of linear elastic fracture mechanics (LEFM) have been successfully applied to brittle solids containing large cracks. However, these "deterministic" relations cannot describe adequately the size effect observed in smooth bar tests, nor the role of rounded notches and other stress concentrations. Experience suggests that a "probabilistic" approach best describes the strength behavior, often based on the strength variations in a number of nominally identical test specimens. By fitting a probability distribution to the strength data, Weibull (1939) developed detailed procedures for predicting the role of specimen geometry and stress distribution. After forty years, his methods continue to be extremely useful and find an increasing number of applications (Finnie, 1977). We review Weibull's procedures and establish a distribution of "equivalent Griffith flaws" based on the assumed strength distribution. Next we consider other formulations which have been proposed for the strength distribution of brittle solids and point out some advantages and limitations.

The Weibull approach has been applied to study many configurations. However, the important problem of the blunt notch in a beam appears to have received little, if any, attention. We present some results of an analysis of this problem which is a demanding test of the validity of the Weibull approach. With sharper notches a transition to LEFM behavior is observed. We discuss the condition under which this

MS247

transition occurs and the related question of the effect of notch radius on the apparent fracture toughness of different brittle solids.

THE WEIBULL APPROACH

Since brittle solids are much stronger in compression than tension, Weibull (1939) chose to ignore any contribution of compressive stresses to failure.

As his basic postulate, Weibull envisioned materials filled with small flaws which he described as "... cracks according to Griffith or dislocations in the atomic grid according to Taylor." Avoiding any physical discussion of these flaws, each was classified by its tensile strength. The flaws were viewed as links of a chain with individual strengths which vary from link to link. Because of the catastrophic nature of tensile fracture in brittle solids, when one link breaks, the entire chain is broken. To predict the behavior of this chain, we consider N links connected in series. For large N, the discrete distribution of strengths is replaced by a continuous distribution. The probability that a random link fails at or below a tensile stress (σ) is given by $f(\sigma)$. For the chain to survive, each of the links must independently survive. Hence, the probability of chain survival is given by the product of all individual link survival probabilities:

$$P_{survive} = \prod_{i=1}^{N} (1-f_i) \tag{1}$$

where $(1-f_i)$ is the survival probability of the ith link. Rearranging yields:

$$P_{survive} = \exp[\Sigma_i^N \ell n\ (1-f_i)]$$

Using the approximation $\ell n(1 + x) \sim x$, and taking the individual failure probabilities to be small gives:

$$P_{survive} = \exp(-\Sigma_i^N f_i)$$

For the large number of links, the summation is replaced by an integral. We also assume the number of links to be proportional to the material volume. The chain survival expression may be written as:

$$P_{survive} = \exp[-\int_{vol} (V_i/V)\ f(\sigma)\ dV] = \exp[-\int g(\sigma)\ dV] \tag{2}$$

Here the individual link failure probability (f) is replaced by a distribution per unit volume (g). An interesting point is that $g(\sigma)$ may be shown to represent the expected number of flaws per unit volume that are weaker than strength (σ). The bracketed portion of Eq. 2 was termed the <u>Risk of Rupture</u> (B) by Weibull.

The question that remains to this day is: what is the mathematical form for $g(\sigma)$ that best describes the character of various materials? With great insight, Weibull chose a simple and workable form:

$$g(\sigma) = (\sigma/\sigma_o)^m \tag{3}$$

The exponent (m) serves as a measure of spread or dispersion of observed values, and is often called the Weibull modulus. The denominator (σ_o) is a scale factor, with units of stress times the mth root of volume.

In problems of interest, the tensile stress is often a function of location within the specimen, rather than a single value. The volume element then becomes the product of length increments and the risk of rupture requires a triple integration over the regions stressed in tension:

$$B = \iiint [\sigma(x,y,z) / \sigma_o]^m \, dx \, dy \, dz \tag{4}$$

Problems of interest may also include multiaxial stresses. Weibull assumed that only the tensile component that is normal to the flaw face affects failure. The flaw orientation is defined by the spherical angles between its face normal and the principal stress directions (Fig. 1).

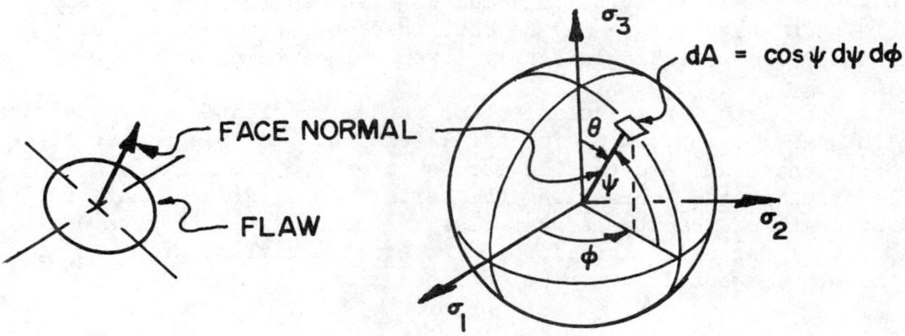

Fig. 1 Spherical coordinate systems.

The magnitude of the tensile stress component tending to open this flaw is:

$$\sigma_n = (\sigma_1 \cos^2 \phi + \sigma_2 \cos^2 \phi) \cos^2\psi + \sigma_3 \sin^2\psi \tag{5}$$

To include randomly oriented flaws and multiaxial stress, Weibull summed the risk of rupture over the range of angles for which σ_n is tensile and adjusted the resulting constants to be in accord with the uniaxial tension results. This summation becomes two integrations, and the risk of rupture is:

$$B = \frac{(2m+1)}{4\pi} \int_{vol} \int_{\phi} \int_{\psi} \left(\frac{\sigma_n}{\sigma_o}\right)^m \cos\psi \, d\psi \, d\phi \, dV \tag{6}$$

These developments are the core of Weibull's description of brittle strength characteristics. It should be pointed out that a general problem may become quite formidable by requiring as many as five successive integrations for the risk of rupture expression. To shorten the analysis with multiaxial problems involving two principal stresses which are perhaps half or less of the maximum principal stress, it is reasonable to use the maximum stress with a uniaxial analysis, since the error is usually a few percent at most.

We next develop a better physical description of the flaws invoked by Weibull using the concepts of LEFM. These flaw size observations are based on the postulate that the material "fracture toughness" is a constant and measurable material property. We assume that the random flaws are dispersed so as to avoid stress field interactions. It also appears acceptable to assume that the flaws are rather flat and sharp edged. In contrast, if the flaws approach spherical shapes, they should all exhibit relatively similar stress concentrations, independent of flaw size, and should fail with little dispersion.

Fracture mechanics solutions describe the strength of a flat flaw having an elliptical perimeter in terms of the semi-minor axis (a) of the ellipse, with a correction to include the length ratio of the major to minor axis. The relation between the normally applied failure stress (σ_{cr}) and the flaw size (a) has the form:

$$\sigma_{cr} = K_{IC} / (\lambda a)^{1/2} \quad (7)$$

where λ is a shape factor. In the two dimensional Griffith (1920) solution for a through crack, the shape factor is π. In three dimensional solutions for the flat elliptical shaped flaw (Tiffany and Masters, 1965), the shape factor is 1.25 for circular flaws and increases with more elliptic shapes, to a maximum value of π. For less than sharp edged flaws, the maximum edge stress follows the same form. Thus we might expect that failure may be described approximately by the same relation with the shape factor taking slightly lower values than for sharp cracks.

To establish the strength - size relation for randomly oriented flaws we turn our attention to the case of uniform equitriaxial tension. In this stress state all flaws would experience the same tensile stress normal to the flaw face, independent of conceivable orientations. Recalling the definition of $g(\sigma)$ we may conclude that the expected strength of the nth largest flaw in the specimen is obtained by equating n to the risk of rupture. Therefore, with equitriaxial tension (σ_{3t}), the strength of the nth largest flaw may be obtained from:

$$n = (2m + 1)\, V\, (\sigma_{3t}/\sigma_o)^m \quad (8)$$

As we have established that the applied stress (σ_{3t}) is the same as the normal stress (σ_{cr}), regardless of individual flaw orientations, Eqs. 7 and 8 may be combined to produce the relation between size and cumulative ranking:

$$a(n) = [K_{IC}^2 / (\sigma_o^2\lambda)]\, [V(2m+1) / n]^{2/m} \quad (9)$$

This relation gives the size (a) of the nth largest flaw in terms of material constants (m, σ_o, K_{IC}) and a shape factor having a value perhaps slightly larger than unity. It is interesting to note the factor $(2m+1)^{2/m}$ which accounts for the random orientation of flaws when strengths are deduced from uniaxial tests.

APPLIED STRESS STATES AND FRACTURE MECHANICS

Weibull's analysis is based on the assumption that only tensile stresses normal to the crack face (mode I fracture) affect failure. Later LEFM developments suggest that it may be desirable to also include mode II and mode III loading in predicting fracture. The basic concepts for this extension were provided in a paper by Fisher and Holloman (1946), which has recently been extended by Batdorf and co-workers (1974, 1975, 1977a, 1977b) in a series of papers.

We consider a simple example to illustrate the approach taken in these papers. The basic LEFM relation for a through crack in an infinite sheet loaded by uniaxial tension normal to the crack plane is:

$$K_{IC} = \sigma_{cr}\, (\pi a)^{1/2} \quad (10)$$

If the crack face normal is at an angle (θ) to the applied stress (σ) the component of this stress normal to, or across the crack is given by:

$$\sigma_n = (\sigma / 2)\, (1 + \cos 2\theta) = \sigma \cos^2 \theta \quad (11)$$

Combining Eqs. 10 and 11 with the assumption that only normal stresses cause failure gives:

$$K_{IC} = (\pi a)^{1/2} \sigma \cos^2 \theta \quad (12)$$

Rearranging this expression and using Eq. 10:

$$\theta = \cos^{-1} (K_{IC}/\sigma(\pi a)^{1/2})^{1/2} = \cos^{-1} (\sigma_{cr}/\sigma)^{1/2} \quad (13)$$

Viewing this expression in the context of many cracks that are randomly oriented provides a second meaning. For a given applied stress (σ) greater than the critical stress (σ_{cr}) all cracks having orientations of θ or less will have failed at or below this stress. As the applied stress increases to infinity, this "critical angle" (θ_{cr}) increases to $\pi/2$. When normalized, θ_{cr} then represents a two dimenal cumulative distribution of failure with respect to orientation:

$$P(\theta_{cr}) = (2/\pi) \cos^{-1} (\sigma_{cr}/\sigma)^{1/2} \qquad (\sigma_{cr} \leq \sigma) \quad (14)$$

In reality, we must consider three dimensional orientations. Angles are defined by the spherical coordinates (θ,ϕ) of Fig. 1 and again refer to the crack face normal. In this context, the previous critical angle now becomes an angle of revolution. The cumulative failure distribution then becomes the ratio of the area of the spherical cap on which normal stresses are greater than σ_{cr} to the hemispheric surface area:

$$P(\sigma,\sigma_{cr}) = 1 - \cos \theta_{cr} = 1 - (\sigma_{cr}/\sigma)^{1/2} \quad (15)$$

Differing strength (or size) flaws may now be included. The flaw strength density function, given equitriaxial tensile loading is defined as $p(\sigma_{cr})$. For failure to occur, a flaw must have both the appropriate strength and also necessary orientation. To establish the failure probability under uniaxial loading, we must survey all possible strengths and their corresponding orientation probabilities, accomplished as follows:

$$P(\sigma) = \int_0^{\sigma} P(\sigma,\sigma_{cr})\, p(\sigma_{cr})\, d\sigma_{cr} \quad (16)$$

Fisher and Hollomon (1946) first presented this development in a compact and somewhat different form. Later, Batdorf (1974) chose to use the cumulative flaw strength distribution $[N(\sigma_{cr}) = \int p(\sigma_{cr})]$ and defined the orientation function as $\Omega/4\pi$. Equation 16 then becomes:

$$P(\sigma) = \int_0^{\sigma} \frac{\Omega(\sigma,\sigma_{cr})}{4\pi} \frac{dN(\sigma_{cr})}{d\sigma_{cr}}\, d\sigma_{cr} \quad (17)$$

The orientation function may be generalized to include triaxial stress states ($\sigma_1,\sigma_2,\sigma_3$) denoted by Σ. In a general state of triaxial stress, the normal component for a given θ,ϕ orientation is:

$$\sigma_n = (\sigma_1 \cos^2 \phi + \sigma_2 \sin^2 \phi) \sin^2 \theta + \sigma_3 \cos^2 \theta \quad (18)$$

Batdorf (1977b) then invokes an operator function to evaluate the orientation function:

$$\Omega(\Sigma,\sigma_{cr}) = \int_0^{2\pi} \int_0^{\pi} H(\sigma_{cr},\sigma_n) \sin\theta\, d\theta\, d\phi \qquad \text{where: } \sigma_{cr} > \sigma: H = 0 \quad (19)$$

$$\sigma_{cr} \leq \sigma: H = 1$$

This expression may be evaluated analytically for some stress states, or numerically for more general cases. It may also be verified that this approach reduces to Weibull's multiaxial formulation. However, this technique may be modified to invoke other flaw failure criteria such as combinations of the shear and normal stress on the crack face. An effective stress σ_e which is some function of the normal and shear stresses on the crack face then replaces the normal stress (σ_n) in the operator (H) of Eq. 19.

This formulation is unique and valuable in that it isolates the basic flaw strength distribution $N(\sigma_{cr})$, the flaw failure criterion, and the effects of triaxial stress. It should be emphasized that the question of what combination of stresses (or stress intensities) best describes the behavior of flaws or cracks in real materials is presently far from being answered. This is evidenced by Batdorf's (1977b) choice of five different criteria for analysis and Evans' (1978) analysis using yet another criterion. We also note that the choice of failure criterion often represents small percentage differences in strength prediction when comparing uniaxial and multiaxial loading states.

THE MATERIAL FLAW DISTRIBUTION

We next turn our attention to the material flaw population $g(\sigma)$. Weibull's choice for this function (Eq. 3) was strictly empirical. He also suggested a three parameter distribution, which has received less use. McClintock (1974) introduced a "simple if inexact model" which provides insight on the nature of the material flaw population. Later works have attempted to extend this model, or directly establish this function $g(\sigma)$ from data analysis.

Rephrasing McClintock's descriptions, his basic premise was that there is a small but finite probability (p) that each material grain is not bonded to its neighbor, thus forming a small crack. The larger cracks which cause specimen failure are then the result of chance accumulations of single grain defects in the large array of grains. To establish the resulting relations, a one dimensional array of units (grains) is defined. From this array, elements are then defined as either a single bonded unit, or a chance series of non-bonded units followed by a single bonded unit which closes the crack. Hence, the conditional probability of a single unit crack is: $p(1 - p)$. Likewise, the chance of an n unit crack is: $p^n(1 - p)$. Performing the appropriate integrations over the range of elements yields a cumulative flaw size (a) distribution from largest to smallest of the form:

$$g(a) = c_3 \exp(-c_4 a) \tag{20}$$

Here (a) is the crack length (multiples of a single unit), while c_3 and c_4 are grouped material constants. Fisher & Hollomon (1946) used this same relation, however, theirs was a strictly empirical choice. McClintock then chose the size, strength relation:

$$\sigma = K_{IC} / (2a)^{1/2} \tag{21}$$

With this, the cumulative strength distribution takes the form:

$$P_{survive} = \exp\left[- \int c_1 \exp(-c_2 / \sigma^2)\, dV\right] \tag{22}$$

Given the form of Eq. 22, it is at best difficult to obtain analytic relations for test specimen geometries other than uniform tension, a test which is difficult to perform accurately. To overcome these difficulties, we resort to an approximate method, assuming that Weibull methods are a reasonable representation of the actual behavior. Using Weibull's methods we may define a tensile strength (σ_t) that is equivalent to the bending strength (σ_b). For three point bend tests, this relation is:

$$\sigma_t / \sigma_b = [1 / (2(m + 1)^2)]^{1/m} \tag{23}$$

By scaling each of the observed bend test strengths, we may then proceed with the data analysis as if the tests were equivalent to tension. Using bend test data with this approach (Pankow, 1978) found that the two distribution forms fit the data equally well. Using these distributions to predict median tensile strength behavior for magnesia gives Fig. 2.

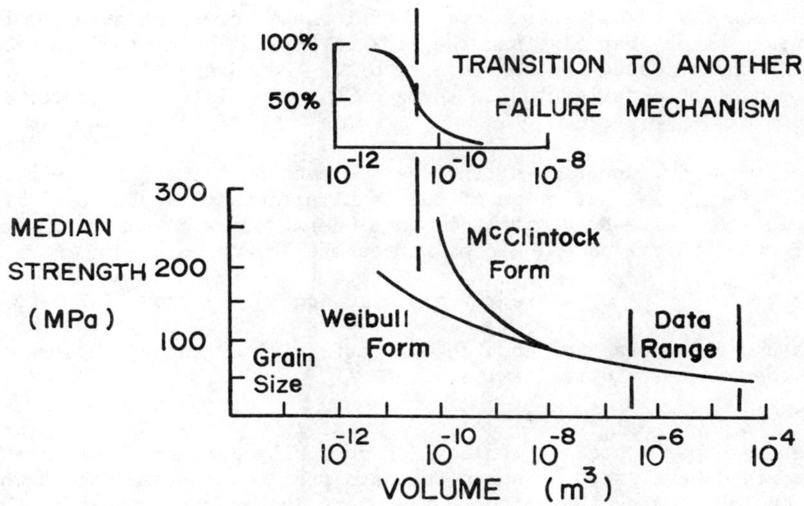

Fig. 2 Predicted Median Tensile Strength Behavior for Magnesia.

We observe that the McClintock form contains an asymptote, which is approximate as his assumptions break down at this point. In the physical sense, this may be viewed as an ever increasing chance of finding no flaws as the region becomes smaller, suggesting that another failure mechanism must become effective.

McClintock's strength criterion (Eq. 21) is inexact as he has considered the major flaw axis rather than the minor axis as the characteristic flaw dimension (Tiffany and Masters, 1965). To model real materials, it would be desirable to consider at least a two dimensional row and column interacting array of units, representing a cubic grain structure. Batdorf (1975) attempted to extend this model by considering each flaw to have a circular perimeter. However, his method still reflects McClintock's one dimensional character.

Taking another approach, Evans and Jones (1978) provided data manipulation techniques for several test geometries wherein $g(\sigma)$ may be obtained without a priori assumptions of its form. This technique is valuable but requires smoothed graphical or numerical derivatives to be established from the discrete data observations.

THE NOTCH STRENGTH OF BRITTLE SOLIDS

LEFM studies have provided an extensive background for the strength prediction of sharp notches in brittle materials based on "fracture toughness" data. However, surprisingly little effort has been given to the strength prediction of blunted notches. Using Weibull's statistical method, we provide a general relation for predicting notch strengths, and discuss the transition from statistical fracture to fracture mechanics as the active failure criterion.

We take the notch to be two dimensional, or a through notch in a specimen of width (w), having a depth (a), and a tip radius (ρ). Three solutions for the stresses in the vicinity of the notch were used: The Inglis (1913) solutions for uniaxial and equibiaxial loading, and the Creager and Paris (1967) solution. In each of these solutions, the stresses can be expressed in the functional form:

$$\sigma_{ij} = K_I \, F(a/\rho) \, f(i,j) \tag{24}$$

where i,j is the notch tip coordinate system. Using this technique, the risk of rupture expression may be expressed in a dimensionless form. Because the high notch tip stresses decay rapidly, and the risk of rupture calls for the mth power of stress, only the stresses in the vicinity of the notch are significant. Thus we are able to predict notch strengths in terms of a stress intensity factor rather than stresses.

Using numerical methods, the risk of rupture for each of the stress field solutions was evaluated. The full description of this analysis was presented by Pankow (1978) and will be published elsewhere with additional data. Viewing numerical solutions, the risk of rupture in all cases, can be represented by the closed form relation:

$$B = V \, (K_I/K_o)^m \qquad \text{where: } V = \rho^2 w/10 \text{ and } K_o = \sigma_o(\pi\rho)^{1/2} \,/\, 1.86 \tag{25}$$

This relation has an accuracy of about 5% or better (with respect to stress) when applied with the non-restrictive limits:

$$m > 6 \;,\; a/\rho > 5 \;,\; ma \,/\, \rho > 100 \tag{26}$$

Figure 3 presents experimental data using a high purity commercial alumina (Al_2O_3) ceramic. The right hand graph presents the data from tests of smooth beams in three point bending. Using the Weibull parameters deduced from these tests with Eq. 25 provides the solid line (center) prediction of notched specimen strength and the dashed line represents the observed strengths. We note that the predicted and measured σ_o values differ by only 10%, whereas the effective stressed volume differs by roughly four orders of magnitude. Finally, the data on the left presents tests of specimens having a notch root radius of 0.02 mm. The lack of data dispersion shows that a basic change in the failure mechanism has occurred.

Equation 25 may also be used to establish the behavior of differing notch radii. A derivation (Pankow, 1978) of the expression for the mean stress intensity at failure (K_{If}) indicates a linear relation when K_{If} to the m/(m - 2) power (~1) is plotted against $\rho^{1/2}$. Likewise, the standard deviation may also be shown to decrease with decreasing ρ. Thus, if all test data are plotted in the form of K_{If} vs $\rho^{1/2}$, an envelope which converges to the fracture toughness (K_{IC}) should result. We have observed this behavior in testing glass and alumina.

McClintock's description, in which flaws are associated with grain boundaries, helps to establish a priori criteria as to which failure criterion is valid. For the Weibull, or statistical fracture mechanism, the notch tip region ($\rho^2 w$) should contain many material grains. For a small notch tip radius, comparable to the grain size, the occasional flaw that occurs might be viewed as an approximation of the LEFM sharp starter crack. A posteriori, the failure mechanism may also be verified by the presence or lack of dispersion in the data.

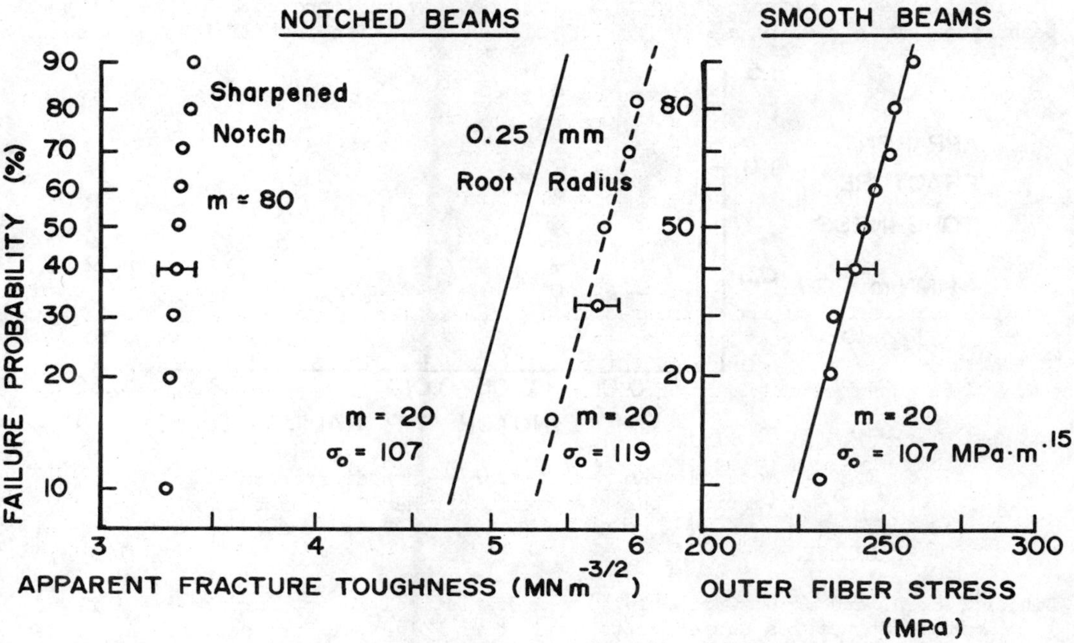

Fig. 3 Test results for alumina. The horizontal bars represent estimated error bounds.

With a larger grain size the transition from LEFM to statistical behavior should occur at a larger notch radius. Also in certain materials, such as rocks and cast iron, it appears to be difficult to prepare notches or cracks without introducing "damage" around the crack tip. For this, or other, reasons the apparent fracture toughness of these materials is insensitive to notch tip radius out to quite large values. In addition one observes the paradox that inherent flaw sizes, calculated from measured values of fracture toughness and smooth bar strength, may be of the same magnitude as the specimen dimensions. For example, Fig. 4 presents data obtained from notched beam tests on a granodiorite rock. This material was quite coarse grained such that each specimen was only a few grains wide. If anything, the fracture toughness increases with decreasing notch tip radius and the inherent flaw sizes deduced for smooth beams exceed the specimen dimensions.

A suggested explanation for this behavior is that fracture is forced through the sound material of individual grains, rather than following the perhaps significantly weaker grain boundaries. This then poses a problem which we cannot yet answer. That is, what fracture toughness does one use to deduce inherent flaw sizes for this type of material?

CONCLUSIONS

We have presented the methods of Weibull in some detail, as they are widely used, and practically proven techniques. Words of caution are due, as the material constants (m, σ_o) are often affected by changes in manufacturing techniques. Indeed, a new population of predominantly surface flaws, that might be termed "production flaws", can be introduced by lackadaisical quality assurance.

As Weibull's analysis is empirical in nature, we have reviewed more recent works, attempting to provide insight to the statistical fracture mechanism. By isolating the relations between the flaw population, the fracture criterion, and multiaxial loading, perhaps statistical prediction methods may be improved. At present these techniques often require a lengthy analysis combined with insufficient test data.

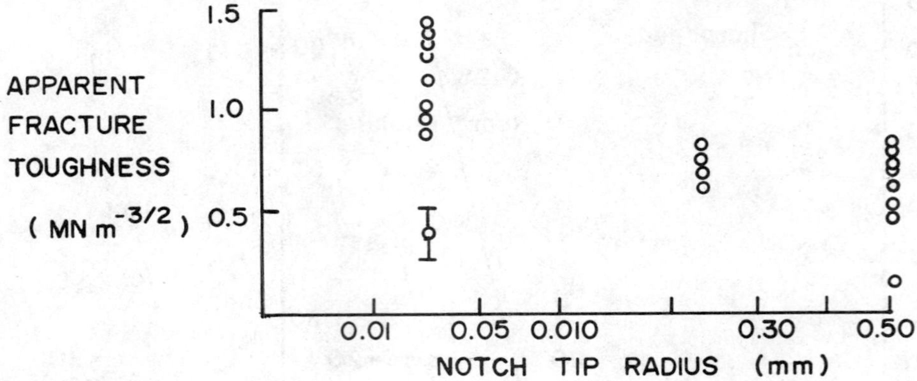

Fig. 4 Notched beam results for a granodiorite rock. Vertical bar represents estimated error bounds.

REFERENCES

Batdorf, S. B. and J. G. Crose (1974). A Statistical Theory for the Fracture of Brittle Structures Subjected to Nonuniform Polyaxial Stresses. J. Appl. Mech., 41, 459-464.

Batdorf, S. B. (1975). Fracture Statistics of Brittle Materials with Intergranular Cracks. Nuc. Eng. Design, 35, 349-360.

Batdorf, S. B. and D. J. Chang (1977a). On the Relation Between the Fracture Statistics of Volume Distributed and Surface Distributed Cracks. UCLA-ENG-7723.

Batdorf, S. B. and H. L. Heinisch (1977b). Weakest Link Theory Reformulated for Arbitrary Fracture Criterion. UCLA-ENG-7728.

Creager, M. and P. C. Paris (1967). Elastic Field Equations for Blunt Cracks with Reference to Stress Corrosion Cracking. Int. J. Fracture, 3, 247-251.

Evans, A. G. (1978). A General Approach for the Statistical Analysis of Multiaxial Fracture. J. Am. Ceram. Soc., 61, 302-308.

Evans, A. G. and R. L. Jones (1978). Evaluation of a Fundamental Approach for the Statistical Analysis of Fracture. J. Am. Ceram. Soc., 61, 156-160.

Finnie, I. (1977) Waloddi Weibull-90 Years Young on June 18, 1977. Trans. ASME, 99H, 193.

Fisher, J. C. and J. H. Hollomon (1946). A Statistical Theory of Fracture. Trans. A. I. Min. Met. Eng., 171, 546-561.

Griffith, A. A. (1920). The Phenomena of Rupture and Flow in Solids. Phil. Trans. Roy. Soc., A221, 163-198.

Inglis, C. E. (1913). Stresses in a Plate due to the Presence of Cracks and Sharp Corners. Trans. Inst. Nav. Arch., 55, 219-230.

McClintock, F. A. (1974). Statistics of Brittle Fracture. In R. C. Bradt (Ed.), The Fracture Mechanics of Ceramics, Vol. I, Plenum Press, New York, 93-114.

Pankow, D. H. (1978). The Size Effect in Statistical Fracture. D. Eng. thesis, Univ. Calif., Berkeley.

Tiffany, C. F., and J. N. Masters (1965). Applied Fracture Mechanics. In Fracture Toughness Testing, ASTM-STP-381, ASTM, Phila., 249-272.

Weibull, W. (1939). A Statistical Theory of the Strength of Materials. Ingvetenskakad. Handl., 151, Stockholm, 5-47.

Volume 3, ICM 3, Cambridge, England, August 1979

DEFORMATION AND FRACTURE OF CONCRETE SUBJECTED TO IMPACT LOAD OR BLAST SHOCK

J. Takeda and H. Tachikawa*

Department of Civil Engineering, National Defence Academy, Yokosuka 239, Japan
***The 4th Laboratory, Technical Research and Development Institute, Defence Agency, Japan**

INTRODUCTION

Concrete is the most commonly used material for buildings, bridges and many other civil engineering structures. However, it has the disadvantage that it is very brittle and that the tensile strength is far less than its compressive strength. These defects may bring about serious damange to the concrete or reinforced concrete structures, especially in the cases of dynamic loading condition, which leads to fracture modes quite different from the static ones. Photo. 1 shows examples of the fracture modes of concrete.

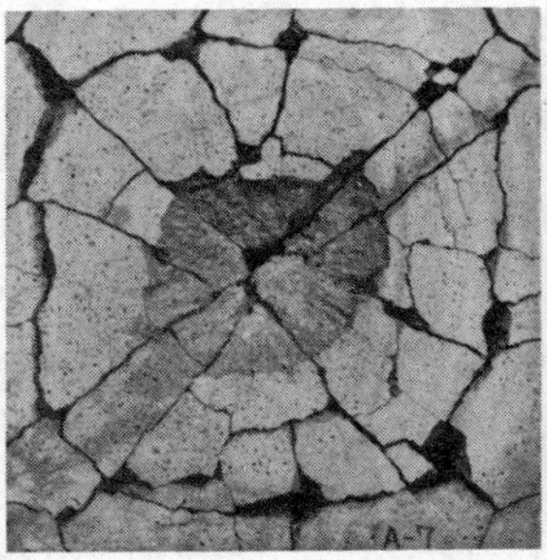

(a)

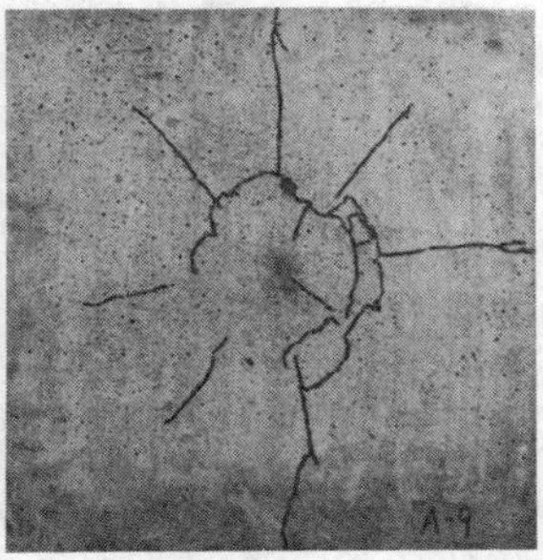

(b)

(c)

Photo.1. Examples of fracture mode of concrete,
(a),(b): fracture of concrete slabs by explosion,
(c): cracks on reinforced concrete retaining wall by low velocity impact.

Though concrete or reinforced concrete walls or slabs loaded statically usually have a fracture mode with several long cracks extending widely (Photo.1c), the fracture by impact of high velocity is usually limited to a zone near to the impact point. Even in the case of an impact by a rolling stone, only a hole is made on a reinforced concrete retaining wall without extending cracks. Not only the variation of the fracture mode, mentioned above, but also the process of concrete is entirely different under dynamic loadings from that in the static condition, as will be explained in this paper later on.

Also there have been two kinds of dynamic experiments carried out generally for a long time, and also in our laboratory; one is a test with high-rate loading and the other is an impact test. Considering the complexity of fracture modes, mentioned above, and the many data of these experiments, the following conclusion was obtained by us: The dynamic behaviour of concrete is superposed by three effects;

(1) transition of mechanical properties and stress-strain relation of concrete brought about by rapid straining(the rate effect on stress-strain relations),
(2) variation of failure criterion of concrete brought about by rapid straining (the rate effect on failure criterion),
(3) contribution of stress waves propagating in concrete.

The method of reaching this conclusion is explained shortly in this paper, though it was reported in ICM-1 and 2 (Takeda and colleagues, 1974,1976).
The main part of this paper is devoted chiefly to the item (3). In order to complete the design method of impact or shock resistance structures, the problems on following subjects should be resolved;

(1) the quantitative determination of stress waves in concrete generated by given impact load,
(2) the propagation characteristics of stress waves in concrete,
(3) the estimation of magnitude of deformation and fracture by impact load,
(4) the estimation of magnitude of stress waves transmitted from other material to concrete,
(5) the estimation of magnitude of fracture inflicted by blast shock caused by explosion.

In this paper, the summarized results of several experiments done are reported, so far as obtained until now.

BEHAVIOUR OF CONCRETE UNDER HIGH RATE LOADING

The behaviour of concrete loaded with high-rate loadings (but without shock[1]) is a part of the whole responses of concrete under impact load or blast shock.
This is understood through the consideration that the elements of concrete under impact or blast shock[2] also cause the deformation with high rate of straining in the same way as in the case of high-rate loadings, which was already ascertained experimentally, as explained later on in this paper.
It is generally known that many matters behave in different ways under high rate loadings from those in static conditions. Such behaviour was observed experimentally in metals by many authorities(Ludwik, 1909; Nadai, 1950) and have been approximately explained theoretically(Freudenthal, 1950). On concrete, the similar behaviour as in metals were ascertained experimentally by many authorities(Evans, 1942; Watstein, 1953) and our group(Takeda, 1962). The experiments on the mechanical behaviour of concrete under high-rate loading in our laboratory were started in 1958 using a new hydraulic testing machine designed by one of our group which enabled the application of loads with the loading rate ranging from that of static tests to about 7×10^3 ton per second within a maximum load of 140 tons, and have been continued

---

[1] The terminology of High-Rate Loading is used in this meaning, i. e. the loading with high rate condition, but without shock in the whole stages of loading.

[2] The behaviour of concrete inflicted by blast shock is usually combined with the effects of generated gas by explosion which is in exceedingly high temperature. The latter is excluded here.

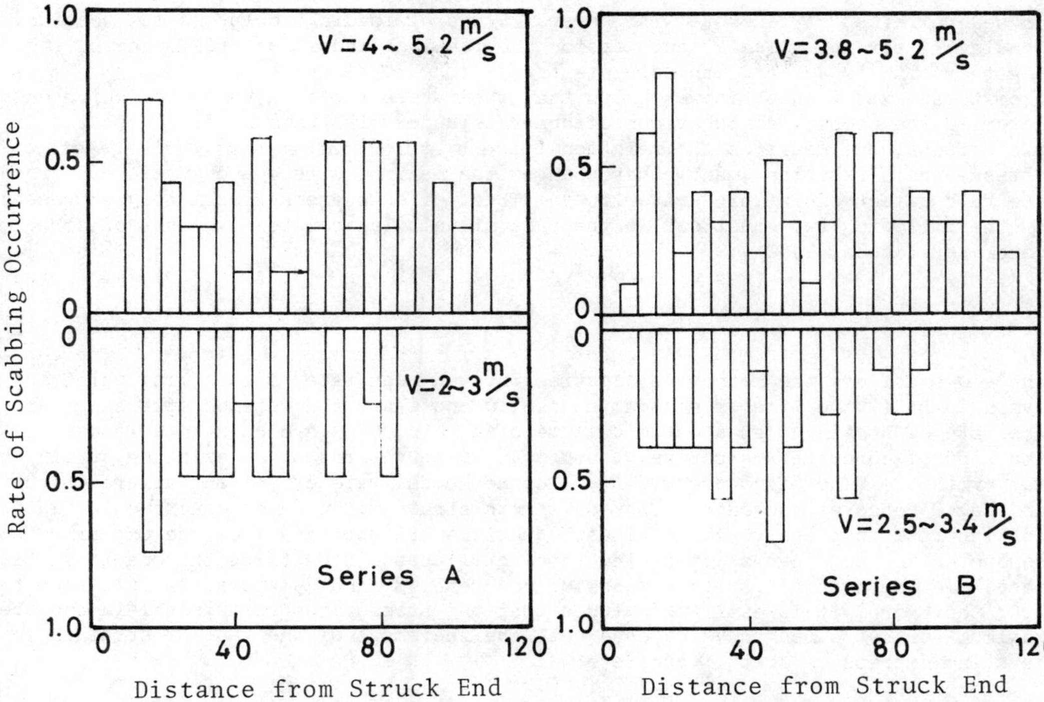

Fig. 1. Histograms of the rate of scabbing occurrence in impact test of mortar prisms.

Examples of the fracture mode of specimens are seen in Photo. 2, which clearly shows the occurrence of so-called scabbings(spallings) which is already given a clear explanation(Rinehart, 1951; Broberg, 1955) that the tensile stress generated in the reflection of compressive stress wave at free end incurs them. Fig. 1 indicates the histograms of the rate of scabbing occurrence in series A and B, calculated from the numbers of scabbings occurred within the range of 5cm wide successively devided on the specimen. It is seen that the numbers of scabbing occurring near the part to the struck end are not always less in low impact velocity than in higher velocity, notwithstanding the total numbers of scabbing occurrence are clearly less in low impact velocity than in higher cases. Of course the scabbing occurrence and its distribution should depend on the shape and type of concerning stress waves in the specimen; therefore, it is most important to know the quantitative data on the stress waves in specimen in order to complete the shock resistance design of structures.

Fig. 2 is an example of equi-strain contour lines on a characteristic plane(x-t plane, x=distance from struck end, t= time), obtained from the strain waves measured at several points in the specimen of series B. It is easily noticed that tensile strain is generated in the wide domain at the time of about 700-800 micro-second after the impact blow which is consistent with the observation that the scabbings occurred almost at the whole part of the specimen, especially in the case of higher impact velocity in Fig. 1. And also seen that the linear parts of contour lines in initial stage of wave propagation do not run parallel with each other in this figure, and hence the velocities of strain wave propagation take different values depending upon the magnitude of strain. This observation is entirely natural because concrete is not elastic. It should be emphasised that plastic wave theory is necessary to analyze and discuss the results of the experiments. To simplify the wave analysis, a 'polygon' stress-strain relation was assumed, as explained in Fig. 3.

on several kinds of concrete with several types of loading(compression, tension, bending, shearing, triaxial compression and tension, and structural members, etc.), together with impact and explosion tests.
From these experiments, it was known that the rate effects on concrete, which characterize the mechanical behaviour of concrete under high rate loading, consist of two effects; one modifies the relation between stress and strain (the rate effect on stress-strain relation) and other changes the maximum stress and strain concerned (the rate effect on failure criterion). These effects are approximately expressed by the following two equations in the case of uniaxial compressive test of normal concrete(Takeda, 1962);

$$\bar{\dot{\sigma}} = \alpha_1 \{ \bar{\dot{\varepsilon}} \}^{\beta_1} \qquad (1)$$

$$\dot{\sigma} = \alpha_2 \{ \dot{\varepsilon} \}^{\beta_2} \qquad (2)$$

where $\bar{\dot{\sigma}}$ and $\bar{\dot{\varepsilon}}$ are the mean rate of stressing and mean rate of straining which are averaged until the time of maximum stress, $\dot{\sigma}$ and $\dot{\varepsilon}$ are the rate of stressing and the rate of straining at a certain value of strain, α_1, β_1 are constants related to the kind of concrete respectively, and α_2, β_2 are functions of an existing strain and vary with the kind of concrete. Eq. 1 concerns the rate effect on failure criterion and Eq. 2 concerns the rate effect on stress-strain relation of concrete subjected to high rate compressive load. These equations are obtained from the experiments conducted in the range of straining rate until about 5×10^5 times the one in static test. In the test of fourteen kinds of concrete, β_1 and β_2 are slightly larger than 1.0; therefore, it is easily concluded that the stress at a certain strain and the maximum stress(strength) increase with the increment of the rate of straining concern in compressive test of concrete.

BEHAVIOUR OF CONCRETE PRISMS SUBJECTED TO LONGITUDINAL IMPACT

The behaviour of concrete prism subjected to longitudinal impact has been studied experimentally and theoretically, because this is the simplest case in the general response under dynamic loadings. Concrete prisms of 4cm x 4cm x 40 - 160cm were tested with several setups and newly designed impact machines. The experiments consist of two series(A and B). In series B, the steel prism(4cm x 4cm x 120cm), which was stuck to one end of the specimen with gypsum, was placed between the impact apparatus and the specimen and impact blow at the head of the steel prism.
The specimen was made of normal concrete and mortar(having the same mix proportion as the mortar matrix in the concrete), with the w/c of 0.5 and 0.7 in series A and of 0.5 in series B. After the curing in water for 28 days, the specimens were placed in air in a basement room, and tested at the age of 90-150 days.
In series A, a steel block(6cm x 6cm x 6cm) was used to make the specimen co-axially compressed, instead of steel prism in series B. The strain waves in the specimen and the steel prism were measured with strain gages stuck to their flank, and with wide-response-range amplifiers(response range of 0-100KHz, NEFF-122) transient recorders(Biomation-805) and memory-scope(Tektronix D13).

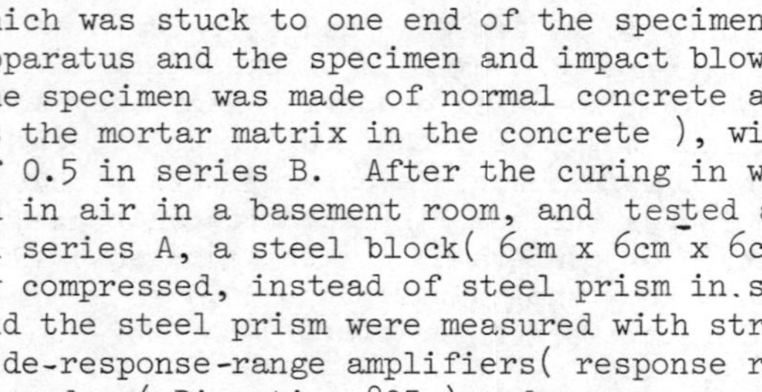

Photo. 2. Examples of scabbings occurred in the impact test of mortar and concrete prisms in series B.

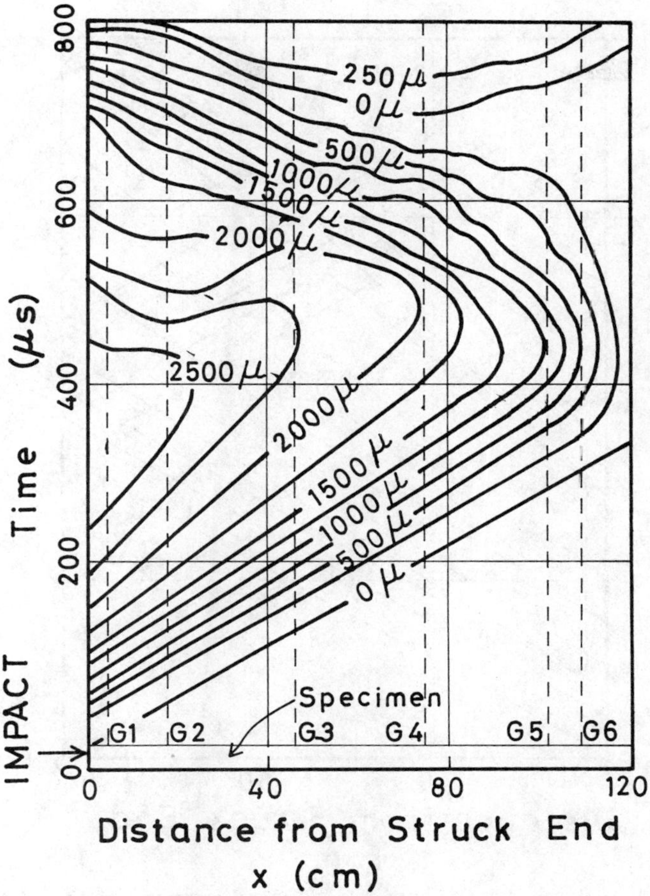

Fig. 2. Example of equi-strain contour lines on characteristic plane, in impact test (mortar prism, 120cm long, in series B).

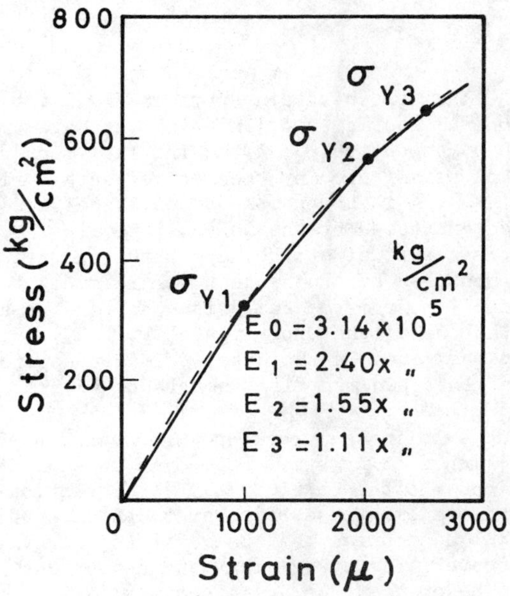

Fig. 3. Stress-strain curve obtained from Fig. 2 (broken line) and polygon stress-strain relation assumed for wave analysis.

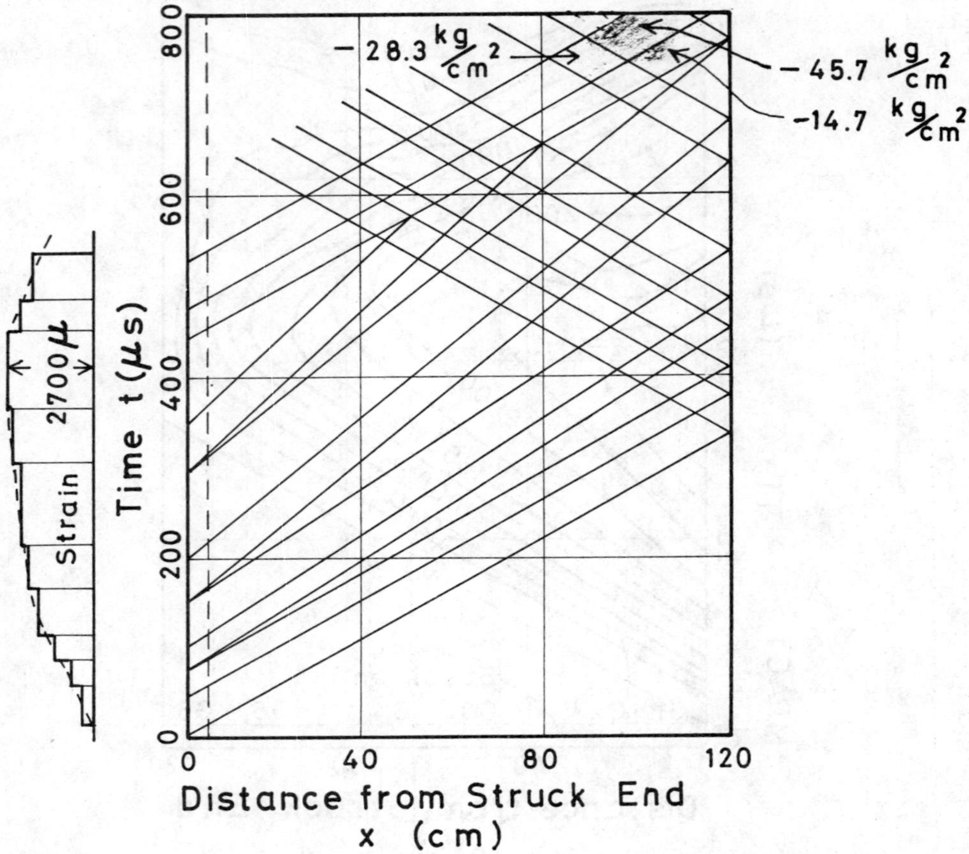

Fig. 4. Example of strain wave analysis of mortar prism, 120 cm long, by plastic wave theory.

The stress-strain curve in Fig. 3(broken line) is completely different from the one in static case, in which the maximum stress and strain are 441 kg/cm^2 and 2550 μ respectively and the initial Young's modulus is 2.2x10^5 kg/cm^2, and this is the one affected by the rate effects during the high rate straining of the specimen. Fig. 4 is an example of wave analysis carried out with the constitutive relation expressed by the polygon(in Fig. 3) and with the assumption that the input strain wave of stair shape shown in the figure which is following the strainwave measured by the nearest strain gages to the struck end(4cm distant from the struck end). From the analysis, the tensile strain emerged in the extremely near domain to the one of the experiments(Fig. 2). This is a very powerful support to the concept that even in the case of impact application, the rate effects by the rapid straining of the specimen also contribute to its behaviour, as the concerning stress waves do.

In the experiment of series B, the strain waves in the specimen were transmitted through the steel prism stuck to the specimen with gypsum. Fig. 5 is an example of the strain waves measured in the specimen and the steel prism. The quantitative relation between the incident and the transmitted strain waves can be obtained by the plastic stress wave theory using the assumption that there is no energy loss in any stages of the wave propagation, however, the energy loss at the boundary of the specimen and steel prism could not be omitted during wave transmission nevertheless the specimen and the steel prism were stuck with gypsum. This is evident in Fig. 6 which shows the comparison of the transmitted and reflected waves at the boundary obtained theoretically (the broken lines) with the experimental results (the solid lines) in the inital stage of wave transmission, using elastic wave theory. The considerable discrepancies between the two waves can be seen, and it is suggested that the appropriate effective ratios of impedance should be obtained and utilized for wave transmission analysis. One more datum, which is necessary for practical wave analysis, is that concerning the wave dissipation during the propagation in concrete. However, it is very difficult to obtain the datum experimentally, although it is encouraging that the approximate analysis

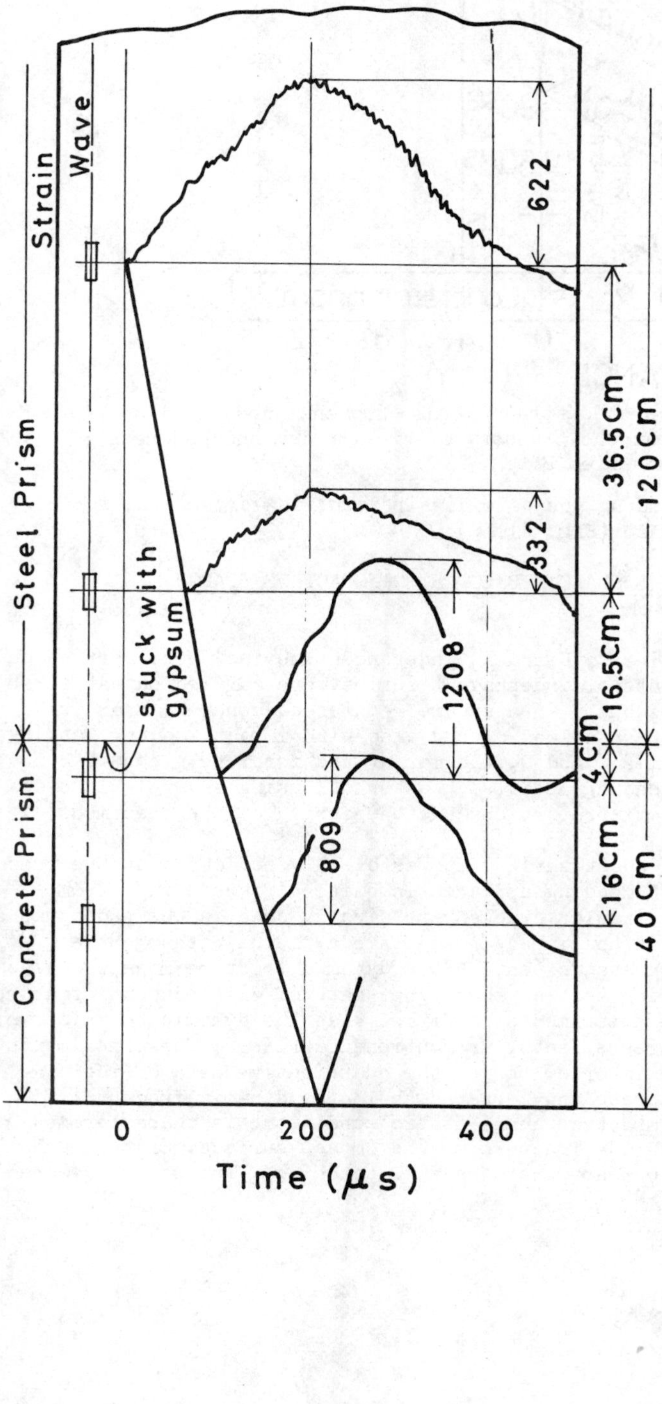

Fig. 5. Example of strain waves measured in steel prism and concrete prism, 40 cm long, in series B.

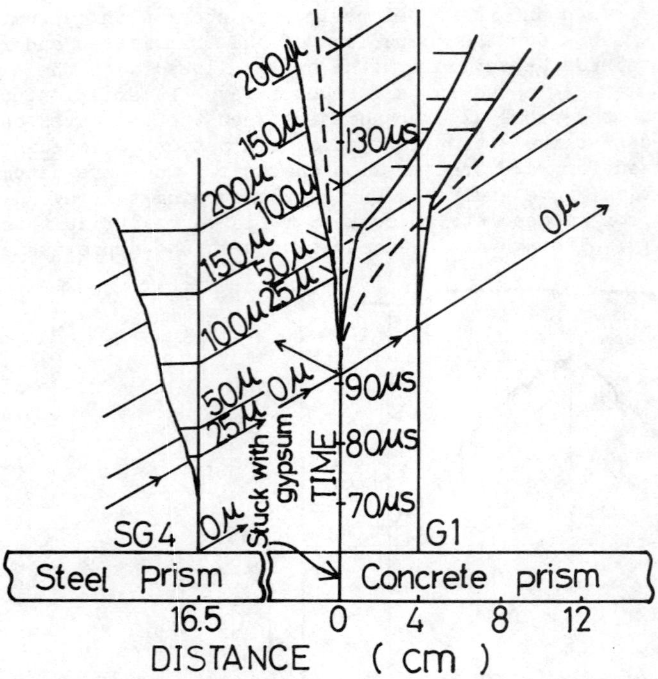

Fig. 6. Comparison of calculated strain waves (broken lines) to measured values (solid lines) at the boundary of steel prism and concrete prism, 40 cm long, in series B.

using plastic wave theory, excluding the wave dissipation, is very useful for estimating the scabbing occurrence (Fig. 4).

BEHAVIOUR OF CONCRETE SLABS SUBJECTED TO IMPACT OR BLAST SHOCK

Three kinds of experiments (series C, D and an independent series) were carried out in this category. The experiments of independent series were carried out at first on the mortar and concrete slabs built in a square reinforced concrete frame with the concentrated static and high-rate loading and also with impact loading, at its center (Tadeda, 1976a). In series C and D, the mortar and concrete slabs with free edges were tested with impact loading and also blast shock using several kinds of explosives at the center. The mix proportions and the curing of the specimens were the same as in series B.

In the experiments of the independent series, it was clearly noticed that the fracture mode of specimens in impact loading appeared in quite different ways from those in static condition; the circular cracks were firstly seen in the photographs taken with a high-speed camera in the impact test (Takeda, 1976b), though the diagonal cracks prevailed in the static test. Such dynamic fracture mode was also seen even in the high-rate loading test at the higher rate of displacement of about 10 cm/sec at the center. These experiments confirmed that the dynamic behaviour of concrete is the synthesis of three effects, as mentioned earlier. The experiments of series C and D were executed in order to get the quantitative data of the deformation and fracture of concrete and mortar slabs. The dynamic fracture of concrete slabs, excluded slow impact or quasi-static condition, is characterized by two modes of fracture; one is the fracture on the impact surface, crater, and the other is the fracture of the rear part of slabs, scabbing. The scabbings are

caused by stress waves apparently, however, analytical estimation of scabbing occurrence by plastic wave theory is very difficult at present.
Therefore, the analysis on the basis of elastic wave theory was done following Norman David(1959). Fig. 7 is a result of the analysis in the case of Lame's constants $\lambda = 0.778 x 10^5$ kg/cm^2, $\mu = 1.167 x 10^5$kg/cm^2, density $\rho = 2.35 x 10^{-6}$kgs^2/cm^4, velocity of propagation of longitudinal wave C_L = 3640m/s, velocity of propagation of transverse wave C_T = 2230m/s, and concentrated impact load at center $P = P_0$ x exp(-bt), b=2.75 micro-sec. The numbers in the figure coincide with the values of stress obtained in the unit of ($1/2\pi P_0$). In the figure, the prevailing tensile strain emerges at the depth of 1.0 - 1.2 cm distant from the back surface and the time of about 16 micro-sec after the impact blow. Phot. 3 is an example of scab, occurred in the test of the series D, in which an apparent crack at the extremely near position to the zone on which the tensile strain wave appeared earliest in Fig. 10, however, many other cracks are also seen in Phot. 3. The latter observation is not helpful for the estimation of depth of scabbings, though it becomes clear that the application of elastic wave theory is useful for the estimation of the depth of initial(presumably) scabbing.
On the crater incurred by explosion, on the other hand, a simple relation that the depth of crater strongly depends upon a parameter λ (Eq. 3), was found and reported in ICM-2(Takeda, 1976a).

$$\lambda = T_1 \times V_0 \times W \qquad (3)$$

where T_1 is explosion temperature, V_0 is specific volume of generated gas and W is weight of explosive.

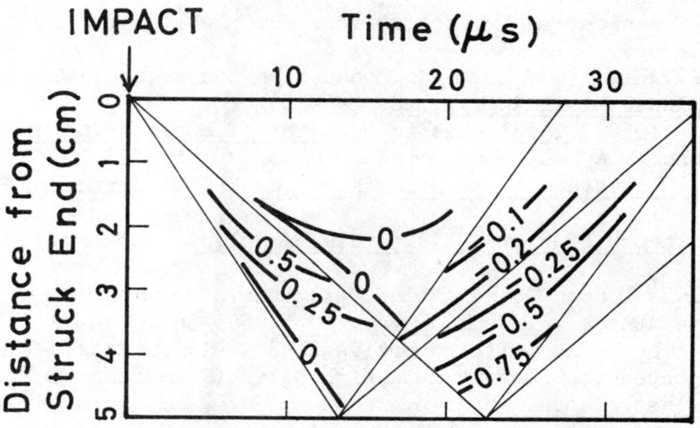

Fig. 7. Stress wave analysis of impacted concrete slab, 5 cm thick, by elastic wave theory.

Phot. 3. Example of scab, in explosion test of mortar slab, 5 cm thick,in series B.

Phot. 1 (a) and (b) , on the first page of this paper, show examples of fracture mode of the specimen of series D carried out for the determination of the depth of scabbing and crater. Results of these will be reported in the near future.

CONCLUSIONS

1. The dynamic behaviour of concrete structures is a superposed result of,

   (1) the rate effect on stress-strain relation of concrete,
   (2) the rate effect on failure criterion of concrete, and
   (3) the contribution of stress waves propagating in concrete.
2. The propagation of stress waves and scabbing occurrence in mortar and concrete prisms can be estimated approximately by the wave analysis using plastic wave theory.
3. The stress wave in concrete prism in the initial stage of impact, transmitted from a steel prism stuck to the concrete prism with gypsum, is considerably less than the value calculated by elastic wave theory.
4. The application of elastic wave theory to the esitmation of the scabbing of concrete slabs under impact or blast shock is not always useful, because the scabbings occur successively in such loading.
5. On the depth of crater in concrete slabs subjected to blast shock by explosion, the relation (Eq. 3) exists.

REFERENCES

Broberg, K. B. (1955). Studies on Scabbing on Solids under Explosive Attack. J. Appl. Mech., 22, 317-323.

David, N. (1959). Transient Analysis of Stress Wave Penetration in Plates. J. Appl. Mech., 26, 651- 660.

Evans, R. H. (1942). Effect of Rate of Loading on the Mechanical Properties of Some Materials. J. Inst. Civil Engrs., 18, 296-306.

Freudenthal, A. M. (1950). The Inelastic Behavior of Engineering Materials and Structures. John Wiley and Sons, London. Chap. 3, pp. 135-140.

Ludwik, P. (1909). Elemente der technologischen Mechanik. Springer, Berlin.pp. 44-47.

Nadai, A. (1950). Theory of Flow and Fracture of Solids. Mc Graw-Hill, N.Y. pp.21-24, 309-316.

Rinehart, J. S. (1951). Some Quantitative Data Bearing on the Scabbing of Metals under Explosive Attack. J. Appl. Phys., 22-5, 555-560.

Takeda, J. (1962). The Mechanical Properties and Their Fundamental Equations of Concretes Subjected to Compressin in High Rates of Loading (in Japanese). Trans. Architectural Inst. Japan, 78, 1-6.

Takeda, J. (1972). Deformation and Fracture of Concrete Subjected to Dynamic Load. ICM-1, IV, 267- 277.

Takeda, J. (1976a). Influences of Straining Rate and Propagating Stress Waves on Deformation and Fracture of Concrete. ICM-2, 1468-1472.

Takeda, J. (1976b). Deformation and Fracture of Concrete under Impact or Blast Shock (in Japanese). Cement and Concrete, 356, 29-36.

Watstein, D. (1953) Effect of Straining Rate on the Compressive Strength and Elastic Properties of Concrete. Proc. Am. Concrete Inst., 49, 729-744.

Volume 3, ICM 3, Cambridge, England, August 1979

EFFECTS OF RADIATION INDUCED POLYMERISATION ON THE MECHANICAL PROPERTIES OF POLYMER IMPREGNATED CONCRETE

S. Ohgishi, H. Ono*, K. Araki and Y. Kasahara****

Nagoya Institute of Technology, Nagoya, Japan
***Chubu Institute of Technology, Kasugai, Japan**
****Japan Atomic Energy Research Institute, Takasaki, Japan**

INTRODUCTION

Polymerisation in making polymer impregnated concrete (PIC) is usually accomplished either by a thermal catalytic or a radiation method. The latter has the merits of increasing the monomer life, reducing the time for polymerising, and preventing evaporation of the monomer. Studies on radiation polymerising of PIC by gamma rays were initiated in 1967 as a joint effort between BNL and USBR. Thereafter, results of many investigations of the gamma ray method were published, however, there are few reports concerning polymerisation by accelerated electrons and X-rays. Each radiation source has a unique energy and dose rate, therefore it requires a different irradiating time; and the characteristics of the radiation affect the polymerisation and mechanical strength of PIC. Electron irradiation has a far stronger dose rate than gamma rays, and electron polymerising is a suitable method for efficient production of PIC with a thin cross section or low density of concrete.

In this programme, effects of irradiation energy on mechanical properties of PIC were examined with regard to the density of the base cement mortar, the total exposure dose, the radiation source and other factors.

EXPERIMENTAL

Preparation of Specimens

The physical properties of each aggregate and the mixing proportions of base cement mortars are shown in Table 1. Three types of test specimen were used. The mortar specimens for the PIC were cured for 14 days in water and dried in an oven for 12 hr at 150°C. The specimens were soaked in monomer under a vacuum of 8×10^2 Pa, and the monomer was pressured by N_2 gas at 3.92×10^5 Pa for one hour. The control specimens were tested after curing 28 days in water (20°C). The kinds of monomer used were as follows.

a) MMA 100% (methyl methacrylate)
b) MMA90%+PEA10%(PEA=polyethylacrylate)
c) MMA90%+Pst10%(Pst=Polystyrene)
d) MMA92%+PMMA8%

Impregnation Method

The characteristics of the irradiation apparatus and testing conditions for the accelerated electron beam are shown in Table 2. In this experiment two types of electron apparatus were used, namely a Cockroft Walton's apparatus with 64.08(pJ) energy and a van de Graaff generator with 32.04 or 19.22(pJ) energy. The specimens, wrapped in aluminum foil, were irradiated on a conveyer belt moving under the scanner and returning as shown in Fig. 1.

Source characteristics of electro magnetic wave and irradiating conditions are shown in Table 3. In the case of X-ray irradiation, the specimens were polymerized in a water vessel (40 × 60 × 8 cm). Fig. 2 shows the X-ray apparatus.

The Gamma ray method used two radiation sources either ^{137}Cs (4800 Ci) in the shape of a cylinder (20 cm ϕ) Fig. 3, or ^{60}Co (120 kCi) with the shape of a plank (120 × 10 cm).

TABLE 1 Physical Properties of Fine Aggregate and Weight Mixing Proportion

Kinds of aggre.	Density of agg. (g/cm)	Water absorp. (%/wt)	Max.[1)] size (mm)	F.M.	2×4×16 cm Mixing proportion[2)] C	S	W	Flow (mm)	ρ g/cm³	0.3×4×16 cm Mixing propotion[3)] C	S	W	Flow (mm)	ρ g/cm³
River sand	2.32	1.15	1.2	1.88	1	2	0.45	188	2.01[4)]	1	2	0.55	198	2.05[4)]
			5.0	2.70					2.16[5)]					2.28[5)]
Light weight	1.85	9.14	1.2	1.90	1	1.47	0.45	220	1.53[4)]	1	1.47	0.55	263	1.60[4)]
			5.0	2.72					1.80[5)]					1.91[5)]
Perlite	1.10	28.2	1.2	2.22	1	0.87	0.45	209	1.09[4)]	1	2	0.55	224	1.19[4)]
			5.0	2.72					1.80[5)]					1.59[5)]

1) Aggregate size 1.2mm was used for 0.3×4×16cm and 5.0mm for 2×4×16cm.
2) Mixing ratio by volume is held constantly as C/S/W=1/2.4/1.42 .
3) Constant volume ratio of c/s/w/=1/2.4/1.73 .
4) Drying density.
5) Values impregnated by MMA monomer.

TABLE 2 Characteristics of Linear Accelerator and Irradiating Conditions.

Accelerator		No.1	No.2	No.3
Capability	Beam energy (pJ)	19.22	32.04	64.08
	Beam current (mA)	2.0	2.0	200.0
Testing	Scanning width(cm)	60.0	60.0	22.0
	Conveyer speed (cm/sec)	2.7	2.7	2.3
	Dose rate (Mrads/sec)	0.125	0.125	1.1
	Time interval	60 sec/pass		
	Total dose(Mrads)	12.5, 25, 50 and 100/ face		

Fig.1
Irradiating apparatus for accelerated electron beam.

TABLE 3 Source Characteristics of Electro Magnetic Wave and Irradiating Conditions.

	Source type	X-ray	^{137}Cs	^{60}Co
Capability	Energy (pJ)	1.28	10.57	20.03
	Radiation source	E=150kVP I=400mA	4800 Ci	120 kCi
Testing	Dose rate (R/h)	5×10^5	4.5×10^4	1.0×10^6
	Irradiation	under water		
	Total dose (MR)	1, 2, 3 and 5	0.45, 1.08, 1.4 and 2.34	2, 3, 5 and 22

Fig.2
Irradiating apparatus for X-ray polymerisation.

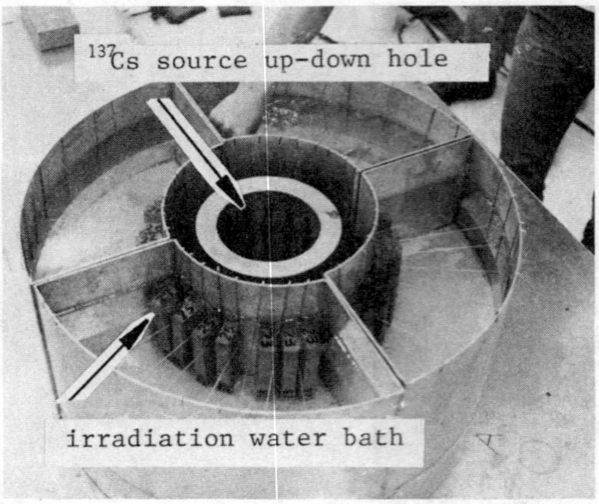

Fig.3
Irradiation water bath for ^{137}Cs γ-rays source.

RESULTS AND DISCUSSION

Polymerisation by Accelerated Electrons

To investigate the polymerising depth from the surface of mortar specimens, specimens of 3 × 40 × 160 mm perlite, light weight and river sand mortar were prepared. The specimens, impregnated with monomer were piled up as shown in Fig. 4, and polymerised by accelerated electrons under the same exposure dose (50 Mrads). Figure 4 shows the relationship between the flexural strength and the thickness of mortar from irradiated face. The results show that when the energy of the electron is smaller, the strength of PIC specimens decreases more abruptly through the thickness. It seems that when the energy intensity of the radiation source is lower, the transmission loss is increased. Such dependency is found similarly in the case of light weight and perlite mortar for the relations between flexural strength and the thickness of specimens. The polymer loading and flexural strength of each lamina were examined for three kinds of mortar with different density, and results are summarised in Table 4. Each value of polymer loading shows that the first lamina has about twice the strength of the fourth lamina.

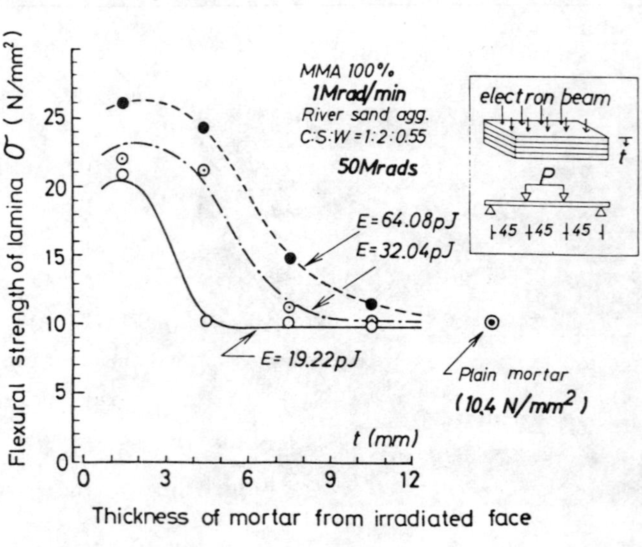

Fig.4 Flexural strength vs. the thickness of mortar from irradiated face.

Test results in Table 4 are plotted in Fig. 5. The strength ratio for the product of thickness and density of mortar (t.ρ) shows the same trend as the relationship between the absorption energy of the accelerated electron beam and the values of (t.ρ). If the energy intensity is changed by degrees from 64.08 to 19.22, we can estimate the desirable conditions for polymerisation by the accelerated method for any kind and thickness of mortar. However, the curves coincide imperfectly because the polymerisation reaction raises the temperature to 80 - 100 °C and a proportion of the electrons which collide with heavy metal atoms (Fe,Al,Ca) in the mortar specimens transform to X-rays.

TABLE 4 Polymer Loading and Flexural Strength of Laminate PIC.

Kinds of mortar	Polymer loading Flexural strength	Plain (N/mm²)	19.22pJ				32.04pJ				64.08pJ				
			1st	2nd	3rd	4th	1st	2nd	3rd	4th	1st	2nd	3rd	4th	
River sand mortar	(%/wt)		7.6	3.7	3.7	2.1	8.9	6.3	4.7	2.8	17.0	15.0	8.0	6.0	1)
	(N/mm²)	10.4	21.0	9.3	10.6	10.0	21.8	21.2	11.5	11.0	25.8	24.2	15.0	14.5	2)
Light weight mortar	(%/wt)		10.4	2.9	2.8	1.8	14.3	10.0	4.7	2.9	22.0	20.0	17.0	8.0	1)
	(N/mm²)	10.34	18.3	11.1	9.3	8.2	18.4	16.1	11.7	10.4	22.1	22.1	19.7	9.8	2)
Perlite mortar	(%/wt)		22.2	5.2	5.4	3.8	24.5	19.6	7.7	6.8	28.5	30.2	26.2	17.2	1)
	(N/mm²)	6.72	12.7	7.9	7.6	7.6	13.1	12.2	7.3	7.7	17.7	16.8	16.8	9.6	2)

Polymer loading = [impregnated polymer weight/PIC Weight] x 100 (%)

1) Polymer loading (%/wt)

2) Flexural strength (N/mm²)

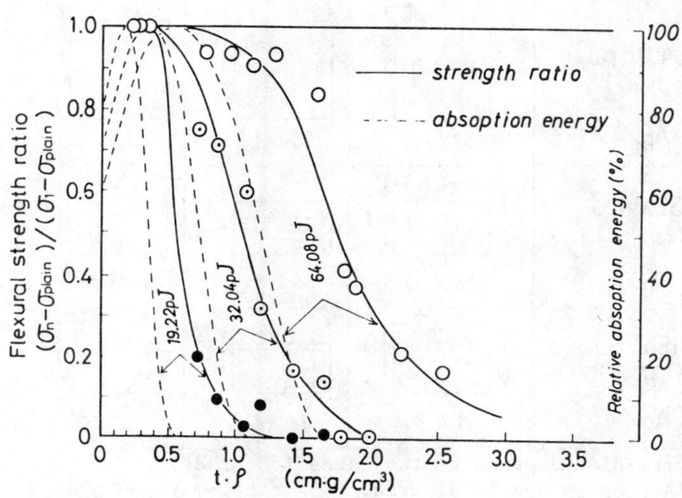

Fig.5 Flexural strength ratio of PIC vs. t·ρ of cement mortar and absorption energy of electrons.

The results on the effect of total exposure dose on flexural strength are shown in Fig. 6. From the accelerated electron method, we can conclude that the flexural strength does not always increase in proportion to the increased exposure dose.

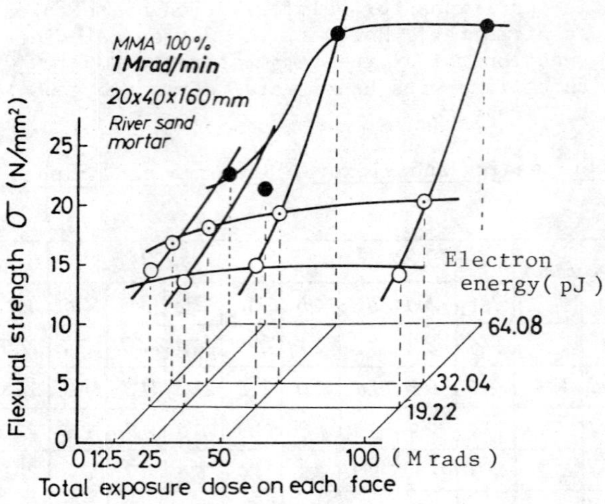

Fig.6 Relation between total exposure dose, beam energy and flexural strength of PIC.

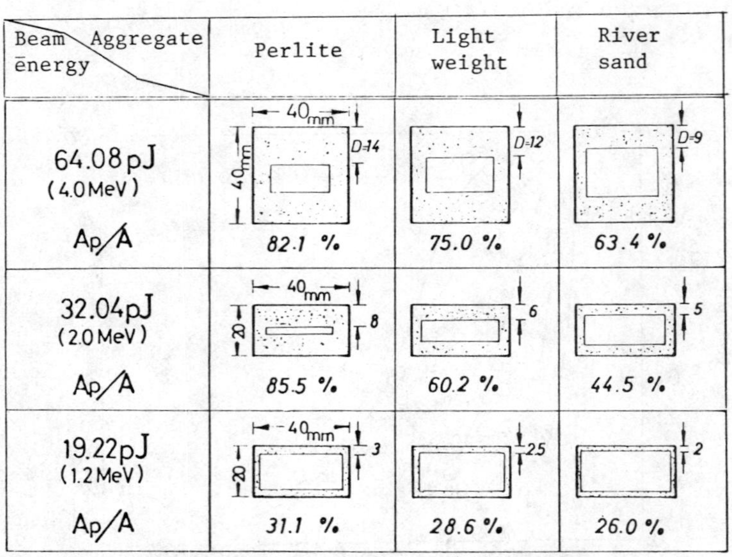

Irradiated opposite two faces (50 Mrads)
A : cross sectional area. A_p : polymerised area.
D : polymerising depth (mm)

Fig.7 Cross sections of PIC prism specimens.

Figure 7 shows schematically cross sections of PIC prism specimens polymerised by accelerated electrons under 50 Mrads. The impregnation depth from the surface of specimens is greater, in proportion to the energy intensity, and the density of mortar is smaller. Of the three beam energies 64.08 pJ energy is most effective in increasing the polymerisation depth.

It has been reported that cross linking agents have an influence on polymer loading, the strength of PIC and resistance to chemical attack. In these experiments agents such as Pst, PEA and PMMA were added in MMA monomer. Pst and PEA improve the polymerising depth and the flexural strength as shown in Fig. 8.

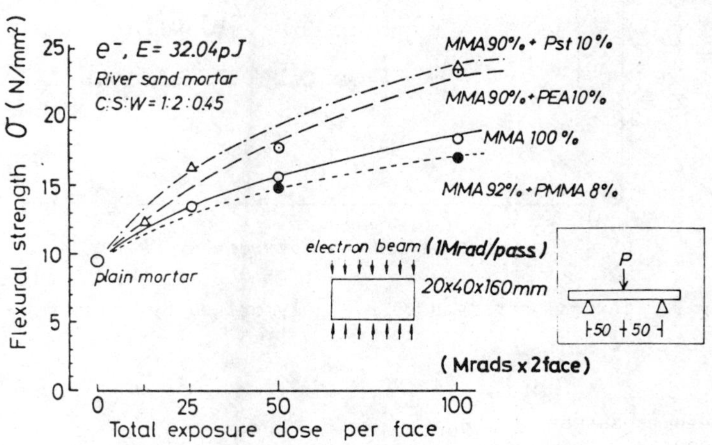

Fig.8 Flexural strength of PIC for different monomers.

Polymerisation by Electro-Magnetic Waves

From the authors previous investigation, it was known that optimum conditions for gamma ray polymerisation depend on the dose rate of the radiation source, total exposure dose and ambient temperature. In this programme, the same trends were obtained, as shown in Fig. 9. From the test results, the optimum exposure dose is estimated to be about 3-4 MR for ^{60}Co and about 1 MR for ^{137}Cs. Because the radiation energy of ^{137}Cs is smaller than that of ^{60}Co, with the dose rate being very much lower, the strength of PIC after irradiation by ^{137}Cs is less than that obtained by using ^{60}Co. However, polymerisation by ^{60}Co gave a slightly higher strength of PIC compared with ^{137}Cs.

X-ray polymerisation. Impregnated mortar specimens were cured by X-rays using a dose rate of 5×10^5 R/h. From the experimental information on strength, it was considered that the desirable total exposure dose was 2 MR, and a flexural strength of 35 N/mm^2 (Fig. 10) and compressive strength of 120 N/mm^2 were obtained. These values are approximately the same as those obtained by ^{60}Co irradiation. Three radiation induced polymerisation methods were studied to determine the effect of energy on MMA-impregnated cement mortar. Figure 11 shows the relationship between the total exposure dose (Mrads), energy density of radiation and flexural strength of PIC.

Each radiation source requires specific polymerisation conditions related to energy and exposure dose. If polymerising is done under the optimum conditions, the strength of PIC is roughly the same regardless of energy and radiation type, but exposure times differ.

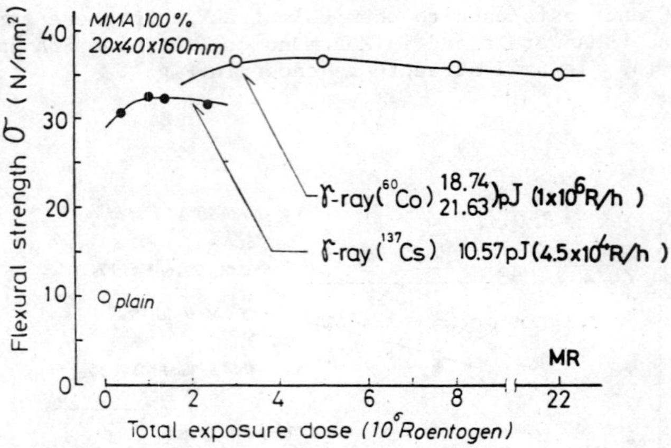

Fig.9 Flexural strength of PIC polymerised by gamma rays.

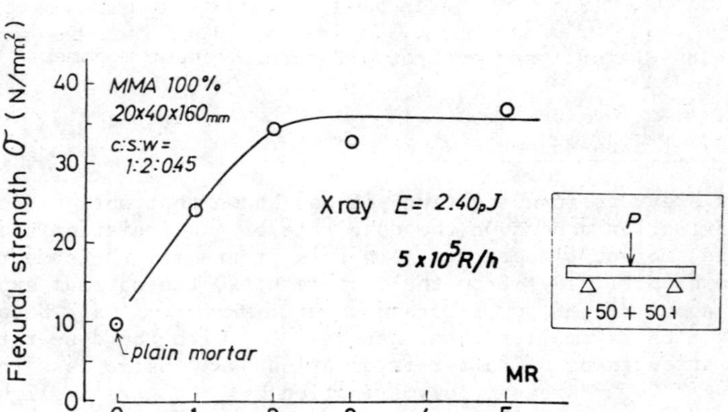

Fig.10 Flexural strength of PIC polymerised by X-rays.

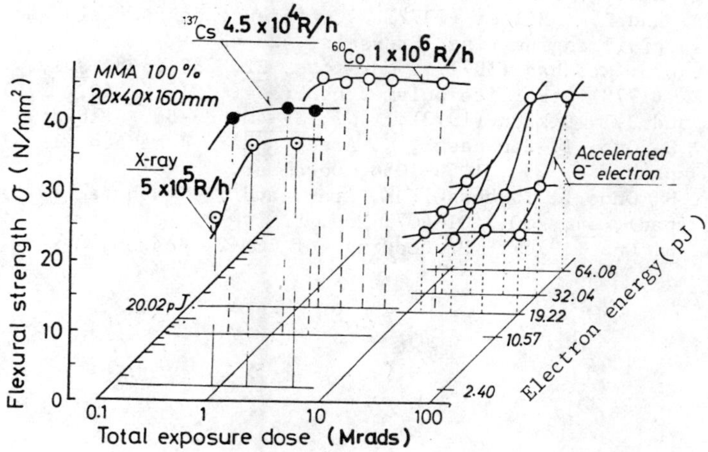

Fig. 11 Relation between total exposure dose, radiation energy and flexural strength of PIC.

CONCLUSIONS

In polymerising by an accelerated electron beam:

1) The strength ratio and relative absorption energy of the electron beam for the product of thickness and density have the same trend (Fig. 5).
2) The maximum energy used (64.08 pJ) is most effective in increasing the polymerisation depth and flexural strength. It was estimated that the impregnation depths of specimens in river sand mortar MMA-PIC were about 9 mm, 5 mm and 2 mm respectively for 64.08, 32.04 and 19.22 pJ under 50 Mrads irradiation (Fig. 7).
3) The polymer loading and strength of PIC are improved by the coexistence of polymers in monomer, and these polymers are found to cross-link in PIC.

In polymerising by electro-magnetic waves

4) It was found that the strength of PIC polymerised by X-rays takes approximately the same values as obtained by the ^{60}Co γ-ray method, and therefore X-rays can be used for polymerisation of PIC (Fig. 10).
5) Optimum total exposure dose was estimated to be about 3 MR for ^{60}Co (1×10^6 R/h) and about 2 MR for ^{137}Cs (4.5×10^4 R/h) (Fig. 9).
6) From the above conclusion, we see that an energy dependency is noticeable only when comparing the same kinds of radiation source.

REFERENCES

1) Steinberg, M., L.E. Kukacka, P. Colombo and others (1968). Concrete polymer material. First topical report *BNL 50134 (T-509) USBR Gen. Rep. 41*. December.
2) Steinberg, M., L.E. Kukacka, J.T. Dikeou and others (1969). Concrete polymer material. Second topical report *USBR REC-OCE-70-1 BNL 50218 (T-560)*. December.
3) Dikeou, J.T., M. Steinberg, W.C. Cowan and others (1971). Concrete polymer material. Third topical report *USBR REC-ERC-71-6 BNL 50275 (T-602)*. January.
4) Kukacka, L.E., W.G. DePuy and others (1972). Concrete polymer material. Fourth topical report *USBR REC-ERC-72-10 BNL 50328*. January.

5) Yuan, H.C., W.W. Hsu, U.P. Wang, C.C. Wu and J.S. Shieh (1969). International Atomic Energy Agency *(IAEA)-SM-123/28*, Munchen, 383-397.
6) Rotariu, G.J. and L.A. Miller (1972). Conference on radiation and isotope techniques in civil engineering, Brussels, 425-436.
7) Hills, P.R., D.J. McGahan (1971). *Ceramics*, 22, 269, 6-8. July.
8) Kukacka, L.E. (1973). Perlite polymer concrete, *ACI*, SP-40, 137-189.
9) DePuy, G.W. and J.T. Dikeou (1973). *ACI*, SP-40, 33-56. (ibid)
10) Ohgishi, S., H. Ono, I. Tanahashi, K. Araki and Y. Kasahara (1974). *J. Soc. Mat. Sci.* (Japan), 23, 255, 1050-1056, December.
11) Ohgishi, S., H. Ono, I. Tanahashi, K. Araki and Y. Kasahara (1975). *J. Soc. Mat. Sci.* (Japan), 24, 260, 401-407, May.
12) Ohgishi, S., H. Ono (1976). Proceedings of ICM-2, Boston, August, 1383-1388.

Volume 3, ICM 3, Cambridge, England, August 1979

FAILURE PROCESS OF CONCRETE: CRACK INITIATION AND PROPAGATION

H. Mihashi, T. Sasaki and M. Izumi

Department of Architecture, Tohoku University, Sendai 980, Japan

1. INTRODUCTION

Concrete fracture behaviour is very complex compared to other homogeneous materials, especially fracture under compressive load which is caused by the accumulation of local failure processes. Because the material contains many defects like pores and aggregates, the internal mechanical behaviour is far from elastic and linear. The mechanical properties and fracture behaviour are also affected by these defects. Therefore investigating the structural behaviour and mechanisms as well as the fracture behaviour needs to be done in order to achieve reasonable designs. There has been much information concerning concrete fracture behaviour (for example, Buyukozturk, Nilson & Slate, 1971; Kosaka & Tanigawa, 1975; Mihashi & Izumi, 1975; Santiago & Hilsdorf, 1973; Shah & Chandra, 1968; Stroven, 1973), but the actual behaviour of cracking inside a concrete system has had little study. Hence many important questions remain unanswered. For example, what is the most important factor that causes microcracking? How does the microcrack develop? What is the mechanism of fracture? The purpose of this paper is to show the property of microcracking in plain concrete under uniaxial compressive load and to discuss the fracture behaviour of concrete. The failure process of model concrete as a two-phase composite material is investigated by means of a photo-elastic coating technique. The coating film is rather thin and very soft. Uniaxial compressive tests are performed on concrete plates in a plane stress condition to obtain information on cracking behaviour at the sub-macroscopic level. Hence it is possible to observe the actual behaviour of crack initiation and propagation continuously, even in the final stage leading to rupture of the specimen.

2. MATERIALS AND PROCEDURES

2.1. Mechanical Properties of Photoelastic Coating Film

In this paper, a photoelastic coating technique is adopted in order to observe the change of local strain caused by microcracking. Previous studies using this technique seem to fail in investigating the actual crack mechanism because of the thickness and the stiffness of the coating film. (Stroven, 1973) Accordingly we made a film of 1 mm in thickness of epoxy resin with Thiokol 50%. The elastic constants of the film were $E \sim 5 \times 10^2$ kg/cm^2, $\nu \doteqdot 0.40$. As Young's modulus of hardened cement paste (w/c = 60%) was $E \sim 1.47 \times 10^5$ kg/cm^2, the reinforcement effect of the film was negligible. It was also assumed that the film was soft enough not to restrict cracking behaviour.

2.2. Model Concrete

An experimental study was carried out with model concretes of Fig. 1. A commercial Portland cement was used and the sand was Abukuma river sand. Only the fine portion under 0.6 mm of the sand dimension was used in order to achieve mortar of a homogeneous and isotropic condition. Series I was to observe the fracture behaviour in the vicinity of a single pore (Fig. 1a). At the same time, the sand content of mortar was varied: 0 , 20 , 50 , 100% (volume percent). Series II was to research the failure process in the vicinity of a single aggregate (Fig. 1b). This series was carried out with several strength of model aggregate (s/c = 100%, w/c = 30 , 40 , 60 , 80% mortar) and mortar matrix (w/c = 60%, s/c = 50 , 100%). In Series III, the failure process of two model concretes was investigated (Fig. 1c). One was r = 1 cm, d = 4 cm; w/c : inc = 30% and the other was r = 1.5 cm, d = 4.5 cm; w/c : inc = 40%. Both of them have s/c : inc = 100% and w/c : mat = 60%, s/c = 100%.

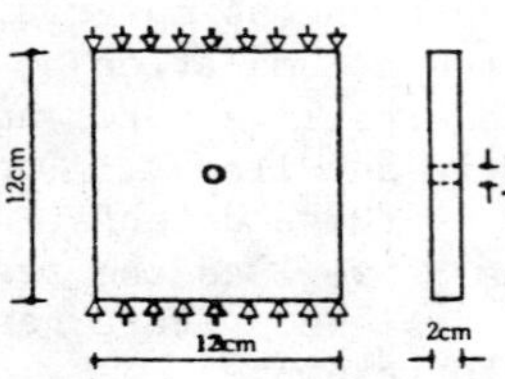

Fig. 1(a)
Model concrete
(Series I)

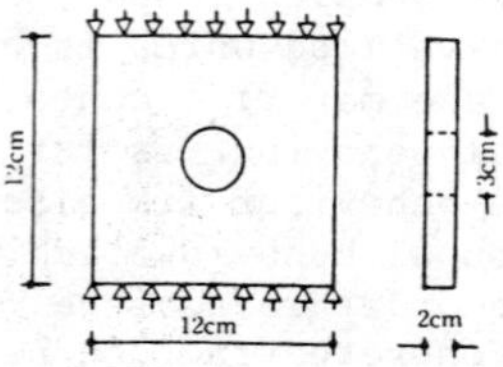

Fig. 1(b)
Model concrete
(Series II)

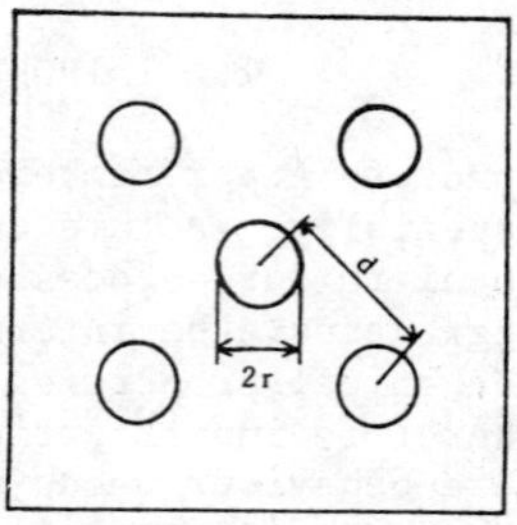

Fig. 1(c)
Model concrete
(Series III)
(12x12x2 cm)

3. RESULTS

3.1. Failure Process in the Vicinity of the Single Hole

The experimental results of Series I are summarized in TABLE 1. Moreover some examples of crack initiation and development around a hole are illustrated in Fig. 2. Here we define the crack on the top of the hole in the loading direction as the first crack and other diagonal cracks as the second crack. As could be expected, the first crack initiation occurred at the top of the hole in the loading direction. The larger the sand content, the higher was the loading level to the first crack. And rupture of the specimen was not caused only by the first crack. In some cases of s/c : mat = 100%, the second crack occurred before any first cracks had been observed. In the case of cement paste, the second cracks were created diagonally in the four directions after the first crack. Then rupture of the specimen occurred accompanied by other cracks in the matrix. Both the first crack and the second cracks formed in a brittle manner and developed in the loading direction. They were very fine cracks. On the other hand, the second cracks in mortar were comparatively wider and the propagation of these cracks was not so rapid as that in cement paste. Therefore the crack development seemed to take place in a zigzag manner or it seemed that there was a failure zone in the vicinity of the crack. Moreover in the case of mortar, rupture of the specimen occurred with only two or three second cracks. Especially it was only two second cracks in a diagonal position that caused rupture of the specimen (Fig. 2c).

TABLE 1 Loading Level (σ/σ_{max}) of Cracking

s/c	0%	20%	50%	100%
First Crack	0000-1 0.24 (12.55t)	0020-1 ? (11. 4t)	0050-1 0.45 ~ 0.67 (11.15t)	00100-1 no (11. 5t)
		0020-2 0.7 (9.95t)	0050-2 0.17 (11. 5t)	00100-2 no (10. 7t)
	0000-2 0. 1 ~ 0. 3 (14.40t)	0020-3 (9.85t)	0050-3 no (11.25t)	00100-3 0.43 ~ 0.65 (10.75t)
Second Crack	0000-1 0.68	0020-1 0.83	0050-1 0.90	00100-1 0.95
		0020-2 0.95	0050-2 0.83	00100-2 0.94
	0000-2 0.76	0020-3 0.74	0050-3 0.96	00100-3 0.88

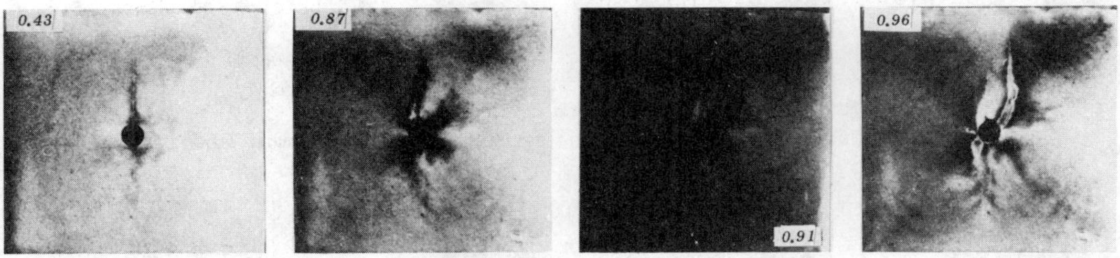

Fig. 2(a) Failure process of cement paste

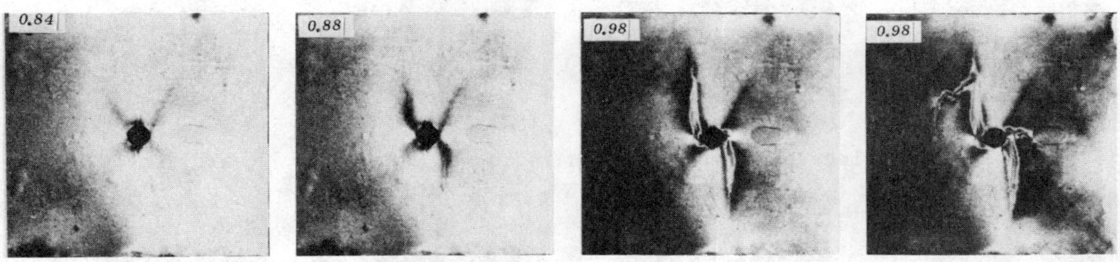

Fig. 2(b) Failure process of mortar (s/c = 50%)

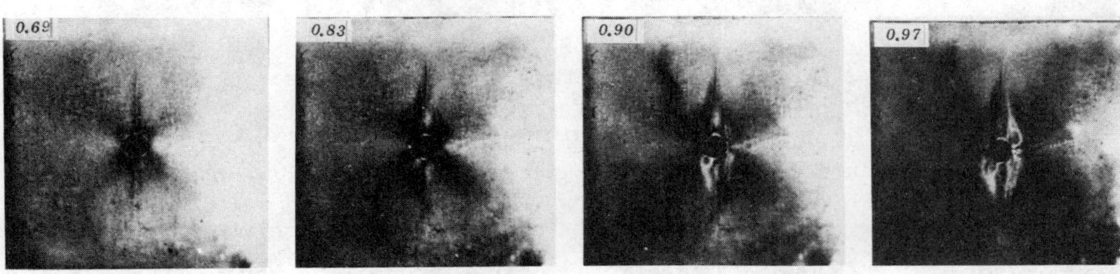

Fig. 2(c) Failure process of mortar (s/c = 100%)

3.2. Failure Process in the Vicinity of a Single Inclusion

Fig. 3 show some examples of failure process around a model aggregate. In Fig. 3a some microcracks perpendicular to the surface of the inclusion are observed. These were considered to be a kind of shrinkage crack, but they did not seem to affect remarkably the crack development process. In all cases, crack initiation was created on the four points of the matrix-inclusion interface first. There were no exceptions even in the case of the same strength between inclusion and matrix. The fluctuation of the position of bond crack initiation is illustrated in Fig. 4a. These bondcracks developed on the interface in the loading direction and linked to each other. The fluctuation of the linking bondcrack length is shown in Fig. 4b. The cracking patterns after this stage were subdivided into the following three types. The first type was to create a triangular deposit portion on the top of the inclusion (Fig. 3a , 3b). This is quite similar to those obtained by previous studies (Perry & Gillott, 1977; Stroven, 1973). The second type was illustrated in Fig. 3c. In this case, bond cracks behaved as only inclined cracks. The third type was that rupture of the specimen occurred immediately after the inclusion was crushed (Fig. 3d). This failure process seems to be similar to that of light weight concrete whose aggregates are weaker than mortar matrix.

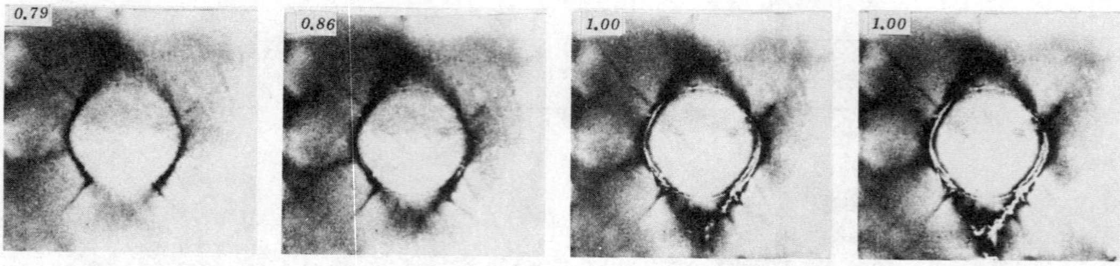

Fig. 3(a) Failure process around an inclusion (s/c:mat=50%; w/c:inc=30%)

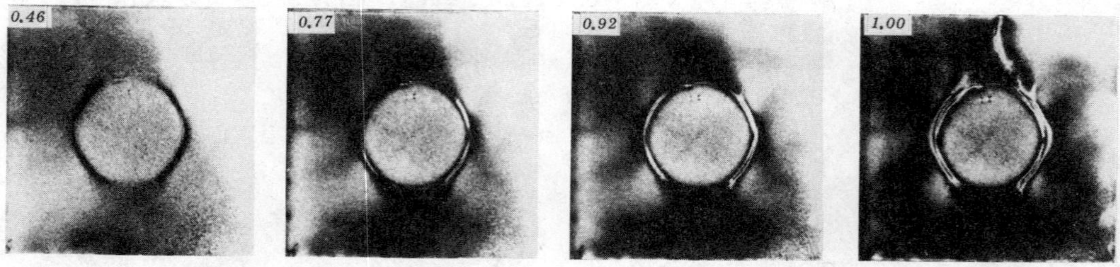

Fig. 3(b) Failure process around an inclusion (s/c:mat=50%; w/c:inc=40)

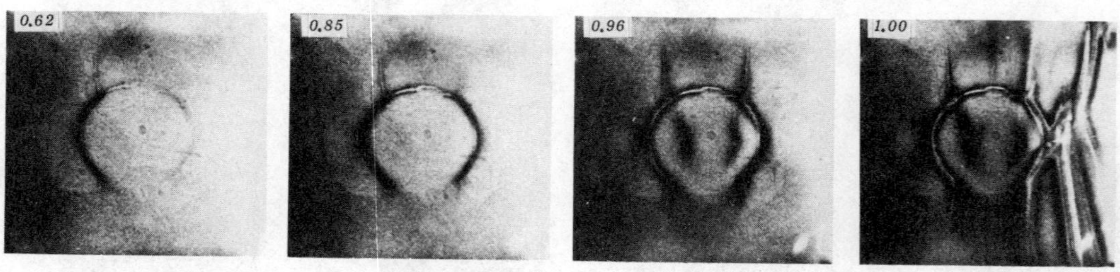

Fig. 3(c) Failure process around an inclusion (s/c:mat=50%; w/c:inc=60%)

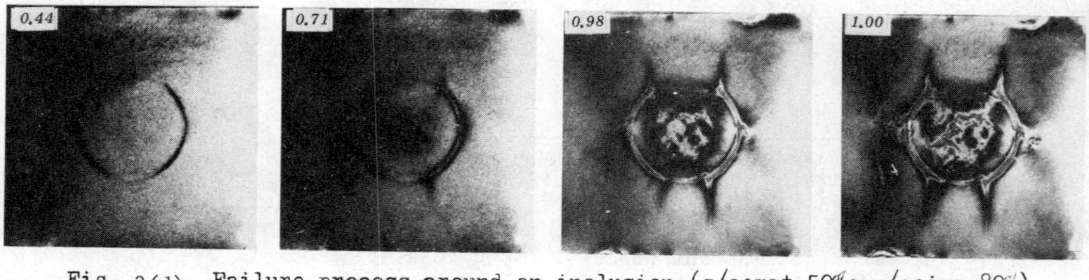

Fig. 3(d) Failure process around an inclusion (s/c:mat=50%; w/c:inc=80%)

3.3. Failure Process of Model Concrete

Examples of failure processes in model concrete with several inclusions are illustrated in Fig. 5. Some shrinkage cracks were observed around inclusions. Until a certain high loading level, almost the same failure process as that in the vicinity of a single inclusion (Series II; Fig. 3a , 3b) was observed around each aggregate: that is, bond crack initiation and triangular deposit portions were created on the top of the inclusions. Then an axial crack was created between inclusions, which seemed to link the peaks of each triangular deposit portion. Moreover another inclined crack was created in the same position and the oblique crack connected bond cracks diagonally (final stage of Fig. 5a , 5b). Rupture of the specimen was caused by the development of this oblique crack into the matrix.

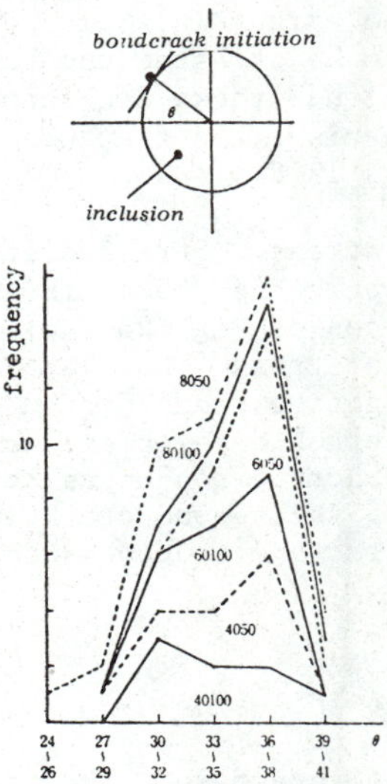

Fig. 4(a) *Position of bond crack initiations*

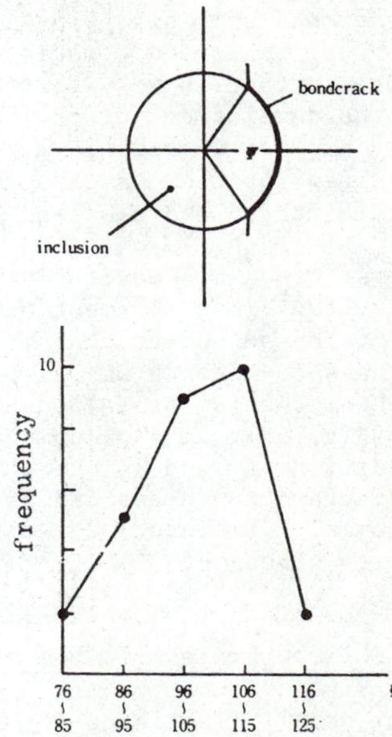

Fig. 4(b)
Length of bond crack

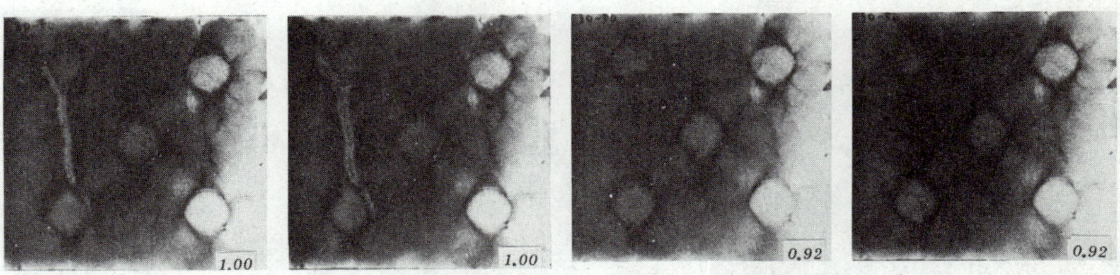

Fig. 5(a) Failure process of model concrete (No. 1)

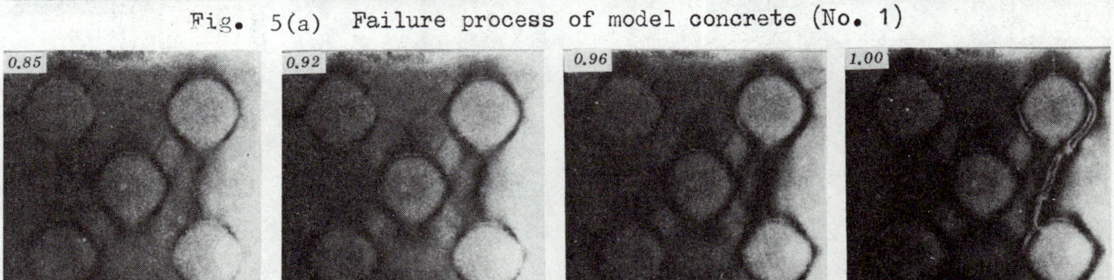

Fig. 5(b). Failure process of model concrete (No.2)

4. DISCUSSION

4.1. Failure Process of Hardened Cement Paste and Mortar

The main material defect of hardened cement paste and mortar (if the sand dimension is small) is the pore. Hence it is reasonable to consider a homogeneous system with randomly distributed pores. In the case of hardened cement paste, the first crack occurs at a rather low loading level. Therefore it is possible that these first cracks around some big pores link up with one another to cause rupture of the speciment before the second cracks are created. As the result of this mechanism, the fracture surface is very flat and parallel to the axis (Fig. 6a). On the other hand, the development of the first crack seems to be arrested and the second crack is very easy to develop in mortar. Hence the second cracks are connected and form an inclined surface. As the result of this mechanism, the fracture surface of mortar is uneven and slanting (Izumi & co-worker, 1978).

The first crack was obviously caused by tensile stress. Stress distribution calculated by the Finite Element Method (FEM) is shown in Fig. 6b to investigate the cause of the second crack initiation. A comparison of Fig. 2a and 6c indicates that the second crack was also created by tensile stress. But in the case of mortar, the mechanism of second crack initiation might be changed from cleavage to shear (Fig. 6d). The phenomenon that the first crack development was delayed by the addition of sand is due to the increase of fracture toughness Kc. But the high fracture toughness did not mean the delay of the second crack initiation nor development. Therefore the material with high fracture toughness does not always have high strength.

4.2. Failure Process in the Vicinity of a Single Inclusion

A crack usually initiates at the matrix-inclusion interface. The fact that the interface is the weakest phase was proved by the experimental results of 60100 test pieces (the strengths of the matrix and the inclusion were almost the same). But the mechanical condition around an inclusion is also readily observable. Tensile

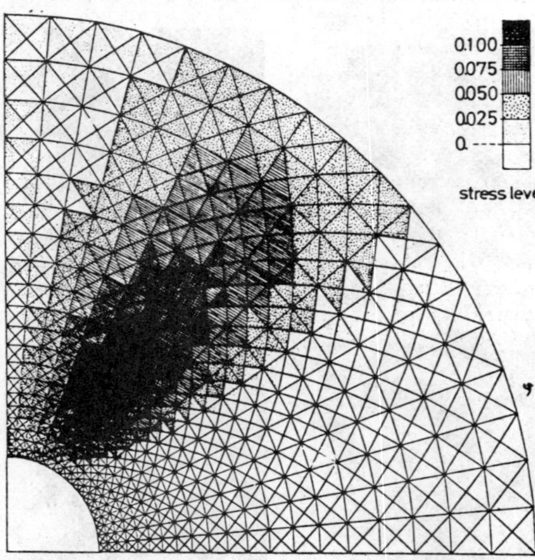

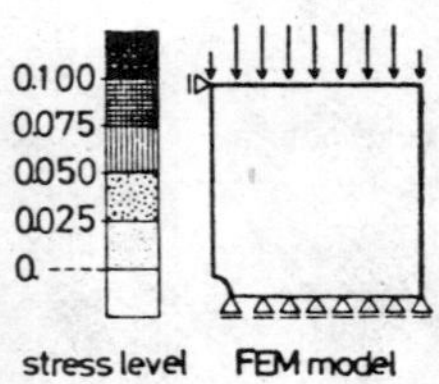

Fig. 6(a) Fracture mode of cement paste

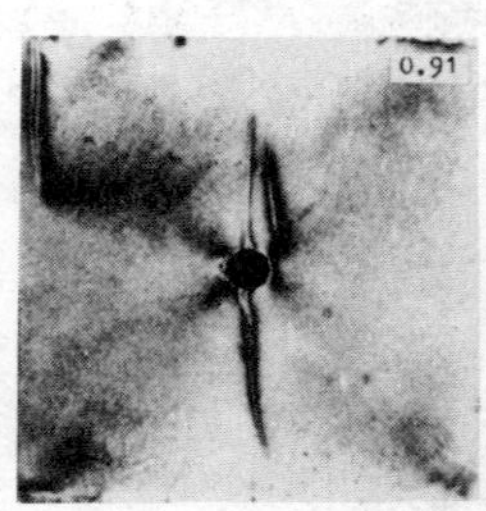

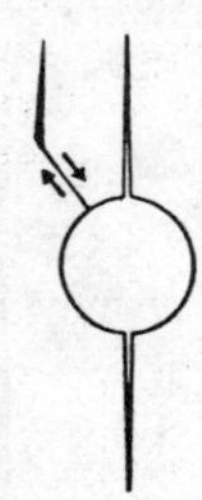

Fig. 6(b) Tensile principal stress (calculated result by FEM)

Fig. 6(c) Second crack in cement paste

Fig. 6(d) Mechanism of second crack in mortar

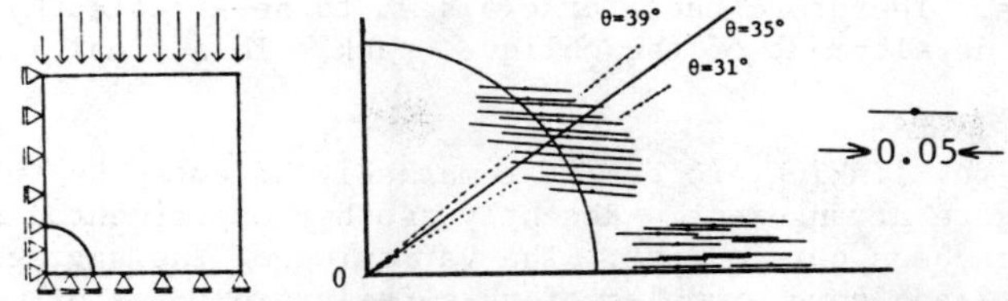

Fig. 7 Tensile principal stress around an inclusion (calculated result by FEM/ E:mat=382540 kg/cm², ν=0.26; E:inc=620460 kg/cm², ν=0.25)

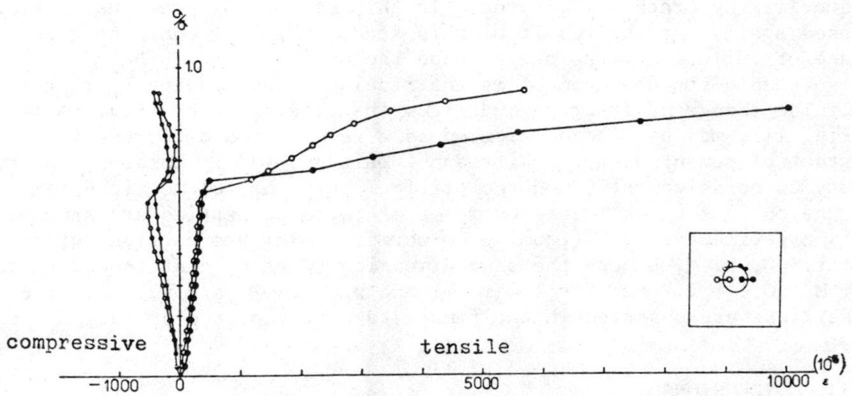

Fig. 8 Failure process obtained by strain gages

principal stress distribution obtained by FEM is illustrated in Fig. 7. Maximum tensile stress are located in the region of $\theta \sim 35^{\circ} \pm 4^{\circ}$ on the interface. This region agrees with the high frequency position of the bondcrack initiation (Fig. 4a). Moreover the area of the region is proportional to the size of inclusion. Therefore the probability of bond cracking increases and the strength of crack initiation decreases when the aggregate size is bigger. Such results were obtained by Liu, Nilson & Slate (1972). The mechanical behaviour in the vicinity of the inclusion is shown in Fig. 8. From the load level 0.6, a violent increase of tensile strain was observed on the matrix-inclusion interface perpendicular to the load axis. At the same level, a temporary relaxation phenomenon was observed in the compressive strain. Accordingly a change of the mechanism to bear the load is considered as follows in the case of concrete with stiffer and stronger aggregate than matrix: (1). Aggregate and matrix bear the load as one body and high stress is conveyed to aggregate through the interface. (2). The bearing system is changed into another one shown in Fig. 9 by bond cracking. In this case, the compressive strength of matrix and the local shear strength and the cleavage strength in the region above the peaks of the triangular deposit portions are compared. Generally the cleavage strength is the weakest and then the shear crack is created. On the other hand, the crushed aggregates act in the same way as holes if the aggregates are weaker than matrix.

4.3. Failure Process of Model Concrete

The linking process in the final stage leading to rupture of the specimen is subdivided into three stages. The first stage is the cleavage cracking between aggregates in the loading direction. The second stage is the changing from a cleavage axial crack into a shear oblique crack. The former cleavage crack ends in the peaks of triangular deposit portions and the latter oblique crack is absorbed in bondcracks at both ends. Therefore these cracks seem to be arrested by aggregates. The third stage is the development of the oblique crack. This final stage causes rupture of the specimen.

It is obvious that the linking process is remarkably affected by the geometrical condition of aggregate arrangement. Recently another experimental study (Izumi & co-worker, 1978) has been carried out. The variation of the aggregate arrangement is listed in TABLE 2 and some examples of this failure process of model concrete are illustrated in Fig. 10. In this study, almost the same phenomena as those mentioned above were observed. Especially in Fig. 10e, a cleavage crack linking bond cracks in the loading direction was observed. But this was not unstable and then another oblique linking crack was created. In this stage, the previous cleavage crack was closed again. As the result of this study, the most dominant crack in the final stage of failure process seems to be the oblique crack. Because the crack is not so stable its development is the final process leading to rupture of the specimen. The change of fracture mode from the cleavage type (Fig. 6a) to the shear type (Fig. 11), may be also considered as a result of a different failure process from that of cement paste. (Wittmann & Zaitsev, 1971) Moreover the non-linear property of poission ratio and the critical point of volumetric strain may be caused by the oblique crack. Some examples of these properties are shown in Fig. 12. This experimental work (Izumi & co-worker, 1978) was carried out on river and sand mortar. O, S, M, L mean the sand dimension (O means neat cement paste and S = ∿ 0.6 mm; M = 0.6 ∿ 1.2 mm; L = 1.2 mm ∿) and w/c = 40%, s/c = 100%. The characteristic properties were observed in the final stage of the case of only L Series.

5. CONCLUSIONS

A coating technique on model concrete is very effective for investigating actual

TABLE 2 List of Test Pieces

Series I′

No.	d (cm)
1	3
2	4
3	5
4	6

Series II′

No.	d_1 (cm)	d_2 (cm)
1	0.5	3
2	1.0	3
3	0.5	4
4	1.0	4
5	1.5	4
6	2.0	4
7	0.5	5
8	1.0	5
9	1.5	5
10	2.0	5
11	0.5	6
12	1.0	6
13	1.5	6
14	2.0	6

Series III′

No.	d_1 (cm)	d_2 (cm)
1	3	3
2	4	3
3	5	3
4	3	4
5	4	4
6	5	4
7	6	4
8	7	4
9	3	5
10	4	5
11	5	5
12	6	5
13	7	5
14	3	6
15	4	6
16	5	6
17	6	6
18	7	6
19	3	7
20	4	7
21	5	7
22	6	7
23	7	7

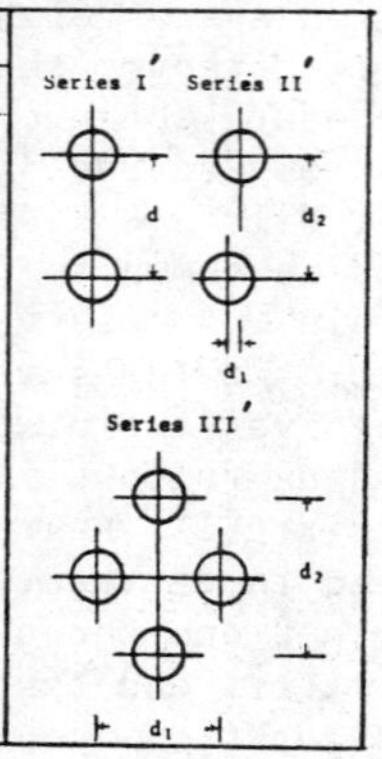

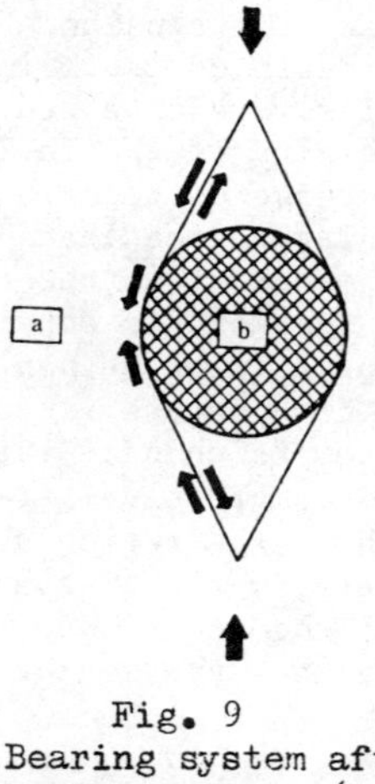

Fig. 9
Bearing system after bond cracking (a: matrix, b:inclusion)

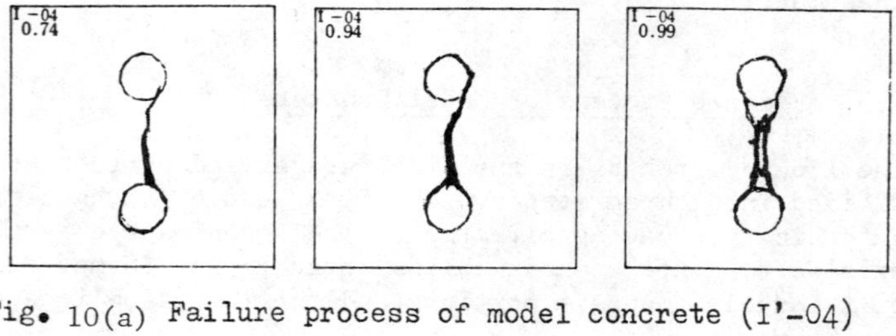

Fig. 10(a) Failure process of model concrete (I'-04)

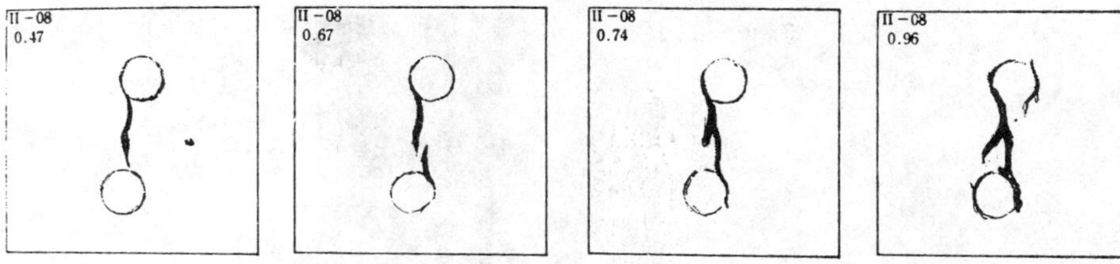

Fig. 10(b) Failure process of model concrete (II'-08)

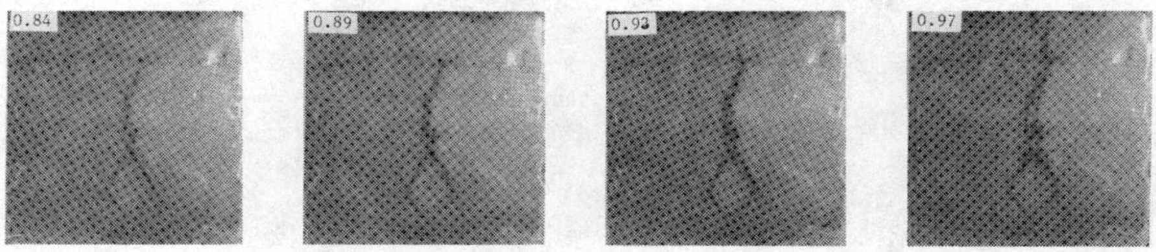

Fig. 10(c) Failure process of model concrete (II'-11)

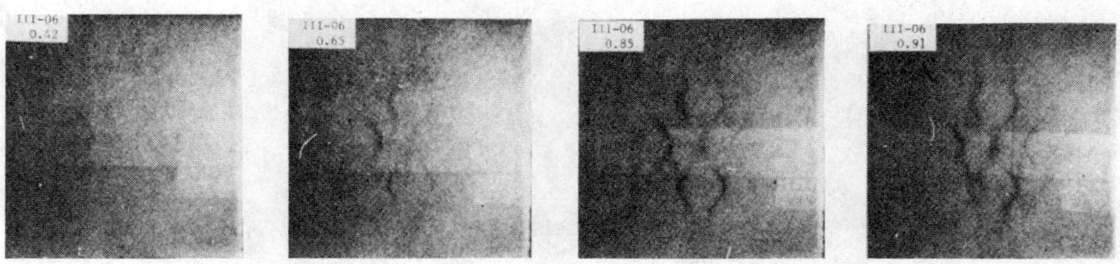

Fig. 10(d) Failure process of model concrete (III'-06)

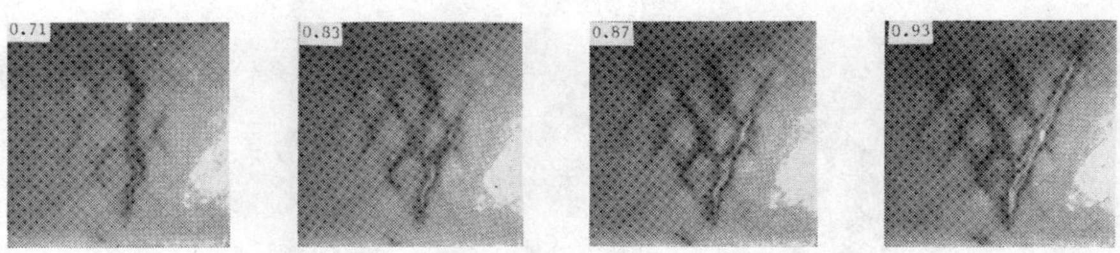

Fig. 10(e) Failure process of model concrete (III'-07)

Fig. 10(f) Failure process of model concrete (III'-14)

Fig. 11 Fracture mode of mortar (40-L)

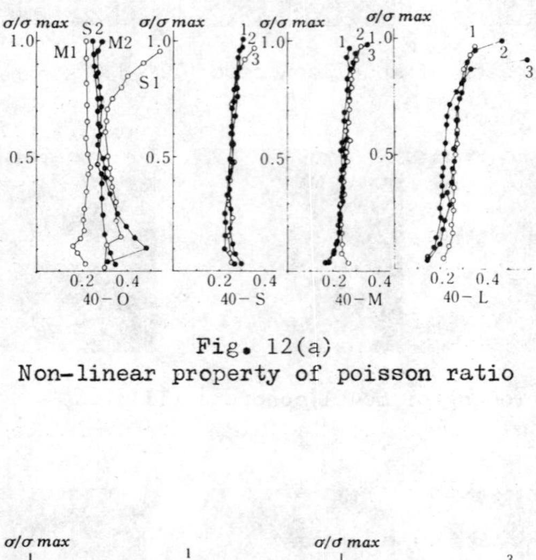

Fig. 12(a)
Non-linear property of poisson ratio

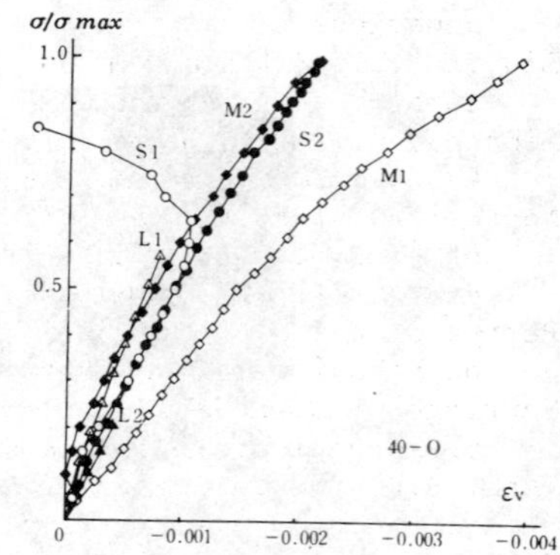

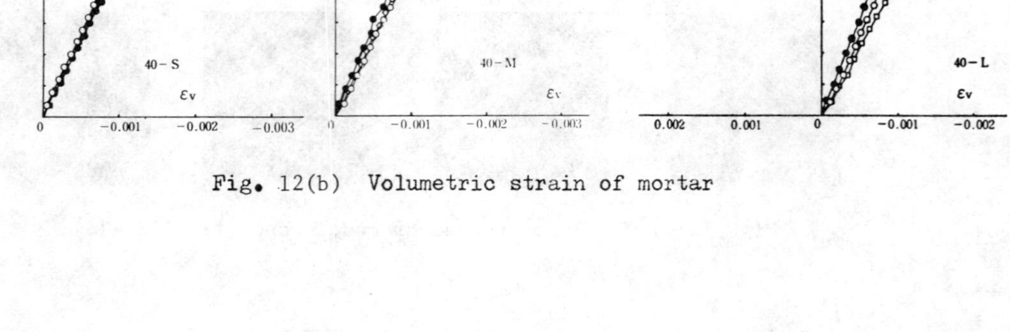

Fig. 12(b) Volumetric strain of mortar

cracking behaviour. The fracture behaviour of concrete under uniaxial compressive load may be considered as a series of cracking process (initiation and propagation), which is created in a local tensile stress field caused by material defects. The failure process of concrete may be subdivided into the following four stages. (1). Bond cracking on the matrix-aggregate interface; (2). Its extension into the mortar matrix, or linking bond cracks by cleavage cracking; (3). Linking bond cracks by an oblique shear crack; (4). Rupture after the unstable development of a dominant oblique crack. In the second stage, cracks develop in the loading direction or link the weakest path rapidly along the interfaces. But these cracks do not on their own cause rupture of the specimen. In the third stage, the initial oblique crack is not brittle nor propagates rapidly, because it is absorbed in the bond cracks at each end. The final oblique cracking pattern may cause some characteristic mechanical properties; that is, the non-linear poisson ratio and the critical point of volumetrical change. If the development of the oblique crack can be recognized, the critical condition leading to rupture of the specimen may be predicted.

6. REFERENCES

Buyukozturk, O., A.H. Nilson and F.O. Slate (1971). Stress-strain response and fracture of a concrete model in biaxial loading. *J. ACI.*, 68, 68-52.

Izumi, M., H. Mihashi and T. Sasaki (1978). Failure process of concrete. *Arch. Rep. of Tohoku Univ.*, 19, 61. (in Japanese)

Kosaka, Y. and Y. Tanigawa (1975). The effect of coarse aggregates on fracture behaviour of concrete. *Trans. of Arch. Inst. Jap.*, 228, 1. (in Japanese)

Liu, T.C.Y., A.H. Nilson and F.O. Slate (1972). Stress-strain response and fracture of concrete in uniaxial and biaxial compression. *J. ACI.*, 69, 69-31.

Mihashi, H. and M. Izumi (1975). Study of fracture behaviour of concrete. *Arch. Rep. of Tohoku Univ.*, 16, 41. (in Japanese)

Nepper-Christensen, P. and T.P.H. Nielsen (1969). Modal determination of the effect of bond between coarse aggregate and mortar on the compressive strength of concrete. *J. ACI.*, 66, 66-7.

Perry, C. and J.E. Gillott (1977). The influence of mortar-aggregate bond strength on behaviour of concrete in uniaxial compression. *Cem. & Concr. Res.*, 7, 5.

Santiago, S.D. and H.K. Hilsdorf (1973). Fracture mechanism of concrete under compressive loads. *Cem. & Concr. Res.*, 3, 363.

Shah, S.P. and S. Chandra (1968). Critical stress, volume change and microcracking of concrete. *J. ACI.*, 65, 65-67.

Stroven, P. (1973). Some aspects of the micromechanics of concrete. Ph.D. Thesis, Delft Univ. of Tech.

Wittmann, F.H. and Ju. Zaitsev (1971). Behaviour of hardened cement paste and concrete under high sustained load. *Proc. Int. Conf. on Mech. Behav. of Mat., Kyoto, Japan*, IV, 84.

Volume 3, ICM 3, Cambridge, England, August 1979

TRIAXIAL COMPRESSIVE CREEP OF CONCRETE AT TEMPERATURES 20°C TO 320°C

S. Ohgishi, M. Wada and H. Ono*

Nagoya Institute of Technology, Nagoya, Japan
***Chuba Institute of Technology, Kasugai Japan**

INTRODUCTION

To improve safety in the design of prestressed concrete pressure vessels (PCPV) for nuclear reactors of the gas cooled type, a study of triaxial creep of concrete at elevated temperature was begun by D. H. Hannant (1) at the end of 1960's. Since then, only a few studies have been carried out (2) (3) (4) (5). The temperature range of all these tests on triaxial creep was below 80°C.

The Authors (6) (7) have developed a triaxial compressive creep test apparatus particularly designed for up to 300° to 500°C. In this apparatus, a cylindrical concrete specimen of 10ϕx20cm is used. Hydraulic pressure is applied to the surface of the sealed concrete specimen by silicon oil at up to 350°C. If it is necessary for a triaxial creep test in the temperature range 350° to 500°C, melted antimonial alloy with a melting point of 62°C can be used as a pressure medium instead of silicon oil.

In the present paper, concrete cylinders sealed by a thin copper mold were tested under hydraulic triaxial pressure and uniaxial compressive stress at temperatures 20° to 320°C. Applying a rheological model to the test results, the temperature dependence of the elastic modulus and the coefficient of viscosity of concrete were analyzed. Also discussed are the activation energy for creep of concrete, triaxial creep Poisson's ratio, a time shift principle (also called "method of reduced variables") and a master curve for creep. The distribution function of retardation time of the test concrete is also analyzed.

EXPERIMENTS

The mix proportion and mechanical properties of the test concrete are shown in Table 1. Two Ailtech strain gauges with AC thermo-couples were placed orthogonally in the center of the copper mold 10ϕx 20 cm. (t_1=0.5 mm, cf. Fig. 1). The gaps between the stainless steel sheath of the gauge lead wire and its hole in the copper mold were welded by silver wax. The copper mold was then filled with concrete. After curing in 20°C water for 15 weeks, all specimens were sealed by welding a copper cap (t_2=1 mm) to the mold with silver wax. But the specimens for the creep test at above 100°C were pre-heated at 150°C for 6 hours in an oven till the concrete attained constant weight. This procedure was intended to separate drying shrinkage from creep strain.

TABLE 1 Mixing and Properties of Test Concrete.

C*1 W S*2 G*3 (kg/m³)	Sl (cm)	Test age	σ_B *4 (N/mm²)	E_i *5	$E_{1/3}$ *5 (kN/mm²)	uni ν_e, *6
322 225 994 813	22	196d.	38.1	31.5	28.4	0.18

*1 Ordinary portland cement, *2 River sand(under 5mm, FM;2.71) *3 River gravel(under 15mm,FM;6.62) *4 Compressive strength under uniaxial load(10ϕx20cm) *5 Initial tangent and secant modulus at $\sigma_B/3$, *6 Uniaxial Poisson's ratio.

The test apparatus for triaxial compressive creep in this investigation consists of the hydraulic cell shown in Fig. 1, oil pumps, automatic control intensifier, temperature and pressure recorders and electric heater. Block diagram of the pressure control system is given in Fig. 2. As the hydraulic pressure medium is required to have a high boiling point and chemical stability at elevated temperature, the silicon oil KF-54 was used in this experiment. Under closed conditions without air, silicon oil can be used at up to 380°C as a pressure medium.

The creep stress level in the test is only one step at $\sigma_B/3$ (= 14N/mm²), and the creep loading consists of two types of uniaxial (${}_a\sigma$= 14N/mm², ${}_r\sigma$=0) and hydrostatic pressure (${}_a\sigma$ = ${}_r\sigma$ = 14N/mm²). In both cases, axial and radial creep strain (${}_a\varepsilon$, ${}_r\varepsilon$) was measured by embedding two Ailtech strain gauges at six temperature steps as 20°, 95° (tri;90°), 140°, 200°, 260° and 320°C (tri;302°) each for 12 hours (= 4.32x10⁴ sec.). The rising rate of temperature was kept at 1°C/min at the center of specimen. The average loading age of the creep test was thirty weeks.

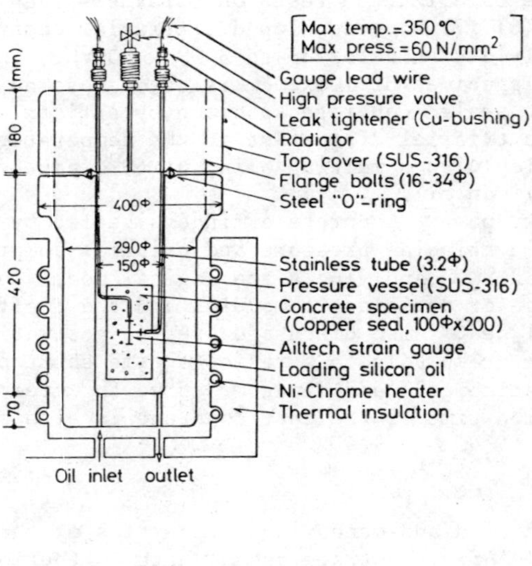

Fig. 1. Hydraulic triaxial cell.

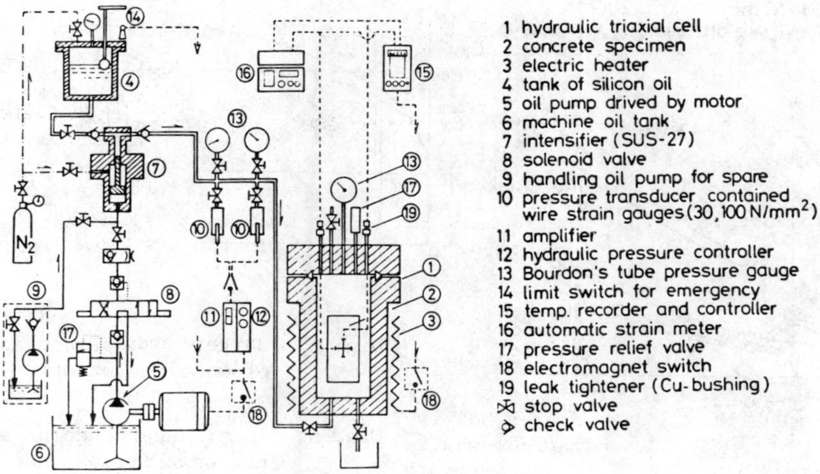

Fig. 2 Hydraulic pressure control system.

TEST RESULT

Radial total strain curves per unit stress under triaxial load are shown in Fig. 3. The figure showing axial strain was omitted. Fig. 4 shows the strains under uniaxial load. From these data, the elastic strain ($_a\varepsilon_e$, $_r\varepsilon_e$) and creep strain ($_a\varepsilon_c$, $_r\varepsilon_c$) at 4.32x10^4 sec. after loading are drawn against temperature in Fig. 5. The large strains at 100°C are a conspicuous feature in this figure. This is assumed to be due to the loss of free water in 140°C and 200°C concrete by preheating.

Poisson's ratio under uniaxial load, $_u\nu_e$, $_u\nu_c$ and strain ratio ($_r\varepsilon_e/_a\varepsilon_e$), ($_r\varepsilon_c/_a\varepsilon_c$) under triaxial load was calculated as listed in Table 2. The values of $_u\nu_e$ and $_u\nu_c$ are constant regardless of temperature. In both cases, the ratio of radial strain to axial strain under triaxial stress is generally taken as 1.1 in specimens with height/diameter ratio (H/D)= 2.

DISCUSSIONS

Dependence of Visco-Elasticity on Temperature

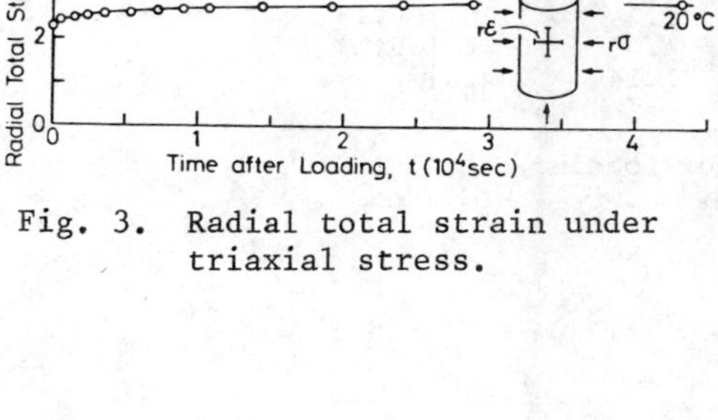

Fig. 3. Radial total strain under triaxial stress.

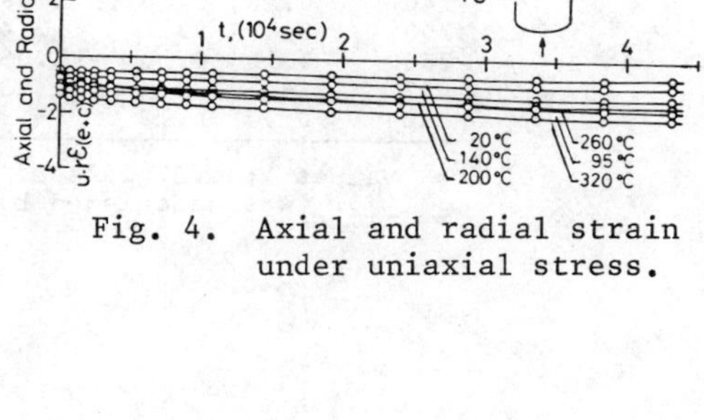

Fig. 4. Axial and radial strain under uniaxial stress.

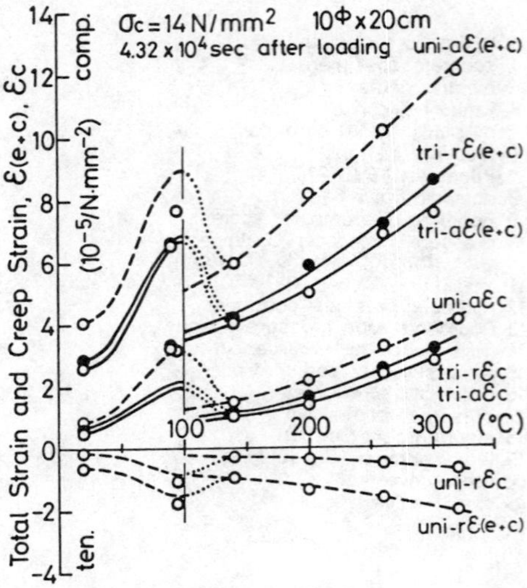

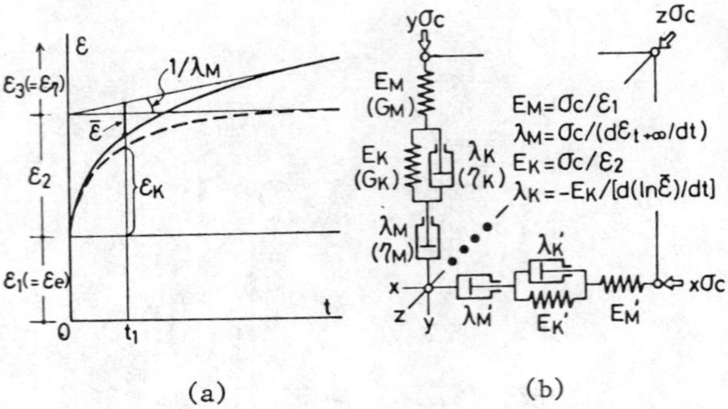

Fig. 6. Burger's model and analysis of model constants.

Fig. 5. Temp. dependence of strain per unit stress under uni- and tri-axial load.

Applying Burger's model in Fig. 6 to the creep curve, the four model constants (E_M, E_K, λ_M and λ_K) were analyzed and plotted against temperature in Fig. 7. Here λ is the coefficient of viscosity under normal stress. Since triaxial load restrict deformation stronger than uniaxial load, apparent E and λ under triaxial load are larger. These four mechanical constants have similar dependence on temperature. Discontinuous change at 100°C is seemed to be related to the moisture condition in the concrete. The values of the bulk modulus K_e of concrete obtained by eq. (1) under hydrostatic stress are K_e = 15kN/mm² at 20°C, 10kN/mm² at 100°C, and 6kN/mm² at 302°C.

Where σ_H is the hydraulic pressure and ${}_v\varepsilon_e$ the bulk elastic strain,

$$K_e = \frac{\sigma_H}{{}_v\varepsilon_e} = \frac{\sigma_H}{({}_a\varepsilon_e + 2\,{}_r\varepsilon_e)} \qquad (1)$$

Triaxial Poisson's Ratio (${}_t\nu_e$, ${}_t\nu_c$)

The time variation of uniaxial creep Poisson's ratio, ${}_u\nu_c$, is shown in Fig. 8. If creep strain conforms to the principle of superposition, creep Poisson's ratio

TABLE 2 Elastic and Creep Poisson's Ratio(ν_e,ν_c) under Uni- and Tri-axial Load.

Creep temp. (°C)	Uniaxial		Triaxial			
	${}_{uni}\nu_e$	${}_{uni}\nu_c$ *1	${}_r\varepsilon_e/{}_a\varepsilon_e$	${}_r\varepsilon_c/{}_a\varepsilon_c$ *1	${}_{tri}\nu_e$	${}_{tri}\nu_c$ *1 *2
20	0.15	0.17	1.09	1.11	0.18	0.18
90	–	–	1.00	1.02		
95	0.15		–	–	–	–
140	0.15	0.18	1.08	1.07	0.17	0.15
200	0.17	0.16	1.14	1.24	0.20	0.18
260	0.16	0.15	1.05	1.06	0.18	0.14
302	–	–	1.13	1.14	0.20*	0 16*
320	0.17	0.17	–	–		

*1 Values at 4.32x10⁴sec. after loading,
*2 ${}_{tri}\nu_c$ was calculated by eq.(2).

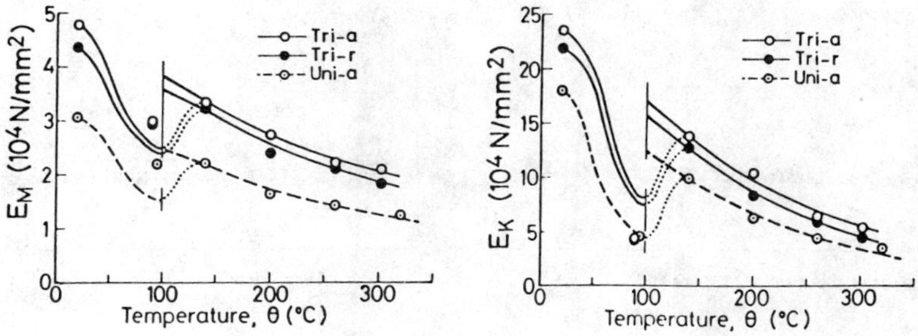

(a) Modulus of Elasticity.

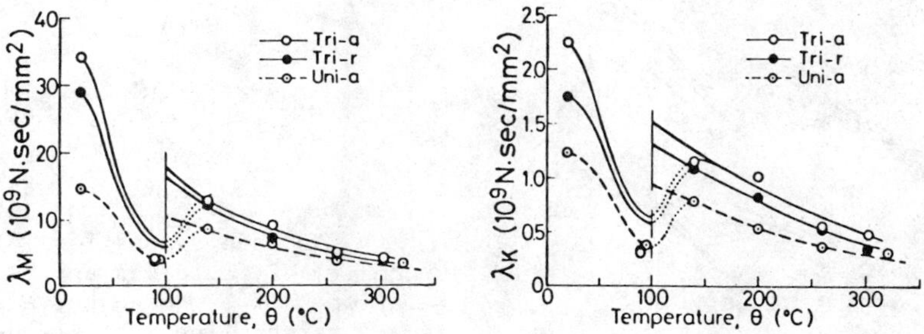

(b) Coefficient of Viscosity.

Fig. 7. Dependence of Visco-elasticity on Temperature.

$_t\nu_c$ under hydrostatic stress is expressed as eq. (2).

$$_t\nu_c = \frac{\sigma_H \cdot {}_{u,a}\varepsilon_c - {}_{t,a}\varepsilon_c}{2\sigma_H \cdot {}_{u,a}\varepsilon_c} \qquad (2)$$

Where $_{u,a}\varepsilon_c$ is the axial creep strain under uniaxial load, and $_{t,a}\varepsilon_c$ that under triaxial load.

The $_t\nu_e$ and $_t\nu_c$ values in Table 2 were calculated by eq. (2) from the data in Fig. 5. It is proved that $_t\nu_c \sim {}_u\nu_c$ (=0.16 ~0.18), and $_t\nu_e > {}_t\nu_c$ according to some previous reports (3) (5) (8). The time variation of triaxial creep Poisson's ratio is shown in Fig. 9. The values of $_t\nu_c$ is known to be constant with time regardless of temperature. The same view is reported by D. J. Hannant (9) and I. J. Jordaan (10).

Activation Energy for Creep

The relation between the temperature and coefficient of viscosity η_M, η_K of concrete which are calculated by Trouton's eq. (3) is shown in Fig. 10.

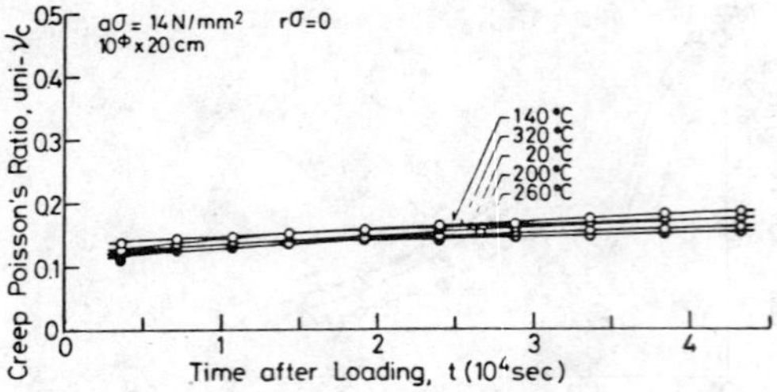

Fig. 8. Time variation of $u\nu c$.

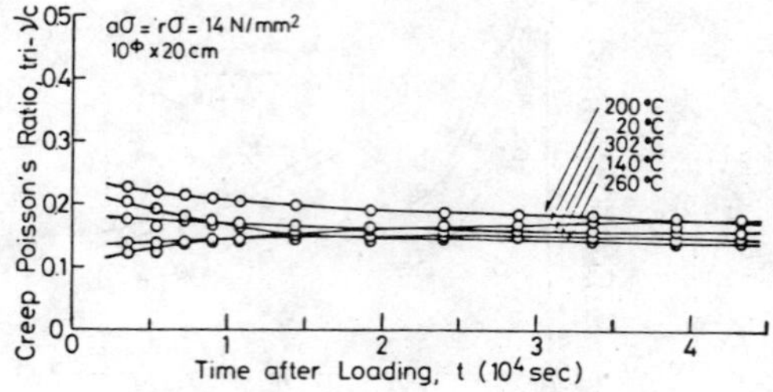

Fig. 9. Time variation of $t\nu c$.

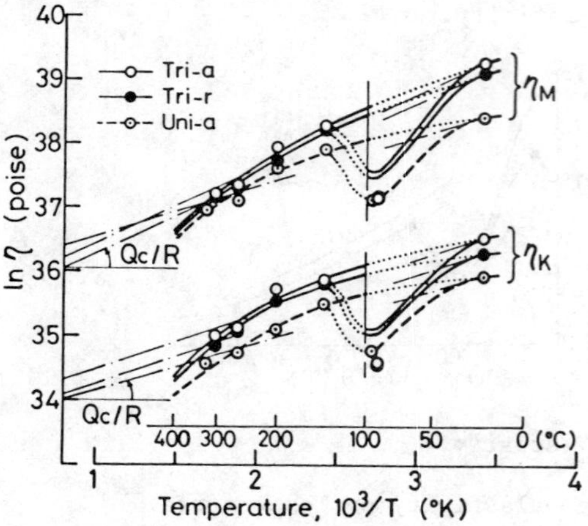

Fig. 10. Coefficient of viscosity η.

$$\eta = \lambda/3 \qquad (3)$$

Andrade's equation (4) on viscosity of fluid was applied to this relation. If each curve in Fig. 10 is a straight line, the viscosity η is conformable to Andrade's equation. Regarding each of these relations as a straight line, the apparent activation energy for creep, $_{app}Q_c$, is calculated as follows: $Q_{c,tri}$ = 10.55, 6.82 (kJ/mol), $Q_{c,uni}$ = 6.23, 5.61 (kJ/mol).

$$\eta = A.e^{Q/RT} \qquad (4)$$

$$\ln \eta = \ln A + B(1/T) \qquad (4')$$

Where A is an experimental const., R the gas const.(= 8.315 J/mol·°K), and B = Q/R. Strictly, activation energy needs to be expressed as a function of temperature. By definition, the coefficient of viscosity, λ, is obtained by eq. (5):

$$\lambda = 1/\dot{\varepsilon}_c(t) \qquad (5)$$

That is to say, λ, is given by the reciprocal of tangential gradient of the creep curve (creep strain rate) in Fig. 3 and Fig. 4. By this method, the coefficient of viscosity (λ in Fig. 3) is drawn in Fig. 11. We fix only the curve at 20°C on Fig. 11, and when other lnλ - curves are shifted to the right and upward parallel with axes, one well superimposed master curve is obtained as the lowest curve in Fig. 12. The quantity of shifting lnλ curves to the right, that is, shift factor, a_T, is listed in Table 3. Like this, when (lnλ~t) curves at any temperature T can be superposed on production of one, at the base temperature T_0, by shifting along the time-axis for a_T, the following relations are obtained

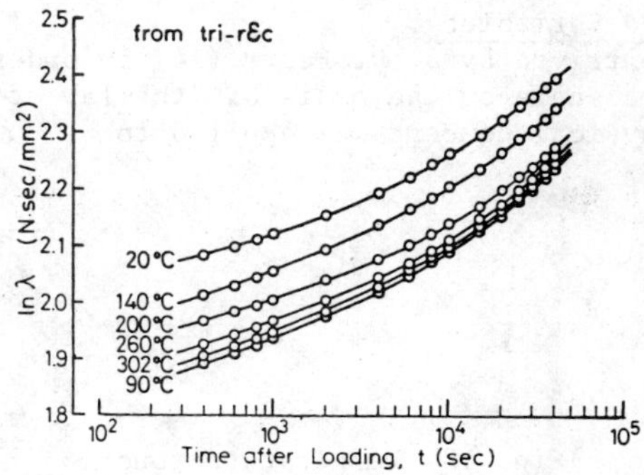

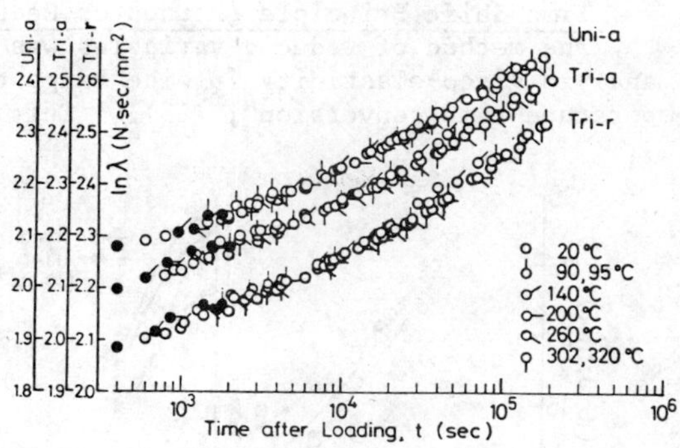

Fig. 11. Relation (lnλ~t) from radial creep strain, $_{t,r}\varepsilon_c$.

Fig. 12. Master curve of viscosity, lnλ.

TABLE 3 Shift Factor a_T and Shift Time S_T.

Base temp.; T_0=20°C

	θ(°C)	20	90	95	140	200	260	302	320
a_T	Tri-axi.	1	5.00	-	1.50	2.05	2.95	3.87	-
	Tri-rad.	1	4.50	-	1.75	2.15	3.50	4.25	-
	Uni-axi.	1	-	3.70	2.25	3.00	3.70	-	4.50
S_T	Tri-axi.	1	100	-	5.60	10.0	36.0	60.0	-
	Tri-rad.	1	90	-	5.60	13.2	36.0	76.0	-
	Uni-axi.	1	-	46	5.79	18.1	54.0	-	100

a_T ;Shift factor in master curves on lnλ shown in Fig.12.
S_T ;Shift time in master curves on ε_c shown in Fig.16.

from Andrade's equation. Using eq. (6)

$$\ln\left(\frac{\eta}{\eta_0}\right) = \frac{Q_c}{R}\cdot\left(\frac{1}{T} - \frac{1}{T_0}\right) \tag{6}$$

and eq. (7), a_T is expressed as eq. (8). Then eq. (8) is rewritten for Q_c in eq. (9), where R·ln10 = 19.14(J/mol·°K).

$$(\eta/\eta_0) = a_T\ (T/T_0) \tag{7}$$

$$\ln a_T = -\frac{Q_c}{19.14T_0}\left(\frac{T-T_0}{T}\right) - \ln\left(\frac{T}{T_0}\right) \tag{8}$$

$$Q_c = -19.14\frac{T.T_0}{T-T_0}\ln\left(a_T\cdot\frac{T}{T_0}\right) \tag{9}$$

As values of a_T were given experimentally in Table 3, Q_c was calculated as shown in Fig. 13 by eq. (9). Activation energy for creep of concrete, Q_c is a function of temperature, which is discontinuous around 100°C. Author has already reported dependence of Q_c on temperature (11) (12) and on stress (13) from uniaxial creep test at elevated temp. P. Klug (18), and G. A. Hirst (19) also discussed Q_c of concrete.

As another method, we can obtain Q_c by eq. (10) using a shift time, S_T which is mentioned in the next paragraph.

$$Q_c = R\frac{d(\ln S_T)}{d(1/T)} = 2.303\ R\ \frac{d(\log S_T)}{d(1/T)} \tag{10}$$

Time Shift Principle (Method of Reduced Variables)

The method of reduced variables was contrived by J. D. Ferry (14) in order to analyze visco-elasticity in wide range time scale on the basis of "the law of temperature-time conversion", (15). This physical concept was applied to

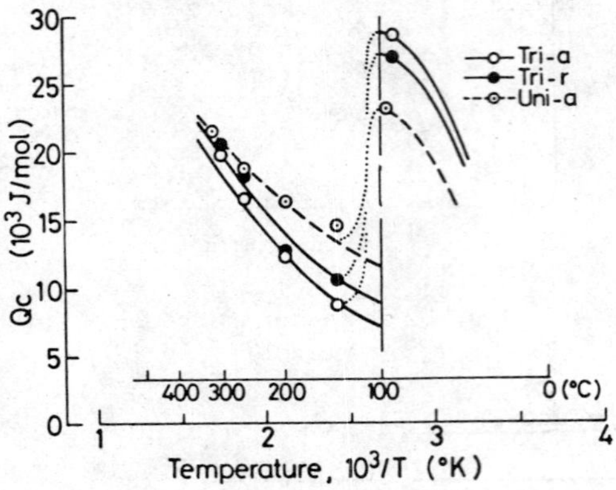

Fig. 13 Temperature function of activation energy.

concrete by the first author (11) (12) (13), M. A. Mukaddam (16) and B. Bresler (17). The law is based on the following three hypotheses; (1) The modulus of elasticity E is proportional to the temperature $T(^oK)$, (2) E is proportional to the density ρ in solid, and (3) when the temperature changes, the relaxation time or retardation time τ_K is uniformly multiplied by a_T, which is a function of $T(^oK)$. The value a_T is called "shift-factor", and it was examined in Table 3.

According to "the law of temperature-time conversion", when the temperature and density changes from (T_0, ρ_0) to (T, ρ), a compliance $J_T(t)$ at temperature T is expressed in eq. (11). If a_T is determined experimentally, creep compliance $J_T(t)$ at any temperature T and any time t can be known by eq. (11).

$$J_T(t) = \frac{T_0\rho_0}{T\rho}\int_0^{\infty} J(\tau_k)[1 - \exp(-\frac{t}{a_T\cdot\tau_k})]d\tau_k \qquad (11)$$

$$= \frac{T_0\rho_0}{T\rho} J_{T_0}\left(\frac{t}{a_T}\right) \qquad (11')$$

Shift time, S_T in this section is induced from creep master curve under tri- and uni-axial stress. S_T is comparable to a_T physically. The first author (12) found experimentally that the above-mentioned hypothesis (1) is not applicable to concrete and that in hypothesis (3) master curve on creep accord with actual long-term creep curve rather than doing temperature-density correction. So, $(T\rho/T_0\rho_0)= 1$ was assumed in the following process. Pure creep strain in Fig. 3 was redrawn by logarithmic time axis in Fig. 14. Similarly, Fig. 15 was redrawn from Fig. 4. Longitudinal axis in the two figures give expression of the creep compliance $J_c(t)$ because of the relation in eq. (12).

$$J_c(t) = 1/E(t) = 1/[1/\varepsilon_c(t)] = \varepsilon_c(t) \qquad (12)$$

Sliding the creep strain curves in Fig. 14 (except for the 20° curve of base temperature) to the right along the time axis so as to be superposed on the extrapolated 20^oC curve, a master curve on creep strain was obtained as the middle line in Fig. 16. In the same way, from Fig. 15, master curve was obtained as top line in Fig. 16.

The quantity of sliding creep curve on time axis, S_T is drawn in Fig. 17 and listed in Table 3. Though a_T and S_T in Table 3 are physically the same idea and their object are the same concrete, they are not equal. It seems to us that concept S_T is more opportune than a_T in view of phenomenalism of concrete creep strain. As another method, activation energy can be given also from S_T by eq. (13). The induction of eq.(13) and calculations on Q_c of concrete was omitted.

$$Q_c = R\,\frac{d(\ln S_T)}{d(1/T)} = 2.303\,R\,\frac{d(\log S_T)}{d(1/T)} \qquad (13)$$

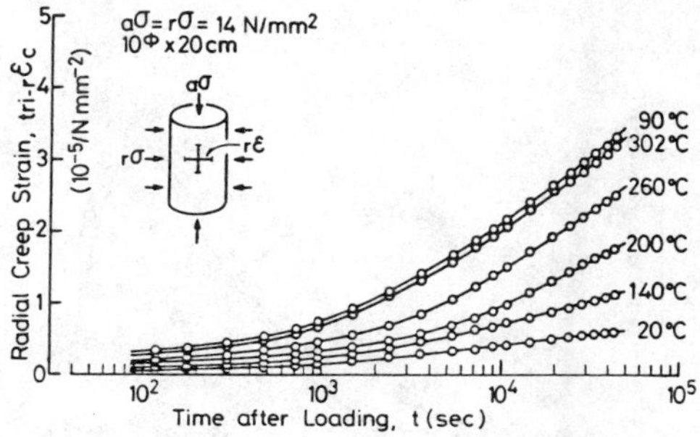

Fig.14. Creep strain curves of $_{tri\text{-}r}\varepsilon_c$.

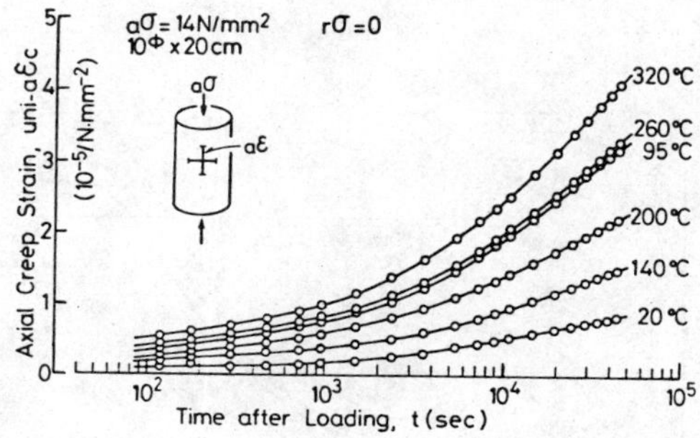

Fig.15. Creep strain curves of $_{uni\text{-}a}\varepsilon_c$.

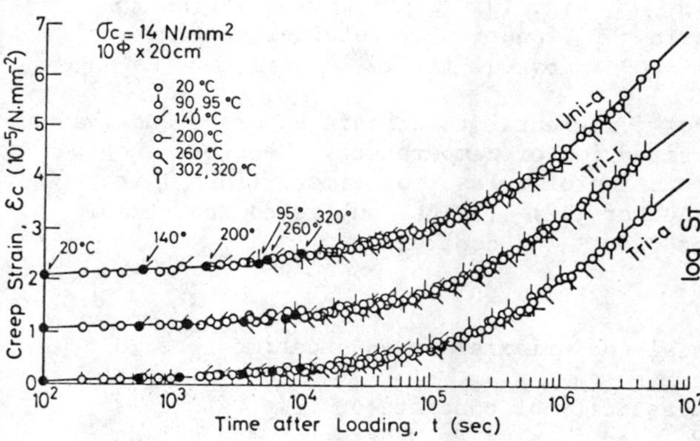

Fig.16. Master curve on creep strain obtained from Fig. 14 and 15.

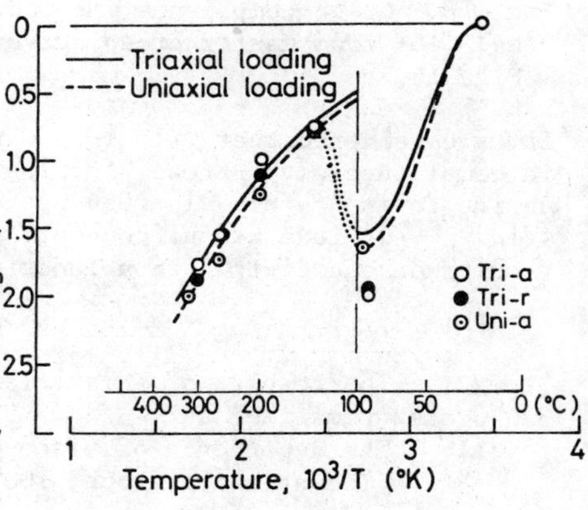

Fig. 17. Dependence of shift time on temperature.

<u>Continuous Distribution Function of Retardation Time of Concrete</u>

Fig. 16 induced by the time shift principle are assumed long-term creep curves. Considering creep strain consists of retardation component, ε_K and pure flow component, ε_λ, the total strain is expressed in eq. (14). It is rewritten in relation of compliance in eq. (15). The creep compliance can be divided between retardation elasticity compliance, $J_c(t)$ and pure flow compliance, t/λ in eq. (16).

$$\varepsilon(t) = \varepsilon_e + [\ \varepsilon_k(t) + \varepsilon\lambda(t)\ ] \qquad (14)$$

$$J(t) = J_e + [\ J_c(t) + (t/\lambda)\ ] = J_e + J_c{}'(t) \qquad (15)$$

$$J_c(t) = J_c{}'(t) - (t/\lambda) \equiv {}_pJ_c(t) \qquad (16)$$

Retardated elasticity compliance, $_{p}J_{c}(t)$ obtained by eq. (16) from master creep curve in Fig. 18. Continuous distribution function, $\Phi(\ln \tau_k)$ of retardation time, τ_k on test concrete was analyzed approximately as shown in Fig. 19 by Alfrey's equation (20)

$$\Phi(\ln \tau_k) = \left. \frac{d({}^{J}c(t))}{d(\ln t)} \right|_{t=\tau_k} \tag{17}$$

The maximum value of the function arises on ln t = 13~14, and the maximum value under triaxial stress is a little larger than that under uniaxial stress.

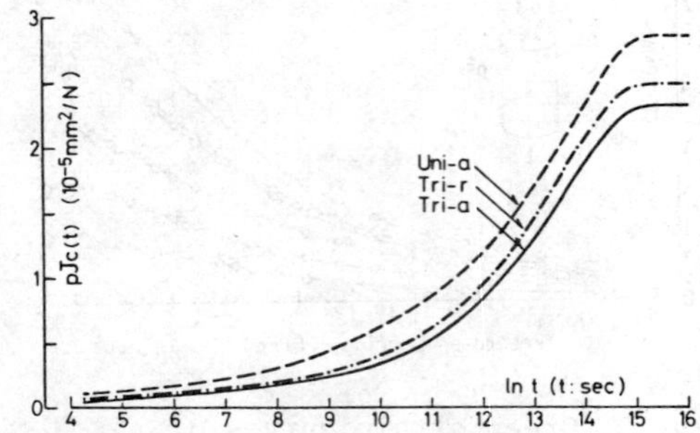

Fig. 18. Creep compliance induced by eq. (16) from master creep curves in Fig. 16.

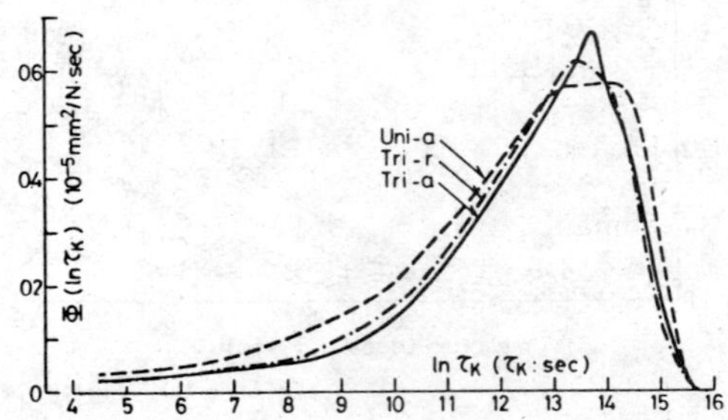

Fig. 19. Continuous distribution function of retardation time analized by eq. (17).

It is considered that $\Phi(\ln \tau_k)$ is not fundamental to certain concrete and changes in compliance with stress, loading situation or temperature. However, when we desire to induce the distribution function of relaxation time, $\Psi(\ln \tau_M)$ from $\Phi(\ln \tau_K)$, it is a useful concept. Author (12) (13) has published analytical results on the distribution function $\Phi(\ln \tau_K$ of concrete.

CONCLUSIONS

The results of this triaxial and uniaxial creep experiment at 20° ~ 320°C are summarized as :-

1) The dependence of visco-elasticity of concrete on temperature is discontinuous around 100°C, and is similar under tri- and uni-axial stress. (Fig. 7).
2) The modulus of elasticity, and the coefficient of viscosity under triaxial stress are larger than the corresponding values under uniaxial stress by 35~60%. (Fig. 7).
3) Poisson's ratio was found to have the following values; $_{uni}\nu_e$ = $_{uni}\nu_c$ = 0.15~0.18, $_{tri}\nu_e$ = 0.186, $_{tri}\nu_c$ = 0.163. (Table 2)
4) The activation energy of concrete is a function of temperature which changes greatly with moisture in concrete. (Fig. 13).
5) Master curves of $\ln\lambda$ and $\varepsilon_c(t)$ comes into existence. Shift factor a_T is not equal to shift time S_T, and S_T is more opportune as a reduced variable in order to predict long-term creep by the temperature-time conversion. (Fig. 12 and Fig. 16), (Table 3), (Fig. 17).

REFERENCES

1) Hannant, D. J. (1967). Conference on prestressed concrete pressure vessel, pp. 177-191, at Church House, Westminster, S.W. 1 13-17, March.
2) Jundi, N., T. W. Kennedy, E. S. Perry, and J. N. Thompson(1972). ACI, SP-34, 2, pp. 641-658. [Semi. on Conc. for Nuclear Reactors, at Berlin, Oct. 5-9, 1970].
3) York, G. P., T. W. Kennedy, E. S. Perry, and J. N. Thompson (1972). ACI, SP-34, 2, pp. 659-697. [ibid]
4) McDonald, J. E. (1972). ACI, SP-34, 2, pp. 735-768. [ibid]
5) Yamazaki, N., and Y. Abe (1972). Annual Technical Report of Kajima Corporation Ltd., 21, pp. 179-185., (1975). 24, pp. 13-18. (In Japanese).
6) Ohgishi, S., K. Kishitani, H. Ohshima, T. Shiire, and M. Wada (1977). Transactions of The Architectural Institute of Japan (AIJ), 262, pp. 1-11, December.
7) Ohgishi, S., and M. Wada (1978). Trans. of AIJ, 269, pp. 39-48, July.
8) Jordaan, I. J., and J. M. Illston (1971). Maga. of Conc. Rese. 23, No.75-76, pp. 79-88, June-Sep.
9) Hannant, D. J. (1969). Proc. of ACI, 66, No.5, pp. 391-394, May.
10) Jordaan, I. J., and J. M. Illston (1969). Maga. of Conc. Rese. 21, No. 69, pp. 195-204, December.
11) Ohgishi, S. (1971). Trans. of AIJ, 187, pp. 9-18, Sept.
12) Ohgishi, S. (1972). Trans. of AIJ, 197, pp. 11-25, July.
13) Ohgishi, S. (1973). Trans. of AIJ, 209, pp. 1-11, July.
14) Marvin, R. S., E. R. Fitzgerald, and J. D. Ferry (1950). Jour. of Appl. Phys., 21, pp. 197.
15) Ferry, J. D. (1950). Jour. of Amer. Chem. Soci., 72, p. 3746.
16) Mukaddam, M. A., and B. Bresler (1972). ACI, SP-34, 2, pp. 771-797.
17) Fahmi, H. M., B. Bresler, and M. Polivka (1973). Jour. of ACI, pp. 709-713, Oct.
18) Klug, P., and F. Wittmann (1974). Materi. Scie. and Engi., 15, pp. 63-66.
19) Hirst, G. A., and A. M. Neville (1977). Maga. of Conc. Rese., 29, 98, pp. 13-18, Mar.
20) Alfrey, T., and P. Doty (1945). Jour. of Appy. Phys., 16, p. 700, (1950) 21, p. 513, (1951) 22, p. 717.

Volume 3, ICM 3, Cambridge, England, August 1979

BEHAVIOR OF CONCRETE IN THE COMPRESSION ZONE OF FLEXURAL MEMBERS; A NEW TESTING METHOD

K. Suzuki and T. Nakatsuka

Department of Architectural Engineering, Faculty of Engineering, Osaka University, Yamada-Kami, Suita-City, Osaka, Japan

ABSTRACT

A very stiff testing apparatus consisting of a conventional testing machine plus simple additional devices including a one point loading system, permitted an experimental investigation into the effects of (i) neutral axis depth of concrete flexural members and, (ii) compressive strength of concrete, on the flexural behavior of concrete in the compression zone. The new test method makes it possible to observe the post-crashing behavior of concrete in flexure without a known section-stress history. The investigation considers the strain re-distribution in the failure zone. Flexural test results are discussed in comparison with the results of concentric compression tests of cylindrical specimens. It is deduced that the stress-strain relation in flexure when the depth of neutral axis is more than about 20cm,is likely to agree with that determined from concentric compression tests.

1. INTRODUCTION

At the ultimate state of failure in concrete flexural members, both flexural moment and rotational capacity are greatly influenced by the mechanical behavior and condition of the concrete within the compression zone. Thus the properties of the concrete, the magnitude of the strain gradient, the distribution of the neutral axis along the longitudinal direction of a member etc., are important. More information on pre- and post-crushing behavior of concrete in flexure is therefore needed to establish a more reasonable ultimate strength design method for flexural members.

In earlier investigations (references 1 to 9 inclusive)concerning the behavior of concrete in the compression zone of flexural members, the following factors have not been taken into consideration sufficiently.

(2) Control of eccentricity of an axial load or adjustment of magni-

tude of two thrusts to keep the strain at one outer fiber constant resulting in an unknown stress distribution and history.

(2) A high stiffness of testing apparatus to prevent the sudden failure of a specimen and so investigate the process of the post-crushing behavior of concrete.

(3) Because of the strain softening characteristics of concrete, a strain redistribution in the compression zone takes place along the axis of a prism during the ultimate state and failure is thereby constrained to occur within a restricted zone of a flexural member.

This paper describes results obtained by a new high stiffness eccentric-compression test followed by a discussion of the effect of neutral axis depth and concrete strength on the mechanical properties and failure process of concrete in the compression zone of flexural members in relation to the properties of concrete under uniaxial compression.

2. EXPERIMENTAL PROGRAM

2-1 Method of Flexural Test

The traditional eccentric-compressive test methods which intend to keep the strain at zero at one surface of the specimen have intrinsic disadvantages. As illustrated in Fig.1, when the extreme fiber strain is beyond the strain given by the minimum value of the stress block coefficient k_2/k_1k_3 (see Fig.5), strain redistribution will occur and the difference of neutral axis depth between failed and un-failed zones will retain the coincidence of the positions of compressive resultants in both zones. The strain redistribution mechanism suggests that a test method is required to consider the movement of neutral axis depth or the occurrence of tensile cracks and the distribution of the neutral axis along the longitudinal axis.

The apparatus and specimens adopted in this study have simple devices to simulate more precisely the compression zone of flexural

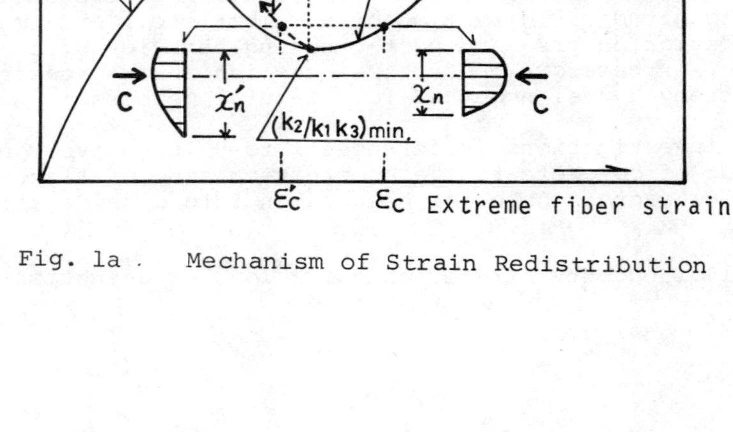

Fig. 1a . Mechanism of Strain Redistribution

members and remove the above mentioned disadvantages involved in earlier methods. In order to give the conventional universal testing machine more stiffness, or to control the release of stored elastic strain energy in the frame of the machine at the stage beyond the maximum load of each specimen, high tensile steel bars are attached to the machine as shown in Fig.2. The stiffness of the apparatus can be changed easily, to relate to the strain softening characteristics of specimens, by varying the length and total cross-sectional area of the high tensile steel bars.

Fig.2 shows also the specimen which is an eccentrically loaded plain concrete prism with a tensile steel plate on the tension side. From equilibrium of both axial load and flexural moment both in the elastic state and ultimate state of a specimen, the amount of cross-sectional area of the tensile plate is determined in relation to the eccentricity of load and the overall depth of specimen so as to keep the neutral axis depth in the failure zone approximately constant throughout the test without any adjustment of load or loading point. In the test method proposed herein, only one load with constant eccentricity is subjected to the specimen, therefore, the unknown stress history in the section, illustrated in Fig.3, can be eliminated.

Specimens are of rectangular cross section with a width of 10cm, a depth determined in combination with the amount of cross-sectional area of steel plate and with a length of about six times the neutral axis depth. The test zone is the middle part of specimens and has a length of about four times the neutral axis depth to avoid the strain restraint from end blocks. The tensile steel plate is fastened rigidly to both loading steel frames and specimen with high tensile steel bars at each end of the specimen prior to a test (see Fig.2b).

The load on the specimen is measured by a load cell, and the tensile force in the tensile steel plate is determined by strain gauges calibrated on the same steel plate specimen under uniaxial tension. The compressive load on the concrete prism is determined from longitudinal equilibrium conditions.

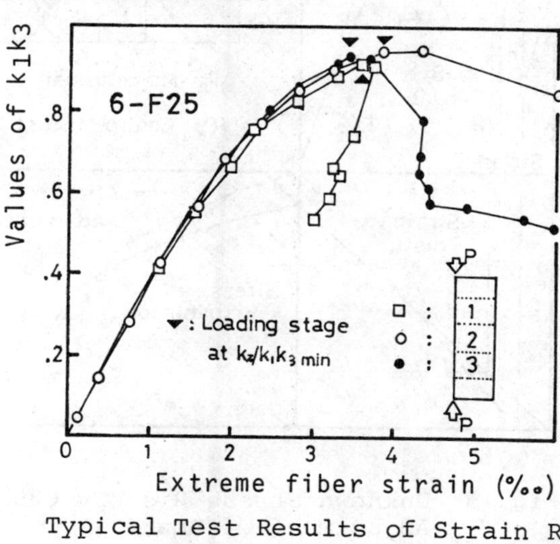

Fig. 1b Typical Test Results of Strain Redistribution

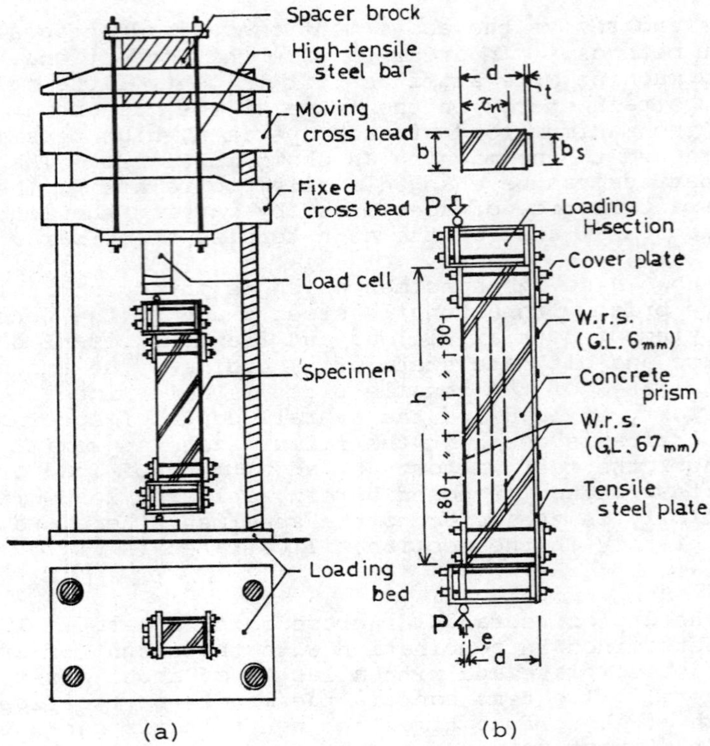

Fig. 2 Testing Apparatus and Concrete Specimen for Flexural Test

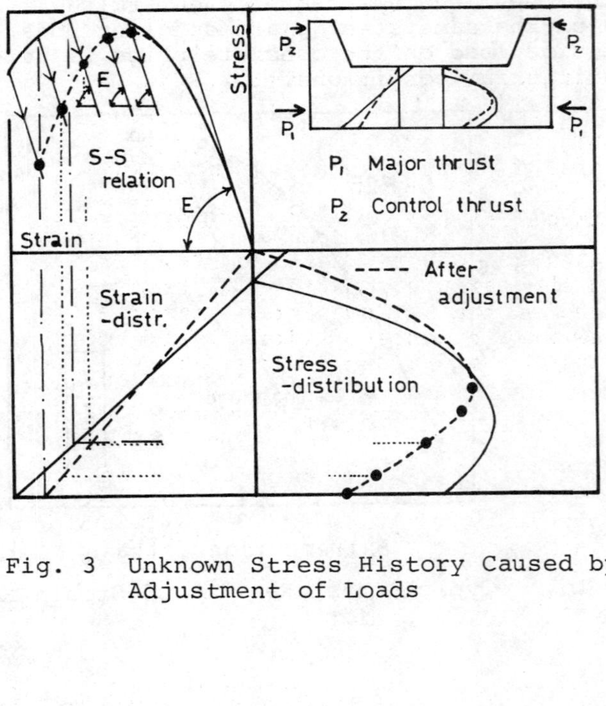

Fig. 3 Unknown Stress History Caused by Adjustment of Loads

Since the local strain behavior is to be measured to investigate the mechanism of strain redistribution in the ultimate state, 67mm wire resistance gauges are attached on the surfaces of prisms with a separation distance of 80mm in the longitudinal direction and at four levels over the depth of prisms in the testing zone as shown in Fig.2b. In order to monitor the extreme fiber strain, the mechanical displacement instruments are used at the same time. The strain measurements of the tensile steel plate are made by 6mm wire resistance gauges.

Strain distributions adjacent to the neutral axis of each specimen are found to be essentially linear even in the failure zone and at high overload, therefore the depth of the neutral axis at any loading stage is determined from strain distributions measured by the wire resistance gauges attached to the concrete prisms.

2-2 Uniaxial Compression Test

For the purpose of comparing the mechanical behavior of concrete in flexure and in uniaxial compression, concentric compression tests are carried out by the following method. In order to prevent sudden failure of specimens, four or six 33mm diameter high strength steel rods and a cylindrical specimen of 10cm diameter and 20cm height are located in parallel between the same top and bottom thick steel plates, and load is applied to these steel plates. In these tests, the strain measurement is made by displacement transducers of 18cm gauge length.

2-3 Materials and Variables

The Cement used is high early strength portland cement. Fine and coarse aggregates are river sand and gravel, and their maximum sizes are 5mm and 20mm, respectively. Mix proportions and properties of the concrete are shown in Table 1, and the yield strength of the steel plate is shown in Table 2.

Variables adopted in this test program are neutral axis depth (Xn) and compressive strength of concrete (f'c). Those are considered to be major factors which extraordinarily influence the ultimate behavior of concrete in the compression zone. Properties of specimens are shown in Table 2.

3. TEST RESULTS AND DISCUSSION

3-1 Length of Failure Zone and Crushing Zone

After the strain redistribution starts, three different zones are observed in the failure zone of the prism and those three should be distinguished in the following discussion. The first is the zone in which the strain concentration, spalling of concrete fragments and powdering of mortar are observed in its final state. This zone is termed "crushing zone" in this paper. The strain in the second zone adjacent to the crushing zone remains approximately constant. In the third zone which is located between the second one and the unfailed zone, slight failures are observed such as cracks and a small decrease of strain.

The length of the crushing zone (L_{cr}) which strongly influences the rotational capacity of concrete flexural members is 1.0 or 1.5

TABLE 1 Properties of Concrete

Designation	W/C (%)	Mixpropertion (KN/m^3)				Test age (days)	f'_c (KN/cm^2)
		Water	Cement	Fine	Coarse		
F-25	60	1.77	2.97	7.90	10.39	45	2.86
F-35	50	1.89	3.80	7.37	9.98	46	3.25
F-45	38	1.82	4.76	6.45	10.00	43	4.20

W/C:water cement ratio, f'_c:concrete cylinder strength

TABLE 2 Properties of Specimens

Designation	Prism		Tensile steel plate			h	e	x_n (planned)
	b	d	t	bs	σy (KN/cm^2)			
6-F25				7.6				
6-F35A	10	10	0.16	9.8	32.3	40	0	6
6-F35B				10				
6-F45			0.32	7.3	33.7		1	
12-F25			0.32	7.6	33.7			
12-F35A								
12-F35B	10	20	0.45	7.1	39.2	65	0	12
12-F35C								
12-F45			0.6	7.1	19.2		1	
18-F25			0.45	8.1	39.2			
18-F35	10	30	0.6	8.0	19.2	95	0	18
18-F45			0.87	7.9	24.2		1	

b, d, t, b_s, h, e, X_n:see Fig.2 (UNIT:cm)

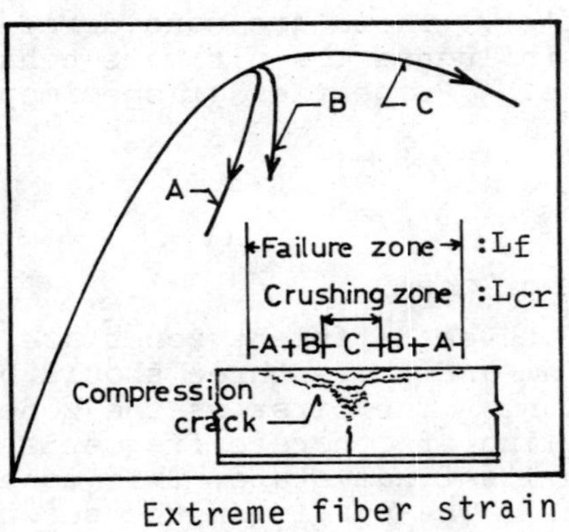

Fig. 4 a) Definitions of Failure Zone and Crushing Zone

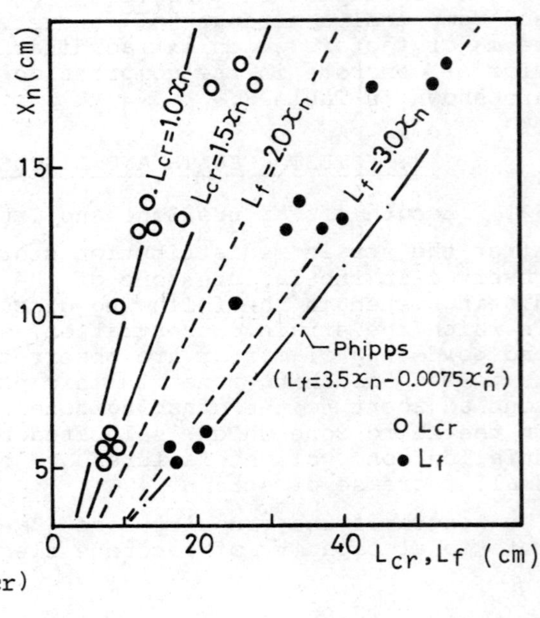

Fig. 4 b) Test Results of Failure Zone Length(L_f) and Crushing Zone Length(L_{cr})

times the neutral axis depth as shown in Fig.4. The measured length of the failure zone (L_f) is two or three times the neutral axis depth. This result agrees approximately with Phipp's results (1976).

3-2 $k_1.k_3$ max and ε_u

Stress block coefficients k_1, k_2 and k_3 defined by Hognestad (1955, see Fig.5),especially the maximum value of $k_1.k_3$ in the crushing zone and the extreme fiber strain corresponding to $k_1.k_3$ max,are important parameters for ultimate strength design of concrete flexural members. Figure 6 in which results obtained herein are plotted against concrete strength indicates that : on the effect of neutral axis depth on $k_1.k_3$ max, the maximum values obtained from prism specimens with small neutral axis depth agree sufficiently with the relation obtained from flexural tests on prestressed concrete beams with small neutral axis depth in the ultimate state (see line in Fig.6(a) by Billet-Appleton (1965), Xn = 3~7 cm), however, the values obtained from prism specimens with larger neutral axis depth are fairly close to the values proposed by Hognestad (1955, see Fig.6(b), Xn=20 cm), and the maximum values of $k_1.k_3$ decreases with the increase of the depth of neutral axis and tend to converge to the values determined from the stress strain relations of cylindrical specimens which are subjected to uniaxial compression (see Fig.6(b)).

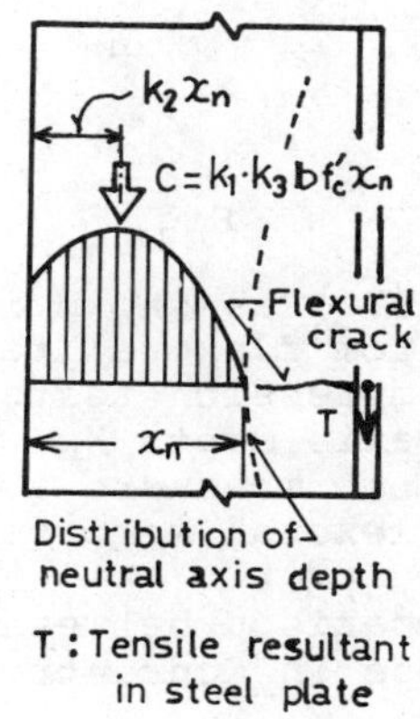

Fig. 5 Definitions of Coefficients of $k_1 \cdot k_3$ and k_2

Similar tendency is found in the relation between extreme fiber strain (ε_u) at $k_1.k_3$ max and neutral axis depth, and this tendency agrees with the experimental results by

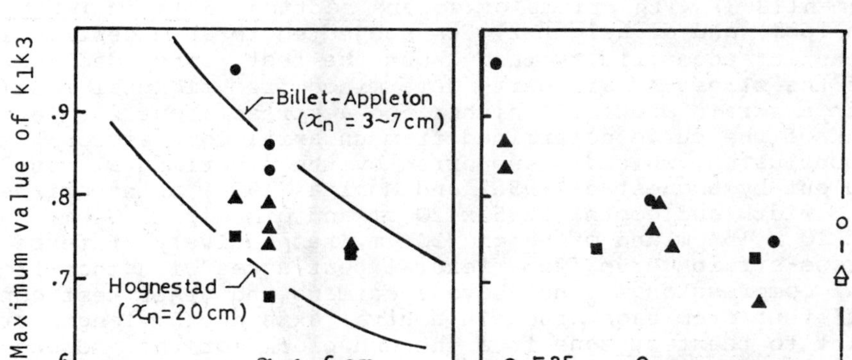

Fig. 6 Maximum Values of Coefficient $k_1 \cdot k_3$

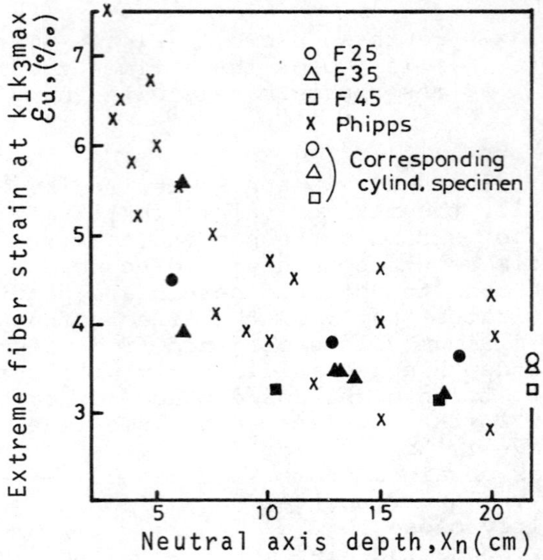

Fig. 7 Extreme Fiber Strains at $k_1 \cdot k_3$max

Phipps(1976) as shown in Fig.7. Since values of $k_1 \cdot k_3$max and ε_u obtained from flexural tests converge to those determined from concentric compression tests of cylindrical specimens with the increase of neutral axis depth X_n, it may be deduced that those values on cylindrical specimens are available as reasonable design parameters to actual flexural members with X_n of more than 15 or 2o cm.

3-3 Comparison between Stress-Strain Relations in Flexure and Those in Concentric Compression

As to the stress-strain relation in the compression zone of flexural members, two different conclusions have been reported. The first one by Struman(1965) with prisms of square section 15x15 cm (with smaller X_n than 15cm) and of height 27.5cm subjected to an eccentric load with constant eccentricity throughout the test, concluded that the peak of the stress-strain curve determined from flexural test was located at a strain about 50% higher and a stress about 20% larger than the peak of the curve determined from uniaxial compression. The other conclusion, which is supported by the experimental results carried out by Hognestad (1955) and Morita (1971) using prisms of section (width and depth) 12.5 x 20 cm and of height 40 cm, and of section 10 x 15 cm and of height 30 cm, respectively, inferred that the stress-strain curve from flexural test agree with that from uniaxial compression. The above specimens and other test conditions are different from each other in neutral axis depth, length or strain restraint to the test zone from the end block (or the end steel plate).

In the present investigation, calculations of the stress-strain relations from the results of flexural tests are carried out bu Eq.(1) (Hognestad, 1955) in which (i) it is assumed that all fibers in a section follow one and the same stress-strain curve and (ii) the strain distribution across a section is linear.

$$\sigma_c = \left(\varepsilon_c \cdot \frac{d(k_1k_3)}{d\varepsilon_c} + k_1k_3 \right) \cdot \sigma_c \qquad (1)$$

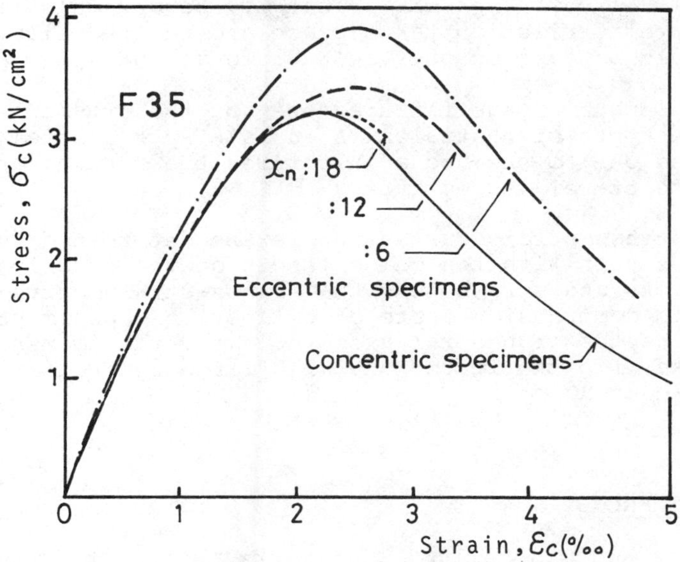

Fig. 8 Flexural Stress-strain Curves in Crushing Zone

Here, σ_c and ε_c are the compressive stress and strain in concrete respectively, $\bar{\sigma}_c$ is the average stress in the concrete of the compression zone. The results obtained for the crushing zone of prisms are shown in Fig. 8. Peaks of curves determined from flexural tests have a tendency to transfer upward and toward larger values in strain from the curve of concentric specimen with decreasing depth of neutral axis. The maximum stress and its corresponding strain of prisms with about 6 cm depth of neutral axis are approximately 1.2 times larger than those of concentric specimens. Results on the prisms with about 18 cm depth of neutral axis, show similar curves to those of cylindrical specimens.

From the above investigations it can be presumed that the previous different conclusions of several researchers on the flexural stress-strain relation of concrete are mainly due to the difference of specimen size and loading condition, i.e. the depth of neutral axis and perhaps the strain restraint from end blocks.

4. CONCLUSIONS

The present experimental investigation carried out by a new method leads to the following conclusions.

1. The test method proposed herein is useful to investigate the pre- and post-crushing behavior of concrete in the compression zone of flexural members and is especially useful for measuring strain distribution without sudden failure of specimens and eliminates the problem of unknown stress histories due to adjustment of loads.

2. Different stress distributions in cross sections along the axis

of a prism can be recognised at the same time after the loading stage at the minimum value of $k_2/k_1.k_3$. That is, beyond this loading stage, strain redistribution takes place around the failure zone and each test should be carried out taking into account this phenomena.

3. Both the length of the failure zone and the crushing zone are related to the depth of neutral axis and are approximately two to three times and one to one and a half times the neutral axis depth in the ultimate state, respectively.

4. Derived stress-strain curves in flexure determined by Eq. (1) and the fiber strain with the gauge length of the crushing zone is greater in stress and strain than the stress-strain curve obtained from concentric compression tests of cylindrical specimens in the case of small depths of neutral axes, however, the former curve is likely to agree with the latter one with large neutral axis depths of more than about 20 cm.

REFERENCES

(1) Billet,D.F. and others(1965). Flexural strength of prestressed concrete beams. ACI Journal. June,
(2) Hatcher, D.S. and others(1975). Post-crushing behavior of unbound concrete beams. J. of Structural Division, ASCE. Nov.
(3) Hognested,E., N.W.Hanson and D.McHenry(1955). Concrete stress distribution in ultimate strength design. ACI Journal,Dec. 445-479.
(4) Karsan,I.D. and J.A.Jirsa(1970). Behavior of concrete under varing strain gradient. J. of Structural Division, ASCE. Aug. 1675-1696.
(5) Morita,S. and N. Adachi(1971). The stress-strain behavior of concrete in the compression zone of flexural members. Proc. of Int. Conf. of Mechanical Behavior of Materials(ICM-1). 162-171.
(6) Phipps,M.E.(1976). The strain capacity of compression-zone concrete subjected to short-term loading. Mag. of Concrete Research, June, 85-100.
(7) Rüsch,H.(1960). Research toward a general flexural theory for structural concrete. ACI Journal, July. 1-28.
(8) Smith, R.G.(1960). The determination of the compressive stress-strain properties of concrete in flexure. Mag. of Concrete Research, Nov.
(9) Sturman,G.M., S.P.Shah and G.Winter(1965). Effect of flexural strain gradients on micro-cracking and stress-strain behavior of concrete. ACI Journal, July. 805-822.

Volume 3, ICM 3, Cambridge, England, August 1979

MECHANICAL PROPERTIES OF MASS-CONCRETE AT ELEVATED TEMPERATURES

H. Aschl

Aschl-Consult, D-8000 Munich 60, F.R.G.
(formerly Technical University of Munich)

INTRODUCTION

At the Institut fuer Massivbau of the Technical University of Munich, tests were carried out to determine the thermal expansion coefficients as well as the modulus of elasticity and to study shrinkage and creep of mass-concrete at elevated temperatures at 353 and 393 K. Therefore, testing machines were built, to allow testing of sealed and unsealed cylinders under uniaxial stress-states at elevated temperatures up to 523 K. The construction of the heat cell, the loading equipment and the test programmes are briefly described. In addition to the evaluation of the test results the residual uniaxial strengths of all specimens were determined. Some specimens showed a decrease in strength up to 60 % for unsealed, i.e. drying specimens. A detailed research report will be published (Aschl, Stoeckl 1979).

The mechanical properties of concrete under uniaxial and multiaxial stress states at elevated temperatures must be known for the design of thick-walled prestressed concrete reactor vessels. These are greatly affected by the humidity of the concrete which is dependent on the level of temperatures and temperature gradients. The temperature gradients inside the thick walls cause water migration. Hundt (1975) and England and Skipper (1977) furnished proof of this fact and showed that pore pressure decreases after heating. To study the real conditions in such a reactor wall, sealed specimens with a vapour pressure outside the sealing have to be used. In addition tests on unsealed specimens, i.e. drying ones, were carried out to study the influence of water migration on the mechanical behaviour.

Figure 1 shows the deformation of concrete under sustained loading at normal temperature (293 K). The isothermal deformation of concrete under different states of stress is divided into 4 components:

- elastic deformation
- shrinkage caused by loss of relative humidity and hydration

- basic creep, which is creep of concrete at constant humidity, and
- drying creep, which is caused by drying under sustained loading.

All these mechanical properties are temperature dependent.

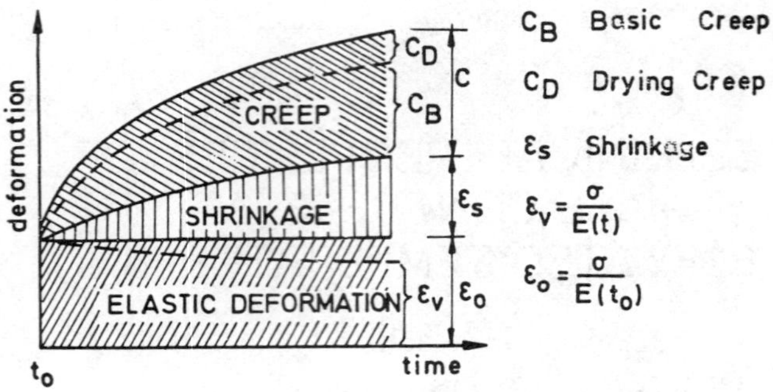

Fig. 1. Schematic sketch of deformation under sustained loading

Also the thermal expansion coefficient can be measured at elevated temperatures. Due to the fact that concrete is a highly disperse system, which means it has a lot of micropores, the relative humidity inside these pores influences all mechanical properties of mass-concrete at elevated temperatures, too.

EXPERIMENTAL DETAILS

In testing the mechanical properties of mass-concrete we had to pay attention to the physical states inside such a thick concrete wall. To avoid drying of the specimens, they were sealed with a copper foil, as usual. At elevated temperatures, above 373 K, the vapour inside the sealing would have caused overpressure which would blow and destroy the sealing. Therefore, we developed a new testing equipment, the essential characteristic of which is the heating of the sealed specimens via a pressure vessel filled with vapour, so the pore pressure inside the sealing is equal to the vapour pressure inside the pressure vessel, i.e. outside the sealing. This test machine corresponds to the physical properties of a thick concrete wall. To study the influence of water migration on the mechanical properties of concrete, unsealed specimens were tested, too. In this case the pressure vessels were opened.

Figure 2 shows two heat cells in one loading frame. The upper part of Figure 2 is a vertical section of such a heat cell with a built-in specimen, the lower one a side view of an isolated heat cell.

Figure 3 gives detailed information of the sealing of the specimen and the heat cell. The specimens were sealed with copper foils to the steel loading platens which were mounted to the top and bottom of the specimens. The space between the sealing and the heat cell was partly filled with water.

The stress states of the specimens were measured by an oil manometer which was part of each loading frame. The longitudinal strains of each specimen were measured via four steel frames which were mounted to the steel loading platens on the top and bottom of the specimen. In each frame a dial gauge measured the deformations with an accuracy of 1 μm. The lateral deformations were not measured. With an additional test programme (Aschl, 1976) biaxial tests were carried out to study the Poisson's creep ratio. A detailed research report will be published (Aschl, Stoeckl, 1979).

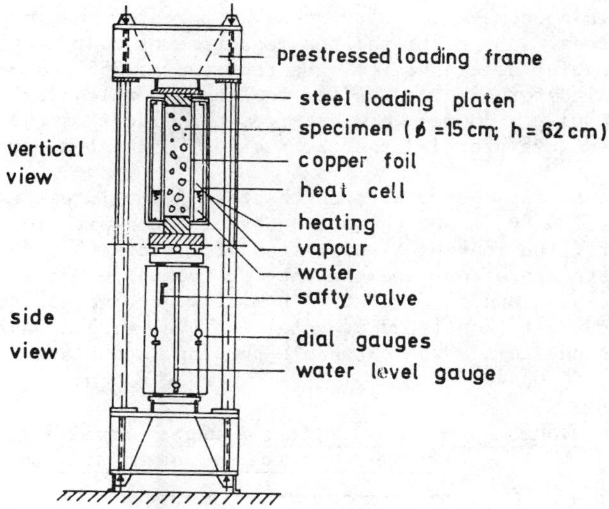

Fig. 2. Testing equipment.

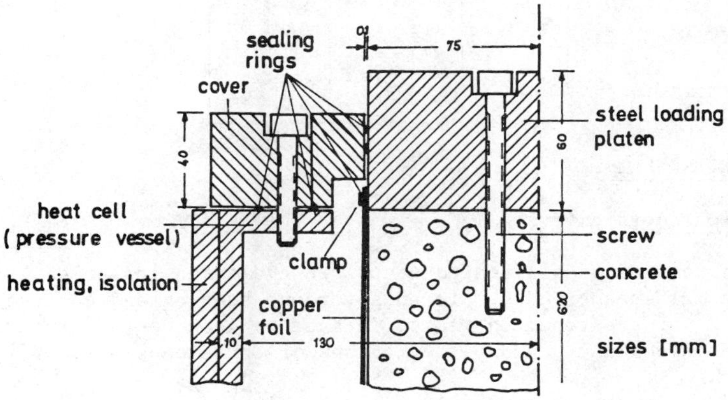

Fig. 3. Details of the heat cell.

Materials and experimental procedure.

36 specimens were cast. One day after casting all specimens were stored in a fog room (293 K, 100 % relative humidity). Afterwards they were treated in 3 different ways:

- after 4 weeks in the fog room, storage at 293 K and 65 % relative humidity (later called "test storage", which was used for unsealed specimens)
- sealage after one week in the fog room (later called "fog storage")
- after 4 weeks in the fog room, 8 weeks of predrying at 378 K (later called "predried storage")

The tests were carried out three months after casting. Additional sealed and unsealed tests were carried out at normal temperature (293 K). The specimens tested at elevated temperatures were heated at 5 K/h up to 353 K. After 12 hours, when a uniform temperature was reached, the cylinders were loaded. Shrinkage was measured at unloaded specimens for nine weeks with heated specimens and for two years with unheated ones. Some tests for the measurements of the thermal expansion coefficients and modulus of elasticity were repeated at 353 and 393 K. Creep measurements were studied for one year. After cooling and reloading of all specimens, cylinders (ø = 150 mm, h = 300mm) were cut from the middle of the specimens, their loading surfaces were grounded and the residual strengths were measured.

The development of the concrete strength after 4 and 13 weeks from casting is shown in Table 1. $ß_{W10}$ and $ß_{W20}$ are the strengths of cubes with 100 or 200 mm side lengths, $ß_C$ for cylinders. The indexes f and v mean "fog storage" and "test storage". No index means a storage of one week in the fog room after casting and then in a climate room with 293 K and 65 % relative humidity, equivalent to German standard practice. The concrete consisted of calcite gravel with a maximum size of 22.4 mm, some quarzite sand and fly ash to enable pumping, and blast furnaced cement. The water-cement ratio was 0.52.

TABLE 1 Mean values and standard deviations of the test for the strength development

Specimen size and storage	Strengths [MN/m²] Mean value and stand. dev. 28 days after casting	91 days after casting
$ß_{W10f}$	62.0 ± 9.8 %	70.1 ± 6.6 %
$ß_{W10}$	64.5 ± 5.9 %	71.4 ± 7.3 %
$ß_{W20}$	59.1 ± 5.2 %	65.3 ± 9.2 %
$ß_C$	54.1 ± 5.6 %	60.5 ± 5.9 %
$ß_{Cv}$	--	61.4 ± 5.9 %

TEST RESULTS

Thermal expansion coefficients

The strain measurements of all heated tests are evaluated in Tables 2 and 3. Table 2 shows the thermal expansion coefficients for one heating and cooling cycle, Table 3 all the results of the repeated tests. In both tables the temperature ranges are shown. The mean values and standard deviations of the expansion coefficients are:

- for predried concrete: $9.5 \times 10^{-6} \pm 0.8 \times 10^{-6}\ K^{-1}$
- for unsealed concrete: $13.2 \times 10^{-6} \pm 0.6 \times 10^{-6}\ K^{-1}$
- for sealed concrete: $13.6 \times 10^{-6} \pm 1.1 \times 10^{-6}\ K^{-1}$.

Table 3 for the repeated tests shows similar results. Due to the limited number of repeated tests no systematic change of the thermal expansion coefficient, caused by heating cycles, can be seen. Also tests up to 393 K show no significant deviations from the results stated above.

But systematic changes of the thermal expansion coefficients can be seen for sealed and unsealed specimens by cooling which shows the influence of humidity. Mean values and standard deviations of the thermal expansion coefficients by cooling are:

- for predried concrete: $-10.1 \times 10^{-6} \pm 1.9 \times 10^{-6}\ K^{-1}$
- for unsealed concrete: $-11.6 \times 10^{-6} \pm 1.4 \times 10^{-6}\ K^{-1}$
- for sealed concrete: $-11.7 \times 10^{-6} \pm 2.0 \times 10^{-6}\ K^{-1}$.

TABLE 2 Thermal expansion coefficients at the first heating cycle

Specimen storage	Specimen number	Thermal expansion coefficients x 10^6 1/K heating T [K]		cooling T [K]	
predried	2b/3	293-353	8.92	353-298	- 8.82
	2b/4	293-361	10.6	353-295	-11.45
unsealed	3c/1	293-353	12.45	353-296	-10.49
	4a/5	293-353	13.75	353-293	-11.22
	4a/16	293-353	19.80	353-305	(-20.42)
	5a/7	293-353	12.68	353-293	-12.30
	5a/8	293-353	13.05	353-293	-12.80
	6a/5	293-353	13.77	353-293	-13.32
	6a/7	293-353	13.48	353-293	- 9.55
sealed	2b/1	293-353	14.50	353-298.5	-13.43
	2b/2	293-353	12.95	353-333.5	- 8.32
	3b/1	293-353	14.75	353-299	-12.22
	3b/2	293-353	14.58	353-299.5	-13.98
	4a/3	293-353	13.40	353-293	-10.15
	4a/4	293-353	12.68	353-293	-10.68
	5a/3	293-353	12.30	353-293	-13.75
	5a/4	293-353	12.03	353-293	-13.63
	6a/1	293-353	13.28	353-293	-10.03
	6a/3	293-353	15.28	353-293	-11.17

TABLE 3 Thermal expansion coefficients for repeated tests

Specimen storage	Specimen number	Thermal expansion coefficients x 10^6 1/K Heating up to 353 K T [K]		Heating up to 393 K T [K]		Cooling T [K]	
predried	2b/3	298-330.5	7.42			330.5-295.5	- 6.40
unsealed	3c/1	296-353	11.58	353-393	13.93	393-301	-10.65
	4a/6	305-353	(+20.58)			353-293	(-17.28)
sealed	2b/1	298.5-353	13.34			353-301	- 8.87
	2b/2			303.5-380	9.96	380-306	- 7.96
	3b/1			299-393	13.72	393-309	-11.74
	3b/2			299.5-383	13.57	383-297.5	-15.92

Modulus of elasticity

All strain measurements for the study of the modulus of elasticity are evaluated in Table 4. Table 5 shows the results of the repeated tests. The mean values and standard deviations are:

- for unsealed concrete at 293 K: 41,770 $\pm$ 930 MN/m²
- for sealed concrete at 293 K: 37,980 $\pm$ 380 MN/m²
- for unsealed concrete at 353 K: 35,385 $\pm$ 290 MN/m²
- for sealed concrete at 353 K: 34,220 $\pm$ 570 MN/m²

During the second heating cycle at 393 K the unsealed specimen 3c/1 showed a modulus of elasticity of 32,500 MN/m², the sealed one 3b/1 32,890 MN/m².

TABLE 4 Modulus of elasticity at the first heating cycle

Specimen storage	Specimen number	T [K]	E [MN/m²]
unsealed	3b/5	293	42,710
	3b/6	293	41,740
	6a/8	293	40,860
sealed	3b/3	293	34,420
	3b/8	293	37,720
	6a/4	293	37,810
unsealed	3c/1	353	35,590
	6a/7	353	35,180
sealed	3b/1	353	34,620
	3b/2	353	34,470
	6a/3	353	33,560

TABLE 5 Modulus of elasticity of repeated tests

Specimen storage	Specimen number	T [K]	E [MN/m²]	Remarks
unsealed	3c/1	353	36,500	2nd heating
		393	32,500	2nd heating
	6a/7	353	35,400	immediately after heating
		353	35,310	2 days after heating
		353	36,100	2 months after heating
		293	37,340	after cooling
		293	36,160	outside the heat cell
sealed	3b/1	393	32,890	
	3b/2	302	41,010	after 2nd cooling
	6a/3	353	33,560	immediately after heating
		353	34,000	2 days after heating
		353	36,960	2 months after heating
		293	40,670	after cooling
		293	40,140	outside the heat cell

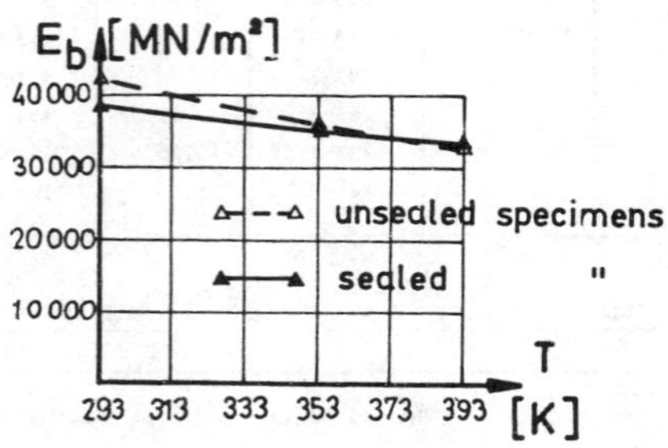

Fig. 4. Modulus of elasticity at elevated temperatures.

Figure 4 shows evaluated test results for the modulus of elasticity as a function of temperature. It is obvious that the decrease of modulus of elasticity is smaller for sealed specimens than for unsealed ones. The decrease is 10 % at 353 K and 13 % at 393 K for sealed specimens, while it is 15 and 22 % for unsealed ones.

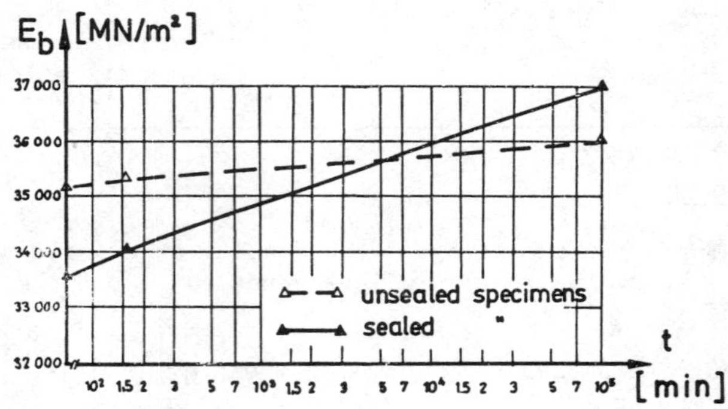

Fig. 5. Modulus of elasticity at sustained temperatures.

For 3 specimens 3b/2, 6a/3 and 6a/7 the modulus of elasticity was measured again after cooling. Here sealed tests showed an increase of about 10 %, unsealed ones of only 4 %.

Shrinkage

The strain measurements for the study of shrinkage are evaluated in Figure 6 for tests at normal temperature of 293 K and Figure 7 for tests with elevated temperatures of 353 K. Shrinkage deformations of all unsealed specimens of the other test programmes and all sealed specimens after unsealing are shown with crosses in Figure 6.

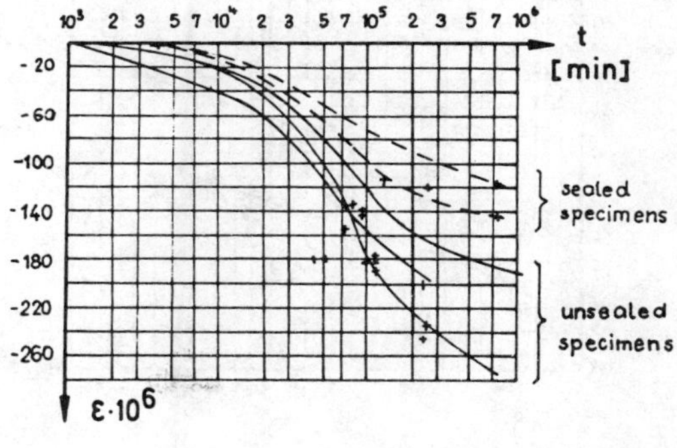

Fig. 6. Shrinkage at 293 K

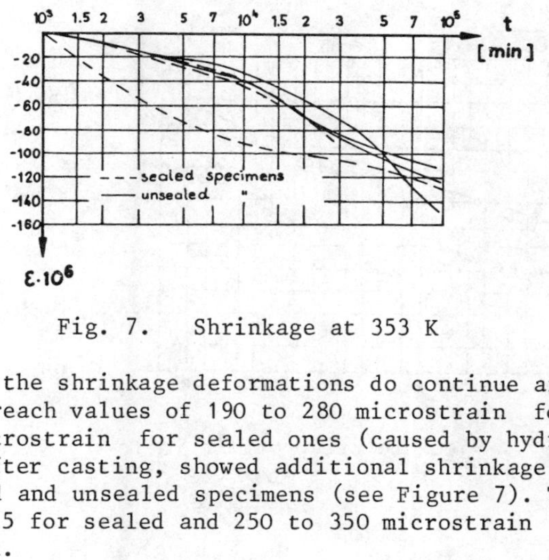

Fig. 7. Shrinkage at 353 K

Both figures show that the shrinkage deformations do continue after testing. After 2 years at 293 K they reach values of 190 to 280 microstrain for unsealed specimens and 120 to 150 microstrain for sealed ones (caused by hydration). The heating, started 90 days after casting, showed additional shrinkage of 110 to 150 microstrain for sealed and unsealed specimens (see Figure 7). Total shrinkage deformations of 190 to 225 for sealed and 250 to 350 microstrain for unsealed specimens were measured.

Creep

Creep deformations of all creep tests are evaluated in Figures 8 and 9. In the first one the creep ratio φ is plotted in logarithmic scale versus time in logarithmic scale, in the second one the specific creep strain ratio c_{sp} is presented in double logarithmic scale versus time.

The creep ratio φ is the ratio between the creep deformations and the elastic deformation at loading. The specific creep strain ratio c_{sp} is the ratio between the creep strain and the sustained stress. Both are related via the modulus of elasticity, whereas the creep ratio φ is the specific creep strain ratio c_{sp} multiplied by the modulus of elasticity at loading. As with the shrinkage the creep deformations do continue after testing. After a sustained stress of one year, the following creep ratios (specific creep strain ratios) were measured for tests at normal temperature of 293 K on sealed specimens of 0.55 (14×10^{-6} m^2/MN) and on unsealed ones of 0.85 (19×10^{-6} m^2/MN), for tests on elevated temperatures of 353 K on sealed specimens of 1.43 (40×10^{-6} m^2/MN) and at unsealed ones of 2.43 (68×10^{-6} m^2/MN). The creep deformations of the heated specimens were always 2.5 to 3.5 times higher than the unheated ones.

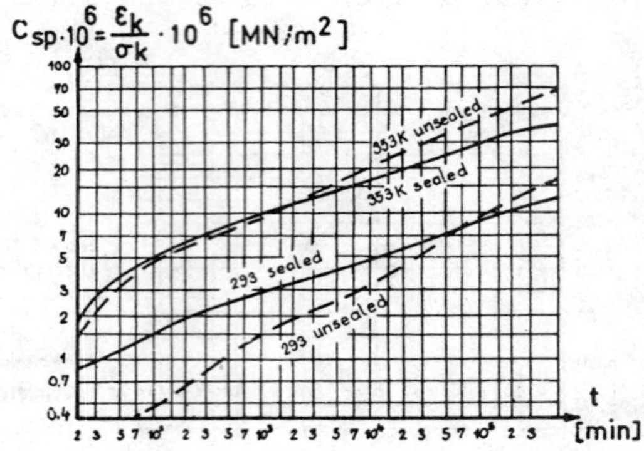

Fig. 8. Creep ratio versus time.

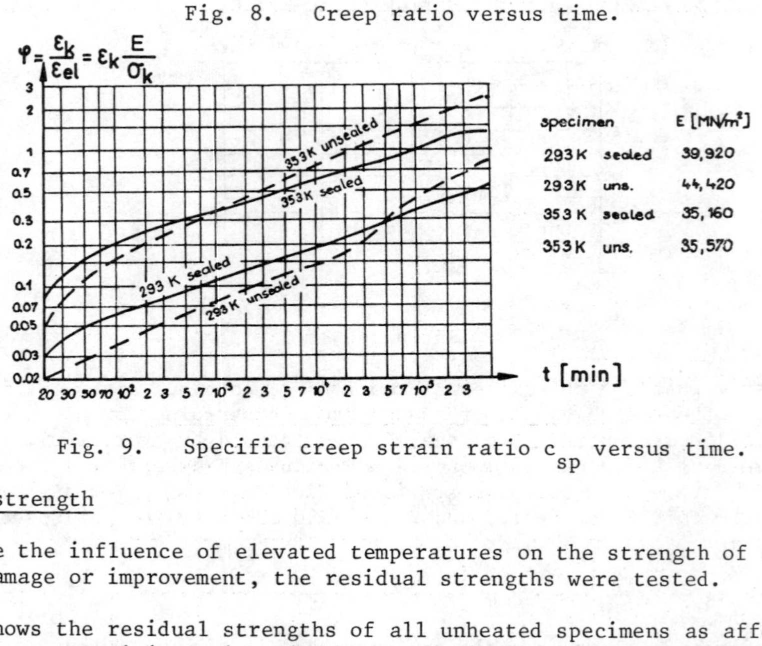

Fig. 9. Specific creep strain ratio c_{sp} versus time.

Residual strength

To measure the influence of elevated temperatures on the strength of concrete, whether damage or improvement, the residual strengths were tested.

Table 6 shows the residual strengths of all unheated specimens as affected by different storage conditions with different relative humidity and with or without sealing. For comparison the strength results of cylinders 3 months (91 days) after casting are shown, too. Unsealed specimens at normal temperature show an increase in residual strength (except specimen 4a/8). So do the sealed specimens except 3b/3, 3b/8, 5a/1 and 61/2. The decreases of the exceptions are caused by shrinkage after sealing, which can be seen on the specimens 4a/1 and 4b/1. They differ in so far as the first was tested immediately after unsealing while the second was used for additional shrinkage measurements for 9 months.

The tests with elevated temperatures (see Table 7) always show decreased residual strengths. The decrease was 5 to 20 % for sealed specimens. The damage caused by elevated temperatures seems to appear immediately after heating while with sustain-

TABLE 6 Residual strengths of unheated specimens

Unsealed specimens:

Specimen number	$\beta_{C,91}$ [MN/m²]	Residual strengths [MN/m²] of the specimens at an age of 120 to 160 days	c 380 days	c 440 days	c 510 days
3b/5	57.5	68.8			
3b/6		60.8			
4a/7	60.3				62.2
4a/8					53.5
5a/5	65.2		68.7		
5a/6			61.4		
6a/6	53.7	63.4			
6a/8				52.1	

Sealed specimens:

Specimen number	$\beta_{Cv,91}$ [MN/m²]	Residual strengths [MN/m²] after different storage conditions at an age of 100 days sealed (+23 days unsealed)	175 days sealed (+270 days unsealed)	370 days sealed (+270 days unsealed)	430 days sealed (+9 days unsealed)	510 days sealed (+9 days unsealed)	510 days sealed (+270 days unsealed)
3b/3	64.1	55.0					
3b/8		58.2					
4a/1	55.8					65.8	
4b/1							59.4
5a/1	64.3			53.7			
5a/2				69.3			
6a/2	60.9				68.9		
61/4			35.5				

ed loading time the residual strengths increased again.

The most significant decreases in strengths were measured in unsealed, i.e. drying specimens. Obviously, the damage caused by elevated temperatures occurred during the heating, as with the sealed specimens. With sustained heating time the residual strengths increased again (compare specimens 3c/1, 3b/7, 4a/5 and 4a/6).

TABLE 7 Residual strengths of heated specimens

Specimen storage	Specimen number	$\beta_{C,91}$ resp. $\beta_{Cv,91}$ MN/m²	Residual strengths after different loading conditions at an age of 2 days 353K	2x2 days 353K	2x2 days 353K +1 day 393K	70 days (353K)	150 days (353K)	360 days (353K)
predried	2b/3	59.0		39.7				
	2b/4		34.7					
sealed	2b/1	59.0		46.6				
	2b/2				42.3			
	3b/1	57.5			45.2			
	3b/2				43.0			
	4a/3	60.3				47.5		
	4a/4					51.7		
	5a/3	65.2				(sealing		51.9
	5a/4					destroyed)		60.2
	6a/1	53.7					46.4	
	6a/3					51.0		
unsealed	3c/1	--			48.5			
	3b/7	64.1			45.0			
	4a/5	55.8				50.5		
	4a/6					54.8		
	5a/7	64.3						25.9
	5a/8							31.7
	6a/5	60.9					37.2	
	6a/7					42.1		

But under sustained stress, specimens at elevated temperatures showed an increase in damage (compare 6a/7 and 4a/5, 4a/6 and 5a/7, 5a/8, 6a/5 with all the others). The longer the duration of sustained stress, the higher was the damage. Compared with the original strengths of the specimens 3 months (91 days) after casting, the residual strengths showed decreases up to 60 %.

All residual strengths of heated specimens were measured after cooling. No statement can be made whether the damage is caused by the elevated temperatures themselves or partly or totally by cooling.

CONCLUSION

Calcite concrete may be used for mass-concrete constructions which are heated up to 353 K, if drying can be avoided. The modulus of elasticity is reduced, the shrinkage deformations increase, the creep deformations are 2.5 to 3.5 times higher than at normal temperatures of 293 K. The decrease in strength is up to 20 %. Further investigations with different concrete mixtures with other gravels have to be carried out and the investigations of combined stress states and elevated temperatures by drying specimens have to be researched carefully.

If drying of concrete cannot be avoided, special attention must be paid to the decrease in strength under sustained elevated temperatures and stresses. Additional research of the mechanical properties of concrete under these conditions is necessary.

ACKNOWLEDGEMENT

The construction of the loading equipment and heat cells and the tests to study thermal expansion coefficients and the modulus of elasticity were financed by Hochtemperatur-Reaktorbau GmbH, Mannheim.

The shrinkage and creep tests were financed by the German Ministry of Research and Technology under the auspices of the Deutscher Ausschuss fuer Stahlbeton.

The author would like to thank these institutions for their financial support. The helpful advice of Professor Kupfer, Professor Grasser, Dr. Linse and Dr. Stoeckl during the experiments is gratefully appreciated.

REFERENCES

Aschl, H., Stoeckl, S. (1979). Waermeausdehnung, Elastizitaetsmodul, Schwinden und Kriechen von Reaktorbeton unter einachsiger Belastung und erhoehten Temperaturen. Volume of the Deutscher Ausschuss fuer Stahlbeton, Berlin (in preparation)

Hundt, J. /1976). Waerme- und Feuchtigkeitsleitung in Beton unter Einwirkung eines Temperaturgefaelles. Volume 256 of the Deutscher Ausschuss fuer Stahlbeton, Berlin.

England, G.R., Skipper, M.E. (1977). On the Prediction of Moisture Movement in Heated Concrete. 4th Int. Conf. on Structural Mechanics in Reactor Technology, Paper H6/2, 8pp.

Aschl, H., Linse, D., Stoeckl, S. (1976). Creep of Mass-Concrete with Biaxial Loading and at Elevated Temperatures. 2nd Int. Conf. on Mechanical Behaviour of Materials, Vol. II, 183-192.

Aschl, H., Stoeckl, S. (1979). Kriechen von Beton unter zweiachsiger Belastung bei erhoehten Temperaturen. Volume of the Deutscher Ausschuss fuer Stahlbeton, Berlin (in preparation).

Volume 3, ICM 3, Cambridge, England, August 1979

MECHANICS OF MICROCRACKING IN CONCRETE SUBJECTED TO FATIGUE LOADING

P. Stroeven

Stevin Laboratory, Delft University of Technology, Stevinweg 4, Delft, The Netherlands

INTRODUCTION

Throughout the many decades of successful use of concrete as a construction material no serious troubles have been experienced from fatigue. A general trend in modern building technology, however, is the erection of more slender constructions in combination with higher allowable working stresses. In addition, we are now facing new applications in off-shore engineering. In a relatively hostile environment these marine constructions have to sustain fluctuating loads due to waves and wind. The varying level of the stored oil and the accompanying changes in the temperature gradients involve equally serious attacks on the integrity of the construction. Since this type of construction is designed for temporary use, one is moreover tempted to raise the working stresses, consequently reducing the strength reserve available to guarantee the serviceability during its planned life-time. Accurate knowledge should therefore be obtained of the mechanical behaviour of concrete subjected to high repeated loadings above the endurance limit.

FATIGUE LOADING OF CONCRETE

Research on fatigue loading of concrete has been carried out for some decades (McLaughlin, 1960). It mainly concentrated, however, on loadings below the endurance limit. Although the more recent publications do not present any evidence for the existence of such a limit state, the phenomenological behaviour is nevertheless different in the stress domains separated by it (Shah and Chandra, 1970; Awad and Hilsdorf, 1974)

As for the mechanics of microcracking, for fluctuating compressive stresses below the endurance limit no differences have been observed as compared to the case of monotonically increasing uniaxial compression (Plähn, 1962; Gaede, 1962; Hansen, 1966). As the upper load level is further increased, the scarce experimental data point towards a change in structural behaviour. Graf and Brenner (1939) report cracks in the loading (axial) as well as in the transverse direction. Although some crack length and orientation measurements have been performed to compare the effects of sustained and cyclic loadings (Bennet and Raju, 1969; Shah and Chandra, 1970), this supplies insufficient information to determine particular mechanisms.

Since morphometrical data are intimately related to the level of the microstructure considered, and the sensitivity of the observational technique is not mentioned, this makes a comparison and hence an evaluation of published results an

 MS216

impossible task.

The data presented by Shah and Chandra (1970) reveal only slight differences in the total crack length at 95% of the time at rupture for sustained and cyclic loadings, at a maximum load of 85% of ultimate (in a short time test). Crack development was more gradual under cyclic loadings, however. Bennet and Raju (1969) demonstrated that only moderate differences can be observed in the orientation distribution of cracks at the surfaces of specimens subjected to static and fatigue loadings, respectively. The sensitivity of the observational technique was low as compared to the one employed in our tests. Further, bond cracking was observed to be more pronounced in the fatigue tests. Stroeven (1975) has shown that all those results depend on the minimum crack length taken into consideration. An additional drawback of the above mentioned investigations is that the image analyses performed were not based on objective procedures. This makes their results somewhat dependent on the applied techniques (see e.g. Stroeven (1973)). The quantitative image analysis techniques and the stereological relationships used to quantify the three-dimensional features of the crack structure have been previously described by this author (Stroeven, 1973, 1979).

Application to the static case has proven to be fruitful in relating phenomenological behaviour to the characteristics of structural loosening, i.e. microfissuration. It also facilitated the construction of a closed and causal model for crack development around the inclusions present in mortar or concrete (Stroeven, 1973, 1976). This model is in agreement with information about crack mechanisms published in the literature (Vile, 1965; Newman, K. and J. B. Newman, 1969; Perry and Gillott, 1977; Stroeven and de Haas, 1974). On the analogy of the static case we will extend this framework for structural loosening to the one for high cyclic loadings.

EXPERIMENTAL APPROACH

The investigations included the influence of the upper stress level σ_t', the relative stress amplitude S, and the frequency f, on the number of cycles to rupture N of concrete specimens subjected to a sinusoidally fluctuating load. By definition $S = (\sigma_t' - \sigma_b')/\sigma_{f28}'$, in which σ_b' is the lower stress level and σ_{f28}' the ultimate compressive strength of prisms loaded after 28 days. For the phenomenological research program a sustained load level of $0.875\sigma_{f28}'$ was selected, whereas for the fatigue tests σ_t' was kept constant during the tests at two different levels, $0.875\,\sigma_{f28}'$ and $0.850\sigma_{f28}'$, respectively. The distribution of the lower stress levels σ_b' was chosen in such a way, that the peak in the S-logN curve, suggested by the results of Awad and Hilsdorf (1974) and of Weigler and Freitag (1975) could be located, the extreme values being $\sigma_b' = 0.875\sigma_{f28}'$ (permanent loading) and $\sigma_b' = 0.4\sigma_{f28}'$. Although no decisive evidence is presented, the assumption of a peak value also conforms to "contradictory" results obtained by Mehmel and Kern (1962), i.e. contradictory in so far as some results conflict with the generally observed trend that the strength is reduced by increasing the stress amplitude. The frequency levels considered are f = 0.175 Hz and f = 17.5 Hz.

Concrete with a maximum grain size of 8 mm and a W/C-ratio of 0.46 was used for the fabrication of prismatic specimens, having dimensions of 100x100x345 mm^3. Portland cement type B was applied to obtain a fairly constant quality at the day of testing (after about 1 month). The specimens were stored under water for 10 days followed by a 7 days storage under climatized conditions (20°C and 50% RH). Finally, the specimens were placed in the testing room (about 20°C and 30% RH). A servo-hydraulic testing machine (Schenck) was utilised. Deformations in two orthogonal directions were recorded by demountable strain gauge transducers.

The tests at an engineering level were complemented by a structural study of the effect of the most interesting factor, the stress amplitude, on the crack structure. From three series of "identical" specimens those were selected whose deformational behaviour nearly as possible approached the group average behaviour. One specimen represented a series subjected to cyclic loading with a relative stress amplitude S = 0.075 for which strength was found to be an optimum. On both sides of the optimum we have taken a specimen of the sustained loading series and one from a series subjected to a cyclic loading with large amplitude (S = 0.35).

The specimens were sliced in the direction of loading, providing three equally spaced sections, 27 mm apart, suitable for image analysis. Due to automatic termination of the test at a predetermined value of the transverse strain rate, the morphological features of the crack structure refer to this state "close to rupture". Visibility of the surface cracks in the sections was enhanced by means of the filtered particle method (Stroeven, 1973; 1974, 1979; de Haas and Stroeven, 1975), implying spraying on fluorescent dye and photographing under illumination with UV-light. The photographic negatives were projected on a semi-transparent screen and all cracks longer than 1 mm were manually copied. An example is shown in Fig. 1. These patterns of lineal features in a plane were analysed by means of Saltikov's directed secants technique. The details of the image analysis and the stereological framework applied have been described elsewhere (Stroeven, 1973, 1974, 1978). It suffices here to remark, that a line grid is superimposed on the crack pattern. Thereupon, the number of intersections P is counted and divided by the total line length L. This quantity, i.e. $P/L = P_L$, determined in two orthogonal directions, suffices to calculate the total crack length per unit area ($L/A = L_A$) or the crack surface area per unit of volume ($S/V = S_V$).

Counting the number of cracks (or, merely, crack sections) facilitates calculating the average crack length $\bar{x}$ ($\bar{x} = L_A/N_A$, where N_A is the number of cracks per unit of area). Since it was shown (Reinhardt, and others, 1978) that the crack size distribution could be approximated by a simple exponential function, the size of the largest crack could be estimated as well. This is a leading parameter in fracture mechanics. Plotting intersection counts for various positions of the line grid, provides a polar figure (rose of the number of intersections) that expresses the

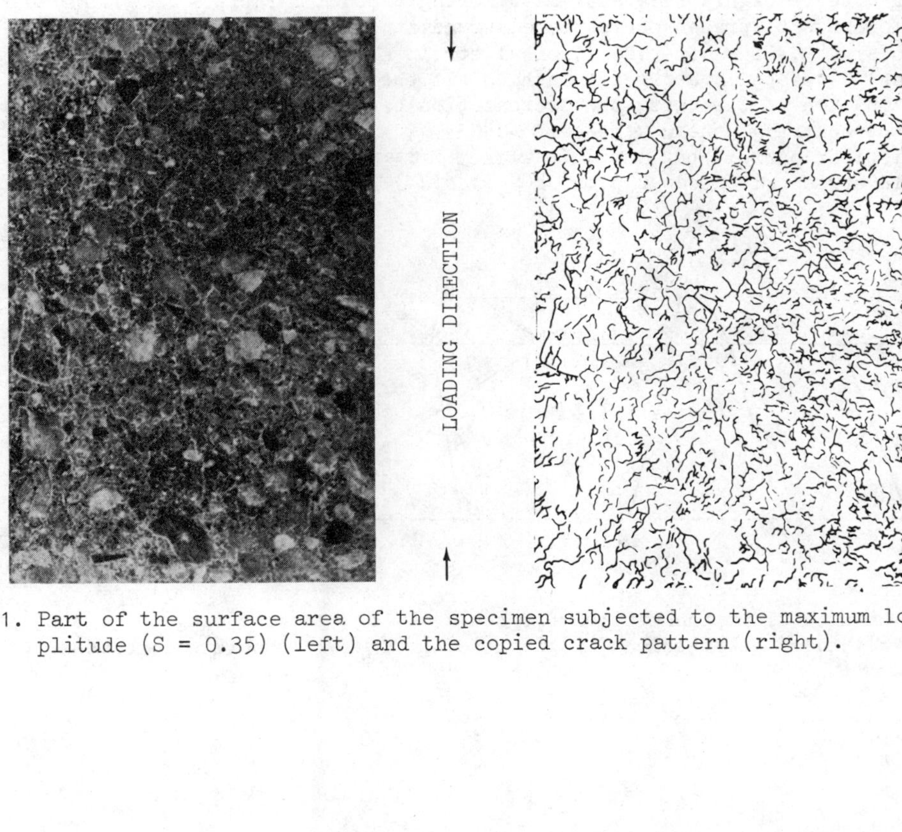

Fig. 1. Part of the surface area of the specimen subjected to the maximum load amplitude (S = 0.35) (left) and the copied crack pattern (right).

planar characteristics of the orientation distribution. These results can be contracted to a single parameter, the degree of orientation ω ($\omega = 0$ for a random and $\omega = 1$ for a unidirectional system).

The accuracy of the results is dominantly influenced by the total number of observations, hence by the sampled area. We have investigated centrally located areas in the section plane with dimensions $A = 51 \times 102\ mm^2$. The analysis of the variance has been discussed by Stroeven (1978). Trends in group averages used as a basis for the construction of the model for crack formation in concrete subjected to cyclic compressive loadings have been demonstrated to be significant.

RESULTS

Some of the basic results relating engineering aspects of mechanical behaviour (cycles to rupture N) to morphological features of the crack structure "close to rupture", are presented in Fig. 2. An optimum can indeed be observed in the stress amplitude (S) v. log cycles to rupture (logN) graph. The changes in morphological parameters with a rising value of the stress amplitude are more complicated. Despite a steady increase in the values of the specific crack length (L_A) and specific surface area (S_V), extreme values are found for the crack density (N_A), the average crack length ($\bar{x}$) and, most pronounced, for the degree of orientation (ω). Also, the histograms for crack size presented (Reinhardt, and others, 1978) showed a reduced peakedness for S = 0.075, hence providing additional evidence for a relatively large maximum crack length. The most interesting parameter in Fig. 2, however, is ω. An increase in the degree of preferred axial orientation is changed for maximum amplitude in a preferred orientation in the tranverse direction (though only 3% of the cracks show this orientation).

In an analogous study of the mechanical behaviour of concrete subjected to a stepwise increasing uniaxial compressive loading (Stroeven, 1979), it was demonstrated that tranverse strains ε_t, Poisson's ratio υ, or volumetric strains e can directly be related to the increment of the specific crack surface area ΔS_V. It is therefore considered to be highly relevant in tailoring a model for crack *development* under cyclic loadings to present here the observed strain measurements. A selection is given in Fig. 3. Since the longitudinal strain ε_1 (after a relatively small number of cycles) develops in a regular fashion all the way up to a point very close to the termination of the test, the presentation of the ε_1-logN-curve could be omitted. On the contrary, the ε_t-(hence, υ- and e-) N-curve manifests in general three quite distinct parts. When the time effect outweighs the cycle effect (above the threshold stress situated at about 80% of σ'_{f28}) we see that the crack development

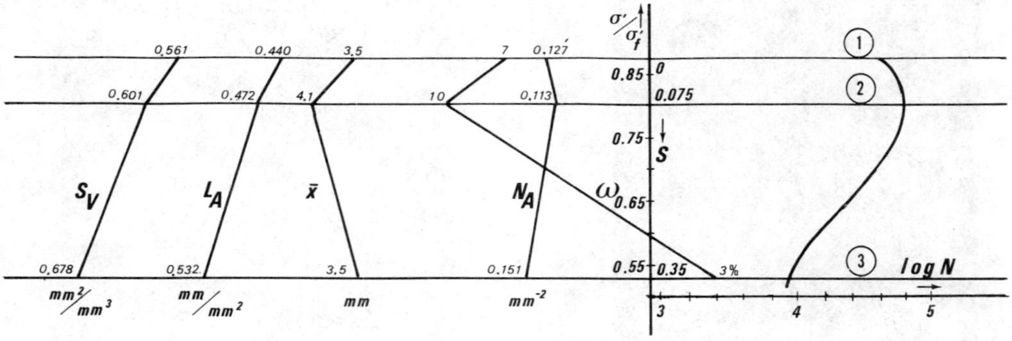

Fig. 2. Results of morphometrical (at the left) and phenomenological tests (at the right) of compression fatigue loading of plain concrete specimens.

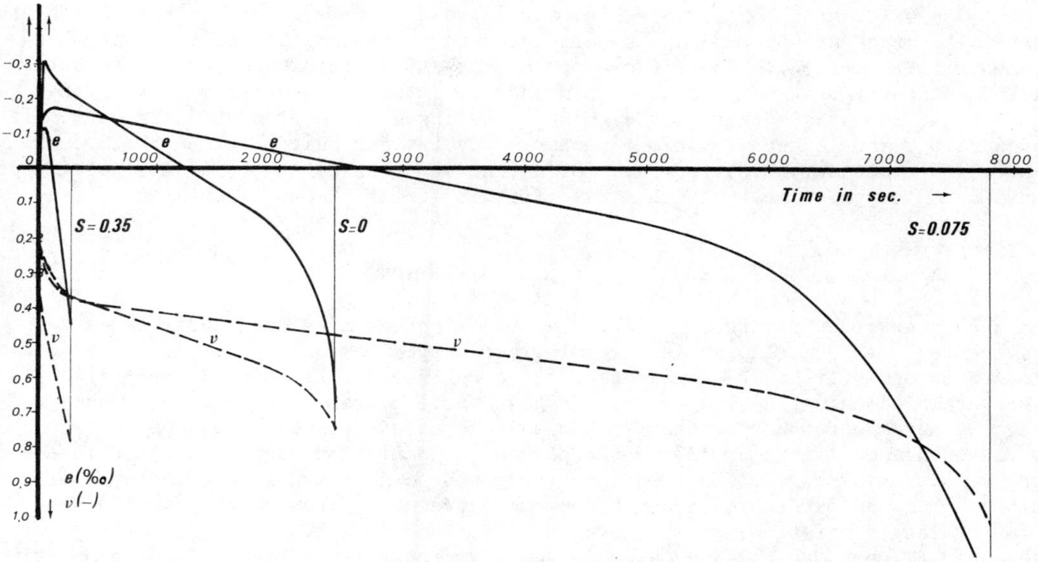

Fig. 3. Development of Poisson's ratio υ and volumetric strain e with time t for the three stress amplitudes S investigated.

process is accelerated at roughly 60% of the time to rupture. For S = 0.35 we see however that no warning can be expected from the deformation measurements. The part of the graph that shows a proportional increase in volumetric strains extends over roughly half the number of cycles to rupture. In Fig. 4 we have presented various strain measurements and crack surface area data as functions of the stress amplitude. In addition, data are presented for N = 2400 cycles (outside the area where a disproportional increase of ε_1 to N can be observed). Comparison of the various curves gives a qualitative insight into crack development under different loading regimes. Worth mentioning is that ε_1 decreases proportional to S, which confirms results of Awad and Hilsdorf (1974)

For practical situations the intermediate part of the e-time (or N)-curve, shown in

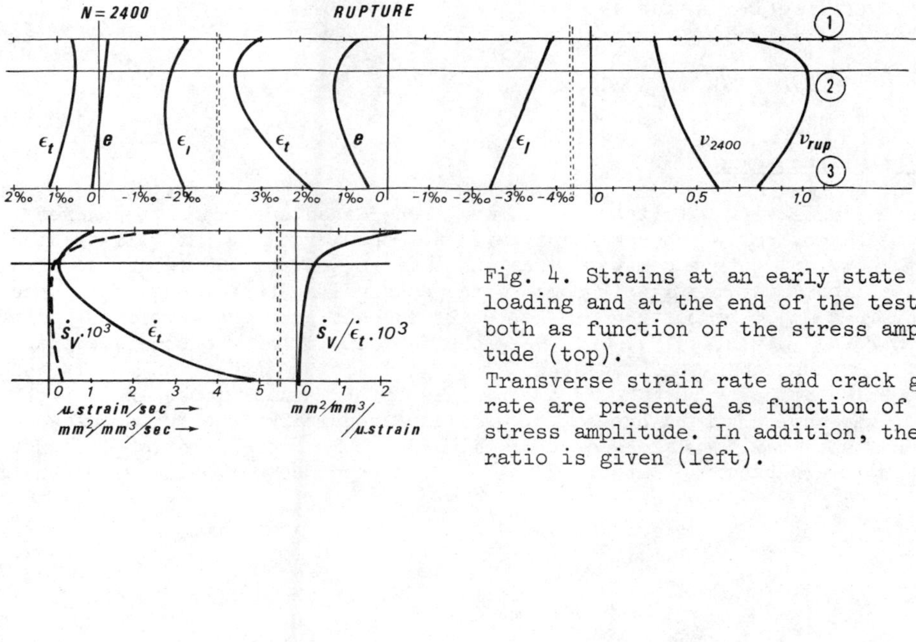

Fig. 4. Strains at an early state of loading and at the end of the test, both as function of the stress amplitude (top).
Transverse strain rate and crack growth rate are presented as function of the stress amplitude. In addition, their ratio is given (left).

Fig. 3, is of paramount relevance. The strain rate $\dot{\varepsilon}_t$ refers to this portion of the curve. The crack growth rate $\dot{S}_V$ is calculated on the assumption of a constant increment in S_V per cycle. The ratio of both quantities yields the rate of structural loosening per unit of lateral strain increment during the stable part of the process. It is obvious that if one thinks of using $\dot{\varepsilon}_t$ as a predictor for structural degradation, knowledge of both the number of cycles per unit of time (that possibly can fairly accurately be determined for marine structures) and the amplitude range (estimation of which may be much more difficult) is of utmost importance.

EVALUATION OF RESULTS

The characteristic features in mechanical behaviour as developed under the three loading regimes, as well as the morphological properties, all outlined in the previous section, will now be integrated into a general, qualitative description of structural loosening under cyclic loadings. Starting from the sustained loading case, we have seen that the drop in the average stress (that may be 20% in respect to the endurance limit, or 50% of the permanent loading strength) resulted in a *more stable crack development*. On the average the cracks are longer, because of joining their number is reduced and they more frequently run in the axial direction. Prolonged lifetime causes, nevertheless, the total crack surface area to be increased. These longer and as a consequence wider cracks produce the disproportional increase in the transverse deformations.

In further reducing the average stress we pass the threshold stress, whereupon the cycle effect starts to dominate the time effect. The transverse strain rate and crack propagation rate are drastically accelerated. However, more and finer cracks are now being created, reducing the possibilities for a proportional increase in transverse strain rate. Additional fine cracks are moreover initiated during stress release. They run on the average in the transverse direction, destroying the load bearing column-like structure usually found in highly stressed uniaxially compressed concrete specimens. Reducing the frequency would offer more favourable conditions for reversed slip, thus eliminating the transverse cracking under stress release. Indeed, for f = 0.175 Hz the S-logN-curve does not show the strong cycle effect. Obviously, the counteracting effects of reduced average stress and cycling are balancing each other in this very case (Reinhardt, and others, 1978). The very effective destruction of the internal structure is reflected by the high transverse strain rate. The small amount of tiny cracks perpendicular to the loading direction contributes only in a minor way to the total crack surface area, however, yet lifetime has drastically decreased.

MECHANISMS OF MICROCRACKING

Stages of Microcracking

On a superficial view with the unaided eye, non-loaded specimens of concrete, though macroscopically heterogeneous, will not show cracking (if well-treated). Working, however, with a somewhat more sensitive observational technique (as in our tests), one must inevitably come to the conclusion, that even virgin specimens can contain a major proportion of the cracks present over the complete loading range (Stroeven, 1973, 1979). Only when the sensitivity of the observational method is low, it makes sense to define the (crack)"initiation stress" as an alternative for the "discontinuity point" (in the process of structural loosening). On the basis of the same argument one can give serious considerations to the usage of additional alternatives like "onset of (micro)cracking" or "onset of more severe (or major) microcracking" (Stroeven, 1973) in dependence on the level of the microstructure observed. Loosely speaking, these terms define a transition stage (contracted for practical purposes to a point) between slow stable crack extension and an accelerated stage in which unstable local crack growth takes place. There is some experimental evidence that the fatigue limit and the discontinuity point fall close together at, say, 50-60% of σ'_f (Newman, K. and J. B. Newman, 1969; Berés, 1969).

A Model for Crack Development

On a sub-macroscopical level, where concrete is found to be a two-phase material, in which the phases are completely bonded, solutions for the stress-strain state around inclusions embedded in a homogeneous matrix can provide the theoretical counterpart for the experimental observations (Stroeven, 1973). On a macroscopical level, where the bond is accepted as incomplete, one can apply fracture mechanics solutions, like that developed by Perlman and Sih (1967) for the stress-strain state around partly debonded inclusions in a homogeneous matrix. For more detailed information see Stroeven (1973). It is demonstrated that in low-stressed uniaxially compressed specimens the interfacial cracks tend to grow due to high shear and low compressive stresses in the interface until a more axially oriented system is obtained.

A rising load causes further crack growth in the interface under local shear stresses. The stress state in the interface, however, becomes gradually more unfavourable because of increasing radially directed compressive stresses. In case the surface roughness of the aggregate resists slip along the interface, it has been shown that the crack can propagate in a less dense layer in the matrix a small distance (20-40 μm) from the aggregate surface (Barnes, Diamond and Dolch, 1978; Stroeven, van Staveren and de Wind, 1979). The surfaces of the aggregate particles, exposed after final rupture of the specimen, will show the debris of the cement matrix still fixed to the surface (Perry and Gillott, 1977).

A stable cone-like zone of triaxially compressed material at the "poles" of the particles prevents further crack extension in the interface. The crack will thereupon branch at penetration in the matrix at an angular orientation of about 50°-60°. The conical zones are surrounded by axially oriented bond cracks of smaller particles, thus creating "en échelon" crack arrays. They were detected by Gramberg (1970) in marble and later by Stroeven (1973) in concrete. These crack arrays constitute areas with low resistance to slip. At discontinuity sand and gravel particles will therefore be debonded from the matrix in a characteristic way. This has been experimentally verified by Stroeven (1973) by means of photoelastic coatings, microscopical observations and holographic interferograms (see also de Haas and Stroeven (1975)). Various researchers (Vile, 1965; Perry and Gillott, 1977; Stroeven, 1973, 1976) have moreover shown that after rupture of the specimen particles provided with conical mortar caps can be peeled off from the fragments.

After passing the "onset of structural loosening", axially oriented cracks are initiated, caused by the oblique cracking in the shear areas. In a following stage, these cracks bridge the isolated microcracks around (especially larger) particles, finally producing continuous cracks that grow into the macroscopical level. Successively these sub-macrocracks grow and stop against barriers, finally splitting up the specimen into "slices" or even "columns" (under conditions of pure uniaxial stress). It goes without saying that the largest grains can be expected to determine the cross-sectional dimensions of the load-bearing column-like structure. Nelissen (1972), indeed, found an average macrocrack spacing in accordance with the maximum grain size .

These features of microcracking in concrete subjected to static loadings also cover

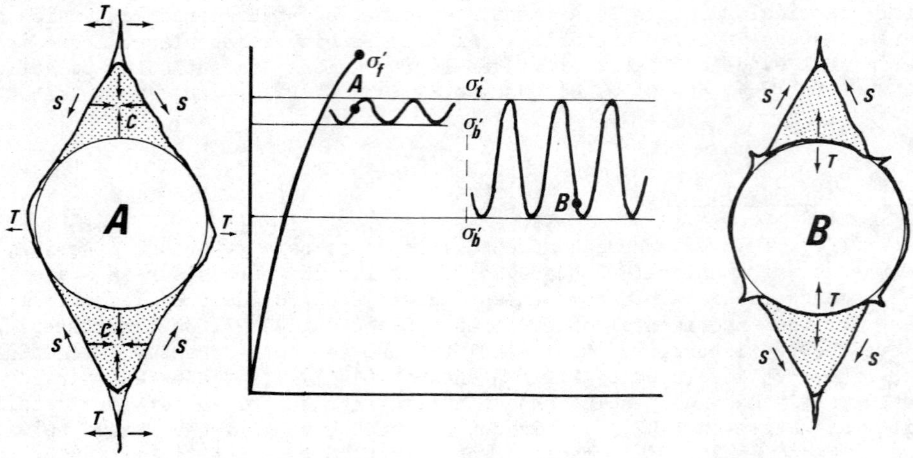

Fig. 5. Models for crack formation around inclusions in concrete specimens subjected to high compressive cyclic loadings. A refers to small load amplitudes. B covers the case of large amplitudes.

the case of high cyclic loadings as long as the amplitude is small. However, when the amplitude is such that on the descending branch of the load-time curve the discontinuity point is passed a new mechanism comes into action. Strain release under reduced stresses implies reversed slip along the shear areas. The rough slip areas prevent, however, complete recovery in the short time interval available (in case of f = 17.5 Hz). The expanding matrix will as a consequence introduce axially oriented tensile stresses in the characteristic elements described. Due to the low strength or the presence of cracks, the interface between particle and conical deposit splits open. This initiates a complete deterioration of the column-like structure that collapses in a rather brittle way at an early stage of cyclic loading. The essential features of both models for high compressive cyclic loadings are presented in Fig. 5. They are supported by microscopical observations as well as quantitative image analysis data (Kooistra and Vrencken, 1975/76; Stroeven, 1973).

CONCLUSIONS

A study of the mechanical behaviour of concrete under high compressive cyclic loadings in combination with a quantitative image analysis, using stereological relationships for the structural evaluation of the crack morphology data, has proven to be fruitful in establishing the relevant features of models encompassing the successively acting crack mechanisms. This can enhance the understanding of the behaviour of concrete in an hostile marine environment, hence offering possibilities for appropriate design.

LITERATURE

Awad, M. E., H. K. Hilsdorf (1974). Strength and deformation characteristics of plain concrete subjected to high compressive repeated and sustained loads. In *Fatigue of concrete, ACI Special Publ.*, SP-41, Am. Concr. Inst., Detroit. pp. 1-13.

Bennet, E. W., N. K. Raju (1969). Cumulative fatigue damage of plain concrete in compression. In M. Te'eni (Ed.), *Proc. Southampton Civ. Engr. Mat. Conf.* Wiley, New York. pp. 1089-1102.

Berés, l. (1969). Relationship of deformational processes and structure changes in concrete. In M. Te'eni (Ed.), *Proc. Southampton Civ. Engr. Mat. Conf.* Wiley, New York. pp. 643-651.

Barnes, B. D., S. Diamond, W. L. Dolch (1978). The contact zone between portland cement paste and glass "aggregate" surface. *Cement and Conc. Res.*, 8, 233-244

Gaede, K. (1962). Versuche über die Festigkeit und die Verformung von Beton unter Druck-Schwellbeanspruchung. *Dtsch. Aussch. für Stahlbet.*, 144. 1-41.

Graf, O., E. Brenner (1939). Versuche zur Ermittlung des Gleitwiderstandes von Eiseneinlagen im Beton bei stetig steigender und oftmals wiederholter Belastung. *Dtsch. Aussch. für Eisenbet.*, 93, 1-23.

Gramberg, J. (1970). *Klastische en kataklastische processen en hun betekenis voor de gesteentemechanica.* Ph. D. Thesis, Delft.

Haas, H. M. de, P. Stroeven (1975). *Strain measurement techniques used in concrete materials research.* Stevin report, 1-75-1, Delft.

Hansen, T. C. (1966). Cracking and fracture of concrete and cement paste. In *Causes, mechanism and control of cracking, ACI special publ.*, SP-20, Am. Concr. Inst., Detroit. pp. 43-66.

Hsu, T. C. (1963). Mathematical analysis of shrinkage stresses in a model of hardened concrete. *J. Am. Concr. Inst.*, 60, 3, 371-390.

Kooistra, T. R., J. H. A. M. Vrencken (1975/76). *Herhaalde belasting van ongewapend beton.* Graduation work, Delft.

McLauglin, J. F., and others (1960). *Fatigue of concrete, annotated.* Am. Concr. Inst. bibliography, Vol. 3, Detroit.

Mehmel, A., E. Kern (1962). Elastische und plastische Stauchungen von Beton infolge Druckschwell- und Standbelastung. *Dtsch. Ausch. für Stahlbet.*, 153.

Nelissen, L. J. M. (1972). Biaxial testing of normal concrete. *Heron*, 18.

Newman, K., J. B. Newman (1969). Failure theories and design criteria for plain concrete. In M. Te'eni (Ed.), *Proc. Southampton Civ. Engr. Mat. Conf.* Wiley, New York. pp. 963-995.

Perlman, A. B., G. C. Sih (1967). Elastostatic problems of curvilinear cracks in bonded dissimilar materials. *Int. J. Engr. Sc.*, 5, II, 845-867.

Perry, C., J. E. Gillott (1977). The influence of mortar-aggregate bond strength on the behaviour of concrete in uniaxial compression. *Cement and conc. Res.*,

Plähn, J. (1969). Zur Frage der bleibenden Verformungen des Betons bei Dauerschwingbeanspruchung, *Bet. und Stahlbet.*, 64, 292-295.

Reinhardt, H. W., and others (1978). Einfluss von Schwingbreite, Belasungshöhe und Frequenz auf die Schwingfestigkeit von Beton bei niedrigen Bruchlastwecksel-zahlen. *Betonw. und Fertigteiltechn.*, 44, 9, 498-503.

Shah, S. P., S. Chandra (1970). Fracture of concrete subjected to cyclic and sustained loadings. *J. Am. Concr. Inst.*, 67, 10, 816-825.

Stroeven, P. (1973). *Some aspects of the micromechanics of concrete.* Ph. D. Thesis, Delft.

Stroeven, P. (1974). Methods for the evaluation of topological characteristics of the pore structure of concrete; application to crack porosity. In S. Modrý (Ed.), *Proc. RILEM/IUPAC Symp. on pore structure and properties of materials.* Academia, Prague. pp. C529-C538.

Stroeven, P., H. M. de Haas (1974). Detection of cracks in concrete by holographic interferometry. In *Proc. RILEM 2-nd Int. Symp. New developments in non-destructive testing of non-metallic materials.* INCERC, Constanta. Vol. II, pp. 19-26.

Stroeven, P. (1975). Stereometric analysis of structural inhomogeneity and anisometry of concrete. In H. E. Exner (Ed.), *Proc. Symp. Quantitative analysis of microstructures, Special Issues Pract. Metallogr.* Dr Riederer Verlag, Stuttgart. Vol. 5, pp. 291-307.

Stroeven, P. (1976). Structural loosening of plain concrete in uniaxial compression. In *Proc. Conf. Mechanics and technology of composite materials.* Varna.

Stroeven, P. (1978). Amplitude effect on fatigue behaviour of concrete qualitatively evaluated by stereological techniques. In J. L. Chermant (Ed.), *Proc. Symp. Quantitative Analysis of microstructures, Special Issues Pract. Metallogr.* Dr. Riederer Verlag, Stuttgart Vol. 8, pp. 264-272.

Stroeven, P. (1979). Geometric probability approach to the examination of microcracking in plain concrete. To be published in *J. Mat. Sc.*

Stroeven, P., R. van Staveren, G. de Wind (1979). Mechanics of debonding in pull-out testing. To be published in *Proc. Int. Kolloq. Festigkeits- und Verformungsverhalten von Beton.* Dresden.

Vile, G. W. D. (1965). *Behaviour of concrete under simple and combined stresses.* Ph. D. Thesis, Univ. London.

Weigler, H., W. Freitag (1975). Dauerschwell- und Betriebsfestigkeit von Konstruktionsleichtbeton. *Dtsch. Aussch. für Stahlbet.*, 247, pp. 47-74.

Volume 3, ICM 3, Cambridge, England, August 1979

STRENGTH PROPERTIES OF STEEL FIBER REINFORCED POLYSTYRENE-IMPREGNATED CONCRETE

Y. Ohama, T. Nishimura and H. Hachisuka

Department of Architecture, College of Engineering, Nihon University, Koriyama, Fukushima-ken, 963, Japan

INTRODUCTION

Fibrous reinforcement was applied for improving the flexural strength, toughness and extensibility of polymer-impregnated mortar (SFRPIM) and concrete (SFRPIC). SFRPIM and SFRPIC were prepared by using six types of steel fibers having different shapes and tensile strengths at various steel fiber contents, and effects of these steel fibers on their mechanical properties such as compressive strength, flexural strength, toughness and extensibility are examined. Bond properties and orientation of steel fibers in cement-polymer matrix are discussed.

The test results indicate that among the six types of steel fibers piano wires are the most effective for increasing the flexural strength, toughness and extensibility. Flexural failure of SFRPIM and SFRPIC containing piano wires occurs from pull-out of wires and that of the others by rupture of steel fibers. It is found that the flexural strength of SFRPIM and SFRPIC is strongly affected by the tensile stress in steel fibers, steel fiber content and orientation of steel fibers, and can be expressed by applying the law of mixtures as follows;

$$\sigma_c = A\,\sigma_m\,(1 - V_f) + B\,\alpha\,\sigma_f\,V_f$$

where σ, V are stress and volume fraction respectively, suffixes c, m, f are composite, cement-polymer matrix and steel fiber respectively, α is orientation factor of steel fibers and A, B are empirical constants. In this study the constants A and B are 0.956 and 0.582 respectively.

Polymer-impregnated concrete is superior to ordinary cement concrete in its strength, watertightness and chemical resistance. But it has a few disadvantages such as lower toughness and lower extensibility. For the purpose of improving

MS225

these properties, the use of fibrous reinforcement can be easily considered. In the case of steel fiber reinforced polymer-impregnated concrete which has high bond between steel fibers and cement-polymer matrix, and high strength of the cement-polymer matrix, the steel fibers are apt to rupture at failure of the concrete. It is therefore anticipated that the beneficial effect of fiber reinforcement will sufficiently be obtained only at high steel fiber content.

The paper describes mechanical properties, i.e., compressive strength, flexural strength and toughness of steel fiber reinforced polystyrene-impregnated mortars (SFRPIM) and concretes (SFRPIC) with several steel fibers. Bond properties and orientation of steel fibers in cement-polymer matrix are also discussed.

MATERIALS

Cement and Aggregates

Normal portland cement and Toyoura standard sand as specified in the JIS (Japanese Industrial Standard) were used for preparation of steel fiber reinforced mortars as base materials. Ohkawa river gravel (size, 5-10 mm) as a coarse aggregate and Abukumagawa river sand (size, 2.5 mm or finer) as a fine aggregate were used for steel fiber reinforced concretes as base materials.

Steel Fibers

Six types of steel fibers used were listed in Fig. 1. These steel fibers had different shapes and were made from steel sheets and wire of the different tensile strengths.

Type I Flat Steel Fiber
(Size, 0.5 x 0.5 x 30 mm)

Type II Flat Steel Fiber
(Size, 0.35 x 0.7 x 30 mm)

Type III Flat Steel Fiber
(Size, 0.5 x 0.5 x 30 mm)

Type IV Deformed Steel Fiber
(Size, 0.25 x 0.5 x 25 mm)

Type V Flat Steel Fiber
(Size, 0.5 x 0.5 x 15 mm)

Type VI Piano Wire
(Size, ϕ 0.35 x 30 mm)

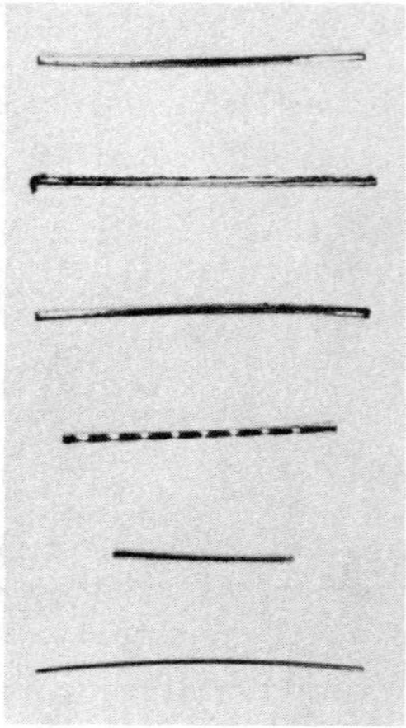

Fig. 1. Six shapes of steel fibers used.

Chemicals for Impregnant

Styrene monomer (St), trimethylolpropane trimethacrylate (TMPTMA) as a crosslinking agent, 2.2'-azobisisobutyronitrile (AIBN) as a catalyst and γ-methacryloxypropyltrimethoxy silane (Silane) as a coupling agent were employed for impregnation.

TESTING PROCEDURES

Pull-Out Test of Steel Fibers (Namman and Shah, 1976; Maage, 1977)

Mortar was mixed as cement : Toyoura standard sand = 1 : 3 (by weight), water-cement ratio, W/C of 80.0% and its flow 170 mm. In the preparation of the pull-out specimens, the briquette molds specified in the ASTM (American Society for Testing and Materials) standard were divided into two parts by two plastics sheets and then mortar was placed in one half of the mold after the steel fibers had been inserted through small holes of the plastics sheet. The number of steel fibers and their total lengths embedded were illustrated in Fig. 2. Then the other half was filled up with mortar and compacted, and the specimens were given 1-day-20°C-80%R.H.-moist, 6-day-20°C-water cure. After curing, they were dried to constant weight at 150°C for about 24 hours and allowed to cool. The dried specimens were evacuated to 0.1 - 0.6 kPa at ambient temperature for 30 minutes, and soaked in an impregnant prepared as St : TMPTMA : AIBN : Silane = 90 : 10 : 1 : 1 (by weight), under pressurized air of 200 kPa for one hour for impregnation. For the preparation of pull-out specimens, the mortar specimens impregnated with the impregnant were placed in 80°C hot water for two hours for thermal polymerization and then allowed to cool.

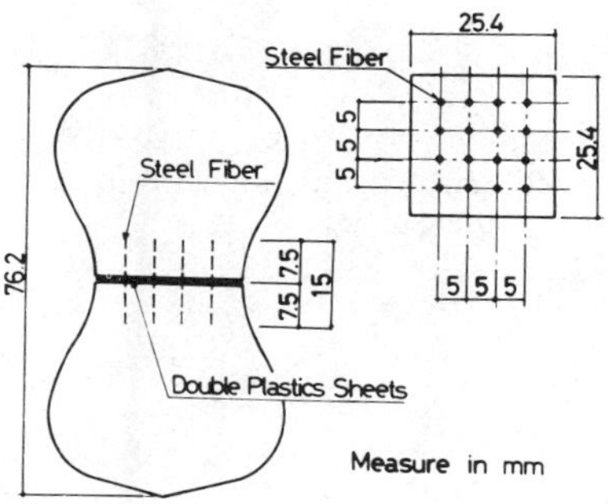

Fig. 2. Preparation of pull-out specimen.

Pull-out specimens were tested in accordance with ASTM C 190-72 (Standard Method of Test for Tensile Strength of Hydraulic Cement Mortars). Tensile stress in the steel fibers was calculated by the following formula.

$$\text{Tensile stress in the steel fibers (N/mm}^2\text{)} = \frac{\text{Maximum tensile load (N)}}{\left(\text{Number of steel fibers}\right) \times \left(\text{Sectional area of steel fiber (mm}^2\text{)}\right)}$$

Strength Tests of SFRPIM

Steel fiber reinforced mortars as base materials were mixed as cement : Toyoura standard sand = 1 : 3 (by weight), water-cement ratio, W/C of 80.0%, steel fiber contents of 0, 1.0, 2.0 and 4.0 vol.%. Mortar specimens 40 x 40 x 160 mm and 20 x 40 x 160 mm were molded. SFRPIM specimens were prepared by the polymer impregnation of the mortar specimens by the same method as described in the above paragraph. SFRPIM specimens were tested for compressive and flexural strengths according to JIS R 5201(Physical Testing Methods of Cement). In the test, the central deflection was measured by means of dial gages. SFRPIM specimens 20 x 40 x 160 mm were not tested for compressive strength.

Strength Tests of SFRPIC

Steel fiber reinforced concretes as base materials were mixed as unit cement content of 400 kg/m^3, water-cement ratio, W/C of 61%, sand content of 60%, maximum aggregate size of 10 mm and steel fiber contents of 0, 1.0, 2.0vol.%. Concrete specimens ϕ75 x 150 mm for compression test and 50 x 100 x 400 mm for flexure test were molded. SFRPIC specimens were prepared by the polymer-impregnation of the concrete specimens by the same method as described in the above paragraph. SFRPIC cylindrical specimens were tested for compressive strength in accordance with JIS A 1182 (Method of Test for Compressive Strength of Polyester Resin Concrete), and SFRPIC prism specimens for flexural strength in accordance with JIS A 1184 (Method of Test for Flexural Strength of Polyester Resin Concrete). The central deflection was measured by means of dial gages in the test.

Observation of Orientation of Steel Fibers in Specimens

Orientation of steel fibers in specimens were observed on the cross sections (40x 40, 20x40 and 50x100mm) of SFRPIM and SFRPIC. The number of steel fibers on their cross sections were calculated. The orientation of steel fibers in SFRPIM and SFRPIC was determined using the following equation (Kobayashi, Izumi and Cho, 1977):

$$\alpha = \rho a / V_f$$

where α represents orientation factor of steel fibers, ρ, a and V_f represent quantity of steel fibers per unit volume of the specimen, sectional area of the steel fiber and steel fiber content, repsectively.

TEST RESULTS AND DISCUSSION

Strengths of SFRPIM

Figs. 3 and 4 show the relationship between steel fiber content and mechanical

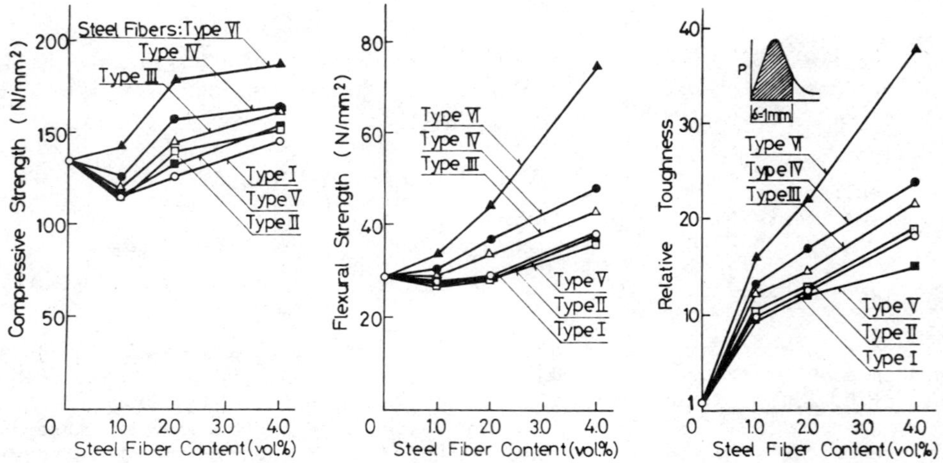

Fig. 3. Steel fiber content vs. compressive strength, flexural strength and relative toughness of SFRPIM 40x40x160mm using various steel fibers.

properties, i.e., compressive strength, flexural strength and relative toughness of SFRPIM 40 x 40 x 160 mm, and flexural strength of SFRPIM 20 x 40 x 160 mm. In this study, the toughness was calculated as the area under load-deflection curve up to 1% deflection of beam span. Generally, the compressive strength, flexural strength and toughness of SFRPIM are elevated with increasing steel fiber content, and particularly, Type VI fibers (piano wires) are the most effective for increasing the flexural strength and toughness of SFRPIM.

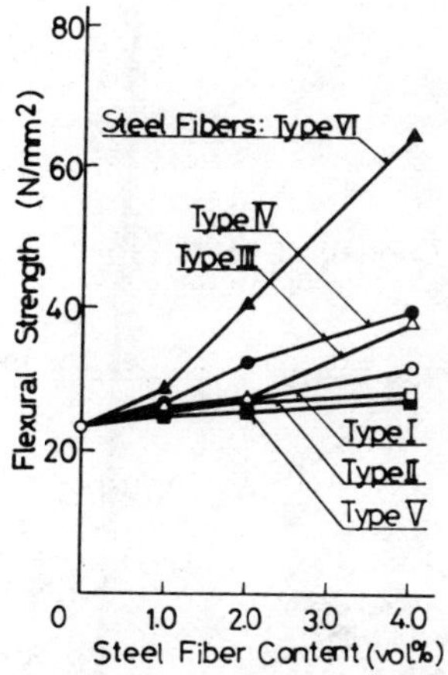

Fig. 4. Steel fiber content vs. flexural strength of SFRPIM 20x40x160mm using various steel fibers.

Strengths of SFRPIC

Fig. 5 gives the effects of type of steel fibers and steel fiber content on mechanical properties, i.e., compressive strength, flexural strength and relative toughness of SFRPIC. The flexural strength and relative toughness of SFRPIC are improved with increasing steel fiber content, and in particular, Type VI steel fibers (piano wires) are the most effective for improving these properties,

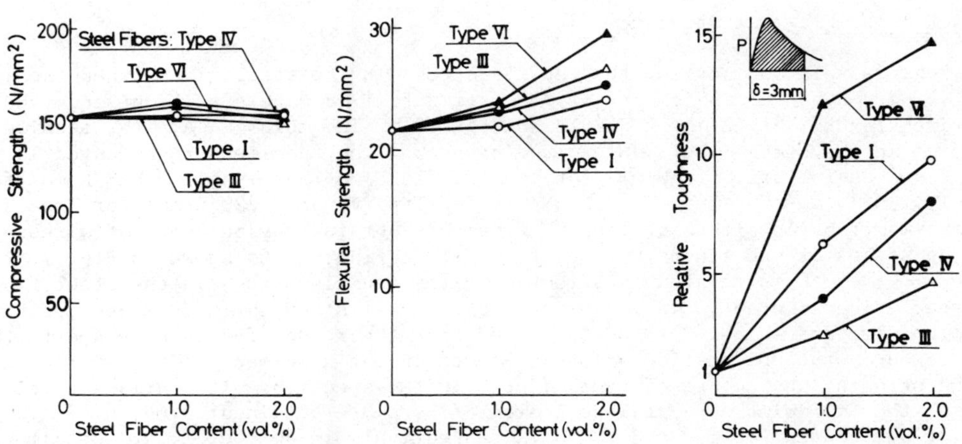

Fig. 5. Steel fiber content vs. compressive strength, flexural strength and relative toughness of SFRPIC 50x100x400mm using various steel fibers.

like the strength of SFRPIM. The compressive strength does not change with variation of steel fiber content and type of steel fibers.

The load-deflection curves for SFRPIC with Type I and Type VI steel fibers are indicated in Fig. 6. Polystyrene-impregnated concrete has high bond between steel fibers and concrete matrix in comparison with ordinary cement concrete. Consequently SFRPIC with steel fibers of low tensile strength is fractured by rupture of steel fibers, the extensibility and toughness are hardly improved by adding the steel fibers.

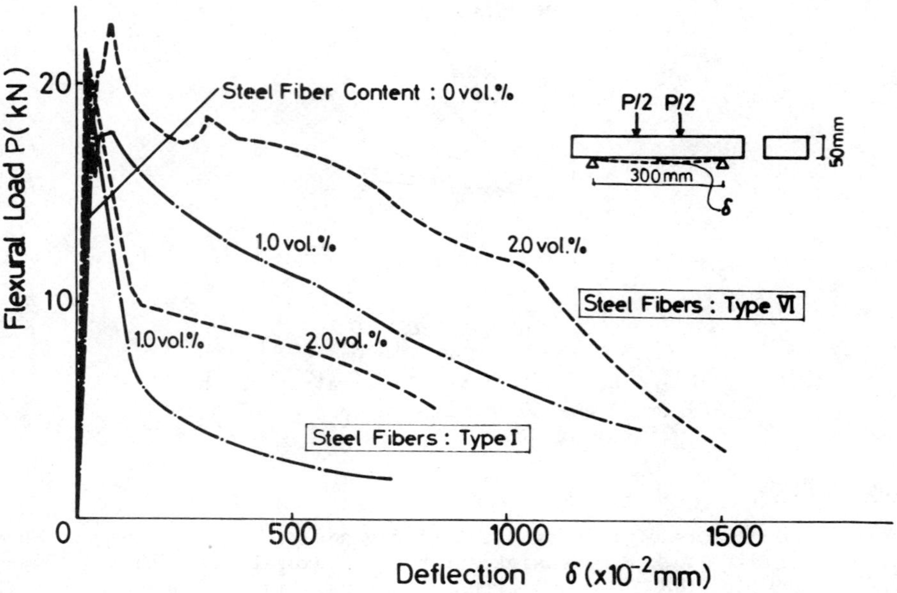

Fig. 6. Flexural load-deflection curves for SFRPIC with Type I and Type VI steel fibers.

Prediction of Flexural Strength

Fig. 7 gives relation between the tensile stress in the steel fibers when the steel fibers embedded in cement-polymer matrix fail or make a pull-out and the relative flexural strength of SFRPIM and SFRPIC. In this case, the tensile stresses in the steel fibers determined by use of the specimens represented in Fig. 2, are as follows; 527 N/mm^2 for Type I, 515 N/mm^2 for Type II, 858 N/mm^2 for Type III, 936 N/mm^2 for Type IV, 524 N/mm^2 for Type V, and 2220 N/mm^2 for Type VI. The maximum tensile stress of Type VI fibers is obtained by pull-out of wires, and the stress of the others by rupture of steel fibers. As shown in Fig. 7, the flexural strength is remarkably affected by the tensile stress in the steel fibers and steel fiber content. In addition, the flexural strength is dependent upon the specimen size and shape. It is considered that the reason for this is due to the difference in the orientation of steel fibers in the specimens. By way of example of the orientation factors of steel fibers in the specimens with various steel fibers, the following data of Type I steel fibers are obtained; 0.68, 0.67 for steel fiber content of 1.0 and 2.0vol.% respectively in 40 x 40 x 160 mm specimen; 0.88, 0.90 for steel fiber content of 1.0 and 2.0vol.% respectively in 20 x 40 x 160 mm specimen, and 0.67, 0.49 for steel fiber content of 1.0 and 2.0vol.% respectively in 50 x 100 x 400 mm specimen.

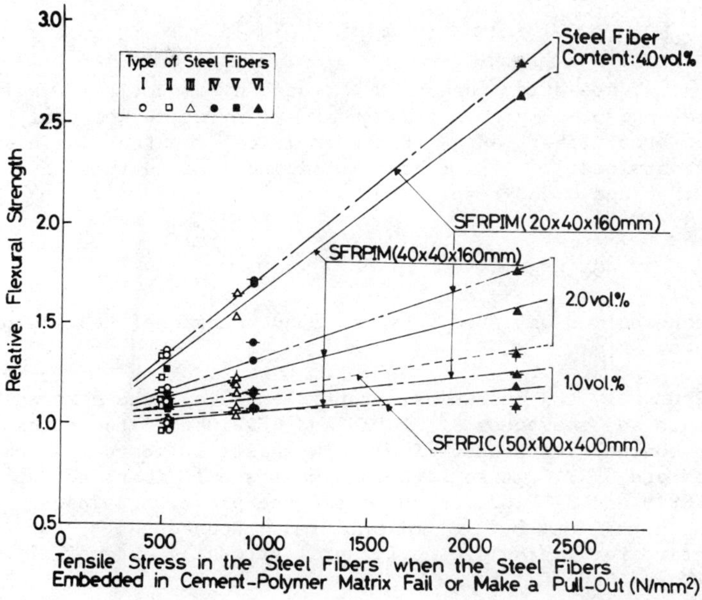

Fig. 7. Maximum tensile stress in the steel fibers vs. relative flexural strength of SFRPIM and SFRPIC.

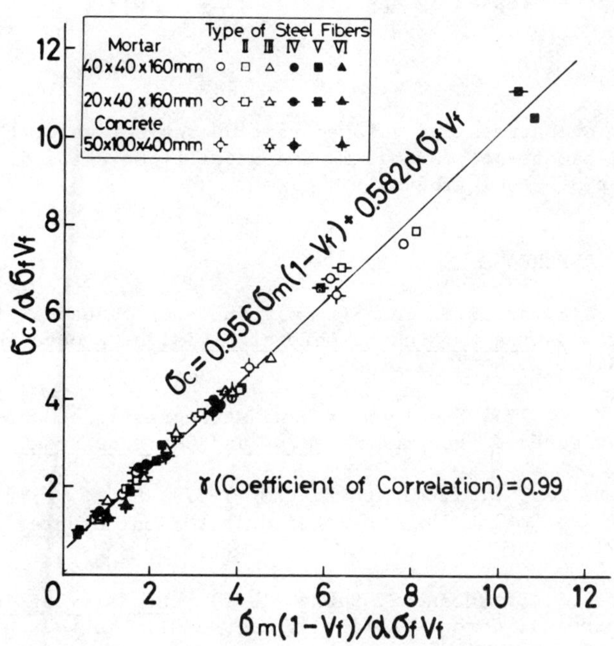

Fig. 8. Prediction of flexural strength of SFRPIM and SFRPIC using various steel fibers.

From the above-mentioned results the flexural strength of SFRPIM and SFRPIC can be expressed by applying the law of mixtures as follows (Noda, Saito and Sakka, 1977).

$$\sigma_c = A\sigma_m\ (1 - V_f) + B\ \alpha\ \sigma_f V_f$$

where σ, V are stress and volume fraction respectively, suffixes c, m, f are composite, cement-polymer matrix and steel fiber respectively, α is orientation factor of steel fibers, and A, B are empirical constants. This relationship empirically obtained is illustrated in Fig. 8, and in this case the constants A and B are 0.956 and 0.582 respectively.

CONCLUSIONS

The conclusions obtained from the above-mentioned test results are summarized as follows:

1. Generally the mechanical properties, i.e., the compressive strength, flexural strength and toughness of SFRPIM and SFRPIC tend to increase slightly as the steel fiber content increases. Bond between steel fibers and cement-polymer matrix in SFRPIM and SFRPIC is so high that most steel fibers are apt to rupture at failure of SFRPIM and SFRPIC. In this case the piano wires having the highest tensile strength pull out but do not rupture. Consequently, the piano wires are the most effective for improving the flexural strength, extensibility and toughness of SFRPIM and SFRPIC.

2. The flexural strength of SFRPIM and SFRPIC is strongly affected by the tensile stress in steel fibers, the steel fiber content and the orientation of steel fibers. The flexural strength of SFRPIM and SFRPIC from the test data of the authors can be expressed by applying the law of mixtures as follows.

$$\sigma_c = 0.956\ \sigma_m\ (1 - V_f) + 0.582\ \alpha\ \sigma_f V_f$$

where σ, V are stress and volume fraction respectively, the suffixes c, m, f are composite, cement-polymer matrix and steel fiber respectively, and α is orientation factor of steel fibers.

REFERENCES

1. Namman, A.E. and S.P. Shah (1976). Pull-out mechanism in steel fiber-reinforced concrete. Journal of the Structural Division, Proceedings of the ASCE, 102, 1537-1548.

2. Maage, M. (1977). Steel fiber bond strengths in cement-based matrixes influenced by surface treatment. Cement and Concrete Research, 7, 703-710.

3. Kobayashi, K., I. Izumi and R. Cho (1977). Steel fiber reinforced concrete -properties, mechanism and fiber orientation. Concrete Journal (in Japanese), 15, 7-21.

4. Noda, T., H. Saito and S. Sakka (1977). Chemistry of Inorganic Materials II (in Japanese). Corona Publishing Co., Ltd., Tokyo. pp.254-262.

Volume 3, ICM 3, Cambridge, England, August 1979

CORRELATION BETWEEN BRITTLE FRACTURE SURFACE FEATURES AND MICROSTRUCTURE IN NODULAR GRAPHITE CAST IRON

J. Pirs

Materials Department, University of Rijeka, Technical Faculty, Rijeka, Yugoslavia

INTRODUCTION

Fully annealed nodular graphite irons with a ferritic matrix were investigated at room temperature, - 46 and - 55 °C in Charpy impact testing. The ductile fracture surfaces obtained at room temperature and the brittle fracture surfaces obtained at low temperatures were observed in a scanning electron microscope (SEM). The SEM, having the advantages of allowing continuous observation from low to high magnifications and a large depth of focus, makes it possible to observe directly any correspondence of fracture surface with microstructure. There have been many studies on the impact toughness of cast iron in relation to microstructure, but few have considered principally the fracture mode based on observations of fracture surfaces. Early investigations of the fracture behaviour of nodular graphite iron at low temperatures (Pirs and Shabaik, 1977) were followed by investigations of fracture behaviour and microstructure. The purpose of the present study was to investigate the direct relationship between the fracture surfaces of fully annealed ferritic nodular cast irons and the structure of these irons at room and subzero temperatures. A further purpose was to investigate the problem of discontinuous crack propagation experimentally by observation of fracture surfaces.

EXPERIMENTAL PROCEDURE

To obtain a fully annealed ferritic matrix structure the bars were heated to 920 °C and maintained at this temperature for 4 hr. Afterwards the bars were furnace cooled to 690 °C and held for 8 hr before cooling to room temperature. The ferrite produced by this treatment was distributed around the graphite nodules. Brittle fracture surfaces were prepared by breaking Charpy impact specimens with a 2 mm V notch at - 46 °C, and with a U notch at - 55 °C. Ductile fracture surfaces were obtained by testing 2 mm V notched specimens of the same cast iron at room temperature. The chemical composition of the cast irons is given in Table 1.

TABLE 1 Chemical Composition

Alloy Type	Test Series	T.C	Si	Mn	S	P
SFF*	C 14	3.41	2.32	0.26	0.004	0.03
SFF	C 15		2.10	0.31	0.010	0.04

*Meehanite Type Ferritic Nodular Cast Iron

The results of impact tests on cast irons are shown in Table 2.

TABLE 2 Impact Test Results

Test Series	Sample	Notch Type	Temperature, °C	Impact Value,* J
C 14	1	V	Room	17.5
	2	V	Room	17.5
	3	V	- 46.0	5.0
	4	V	- 46.0	5.5
	5	U	- 55.0	5.0
	6	U	- 55.0	5.0
C 15	7	V	Room	17.0
	8	V	Room	19.0
	9	V	- 46.0	6.5
	10	V	- 46.0	6.0
	11	U	- 55.0	9.0
	12	U	- 55.0	8.5

*Impact values are given in N.m

It is considered that the impact test is the most suitable test of the quality of nodular cast iron, both ferritic and pearlitic. The impact testing of nodular irons is complicated by the many different designs of test bars in use. Impact values are quoted for both notched and unnotched specimens and it is evident that the notched specimens are preferred. Three types of notched bars are commonly used in Europe, namely the Izod, Charpy V and D.V.M. testpieces, the latter in accordance with German Standard DIN 50 115. It is accepted that Izod and Charpy specimens give comparable results and specific values quoted for both are identical. The relationship between values obtained from D.V.M. testpieces and Izod and Charpy specimens for Meehanite nodular cast irons has been shown (Attenborough, 1969), and the Unified Standards for these irons give the values of impact strength obtained on D.V.M. testpieces and Charpy specimens. The results at room temperature, given in Table 2, are in accordance with values for nodular irons given in the U.S.. Conventional metallographic examinations of the fractured samples by optical microscopy were performed, and the results obtained on sample 12 are shown in Fig. 1.

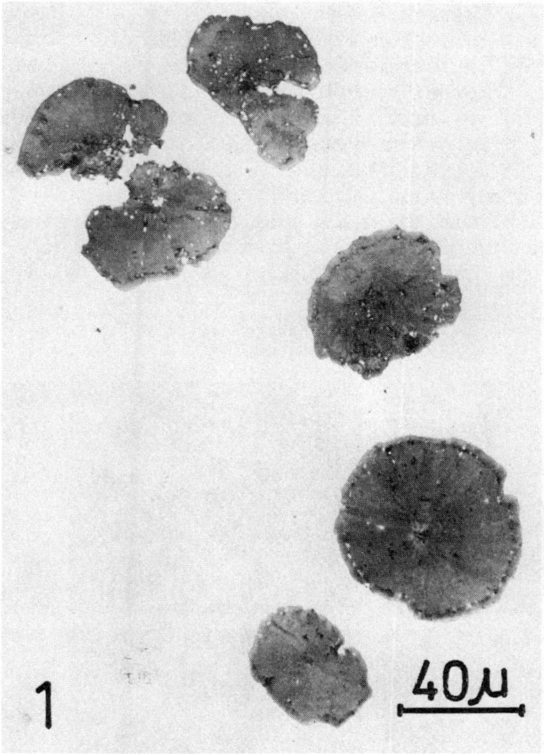

Fig. 1. Graphite nodules in nodular iron

Further metallographic studies were performed on nickel plated fracture surfaces of broken impact specimens. The section examined was parallel to the direction of crack propagation and nearly orthogonal to the fracture surface. It was etched with facet pit reagents. After removing the nickel coating, the fracture surface and the corresponding section were examined in the SEM. To make it easy to strip the nickel coating from the fracture surfaces the specimens were immersed in a 5% CrO_3 aqueous solution for 30 s before plating. The facet pits were revealed using the following reagents:

$$15 \text{ cc } H_2O_2 + 2 \text{ cc } HCl + 100 \text{ cc } H_2O$$

$$1 \text{ cc } HC + 10 \text{ cc Sat.aq. } FeCl_3 \; 6\, H_2O + 100 \text{ cc } H_2O$$

$$50 \text{ cc } C_2H_5OH + 50 \text{ cc } CH_3COOH \text{ or } HCOOH$$

The facet pit faces were mainly {110} and secondarily {100}. The crystallographic orientations of cleavage plane and facet pit were measured by a gonio-microscope.

EXPERIMENTAL RESULTS

The present study has considered the change from ductile to brittle behaviour and the mechanisms of ductile and brittle fracture in annealed ferritic nodular graphite irons. In impact testing the fracture surfaces change from dark and ductile at room temperature to a zoned dark/bright appearance, as the ductility decreases with falling temperature. This is consistent with the increased resistance to plastic deformation with falling temperature, and the start of a ductile to brittle transition is indicated by a rapid decrease in impact toughness. The temperature for the start of this transition is below zero in the irons examined. Recently, similar results

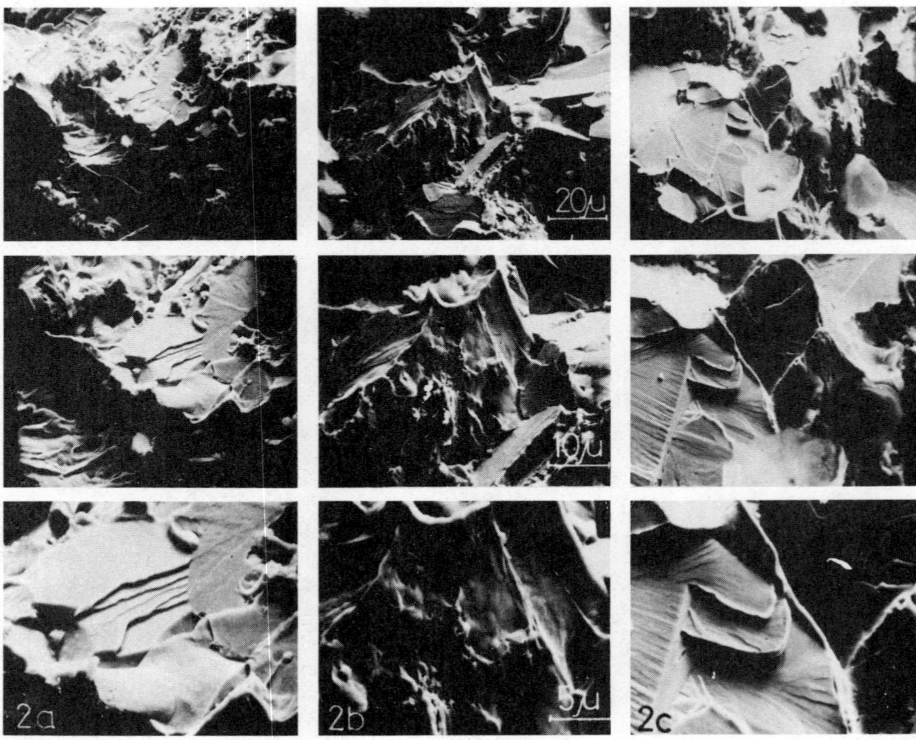

Fig. 2. Scanning electron micrographs of ductile and brittle fractures in the ferritic nodular graphite irons.

(a) Brittle;
(b) Ductile;
(c) Brittle

(Rickards, 1971) were obtained during investigations of variations of tensile properties of ferritic nodular irons with temperature.

The scanning micrographs in Fig. 2 show primarily the ferrite bridges between the graphite nodules. When these bridges fail in ductile fracture the rupture takes place by tearing along grain boundaries. The resulting ductile fracture appears dark because it is deeply indented and the graphite nodules are exposed on the fracture surface. Brittle fracture occurs primarily by cleavage through the ferritic grains. Cleavage fracture initiation occurs at some critical stress which may be sufficiently high to cause considerable prior plastic deformation. Brittle cleavage occurred discontinuously, resulting in the formation of steps on the fracture surface. Cleavage steps may also be formed at a microcrack/twin intersection or when a propagating brittle microcrack crosses a grain boundary. The stepped nature of the brittle cleavage surface obtained at - 46 ^{o}C is shown in Fig. 2.

On subjecting a ductile ferritic nodular graphite iron to an impact stress, the ferritic grains become distorted in the direction of applied stress and the graphite nodules become elongated longitudinally and compressed laterally. The nodules fracture along the weak interfaces between the primary graphite core and the peripheral secondary graphite which was deposited during the ferritising heat treatment. The volumes occupied by the graphite become enlarged in the direction of applied stress, the cores of the nodules remain within the space, while the secondary graphite continues to adhere to the surrounding ferrite grains.

Further investigations were made of the fracture surfaces obtained at - 55 ^{o}C and the secondary electron image of a crack initiation region of an impact fracture surface is shown in Fig. 3. In determination of the fracture characteristics at - 55 ^{o}C it was observed that the cleavage crack initiated in the region 0.15 to 0.20 mm in from the tip of the notch. A small amount of ductile fracture was observed locally not only in the vicinity of the notch but in the entirely brittle fracture surface.

Observation of the fracture surface revealed that the fracture propagated by repeated nucleation, propagation and arrest of microcracks. This microcrack propagation unit is considered to be the fracture facet, as shown by the continuous line in Fig. 3, each unit having one microcrack nucleus. A unit joins an adjacent unit discontinuously by the nucleation of another microcrack in that unit. As the test temperature was raised ductile fracture was observed also at the fracture facet boundary. Similar results were obtained by Terasaki (1967) during investigations of pure iron.

The dimensions of the facets were somewhat the same as the ferrite dimensions but were generally larger than the ferrite grain size. This might be due to the small orientation difference about the cleavage fracture surface {100}, because the fracture surfaces determined from the facet pit and the fracture surface were {100} planes.

A schematic illustration of a facet pit observed in the direction parallel to the [001] axis is shown in Fig. 4.

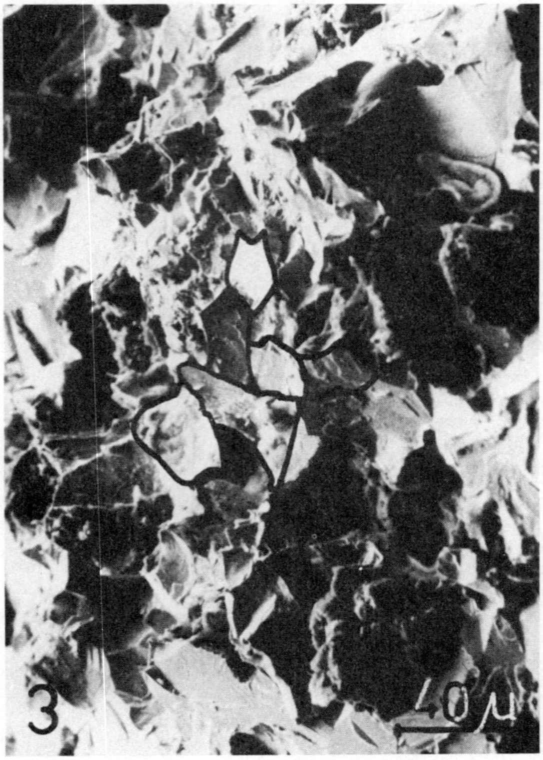

Fig. 3. Fractograph of nodular cast iron broken at - 55 °C. Continuous lines indicate microfracture units.

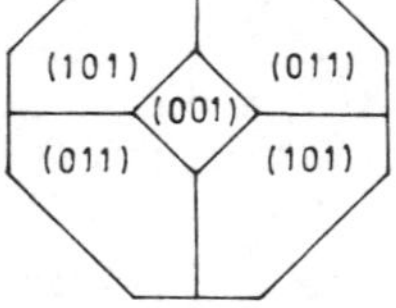

Fig. 4. Schematic illustration of a facet pit observed in the direction parallel to the [001] axis.

Figure 5 shows the direct correspondence of microstructure with the fracture surface obtained at - 55 °C. The directions of observation are illustrated in Fig. 5. The fracture surface was observed in the direction normal to it and the microstructure was observed in cross-section parallel to the direction of crack propagation and nearly orthogonal to the fracture surface, and also from the 45 ° direction to both planes. The crystallographic orientations of fracture surfaces determined by the facet pits and the fracture surface were {100}.

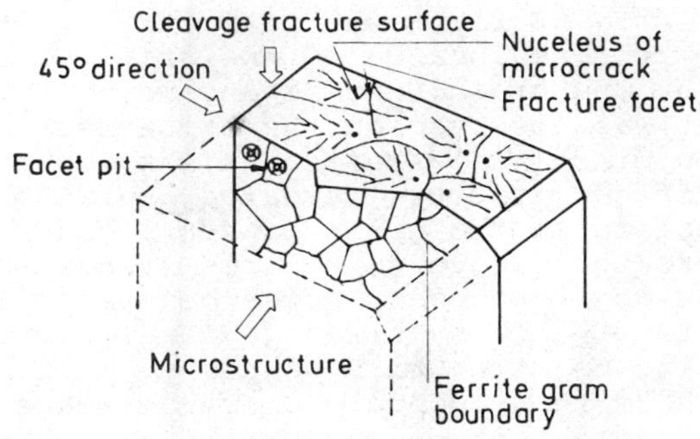

Fig. 5. Observation of direct correspondence of microstructure to fracture surface.

Finally in Fig. 6 the crystallographic orientation of cleavage planes observed by a gonio-microscope in comparison with that by facet pit measurements is shown. The directly measured orientation

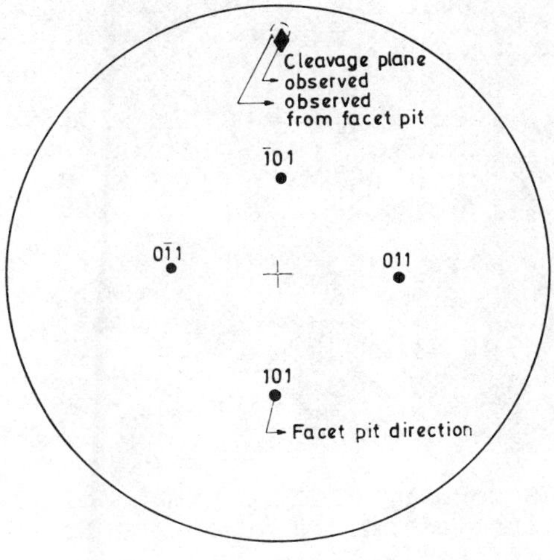

Fig. 6. Crystallographic orientation of cleavage plane observed by gonio-microscope in comparison with that determined by facet pit measurement.

of fracture surfaces showed them to be on the (010) plane, which was determined independently from the metallographic orientation measurements of facet pits on the cross section, as shown in Fig. 5.

Further investigation concentrated on the fracture facet in the ferrite. In ferrite, as is expressed in the well known Petch Relation (Henslop and Petch, 1958) the grain size has an important effect on toughness. Early results (Cottrell, 1958) showed that this is caused by the resistance of the grain boundary to crack propagation and by the grain size dependency of yield stress. In many cases, it was observed on the cross-section of the fracture surface, that the crack was arrested at the ferrite grain boundary. The observation of the fracture revealed that the fracture path changed its direction at the grain boundary and steps called a river pattern were formed due to the resistance to fracture. The resistance of the grain boundary arose because brittle fracture at low temperature took place by cleavage on {100}, and the cleavage plane is discontinuous at a grain boundary. This can be clearly seen in Fig. 7 where the secondary electron image of the crack initiation region on the fracture surface of the sample No. 11 is shown. Besides ferritic structures in the quenched and tempered structures, the grain size corresponding to the ferrite grain is not clear. The austenite grain size and martensite laths have been taken as the metallurgical factors controlling strength and toughness (Cina and Jubb, 1959).

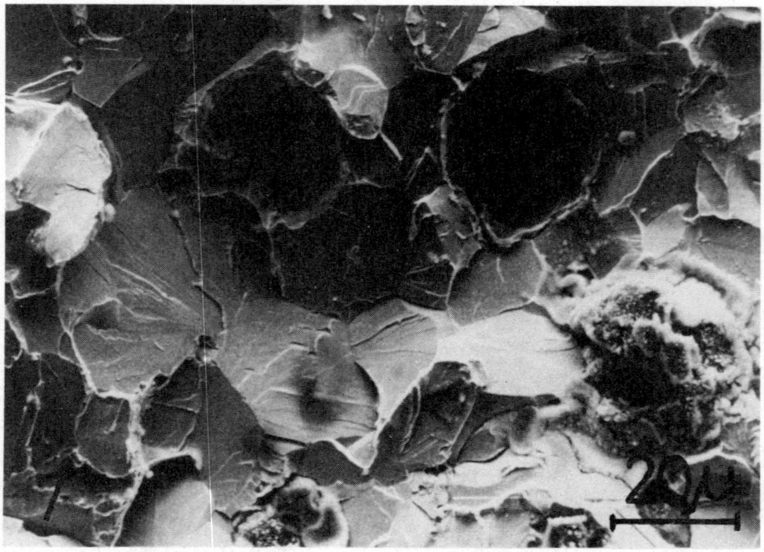

Fig. 7. Secondary electron image of the crack initiation region on the fracture surface of sample No.11.

From Fig. 7 it can be concluded that the fracture facet is generally larger than the ferrite grain size. When the orientation difference of {100} planes between adjacent grains is small, the ferrite grain boundary may be considered as if it is sub-boundary, and cannot be an effective barrier to crack propagation. These possibilities are illustrated in Fig. 8. The fracture appearance depends on the kind of small angle boundary, i.e. tilt boundary or twist boundary. These results are in accordance with previous results (Turkalo, 1960) where it was reported that the fracture facet in ferrite-pearlite was a pearlite colony at least and a number of pearlite colonies, up to an austenite grain, at most. These facet sizes are considered to correspond to the area having the same crystallographic orientation of {100} planes, and it can be explained from this study.

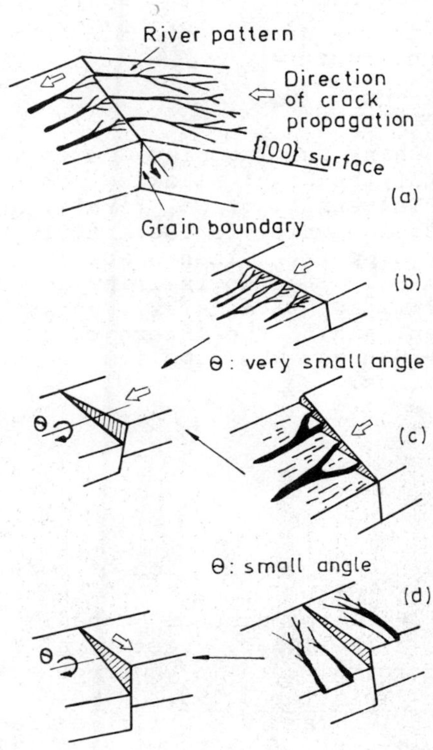

Fig. 8. Schematic representation of the changes in cleavage fracture surface at various kinds of small angle grain boundaries.

Further studies have been devoted to the influence of the interparticle spacing of nodular graphite on the flow stress and also of the dispersion of graphite particles, following previous investigations in this field (Mogford and co-workers, 1967). The results obtained will be reported later.

SUMMARY

Ductile and brittle fracture surfaces of ferritic nodular graphite irons were investigated in the SEM. The correspondence of the fracture surface with the microstructure was directly investigated and the fracture mode was studied from the view point of crystallographic orientation relationships.

The fracture surfaces obtained by Charpy impact tests revealed a discontinuous process during testing of brittle specimens. The fracture surface was formed by repeated nucleation, propagation and arrest of microcracks, and the cleavage plane was determined as {100} by orientation measurements of the facet pit and the fracture surface using a gonio-microscope. The microcrack propagation unit could be considered to be the fracture facet, and it was composed of adjacent grains in which the crystallographic orientations about {100} were nearly equal. The fracture facet was generally larger than the ferrite grain size.

ACKNOWLEDGEMENTS

The author wishes to thank the Yugoslav Commission for Exchange of Scholars in Beograd for the Grant allowing him to perform his postdoctoral research at University of California, and also the Republic Selfgoverning Interest Community for Scientific Work in Zagreb, SIZ-1, for financial support. Thanks are due to the Chairman of Materials Department, School of Engineering and Applied Science, University of California at Los Angeles, Professor Cris N. Wagner for helpful discussion, and to the International Meehanite Metal Company Ltd, Reigate, Surrey, who supplied the nodular cast iron used in the investigation.

REFERENCES

Attenborough, P. (1969). Specification of nodular Meehanite. Report No. E 1161. Group No. 2.D. 37 th Annual Conference of Licensees of The International Meehanite Metal Company.

Cina, B. B. and P. Jubb (1959). Factors affecting the transition temperature of forgings used in power-generating equipment. J. Iron and Steel Institute, 193, 329-349.

Cottrell, A. H. (1958). Theory of brittle fracture in steel and similar metals. Transactions AIME, 212, 192-203.

Henslop, J. and N.J. Petch (1958). The ductile-brittle transition in the fracture of α-iron. II. The Philosophical Magazine, 3, 1128-1136.

Mogford, I. L., I. L. Brown, and D. Hull (1967). Fracture of nodular cast iron. J. Iron and Steel Institute, 201, 729-732

Pirs, J. and A.H. Shabaik (1977). Fracture behaviour of nodular cast iron. Fracture 1977. Vol. 3, University of Waterloo Press, Waterloo. pp. 819-827.

Rickards, P.J. (1971). Ductile and brittle fracture in ferritic nodular-graphite irons. J. Iron and Steel Institute, 205, 190-196.

Terasaki, F. Etude des proprietes mécaniques a 77 °K et de la rupture par clivage des monocristaux de fer pur. Acta Met, 15, 1057-1072.

Turkalo, A.M. (1960). The morphology of brittle fracture in pearlite, bainite and martensite. Trans. AIME, 218, 24-30.

Volume 3, ICM 3, Cambridge, England, August 1979

DISPERSION-STRENGTHENED ZIRCONIUM BASED ALLOYS

F. D. S. Marques

Universidade do Minho, Palácio Vila Flor, Guimarães, Portugal

INTRODUCTION

Since the strength of zirconium based alloys decreases rapidly with increasing temperature it is important, for nuclear applications, to increase the elevated temperature strength of these alloys. This type of increase has been achieved in other materials by dispersion-strengthening, the mechanisms of which have been reviewed by Ansel (1968).

For low temperature applications (e.g. CANDU-PHW reactors which operate at ~566 K alloys which incorporate both solid solution and intermetallic dispersion strengthening are practically more attractive than inert particle-dispersion strengthened alloys. In zirconium the effects of intermetallic dispersion strengthening appear to be greater than those of solid solution strengthening. These effects have been reviewed by Douglass (1971).

However, the solubility of the intermetallics in the zirconium matrix increases rapidly with increasing temperature. Thus for applications at elevated temperatures (e.g. superheated steam reactors which operate in the temperature range 723-773 K), or at temperatures near the α/β transus, only inert particle dispersion strengthened alloys can be potentially useful, since the intermetallics cease to be effective barriers to dislocation motion. These inert particles can function as primary strengthening agents and/or as stabilising agents to the other structure. Dispersion strengthening with inert particles also offers the advantage of good corrosion properties, which generally limit the volume fraction of the intermetallics that can be present in a given alloy. Dispersion strengthening with stable oxides has previously been attempted in zirconium based alloys using either internal oxidation or powder-metallurgy techniques (Antony and Keepler, 1965; Parsons and Adolph, 1972; Wilkins and Chater, 1969), and its effects have recently been reviewed (Antony and Keepler, 1965).

Various theories and models have been proposed in the literature for calculating the strength of dispersion-strengthened materials, but no single theory has yet satisfactorily explained all the experimental results (Ansel, 1968). However, most theories assume an inverse relationship between the yield strength and the inter-particle spacing. The interparticle spacing of a dispersion may be defined in many different ways, to suit a particular dispersion strengthening model (Antony and Keepler, 1965).

An ideal dispersion alloy must have a minimum inter-particle spacing and the dispersant must be in the form of hard and stable particles. The dispersion strengthening effect appears to be considerable for inter-particle spacings of about 1μm(Douglass, 1971).

The development target for a suitable pressure tube material for superheated steam reactors has been arbitrarily defined by (Mc Donald, 1971; Williams, 1970; Williams, Ells and Dixon, 1972).

This leads to the conclusion (Douglass, 1971; Parsons and Adolph, 1972) that dispersed particles of approximately 1000 Å in size with an inter-particle spacing about 3000 Å may be required. However, this target has not yet been achieved.

The selection of a suitable inert dispersant for zirconium based alloys is difficult since it would have to be more stable than ZrO_2, have a minimum specific surface area, be hard, and have a low neutron capture-cross section. Table 1 lists various oxides with a free energy of formation negatively greater than that of ZrO_2. Although the solubility and diffusivity of oxygen in the zirconium lattice will influence the stability of the oxide, La_2O_3 is expected to be quite stable.

TABLE 1 Heats, Entropies and Free Energies of Formation (k Joule $mole^{-1}$)

Oxide	$-\Delta H_{298\ K}$	S_{298}	$-\Delta G_{300\ K}$	$-\Delta G_{500\ K}$	$-\Delta G_{1000\ K}$	$-\Delta G_{1500\ K}$	$-\Delta G_{2000\ K}$	±kJ
ZrO_2	1086.5	50.7	1027.9	988.9	895.1	803.5	714.7	16
ThO_2	1227.6	65.3	1173.6	1135.0	1048.8	960.9	—	20
Al_2O_3	1674.7	51.1	1580.5	1517.3	1359.8	1143.4	—	16
Ce_2O_3	1820.4	150.7	1731.2	1674.7	—	—	—	16
La_2O_3	1867.3	128.1	1779.4	1721.6	1582.2	—	—	16
Er_2O_3	1899.1	153.2	1808.7	1748.4	—	—	—	8
Y_2O_3	1906.7	99.2	1817.9	1759.3	1618.6	1482.0	1348.1	20

EXPERIMENTAL PROCEDURE

Materials and Alloy Preparation

Typical analyses of the materials used are given in Table 2. To prevent fluctuations in the alloy composition due to losses by sputtering and to eliminate gas porosity, zirconium sponge was melted into rod-shaped ingots prior to use. Oxgen was added in the form of zirconium dioxide " grade S " by filling cavities previously drilled in the zirconium ingot, which were closed with previously machined zirconium stoppers. A Zr-15.5 at% O master alloy was initially made. The alloys were made as 50 g ingots in an electric arc melting furnace equipped with a non-consumable high purity tungsten electrode and a water-cooled copper hearth. A protective atmosphere of high purity argon and a zirconium getter were also employed. To ensure homogeneity, the alloys were generally remelted ten times. Weight changes resulting from melting were small (±0.05wt%) and the compositions of the alloys were thus calculated from the weights of the components.

TABLE 2 Purity of Materials

Zirconium (ppm)

C 170 H 50 N 16 O 416 Cl 150 Fe 240 Hf 200 Al 65 Na 50

Silicon (ppm)

Al <100 Fe <100 Ti <50 B <50 Mn <100

Zirconia (ppm)

TiO_2 680 SiO_2 200 Fe_2O_3 30 L.O.I. 3100 SO_3 2600

Molybdenum 99.9 % Mo; Aluminium 99.999 % Al; Tin 99.99 % Sn

Heat Treatment and Electron Metallography Techniques

Details of the heat treatment techniques and of the electron metallography procedure have been described elsewhere (Bradbrook and co-workers, 1971; Marques, 1976, 1977, 1978).

X-Ray Diffraction Analysis

X-ray diffraction analysis was carried out using the Debye-Scherrer technique with characteristic copper K_α radiation.

Mechanical Test Methods

Compression tests were carried out on specimens with nominal dimensions 6x2.5x2.5 mm using an Instron Universal Testing Machine. The test temperature ranged from 300 to 1000 K with an initial strain rate of $1.39 \times 10^{-4} s^{-1}$. A protective dynamic atmosphere of high purity argon was used above $400^{\circ}K$. The strain rate was increased by a factor of ten during strain rate cycling tests to determine the strain rate sensitivity. The movement of the cross head was stopped during stres relaxation tests to determine the relaxation stress. The strain rate sensitivity was calculated by the equation

$$\lambda = \Delta\sigma / \ln\dot{\varepsilon}$$

where $\Delta\sigma$ is the increase in the flow stress correspondent to an increase in the strain rate $\dot{\varepsilon}$. The relaxation stress was taken as the drop in the stress during a relaxation period of ten minutes.

RESULTS AND DISCUSSION

In order to investigate the feasibility of combined intermetallic and oxide dispersion strengthening and to evaluate the relative contributions of the various strengthening mechanisms, the following zirconium based alloys: Zr-3Mo-3Al-1.07O-0.3La, Zr-3Mo-1.5Al-1.5Sn-0.5Si-1.07O-0.5La, Zr-0.25Si, Zr-4Al-560ppmO, Zr-1.07O and Zr-1Mo-1250ppmO were prepared and studied. Table 3 shows the thermal and mechanical treatments carried out and the hardness results obtained. A summary of the results of the X-ray and electron diffraction analysis is also shown in this table. Figures 1 to 5 show characteristic microstructures and electron diffraction analysis of these alloys. The temperature dependence of the 0.2% proof stress, strain rate sensitivity and relaxation stress is shown respectively in Figs. 6, 7 and 8.

TABLE 3 Thermal and Mechanical Treatments, Hardness, Oxygen Analysis and Phase Nature

Alloy Composition at%	Treatments	Hardness DPH	Oxygen ppm	Phase Nature
Zr-3Mo-1.5Al-1.5Sn-0.5Si-1.07O-0.5La	1293 K/1h/OQ	374	1900	La_2O_3, Zr_3Si,
"	1293 K/10h/OQ	405	"	Zr_4Sn, β, α,
"	1293 K/100h/OQ	385	"	ω traces
Zr-3Mo-3Al-1.07O-0.3La	1293 K/1h/OQ	-	1900	
"	1293 K/10h/OQ	396	"	La_2O_3, β,
"	1293 K/100h/OQ	380	"	α, ω
Zr-0.25Si : 60% C.R.	1473 K/15m/WQ	157	-	α
Zr-4Al-560ppmO	1373 K/4h/WQ	256	560	α
Zr-1.07O	1573 K/24h/WQ	230	1900	α
Zr-1Mo-1250ppmO : 60% H.R. + 25% C.R.	1473 K/15m/WQ	336	1250	α

The Microstructure

Specimens of the alloys containing lanthanum were examined by optical reflected light microscopy (LM) and transmission electron microscopy (TEM), with respect to uniformity of dispersion, grain size, dispersant particle size and nature, inter particle spacing and thermal stability.

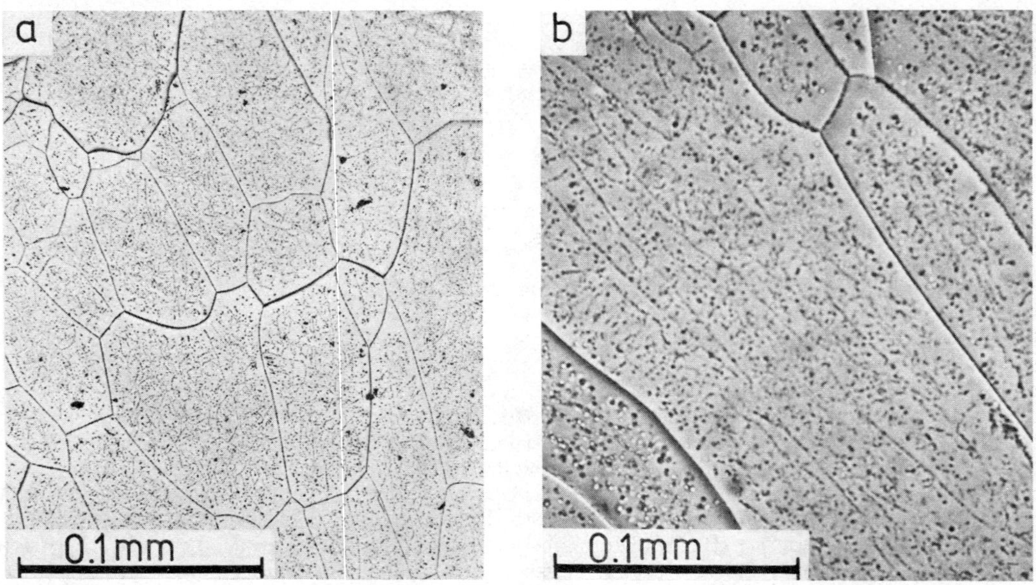

Fig. 1 Fine dispersions of particles uniformly distributed in a Zr-3Mo-1.07O-0.3La alloy : 1293°K/10h/OQ. (a) The original "β" grain boundaries and dislocation arrays (which appear as sub-boundaries) are pinned by dispersant particles. (b) uniform particle distribution within a typical columnar shapped grain.

Uniform distributions of various types of dispersant particles were observed in both alloys (Figs. 1 and 2). It should be noticed that due to the etchant "eroding action" the size of the particles observed by light microscopy is always larger than the actual size. Although LM is ideal to study the uniformity of the dispersion, TEM allows a more accurate determination of the particle size and inter-particle spacing.

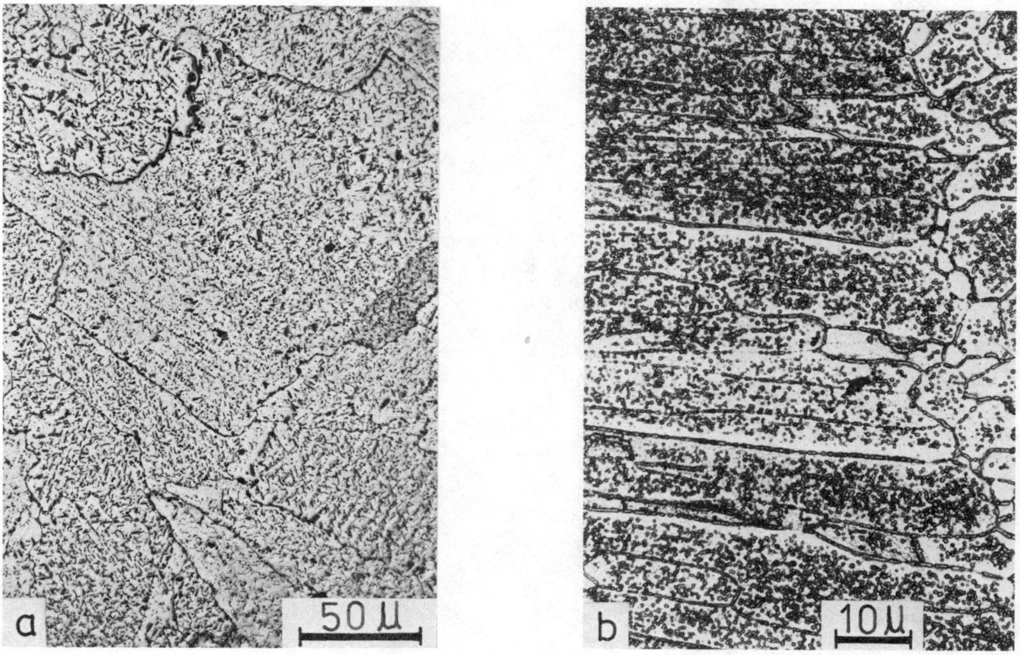

Fig. 2 Zr-3Mo-1.5Al-1.5Sn-0.5Si-1.07O-0.5La alloy: (a) 1293 K /100h/OQ showing irregular shaped grains with a uniform dispersion of particles, of which some have an acicular morphology; (b) 1293 K/10h/OQ: columnar grains exhibiting a fine dispersion of particles both at the original "β" grain boundaries and within the grains, which form subgrains.

An important effect of dispersed particles is to impede the movement of grain boundaries and thus reduce the average grain size. Figures 1 and 2 show how particles pinned original "β" boundaries. After 10 hours at 1293°K the average original "β" grain sizes were 50 and 5μ respectively for the Zr-3Mo-3Al-1.07O-0.3La and for the Zr-3Mo-1.5Al-1.5Sn-0.5Si-1.07O-0.5La alloys. These grains were columnar and not equiaxed which appears to confirm that grain boundaries were indeed pinned by dispersant particles. The pinning of the boundaries can explain the fact that in the Zr-3Mo-1.5Al-1.5Sn-0.5Si-1.07O-0.5La alloy the grain size is smaller ($5\mu m$) than in the Zr-3Mo-3Al-1.07O-0.3La alloy ($50\mu m$) since only a smaller number of dispersant particles are available to pin grain boundaries in the latter alloy (Figs. 1 and 2). Particles of both La_2O_3 and Zr_3Si were observed at grain boundaries.

The original grains contained dense arrays of dislocations forming large numbers of sub-boundaries which were pinned by dispersant particles (Figs. 1 and 2). Although no quantitative measurements of the size of the subgrains were made, they were uniformly distributed within each grain (Figs. 1 and 2). This is believed to have reduced the effective grain size in these alloys considerably.

Particles of various dispersants were observed in these alloys and their sizes ranged from 20 Å to 5000 Å. Clusters of two to ten dispersant particles were also observed. Although the particles of the same dispersant were usualy uniformly distributed (Figs. 1 to 4), discontinuities in the inter-particle spacings resulted from the interpenetration of the various dispersions.

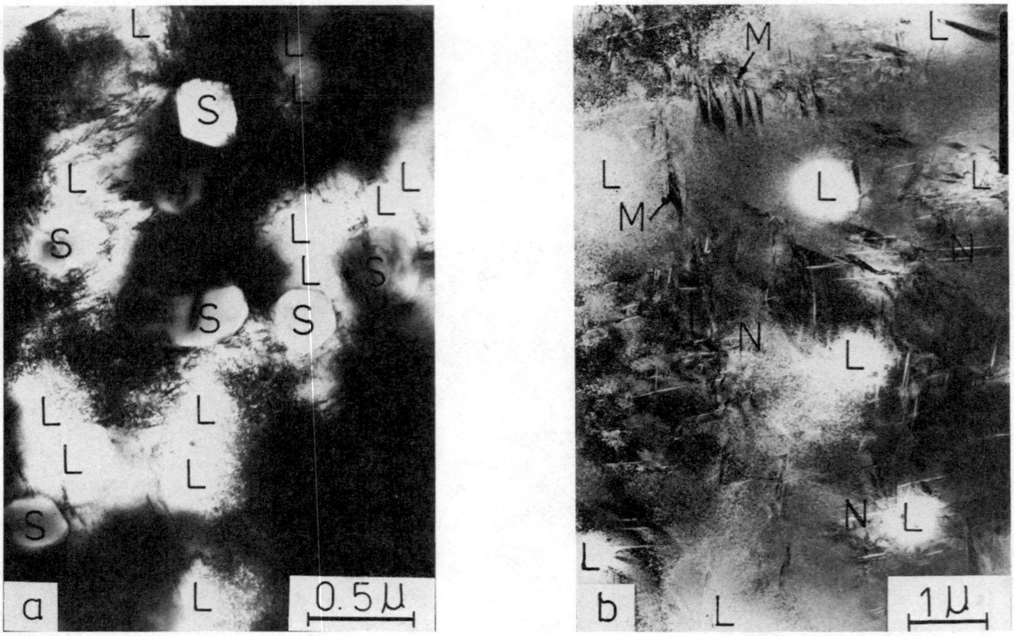

Fig.3 Dispersions of various types of particles, uniformly distributed in a Zr-3Mo-1.5Al-1.5Sn-0.5Si-1.07O-0.5La alloy: 1293 K/10h/OQ: (a) Particles marked as "L" are La_2O_3 and particles marked as "S" are Zr_3Si; (b) Particles with a plate-like morphology and marked as "M" are $ZrMo_2$; particles with an acicular morphology and marked as "N" were not identified.

In the Zr-3Mo-3Al-1.07O-0.3La alloy the dispersions consisted of uniformly distributed La_2O_3 particles, very fine plates of internally twinned martensite, small amounts of ω-phase particles and very fine coherent ordered particles.

In the Zr-3Mo-1.5Al-1.5Sn-0.5Si-1.07O-0.5La alloy at least five different types of dispersant particles were observed uniformly distributed throughout as shown in Figs 3 and 4. Among these particles those of La_2O_3 (marked as "L") and of Zr_3Si (marked as "S") were identified and are shown in Figs. 3, 4 and 5(b). Dislocations were observed at interfaces of some of the Zr_3Si particles which were beginning to loose coherency, as shown in Fig. 5(a). Dispersions of three other types of particles were also observed. Those with a plate-like morphology and marked as "M" (Figs. 3 and 4) were indexed as $ZrMo_2$. These particles averaged 3000 Å in length. Particles with an acicular morphology and an average length of about 5000 Å were also observed and are marked as "N" (Figs. 3 and 4); but were not identified. Particles which could be indexed as Zr_4Sn by X-ray diffraction analysis, were also observed.

Another type of dispersion, observed in both alloys containing lanthanum, consisted of very fine coherent ordered particles, approximately 20 Å in size and with an inter-particle spacing of about 100 Å, as shown in Fig. 4. Figure 5(b) shows the superlattice reflections due to these coherent ordered particles.

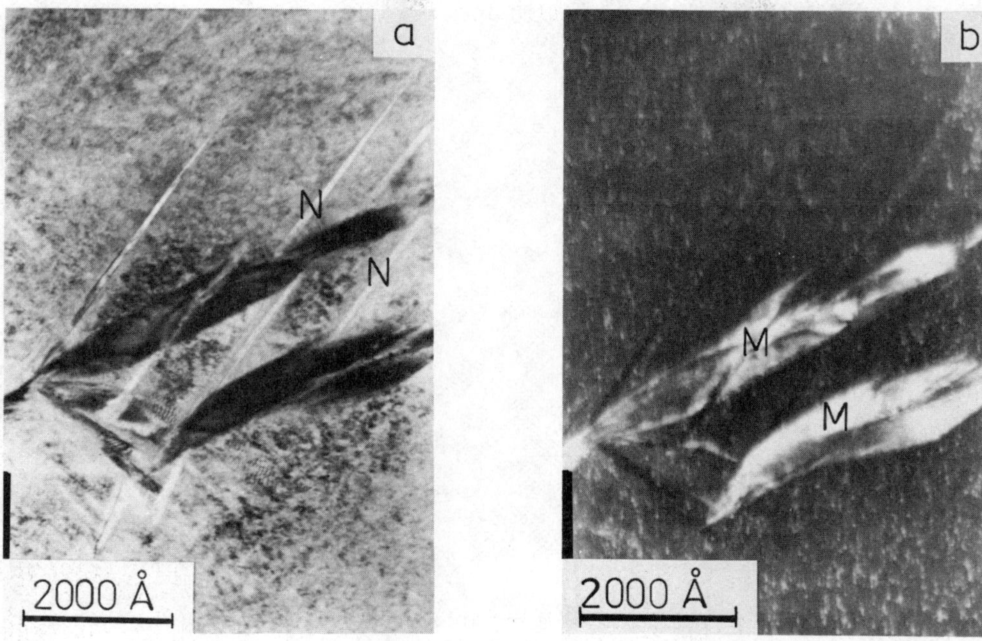

Fig. 4. Three types of dispersant particles in a Zr-3Mo-1.5Al-1.5Sn-0.5Si-1.07O-0.5La alloy: 1293 K/10h/OQ: (a) Particles with plate-like (marked as "M") and acicular (marked as "N") morphologies; (b) High resolution dark field of (a) showing $ZrMo_2$ particles and a very fine dispersion of ordered particles.

The interpenetration of the various types of dispersions and the fact that it was not always possible to differentiate between the various dispersant particles, made the measurement of the inter-particle spacing difficult. However, some average values for the inter-particle spacing of the same dispersant were obtained. The coarser dispersant had an inter-particle spacing of less than 1μm and the finer dispersant had an inter-particle spacing of less than 100 Å. The inter-particle spacing for the other dispersants varied between these values.

To investigate the thermal stability of these alloys they were heat treated at 1293°K for 100 hours. They were then examined with respect to grain growth, particle stability and hardness. The grain boundaries exhibited a very irregular wavy configuration (Fig. 2) which is attributed to the fact that these boundaries were pinned. The extent of the grain growth observed was very small (5 to 20μm max). Some of the particles had coarsened to a maximum of twice their original size. Other particles remained relatively stable in size. The small decreases of 15 to 20 points observed in the hardness of these alloys (Table 2) are in agreement with the metallographic observations.

It is therefore reasonable to conclude that at the temperatures at which the reactors operate (723-773 K) these particles should be very stable.

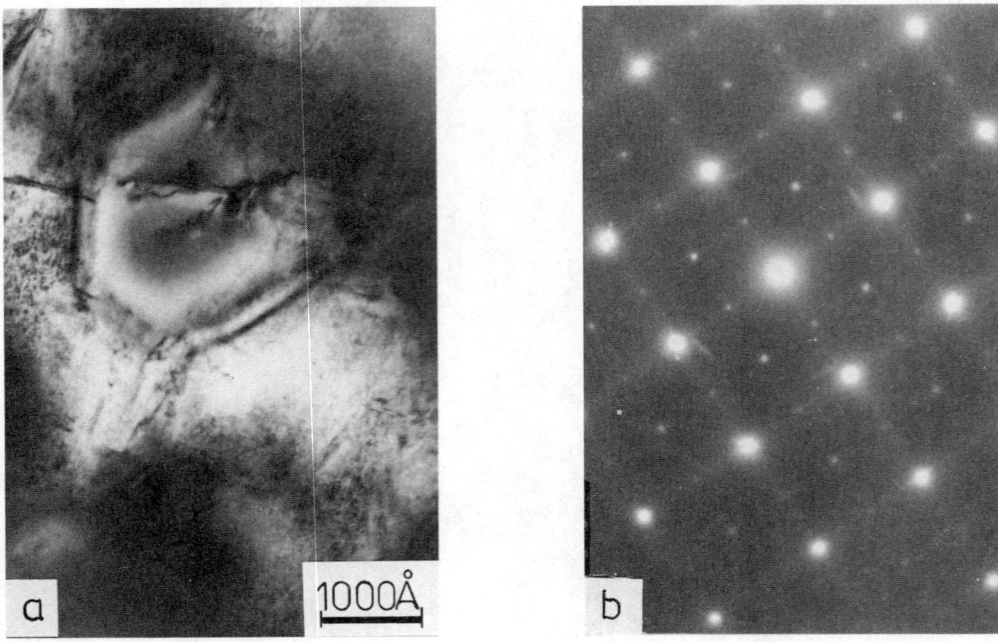

Fig. 5. (a) A Zr_3Si particle exhibiting interface dislocations; (b) Selected area diffraction pattern of Fig. 4, showing superlattice reflections due to ordered particles.

Mechanical Properties Measurements

The 0.2% proof stress, strain rate sensitivity and relaxation stress of these alloys were measured in the temperature range 293-900 K and are shown respectively in Figs. 6, 7 and 8. The 0.2% proof stress of the Zr-3Mo-1.5Al-1.5Sn-0.5Si-1.07O-0.5La alloy was 1420 MN m^{-2} at 293 K and decreased approximately linearly with increasing temperature, to reach the value of 1000 MN m^{-2} at about 700 K. It then increased to reach a value below 1250 MN m^{-2} near 800 K and then decreased again to a value of 1000 MN m^{-2} at about 880 K.

Strengthening Mechanisms

The main effect of the dispersant particles on the microstructure of these alloys was to pin grain boundaries, which reduced grain boundary sliding at elevated temperatures. The formation of a high density of sub-boundaries also pinned by dispersant particles, restricted the dislocation movement considerable and reduced the average grain size operating in these alloys significantly.

The interaction of dislocations with dispersant particles was observed by TEM, but the resulting strengthening effect is believed to be smaller than that due to the pinning of grain boundaries and sub-boundaries. The relative contributions of grain size and dispersion strengthening are difficult to determine due to the high density of sub-boundaries and the various types of dispersants present.

The peak in the 0.2% proof stress-temperature curves is attributed to the interaction between moving dislocations and dispersant particles, and to dynamic strain ageing effects (Fig. 7 and 8). These dynamic strain ageing effects are further suported by the following observations:

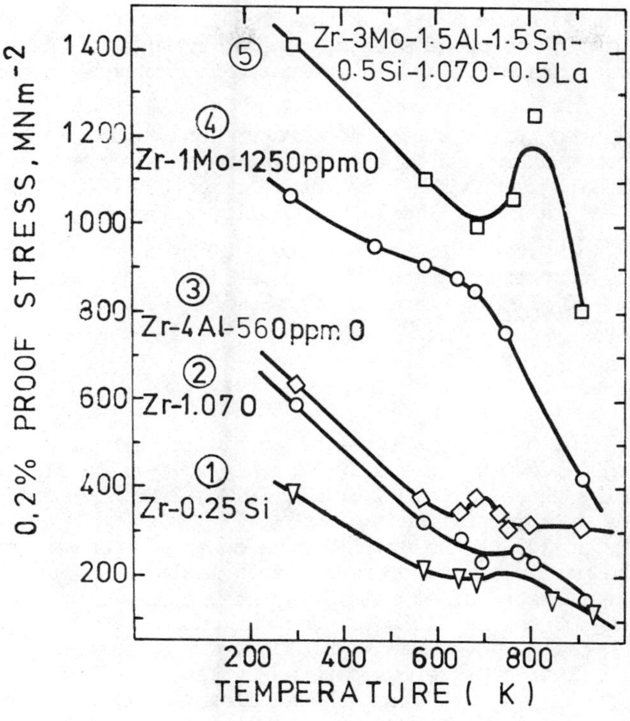

Fig. 6 Proof stress data

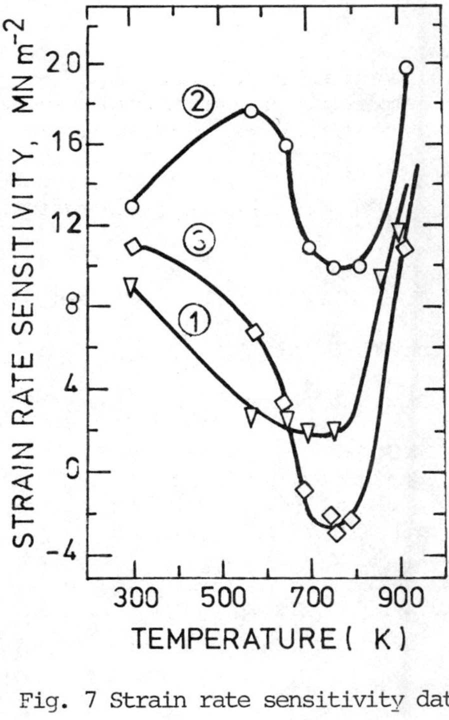

Fig. 7 Strain rate sensitivity data

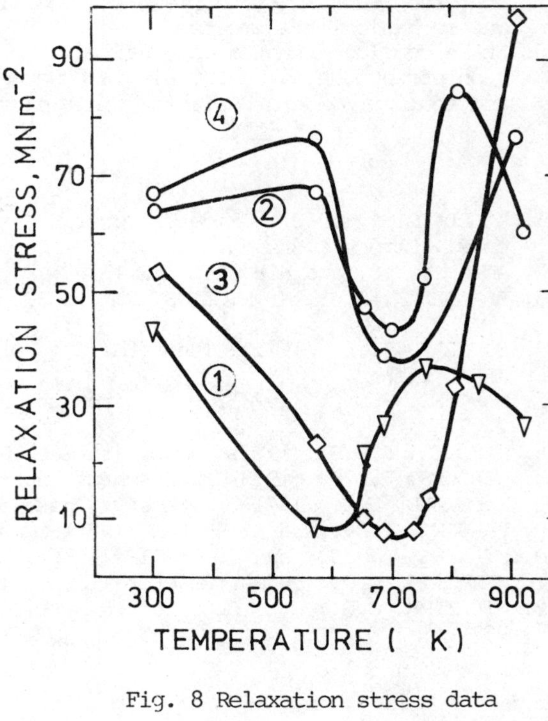

Fig. 8 Relaxation stress data

(a) In the temperature range 650-750 K the Zr-4Al-560ppmO alloy shows a rise of about 50 MN m^{-2} in the proof stress, a negative minimum in the strain rate sensitivity-temperature curve and in the relaxation stress-temperature curve (Fig. 6, 7 and 8)

(b) The plateaux in the proof stress-temperature and minima in both the strain rate sensitivity-temperature and in the relaxation stress-temperature curves found in the Zr-0.25Si alloy in the temperature range 600-850 K; in the Zr-1.07O alloy at 700-825°K and in the Zr-1Mo-1250ppmO alloy at 550-650 K (Fig. 6, 7 and 8).

These results are consistent with a Snoek type of ordering of O-O, Al-O and Si-O atom pairs and are further discussed by Marques (1977). The solution hardening effects due to atoms of Si, Al, O and Mo are shown in Fig. 6 and are further discussed by Marques (1976, 1977).

SUMMARY

1. Uniform dispersions of various types of particles were produced in the Zr-3Mo-1.5Al-1.5Sn-0.5Si-1.07O-0.5La and Zr-3Mo-3Al-1.07O-0.3La alloys. Among the dispersant particles those of La_2O_3, Zr_3Si, Zr_4Sn, $ZrMo_2$, ω-phase and coherent ordered particles were identified.
2. After 100 hours at 1293 K the grain growth observed was very small (5 to 20μm max.) and some of the particles had coarsened to a maximum of twice their original size. Other particles remained relatively stable in size.
3. The average size of the particles of the various dispersants ranged between 20 Å and 3000 Å. The coarser dispersants had an average inter-particle spacing of less than 1μm and the finer ones were separated by less than 100 Å.
4. This resulted in alloys with strengths at elevated temperatures, well in excess of the targets previously fixed. In particular the Zr-3Mo-1.5Al-1.5Sn-0.5Si-1.07O-0.5La alloy has 0.2% proof stress values above 1000 MN m^{-2} up to 880 K. Further work will be of interest to investigate the ductility, creep strength and creep rupture properties of these alloys and their response to other thermal and mechanical treatments.
5. It is concluded that a combination of both intermetallic and inert oxide dispersion strengthening can result in considerable improvement of zirconium alloys and could provide a basis for their elevated temperature high strength.

REFERENCES

1. ANSEL, G.S. (1968). Oxide Dispersion Strengthening. Metall. Soc. Conf., vol. 47, pp.61. Gordon & Breach.
2. ANTONY, K.C. and H.H. KEEPLER (1965). J. Less-Common Metals, 8, 36.
3. BRADBROOK, J.S., J. HIGGINS, R.M. HOBBS, and J.E. MILLER (1971). Pract. Metallog. 6, 359.
4. DOUGLASS, D.L. (1971). The Metallurgy of Zirconium. Suppl. Atom. Energ. Rev., IAEA, Vienna.
5. MARQUES, F.D.S. (1976). Proc. 4th Int. Conf. Strength of Metals & Alloys, Nancy, 2, 736.
6. MARQUES, F.D.S. (1977). Proc. 4th Gen. Ass. CENIM, Madrid, 1, 59.
7. MARQUES, F.D.S. (1978). Zeitschrift fur Metallkunde, 69, H. 3, 167.
8. McDONALD, N.R. (1971). J. Austr. Inst. Metals, 16, 4, 794.
9. PARSONS, P.D., and E. ADOLPH (1972). Can. Metall. Quart., 11, 1, 223.
10. WILKINS, B.J.S., and J. CHATER (1969). J. Nucl. Mater., 31, 288.
11. WILLIAMS, C.D. (1970). Reactor Techn., 13, 2, 147.
12. WILLIAMS, C.D., C.E. ELLS, and P.R. DIXON (1972). Can. Metall. Quart., 11, 1, 257.

Volume 3, ICM 3, Cambridge, England, August 1979

THE INFLUENCE OF GRAPHITE SHAPE ON MECHANICAL PROPERTIES OF FERRITIC CAST IRON

D. Löhe, O. Vöhringer and E. Macherauch

Institut für Werkstoffkunde I, Universität Karlsruhe (TH), Karlsruhe, F.R.G.

INTRODUCTION

Grey cast iron shows quite different mechanical behaviour according to chemical composition, shape of the graphite inclusions, and heat treatment. A great number of publications exists concerning the technological properties of flake graphite cast irons, vermicular cast irons, and nodular cast irons (Gilbert, 1961, 1964; Holzmüller, 1976; Matting and Krüger, 1970; Nechtelberger, 1977; Rickards, 1969a, 1969b; Riemer, 1973; Ruff and Wallace, 1976; Sissener and others, 1972). However, only a few investigations (Emerson and Simmons, 1976; Hüttebräucker, Vöhringer and Macherauch, 1978; Riegger, Vöhringer and Macherauch, 1971) have been performed regarding the influence of graphite shape on the uniaxial deformation behaviour of these materials, whilst holding all other parameters constant. It is well known that the behaviour of cast iron under tensile load is different from that under compressive load. Therefore, this paper deals with the results of tensile and compressive tests in the temperature range 77 K $<$ T $<$ 873 K on ferritic cast iron specimens with nearly the same chemical composition but various graphite shapes.

MATERIALS AND TEST PROCEDURE

Five cast irons have been investigated: Nodular cast iron GGG (Y1), vermicular cast iron GGV (Y2), and three flake graphite cast irons GG (Rst), GG (Y1), and GG (Y2). The abbreviations in brackets characterize the moulds described in Table 1. As can be seen from Table 1, the chemical composition of all cast irons was nearly the same. The fraction of unspecified elements was below 0.2 wt.%. The graphite morphologies have been changed by suitable inoculation-treatments and different cooling-rates. For reasons of comparison, an iron-silicon alloy (FeSi 2.5) with the same chemical composition as the metallic matrix of the cast irons has been included in the investigation. All materials were annealed to fully ferritic condition according to the heat-treatments listed in Table 1. Figure 1 shows the corresponding micrographs. As characteristic structural features of the graphite morphology, the mean ferrite grain diameter d and the mean aspect ratio a/b (a length, b width of the graphite particles) are also included in Table 1.

The mechanical behaviour of the materials has been determined in the temperature range 77 K $<$ T $<$ 873 K under tensile loading and the range 77 K $<$ T

TABLE 1 Chemical Composition, Treatments and Structural Features of Investigated Materials

Material	GGG (Y1)	GGV (Y2)	GG (Rst)	GG (Y1)	GG (Y2)	FeSi 2.5
wt.% C	3.47	3.46	3.34	3.46	3.54	0.025
wt.% Si	2.69	2.55	2.82	2.79	2.84	2.59
Inoculant	0.9 % SiMg (31.1 % Mg) + 0.5 % FeSi (74 %)	0.145 % Cer-mischmetal + 0.5 % FeSi (75 %)	CaC_2 + CaF_2	0.5 % FeSi (76 %)	0.5 % FeSi (74 %)	--
Mould	Y-shaped 10 x 20 x 170 mm^3 *)	Y-shaped 25 x 40 x 300 mm^3 *)	Ø 18 mm x 420 mm	Y-shaped 10 x 20 x 180 mm^3 *)	Y-shaped 25 x 40 x 260 mm^3 *)	--
Heat treatment	3h/920°C + 16h/700°C furnace cooled	3h/900°C + 16h/690°C furnace cooled	3h/920°C + 16h/690°C furnace cooled	3h/920°C + 18h/700°C furnace cooled	3h/920°C + 6h/700°C furnace cooled	vacuum 2h/850°C furnace cooled
d (μm)	35	44	≈ 14	≈ 20	≈ 17	45
Ratio a/b	1.2	4.0	7.2	22.2	34.0	--

*) Size without feeder head

< 623 K under compressive loading. The deformation rate in all cases was $\dot{\epsilon} = 3.3 \cdot 10^{-4}\ s^{-1}$. The temperature range in the compressive tests was limited by the testing equipment.

RESULTS

Figure 2 shows the temperature dependence of the 0.2 % proof stress of the materials under investigation. It clearly demonstrates the influence of the graphite shape, which changes from nodular form (GGG (Y1)) to lamellar form (GG (Y2)), on this mechanical quantity and its temperature dependence. In the temperature range between 77 K and 400 K the tensile 0.2 % proof stress decreases in the sequence FeSi 2.5, GGG (Y1), GGV (Y2), GG (Rst), GG (Y1), and GG (Y2). Below 350 K the measured values increase with decreasing temperature to a different extent. Nodular cast iron shows a much stronger temperature dependence than flake graphite cast iron. Between 350 K and 600 K nearly constant values are obtained for all materials except FeSi 2.5 . In this temperature range the stress - strain curves are serrated due to dynamic strain ageing. Beyond 650 K a decrease of the proof stress is seen. This decrease is less pronounced for GG (Y1) and GG (Y2) than for the other materials. The proof stress / temperature curve of GG (Rst) intersects at higher temperatures the corresponding curves of GGV (Y2), GGG (Y1), and FeSi 2.5. The compressive 0.2 % proof stress is plotted as a function of temperature in Fig. 3. As can be seen from a comparison with Fig. 2, the measured quantities, apart from those of FeSi 2.5, are greater than the respective tensile proof stresses. Also the sequence of the materials with respect to the

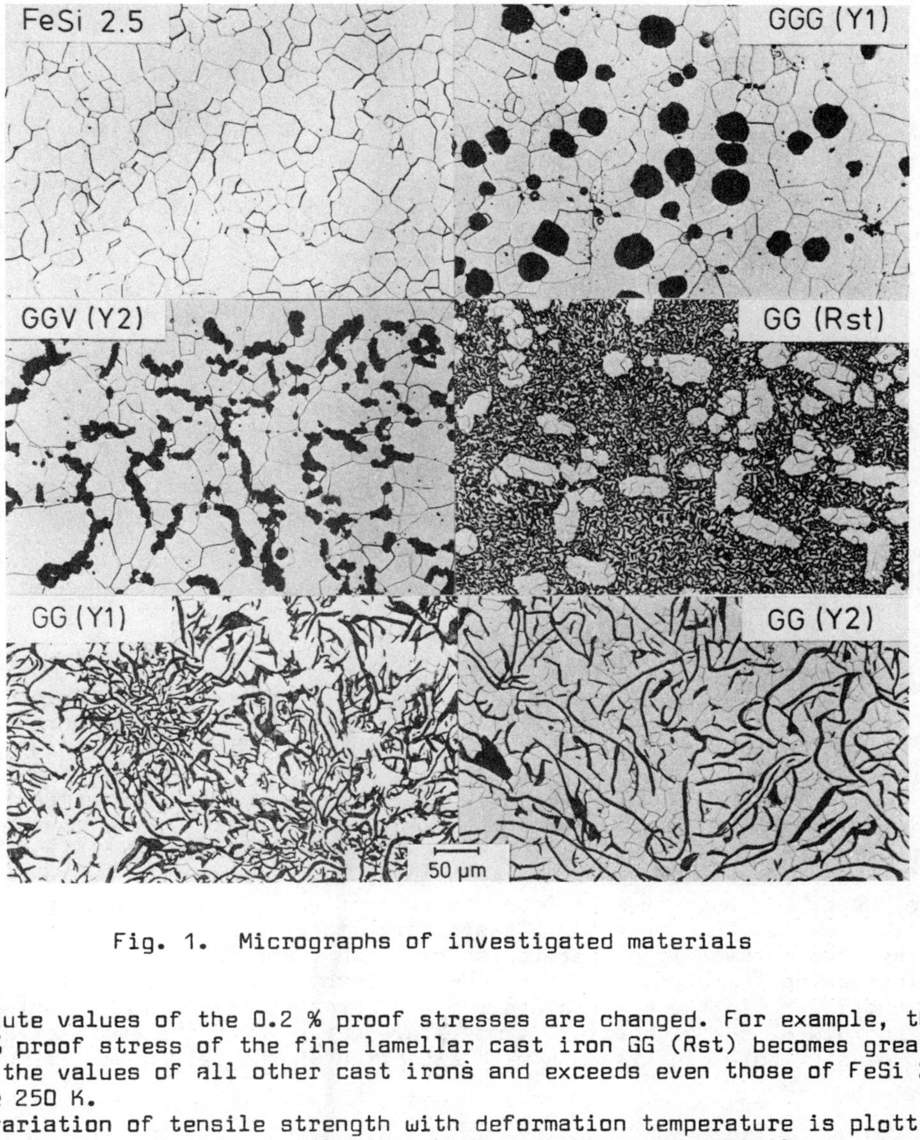

Fig. 1. Micrographs of investigated materials

absolute values of the 0.2 % proof stresses are changed. For example, the 0.2 % proof stress of the fine lamellar cast iron GG (Rst) becomes greater than the values of all other cast irons and exceeds even those of FeSi 2.5 above 250 K.
The variation of tensile strength with deformation temperature is plotted in Fig. 4. The graphs for the flake graphite cast irons are similar to the temperature dependence of the tensile 0.2 % proof stress in Fig. 2. For all other materials, the tensile strength / temperature curves are much more separated than the tensile 0.2 % proof-stress / temperature curves up to 600 K (compare for example the vermicular cast iron GGV (Y2) with the fine flake graphite cast iron GG (Rst)). Further, the temperature-independent stress plateaus in the range of dynamic strain ageing are more distinct for the tensile strength than for the tensile 0.2 % proof stress. Obviously, the increase of tensile strength of FeSi 2.5 and nodular cast iron with decreasing temperature diminishes below 100 K.
Fig. 5 shows the percentage elongation before reduction (A_g) as a function of the deformation temperature. In the whole temperature range GG (Rst), GG (Y1), and GG (Y2) show very small A_g - values, which do not exceed 0.5 %.

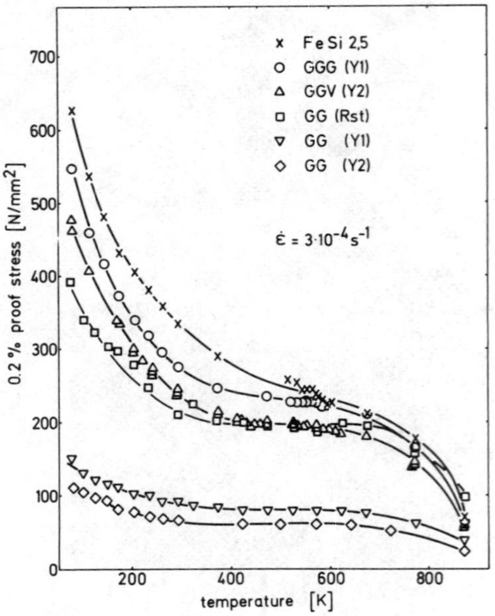

Fig. 2. Tensile 0.2 % proof stress vs temperature

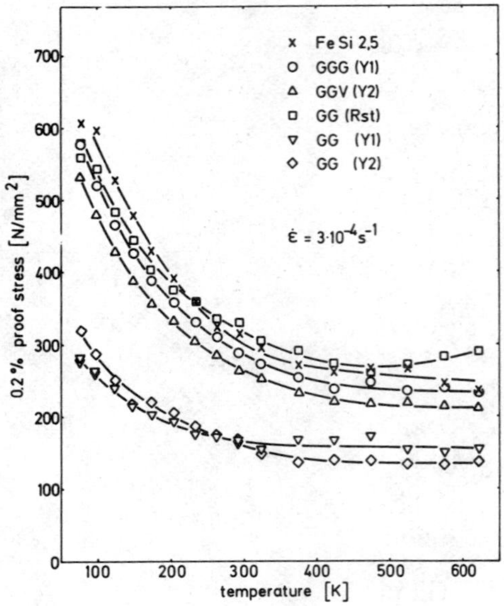

Fig. 3. Compression 0.2 % proof stress vs temperature

Apart from FeSi 2.5, the greatest values are measured with nodular cast iron (GGG (Y1)). A sharp drop is seen in the value of A_g for this material in the temperature range 500 $< T <$ 600 K. The A_g - values of vermicular cast iron are significantly lower than those of the nodular cast iron in the whole temperature range, but much greater than the respective values for flake graphite cast iron. Below 100 K relatively small percentage elongations before reduction for nodular cast iron, vermicular cast iron, and FeSi 2.5 are measured. With increasing temperature, the A_g - values reach relative maxima between 150 K and 300 K. In the temperature range of dynamic strain ageing between 350 K and 600 K, the elongation / temperature curve shows the smallest A_g - values of the whole measurements with exception of the values determined at 77 K.

DISCUSSION

It is well established (Hüttebräucker, Vöhringer and Macherauch, 1978) that the flow stress of a material can be split into several components due to different hardening processes. For the cast iron types under investigation, the following flow stress components have to be considered:
The thermal flow stress component (σ^*) which can be expressed by the relationship

$$\sigma^* = \sigma_0^* \left[1 - \frac{T}{T_0}\right]^2 \tag{1}$$

σ_0^* is the thermal flow stress at 0 K, and T_0 is the characteristic temperature at which σ^* becomes zero.
The flow stress component due to dislocation hardening is given by

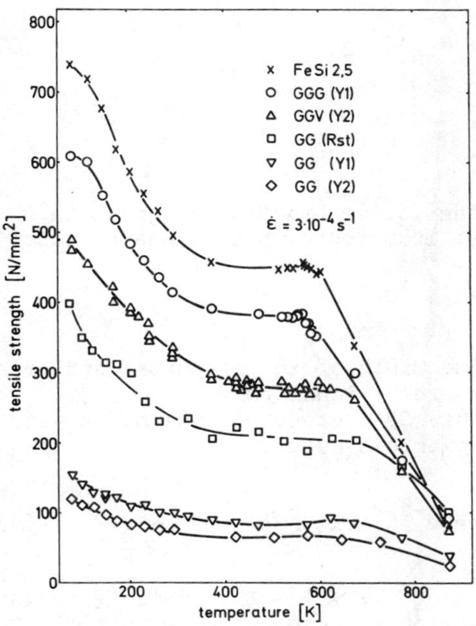

Fig. 4. Tensile strength vs temperature

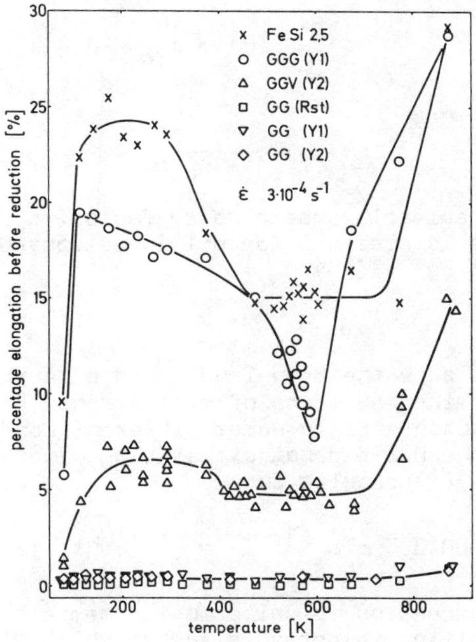

Fig. 5. Percentage elongation before reduction vs temperature

$$\sigma_{dh} = \alpha_1 \cdot \sqrt{\rho_t} \tag{2}$$

where α_1 is constant and ρ_t is the total dislocation density.
Solid solution hardening leads to a flow stress component

$$\sigma_{sh} = \alpha_2 \cdot c^n \tag{3}$$

with α_2 constant and c the concentration of solute atoms. The exponent takes values $0.5 \leq n \leq 1$. According to the Hall-Petch-relationship, grain boundary hardening results in a flow stress component

$$\sigma_{gh} = K \cdot d^{-1/2} \tag{4}$$

K is a material-dependent quantity and d is the mean grain size.
To take into account the influence of the graphite particles on the flow stress, two aspects have to be regarded. Firstly, the graphite inclusions may lead to a hardening effect like second phase particles. If this were true, they should produce a flow stress component

$$\sigma_{ph} = \alpha_3 \cdot \lambda^{-m} \tag{5}$$

where α_3 is a constant and λ is the mean spacing between the graphite particles. Appropriate values of m should be 0.5 or 1 (Liu and Gurland, 1968; Marder and Bramfitt, 1976).
Secondly, the local stress concentrations near the graphite particles should be considered. These influence the above mentioned stress components to nearly the same extent. To a first approximation, the total flow stress can be written as

$$\sigma = (\sigma^* + \sigma_{dh} + \sigma_{sh} + \sigma_{gh} + \sigma_{ph}) \cdot F_n \quad (6)$$

where

$$F_n = \frac{1}{1 + c_G \cdot \eta} \quad (7)$$

is a suitable chosen notch factor. c_G is the volume-fraction of graphite, which is constant for all investigated cast iron types. η is a shape factor given by

$$\eta = \beta \cdot \frac{a}{b} \quad (8)$$

where a is the mean length and b is the mean width of the graphite particles determined by means of micrographs. β is a proportional factor including the notch effects under different conditions. Of course, eq. (6) holds also for the 0.2 % proof stress ($R_{p0.2}$) of the examined materials. Summarizing eqs. (1) - (8) results in

$$R_{p0.2} = \left[\sigma_0^* \left(1 - \frac{T}{T_0}\right)^2 + \alpha_1 \cdot \sqrt{\rho_t} + \alpha_2 \cdot c^n + K \cdot d^{-1/2} + \alpha_3 \cdot \lambda^{-m}\right] \frac{1}{1 + c_G \beta \frac{a}{b}} \quad (9)$$

This equation is valid in the temperature range $T_t \leq T \leq T_0$, where T_t is the transition temperature, above which dislocation movements induce sufficient plastic deformation before fracture occurs. For more details with respect to the considered flow stress components see Hüttebräucker, Vöhringer, and Macherauch (1978). Since all materials in this investigation have the same chemical composition and the same ferritic matrix, it seems reasonable to assume that the thermal flow stress and the flow stress components due to dislocation hardening and solid solution hardening are also equal in all cases. However, there will be differences in the grain boundary hardening and the second phase hardening as well as in the notch factor F_n for the different materials. Consider first the notch factor F_n, which is determined by the shape factor η. The assumed form of η in this investigation diverges from other proposals to take into account the notch effect of the graphite particles (Capaletti and Hornaday, 1974; Hornbogen and Motz, 1977; Pohl, 1977; Stroppe, 1966). For example, Stroppe has derived an expression similar to eq. (8) but with a constant value $\beta = 0.75$. On the other hand, Hornbogen and Motz (1977) have used a notch factor

$$F_n' = c_M + c_G \cdot \beta' \cdot \left(1 + 2 \cdot \sqrt{\frac{l}{r}}\right) \quad (10)$$

to determine the influence of the shape of the graphite particles on the fracture toughness of grey cast irons. c_M is the volume fraction of the metallic matrix, β' is constant, l is the mean length and r the mean radius of curvature at the end of the graphite lamella. There is no recommendation, however, as to how r should be measured accurately. Furthermore, assuming elliptical shaped graphite particles and taking $\beta = 1.70$, eq. (10) approximately equals eqs.(7) and (8). So it has to be proved whether or not the proposed notch factor is correct. The simplest procedure to check the validity of F_n includes the measurement of Young's Modulus under similar conditions for all materials listed in Table 1. There should be a strong dependence of it on the notch effect of graphite particles with different shapes. Fig. 6 shows the results of corresponding measurements at room temperature. In the figure the ratio of Young's Modulus $\frac{E^{(\text{cast iron})}}{E^{(\text{FeSi } 2.5)}}$

is plotted as a function of the aspect ratio a/b. a/b = 1.2 belongs to nodular cast iron, a/b = 34 to the coarse flake graphite cast iron GG (Y2). The measured values are well described by

$$F_n = \frac{1}{1 + 0.75 \cdot \frac{a}{b}} \tag{11}$$

This shows that the assumed form of the notch factor reproduces the experimental values in a quantitatively correct manner.
If all hardening effects summarized in eqs. (6) and (9) would be the same for all measured materials the relationship

$$\frac{R_{p0.2}^{(\text{cast iron})}}{R_{p0.2}^{(\text{FeSi } 2.5)}} = F_n^{(\text{cast iron})} = \frac{1}{(1 + c_G \cdot \beta \cdot \frac{a}{b})} \tag{12}$$

should be valid for the tensile 0.2 % proof stress. Actually, as shown in Fig. 7, eq. (12) is well fulfilled if β is 0.86. This β-value, however, is nearly the same as that in eq. (11). So one can conclude that the notch effect is the dominating factor in determining the tensile 0.2 % proof stress of grey cast irons of the same chemical composition but with different shapes of graphite particles. Nevertheless, the hardening mechanisms are effective to different degrees for the materials with different graphite shapes. During tensile loading the matrix deformation is the more inhomogeneous the greater the aspect-ratio a/b. At high η-values the plastic deformation is concentra-

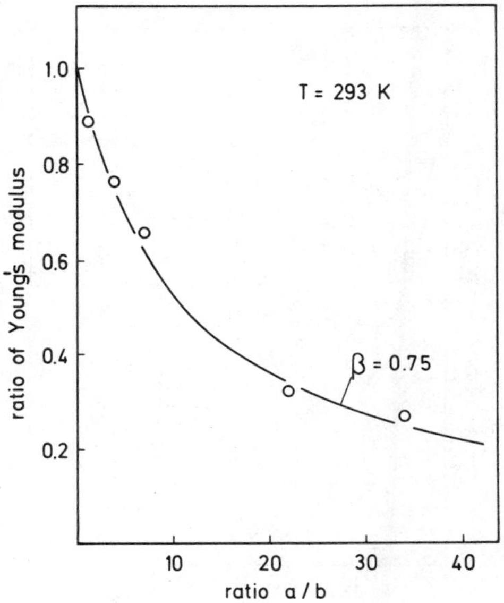

Fig. 6. Ratio of Young's Modulus vs ratio a/b

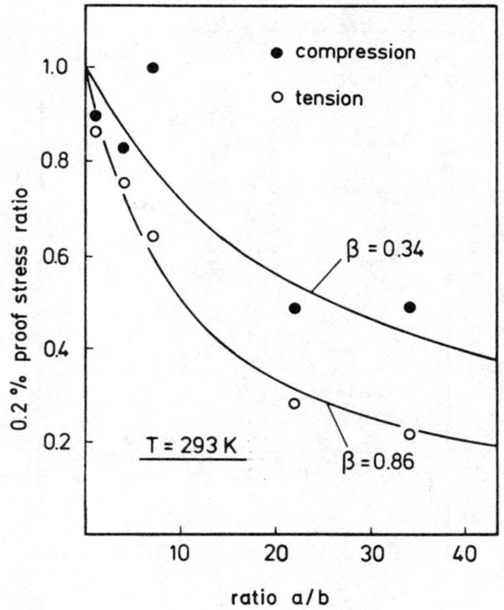

Fig. 7. 0.2 % proof stress ratio vs ratio a/b

ted in the very vicinity of the graphite particles. Since the stress peaks are reduced to the nominal stress along a distance of several grain diameters, no grain boundary hardening and no second phase hardening are effective. On the contrary, it is well known that in the case of nodular and vermicular cast iron (small aspect ratios a/b!) grain boundary hardening influences the tensile 0.2 % proof stress (Gilbert, 1956; Hüttebräucker, Vöhringer, and Macherauch, 1978; Parent-Simonin, 1974). In as much as both materials reveal nearly the same grain sizes according to Table 1, in this investigation only a small influence of grain size should be expected. The comparison of Figs. 2 and 3 shows pronounced differences in the tensile and compressive 0.2 % proof stress of the different cast iron types. With respect to the notch factor F_n, this behaviour is not surprising. According to Shibata and Ono (1978) an oblate spheroidal inclusion shows under tensile loading a much greater stress concentration than under compressive loading. Therefore the compressive 0.2 % proof stress ratios plotted as a function of a/b should decrease with increasing a/b to a smaller extent than the tensile 0.2 % proof stress ratios. This expectation is fulfilled, as can be seen from Fig. 7. The curve with the best fit produces $\beta = 0.34$. However, there are marked deviations between calculated and measured compressive 0.2 % proof stresses. For example, the 0.2 % proof stress of GG (Rst) with very fine graphite flakes is much higher than expected. This may be caused by grain boundary- and second phase hardening, which obviously are more effective in compressive loading, because the deformation is much more homogeneous. As a consequence of the transmission of stress by the graphite particles, differences in grain sizes and interlamellar spacings result in differences in the flow stress components of the respective hardening mechanisms.

As mentioned above, eq. (9) holds for $T_t < T < T_0$. Since the notch effect is an athermal effect, it is expected that at different temperatures the same

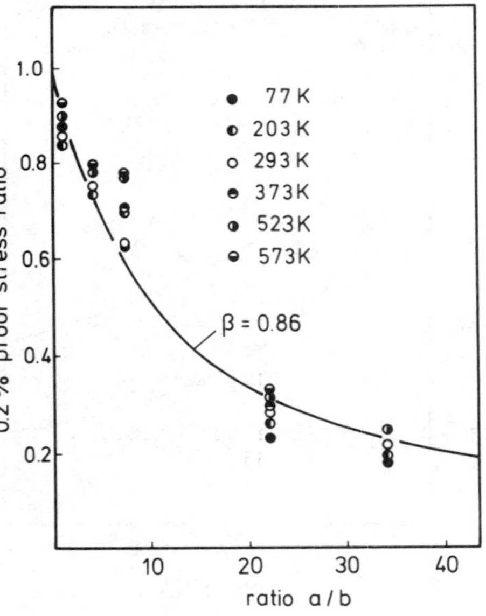

Fig. 8. 0.2 % proof stress ratio (tension) vs ratio a/b at different temperatures

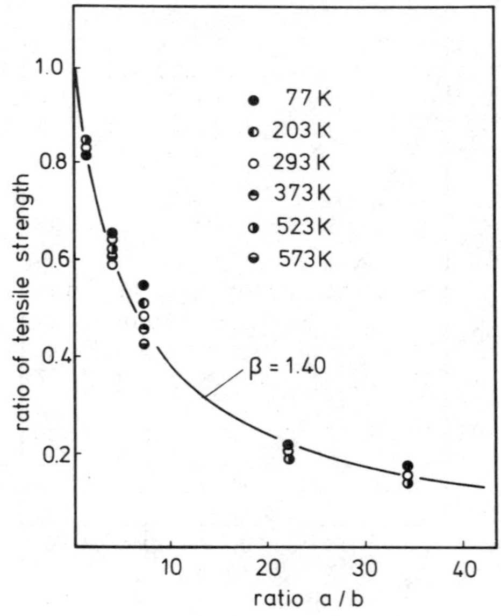

Fig. 9. Ratio of tensile strength vs ratio a/b at different temperatures

dependence of the tensile 0.2 % proof stress ratio on the aspect ratio a/b occurs. As can be seen from Fig. 8, qualitatively similar relations exist nearly independent of temperature. At low temperature, a slightly higher, at high temperatures, a slightly lower notch effect appears than at room temperature. The calculated values (full line) are valid for ß = 0.86. The high temperature values of GG (Rst) diverge from this curve more than all the other data. This is probably caused by the variation of dynamic strain ageing processes through the fine graphite morphology of GG (Rst), which is not considered by eq. (9).

The tensile strength / temperature curves given in Fig. 4 are even more influenced by variations of the graphite shape than the tensile 0.2 % proof stress / temperature curves plotted in Fig. 2. This result is easy to understand by considering the percentage elongation before reduction given in Fig. 5. The room temperature A_g - values of nodular cast iron and FeSi 2.5 are about three times higher than those of vermicular cast iron, while the elongation A_g of flake graphite cast iron does not exceed 0.5 %. The marked differences in tensile strength compared with tensile 0.2 % proof stress, are caused by this highly different elongation before reduction, which involves different dislocation hardening until fracture.

This relationship causes a strong dependence of ratio of tensile strength on aspect ratio a/b, given in Fig. 9. The calculated values (full line) are valid for ß = 1.40. The ratio of tensile strength shows no clear dependence on temperature for all aspect ratios a/b.

SUMMARY

Tensile and compressive tests have been performed at different temperatures on ferritic cast iron with nearly constant chemical composition, but with various graphite shapes. The results show that the 0.2 % proof stress of flake graphite cast iron in tensile deformation is mainly influenced by the notch effect of graphite particles. The dependence of tensile strength on this notch effect is even greater. In compressive deformation, the 0.2 % proof stress is less influenced by the stress concentrations due to graphite particles. In this case, the flow stress components of grain boundary hardening and second phase hardening cannot be neglected.

REFERENCES

Capaletti, T. L., and J. R. Hornaday (1974). Nodular iron shape factor - a new approach to quantifying graphite morphology. AFS Transactions, 82, 59 - 64.

Emerson, P. J., and W. Simmons (1976). Final report on the evaluation of the graphite form in ferritic ductile iron by ultrasonic and sonic testing, and of the effect of graphite form on mechanical properties. AFS Transactions, 84, 109 - 128.

Gilbert, G. N. J. (1956). The effect of grain size on the mechanical properties of ferritic nodular irons. BCIRA J., 4, 430 - 435.

Gilbert, G. N. J. (1961). Stress/strain properties of cast iron and Poisson's ratio in tension and compression. BCIRA J., 9, 347 - 363.

Gilbert, G. N. J. (1964). The stress/strain properties of nodular cast irons in tension and compression. BCIRA J., 12, 170 - 193.

Holzmüller, A. (1976). Gußeisen mit Vermiculargraphit - ein Spezialwerkstoff für den Motorenbau. Giesserei, 63, 281 - 291.

Hornbogen, E., and J. M. Motz (1977). Über die Bruchzähigkeit von graphithaltigen Eisen - Kohlenstoff - Gußwerkstoffen. Giess.-Forsch., 29, 115 - 120.

Hüttebräucker, K., O. Vöhringer, and E. Macherauch (1978). Verformungsverhalten von ferritischem Gußeisen mit Vermiculargraphit. Giess.-Forsch., 30, 39 - 45.
Liu, C. T., and J. Gurland (1968). The strengthening mechanism in spheroidized carbon steels. Trans. Met. Soc. AIME, 242, 1535 - 1542.
Marder, A. R., and B. L. Bramfitt (1976). The effect of morphology on the strength of pearlite. Metallurg. Trans., 7, 365 - 372.
Matting, A., and U. Krüger (1970). Ermittlung der mechanischen Eigenschaften von Gußeisenwerkstoffen bei tiefen Temperaturen. Giess.-Forsch., 22, 29 - 37.
Nechtelberger, E. (1977). Gußeisenwerkstoffe für Zylinderdeckel. Forschungsberichte Verbrennungskraftmaschinen, Heft 230.
Parent - Simonin, S. (1974). Ferrite grain: application to malleable cast iron. AFS Transactions, 82, 209 - 216.
Pohl, D. (1977). Die Bedeutung der Graphitform für die mechanischen Eigenschaften des Gußeisens mit Kugelgraphit. Giesserei, 64, 629 - 633.
Rickards, P. J. (1969a). Tensile properties of grey irons at low temperatures. Cryogenics, 9, 186 - 193.
Rickards, P. J. (1969b). The low temperature tensile properties of nodular graphite cast irons. Cryogenics, 9, 465 - 474.
Riegger, S., O. Vöhringer, and E. Macherauch (1971). Untersuchungen zum Verformungsverhalten von Gußeisen mit Kugelgraphit GGG-38/42 im Temperaturbereich zwischen 77 K und 873 K. Giess.-Forsch., 23, 35 - 41.
Ruff, G. F., and J. F. Wallace (1976). Control of graphite structure and its effect on mechanical properties of gray irons. AFS Transactions, 84, 705 - 728.
Shibata, M., and K. Ono (1978). Stress concentration due to an oblate spheroidal inclusion. Mat. Sci. Eng., 34, 131 - 137.
Sissener, J., and others (1972). Cast iron with vermicular graphite. AFS - Cast Met. Res. J., 8, 178 - 181.
Stroppe, H. (1966). Untersuchungen zum Elastizitätsverhalten von Grauguß. II. Theoretischer Teil. Wiss. Z. TH Magdeburg, 10, 159 - 173.

THEME V

DESIGNING WITH NEW MATERIALS

Volume 3, ICM 3, Cambridge, England, August 1979

THE MECHANICAL PROPERTIES OF JUTE FIBERS AND POLYESTER/JUTE COMPOSITES

K. K. Chawla and A. C. Bastos

Centro de Pesquisas de Materiais and Seção de Engenharia e Ciência dos Materiais, Instituto Militar de Engenharia, Urca 22290, Rio de Janeiro, RJ, Brazil

INTRODUCTION

Jute and sisal fibers are vegetable fibers traditionally used for making cords and gunny sacks. Brazil is a major producer of these fibers. These vegetable fibers are different from man-made fibers in that they are products of the vegetable kingdom and thus constitute a renewable engineering material. A research program has been going on, here at IME, since 1975 involving study of the mechanical properties of these fibers and exploring an unconventional use of these fibers as reinforcement for unsaturated polyester resin (Chawla, 1976). The main driving force for this program is the following: these fibers are extremely cheap, abundantly available in Brazil and, thus, appear promising as a substitute for the conventional glass fibers in many situations. This work describes the mechanical properties of jute fibers and polyester/jute fiber composites.

JUTE FIBERS

Chemically, all vegetable fibers have cellulose as the main constitutent plus various amounts of lignin, hemicellulose, pectins and waxes. Jute, specifically, has 58-63% α - cellulose, 21-24% hemicellulose, 12-14% lignin, and small amounts of wax, pectin etc. (The New Encyclopaedia Britannica, 1975, Barkakaty, 1976). Jute is a bast fiber as it is obtained from the inner bast tissue of the plant stem. The fibers are bound together by gummy materials in the tissues under the bark. These gummy tissues are softened, dissolved and washed away by steeping the stems in the water (the retting process) and the fibers extracted. Micro-organisms decompose the gums and soften the tissues in 5 to 30 days depending upon the temperature and type of water (The New Encyclopaedia Britannica, 1975). After retting, the fibers are removed from the stalk, washed, and dried.
The as received jute fibers from Paraiba, Brazil, showed a very irregular transverse section (Fig.1). The transverse section, Fig.1, showed that the jute fibers, like sisal fibers (Chawla, 1976, Chawla et al., 1978), had a multicellular structure

composed of microfibrils. The fiber cross-section was highly non-uniform. The fiber cross-sectional area (average value of 100 measurements) was 1 x 10^{-3} mm^2 (with a standard deviation of 0.7 x 10^{-3} mm^2). Thus, one notes that the biological origin of these fibers leads to a disadvantage vis a vis man made fibers, viz. that one has very little control over their properties and characteristics. Variability in properties, due to geographic, climatic and growth conditions, and processing is the rule and structural consistency is conspicous by its absence. The color, lustre, and strength are very much influenced by the retting process, especially, overretting leads to weak fibers.

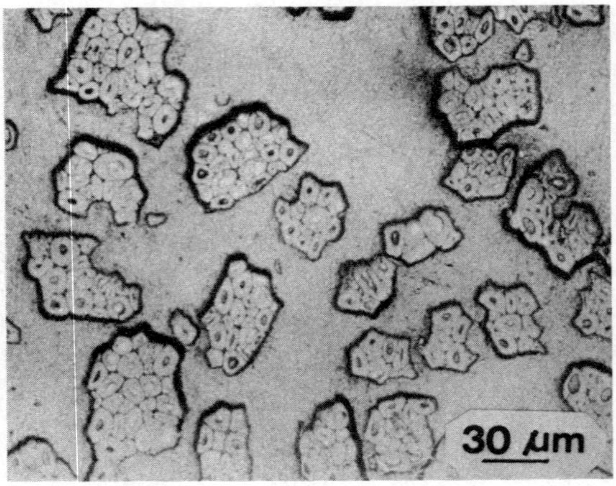

Fig.1 - The distribution of jute fibers in polyester resin. Notice the transverse section of the jute fibers consisting of microfibrils.

The jute fibers were tested in tension in an Instron machine, using the technique described in (Chawla, 1976, Chawla et al., 1978). The results given below are the average of 22 tests and with a gage length of 50 mm. The as-received fibers were washed with a detergent to remove any grease, emulsifers etc. on the fiber surface before testing and incorporating in the unsaturated polyester resin. The average elastic modulus of jute fibers was 21.78 GPa (with a standard deviation of 3.12 GPa) while the average tensile strength was 270.5 MPa (with a standard deviation of 66.6 MPa). Practically no plastic deformation was observed and the average deformation at fracture was 1.28% (standard deviation = 0.30%). This value seems to be at the lower end of the range generally associated with vegetable fibers.

POLYESTER/JUTE COMPOSITES

Nominally unidirectional composites with varying volume fractions

(V_f) of jute fibers in unsaturated polyester resin were made by leaky mold technique. The longitudinal distribution of the fibers was such that 90% of the fibers lay within ± 10° of the axis of the specimen. The non-uniformity of the fiber cross-section within a given fiber was quite random and the volume fraction of jute fibers did not vary more then 10% along the length of the composite specimen. Tensile specimens cut out from the composite sheets were tested in an Instron testing machine. A clip-on type extensometer was used for measuring the specimen elongation. Some three point bending tests were also done in an Amsler machine. The composites showed a more or less linear relationship between the tensile modulus of elasticity, E, and the jute fiber volume fraction, V_f, as well as between the ultimate tensile strength, σ_{max} and V_f (Figs.2 & 3). A similar linear relationship was obtained between σ_{max} and V_f in the bending test (Fig.4). The linear increase in strength and modulus of the polyester/jute fiber composites with fiber volume fraction demonstrates the phenomenon of reinforcement of the expensive unsaturated polyester resin by the cheap jute fibers.

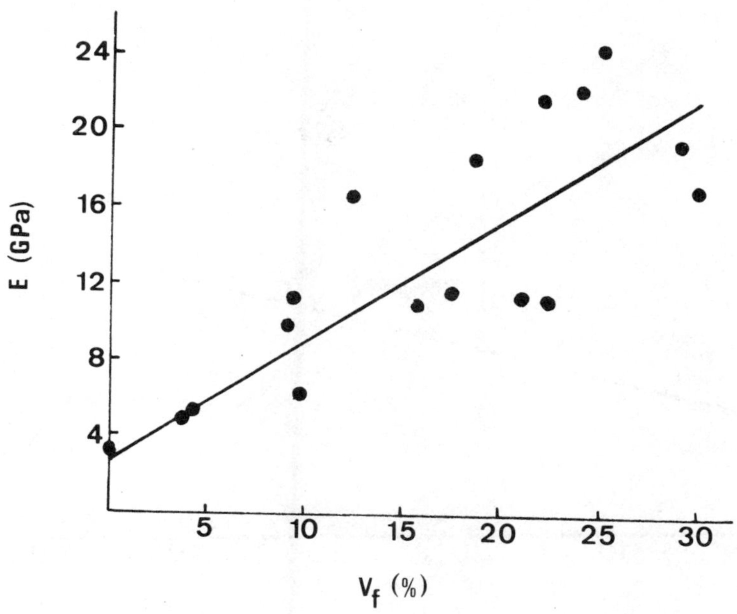

Fig.2 - Variation of the composite Young's modulus, E with jute fiber volume fraction, V_f. Tension.

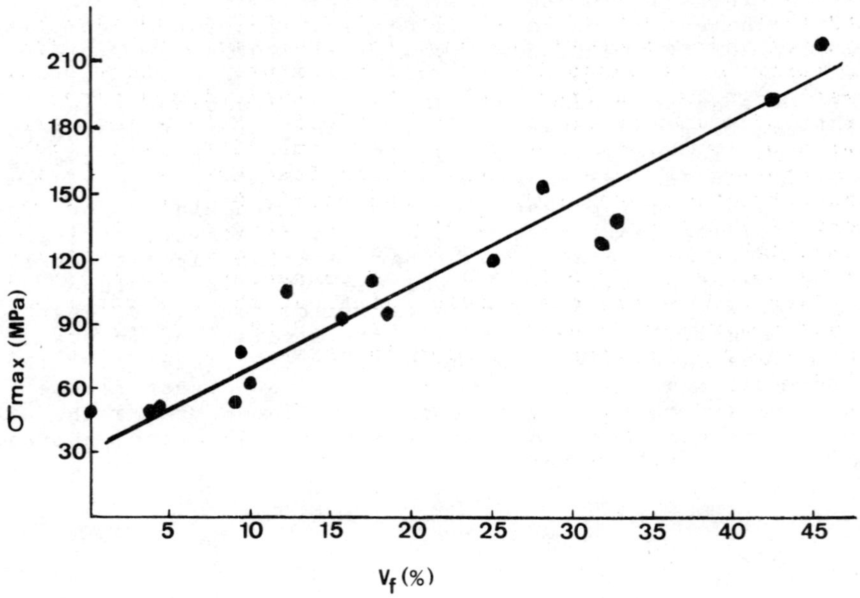

Fig.3 - Variation of the composite maximum strength, σ_{max}, with jute fiber volume fraction, V_f. Tension.

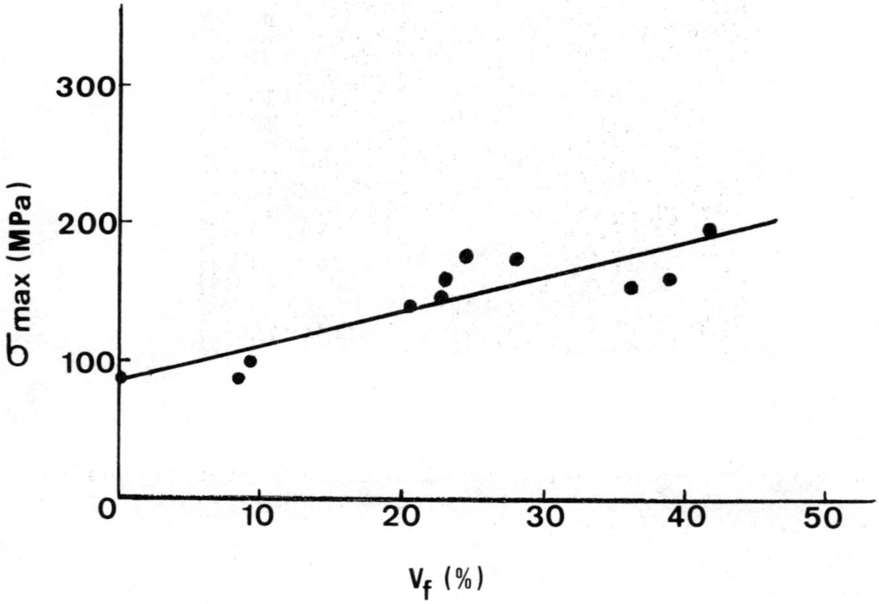

Fig.4 - Variation of the composite maximum strength, σ_{max}, with jute fiber volume fraction, V_f. Bending.

The fracture surfaces of composites were examined in an SEM in the secondary electron mode at 10 kV and after sputter-depositing a thin layer of gold to avoid the charging problem in SEM. The fracture surfaces of composites showed a mixture of fiber breakage in the matrix plane and fiber pull out. A typical scanning fractograph of a composite is shown in Fig.5. Notice the form of the fiber cross-section. One notes that some fibers are projecting out of, or have left holes in, the matrix fracture surface while others have broken more or less in the matrix fracture surface. The amount of fiber pull out increased with increasing V_f. Presumably, this was due to the difficulty of adequate penetration of resin at larger V_f's (V_f > 25%) and the consequent poor bonding at the interface. However, one should point out that as compared to sisal fiber/polyester composites (Chawla, 1976) there was a smaller incidence of fiber pull out in these jute/polyester composites.

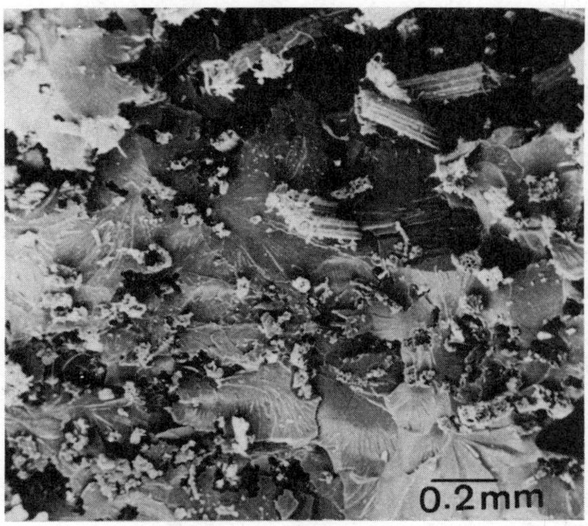

Fig.5 - A typical scanning electron fractograph of a composite showing fiber pull out and some fiber breakage in the matrix fracture surface.

CONCLUSION

In conclusion one can say that the incorporation of jute fibers in the polyester matrix leads to a linear increase in strength and modulus as a function of the fiber volume fraction. The fracture surface showed a mixture of fiber pull out and fiber breakage in the matrix fracture plane indicating that the adhesion between jute/polyester was better than that between sisal/polyester (cf. Chawla, 1976). Promising through the results are, one must recognise that other tests, particularly with respect to microbial decomposition of any kind of vegetable fibers and their weathering resistance, need to be carried out.

ACKNOWLEDGEMENTS

This work was supported by the Brazilian Ministry of Army, CNPq and FINEP. We thank Eng. J.R. Moraes d'Almeida for assistance in fiber testing.

REFERENCES

Barkakaty, B.C. (1976). J.App.Polymer Sci., 20, 2921.

Chawla, K.K. (1976). Proc.Intl.Conf.Mech. Behavior of Materials II, Amer.Soc.Metals, Metals Park, Ohio, p.1920.

Chawla, K.K., R.R.Avillez, R.R.Rodrigues, A.C.M.Sá, R.G.Serra and L.G.P.L.Cavadas (1978), Revista Brasileira Tecnologia, in press.

The New Encyclopaedia Britannica. (1975). Macropaedia, vol.7, Encyclopaedia Britannica Inc., Chicago, p.271.

Volume 3, ICM 3, Cambridge, England, August 1979

MECHANICAL BEHAVIOUR OF FIBRE-STRENGTHENED MATERIALS CONTAINING A HOLE

T. Hirai and Teruo Kimura

Department of Mechanical Engineering, Doshisha University, Kyoto, Japan

INTRODUCTION

F.R.P. plate reinforced by roving glass cloth is widely used for structural elements because of its superior mechanical properties. There have been many reports on analytical researches on the mechanical behaviour of composite materials, but none of these investigations has succeeded in solving problems concerning the structural effects of reinforcement, such as stress concentration and interlayer separation, because of a too macroscopic or too simple assumption of the homogeneous and anisotropic characteristics of these materials derived from their fibre orientation. The assumed representation of their homogeneous constitutive behaviour simplifies analysis of their mechanical behaviour. However, it is an error to fail to recognise the essentially heterogeneous characteristics of composite materials. Moreover the non-linear mechanical behaviour of the F.R.P. plate is characterized by a "knee" in the stress-strain or load-strain curve. Various models to account for the "knee" have been devised (1) (2), but they have not yet gone beyond the assumption of homogeneous characteristics. Therefore it is necessary to develop a mechanical model which clearly represents these heterogeneous characteristics.
In this report, a model is constructed by beam elements and plane elements according to the arrangement of reinforcement and matrix respectively so as to make it amenable to finite element methods (F.E.M.). An attempt is also made to explain the mechanism of stress transmission and the interlayer separation mechanism encountered in F.R.P. plate reinforced by roving glass cloth, and having a circular hole under uniaxial loading.

FINITE ELEMENT MODEL AND FORMULATION

Figure 1 shows the unit cross section of the F.R.P. plate reinforced by roving glass cloth. In this figure, the warp, the weft and the resin parts are shown by '1', '2' and '3' respectively, and it should be noted that the three parts constitute a three dimensional cross linking structure. In this report, a finite element model was used to analyse this structure. The fabric structure of the F.R.P. plate is treated as a space rahmen element having a three dimensional structure, and the resin is treated as a triangular plane element having a two dimensional structure to avoid computational inefficiency. Moreover, an imaginary spring element is established to represent a linking mechanism for transmission of the compressive force between the warp and the weft, as shown in the unit model illustrated in Fig. 2.

As shown in Fig.3, the finite element mesh for the F.R.P. plate with a circular hole can be given by assembling the above unit model. One quarter of the plate was analysed, with appropriate symmetry conditions along the centerline boundaries of the model. The material constants for each element will be explained in the next section. The following equations show the finite element formulation in brief: The strain energy for the space rahmen element U_1 can be formulated as follows:

$$U_1 = \frac{1}{2}\int_V [(E^t\varepsilon^t + E^b\varepsilon^b)(\varepsilon^t + \varepsilon^b) + G\gamma_{yx}^2 + G\gamma_{zx}^2]. \qquad (1)$$

Considering the difference between tensile and bending properties for composite materials, the strain is separated into axial strain ε^t and bending strain ε^b in the above equation. The strain energy for the plane element U_2 can be formulated as follows:

$$U_2 = \frac{1}{2}\iint \{\varepsilon\}^T\{\sigma\}\cdot t\cdot dx\cdot dy \qquad (2)$$

where 't' represents the thickness of the element.
The total strain energy U can be obtained by summing the U_1 and U_2 as:

$$U = U_1 + U_2 . \qquad (3)$$

Minimizing the total potential energy taking account of the finite strain, the force equilibrium of the system of the finite element can be formulated as follows:

$$([K_1]_o + [K_2]_o + [K_1]_G + [K_2]_G)\cdot\{\phi\}^e - \{F\}^e = 0 \qquad (4)$$

where $\{\phi\}^e$ denotes the nodal point displacement, $\{F\}^e$ denotes the nodal point force, $[K_1]_o$ and $[K_2]_o$ are the respective elastic stiffness matrices for the space rahmen element and the plane element, and $[K_1]_G$ and $[K_2]_G$ are the respective geometrical stiffness matrices for the space rahmen element and the plane element. The compressive property of the spring is superposed on the $[K_1]$.

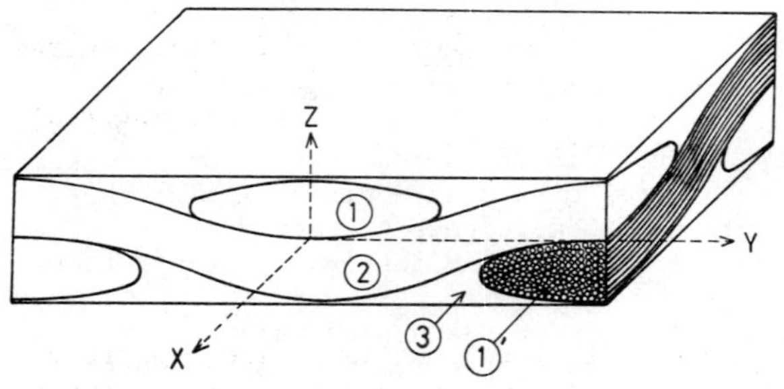

Fig. 1. Unit cross section of the F.R.P. plate reinforced by roving glass cloth.

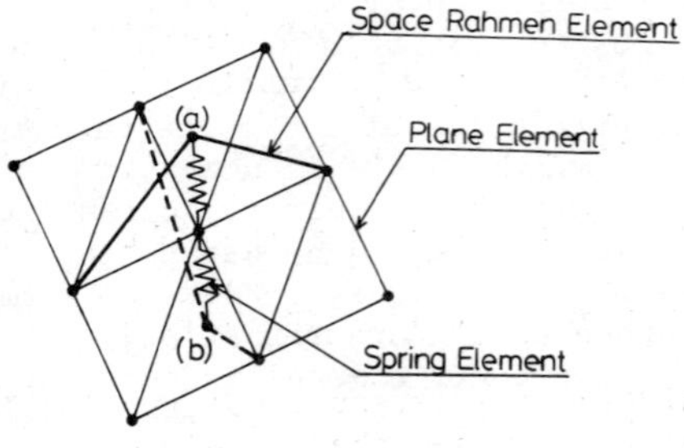

Fig. 2. Unit model.

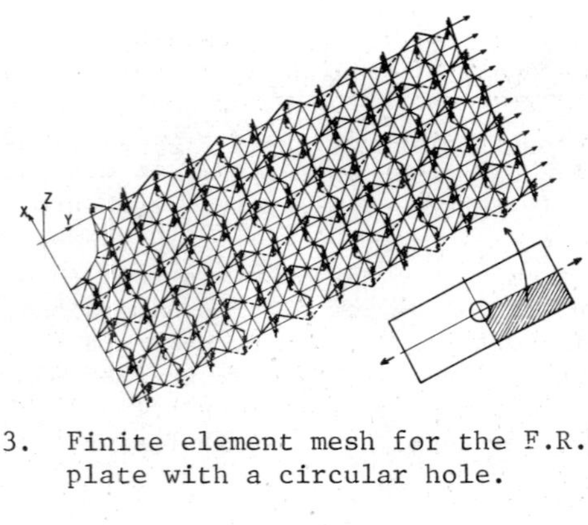

Fig. 3. Finite element mesh for the F.R.P. plate with a circular hole.

MATERIAL CONSTANTS AND MATERIAL NONLINEARITY FOR CALCULATION

The width of bundle of the glass fibre (EWR 55) used in this experiment is about 4 mm. The bundle is treated as a space rahmen element in the F.E.M. Therefore, the material properties used in the F.E.M. are measured by bending and tension tests for a bundle of the glass fibres which is saturated by resin. The property of the spring element which represents the transmission of the compressive force between the warp and the weft is measured by a compression test for the bundle descrived above. In this case, the direction of the load is vertical to the fibre orientation. The material properties measured by the tests are as follows:

warp part: E(tension)=19.6 GPa, E(bending)=10.3 GPa, E(spring)=660 N/mm,

weft part: E(tension)=21.6 GPa, E(bending)=11.2 GPa, E(spring)=765 N/mm,

resin part: E=1.55 GPa, ν=0.4, G=0.55 GPa.

The thickness of the plane element (i.e. resin part) used in the F.E.M. is decided by the following equation:

$$\text{Thickness} = \frac{A\times(1-\alpha)-C\times(1-\beta)(1-0.5)}{B} \qquad (5)$$

where A: volume of F.R.P. plate,
B: area of F.R.P. plate,
C: volume of space rahmen element,
α: volume fraction of the fibre in the F.R.P. plate and
β: volume fraction of the fibre in the space rahmen element.

Thus the resin saturated in the interstices of a fibre bundle will also affect the behaviour of the perpendicular direction to the axis of fibre orientation, but the effect cannot be represented fully by the space rahmen element. Therefore the thickness which is equivalent to the effect of the above resin is added to the thickness of the resin part consisting of plane element shown by '3' in Fig.1.
A similar numerical analysis is also done on a homogeneous orthotropic material, and the mechanical properties used in this case are as follows:

warp direction: E=8.50 GPa, ν=0.241, G=2.94 GPa,

weft direction: E=7.85 GPa, ν=0.221, G=2.94 GPa.

Analysing the stress concentration for the F.R.P. plate, the mechanism of stress transmission and the place where the crack occurs are considered, but the fracture of the F.R.P. plate can not be represented by the analysis. Thus the phenomenon of interlayer separation is analyzed by making the following suppositions in the F.E.M. calculation.

Interlayer separation is supposed to be caused by the distortion between the warp and the weft. In the analysis of the interlayer separation, equation (4) is calculated by using the incremental method, and the progress of interlayer separation is investigated. It is always difficult to judge when interlayer separation occurs, but in the present investigation, its estimation is carried out by calculating the distance between the top point of the warp (a) and the mating point of the weft (b) in each incremental step. If the distance exceeds the constant value decided by the experimental results, it is assumed that interlayer separation has occurred in the element examined. The point (a) and (b) are shown in the unit model in Fig.2. In the succeeding incremental step, the value of the shearing modulus of the resin part is introduced by considering the decreasing value caused by the interlayer separation. The effect of the resin in the weft part on the stiffness in the tensile direction seems to be influenced by the occurrence of interlayer separation. If interlayer separation has occurred, the effect of the resin in the weft part is removed from the stiffness of the resin element, as represented by the plane element in the calculation. This effect is anisotropic with respect to the stress, but is negligible for the compressive stress field along the loading direction. Moreover, the bending modulus for the rahmen elements is varied at the place where the crack occurs.

EXPERIMENTAL METHOD

Observation of the stress concentration was by the photoelastic technique. To compare the effects of the textile structure of the roving glass cloth and the effects of the discontinuity of the glass fibre bundle with the above numerical results, specimens consisting of one layer of roving glass cloth were selected. The diameter of the hole was made an integral number of bundle widths, that is, 4, 8, and 12 mm. The orientation of the warp coincides with the direction of the uniaxial load. The dimensions of the specimen for the tensile test are as follows:

width=80 mm, gauge length=80 mm and thickness=1 mm.

The volume fraction of the reinforcement is about 23.2%. For observation of the interlayer separation, the specimen with a hole of 12 mm diameter was selected for the tensile test. In this case, the dimensions of the specimen were as follows:

width=40 mm, gauge length=100 mm and thickness=1 mm.

RESULTS OF ANALYSIS AND COMPARISON WITH EXPERIMENT

In the F.R.P. plate with a hole, the mechanism of stress transfer becomes complex because the bundle of glass fibres is cut off. Figure 4 and Fig. 5 show the axial stress in the warp and the weft, respectively, which are obtained by the F.E.M. based on assuming heterogeneous characteristics of the F.R.P. plate. From these figures, the axial stress in the distontinuous fibre bundle is smaller than that of the continuous one, and the tendency to this phenomenon depends on the size of the hole relative to the dimension of fibre bundles in the F.R.P. plate Then the value of stress around the hole becomes smaller than in homogeneous materials.

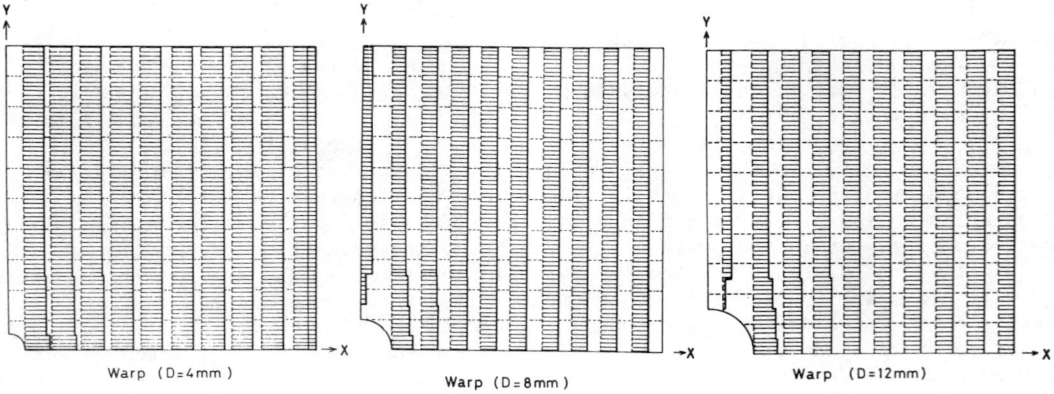

Fig. 4. Axial stress transmitted in the warp.

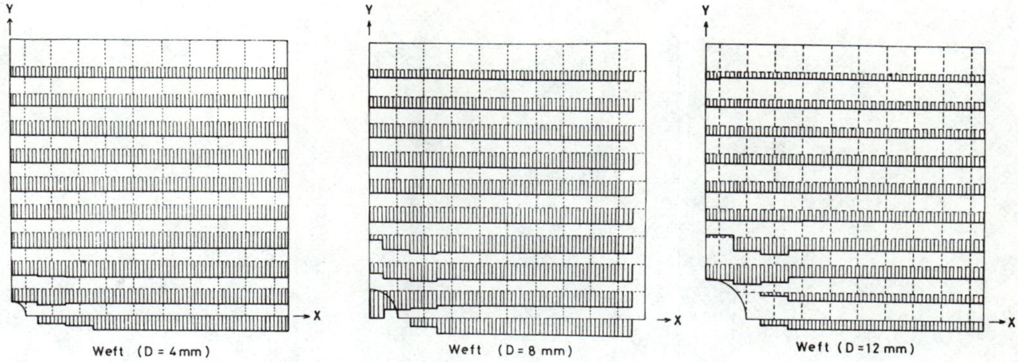

Fig. 5. Axial stress transmitted in the weft.

Figure 6 shows the relation between the form factor α_{max} and the diameter of a hole. The value of α_{max} obtained by F.E.M. assuming homogeneous characteristics for the composites is larger than that obtained by experiment because the stress transmission mechanism in the discontinuous fibres is neglected. The results obtained by F.E.M. based on the heterogeneity of the composite agree with the experimental results. In this figure, the stress concentration factor α is defined by σ_n / σ_o , where σ_n is the normal stress with respect to the tangential direction around the circular hole and σ_o is the uniform tensile stress remote from the hole. As discussed above, the mechanism of stress transfer around the hole was made clear, but it is necessary to analyse the problem of interlayer separation The fracture mechanism of the F.R.P. plate reinforced by roving glass cloth depends on this phenomenon. The stress-strain response obtained in the tensile test is shown in Fig. 7, and Fig. 8 shows the state of the interlayer separation in three progressive loading steps. As shown in case 1, the first stage of interlayer separation occurred at the corner of the circular hole near the horizontal centre line of the specimen. Prior to this however, minute cracks occurred in the weft fibre bundles and these hair cracks could be observed as a white thin line in the spots. On developing the cracks over the unit element, the Young's modulus of the spot decreased. In the model used in the present investigation, this effect was treated by considering varied characteristics of the resin equivalent to the spots. In the large stress field (case 3), the interlayer separation extended to both edges of the specimen shown in Fig. 8.

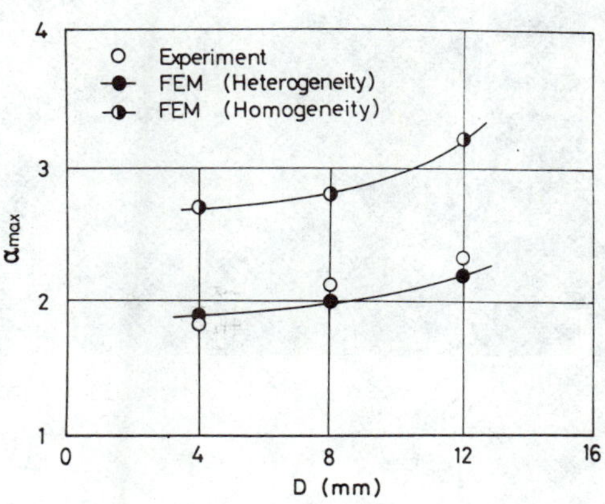

Fig. 6. Effect of a hole diameter to the maximum stress concentration factor.

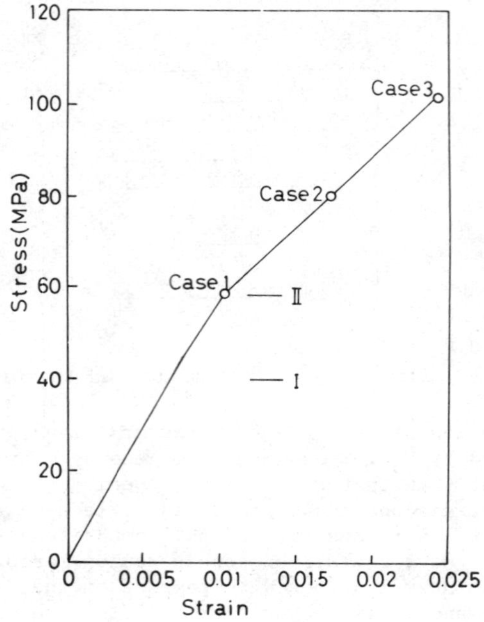

Fig. 7. Stress-strain curve in the case of 0° orientation of roving glass cloth.

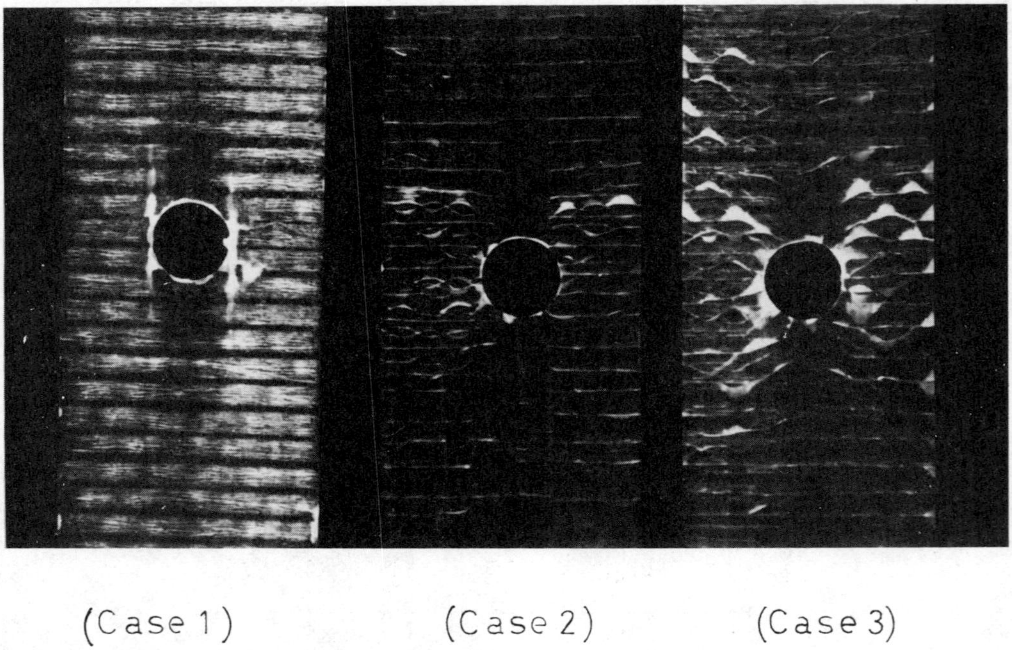

Fig. 8. Aspects of interlayer separation in the case of 0° orientation of roving glass cloth.

The calculated load-displacement response is shown in Fig. 9 with experimental results for comparison. They agree reasonably well. Figure 10 shows the progress of interlayer separation calculated by F.E.M., and good agreement was obtained with the experimental results shown in Fig. 8.

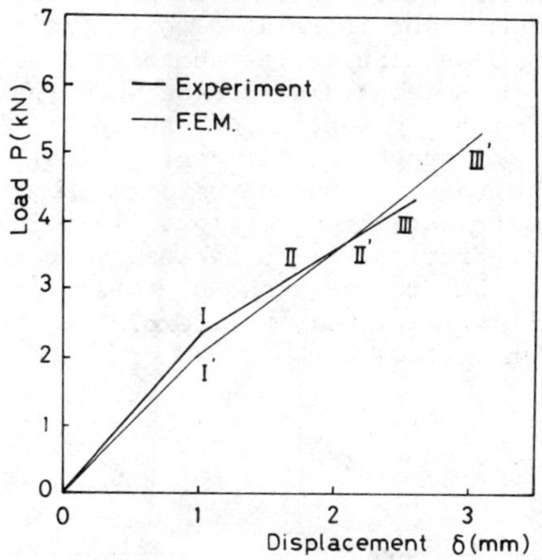

Fig. 9. Load-displacement curve in the case of 0° orientation of roving glass cloth.

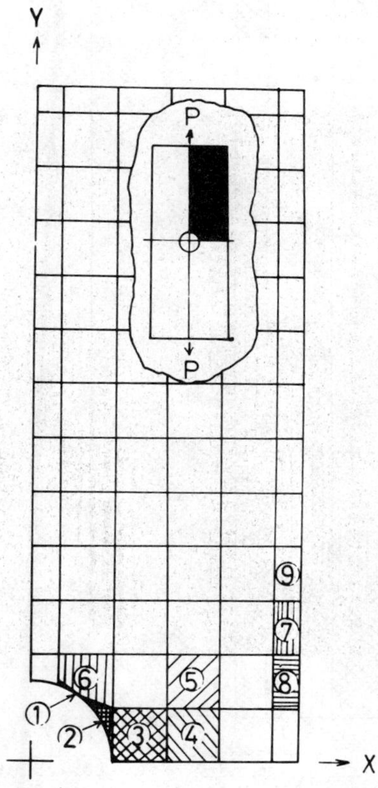

Fig. 10. Progress of interlayer separation calculated by F.E.M.

The distribution of the varying distance between the top point of the warp and the mating point of the weft at the final stage of the calculation is shown in Fig. 11. In this figure, for example, type 1 and type 4 indicate the regions of largest and smallest separation between these two points, respectively. This figure shows the possibility of interlayer separation but, in the experiment, no such separation could be found along the centre line parallel to the loading axis as shown in Fig. 8. It seems that interlayer separation is related to the hair crack orientation with respect to the loading direction by its anisotropic characteristics for the stresses. The regions having a large distance between the top point of warp and the mating point of the weft in the x-direction are shown in Fig. 12. In this figure, interlayer separation does not appear at the centre line. Then it is understood that the interlayer separation depends on the relative distance parallel to the crack orientation in relation to the loading direction. In this way, to gain an exact result for the interlayer separation, it is necessary to consider the direction of the displacement of bundles, but in the scope of this report, the development of interlayer separation can be sufficiently well explained by relative distance no matter what the direction of the displacement.

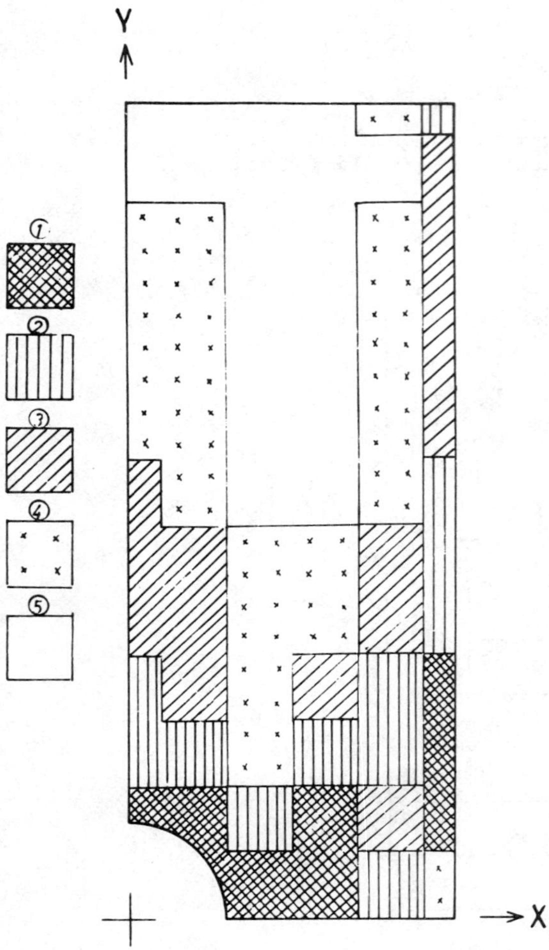

Fig. 11. Distribution of the varying distance between the top point of the warp and the mating point of the weft.

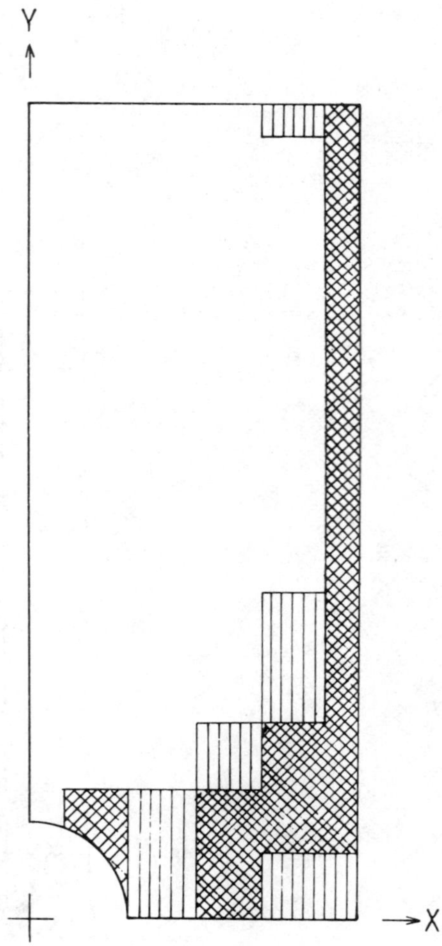

Fig. 12. Distribution of large distance between the top point of the warp and the mating point of the weft in the x-direction.

SUMMARY AND CONCLUSIONS

In this report a model based on the heterogeneity of F.R.P. plate reinforced by roving glass cloth was developed to explain the mechanism of the stress transfer phenomenon and the phenomenon of interlayer separation. The model made it possible to explain successfully the complex phenomena of heterogeneous materials such as composites. In the F.R.P. plate with a hole the stress concentration factor α around the hole is smaller than that of a homogeneous material because of its bundle being cut off. The value of α_{max} nearly equals 2 for the specimen with a hole whose diameter is twice the width of the fibre bundle. By calculating the relative distance between the top point of the warp and the mating point of the weft, the progress of interlayer separation was simulated, to allow a valid explanation of the experimental results. It was also confirmed that the separation extended from the edge of the circular hole to the outer edge of the specimen. The simulation model proposed for the F.R.P. plate reinforced by roving glass cloth is found both practical and sensible, and the results obtained from this model are adequate for engineering purposes.

ACKNOWLEDGMENT

The authors' thanks are due to Mr. T. Katayama, Osaka Prefectural Industrial Research Institute, and Mr. T. Nishikawa, Osaka Gas Co.,Ltd. for their helpful discussions.

REFERENCES

1. S.C.Chou, O.Orringer and J.H.Rainey (1976). Post-Failure Behavior of Laminates: I-No Stress Concentration. J.Composite Materials,10, October, 371-381.
2. S.C.Chou, O.Orringer and J.H.Rainey (1977). Post-Failure Behavior of Laminates: II-Stress Concentration. J.Composite Materials,11, January, 71-78.

Volume 3, ICM 3, Cambridge, England, August 1979

A RELIABILITY ANALYSIS APPROACH TO FATIGUE LIFE DISPERSION OF LAMINATED GLASS FIBER COMPOSITE MATERIALS

T. Tanimoto, S. Amijima and H. Ishikawa*

Department of Mechanical Engineering, Doshisha University, Kyoto, Japan
***Department of Statistics, School of Business, University of Kagawa, Kagawa, Japan**

INTRODUCTION

An investigation was conducted to study the statistical nature of fatigue life of laminated glass fiber composite materials, hereinafter referred to as "FRP". The materials used in experiments were four different kinds of FRP laminates, satin woven FRP, plain woven FRP, roving woven FRP and SMC laminated FRP. Additionally unidirectional carbon fiber reinforced plastic was used for comparison purposes. Sufficient data were accumulated to produce the P-S-N diagram for each type of laminate. A statistical approach based on Weibull distribution was applied to the test data to evaluate the dispersion in the fatigue life of the material. The fatigue life and its dispersion at varying applied stress level, specimen size and glass fiber content in the specimen was discussed. Test results showed that the widest scatter was observed at the life range of 10^5 to 10^6 cycles for every kind of FRP. Consequently sufficient attention should be paid to this for the design and application of FRP.

Glass fiber reinforced plastics are becoming more extensively used as structural materials, machine parts and other material parts. Considerable design data for this material has been generated. However the majority of published fatigue results for reinforced plastics do not include replicate tests so that the true extent of scatter has been suppressed. Generally the result of a fatigue test is considered to have scatter inherently. A good understanding of statistical aspects of fatigue properties is essential for the successful application of such materials because these composites possessless material uniformity as compared with conventional metallic materials.
The purpose of this paper is thus to report the experimental results for fatigue life of various types of laminated FRP and to draw some statistically meaningful inference from these results.

SPECIMEN PREPARATION AND EXPERIMENTAL METHOD

The materials used were four kinds of laminate consisting of the same matrix and different types of reinforcement. The composition of the materials used is shown in Table 1, while that of carbon fiber reinforced plastics is shown in Table 2. All the specimens were prepared from laminates produced in the university laboratories by a hand-lay-up method. The plies of the glass cloth of the specimen were changed for the test in order to obtain different glass content. The glass fiber contents for each laminate are shown in Tables 3 and 4. Those laminate were cured for four hours at 80°C. After being removed from the oven, it was cut in the direction parallel to the warp of glass fabric to the dimensions shown in Fig. 1. The pieces were then kept for at least two days prior to testing under the constant conditions of 23°C and 65% humidity.

Fatigue tests with repeated tension and compression were conducted at a speed of 2000 r.p.m. and the mean stress values were changed at time of testing in order to examine the effect of mean stress on the fatigue life. The frequency of rotating bending fatigue tests (Cantilever type) was 3600 r.p.m..

TABLE 1 Composition of FRP Specimens

Matrix	Polyester resin (Polylite FG-284)
Reinforcement	Satin woven glass cloth (SLS-212)
	Plain woven glass cloth (MG-252)
	Woven roving glass cloth (RH-402)
Catalyst	Methyl ethyl ketone peroxide
Hardener	Cobalt naphthenate 0.6% solution

TABLE 2 Composition of Carbon Fiber Reinforced Plastics

Matrix	Epoxy resin (EPIKOTE 828)
Reinforcement	TORAY carbon fiber (T 300)
Catalyst	BF_3MEA

TABLE 3 Volume Content of Glass Fiber for Each Type of F.R.P.

Kind of F.R.P.	Weight content W_f(%)	Volume content V_f(%)	Tensile strength σ_t(MPa)	Bending strength σ_b(MPa)
Satin woven F.R.P.	54.47	35.78	294	394
Plain woven F.R.P.	51.72	33.28	223	244
Roving woven F.R.P.	57.14	38.30	319	270
	62.63	43.83	326	294
	66.23	47.73	347	334

TABLE 4 Composition of SMC Specimen (Weight Content)

Matrix	Polylite	34%
Reinforcement	Glass roving (1 in.)	30%
Filler	$CaCO_3$	34%
Hardener etc.		1.6%
	MgO	0.4%

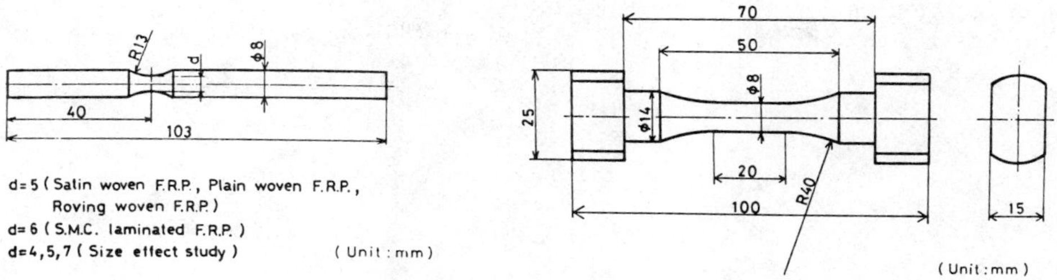

(a) Rotating bending fatigue test specimen.

(b) Repeated tension and compression fatigue test specimen.

Fig. 1. Fatigue test specimens.

TABLE 5 Statistical Analysis of Fatigue Life of Plain Woven F.R.P.

σ(MPa)	69	78	98	118
T(i) 1	21.047×10^5	16.92×10^4	2.00×10^4	0.68×10^4
2	59.744×10^5	23.70×10^4	2.40×10^4	0.90×10^4
3	78.687×10^5	28.40×10^4	2.70×10^4	0.97×10^4
4	135.041×10^5	94.00×10^4	5.00×10^4	0.97×10^4
5	156.271×10^5	100.97×10^4	5.06×10^4	1.02×10^4
6	178.713×10^5	148.46×10^4	5.09×10^4	1.12×10^4
7	222.349×10^5	176.81×10^4	5.30×10^4	1.15×10^4
8	379.395×10^5	233.16×10^4	6.73×10^4	1.37×10^4
9	454.994×10^5	725.73×10^4	28.62×10^4	1.39×10^4
α (MLE)	1.37	0.915	1.13	5.61
β (MLE)	2.05×10^7	1.64×10^6	7.39×10^4	1.15×10^4
α (LSM)	1.04	0.839	1.46	4.59
β (LSM)	2.15×10^7	1.65×10^6	6.89×10^4	1.16×10^4
E[T] (LSM)	2.12×10^7	1.81×10^6	6.25×10^4	1.06×10^4
(MLE)	1.87×10^7	1.71×10^6	7.06×10^4	1.07×10^4
V[T] (LSM)	4.15×10^{14}	4.72×10^{12}	1.90×10^9	6.86×10^6
(MLE)	4.19×10^{14}	3.49×10^{12}	4.56×10^9	4.05×10^6
$\sqrt{V[T]}$ (LSM)	2.04×10^7	2.17×10^6	4.36×10^4	2.62×10^3
(MLE)	2.05×10^7	1.87×10^6	6.75×10^4	2.01×10^3
CV (LSM)	0.962	1.20	0.698	0.247
(MLE)	1.10	1.09	0.956	0.188

RESULTS AND DISCUSSIONS

Scatter of the data in this paper was analyzed by using a two-parameter Weibull (1951) distribution. It is characterized by a probability density function f(t) and the associated cumulative distribution function F(t) as follows.

$$f(t)=\frac{\alpha}{\beta}\left(\frac{t}{\beta}\right)^{\alpha-1}\exp\left\{-\left(\frac{t}{\beta}\right)^{\alpha}\right\} \qquad (1)$$

$$F(t)=1-\exp\left\{-\left(\frac{t}{\beta}\right)^{\alpha}\right\} \qquad (2)$$

Where α and β are shape parameter and scale parameter respectively.

Mean life, variance and coefficient of variation of the two- parameter Weibull distribution are denoted here by E[T], V[T] and CV repectively and are shown below.

$$E[T]=\int_0^\infty t\cdot f(t)dt=\beta\cdot\Gamma(1+\frac{1}{\alpha}) \qquad (3)$$

$$V[T]=E[T^2]-(E[T])^2$$
$$=\beta^2[\Gamma(1+\frac{2}{\alpha})-\Gamma^2(1+\frac{1}{\alpha})] \qquad (4)$$

$$CV=\sqrt{V[T]}/E[T]$$
$$=\sqrt{\Gamma(1+\frac{2}{\alpha})-\Gamma^2(1+\frac{1}{\alpha})}/\Gamma(1+\frac{1}{\alpha}) \qquad (5)$$

The results of statistical analysis on the actual fatigue life data for plain woven FRP are shown in Table 5 as an example. Estimation of parameters, α and β in the table was performed by the Least Squares Method (LSM) and Maximum Likelihood Estimation (MLE). No significant difference was found in the estimated values by either method as expected. Accordingly the discussion hereinafter is carried out based on the estimate of parameters by LSM.
Fig. 2 shows a representative fatigue life distribution in rotating bending fatigue tests on plain woven FRP plotted on Weibull probability paper.
The mean rank defined by

$$F(t)=i/(n+1) \qquad (6)$$

was used to plot the distributions. Here i is the failure order number and n is the total number of samples in each test.
The linearity of the test data for stress amplitudes of 118 MPa and 69 MPa supports the choice of the Weibull distribution. However the results for the stress amplitudes of 98 MPa and 78 MPa imply the possibility of a bimodal life distribution. It is well known that FRP materials exhibit complex fracture mechanisms, being a combination of fiber failure and delamination at the fiber-resin interface (Tanimoto, 1975). At lifetimes of 10^5 to 10^6 cycles, the fatigue life of FRP would tend to separate into two populations caused by two typical cases where either fiber failure or delamination at the fiber-resin interface mainly controls the final fracture of the specimen. This may partly contribute to the greater variance at 10^5 to 10^6 cycles as will be discussed later.

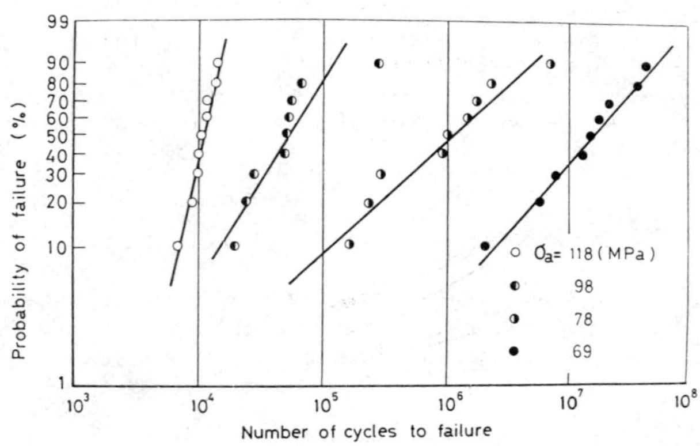

Fig. 2. Fatigue life distribution of plain woven F.R.P.

P-S-N Diagram

The P-S-N diagrams for plain woven FRP for the different probabilities of failure, 5, 10, 50, 90 and 95 percent is shown in Fig. 3 as an example. It is generally believed that the fatigue dispersion at lifetimes near the endurance limit is widest in the common structural metals. However according to our experimental results on FRP, the widest scatter was observed at the life range of 10^5 to 10^6 cycles. The trend in the P-S-N diagram was the same for every kind of FRP as shown in Figs. 4 through 7. It is actually difficult to produce the P-S-N diagram for carbon fiber reinforced plastics due to excessive experimental scatter so that the mean life and its dispersion are shown in the S-N diagram as a substitute for it. It was also revealed that no rigorous fatigue limit within the cycle range of the present test was observed and the slope of S-N curves is not flattened out even over 10^7 cycles.

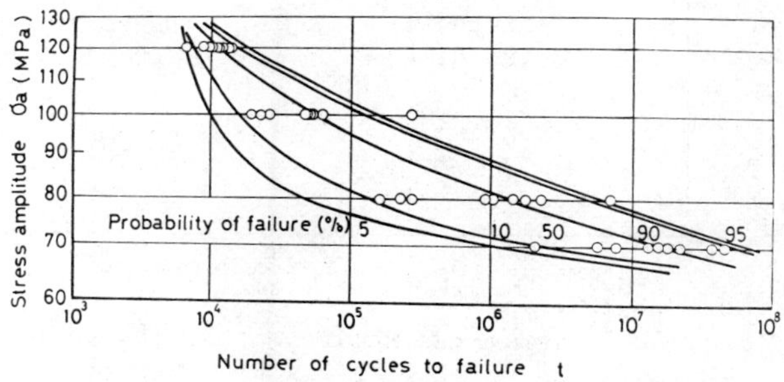

Fig. 3. P-S-N diagram of plain woven F.R.P.

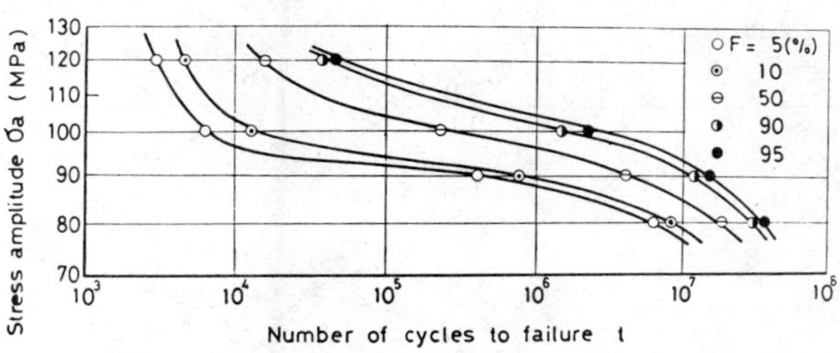

Fig. 4. P-S-N diagram of satin woven F.R.P.

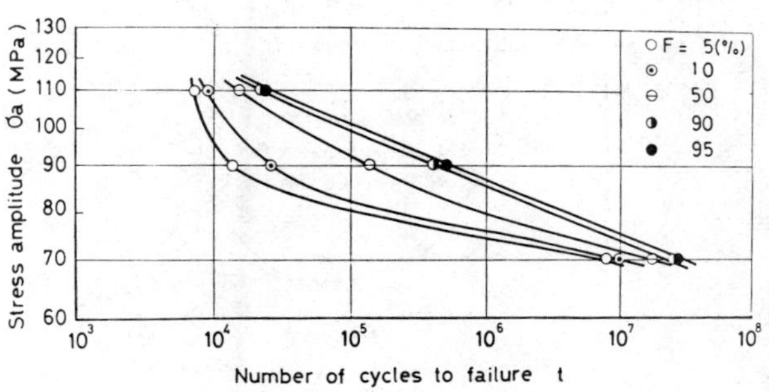

Fig. 5. P-S-N diagram of roving woven F.R.P. (V_f=38.3%).

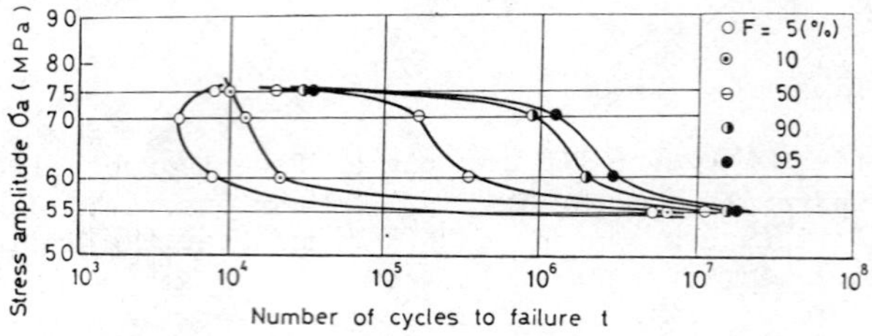

Fig. 6. P-S-N diagram of SMC laminate.

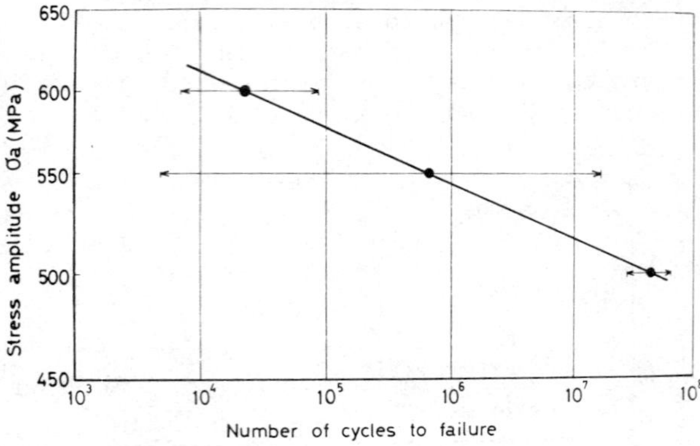

Fig. 7. S-N diagram of unidirectional carbon fiber reinforced plastics (Vf=56.0%).

Fig. 8 shows the Weibull shape parameter α and coefficient of variation C.V. versus expected fatigue life E[T]. It is more evident from this figure that the scatter in the fatigue life becomes widest at 10^5 to 10^6 cycles for every kind of FRP. This tendency in the dispersion of fatigue life at varying stress levels is extremely important and deserves much attention for the design and application of FRP.

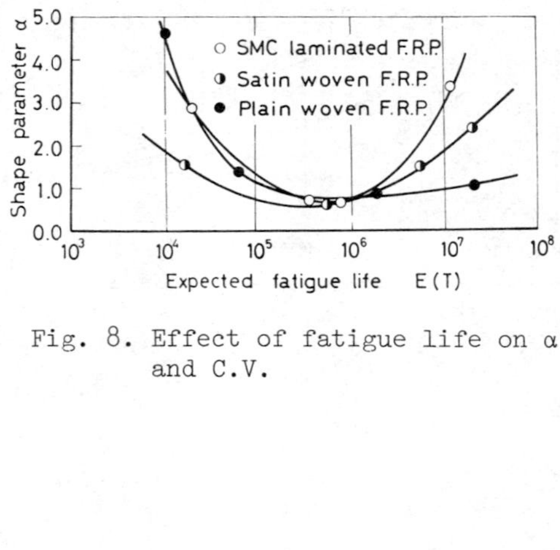

Fig. 8. Effect of fatigue life on α and C.V.

S-N Diagram

Fig. 9 shows a series of S-N diagrams for various types of FRP used in our experiments. The experimental equations showing the relation between expected life and applied stress amplitude(in MPa) according to the least square linear estimation are shown below for each type of FRP.

Satin woven FRP :	$\sigma T^{0.0552}=223.90$	(7)
Plain woven FRP :	$\sigma T^{0.0697}=218.11$	(8)
Roving woven FRP :	$\sigma T^{0.0646}=199.38$	(9)
SMC laminated FRP :	$\sigma T^{0.0510}=123.67$	(10)

No rigorous fatigue limit was observed for any kind of FRP within 10^7 cycles. S-N diagrams of roving woven FRP for the various glass contents are shown in Fig. 10. In the figure estimated values of shape parameter for each stress amplitude are indicated for reference. The dispersion in the fatigue life is widest at the medium stress level, while those at high and low stress levels are comparatively small.

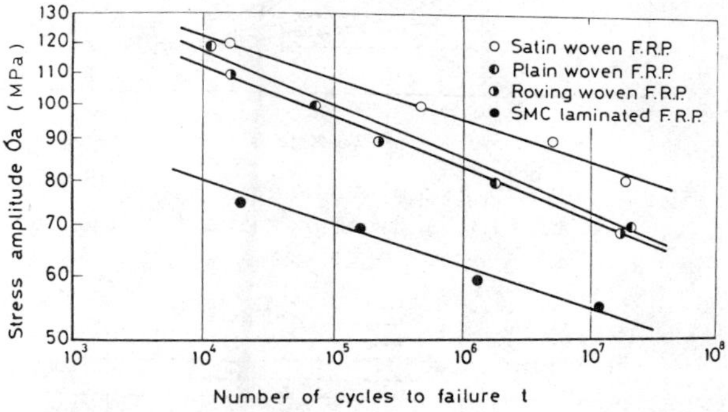

Fig. 9. S-N diagram for various kinds of F.R.P.

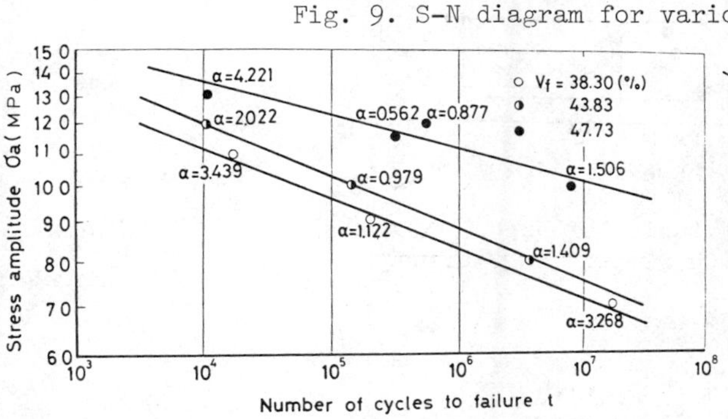

Fig. 10. S-N diagram of roving woven F.R.P. at various V_f.

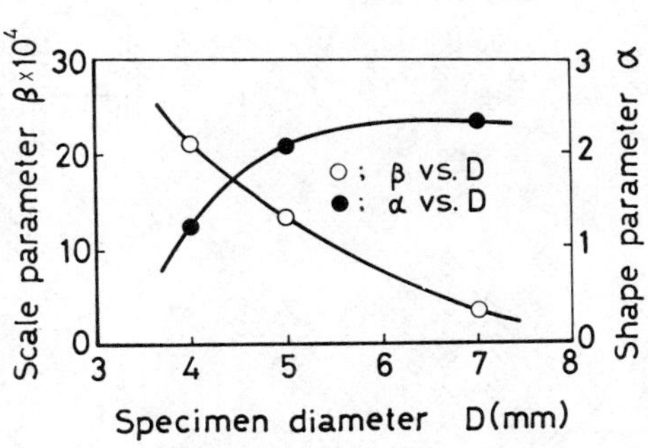

Fig. 11. Size effect on Weibull parameters α and β.

Effect of Specimen Size

Effect of specimen diameter on the fatigue life and its dispersion is shown in Fig. 11. It can be seen from the figure that an increase in the specimen diameter causes a reduction in the fatigue life of the composite, on the other hand the value of α, which is an inverse measure of dispersion, increases with increasing specimen diameter.

Effect of Mean Stress

Experimental results of repeated tension and compression fatigue tests at zero mean stress were plotted on the P-S-N diagram as shown in Fig. 12. The trend was obtained in this P-S-N diagram as was seen in the rotating bending fatigue test. In other words, variation of the lives is comparatively small at the higher stress levels and the lower stress levels, while greater variances appeared at the medium stress levels.
Fig. 13 shows the variation of Weibull shape parameter with respect to expected fatigue life at varying mean stress. The value of α tends to increase with an increase of mean stress. Clearly fatigue life dispersion is widest under zero mean stress.

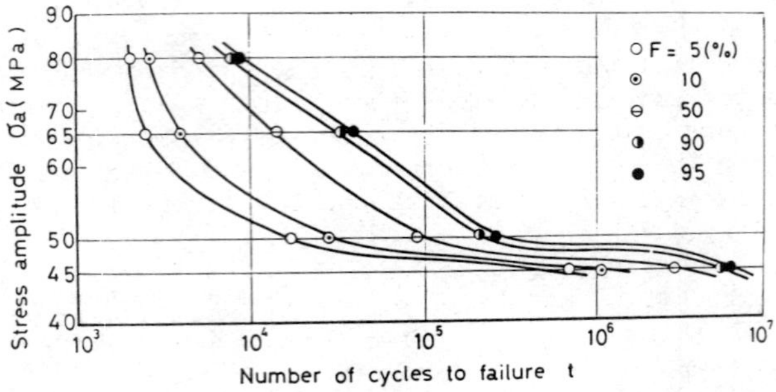

Fig. 12. P-S-N diagram of satin woven F.R.P. in tension and compression fatigue test at zero mean stress.

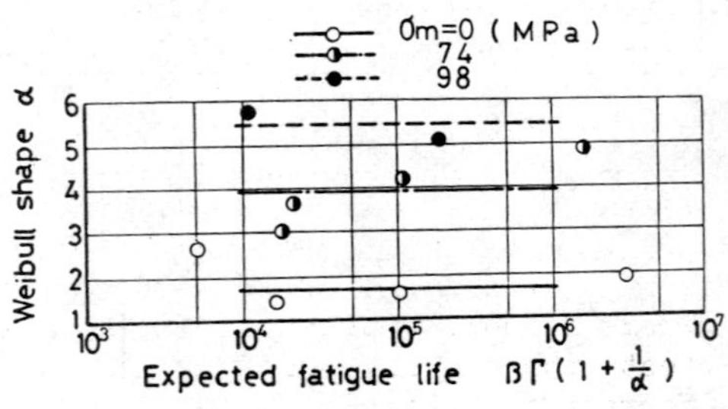

Fig. 13. Variation of Weibull shape α with respect to fatigue life.

Reliability Based Design Safe Life

Safe design life (Hahn, 1970; Knight, 1975) based on the concept of TTFF (time to first failure) in statistics of extremes is now considered. If the shape parameter α is known, the scale parameter β can be estimated by calculating the maximum likelihood estimator $\hat{\beta}$ from the available test data. A lower bound estimate $\breve{\beta}_\gamma$ with confidence level γ is then related to $\hat{\beta}$ by

$$\breve{\beta}_\gamma = \frac{\hat{\beta}}{\left\{\frac{1}{2n}\chi_\gamma{}^2(2n)\right\}^{1/\alpha}} \tag{11}$$

Where $\chi_\gamma^2(2n)$ is the γ-percentile of the chi-square distribution with 2n degrees of freedom.
Now consider a fleet size N, each member of which has the same Weibull distribution of fatigue life with the lower bound scale parameter $\breve{\beta}_\gamma$. The reliability function based on the TTFF concept is represented as $R=\bar{R}^N$, where R is the reliability of entire fleet (weakest member).
Then the design or safe life of one member fleet $\breve{T}_R$ with the desired reliability R is written as

$$\breve{T}_R=\breve{\beta}_\gamma\{\ln(1/R)\}^{1/\alpha}=\breve{\beta}_\gamma\{\tfrac{1}{N}\ln(1/\bar{R})\}^{1/\alpha}$$

$$=\hat{\beta}\left\{\frac{2n}{N}\,\frac{\ln(1/\bar{R})}{\chi_\gamma^2(2n)}\right\}^{1/\alpha}=\hat{\beta}/S \qquad (12)$$

Scatter factor S can be subdivided into three factors, S_n, S_N and $S_{\bar{R}}$.

$$S=S_n\cdot S_N\cdot S_{\bar{R}} \qquad (13)$$

Where $S_N=N^{1/\alpha}$, $S_n=[\chi_\gamma(2n)/(2n)]^{1/\alpha}$, $S_{\bar{R}}=(\ln(1/\bar{R}))^{-1/\alpha}$

S_N, S_n and $S_{\bar{R}}$ are called fleet size factor, sample size factor and reliability factor respectively. All the scatter factors depend on the shape parameter.
Figs. 14 through 16 show S_n, $S_{\bar{R}}$ and S_N based on the estimated values of α for satin woven FRP. Shaded areas represent the changing band of scatter factors for various stress levels. A 95 percent lower bound confidence level was used with equation (12) to find the lower bound life. In the figures each value for low carbon steel (Silicon killed steel, CH10K) is shown for comparison purposes.
The sample size factor which depends on the number of samples and selected confidence level decreases slowly as the number of samples increases.
The reliability factor increases rapidly as high reliabilities are imposed on the weakest member.
The fleet size factor is extremely significant for larger fleet sizes.
The shape parameter for FRP is much lower than those for low carbon steel CH10K indicating wider scatter in fatigue life. Consequently, a higher safety factor should be chosen when designing with FRP than other materials and larger reduction of design life is necessary to assure the same level of confidence and reliability. Furthermore, a considerable reduction of design life is required for FRP in case of designing so as to protect the weakest member of a larger fleet size and this may prove to be a major deterrent to using FRP to the structures of a larger fleet size. It would be essential to pay sufficient attention to this point for the application of FRP as structural material.

CONCLUDING REMARKS

An experimental investigation was conducted to study the statistical fatigue properties of laminated composites under rotating bending and repeated tension and compression. The dependences of fatigue life and its dispersion on the applied stress level, specimen size and fiber content in the specimen were discussed.
The test results showed a wider experimental scatter particularly at the life range of 10^5 to 10^6 cycles than that observed in conventional metallic materials. Thus it is emphasized that a good understanding of such statistical aspects of fatigue properties is essential for the optimum utilization of this material.

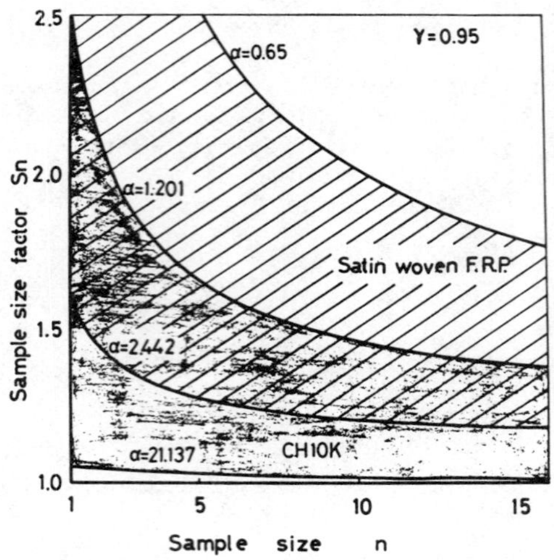

Fig. 14. Domain of sample size factor S_n of satin woven F.R.P. and CH10K as functions of sample size n and Weibull shape parameter α.

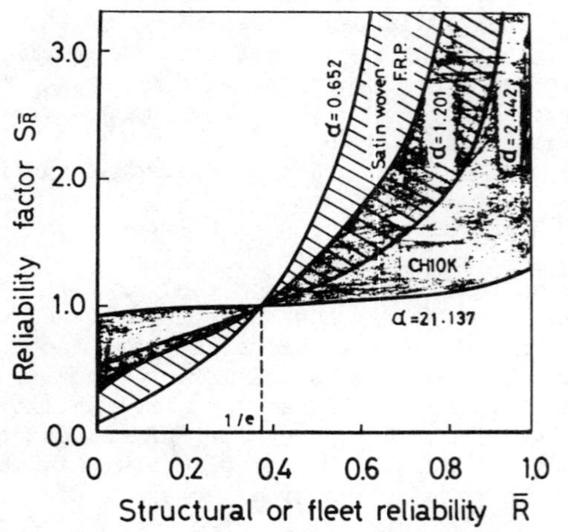

Fig. 15. Domain of reliability factor $S_{\bar{R}}$ of satin woven F.R.P. and CH10K based on TTFF as functions of structural or fleet reliability $\bar{R}$ and Weibull shape parameter α.

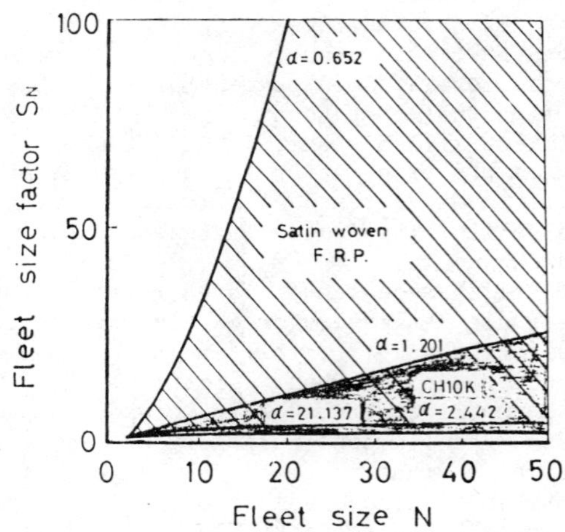

Fig. 16. Domain of fleet size factor S_N of satin woven F.R.P. and CH10K based on TTFF as functions of fleet size N and Weibull shape parameter α.

REFERENCES

[1] Weibull, W. J. (1951). J. Applied Mechanics, 18, 293.
[2] Tanimoto, T., and S. Amijima (1975). J. Composite Materials, 9, 380.
[3] Knight, M., and H. T. Hahn (1975). J. Composite Materials, 9, 77.
[4] Hahn, J. C., J. R. Kopf, and W. Goldberg (1970). J. Composite Materials, 4, 462.

Volume 3, ICM 3, Cambridge, England, August 1979

ASSESSMENT OF FATIGUE DAMAGE IN THE CYCLIC LOADING OF A GLASS-FIBRE POLYESTER COMPOSITE

E. W. Smith and K. J. Pascoe

University Engineering Department, Trumpington Street, Cambridge CB2 1PZ, U.K.

INTRODUCTION

In metals, once a fatigue crack has initiated, the progress of the propagating crack can be monitored and empirical laws exist to predict fatigue lives. In contrast, the fatigue failure of fibre composite materials is not dominated by the propagation of a single crack: progressive damage will occur in the bulk of any highly strained area and realistic quantitative assessment of damage can be made only by monitoring some material property. Further, fatigue failure in glass-fibre reinforced plastics (g.r.p.) is not confined to one mode but may comprise debonding of fibres transverse to the loading direction, matrix cracking, matrix void growth, delamination, fibre pull-out, fibre buckling and fibre failure. The presence of any mode will depend on the reinforcement, the matrix, the fibre/matrix bond and the loading condition. Early laboratory research on the fatigue of g.r.p. (dating from the work of Boller (1952)) looked upon specimen separation as failure. Smith and Owen (1969) felt that specimen separation was not an adequate design criterion and split fatigue failure into three phases: fibre debonding, matrix cracking and specimen separation. Later workers quantitatively measured progressive damage by crack density counts (Broutman and Sahu, 1969), fall in material modulus (Agarwal and Dally, 1975), increase in material opacity, (Endo and Watanabe, 1971), ability to absorb water, (McGarry and Desai, 1959), acoustic emissions, (Mehan and Mullin, 1971), and temperature rise (Stinchcomb and others, 1975).

In the references cited little or no attempt is made to distinguish between modes of damage. Are the progressive trends of different techniques, loading modes or materials the same or does each mode of failure exhibit its own characteristics? This paper will attempt to shed some light on the first two of these queries by comparing the physical appearance of progressive macrodamage with cyclic stress/strain behaviour, extent of hysteretic heating and specimen opacity for a woven roving glass-fibre polyester composite subject to uniaxial and biaxial loading.

EXPERIMENTAL BACKGROUND

All the fatigue tests discussed in this presentation, apart from one, were carried out under sinusoidal loading with maximum stress levels causing permanent damage in the first quarter cycle. A biaxial hydraulic servo-controlled rig was used which had been developed at the Cambridge University Engineering Department primarily for the high strain fatigue testing of steels (Parsons and Pascoe, 1975).

Fully reversed cyclic loading was limited to frequencies of 1 Hz and less but as will be noted later this limitation was no detriment. The test material was made up of 13 laminations giving a specimen thickness of 12 mm and the weave laid orthogonally. Uniaxial specimens were 50 mm wide and biaxial specimens of cruciform shape had a 55 mm square working section. Both employed duralumin plates bonded to the loading arms to prevent premature grip fracture. One low stress uniaxial fatigue test at 16.7 Hz was made using a Losenhausen fatigue machine. Specimen strain was monitored by means of a special strain-gauged extensometer of 25 mm gauge length bonded to the specimen surface (Smith and Pascoe, 1979). Specimen temperature was measured using a Chromel T1/Alumel T2 thermocouple bonded to the specimen surface. Specimen light transmissibility was measured using an unexpanded 2 mW laser beam normal to the surface. A light-sensitive cell was placed on the other side of the specimen and directly opposite. The progressive nature of fatigue damage was observed through a lower power microscope.

Tests were made with twelve states of biaxial stress of which three are of particular importance: equibiaxial loading where there is no in-plane shear; uniaxial loading with the fibres at 45° to the loading direction, in this case equal levels of normal and shear stress are acting on the fibre plane; and in-plane shear loading with the applied loads at 45° to the fibre weave, i.e. a state of pure shear acting on the fibres with no component of normal stress.

ASSESSMENT OF PROGRESSIVE FATIGUE DAMAGE

Physical Observations

Three mechanisms or progressive modes of fatigue damage were recognised under the twelve different loading states:

(i) Rectilinear cracking, delamination and finally lamina instability causing failure in compression.
(ii) Shear deformation along the plane of the fibres causing breakdown of the fibre-matrix bond.
(iii) Combined rectilinear cracking and matrix shear deformation on the fibre plane.

When the fibre axes and the principal stress axes coincide only the rectilinear cracking mechanism of failure is evident. Cracking is perpendicular and to a lesser degree parallel to the major principal stress directions. For in-plane shear with the applied loads at 45° to the fibre weave the failure mechanism is one of pure shear along the fibre plane. For all other off-axis loading cases there appears to be a combined mechanism of rectilinear cracking and matrix shear.

Cyclic Stress/Strain Behaviour

At low stress levels little physical damage appears in the first cycle, but with cycling the accumulation of rectilinear cracking or shear degradation and whitening can be identified easily. However, more direct means of corroborating the failure mode are evident from the cyclic stress/strain behaviour. The rectilinear cracking/delamination mode of failure is characterised by:

(1) A distinct 'knee' in the tensile half cycle of the first cycle stress/strain loop which disappears on subsequent loading.

(2) A linear stress/strain relationship in the compressive half cycle of the stress/strain loop.

(3) Little or no hysteresis in the cycling stress/strain loop even at very high

stress levels.

(4) A progressive build up in strain range (load cycling) in zero-mean-stress loading which is almost entirely confined to the tensile range. The consequence is a stress/strain loop which is distinctly crooked.

(5) A fatigue life and stress/strain behaviour which are not affected by frequency or rest periods at least in the frequency range 0.1 to 1 Hz.

The shear deformation mode is characterised by:

(1) A stress/strain curve showing a low yield point irrespective of the sign of the principal stresses.

(2) A true yield point which does not disappear in the second or subsequent cycles so that the stress/strain loop shows marked hysteresis.

(3) A progressive build up in principal strain range under zero-mean-stress loading which accumulates equally in both the tensile and compressive half cycles.

(4) A fatigue life and stress/strain behaviour which are markedly dependent on cyclic frequency and rest periods in the frequency range of 0.005 to 1 Hz.

Figure 1 compares the progressive stress/strain loops for these two distinctive modes. It is often convenient when comparing the increase in strain range ($\Delta\varepsilon$) with fatigue cycles (N) for different stress levels to plot $\Delta\varepsilon$ v. log N. But use of a log scale masks three prevalent phases of strain accumulation (Fig. 2(a)):

(i) an early phase in which the strain range, particularly in tension, increases rapidly: the rate of increase gradually falling to lead into
(ii) the longest phase where the strain range increases at a steady rate, and
(iii) the final stage in which the strain range (both tension and compression) increases exponentially to final failure.

These phases are characteristic of a creep curve and will be referred to as primary, secondary and tertiary phases respectively, as in the creep of metals.

Hysteretic Heating

Because of the restrictions imposed by the biaxial test rig on the maximum fre-

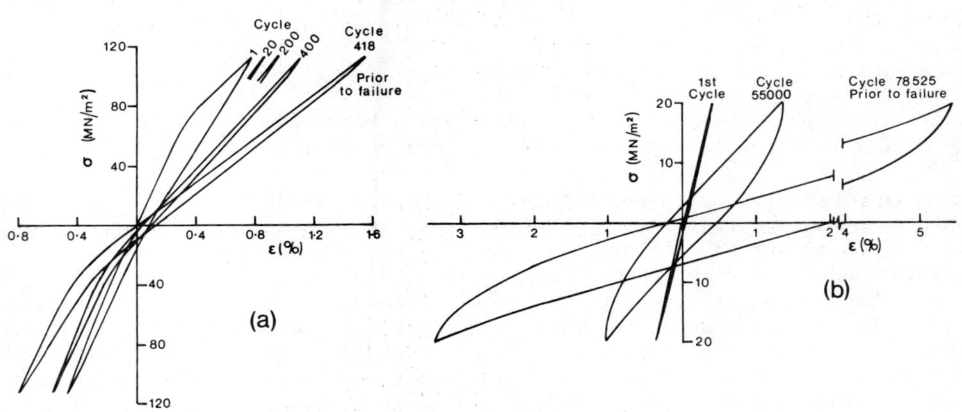

Fig. 1 Progressive stress strain loops for (a) equibiaxial loading and (b) shear loading with the fibres aligned with the maximum shear plane.

quency all the fatigue tests were carried out at frequencies between 0.005 and 0.7 Hz (an exception was a high cycle uniaxial test made on a different rig). This is in contrast to many of the test programmes which can be cited in the literature where frequencies are much higher. For instance, Boller (1952) used a test frequency of 15 Hz with forced air cooling and Smith and Owen (1969) tested at 1.1 Hz. Though the number of high cycle tests has been restricted, the relatively low test frequencies have led to a better understanding of hysteretic heating effects in relation to stress biaxiality and fibre direction.

When loading along the fibre axes, no hysteretic heating effects were apparent in the frequency ranges 0.02 to 0.7 Hz. In contrast, hysteretic heating effects were found to be of prime importance when carrying out off-axis fatigue tests. The effects were manifest as cyclic heating for high frequencies and high stress levels and cyclic creep for very low frequencies, so that in a range of test frequencies from 0.005 to 0.6 Hz there appeared to be an optimum frequency where the fatigue strength could be a maximum. Monitoring of surface temperature showed that the temperature at failure could be of the order of 20 K or more above ambient even though test frequencies were normally less than 0.3 Hz. The rise in surface temperature with fatigue cycles was not linear and did not show a steady state condition but showed the same characteristic phases as the accumulation in strain range (compare Figs 2(a) and 2(b)).

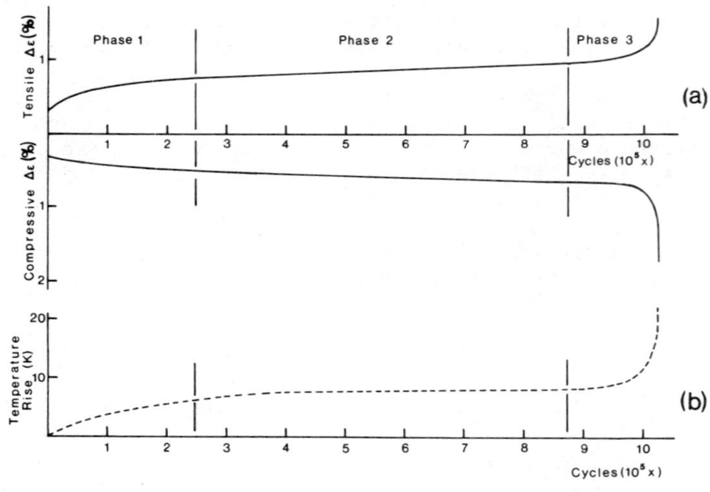

Fig. 2 Three phase pattern of increase in strain range and surface temperature for uniaxial cyclic loading (± 30 MN/m^2 and 0.4 Hz at 45° to the weave).

Only one comprehensive series of fatigue tests was made where different specimens were subjected to the same stress level, but at various frequencies. A stress level of 50 MN/m^2 and 45° off-axis loading gives a nominal life of the order of 3000 cycles and tests with frequencies as low as 0.005 Hz could be completed in one week. Thus seven different frequencies were considered at this stress level from 0.005 Hz up to 0.4 Hz where hysteretic heating caused a rapid degeneration in strength. The results are detailed in Table 1. Fatigue lives varied from 698 to 4738 cycles and the shortest, at the highest test frequency, showed the highest temperature rise, but the longest life was not at the lowest frequency and did not show the smallest temperature rise. The change in life due to cyclic creep is not as marked as for cyclic heating. The appearance of fracture is also affected more by cyclic heating than by cyclic creep. The optimum frequency for this loading situation and stress level is very near to 0.05 Hz when fracture appears on a plane at 45° to the fibres and along a through-thickness shear plane. This mode is explained by the simultaneous growth of rectilinear cracking and fibre/matrix shear degradation; both tend to cause fracture to be aligned with the fibre plane but the former promotes delamination and hence a through-thickness shear component of fracture. Increasing the frequency causes the shear dependent damage mode to predominate; the extent is cumulative as both the modulus and flow stress, though

TABLE 1 Hysteretic Heating Effects Due to Different Test Frequencies for Uniaxial 45° Off-Axis Loading and 50 MN/m^2 Semi-Stress Range

Frequency (Hz)	Strain Range (%) at Failure		Cycles to Failure	Surface Temp. at Failure (K above ambient)
	Tension	Compression		
0.005	4.02	2.98	2,754	0
0.01	4.11	2.86	3,165	0.9
0.02	3.25	2.55	3,480	7.8
0.05	3.89	3.14	4,738	10.5
0.05	3.95	3.28	2,172	17.5
0.1	4.12	3.34	2,016	28.8
0.2	4.18	3.55	798	31.7
0.4	4.39	3.72	698	43.5

essentially the latter, decrease with rising temperature, thus causing an increase in strain range, a further increase in matrix damage and so more heat generation. The mode of fracture becomes more defined as the proportion of rectilinear cracking to matrix/fibre shear degradation drops and at a frequency of 0.4 Hz any rectilinear cracking is barely discernible even at failure, and fracture appears as a distinct shear plane resembling in concept Lüders lines in the plastic deformation of metals with apparently undamaged material above and below the plane of fracture. Conversely, very low frequencies cause a complete break-up of the specimen centre-section, rectilinear cracking will progress at the normal rate whereas the time dependent deformation of the fibre/matrix interface will be exaggerated by the combined effects of creep and the presence of cracks aligned along the shear plane.

Material Opacity

The accumulation of macro-damage can be seen with the naked eye as a gradual increase in specimen opaqueness and the measurement of this opaqueness showed the specimen light transmissibility to drop markedly in the first cycle and then to a lesser extent in successive cycles. This decay, as shown in Fig. 3, was approximately exponential and prior to failure the amount of light transmitted was only a very small fraction of that for an undamaged specimen.

PROGRESSIVE DAMAGE AND PREDICTION OF FATIGUE

Both Broutman and Sahu (1969) and Owen and Howe (1972) have measured the density of cracking and fibre debonding in cyclic uniaxially loaded g.r.p. Both references report a three-phase increase in crack density of similar trend to the three-phase increase in strain range and hysteretic heating noted for the test g.r.p. Whether or not the increase in strain range under load-controlled fatigue testing is due solely to the accumulation of cracks is open to conjecture; the part played by delamination can only be surmised. Cracks running perpendicular to

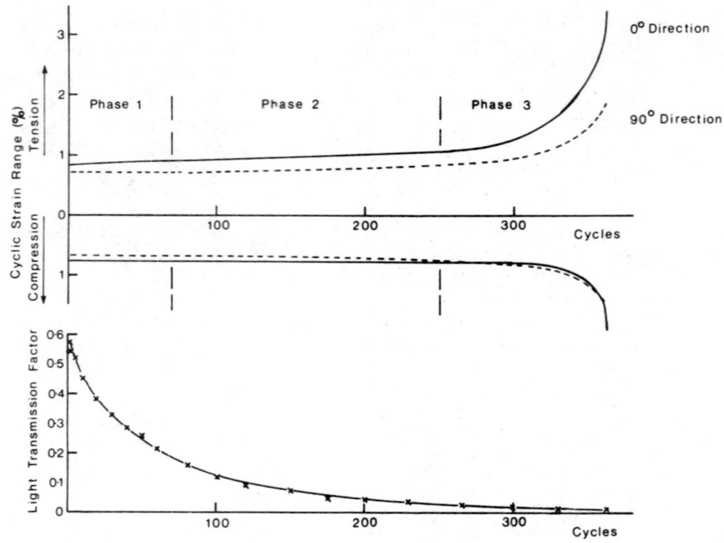

Fig. 3 Comparison of light transmissibility and cyclic strain range (cyclic shear loading of ±100 MN/m^2 and 0.02 Hz with principal stresses aligned with fibres

a principal stress will mainly affect the tensile strain range (due to crack opening) and cause only a little change (if any) in the compressive strain range, whereas fibre/matrix shear degradation will be unaffected by the sign of the principal stress, only the level of the in-plane shear stress along the fibre plane affecting the strain range. In the initial and secondary phases of strain accumulation the compressive strain range for loading along the fibre axes increases only slightly, but in the tertiary phase a rapid increase in the compressive strain range as well as the tensile strain range is witnessed. An obvious explanation for this exponential increase of strain range is the accumulation of delamination, but the beginning of the tertiary phase never appeared to coincide with the early signs of delamination.

The long secondary phase of strain range accumulation shows a linear rate of increase with fatigue cycles and is a minimum for any cyclic stress level. This pattern is similar to a creep curve for metals and Monkman and Grant (1956) showed for a large number of alloys that there is a reasonable correlation between the secondary or minimum creep rate and the creep rupture life. A similar type of relationship was considered for the fatigue life (N_f) of the g.r.p. and the minimum or secondary strain range accumulation rate ($\dot{\varepsilon}_{min}$), that is

$$\log N_f = p \log \dot{\varepsilon}_{min} + c$$

where p and c are constants.

The relationship was applied to the sum of the tensile and compressive strain range accumulation rates for uniaxial loading and the extremes of biaxial loading; equibiaxial loading and shear with the principal load axes at 45° to the fibres. The 'least square fit' lines are shown in Fig. 4. The straight line log-log relationship is shifted in the direction of increasing $\dot{\varepsilon}_{min}$ as the off-axis angle is increased but the gradient, i.e. the constant p, is reasonably constant for all loading conditions. The amount of test data backing up this relationship is small but its use in predicting high cycle fatigue strengths or the probability of a fatigue limit ought to be considered. The fatigue strength is dependent on cyclic test frequency, particularly for off-axis loading and negative load ratios, and an optimum frequency for 45° uniaxial loading to minimise cyclic heating and cyclic creep is estimated to be of the order of 1 Hz for a stress level giving a life of 10^8 cycles, i.e. a test lasting $3\frac{1}{4}$ years! If the constant p is established from short life tests, then high cycle fatigue lives can be estimated once a test has progressed into the secondary phase.

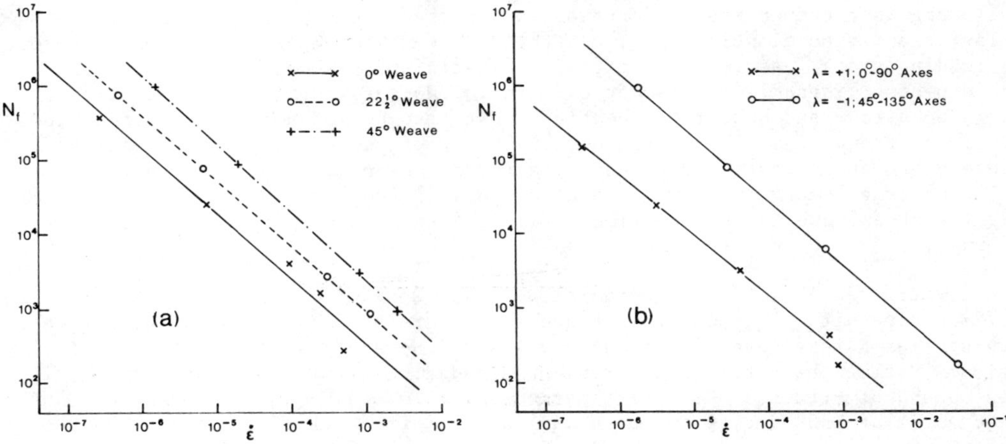

Fig. 4 Logarithmic plots of fatigue life versus minimum rate of strain range increase (a) uniaxial, (b) biaxial

COMPONENTS OF DEFORMATION

So far, only the modes of permanent progressive damage have been mentioned, but it is also possible to break down *deformation* into five distinct components:

(1) Elastic deformation of both fibres and matrix, rate independent and recoverable.

(2) Plastic deformation of the matrix (micro-damage), rate independent and non-recoverable.

(3) Deformation due to macro-damage, rate independent and non-recoverable:

(i) rectilinear cracking (combined resin cracking and fibre/matrix debonding),
(ii) shear degradation of the fibre/matrix interface,
(iii) delamination.

(4) Anelastic deformation of the matrix, rate dependent and recoverable.

(5) Linear and non-linear viscoelastic deformation of the matrix, rate dependent and part recoverable.

The presence of each depends on the loading state and weave direction. Loading along the fibre axes results in the elastic component, the macro-damage components (i) and (iii) and a linear component of viscoelastic deformation. A pure shear stress along the fibre axes involves all the above components except the macro-damage (i). The general off-axis loading case results in all five components of deformation. Quantification of each component is extremely difficult apart from the elastic component as the other four components are dependent on the cyclic stress history and the anelastic and viscoelastic components are time or rate dependent.

DISCUSSION

Fatigue damage in composites has a deleterious effect on the material's physical properties well before final failure. It is quite possible to make some assessment of this damage but to use this assessment to predict the progress of further damage and failure is not easy due to the involved nature of deformation. Of the five recognised components of deformation two are rate dependent, they are the anelastic and viscoelastic parts, and their extent is related to the weave direction and loading state. As recovery of the anelastic and viscoelastic deformation is time dependent, dynamic assessment of material damage, that is, cyclic stress/strain behaviour and hysteretic heating, must also be influenced by rest periods as well as frequency. At low frequencies the rest period will cause complete recovery of any anelastic component of deformation and at high frequencies where cyclic heating occurs there will be a recovery in the material's elastic modulus and flow stress due to the specimen's temperature returning to that of the surroundings.

An uninterrupted 45° off-axis uniaxial test at 50 MN/m^2 and 0.2 Hz showed a 'life' of 798 cycles with a total strain range at failure of 7.73%. But a further test with similar stress level and frequency was deliberately subjected to a rest period each time the total strain range reached 3.6% so allowing recovery of all time dependent strains. Four rest periods from 10 minutes to 24 hours were introduced until at 1350 cycles the test was run to failure (1531 cycles). The resultant behaviour is outlined in Fig. 5. Stopping the test to allow recovery of time dependent strains would appear to cause a substantial increase in fatigue life, in fact if the practice of stopping the test each time the strain range accumulated to a level of 3.6% was continued there is little reason to doubt that the fatigue life for this particular condition would be increased ten-fold or more. In comparison Riddell, Koo and O'Toole (1966) in the testing of unreinforced thermoplastics showed that high frequency cyclic stressing did not cause a true fatigue failure and that after cooling and annealing much of the material's strength and stiffness was regained.

Prolonged testing in order to simulate actual material working conditions is time consuming and consequently expensive. As long as working conditions give slightly longer lives than would be expected from laboratory testing it seems somewhat futile to carry out sustained tests in order to represent a specific condition. Though, conversely, highly accelerated testing may prove unrepresentative and so inefficient due to premature failure. When loading along the fibre axes no frequency effects in the range 0.02-0.7 Hz were apparent. However, in an attempt to carry out an accelerated high cycle fatigue test a uniaxial specimen was cycled at 16.7 Hz. The specimen failed prematurely through cyclic thermal softening. No rectilinear cracking was evident and, unlike the appearance of failure at low frequencies, all damage was localised to a small area around a defined fracture plane. Boller (1952), amongst others, found that the fatigue failure of g.r.p. was affected by heat generation when the test frequency was high (15 Hz) and overcame the problem by forced air cooling of the specimen. This may be a possible remedy whilst recognising only cyclic thermal softening due to linear viscoelastic effects. However, for off-axis loading the stress/strain behaviour is no longer linear and non-linear viscoelastic behaviour will play an increasing part as the off-axis angle is increased or the in-plane shear component made greater. Then the rate-dependent cyclic stress/strain behaviour and the fatigue life are not only affected by heat generation but cyclic creep and stress relaxation will play a part in failure: the latter will not be present under forced cooled high frequency testing and may therefore show over-conservative results.

It has been shown that progressive damage can be estimated by continuous measurement of the cyclic stress/strain behaviour and the fatigue life predicted from the rate of strain range accumulation. However, from the discussion above any prediction ought to be treated with caution. Fatigue life is consequent on stress history: stress level and state, fibre angle, frequency and rest periods. The constant p of the empirical relationship $\log N_f = p \log \dot{\varepsilon}_{min} + c$ is seen to vary with stress state and fibre angle, but it is also frequency dependent and probably influenced by rest periods.

One real purpose of measuring progressive damage is for application to the in-service inspection of structures. Any method must be non-destructive; hence crack density counts by material sectioning can be ruled out. Further, crack density counts are only reliable for assessment of damage due to loading normal to the fibres: they do not quantify the extent of shear degradation. On the other hand, measurement of material stiffness or flow stress, or material opacity does not interfere structurally with the material. The fall in material stiffness or flow stress shows the same three-phase pattern as crack density counts and hysteretic heat build-up. In most cases load/extension measurement of in-service structures is relatively simple, though in the present test programme the direct bonding of strain gauges to a g.r.p. surface was never found reliable, even when only looking at monotonic loading (see Smith and Pascoe, (1979)). In comparing the rate of increase in opaqueness with other damage phenomena the validity of such a technique becomes questionable. The rate of increase in opaqueness with fatigue cycles does not show a three-phase pattern. Instead, as damage progresses so the rate of decay in light transmissibility decreases until in the last 10% of life the change is very small. It would appear that the light scattering properties of the damaged material become saturated as the crack density increases, so that, as a means of assessing fatigue damage the technique loses sensitivity at the most critical part in the fatigue life history.

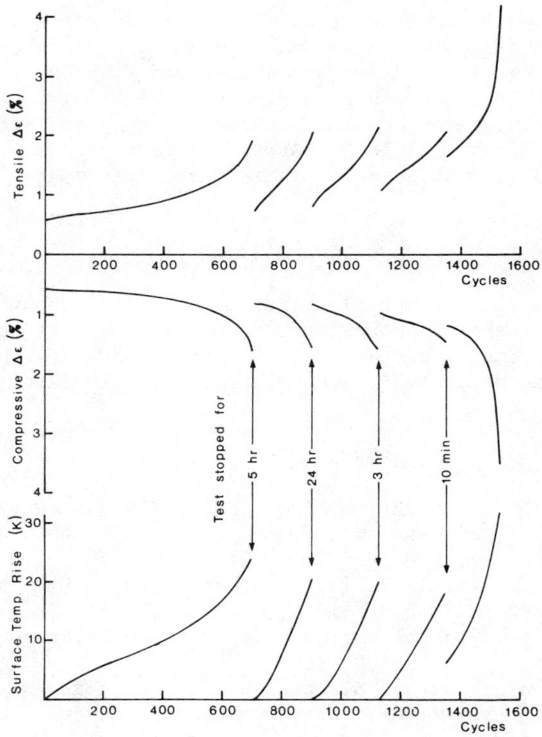

Fig. 5 Influence of rest periods on fatigue life: cyclic strain range and surface temperature rise versus cycles to failure (uniaxial stress 50 MN/m^2, at 45° to weave and frequency of 0.2 Hz).

CONCLUSIONS

Progressive damage and deformation in a woven roving glass-fibre polyester composite subject to uniaxial and biaxial fatigue loading is seen to depend on stress state, weave angle, cyclic frequency and loading history. Damage has been assessed by measuring material stress/strain behaviour, hysteretic heating and material opacity. Accumulation of damage appears in three phases and the dominant phase, the second phase, can be used to predict fatigue life. However, it is found that the acceleration of fatigue tests by using higher frequencies than would be used in service can give rise to misleading data. Frequency effects not only arise from hysteretic heating but cyclic creep and stress relaxation may introduce complications. It is felt that monitoring of material stiffness and flow stress is the best procedure for in-service assessment of progressive damage.

ACKNOWLEDGEMENTS

The experimental programme was carried out at Cambridge University, U.K., with the financial support of the Ministry of Defence. We wish to thank Mr P.R. Christopher, formerly of the Naval Construction Research Establishment, for his help and encouragement; and Mr R.J. Brand for assistance in carrying out the experiments.

REFERENCES

Agarwal, B.B., and J.W. Dally (1975). Prediction of low-cycle fatigue behaviour of GFRP: An experimental approach. *J. Mater. Sci.*, 10, 193-199.

Boller, K.H. (1952). Fatigue tests of glass-fibre base laminates subjected to axial loading. Forest Products Lab. Report No. 1823.

Broutman, L.J. and S.H. Sahu (1969). Progressive damage of a glass reinforced plastic during fatigue. Proc. 24th Conf. SPI Reinforced Plastics Civ. Sect. 11-D.

Endo, K., and M. Watanabe (1971). Fatigue properties of fibrous reinforced plastics. Proc. 14th Japan Congress on Materials Research, Kyoto, 120-123.

McGarry, F.J., and M.B. Desai (1959). Failure mechanisms in fibreglass reinforced plastics. Proc. 14th Conf. SPI Reinforced Plastics Div. Sect. 16-E.

Mehan, R.L. , and J.V. Mullin (1971). Analysis of composite failure mechanisms using Acoustic Emissions. *J. Composite Mat.*, 5, 266-269.

Monkman, F.C. and N.J. Grant (1956). An empirical relationship between rupture life and minimum creep rate in creep-rupture tests. *Proc. A.S.T.M.*, 56, 593-620.

Owen, M.J., and R.J. Howe (1972). The accumulation of damage in a glass reinforced plastic under tensile and fatigue loading. *J. Phys. D: Appl. Phys.*, 5, 1637-1649.

Parsons, M.W. and K.J. Pascoe (1975). Development of a biaxial fatigue testing rig. *J. Strain Analysis*, 10, 1-10.

Kiddell, M.N., G.P. Koo, and J.L. O'Toole (1966). Fatigue mechanisms of thermoplastics. *Polymer Eng. and Sci.*, 6, 363-368.

Smith, E.W., and K.J. Pascoe (1979). To be published.

Smith, T.R., and M.J. Owen (1969). Progressive nature of fatigue damage in reinforced plastics. *Modern Plastics*, 46, 128- .

Stinchcomb, W.W., K.L. Reifsnider, L.A. Marcus, and R.S. Williams (1975). Effects of frequency on the mechanical response of two composite materials to fatigue loads. Fatigue of Composite Materials, ASTM STP 569, 171-188.

Volume 3, ICM 3, Cambridge, England, August 1979

THE MECHANISMS OF INITIATION AND DEVELOPMENT OF DAMAGE IN MULTI-AXIAL FIBRE-REINFORCED PLASTICS LAMINATES

M. G. Bader, J. E. Bailey, P. T. Curtis* and A. Parvizi

Department of Metallurgy and Materials Technology, University of Surrey, Guildford, U.K.
***Materials Department, Royal Aircraft Establishment, Farnborough, Hampshire, U.K.**

INTRODUCTION

Design with fibre-composite materials must be based on an estimation of the properties of the composite from those of the constituent fibrous and continuous (matrix) phases. Current practice is based on the measured properties of uniaxial laminae, which are assumed to be orthotropic so that their elastic properties may be characterised by 5 constants: E_x, E_y, ν_{xy}, ν_{yz}, and G_{xy}. Likewise the lamina strength may be described by its axial and transverse, tensile and compressive strengths and a shear strength. These parameters may alternatively be estimated from the properties of the fibre and matrix by using the more sophisticated semi-empirical approaches of Tsai (1968) and Tsai and Wu (1971), for example.

The properties of a laminate made up from several laminae are then estimated from those of the lamina by the now classical laminate theories, see Ashton, Halpin and Petit (1969) for example. This approach is extremely powerful in that laminate design may be optimised for all possible combinations of service stress conditions. The method, however, suffers from a significant limitation in that the lamina is treated as an anisotropic continuum, and whilst interactions between laminae are considered, interactions <u>within</u> the lamina are ignored ! The present work is concerned with the observation of failure mechanisms in simple laminates and indicates that fibre/interface/matrix interactions have a dominant effect on the early stages of mechanical failure and also that interactions between laminae occur which are not predicted by the classical laminate theory. We conclude that optimum design must acknowledge the significance of these micro-mechanical effects.

In the present work we have manufactured simple laminates with 0° and 90° oriented laminae. These have been loaded in tension and the mechanisms of failure studied. We have identified the mechanisms of failure initiation and propagation, and have established the existence of a number of interactions between laminae which control the first stages of failure. From this work it is possible to formulate some simple <u>rules</u> for the construction of laminates in which failure initiation is significantly suppressed.

EXPERIMENTAL

Laminate Fabrication

Laminates were fabricated from glass-fibre rovings and an epoxy-resin by a wet lay-up process and from a carbon-fibre/epoxy-resin pre-preg which was autoclave cured.

The glass/epoxy (GRP) laminates were made from a 1200 tex, E-glass fibre roving which was wound on to square, open metal frames to produce unidirectional arrays of fibres. These frames were then stacked, one-by-one, on to an aluminium plate (which was smaller than the aperture in the frame), wetted out with the liquid resin at each stage until the required number of layers had been formed. Any air bubbles were then carefully expelled, using a hot air jet to heat the laminate locally to assist the process, and a top glass plate applied. Excess resin was then squeezed out and the laminate consolidated to the required thickness by application of pressure, either by weights or in a hand press. The laminate was cured for 3 to 4 hours at 100°C followed by a post cure of 3 hours at 150°C. The resin used was an anhydride cured diglycidyl ether of bisphenol A with an amide accelerator (Shell Epikote 828 with 80 PHR Epikure NMA and 0.5 PHR BDMA). This formulation was chosen on account of its low viscosity which aided the production of high quality, void-free laminates of good transparency. These glass/epoxy laminates were unidirectional or of a simple cross-ply construction (i.e. 0°-90°-0°) with a fibre volume fraction (V_f) of 0.55, the principal variables were the relative thicknesses of the 0° and 90° layers. The laminate plates were approximately 220 mm square, and parallel sided coupon test pieces, 20 mm wide were cut from these using a water lubricated diamond saw. The ends of the test pieces were reinforced with GRP tabs so that they could be tested without failure at the grips.

The carbon-fibre (CFRP) laminates were fabricated from a commercially available pre-preg of high-strength carbon-fibres in a proprietary poly-functional epoxy resin cured with diaminodiphenylsuphone (DDS), and a boron trifluoride based initiator. The pre-preg was in the form of sheets approximately 1000 x 300 x 0.125 mm with the fibres aligned along the length of the sheet. 300 mm squares were cut from the sheets and stacked to give the appropriate sequence of 0° and 90° layers. The sheets were then cured in an autoclave using a combination of vacuum applied to the pack of laminae and a controlled sequence of external pressure as the temperature was brought up to 170°C. This was designed to produce a void-free laminate with a fibre volume fraction (V_f) of 0.6. Coupon test-pieces, 20mm wide were cut from the laminates and in this case aluminium-alloy end tabs were applied.

In addition to the 0°, 90° series, laminates with all unidirectional sequences were produced.

Testing Procedure

An electrical resistance strain gauge was bonded to the central portion of the gauge length of each test-piece to measure the longitudinal strain (and in some cases the transverse strain) and the test-piece was then extended in tension in an Instron testing machine at a strain rate of about $0.5 \times 10^{-4} s^{-1}$. Load and strain were recorded autographically. An acoustic-emission recording apparatus was also used for some tests, (mainly the CFRP). to monitor the stress waves produced by damage events during the test. A transducer was attached to the central portion of the test-piece and the acoustic-emission rate was recorded by a second pen on the recorder used to monitor load and strain.

The test-piece was also carefully observed during the progress of the test. The GRP specimens were transparent so that whitening and cracking could be detected by eye and correlated with the load/strain and acoustic-emission traces. Some tests were stopped at selected points so that further examination of the test-piece could be carried out using optical or electron-microscopy.

RESULTS

Unidirectional Laminates

The principal mechanical properties of the materials studied, derived from tensile tests are given in Table 1 below:

TABLE 1 Mechanical Properties of Unidirectional Laminates

Property	0° CFRP	0° GRP	90° CFRP	90° GRP	Units
Low-strain Young's modulus (E)	127	42	8.3	14	GPa
Fracture stress (σ_f)	1.7	0.92	0.039	0.056	GPa
Fracture strain (ε_f)	1.2	2.2	0.48	0.50	%
Poisson's ratio (ν)	0.29	0.27	0.02	0.09	-

The GRP 0° test curves were essentially linear elastic to fracture but the 90° specimens showed a pronounced knee at a strain of about 0.3% after which a whitening effect (Parvizi and Bailey 1978) could be observed in the specimen. The 0° CFRP was elastic to failure but the curve was not linear, there being an increase in the modulus with increasing strain so that whilst the initial modulus was 127 GPa, the secant modulus at fracture was ≈150 GPa. This effect is attributable to the fibres and is believed to be due to a reorientation of the fibre structure on straining (Reynolds 1970). The 90° CFRP was linear to failure with no knee or acoustic emission prior to failure. The failure strain in both 90° GRP and CFRP is characteristically low at ≈0.5% and may be explained by strain concentrations in the resin matrix produced by the fibres (Kies (1962) and Herrman and Pister (1963)).

The other features of note are the large differences between the longitudinal (0°) and transverse (90°) Young's moduli and between the two Poisson's ratios.

Cross-Ply Sandwich Laminates (0°-90°-0°)

Figure 1 shows a schematic section of a 0°-90°-0° cross-ply sandwich laminate. The two outside 0° plies are of equal thickness, "b", and the central 90° ply of thickness "2d", the specimen width is "c". When extended in tension initial failure is usually in the central 90° ply which cracks in a direction normal to the applied tension and parallel to the fibres in that layer.

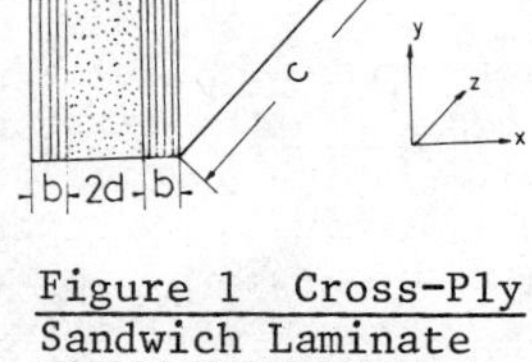

Figure 1 Cross-Ply Sandwich Laminate

The properties of the cross-ply sandwich laminates are summarised in Table 2.

TABLE 2 Mechanical Properties of 0°-90°-0° Cross-Ply Sandwich Laminates

Laminate	Material	b (mm)	d (mm)	E (low strain) (GPa)	Strain at First Transverse crack ε_c (%)	Strain at Onset of Acoustic Emission ε_{ae} (%)	Fracture Stress σ_f (GPa)	Fracture Strain ε_f (%)	Longitudinal Splitting Strain $\varepsilon_{\ell s}$
C1	CFRP	0.5	0.062	115	-	0.66	1.46	1.11	DNS*
C2	"	0.5	0.125	104	-	0.49	1.29	1.09	DNS
C3	"	0.5	0.25	90	-	0.36	1.12	1.11	DNS
C4	"	0.5	0.5	71	-	0.33	0.81	1.10	DNS
G1	GRP	0.5	0.05	39	1.8	-	-	1.8	DNS
G2	"	0.5	0.125	35	1.0	-	0.675	2.2	2.2
G3	"	0.5	0.25	29	0.5	-	0.469	2.1	1.7
G4	"	0.5	0.5	27	0.55	-	0.475	2.3	1.8
G5	"	0.5	1.0	19	0.65	-	0.233	2.0	-
G6	"	0.5	1.33	22	0.55	-	0.247	2.2	-
G7	"	0.5	2.0	17	0.52	-	0.162	2.2	1.4
G8	GRP	0.15	0.5	-	0.5	-	-	2.1	DNS
G9	"	0.25	0.5	22	0.45	-	0.316	2.2	DNS
G10	"	1.0	0.5	33	0.5	-	0.646	2.2	1.5
G11	"	1.5	0.5	42	0.55	-	0.720	2.3	1.3

*Did not split

In general the failure sequence in both sets of laminates follows a similar pattern. The initial obvious failure event is the formation of the first crack in the middle 90° layer. This was detected visually in the GRP and by acoustic-emission and microscopy in the opaque CFRP. This cracking produces a knee on the load/strain curve of the GRP but this is not apparent in the CFRP. Typical load/strain and acoustic emission traces are shown in Figs. 2a-c. On further extension more cracks are formed until the whole gauge portion of the test-piece is filled with a regular array of cracks as shown in Fig. 3 and Fig. 4 for the GRP and CFRP respectively.

The strain at which the first crack occurs increases as the thickness (2d) of the middle 90° layer is reduced and at the same time the crack spacing tends to become smaller. These effects have been described in more detail by Garrett and Bailey (1977a), Parvizi, Garrett and Bailey (1978) and Parvizi and Bailey (1978).

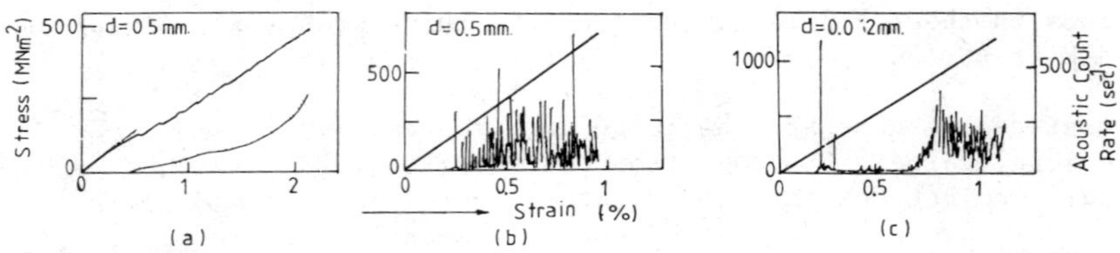

Figure 2 Typical stress-strain curves and acoustic count for (a) GRP, and (b) and (c) CFRP cross ply laminates with different transverse ply thicknesses.

Closer study of the load/extension curve for the GRP Fig. 2a reveals that there are two knees the first at ≃0.3% strain is associated with a visual whitening effect, (which partially disappears if the specimen is unloaded), the second at ≃0.55% strain is associated with both transverse-cracking and the onset of acoustic emission. In the GRP laminates where d>0.25 mm the transverse cracks appear to form instantaneously and always extend right across the section, but in laminates with a thinner middle layer, the cracks are observed to initiate at the edge of the specimen and to propagate in a controlled manner as the load is

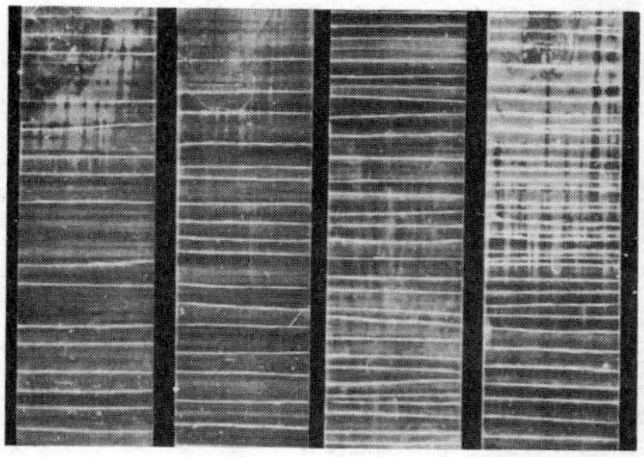

Figure 3 Variation of transverse crack spacing with the applied stress in a GRP laminate with an inner ply thickness, 2d of ≈1.0 mm.

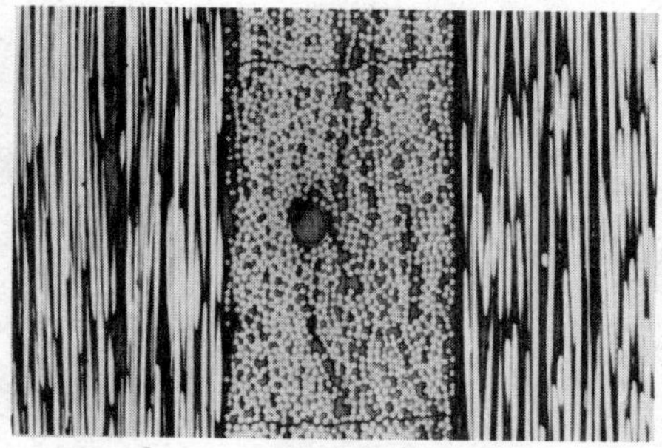

Figure 4 Photo-micrograph showing the transverse cracks in a CFRP sandwich laminate.

increased. The whitening phenomenon precedes the cracking, and in the case of the thinnest transverse layers transverse cracking is not observed at all before the final catastrophic failure of the test piece. The whitening effect cannot be observed in the CFRP but similar trends are observed for crack spacing and crack supression as the thickness of the transverse layer is reduced and transverse cracking is entirely constrained if the layer is sufficiently thin.

Examination of sections cut from transverse cracked laminates under the microscope has shown that in most cases the cracks extend right across the transverse lamina but do not cause delamination at the boundary between the 0° and 90° laminae. The exception is in the case of the thickest 90° laminae (i.e. laminates G6 and G7) shown in Figs. 5a and 5b and a crack with delamination in Fig. 6. It is interesting to note that in the CFRP (Fig. 5b) the crack actually runs through some fibres. This was commonly observed in laminates made from a particular batch of pre-preg but in a subsequent batch made from fibre with a modified surface treatment, designed to reduce the interface strength, the effect was much less common.

Further microscopy has shown that the earliest indications of failure are debonds at or near the fibre/resin interface. These occur at strains even lower than those at which the whitening is observed in the GRP. The next stage is a coalescence of a number of debonds to form a micro-crack. These micro-cracks then appear to grow rapidly when they reach a critical size which is probably about 3-4 fibre diameters, i.e. 50-100μm. Figures 7a and 7b show the development of these cracks for GRP and CFRP respectively.

Multiple-ply Laminates

Two multi-ply laminates were fabricated, one in GRP and one in CFRP, to test whether the cracking constraint was operative. The results, given in Table 3, show that the first transverse cracks are formed at similar strains to those observed in the cross-ply sandwich laminates with similar 90° ply thicknesses. Final failure in these laminates occurred by tensile failure of the 0° laminae and separation of the 90° laminae at the pre-existing transverse cracks. In the case of the CFRP the general effect was that of a "brittle" failure with rather less splitting and delamination than would be observed in the simple sandwich laminates (Fig. 8).

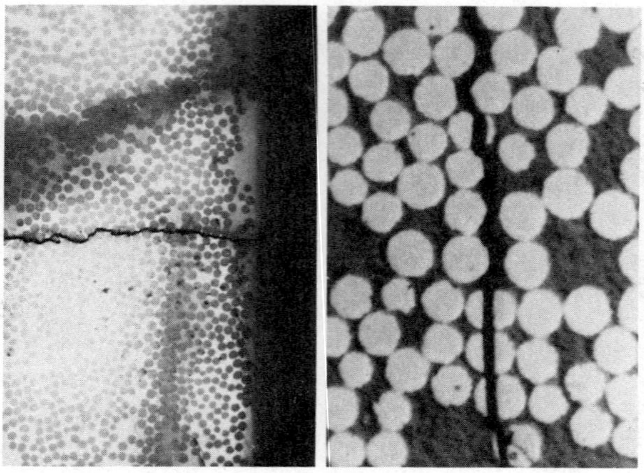

Figure 5 Typical transverse cracks in (a) GRP and (b) CFRP sandwich laminates, showing no delamination between the plies.

Figure 6 Transverse crack in a GRP cross-ply laminate with an inner ply thickness of 4 mm, with some delamination between the plies.

Longitudinal Splitting

Longitudinal splitting was observed to occur in the 0° laminae of the cross-ply laminates at strains intermediate between the transverse cracking strain for the 90° lamina and final failure. The effect is illustrated in Fig. 9 and the strains at which splitting occurred are indicated in Table 2. In this case the strain to initiate splitting increases as the thickness of the longitudinal plies is decreased. In the case of the thicker longitudinal plies ($b>0.5$mm) the splits formed instantaneously whilst for thinner plies the cracks propagated slowly as the strain was increased. Splitting was not observed at all in the thinnest plies ($b<0.25$). Longitudinal splitting has not been observed in the CFRP cross-ply sandwich laminates.

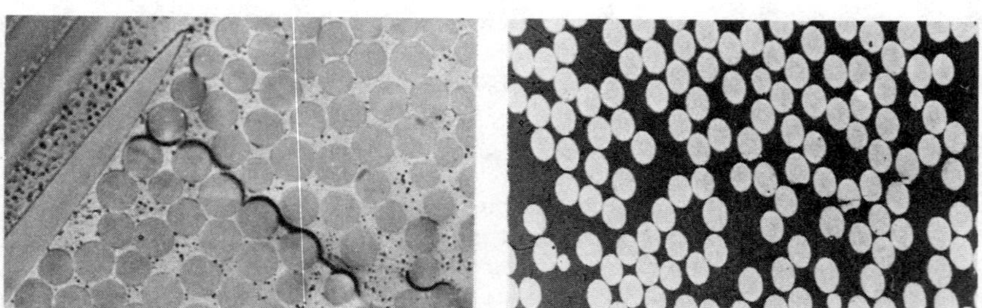

Figure 7 Development of transverse micro-cracks by fibre-matrix debonding in (a) GRP and (b) CFRP cross-ply laminates.

TABLE 3 Mechanical Properties of Multiple-Ply Laminates

Material	Construction	Low-Strain Modulus E (GPa)	Fracture Stress σ_f GPa	Fracture Strain ε_f	Transverse Cracking Strain
CFRP	15 x 0.25 mm plies alternate 0° and 90°	70	0.91	1.14	0.64*
GRP	5 x 0.25 mm plies alternate 0° and 90°	24.8	350	2.1	0.9

*Onset of acoustic emission.

DISCUSSION

Transverse Cracking

Transverse cracking occurs in the 90° plies of 0°-90°-0° and multiple-ply laminates. In general it is the first obvious failure event and can occur at strains of less than 0.5% in both GRP and CFRP. The macroscopic cracks are preceded by microscopic debonding at, or near, the fibre/matrix interface. As the strain is increased these debonds coalesce until a micro-crack of sufficient size to propagate is formed. Debonds have been observed at strains as low as 0.1% and are associated with the stress-whitening effect observed in the transparent GRP laminates. The strain at which macroscopic transverse cracks propagate is dependent on the laminate geometry and in particular on the thickness of the transverse ply. This is clearly illustrated in Fig. 10.

This low strain failure may be explained by the strain magnification effect first proposed by Kies (1962). He proposed that for a square array of fibres, strained in a direction normal to the fibre axis the strain magnification in the matrix (SMF) was given by:

$$SMF = \frac{2 + \frac{\Delta}{r_f}}{\frac{\Delta}{r_f} + 2\left(\frac{E_m}{E_f}\right)} \qquad (1)$$

(where Δ is the inter-fibre separation, r_f the fibre radius and E_m and E_f the Young's moduli of matrix and fibre respectively).

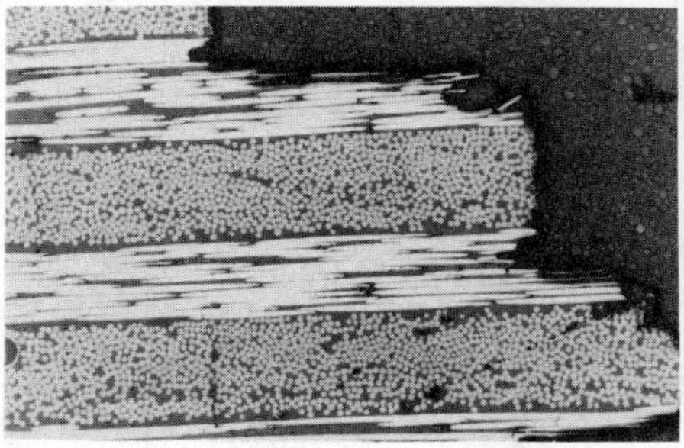

Figure 8 Photo-micrograph showing the brittle fracture of the CFRP multi-ply laminate.

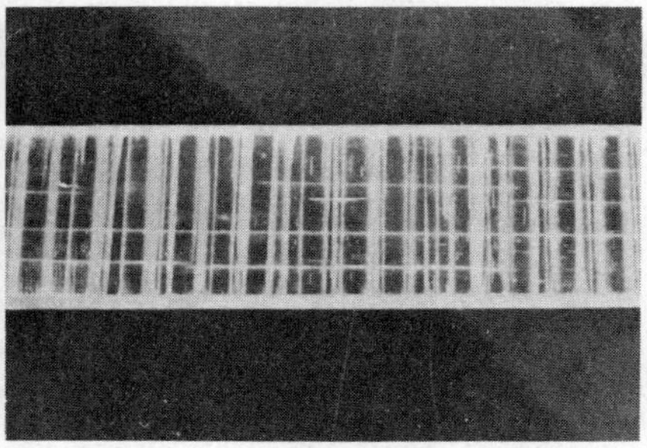

Figure 9 Longitudinal splits in a GRP cross-ply laminate.

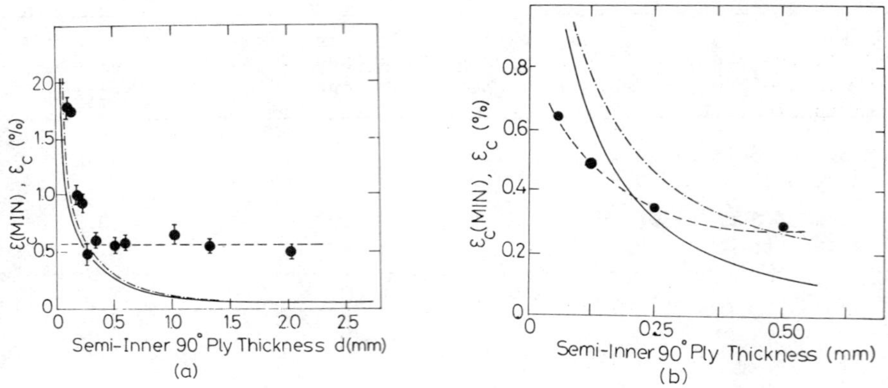

Figure 10 Plot of the theoretical and experimental transverse cracking strain, ε_c, as a function of the inner ply thickness, d, for (a) GRP, and (b) CFRP sandwich laminates. The outer ply thickness in both laminates is 0.5 mm. (——) equation (9), (- - -) equation (8), (-•-) experiment.

This relationship reduces to:

$$SMF = \frac{E_f}{E_m} \qquad (2)$$

as Δ approaches zero, (i.e. the fibres almost touch). Thus SMF would appear to be dependent on the fibre volume fraction but, in fact, even at comparatively low V_f there are invariably regions in the lamina where groups of fibres almost touch, (see Figs. 7a and 7b), so that the effective SMF will tend to approach the limit given by Equation (2). This gives values for SMF of 8-10 for the glass/ epoxy system. (The glass fibres are approximately isotropic, but the transverse Young's modulus of carbon is much lower than its longitudinal modulus and it is this modulus which should be used in Equations (1) and (2). This would put the SMF for carbon/epoxy at about 6.) Our experimental observations indicate that the first cracks form between fibres which are touching or nearly touching along a direction parallel to the principal tension axis. Cracking also appears to occur more readily between groups of fibres concentrated in a region of lower than average V_f, (i.e. fibre rich zones in resin rich regions). This implies that the SMF is more intense when the material around the critical zone is of lower stiffness.

Transverse crack-spacing

In both GRP and CFRP cross-ply sandwich laminates the transverse cracking occurs progressively as the strain is increased after the first crack has formed. The crack-density, and hence the crack-spacing, is related to the geometry of the laminate. This has been discussed by Garrett and Bailey (1977a) and Parvizi and Bailey (1978). They explained the crack-spacing on the basis of a shear-lag model. The tensile load in the central (transverse) ply is zero at the crack but builds up by shear transfer from the outer (longitudinal) plies according to an exponential law; so that at a distance, y, from the crack the load F in the inner ply is given by:

$$F = 2bc\ \Delta\sigma_o\ (1 - \exp(-\phi^{\frac{1}{2}}y)) \qquad (3)$$

$$\text{where } \phi = \frac{E_c\, G_t}{E_\ell\, E_t} \cdot \frac{(b + d)}{bd^2} \qquad (4)$$

These relationships are used to predict the crack spacing for any applied stress. The theoretical and experimental values for the GRP and CFRP laminates studied are compared in Figs. 11 and 12. These show an excellent fit.

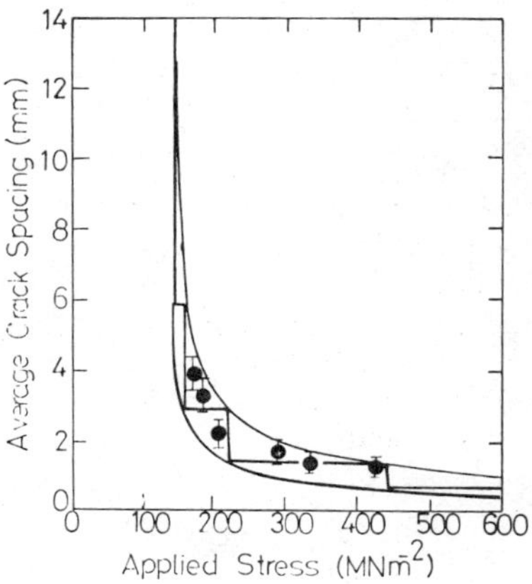

Figure 11 Theoretical and observed transverse crack spacings as a function of the applied stress for a GRP sandwich laminate with d = 0.5 mm and b = 0.5 mm. The stepped curve is the theory for the particular specimen length used. When the first crack forms in the mid-length of the specimen. The continuous curves are for any specimen length and the first crack position (●) experiment.

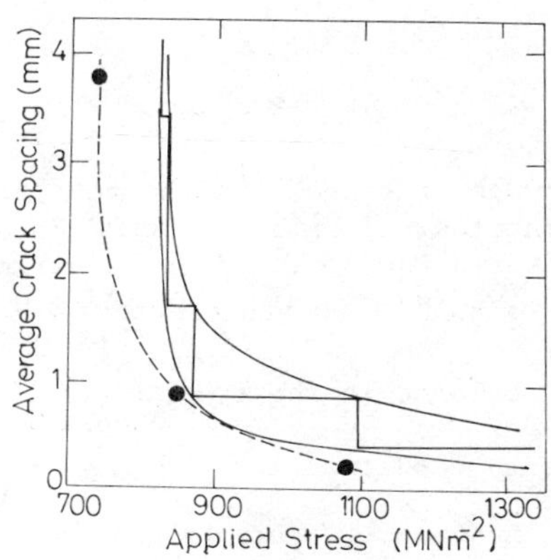

Figure 12 Theoretical and observed transverse crack spacings as a function of the applied stress for a CFRP cross-ply laminate with d = 0.125 mm and b = 0.5 mm. (——) theory, (-●-) experiment.

Transverse Cracking Constraint

This is one of the more significant effects observed. The strain required to initiate transverse cracks is greater when the transverse lamina is thinner, and in some cases cracking is constrained completely up to the strain at which the longitudinal laminae fail catastrophically. The mechanism of constraint has been discussed by Aveston and Kelly (1973), Aveston, Mercer and Sillwood (1974), Parvizi, Garrett and Bailey (1978), and more recently by Aveston and Kelly (1978) and Bailey, Curtis and Parvizi (1978). The basis of our present understanding is that in order for a crack to form it must be both mechanistically possible and energetically favourable. The former requirement is satisfied in the case of the cross-ply sandwich laminates by the strain-magnification concepts of Kies discussed above. This however owes no dependence on the geometry of the laminate and all transverse laminae of similar constitution would be expected to crack at similar strains regardless of the thickness of either the transverse lamina or of the adjacent longitudinal (or any other) laminae. We may, thus, suppose that there is a threshold strain, characteristic of a particular combination of fibre, matrix and interface, at which microscopic cracks will form in all cases. Whether these develop into mini- or macro-cracks is determined by

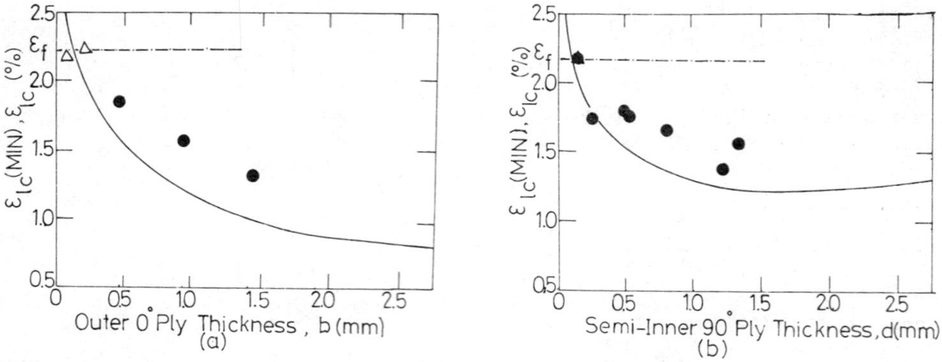

Figure 13 Comparison of the theoretical and the experimental composite strain for the longitudinal splitting of GRP laminates as a function of (a) outer ply thickness for d = 0.5mm, and (b) semi-inner ply thickness for b = 0.5mm. (——) theory, (•) experiment (--Δ--) ultimate fracture.

the energy balance in the system.

The necessary condition for a crack to form is given by:

$$\Delta w > \Delta u + 4\gamma_t dc \qquad (5)$$

where Δw is the work done by the applied stress during the formations of the transverse crack, Δu is the increase in stored elastic energy, and γ_t is the fracture surface energy per unit area of the inner ply. In the case of no pre-existing strains in the composite, Δw is twice Δu and the calculations are thus very simple. Using equation (5), the minimum value of the first cracking strain ε_c(min), is given by:

$$\varepsilon_c(\min) = \left[\frac{2\gamma_t b E_\ell \phi^{\frac{1}{2}}}{(b+d)E_t E_c}\right]^{\frac{1}{2}} \qquad (6)$$

However, in these sandwich laminates, the thermal mismatch between the longitudinal and transverse plies results in thermal strains being developed during the fabrication process. These strains are tensile along the fibre direction and compressive in the direction perpendicular to the fibres. The thermal strains are found to be small in the GRP composites, but in the case of CFRP, these strains are considerably greater and can cause substantial changes in the energetics of the system during crack formation. Equation (6) should therefore be modified to take this thermal effect into account. Upon including the thermal strains, the simple criterion of $\Delta w = 2\Delta u$ is no longer applicable and after calculating Δw and Δu individually and inserting them in Equation (5), the following equation is deduced for ε_c^1(min):

$$\varepsilon_c^1(\min) = -\varepsilon^{th} + \left[\frac{2\gamma_t b E_\ell \phi^{\frac{1}{2}}}{(b+d)E_t\ E_c}\right]^{\frac{1}{2}} \qquad (7)$$

when ε^{th} is the thermal strain in the transverse ply and is tensile. The theoretical values of the minimum cracking strain have been calculated from Equations (6) and (7) as a function of d and are compared with the experimental results in Fig. 10a for GRP laminates. Close agreement is observed between the theory and experiment in the region where d<0.25 mm, indicating an energy controlled crack propagation. For the thicker laminates, however, this theory does not apply and cracking occurs at a constant strain of 0.5% which is close to the cracking strain of the unidirectional 90° lamina. In this region the mechanism of transverse cracking is considered to control the process.

Reasonable agreement between the theory and experiment is also obtained for CFRP composites as shown in Fig. 10b.

Longitudinal Splitting

In GRP laminates, the longitudinal splitting of the outer plies is observed to occur prior to the ultimate fracture of the composite. This phenomenon is considered to be due to the Poisson effect and has been fully discussed by Bailey, Curtis and Parvizi (1978). The Poisson's ratio of the longitudinal lamina is almost three times greater than that of the transverse. Hence,in the sandwich laminate, the differential Poisson contractions of the laminae generate a tensile strain in the outer and a compressive strain in the inner plies in the transverse direction. These strains are increased when the transverse ply cracks. Under some geometrical conditions, the magnitude of the tensile strain in the outer plies, when summed with the existing thermal strain, may be large enough to cause the ply to split. This type of cracking is essentially similar to the transverse cracking and can be treated in the same way. Using an energy based model, the minimum composite strain at which the longitudinal splitting is energetically possible has been calculated and is given by:

$$\varepsilon_{\ell c}(\min) = \frac{1}{\nu_\ell}\left\{\frac{E_c^*\,(b+d)}{E_\ell d}\left[-\tfrac{3}{4}\,\varepsilon^{th} + \left(\frac{1}{16}(\varepsilon)^{th^2} + \frac{\gamma_t d E_\ell \phi^{\frac{1}{2}}}{(b+d)E_c^*\,E_t}\right)^{\frac{1}{2}}\right] + \nu_t\varepsilon_c\right\} \qquad (8)$$

where E_c^* denotes the composite Young's modulus in the transverse direction. Comparison of the above equation with experiment shows a good fit in Fig. 13.

Longitudinal splitting has not been observed in the CFRP laminates although this system shows a greater mismatch between the two Poisson's ratios and the moduli. The reason for this is that the ultimate failure strain of the CFRP composites studied is small and the laminate fractures before large Poisson strains have been developed.

Relevance to Design of Laminates

It has been generally appreciated for many years that GRP and CFRP laminates containing crossed unidirectional plies or manufactured from cloths, woven-rovings or random chopped-strand mats develop microscopic cracks, generally in regions where fibres lie normal to the principle tension axis, at strains which are, at the most, only 30% of the final failure strain. The presence of these micro-cracks can sometimes result in technical failure, e.g. weeping or leakage of GRP pipes and pressure vessels, but in many cases the mere presence of cracks does not appear to be detrimental to the short term mechanical properties of the laminate. However, the fact of transverse cracking presents the designer with a dilemma: whether to base his design on strains below the cracking threshold (typically 0.5% for GRP) or to work to the catastrophic failure strain, which might be 1.5% or more. Clearly the latter criterion is the more attractive if the cracks are truly "inocuous". Unfortunately there is an accumulation of evidence which suggests that micro-cracks may act as nuclei for further local damage leading to ultimate failure under conditions of cyclic or repeated loading and also there is a fear that any cracks would render the laminate more prone to attack and degradation in a hostile environment.

The present work suggests that the formation of "visible" transverse cracks can be constrained or inhibited by constructing the laminate from thinner individual plies. It appears that the absolute thickness of the transverse plies must be minimised. Multi-ply laminates appear to crack at strains comparable to those of the sample cross-ply sandwiches with similar ply thicknesses. It is therefore suggested that multiple ply construction is, in general, preferable especially for the transverse plies. Likewise the longitudinal splitting effect, observed in some GRP laminates, is delayed if the longitudinal ply thickness is reduced - a further argument for multi-ply construction.

Garrett and Bailey (1977b) have shown that the threshold strain for crack initiation, in glass/polyester laminates, is raised if a more flexible resin is used. This effect might be expected to apply also to epoxy systems. However, in most cases, "flexibilising" also reduces the resin stiffness and renders it less resistant to elevated temperatures and hostile environments. Thus the potential usefulness of flexible resins is doubtful.

Another cautionary note is that we have observed that failure in multi-ply laminates tends to be more "brittle", in that the relative volume of material subjected to delamination and "pull out" is reduced, (see Fig. 8).

A final reason for caution is that although the development of visible, or mini-cracks, can undoubtedly be inhibited by the proposed energetic constraint mechanism, so that the threshold strain for cracking is substantially raised, it is also true that the basic strain-magnification mechanism for crack initiation is still operative. In comparing two laminates, one unconstrained and the other constrained, but otherwise similar, we may expect debonding and stress-whitening to occur at similar strains but the coalescence of debonds and the growth of mini-cracks will be delayed to higher strains in the constrained laminate. It is clear that mini-cracks of perhaps >1 mm in length do constitute nuclei for the development of fatigue damage and points of ingress for hostile fluids, but it is not yet known whether the constrained debonds and micro-cracks of perhaps 50μm constitute a similar or a reduced hazard.

ACKNOWLEDGEMENTS

The work discussed was carried out in the laboratories of the Department of Metallurgy and Materials Technology of the University of Surrey. We would like to acknowledge the assistance rendered by our colleagues and also the support of the Ministry of Defence who funded the carbon fibre aspect of the work.

REFERENCES

Ashton, J.E., J.G. Halpin and P.H. Petit (1969). Primer on Composite Materials: Analysis. Technomic Publ. Corp. West Post, Conn.

Aveston, J., and A. Kelly (1973). Theory of Multiple Fracture of Fibrous Composites. J. Mater. Sci., 8, pp. 352.

Aveston, J., and A. Kelly (1978). Tensile First Cracking Strain and Strength of Hybrid Composites and Laminates. To be published. Proc. Roy. Soc.

Aveston, J., R.A. Mercer and J.M. Sillwood (1974), in Composites Standards, Testing and Design. pp. 93. Conf. Proc. National Physical Laboratory I.P.C. London.

Bailey, J.E., P.T. Curtis and A. Parvizi (1978). Cracking and Splitting in Fibre Composites, to be published. Proc. Roy. Soc.

Garrett, K.W. and J. E. Bailey (1977a). Multiple Transverse Failure in 90° Cross Ply Laminates. J. Mater. Sci. 12, pp. 157.

Garrett, K.W. and J. E. Bailey (1977b). The Effect of Resin Failure Strain on the Tensile Properties of Glass-Fibre Reinforced Polyester Cross Ply Laminates. J. Mater. Sci. 12, pp. 2189.

Herrmann, L.R. and K. S. Pister (1963). Composite Properties of Filament Resin Systems. ASME papers 63-WA-239.

Kies, J.A. (1962). Maximum Strains in the Resins of Fibre Glass Composites. U.S. Naval Research Lab. Report No. 5752.

Parvizi, A., and J. E. Bailey (1978). On Multiple Transverse Cracking in Glass Fibre Epoxy Cross-Ply Laminates. J. Mater. Sci. 13, pp. 2131.

Parvizi, A., K. W. Garrett and J. E. Bailey (1978). Constraint Cracking in Glass Fibre Reinforced Epoxy Cross-Ply Laminates. J. Mater. Sci. 13, pp. 195.

Reynolds, W. N., (1970). Structure and Mechanical Properties of Carbon Fibres, Proc. 3rd Conf. on Carbons and Graphite. Soc. Chem. Ind.

Tsai, S.W. (1968) in Fundamental Aspects of Fibre Reinforced Plastics Composites. Ed. by Schwartz, R.T., and H.S. Schwartz, Interscience.

Tsai, S.W., and E. M. Wu (1971). A General Theory of Strength for Anisotropic Materials. J. Comp. Mat. 5, pp. 58.

Volume 3, ICM 3, Cambridge, England, August 1979

DELAMINATION OF GLASS FIBER REINFORCED POLYESTER, AN ACOUSTIC EMISSION STUDY

F. X. de Charentenay, M. Bethmont, M. Benzeggagh and J. F. Chrétien

Département de Génie Mécanique, Université de Technologie de Compiégne, B.P. 233, 60206, Compiègne, France

INTRODUCTION

Fracture in materials is a very important problem regarding safety and durability. In composite materials these aspects of reliability are somewhat confused through lack of full knowledge of the fracture process. Increasing numbers of studies are devoted to define some fracture mechanics for the composites but there are so many different kind of materials that a general theory like linear elastic fracture mechanics for the isotropic and homogeneous materials, can not be built up. However some general trends in the mode of fracture of several composites materials are now better known (Cooper, 1977).

Delamination is one of the more common modes of early failure in laminated fibrous composites materials and the delamination mechanism has been the subject of many studies (Pipes, 1970, Pagano, 1971, Reifsnider, 1977, Rybicki, 1977). However in a large number of studies the stress distribution in the laminate is very difficult to define (short beam test for example) and moreover, the failure initiates at some sample boundary (delamination in angle ply composites in tension). To avoid this lack of mechanical definition and according to the fact that defects are the failure initiation in real structure, we design a simple test in which delamination is localized at the tip of an articifial defect introduced between the reinforcement layers. This study shows in a first part how this simple test may be used for the delamination fracture energy determination and in a second part the use of acoustic emission to monitor the delamination crack propagation.

DESIGN OF TESTS AND EXPERIMENTAL

The simplest crack propagation experiment is the mode I or cleavage mode. This is applied to the laminates by pulling apart the two parts of a partially delaminated composite (Fig. 1). The artificial delamination is introduced during processing by interposition in the resin layer between two reinforcements of a non-adhesive thin film (paraffin or polyethylene). In this test, the plane of the initial crack is thus parallel to the plane of the reinforcements and the loading is perpendicular to it. This cleavage mode was used by some authors in studies on adhesives (Ripling, 1960, Chang, 1976). The mechanics of such joints has been described by Kanninen (1973) and is similar to the DCB sample in linear elastic fracture mechanics.

The shear mode or mode II has also been studied with a four point bending experiment, one of the crack tips being in the shear zone (Fig. 2). Composite laminates are made from 6 to 12 layers of fibre glass reinforcements in the form of woven fabric (Roving satin) of mat (two dimensional random fiber). Resins are various standard polyester from different origin. Samples, typically 30 mm x 120 mm, were cut into 240 mm x 240 mm plates with a 100 mm wide defect, introduced in the plane of symmetry and located at equal distance from both sides.

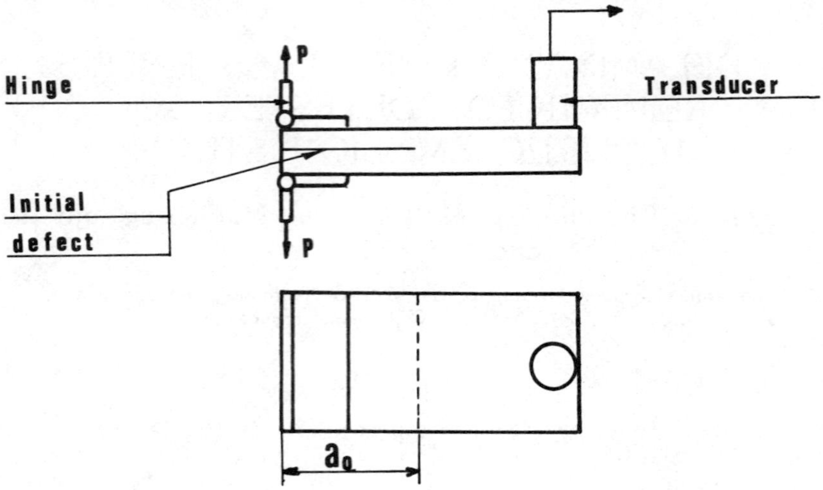

Fig. 1. Cleavage (mode I) experiment

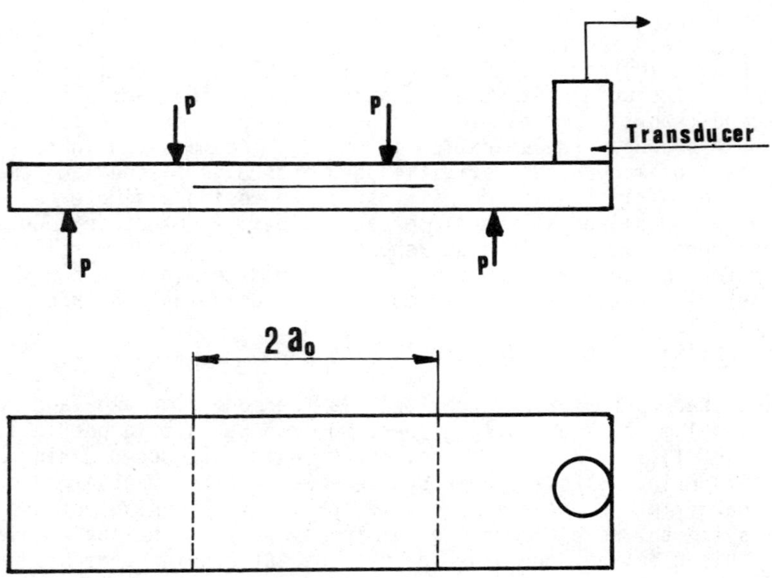

Fig. 2. Shear (mode II) experiment

The load displacement curves are determined on an Instron TTDM testing machine at displacement rate of 2 mm/min. Acoustic emission was used to monitor the stress waves emitted in the microscopic process of fracture. The signal of a piezo-electric transducer is amplified, filtered to eliminate the machine noise and fed to a discriminator which transforms burst signal into counts (Fig. 3). The acoustic emission is evaluated either as count rate or as total count number.

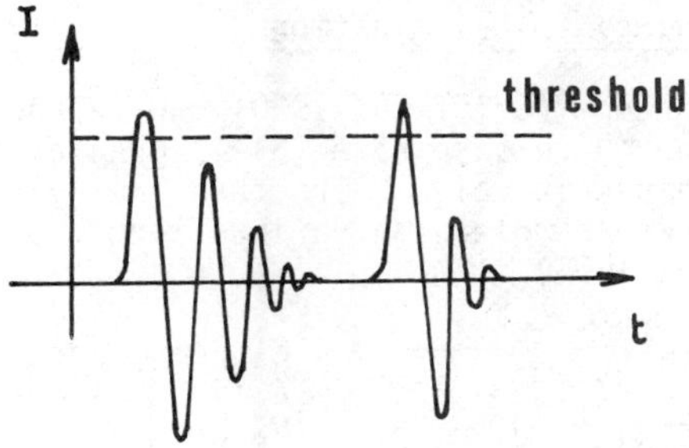

Fig. 3. Acoustic emission counting

RESULTS AND DISCUSSION

A typical load displacement curve together with count rate of acoustic emission is presented in Figure 4. Three stages appear on this curve : first, an elastic

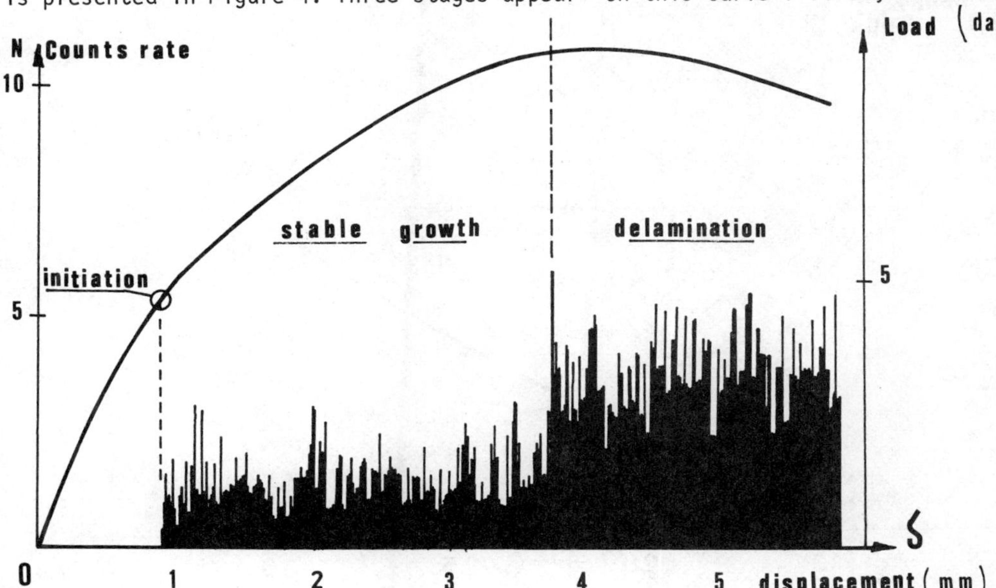

Fig. 4. Load displacement curve and count rate of acoustic emission (cleavage)

deformation of the partially delaminated composite ; second, initiation and stable propagation as indicated by acoustic emission ; third, further propagation with load decreasing. Visual observation confirms that the initiation point coincides with the onset of whitening at the tip of the defect. These tests are discussed in two directions : the initiation point is used for the determination of delamination fracture energy ; the acoustic emission during propagation will afford interpretation of delamination mechanism.

Determination of Fracture Energy in Delamination

The strain energy release concept (Griffith, 1920) can be applied to this cracked sample. If the strain energy release rate (G) is greater than a limit value G_C, the crack propagates. This critical G_C, fracture energy, may be measured with mechanical tests on cracked samples with various length of crack, known as the compliance method (Irwin, 1954).

$$G_c = \frac{P_c^2}{2B}\frac{dC}{da_o}$$

C Compliance
a_0 initial crack length
P_C critical load
B width of specimen

The initiation point on the load displacement curve (Fig. 4) may be determined either by visual observation or by the onset of acoustic emission. Inflexion or slight drop on the load displacement curve may also indicate this initiation point. Figures 5 and 6 show an example of the compliance variation as a function of a_0, and the G_C values calculated for one particulate glass fiber laminate. Results from visual observation and from acoustic emission are very close. Thus

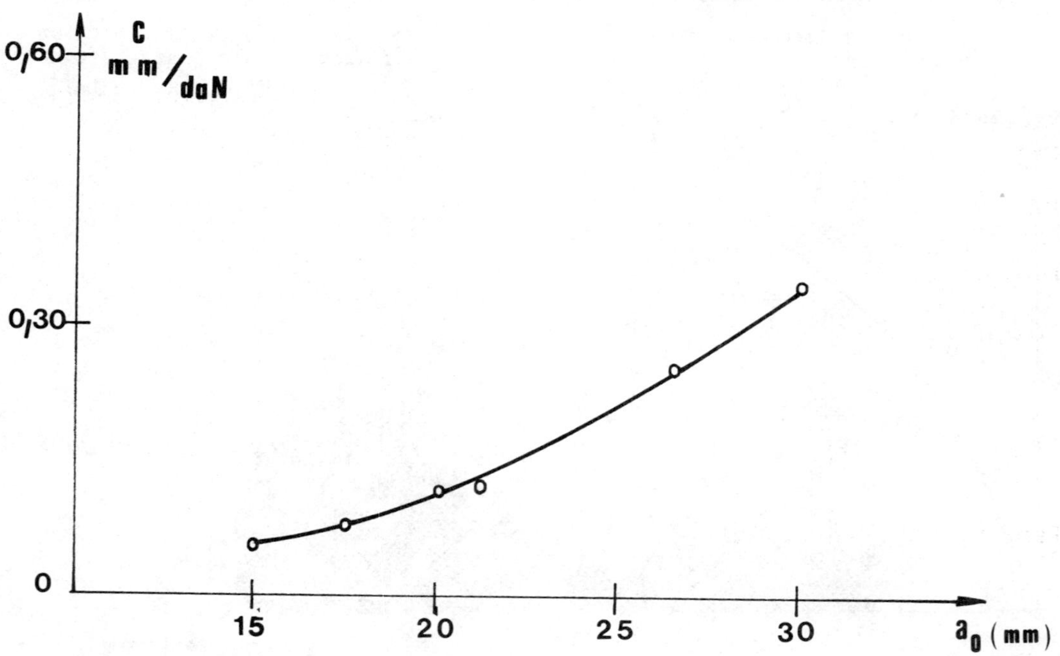

Fig. 5. Variation of compliance with length a_0 of initial defect (1507 LS and mat)

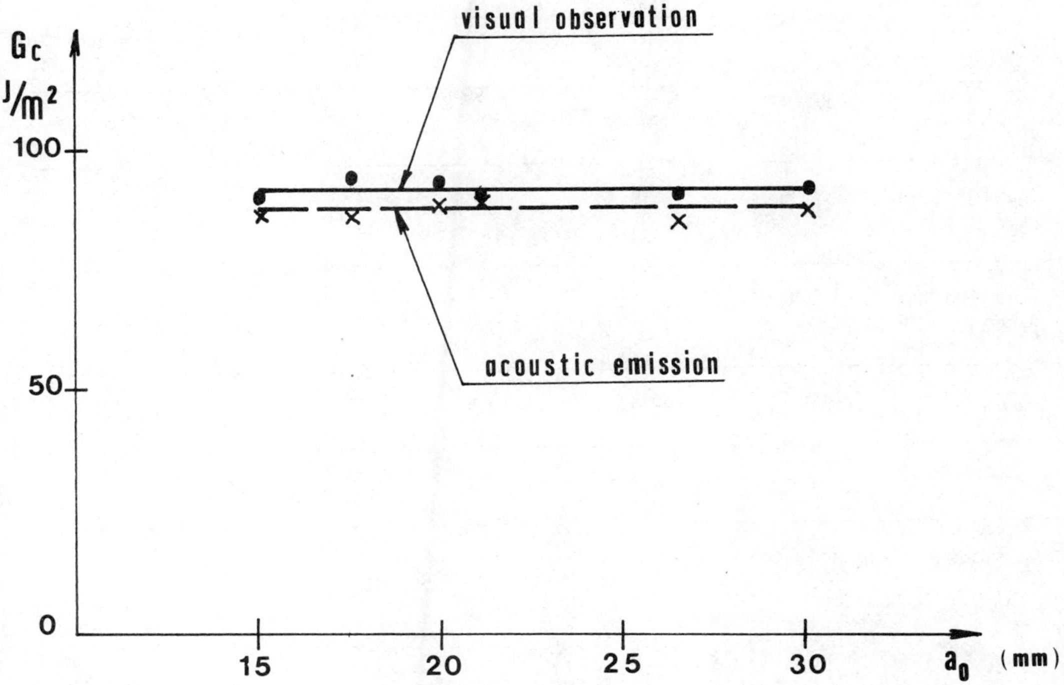

Fig. 6. Variation of fracture energy G_c with length a_0 of initial defect (1507 LS and mat)

acoustic emission may be used for the initiation detection in such experiments on non-transparent samples like CFRP.

The method is restricted by the reproducibility of the different predelaminated samples. Although they are taken from the same plates, it has been found, that there is a large scatter in G_c values as a function of a_0. Slight variation in thickness or in glass percentage within the plates introduces errors in the $\frac{dc}{da}$ term. In that case it is better to use the procedure of Berry (1963) which described the compliance and the G_C value as a function of a_0 :

$$C = \frac{a_0^n}{h} \qquad G_c = \frac{nP\delta}{2a_0B}$$

a_0 length of defect
h constant
P critical load
δ critical displacement
B width of specimen

Results are summarized in Table 1. Although precise comparisons are not yet available the measured G_C values are of the order of the pure polyester resin. This parameter is the critical fracture energy of the initiation of the first fracture event at the tip of the crack, which arises in the resin or at the fiber-resin interface. Thus, the method gives a value which defines delamination resistance of a given laminate system. Systematic determination as a function of reinforcement, resin, cure conditions, etc.. must give interesting information for the comprehension of the mechanical behaviour of laminates.

TABLE 1 - Fracture Energy for Delamination of Glass Fiber Composites

Resin	1507 LS*	544 S **	302 **	302	302	302
Reinforcement (layers number)	mat (6)	mat (8)	mat (6)	woven fabric (10)	mat (6)	woven fabric (10)
Glass volume percentage		25	15	30	16	30
G (J/m^2)	90 (1)	199 (2)	110 (2)	215 (2)	153 (1)	165 (1)
σ	2,5	16,6	19,5	20	20	11

* Rhone Poulenc Polymères

** Synrès S.A.

(1) Compliance method, (2) Berry's method

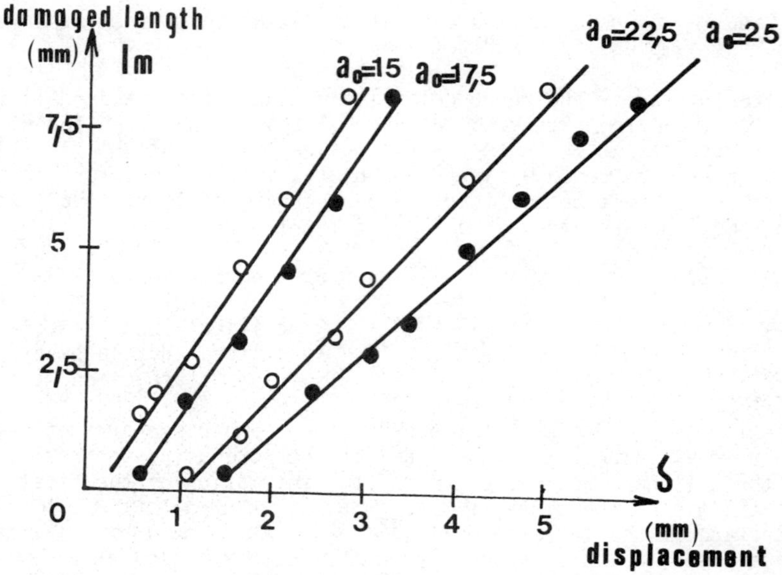

Fig. 7. Propagation of damaged length in a cleavage experiment (1507 LS and mat)

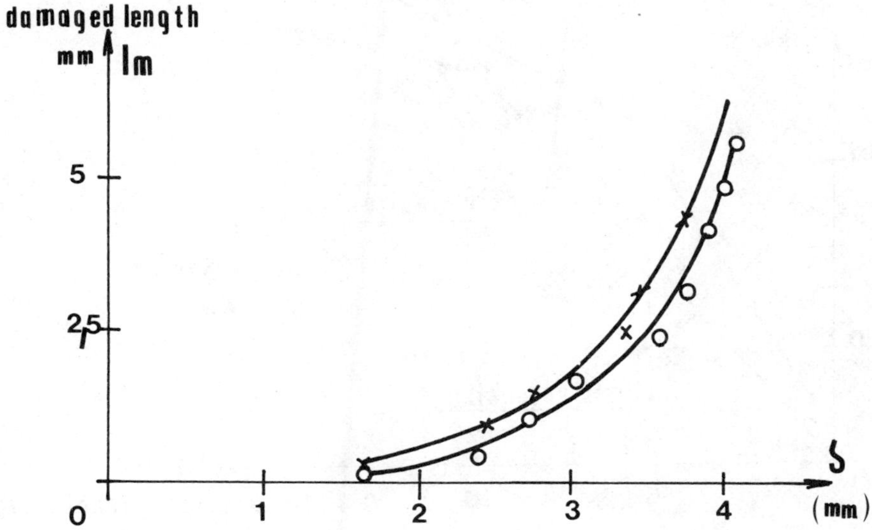

Fig. 8. Propagation of damaged length in a shear experiment (1507 LS and mat)

The Propagation Phase

After the initiation point, the growth of damage at the crack tip is easily measured. The kinetics of growth in a constant crosshead speed experiment are shown in Figure 7 for the cleavage mode and in Figure 8 for the shear mode in the form of the variation of the length l_m of damaged zone as a function of the displacement of crosshead. The cleavage cracking experiment shows stable propagation with a constant rate which depends on the initial delamination crack length a_0. In the shear experiment, damage propagation rate is continuously increasing without onset of instability. In a homogeneous brittle material, such sample design shows unstable propagation unless a_0 is sufficiently large. In this delamination experiment the mode of damaging at the growing tip appears to stabilize and slow down the propagation.

In order to get more understanding about this propagation mechanism, the acoustic emission was recorded during the entire delamination experiment. The logarithms total counts number were plotted against the length l_m (or a - a_0) of visible damage (Fig.9, Fig. 10). These plots present, for all experiments, two straight lines indicating that there are two different parts in the propagation phase, and that the number of acoustic emission events is an exponential function of the length of damage. Further, the slope of the first straight line is independent of the initial delamination length a_0. Different samples (with different a_0) of the same laminates show only slight vertical shift on the log N-l_m plots.

Very few examples of quantitative acoustic emission results have been published. Williams (1974) gave a relationship between counts number and compliance evolution during fatigue of Boron Epoxy laminates. Fitz Randolph (1972) presented also parallel evolution of total counts in acoustic emission and fracture energy in a

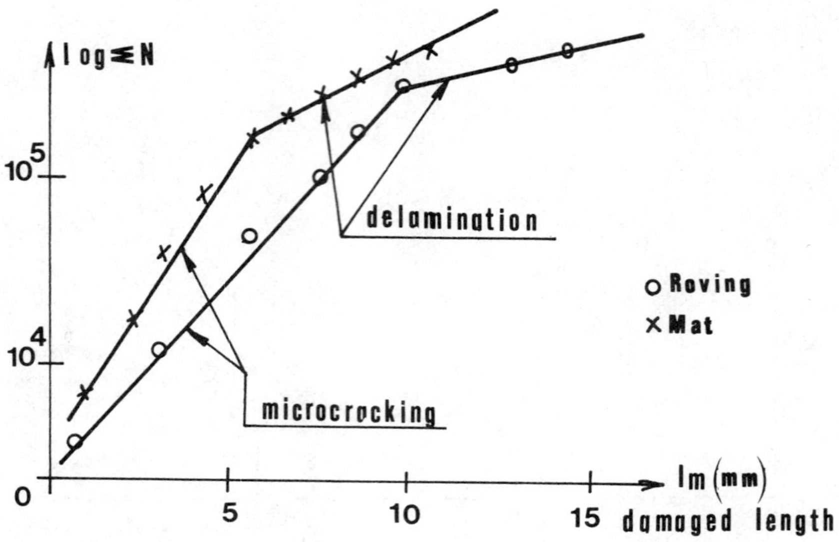

Fig. 9. Variation of logarithm of A.E. total counts number as a function of damaged length (cleavage experiment)

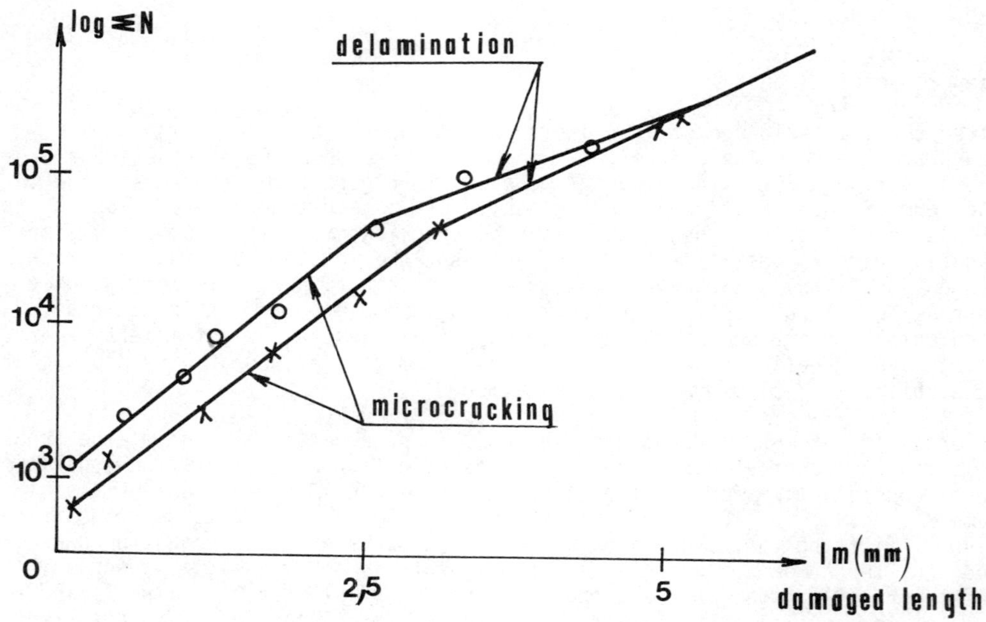

Fig. 10. Variation of logarithm of A.E. total counts number as a function of damaged length (shear experiment, different a_0)

Boron Epoxy laminate.

Close examination of the sample during experiment shows that true delamination (macroscopic opening displacement at the crack tip) occurs only at the cross point of the two straight lines. The first straight line corresponds to micro-crack propagation. Delamination is preceded by a few millimeters (2 to 10) of a microcracked zone, roughly like a crazed zone in some thermoplastics cracking experiment. This is confirmed by plotting on the log N. l_m plot, the load as a function of the length of damaged zone, (Fig. 11). The transition point corresponds to the onset of load dropping indicating opening at the crack tip.

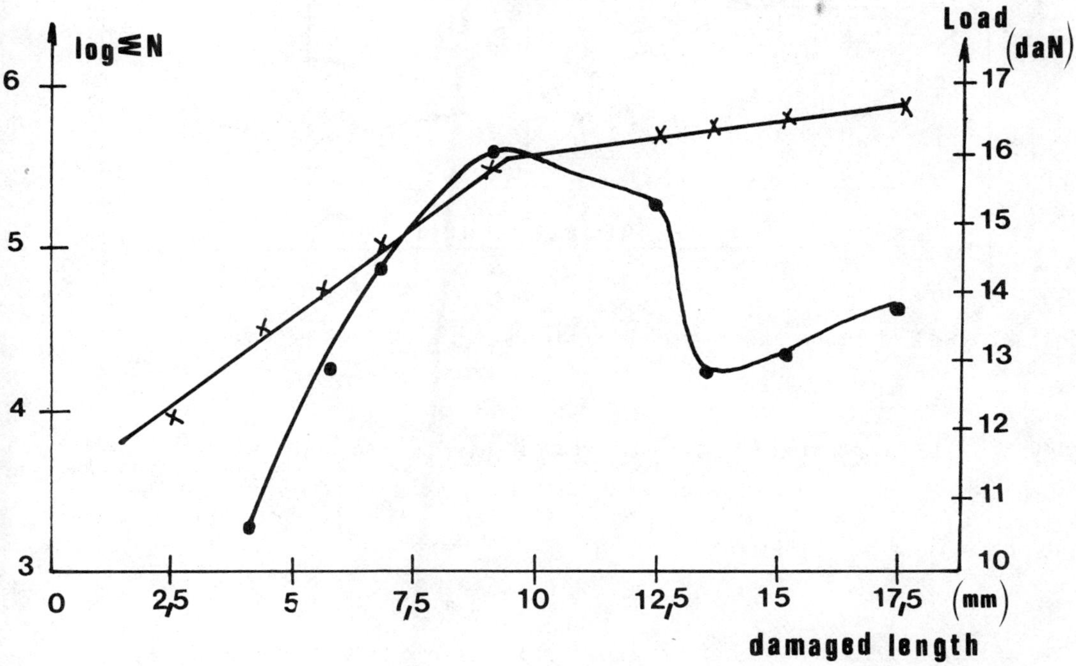

Fig. 11. Comparison of A.E. and load variation with damaged length

Model of microcrack propagation. A simple model is proposed to explain the exponential increase of acoustic emission. The microcrack growth is supposed to follow a chain reaction scheme (Fig. 12). Making the hypothesis that microcracks grow by microscopic steps, each giving an acoustic event, the total counts number is proportional to the total length of microcrack

$$N = k\ l_R \qquad (1)$$

if l_a is the length between points of branching in the reaction scheme, the total length of microcracks l_R is at the step p :

$$l_R = l_a\ (2^P - 1) \qquad (2)$$

Thus $$N = k\ l_a\ (2^P - 1) \qquad (3)$$

for p $\geqslant$ 3 : $$\log N = \log k.l_a + p \log 2 \qquad (4)$$

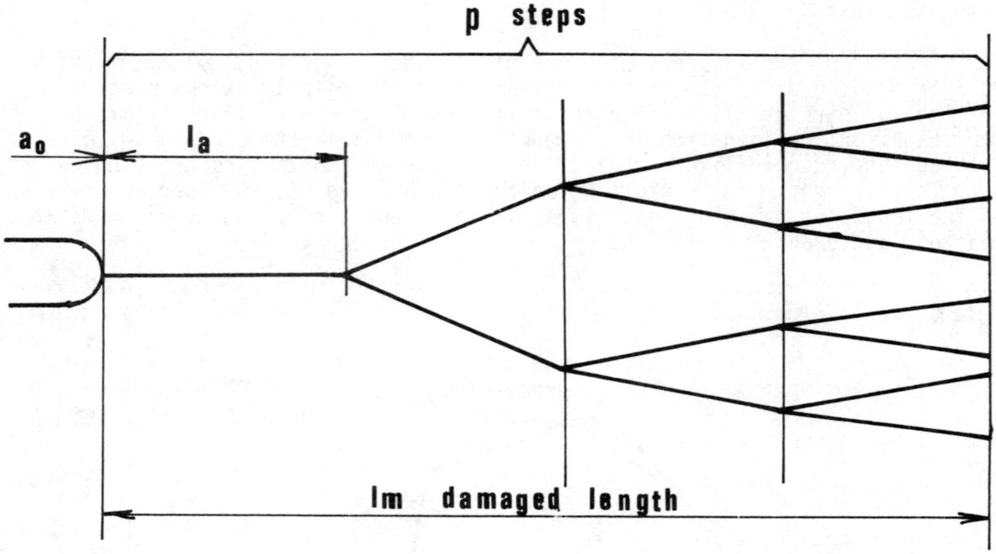

Fig. 12. Model of propagation with branching as in a chain reaction

Figure 12 shows clearly that the observed microcrack length l_m is :

$$l_m = p \; l_a \qquad (5)$$

introducing p from (5) in (4) :

$$\log N = \frac{0.3}{l_a} \; l_m + \log k \; l_a \qquad (6)$$

This relation described the experimental relationship between log N and microcrack length l_m with a slope of $\frac{0.3}{l_a}$. In this expression the slope is a pure material parameter, independent of the experimental conditions of acoustic emission recording. This is confirmed by the reproducibility of the slope from one sample to another (with different a_0) in the same plate.

The length l_a between points of branching, appears to be a material parameter : a characteristic microcrack length which depends on resin, reinforcements, defects etc... This length has been evaluated for two composites with polyester resin : 1.1 mm for the mat (random fiber) reinforcement and 2.0 mm for the woven fabric reinforcement. It is interesting to notice that both materials undergo true delamination when five steps of branching have occured (5.5 mm of microcracking for the mat and 10 mm for the woven fabric). In shear mode, for the same material, the characteristic microcrack length is only 0.5 mm and delamination occurs after five to six steps.

In this exponential microcracking mechanism, the fracture energy must increase drastically during the first propagation part. Development of this study shows that G may increase by factor of 3 to 5 and becomes approximately constant at the delamination point.

CONCLUSION

Simple delamination experiments on predelaminated samples and monitoring of acoustic emission give interesting results on the behaviour of glass fiber reinforced polyester laminates. Delamination initiation may be precisely defined with the onset of acoustic emission. Two parts are distinguished in propagation phase. The microcracking part is shown by acoustic emission to follow a chain reaction with a branching point. Macroscopic delamination then occurs.

This study allows us to define three characteristic parameters of a composite laminate. An initiation fracture energy G_c, a characteristic microcracking length l_a which is the mean propagation length between two points of branching, and a critical microcracking length l_c at which delamination occurs.

These parameters may be very useful for the study of the influence of chemical composition, type of reinforcement and processing mode on the delamination behaviour of composite laminates. Further studies are now in progress on glass fiber and carbon fiber reinforced laminate composites.

REFERENCES

Berry, J.P. (1963). J. Appl. Phys., 34, 62
Chang, D.J., R. Muki and R.W. Westmann (1976). Int. J. Solids. Structures, 12, 13
Cooper, G.A., and M.R. Pigott (1977). I.C.F. 4 (Waterloo, Canada) Proceedings, 1, 557
Fitz Randolph J., D.C. Phillips, P.W.R. Beaumont and A.S. Tetelman (1972) J. Mat. Sci., 7, 289
Griffith, A.A., (1920) Phil. Trans. R. Soc., A. 221, 163
Irwin, G.R. and J.A. Kies (1954). Welding Research Suppl., p. 193
Kanninen, M.F. (1973). Int. J. of Fract., 9, 83
Pagano, N.J. and R.B. Pipes (1971). J. Compos. Mat., 5, 50
Pipes, R.B. and N.J. Pagano (1970). J. Compos. Mat., 4, 538
Reifsnider, K.L., et al (1977), A.S.T.M., S.T.P., 617, 93
Ripling, E.J., S. Mostovoy and R.L. Patrick (1960). Mat. Res. Std., 4, 129
Williams, R.S. and K.L. Reifsnider (1974). J. Compos. Mat., 8, 340

Volume 3, ICM 3, Cambridge, England, August 1979

FRACTURE MECHANISMS AND TOUGHENING OF FIBRE COMPOSITES

M. Munro and P. W. R. Beaumont*

Department of Mechanical Engineering, University of Ottawa, Ottawa, Canada
***Department of Engineering, University of Cambridge, Trumpington Street, Cambridge CB2 1PZ, U.K.**

INTRODUCTION

This paper describes macroscopic observations of the debonding, fracture and pull-out of glass fibres in an epoxy resin, and analyses work of fracture in terms of fibre-matrix sliding processes and the dissipation of energy when a debonded fibre snaps.

When a brittle fibre composite is monotonically loaded in tension, glass fibres in epoxy, for example, it can fail in a number of ways; by matrix cracking, by fibre-matrix debonding, by fibre fracture, by fibre-matrix sliding and by fibre pull-out. Models to describe the more important failure processes have been proposed (Cottrell, 1964, Outwater and Murphy, 1970, Kelly, 1970, for example), and equations derived to estimate the work done in each fracture event (Table I). Theory gives the form of the equation; we then have to resort to experimental data to obtain values for the parameters which enter it.

Beaumont and Harris (1972) measured the work to fracture a carbon fibre-epoxy composite and analysed the results in terms of the work to extract broken fibres out of their matrix sockets. A similar explanation was given by Gershon and Marom (1975) for the work to fracture an E glass fibre-epoxy composite. In continued crack growth experiments on a model glass fibre-polyester composite loaded in tension, Harris et al (1975) equated fracture energy to the work done against friction during sliding of a fibre in its socket and the work to pull a broken fibre from a cracked matrix.

Variations in interpretation of the fracture of similar fibre composites are confusing, perhaps it is due to the difficulty in making precise measurements of parameters which appear in the models, frictional interfacial shear stress developed soon after a fibre debonds, for example. Also the determination of fibre debond length and fibre pull-out length which depend upon the statistical aspects of fibre fracture, cannot easily be evaluated with any degree of confidence. Recently, Kirk et al (1978) described a technique for the fractographic analysis of glass fibre and carbon fibre-epoxy composites. It involved the manufacture of model fibre composites and the use of an optical microscope technique for observing fibre-matrix debonding and fibre pull-out processes. Flexural beam experiments

MS238

were carried out on prismatic bars of epoxy resin containing a single layer of glass fibre rovings and/or carbon fibre tows. The transparent matrix permitted easy recognition of fibre-matrix debonding and the debonded lengths and pulled out lengths of fibres could be measured accurately.

Estimations of the energy dissipated in the various failure processes, fibre-matrix sliding (post-debond sliding energy), fracture of a debonded fibre (fibre deformational energy), and pulling out of broken fibres (fibre pull-out energy), can be made by combining values of fibre debond length and fibre pull-out length, together with expressions for each fracture energy term, assuming a constant value of fibre strength and stiffness, and frictional interfacial shear stress.

Based on the work of Kirk et al in which the post-debond fibre sliding mechanism was proposed as the principal source of energy dissipation, we decided to look at the debonding of glass fibres in epoxy, observing the progressive nature of the debonding process with increasing tensile load (deflexion). It involved determining the work to fracture a glass fibre-epoxy composite as a function of the number of glass fibre rovings, measuring the extent of fibre-matrix debonding with increasing load (deflexion), and measuring the lengths of pulled out broken fibres. The data, together with the fracture models, were used to determine the contribution of each microfracture process to the total work to fracture the composite. Finally, comparisons are made with earlier work of Kirk et al.

EXPERIMENTAL

Work of Fracture

The specimen design and fractographic techniques used have been described previously (Kirk et al). Briefly, a rectangular beam of epoxy resin containing a single layer of glass fibre rovings close to the tensile face is loaded in three-point bending. The dimensions of the beam are 30 mm × 10 mm × 2.5 mm. The work to fracture the specimen is determined by dividing the area under the load/deflexion curve by twice the cross-sectional area of the beam. Average values for the length of debonded fibre and pulled out fibre are evaluated using the optical microscop technique. An average fibre debond length and fibre pull-out length were calculated by summing the measurements of debonding and pull-out for each roving in a number of specimens and dividing by the total number of rovings. The number of glass fibre rovings in each specimen was varied between one and five with 1600 fibres in each roving. At least ten duplicate fracture tests were carried out for each group of specimens and in some instances, average values of fibre debonded and pull-out lengths were obtained from about 100 measurements.

Fibre-Matrix Debonding

In some cases, specimens containing five glass fibre rovings were loaded in flexure and the extent of fibre-matrix debonding was determined using an apparatus shown schematically in Figure 1. The tensile side of each specimen was illuminated and a 35 mm camera used to photograph an image of a debonded glass fibre roving in a mirror located beneath the specimen. The mirror was held parallel to the plane of fibres and since this plane moved as the specimen deflected, the mirror and camera were rotated also. So that the debonding process could be viewed on both sides of the matrix fracture plane, the mirror-camera assembly was pivoted from side to side as the test progressed. As each picture was taken, the load and deflexion of the specimen were noted. The photographs were enlarged and from them an average fibre debonded length for each roving in each specimen was deter-

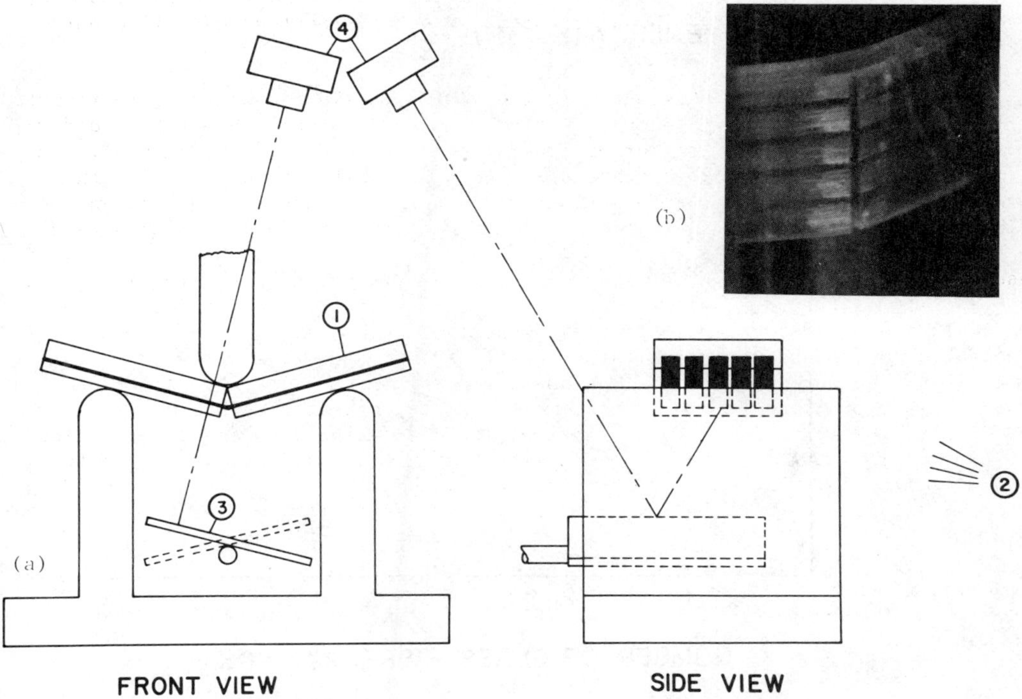

Fig. 1 (a) Experimental rig to photograph the extent of fibre-matrix debonding during a three-point bend test. (1 = specimen; 2 = light source; 3 = pivoting mirror; 4 = 35 mm camera).

(b) A typical photograph of one side of a specimen showing the extent of fibre-matrix debonding in each roving.

mined as a function of load (deflexion).

All tests were carried out at 20°C (± 2°C) using a 10,000 kg Instron tensile testing machine and a displacement-rate of 0.05 mm/s.

RESULTS AND DISCUSSION

Fracture Mechanisms and Toughness

The work of fracture (toughness) as a function of the number of glass fibre rovings is approximately linear between one and four rovings and increases at a higher rate for a specimen containing five rovings (Figure 2). Extrapolating the linear portion of the curve to intersect the ordinate shows a fracture energy for the epoxy resin of 0.5 kJ/m^2 which is reasonable.

The individual fracture energy terms for post-debond fibre sliding energy, fibre debond (deformational) energy and fibre pull-out energy, have been calculated using average values for fibre debond length and fibre pull-out length, together with the expressions for the three energy terms. The distribution of fibre lengths

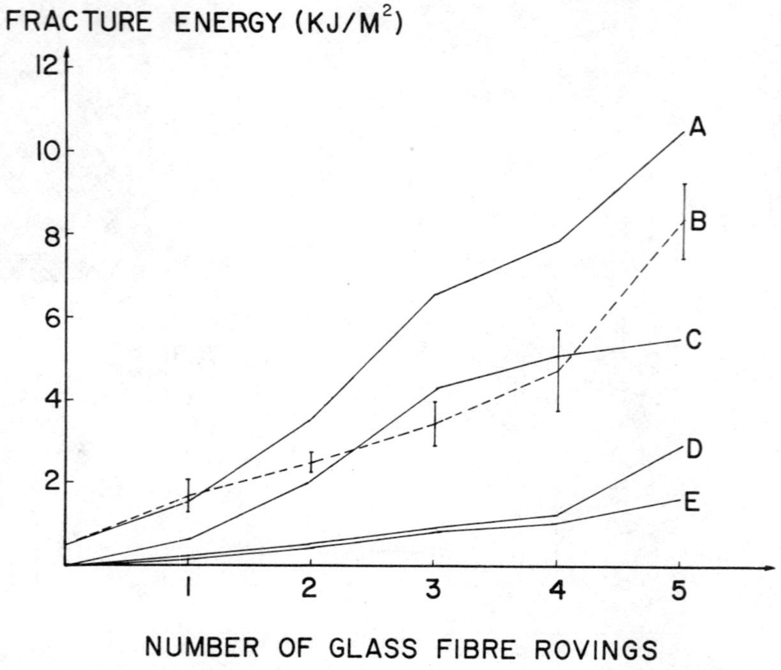

Fig. 2 The individual and total energy terms as a function of the number of glass fibre rovings. The standard deviations are shown by the bars.

A - total theoretical energy,
B - experimental work of fracture,
C - fibre post debond friction (sliding) energy,
D - fibre pull-out energy,
E - fibre debonding energy.

is shown in the form of histograms (Figure 3). The difference between failure strain of fibre and matrix, $\Delta\varepsilon$, has been taken as 0.01. The frictional interfacial shear stress was estimated using the expression $\tau = \sigma_f \, d/8\ell_p$, where ℓ_p is the average fibre pull-out length (Kirk et al). The principal origin of the work to fracture the composite is the post-debonding fibre sliding mechanism; the energy released when a debonded fibre snaps and the energy dissipated when a broken fibre is pulled out of a cracked matrix are comparable in magnitude and together contribute less than one-half of the observed work of fracture. It is interesting to note that the post-debond fibre sliding energy increased slightly for the specimen containing five glass fibre roving, while the fibre pull-out energy term increased by an amount greater than expected from the initial linear portion of the curve. This reflects a larger average length of pulled out fibre for the specimen containing five rovings than the trend indicated. The important point to notice, however, is the total theoretical energy curve (curve A), the sum of curves C, D and E, together with the matrix contribution, continues to increase in agreement with the work of fracture data.

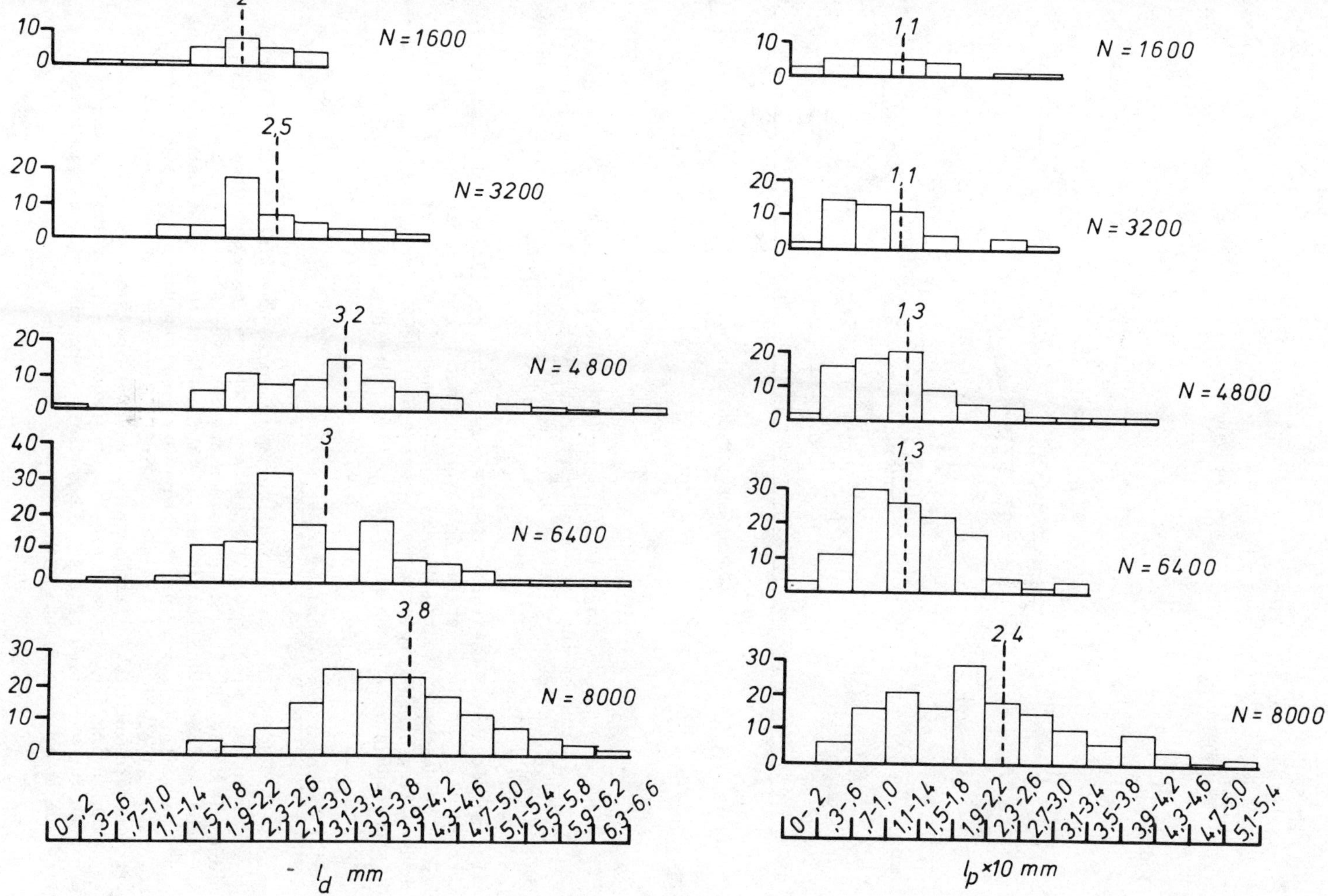

Fig. 3 Histograms showing the distribution of fibre debond lengths (a) and fibre pull-out lengths (b). The average values are shown by the dotted lines. (Each roving contains 1600 glass fibres.)

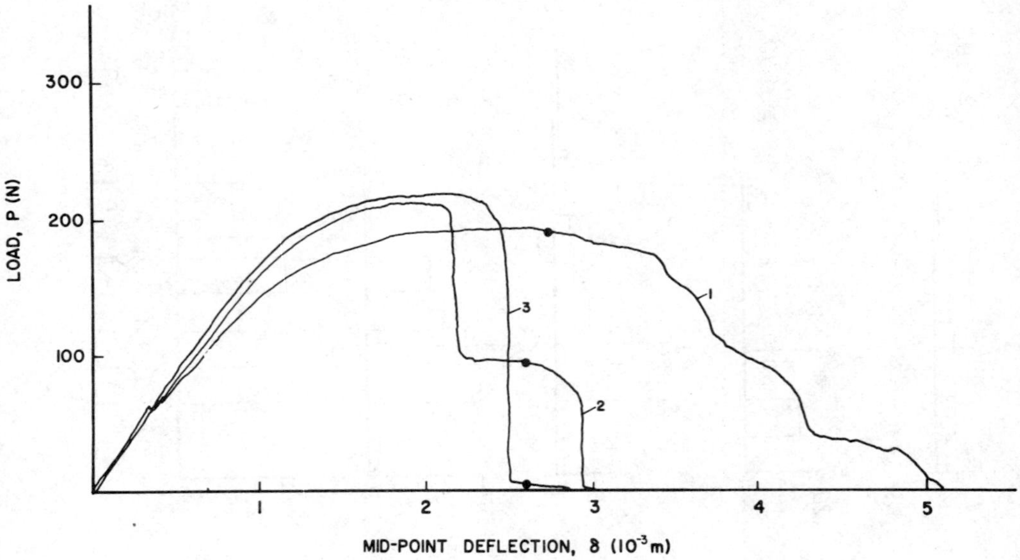

Fig. 4 Load/deflexion curves for the three specimens used in the debonding-deflexion correlation study. The solid points indicate the load and deflexion at maximum fibre-matrix debonding.

The relative magnitudes of the theoretical energy terms are similar to those observed by Kirk et al. The dominant energy absorbing mechanism is post-debond fibre sliding and the important parameter in the model is ℓ_d^2, where ℓ_d is the total debonded length of fibre. In this glass fibre-epoxy system, ℓ_d is an orders of magnitude greater than the fibre pull-out length, ℓ_p.

Dependence of Fibre-Matrix Debonding on Deflexion

Three typical load/deflexion curves for specimens containing five fibre rovings are shown in Figure 4. Even though the shapes of the curves vary, similar values of mid-point deflexion for each specimen are observed for the maximum length of debonded fibre. An instantaneous fibre debond length was determined for each roving in three specimens, together with corresponding load and mid-point deflexion. Typical results are shown in Figure 5 for each glass fibre roving with measurements taken from both halves of a specimen. The data points for each roving are plotted using a least squares linear regression analysis. With the exception of a single roving on one side of a matrix crack, all curves have approximately the same slope.

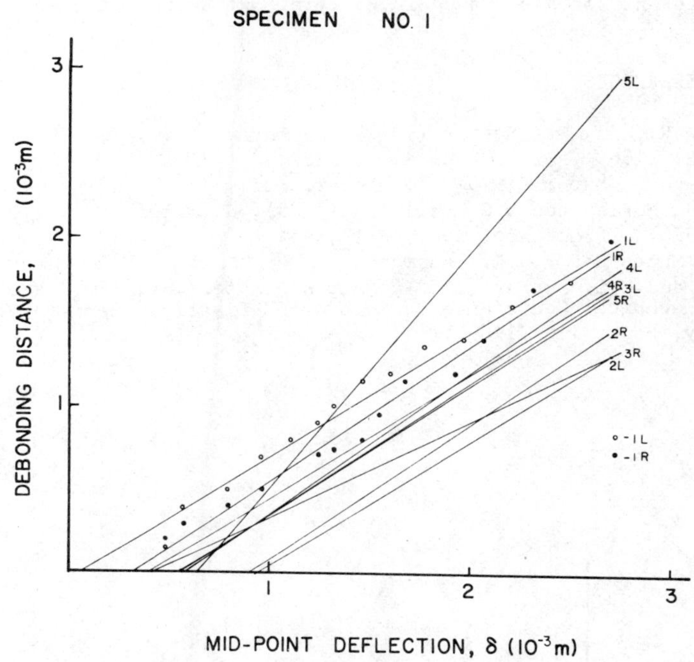

Fig. 5 Fibre debond length as a function of mid-point deflexion for each roving on both sides of the matrix fracture plane. The experimental points are for roving 1 only.

The average debonded length, $\ell_d/2$, for 5 tows in a specimen was determined as a function of mid-point deflexion. Figure 6 shows a least squares linear regression of the average debonded length for one-half of three specimens tested. In each case, there is a linear relationship between fibre debonded length and deflexion. One point worth highlighting is the critical or threshold deflexion that must be exceeded for the debonding process to begin.

CONCLUSIONS

Post-debond fibre sliding appears to be the principal failure mechanism for the dissipation of energy in a glass fibre-epoxy composite. The extent of fibre debonding is directly proportional to the mid-point deflexion of the beam and a threshold value of deflexion must be exceeded for debonding to begin. The work done in the post-debond fibre sliding process contributes more than one-half to the work to fracture the composite, while the energy released when a debonded fibre snaps and the energy dissipated when a broken fibre pulls out are comparable in magnitude.

ACKNOWLEDGEMENTS

We would like to thank the Air Force Office of Scientific Research and the

National Research Council of Canada (Grant No. A0440) for their financial support. Also A. St-Laurent for assistance in carrying out some of the experiments.

REFERENCES

Beaumont, P.W.R., and B. Harris (1972), *J. Mater. Sci.*, 7, 1265.
Cottrell, A.H. (1964), *Proc. Roy. Soc.*, A282, 2.
Gershon, B. and G. Marom (1975), *J. Mater. Sci.*, 10, 1549.
Harris, B., J. Morley and D.C. Phillips (1975), *J. Mater. Sci.*, 10, 2050.
Kelly, A. (1970), *Proc. Roy. Soc.*, A319, 95.
Kirk, J., M. Munro and P.W.R. Beaumont, (1978), *J. Mater. Sci.*, 13, 2187.
Outwater, J. and M.C. Murphy (1970), *Modern Plastics*, 7, 160; see also 24th Annual Technical Conference on Reinforced Plastics, Society of Plastics Industry, Paper 11C (1969).

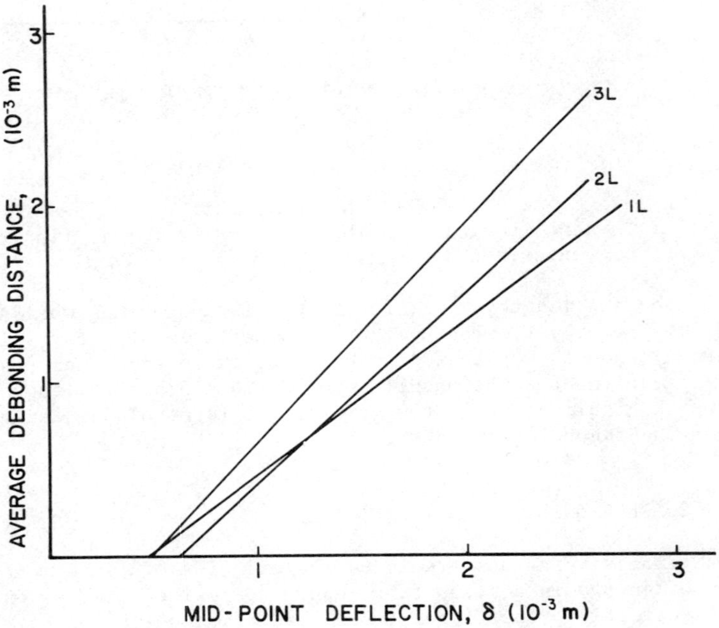

Fig. 6 Average fibre debond length for the left side only of three specimens as a function of mid-point deflexion.

TABLE I Theoretical Models of Fracture

Model	Equation (Kirk et al)	Reference
Fibre debonding (deformational) energy	$W_d = \pi d^2 \sigma_f^2 \ell_d / 8E_f$	Outwater and Murphy (1970)
Post-debond fibre sliding energy	$W_{pdf} = \pi d \tau \ell_d^2 \Delta\varepsilon / 2$	Kelly (1970) Harris et al (1975)
Fibre pull-out energy	$W_p = 2\pi d \tau \ell_p^2 / 3$	Cottrell (1964) Kelly (1970)

d = fibre diameter
$\Delta\varepsilon$ = difference between fibre and matrix failure strains
E = Young's modulus of fibre
ℓ_d = fibre debond length
ℓ_p = fibre pull-out length
σ_f = tensile strength of fibre
τ = frictional interfacial shear stress

Volume 3, ICM 3, Cambridge, England, August 1979

THERMOELASTICITY AND PLASTICITY OF COMPOSITE MATERIALS

O. B. Pedersen

Metallurgy Department, Risø National Laboratory, DK-4000 Roskilde, Denmark

INTRODUCTION

Thermoelasticity represents perhaps the best understood aspect of the behaviour of composite materials. Here the incomplete structural characterisation of the materials typical of the practical situation is faced and taken rigorously into account in theoretical predictions. The linear theory based on Hashin and Shtrikman's (1963) variational principle provides the closest possible upper and lower bounds for the thermoelastic moduli in terms of phase volume fractions. The variational theory has quite recently been formulated as a simple mean field theory (Pedersen 1978) based on the principles of the Brown and Stobbs (1971) theory of the plasticity of dispersion hardened alloys. The mean field theory owes its simplicity to the fact that functionals which must be explicitly minimised in the variational theory are automatically at their minimum in the mean field theory. General matrix equations bounding the effective linear equilibrium - and steady - state transport properties of composites can then be fairly easily obtained.

For thermoelastic and plastic deformation the mean field theory is derived from a simple physical picture: When a force is applied to the composite through external grips (or when its temperature is changed) every inclusion responds like an internal "microgrip" and applies a self-equilibrating distribution of force to the composite. According to the Saint Venant principle this produces substantial local stresses at the microgrip but has negligible effects on the stresses at distances which are large compared with its linear dimensions. However, for a finite composite body Brown and Stobbs emphasise that in addition to the non-uniform local stresses the internal grips must produce a quasi-uniform image stress throughout the material to ensure mechanical equilibrium.

This short communication gives an abbreviated but concise presentation of the mean field theory in its version for the thermoelastic and plastic deformation of 2-phase alloys. It shows that the theory enables one to calculate the internal stress in an alloy with a large volume fraction of hard phase. Further progress can then be made in efforts to develop a quantitative dislocation theory of plasticity which may be of use to engineers and materials scientists concerned with the design and practical applications of composite materials.

MEAN FIELD THEORY

The idea that an inclusion can be thought of as essentially a grip is the basis of Eshelby's (1957) transformation theory which is now a standard method in studies of internal stresses. Consider an infinite homogeneous linear elastic material containing a finite ellipsoidal inclusion which would undergo a prescribed uniform strain (the transformation strain $\underset{\sim}{\varepsilon}^{T}$) if it were free. Because of the constraint of the surrounding material (the matrix) there is a uniform stress

$$\underset{\sim}{\sigma}^{\infty I} = \underset{\sim}{L}(\underset{\sim}{S} - \underset{\sim}{I})\underset{\sim}{\varepsilon}^{T}$$

inside the inclusion. The stress outside the inclusion is highly nonuniform but is at this stage of no interest. Our notatation is that of Nye (1960) and Hill (1965): Latin index pairs in the stress, strain and stiffness tensors (σ_{ij}, ε_{ij} and L_{ijkl}) are replaced by single Greek indices according to the scheme

ij	11	22	33	23/32	13/31	12/21
α	1	2	3	4	5	6

and the resulting matrices (σ_{α}, ε_{α} and $L_{\alpha\beta}$) are represented in the unindexed form ($\underset{\sim}{\sigma}$, $\underset{\sim}{\varepsilon}$ and $\underset{\sim}{L}$). The $\underset{\sim}{I}$ is the 6 x 6 unit matrix. The $\underset{\sim}{L}$ is for isotropic elasticity of the form

$\underset{\sim}{L}$ = 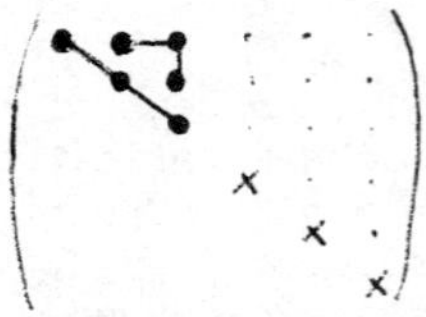

● non-zero element

· zero element

× $\frac{1}{2}(L_{11}-L_{12})$

where the lower half of the array is left out to indicate symmetry about the diagonal. Eshelby's matrix, $\underset{\sim}{S}$, is for an isotropically elastic sphere

$\underset{\sim}{S}$ = 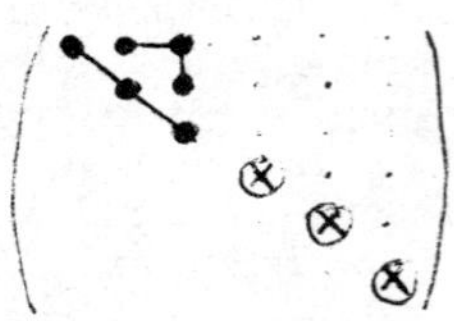

$$S_{11} = \frac{7-5\nu}{15(1-\nu)}$$

$$S_{12} = \frac{5\nu-1}{15(1-\nu)}$$

$$\otimes = S_{11}-S_{12}$$

where ν is Poisson's ratio. For an isotropically elastic circular cylindrical fibre along the x_3-axis we have

$\underset{\sim}{S}$ = 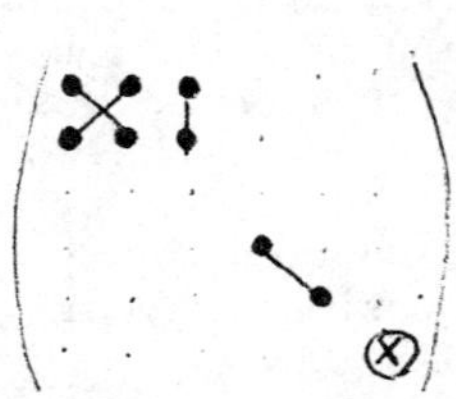

$$S_{11} = \frac{5-4\nu}{8(1-\nu)}$$

$$S_{12} = \frac{4\nu-1}{8(1-\nu)}$$

$$S_{13} = \frac{\nu}{2(1-\nu)}$$

$$S_{44} = \frac{1}{4}$$

where the lower half of the array must be retained since for the fibre $\underset{\sim}{S}$ is not symmetrical about the diagonal.

Let us now imagine that the inclusion is a second phase region whose stiffness $\underset{\sim}{L}_2$ differs from the matrix stiffness $\underset{\sim}{L}_1$. In the transformation theory the stiffness of the inclusion is set equal to that of the matrix. The value of this striking simplification is that an "equivalent transformation strain" $\underset{\sim}{\varepsilon}^{\textcircled{T}}$ must be assigned to the inclusion. It will be shown in a forthcoming detailed paper how $\underset{\sim}{\varepsilon}^{\textcircled{T}}$ can be written as

$$\underset{\sim}{\varepsilon}^{\textcircled{T}} = \underset{\sim}{P}_{12}(\underset{\sim}{L}_{12}\underset{\sim}{\varepsilon}^{U} + \underset{\sim}{L}_{2}\underset{\sim}{\varepsilon}^{T})$$

with

$$\underset{\sim}{L}_{12} = \underset{\sim}{L}_1 - \underset{\sim}{L}_2$$

and

$$\underset{\sim}{P}_{12} = (\underset{\sim}{L}_1 - \underset{\sim}{L}_{12}\underset{\sim}{S}_{12})^{-1}$$

Eshelby's matrix $\underset{\sim}{S}_{12}$ is given indices 1 and 2 to indicate that it is to be calculated on the basis of the elastic constants of phase 1 and the shape of the phase 2 regions.

To describe the thermal and plastic deformation of the composite we write the transformation strain as

$$\underset{\sim}{\varepsilon}^T = -\underset{\sim}{\varepsilon}^P_{12} - \underset{\sim}{\alpha}_{12}\,\Delta T$$

with

$$\underset{\sim}{\varepsilon}^P_{12} = \underset{\sim}{\varepsilon}^P_1 - \underset{\sim}{\varepsilon}^P_2 \quad \text{and} \quad \underset{\sim}{\alpha}_{12} = \underset{\sim}{\alpha}_1 - \underset{\sim}{\alpha}_2$$

where ε^P_1 and ε^P_2 are the uniform plastic strains in phases 1 and 2, $\underset{\sim}{\alpha}_1$ and $\underset{\sim}{\alpha}_2$ are the thermal expansion coefficients and ΔT is the temperature increase.

The $\underset{\sim}{\varepsilon}^U$ is a uniform elastic strain extending throughout matrix and inclusions. In the composite subjected to an applied stress $\underset{\sim}{\sigma}^A$ and differences in the thermal and plastic strains in the phases one can identify such a "surrounding strain", $\underset{\sim}{\varepsilon}^S$; but $\underset{\sim}{\varepsilon}^S$ is highly non-uniform and difficult to calculate. Its mean value is nevertheless easily found: mechanical equilibrium requires that the mean value of the total stress (internal plus applied) over the volume V of the composite equals the applied stress. That is

$$\langle\underset{\sim}{\sigma}\rangle = \underset{\sim}{\sigma}^A$$

where

$$\langle\underset{\sim}{\sigma}\rangle = \frac{1}{V}\int_V \underset{\sim}{\sigma}\, dV$$

The total stress in the matrix differs from that in the inclusions: In the matrix $\underset{\sim}{\sigma}$ simply equals $\underset{\sim}{\sigma}^S$ but in the inclusions it is the sum of $\underset{\sim}{\sigma}^S$ and the inclusion stress $\underset{\sim}{\sigma}^{\infty I}$. Consequently the condition for mechanical equilibrium becomes

$$\langle\underset{\sim}{\sigma}^S\rangle = \underset{\sim}{\sigma}^A - f_2\,\underset{\sim}{\sigma}^{\infty I}$$

where f_2 is the volume fraction of inclusions.

The radical step taken in the variational theory is in effect to represent the surrounding stress as a uniform stress whose value equals the actual mean value of $\underset{\sim}{\sigma}^S$. In other words

$$\underset{\sim}{\sigma}^S \equiv \underset{\sim}{\sigma}^U = \langle\underset{\sim}{\sigma}^S\rangle$$

and, of course,

$$\underset{\sim}{\varepsilon}^S \equiv \underset{\sim}{\varepsilon}^U = \langle\underset{\sim}{\varepsilon}^S\rangle$$

since $\underset{\sim}{\sigma}^U$ and $\underset{\sim}{\varepsilon}^U$ are related by Hooke's law for the matrix

$$\underset{\sim}{\sigma}^U = \underset{\sim}{L}_1\underset{\sim}{\varepsilon}^U$$

By straightforward combination of these equations one then finds the equivalent transformation strain in the form

$$\varepsilon^{(T)} = \underset{\sim}{Q}_{12}\underset{\sim}{P}_{12}\,[\underset{\sim}{L}_{12}\underset{\sim}{\varepsilon}^{A} - \underset{\sim}{L}_{2}(\underset{\sim}{\varepsilon}^{P}_{12} + \underset{\sim}{\alpha}_{12}\,\Delta T)]$$

where

$$\underset{\sim}{Q}_{12} = [\underset{\sim}{I} + f_2\,\underset{\sim}{P}_{12}\underset{\sim}{L}_{12}(\underset{\sim}{S}_{12} - \underset{\sim}{I})]^{-1}$$

The $\underset{\sim}{\varepsilon}^{(T)}$ is clearly related to the mean value of the strain of the composite, as it would be calculated from average displacements on its external surface. The mean value of the internal stress, $\underset{\sim}{\sigma}-\underset{\sim}{\sigma}^{A}$, caused by $\underset{\sim}{\varepsilon}^{(T)}$ is zero, according to the equilibrium condition. The corresponding mean elastic strain then also vanishes, so the contribution of $\underset{\sim}{\varepsilon}^{(T)}$ to the total mean strain is

$$f_2\,\underset{\sim}{\varepsilon}^{(T)}$$

In addition to this strain the composite suffers the uniform strain

$$\underset{\sim}{M}_1\underset{\sim}{\sigma}^{A} + \underset{\sim}{\varepsilon}^{P}_1 + \underset{\sim}{\alpha}_1\,\Delta T$$

where the compliance matrix $\underset{\sim}{M}$ is the inverse of the stiffness matrix

$$\underset{\sim}{M} = \underset{\sim}{L}^{-1}$$

The total mean strain then is the sum

$$<\underset{\sim}{\varepsilon}> = \underset{\sim}{M}_1\underset{\sim}{\sigma}^{A} + \underset{\sim}{\varepsilon}^{P}_1 + \underset{\sim}{\alpha}_1\,\Delta T + f_2\,\underset{\sim}{\varepsilon}^{(T)}$$

which becomes

$$\begin{aligned}<\underset{\sim}{\varepsilon}> &= (\underset{\sim}{I} + f_2\,\underset{\sim}{Q}_{12}\underset{\sim}{P}_{12}\underset{\sim}{L}_{12})\underset{\sim}{M}_1\underset{\sim}{\sigma}^{A} \\ &+ (\underset{\sim}{\alpha}_1 - f_2\underset{\sim}{Q}_{12}\underset{\sim}{P}_{12}\underset{\sim}{L}_2\underset{\sim}{\alpha}_{12})\Delta T \\ &+ (\underset{\sim}{\varepsilon}^{P}_1 - f_2\underset{\sim}{Q}_{12}\underset{\sim}{P}_{12}\underset{\sim}{L}_2\varepsilon^{P}_{12})\end{aligned}$$

when the $\underset{\sim}{\varepsilon}^{(T)}$ expression is inserted. The usual definition of effective properties provides $<\underset{\sim}{\varepsilon}>$ in the form

$$<\underset{\sim}{\varepsilon}> = \underset{\sim}{M}^{C}\underset{\sim}{\sigma}^{A} + \underset{\sim}{\alpha}^{C}\,\Delta T + \underset{\sim}{\varepsilon}^{PC}$$

which by comparison shows that the effective compliance of the composite is

$$\underset{\sim}{M}^{C} = (\underset{\sim}{I} + f_2\underset{\sim}{Q}_{12}\underset{\sim}{P}_{12}\underset{\sim}{L}_{12})\underset{\sim}{M}_1$$

the effective thermal expansion coefficient matrix is

$$\underset{\sim}{\alpha}^{C} = \underset{\sim}{\alpha}_1 - f_2\underset{\sim}{Q}_{12}\underset{\sim}{P}_{12}\underset{\sim}{L}_2\underset{\sim}{\alpha}_{12}$$

and the plastic deformation of the composite is

$$\underset{\sim}{\varepsilon}^{PC} = \underset{\sim}{\varepsilon}^{P}_1 - f_2\underset{\sim}{Q}_{12}\underset{\sim}{P}_{12}\underset{\sim}{L}_2\underset{\sim}{\varepsilon}^{P}_{12}$$

The mean surrounding stress in the composite is important for understanding its ductile strength. The $<\underset{\sim}{\sigma}^{S}>$ is obtained by combining the $\underset{\sim}{\sigma}_{\infty 1}$ expression with the equilibrium condition. The result is

$$<\underset{\sim}{\sigma}^{S}> = \underset{\sim}{\sigma}^{A} - f_2\underset{\sim}{L}_1(\underset{\sim}{S}_{12} - \underset{\sim}{I})\underset{\sim}{\varepsilon}^{(T)}$$

or

$$\langle \sigma^S \rangle = \sigma^A - f_2 L_1 (S_{12} - I) Q_{12} P_{12} (L_{12} M_1 \sigma^A - L_2 \varepsilon^P_{12} - L_2 \alpha_{12} \Delta T)$$

If the inclusions are randomly positioned this will also be the mean stress in the matrix (phase 1). The stress acting on a dislocation in the matrix is then not just the applied stress: it is the applied stress offset by the internal stresses due to inhomogeneous thermal, plastic and elastic deformation.

THERMOELASTICITY

It is essential to make the right interpretation of the mean field equations for M^C, α^C, ε^{PC} and $\langle \sigma^S \rangle$. Clearly, the equations will provide good approximations whenever the inclusions are sufficiently far apart to interact mainly via the image field. This is the situation of a low volume fraction of evenly distributed inclusions. On the other hand, the approximation is likely to break down if the inclusions cluster into close-packed types of configurations, and this must happen when their volume fraction becomes too large. In such situations the inclusions will interact both via the image field and via their local fields. This complex effect is not directly dealt with in the present version of the theory. It is therefore pertinent to ask in what sense we have in fact obtained a valid approximation even for large inclusion volume fractions.

The nature of the mean field approximation is best illustrated through comparisons with other results in the literature on thermomechanical properties of composite materials. This literature is now very extensive, as can be seen in the recent reviews given by Hashin (1970), Rosen (1970), Kelly (1973) and Hale (1976). First, it is interesting to note that the mean field approximation does not seem inconsistent with Hill's (1965) variant of the self-consistent scheme. The equations for M^C and α^C may be written as

$$(M^C - M_1) L_1 L_{12}^{-1} = f_2 Q_{12} P_{12}$$

and

$$(\alpha^C - \alpha_1) \alpha_{12}^{-1} L_2^{-1} = - f_2 Q_{12} P_{12}$$

from which the $f_2 Q_{12} P_{12}$ can be eliminated to provide

$$\alpha^C = \alpha_1 - (M^C - M_1) L_1 L_{12}^{-1} L_2 \alpha_{12}$$

This is Levin's relation, as reviewed and extended by Rosen and Hashin (1970). Levin makes an elegant extension of Hill's method in which he introduces and then eliminates a set of unknown influence coefficients. In the present derivation the eliminated quantity is known (at least approximately), and the two approaches are clearly quite different.

Another basic approach uses Hashin and Shtrikman's (1963) variational principle. In the thermoelastic case this involves minimising free energy with mechanical equilibrium as a subsidiary condition. In essence Pedersen and Brown (1977) showed that for ellipsoidal inclusions the composite is in thermodynamical equilibrium with an applied stress whenever it is in mechanical equilibrium. This suggests that a shift from Hashin and Shtrikman's energy view to the simple force view of Brown and Stobbs (1971) should lead to a simplified formulation of the variational theory for ellipsoid phase structures. This approach is the mean field theory and a detailed proof of its equivalence to Hashin and Shtrikman's approach is given elsewhere (Pedersen 1979). In the variational theory the phase structure is characterised by the phase volume fractions only. Otherwise the structure is left completely unrestricted to cover the ellipsoid structure as one special case. The equivalence of the two formulations then shows that the ellipsoid structure is an extreme one which provides a variational bound rather than any other approximation.

Let us finally apply the mean field theory to predict the stiffness of the technologically important glass fibre reinforced plastics. Kelly (1973) provides the elastic constants of the isotropic phases: Poisson's ratio, ν, and Young's modulus, E, is for epoxy ($\nu = 0.3$, $E = 3GPa$) and for glass ($\nu = 0.25$, $E = 72$ GPa). For 70 vol.pct. long glass fibres along the X_3 axis in epoxy our $\underset{\sim}{M}^C$ expression from the last section then provides (in TPa^{-1})

$$\underset{\sim}{M}^C = \begin{pmatrix} 74.7 & -34.7 & -5.47 & 0 & 0 & 0 \\ & 74.7 & -5.47 & 0 & 0 & 0 \\ & & 19.5 & 0 & 0 & 0 \\ & & & 194.6 & 0 & 0 \\ & & & & 194.6 & 0 \\ & & & & & 218.8 \end{pmatrix}$$

This is one variational bound. In the variational theory (and hence in the mean field equations) the phases are completely interchangeable. Let us calculate $\underset{\sim}{M}^C$ for epoxy fibres in glass; at 70 vol.pct. glass this is in fact not a bad description of the phase structure. The result

$$M^C = \begin{pmatrix} 28.8 & -8.4 & -5.0 & 0 & 0 & 0 \\ & 28.8 & -5.0 & 0 & 0 & 0 \\ & & 19.5 & 0 & 0 & 0 \\ & & & 61.5 & 0 & 0 \\ & & & & 61.5 & 0 \\ & & & & & 72.9 \end{pmatrix}$$

is the other variational bound. The bounds compare favourably with the recent experimental results of Clemens and Moore (1978)

$$M^C_{33} = 17.5 - 18.1 \;\; ; \;\; -M^C_{13} = 3.7 - 4.2$$

$$M^C_{11} = 54.4 - 62.2 \;\; ; \;\; M^C_{55} = 127 - 145$$

which should be 95 per cent confidence limits for nominally the same composite. The compliances are perhaps better understood in terms of engineering moduli:

$$\underset{\sim}{M} = \begin{pmatrix} 1/E_{11} & -\nu_{12}/E_{22} & -\nu_{13}/E_{33} & 0 & 0 & 0 \\ & 1/E_{22} & -\nu_{23}/E_{33} & 0 & 0 & 0 \\ & & 1/E_{33} & 0 & 0 & 0 \\ & & & 1/G_{23} & 0 & 0 \\ & & & & 1/G_{13} & 0 \\ & & & & & 1/G_{12} \end{pmatrix}$$

It is worth noting that for glass-epoxy only the longitudinal constants, E_{33} and ν_{13}, are in fact predicted with good precision. For the other constants the bound separation is typically 60-70 per cent of the value of the upper bound. The self-cocsistent scheme will provide a single value for each of the constants; but it is difficult from the physical picture of this method to say whether the value is more realistic than any other approximate value one could find between the bounds. An examination of this question based on Tanaka and Mori's (1972) and Yoffe's (1974) generalisation of Brown and Stobbs (1971) mean value theorem is in progress. This theorem enables one to calculate a good approximation to the quasi-uniform image field and then take local field interactions approximately into account for various inclusion lattices.

PLASTICITY

The energy expended in reversible thermoelastic deformation is stored as free energy and can be completely recovered by taking the alloy back to its initial state. In contrast, plastic deformation involves internal friction by which free energy is irreversibly lost, mainly as heat. Recent theoretical work (Foreman and Makin, 1966; Ashby, 1966; Brown and Stobbs, 1971; Tanaka and Mori, 1970; Hirsch and Humphreys, 1971) suggests a dislocation picture of the plastic deformation of 2-phase alloys in which a soft phase flows plastically around inclusions of a hard non-plastic phase. The flow stress of the alloy is understood basically as the threshold value of the applied stress above which a (single) dislocation will sweep out a large area in the alloy. In the energy view the hard phase stores free energy which may be recovered, for example in reverse deformation, as in the Bauschinger experiment. There are then two major types of contributions to the workhardening of these alloys: First, the reversible kinematical hardening reflecting path independent storage of free energy; second, the irreversible frictional hardening reflecting the path dependent loss of free energy occurring in dislocation motions. In the force view the kinematical hardening is caused by the internal grips of hard phase inclusions. The grips give rise to the mean stress, which, by opposing further forward motion and assisting reverse motion of the dislocations, gives rise to the permanent softening measured in Bauschinger experiments. The internal grips also produce local stresses which make the dislocation line bend and liberate heat. This friction mechanism leads to the so-called source shortening stress, which, like other friction stresses, is supposed to oppose forward and reverse dislocation motion equally.

For composites with a few volume per cent hard inclusions this picture is quantitatively consistent with a wide range of experimental data (Wilson, 1965; Gould, Hirsch and Humphreys, 1974; Lasalmonie and Martin, 1974; Atkinson, Brown and Stobbs, 1974; Pedersen, 1976; Lilholt, 1977 a). At such dilution it is possible to ignore elastic interactions among the inclusions when calculating the internal stress. However, for large volume fractions, greater than about 10 vol.pct., this approximation fails; Brown and Clarke (1977) and Lilholt (1977 b) mention the considerable difficulties met when attempting to compare theory and experiment at large volume fractions of hard phase. The mean field theory was developed with an eye to these difficulties and it gives quite unambiguous predictions of the internal stress at any volume fraction.

A good illustration of the mechanical behaviour of composites with large volume fractions of elastic inclusions is provided by Kelly and Lilholt (1969). They measured the flow stress at copper containing up to 38 vol.pct. very long tungsten fibres.

Figure 1 shows the stress-strain curve for a composite which is pulled along the direction of the fibre axes (0-X_3). The curve displays three stages: In stage I both phases stay elastic, is stage II the matrix flows around the fibres which do not yet flow, and in stage III both phases flow. The unloading curve demonstrates that a fairly large part of the total strain $\langle\varepsilon_{33}\rangle$ is elastic. This contrasts with the behaviour of dispersion hardened alloys, where the elastic part of $\langle\varepsilon_{33}\rangle$ is usually small. A small amount of the plastic strain is recovered as the composite is unloaded.

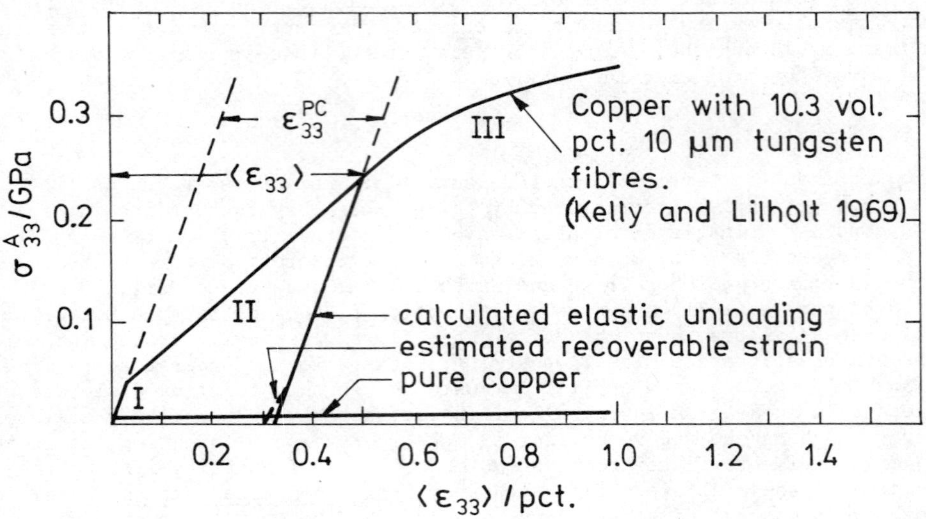

Figure 1. Stress-strain curves for pure copper and for copper with tungsten fibres along X_3, measured by Kelly and Lilholt (1969).

The workhardening rate in stage II can be obtained from Figure 1 as

$$\frac{d\sigma_{33}^{A}}{d\varepsilon_{33}^{PC}} = 61.5 \text{ GPa}$$

and Figure 2 shows results from the whole series of measurements.

The theoretically predicted workhardening rates are also shown. The mean stress contributions were calculated on the basis of Foreman's (1955) effective moduli for copper

$$G_{Cu} = 42 \text{ GPa} \quad \text{and} \quad \nu_{Cu} = 0.45$$

and Lowrie and Gonas' (1965) moduli for tungsten

$$G_{W} = 162 \text{ GPa} \quad \text{and} \quad \nu_{W} = 0.28$$

The bounds for the total mean stress are for all practical purposes equal to the value suggested by Brown and Clarke's (1977) calculations. However, the mean field theory provides two mean stress contributions: <u>The plastic mean stress</u> is the mean surrounding stress in the unloaded composite. It is given in the form

$$f_2 \underset{\sim}{L}_1 (\underset{\sim}{S}_{12} - \underset{\sim}{I}) \underset{\sim}{Q}_{12} \underset{\sim}{P}_{12} \underset{\sim}{L}_2 \underset{\sim}{\varepsilon}_{12}^{P}$$

where ε_{12}^{P} is the plastic strain produced in the matrix during stage II. In the loaded composite there is an additional mean stress contribution, presented here for the first time. It is <u>the elastic mean stress</u> given as

$$- f_2 \underset{\sim}{L}_1 (\underset{\sim}{S}_{12} - \underset{\sim}{I}) \underset{\sim}{Q}_{12} \underset{\sim}{P}_{12} \underset{\sim}{L}_{12} \underset{\sim}{\varepsilon}^{A}$$

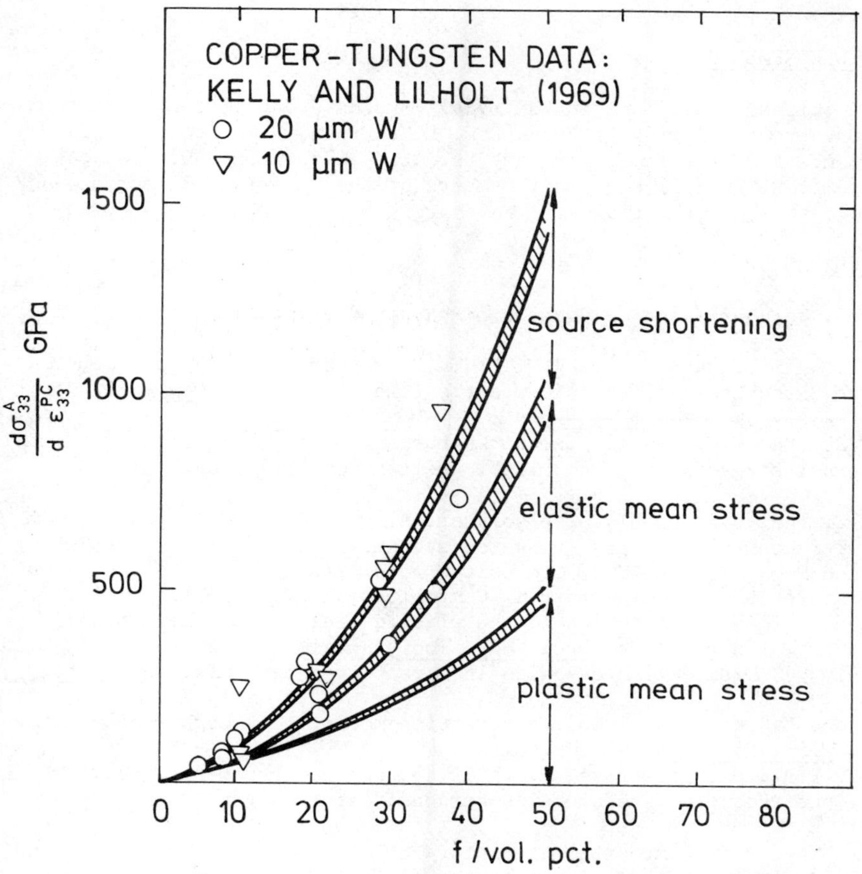

Figure 2. Stage II workhardening of copper-tungsten, measured by Kelly and Lilholt (1969).

At small volume fractions of fibres the elastic mean stress is negligible in comparison with the plastic mean stress; but at large volume fractions the two contributions become roughly equal for copper-tungsten.

The total mean stress alone clearly cannot account for all the hardening. The mean stress looks more like a lower limit to the hardening, as suggested by Brown and Clarke (1977). The extra hardening observed in Figure 2 seems to be due to the internal friction associated with the local stresses: Brown and Clarke's (1977) estimate of the source shortening stress leads to excellent agreement between theory and experiment. Moreover, the source shortening stress is roughly equal to the plastic mean stress, thus providing a natural explanation of the dimensional stability of the plastically deformed composite.

Bauschinger experiments on copper with up to 4 vol. pct. tungsten fibres (Lilholt 1977 a) show that the kinematical hardening and the frictional hardening in copper-tungsten can be separated experimentally. The low volume fraction experiments also cast light on the room temperature processes of plastic relaxation in which dis-location rearrangements driven by stored energy relieve internal stresses. Further

Bauschinger experiments on copper with larger contents of tungsten fibres are in progress at Risø.

ACKNOWLEDGEMENTS

The early part of this work was done in the Metal Physics Group, Cavendish Laboratory, Cambridge University. I wish to thank Drs. L.M. Brown, W.M. Stobbs, E.H. Yoffe and other members of the group for many valuable discussions. Discussions with Drs. H. Lilholt, R. Talreja and other members of the Materials Research Group at Risø have also been a great stimulation and help.

REFERENCES

Ashby, M.F. (1966). Workhardening of dispersion-hardened crystals. Phil.Mag., 14, 1157-1178.

Atkinson, J.D., Brown L.M., and Stobbs W.M. (1974). The workhardening of copper-silica IV. The Bauschinger effect and plastic relaxation. Phil.Mag., 30, 1247-1279.

Brown, L.M., and Stobbs, W.M. (1971). The workhardening of copper-silica I. A model based on internal stresses, with no plastic relaxation. Phil.Mag, 23, 1185-1199.

Brown, L.M., and Clarke, D.R. (1977). The workhardening of fibrous composites with particular reference to the copper-tungsten system. Acta Met., 25, 563-570.

Clements, L.L., and More, R.L. (1978). Composite properties for E-glass fibres in a room temperature curable epoxy matrix, Composites, April, 93-99.

Eshelby, J.D. (1957). The determination of the elastic field of an ellipsoidal inclusion, and related problems. Proc. Roy. Soc., A 241, 376-396.

Foreman, A.J.E. (1955). Dislocation energies in anisotropic crystals. Acta Met., 3, 322-330.

Foreman, A.J.E., and Makin, M.J. (1966). Dislocation movement through random arrays of obstacles. Phil. Mag., 14, 911-924.

Gould, D., Hirsch, P.B., and Humphreys, F.J. (1974). The Bauschinger effect, workhardening and recovery in dispersion-hardened copper crystals. Phil. Mag., 30, 1353-1377.

Hashin, Z. (1970). Theory of composite materials. In F.W. Wendt, H. Liebowitz, and N. Perrone (Eds)., Mechanics of composite materials, Pergamon Press, Oxford. pp. 201-242.

Hashin, Z., and Shtrikman, S. (1963). A variational approach to the theory of the elastic behaviour of multiphase materials. J. Mech. Phys. Solids, 11, 127-140.

Hale, D.K. (1976). The physical properties of composite materials. J. Mat. Sci., 11, 2105-2141.

Hill, R. (1965). A self-consistent mechanics of composite materials. J. Mech. Phys. Solids, 13, 213-222.

Hirsch, P.B., and Humphreys, F.J. (1970). The deformation of single crystals of copper and copper-zinc alloys containing alumina particles. I Macroscopic properties and workhardening theory. Proc. Roy. Soc., A 318, 45-72.

Kelly, A. (1973). Strong solids. Clarendon Press, Oxford.

Kelly, A., and Lilholt, H. (1969). Stress-strain curve of a fibre-reinforced composite. Phil. Mag., 20, 311-328.

Lasalmonie, A., and Martin, J.W. (1974). On the Bauschinger effect in unidirectionally solidified eutectic Al-Al_3Ni crystals. Scripta Met., 8, 377-382.

Lowrie, R., and Gonas, A.M. (1965). Dynamic elastic properties of polycrystalline tungsten, 24°-180°C. J. Appl. Phys., 36, 2189-2192.

Lilholt, H. (1977 a). Hardening in two-phase materials-I. Strength contributions in fibre-reinforced copper-tungsten. Acta Met., 25, 571-585.

Lilholt, H. (1977 b). Hardening in two-phase materials-II. Plastic strain and mean stress hardening rate. Acta Met., 25, 587-593.

Nye, J.F. (1960). Physical properties of crystals. Oxford University Press, London.

Pedersen, O.B. (1976). Bauschinger effect in single phase crystals, Dissertation, University of Cambridge.
Pedersen, O.B. (1978). Transformation theory for composites. Z. Ang. Math. Mech., 58, 227-228.
Pedersen, O.B., and Brown, L.M. (1977). Equivalence of stress and energy calculations of mean stress. Acta Met., 25, 1303-1305.
Rosen, B.W. (1970). Thermodynamical properties of fibrous composites. Proc. Roy. Soc., A 319, 79-94.
Rosen, B.W., and Hashin, Z. (1970). Effective thermal expansion coefficients and specific heats of composite materials. Int. J. Engng. Sci., 8, 157-173.
Tanaka, K., and Mori, T. (1970). The hardening of crystals by non-deforming particles and fibres. Acta Met., 18, 931-941.
Tanaka, K., and Mori, T. (1972). Note on volume integrals of the elastic field around an ellipsoidal inclusion. J. Elasticity, 2, 199-200.
Wilson, D.V. (1965). Reversible work hardening in alloys of cubic metals. Acta Met., 13, 807-814.
Yoffe, E.H. (1974). Calculation of elastic strain: Spherical particle in cubic material. Phil. Mag., 30, 923-933.

Volume 3, ICM 3, Cambridge, England, August 1979

MECHANISMS OF IMPACT DAMAGE IN FILAMENT-WOUND GLASS-FIBRE/ EPOXY-RESIN TUBES

P. W. Manders, M. G. Bader, M. J. Hinton* and P. Q. Flower*

Department of Metallurgy and Materials Technology, University of Surrey, Guildford, U.K.
***Materials Department, Royal Armament Research and Development Establishment, Sevenoaks, Kent, U.K.**

INTRODUCTION

Filament-wound glass-fibre/epoxy tubes are increasingly being used for high pressure containment in chemical plant and aerospace applications. Tube designs optimized primarily for containment of internal pressure are frequently very thin-walled, with excellent cost and weight effective utilization of materials. In practice, however, the final design must also be sufficiently rugged to withstand normal handling and fabricating situations. This paper outlines an experimental approach to determine those factors, such as helix angle and winding pattern, which affect the resistance of a tube to side impacts from blunt objects. A better understanding of these factors and of the types of damage to be expected from accidental impact will enable the designer to specify a construction capable of satisfying both the pressure containment and impact requirements of service conditions.

Low velocity ($\approx$10 ms^{-1}) impacts to the sides of tubes may be classified into three categories, broadly as follows:

i) Diametral crushing between large flat surfaces which causes the tube to fold along its length.

ii) Impacts by pointed objects which penetrate the tube wall at the point of impact.

and

iii) Impacts by blunt objects which do not penetrate but which create regions of intense deformation and localised damage, sometimes away from the point of impact.

Impacts in the first two categories usually give a clear indication of the degree of damage (for example a hole in the tube). However, impacts in the third category are of considerable concern as it is possible for the ultimate burst pressure of the tube to be severely reduced by relatively low energy impacts, which neither penetrate the tube nor give a clear visual indication of the significance of the damage. Figure 1 (from Cawthorne) shows burst pressure test results for 100mm dia., 1mm wall thickness, $\pm 75^{o}$ helix angle tubes which had been impacted by a 50mm diameter hemisphere. At impact energies above a critical value of 6J the burst pressure is progressively reduced from about 24 MPa to about 8 MPa at an impact energy of 15 J. The tubes were all impacted at 1 ms^{-1} in a drop weight machine, and subsequently pressurised with water, (no liner was used and there was no end constraint).

Impact/residual strength behaviour is not in general as simple as these results suggest. There is a spectrum of damage ranging from complete elastic recovery, through resin cracking, extensive microcracking, fibre breakage in concentrated damage zones, and ultimately to complete crushing of the tube as the impact energy is increased. The energy at which impact damage becomes significant depends very much on how the residual properties are assessed. Using the pressure test described above, well-defined bursts which involve rupture of the wall due to fibre breakage can only be obtained with helix angles higher than 65°. Below this angle there is extensive cracking and weepage as the fibres become decoupled from the resin (see Hull 1978 & Spencer 1978). In the present work the test is terminated when the rate of weepage equals the pressurising input. For this reason the residual burst pressure test does not provide a realistic estimate of residual strength which might otherwise be adequate in a suitably lined tube. In practice residual properties need to be assessed appropriately for the application, taking account of any measures taken to prevent weepage.

To avoid this problem our approach has concentrated on the micro-mechanical aspects of damage. We have anticipated that any mechanism which results in fibre breakage will significantly reduce the burst pressure. We have adopted this as a *critical damage* criterion in our preliminary work but recognise that *micro-cracking* and/or *delamination* might in some circumstances also result in a reduction in the burst pressure. Multiple micro-cracking in the resin occurs extensively before fibre breakage, but can probably be tolerated in some applications.

EXPERIMENTAL

Tube fabrication and structure

All the tubes were filament wound on 100mm diameter cylindrical steel mandrels using 25mm wide *tapes* of glass fibre. These tapes comprised tows of 1200 tex (Silenka 051L) glass-fibre which were passed side by side through a bath of liquid epoxy-resin (Ciba-Geigy MY 750 cured with HY917 hardener and DY063 accelerator). After winding the mandrels were placed vertically in an air circulation oven to gel at 90°C and allow excess resin to drain, and cured for 1 hour at 130°C followed by 1½ hours at 150°C. The tubes were then cooled and removed from the mandrels.

The tape width was similar for all tubes and set an upper limit to the practicable helix angle (θ) of 85°, allowing for some overlap at the tape edges. In all tubes of 5 different helix angles were wound at 10° intervals from ±45° to ±85°, the lower limit of 45° being determined by slippage of the tape when reversing the winding direction. Volume fraction averaged 0.54 for all tubes with a variation of ±0.03. To achieve the nominal wall thickness of 1mm the winding process was continued until the mandrel was everywhere covered with four thicknesses of tape. The interweaving of tapes is determined by the order in which they are laid down, in this case, tape-laying passes in the same direction placed tapes adjacent to each other, and at the end of each pass the winding machine carriage reversed and continued laying tape in the opposite direction. This sequence results in a characteristic winding pattern shown in Fig. 2, in which the mandrel is completely and evenly covered when the tube is everywhere two tapes thick ($\pm\theta$), constituting one *cover*, two of which complete a tube.

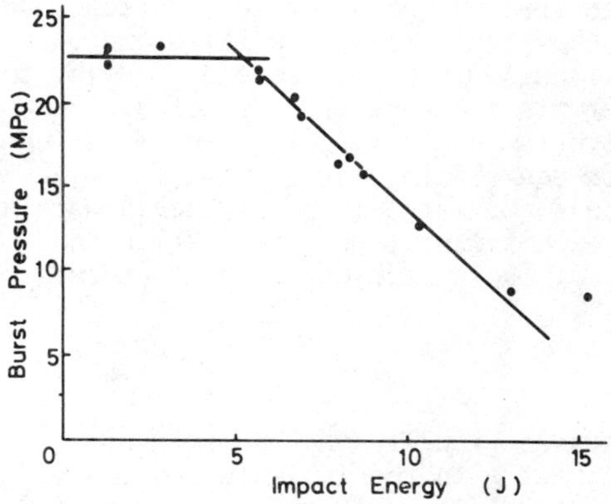

Fig. 1. Reduction of residual burst pressure with impact energy

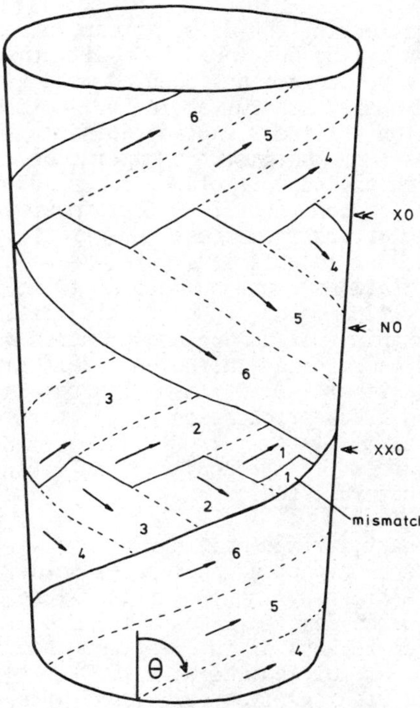

Fig. 2 Single cover showing tape orientation and winding order, XO & NO indicate indentation positions on & off line of cross-overs respectively.

One cover of a ±85° tube is wound with a single carriage pass in each direction and is exceptional in that there is no crossing over of tapes as with each revolution of the mandrel the tape butts against itself. To wind lower angle tubes more passes must be made in each direction and this results in crossing of tapes in the +θ and -θ directions, Fig. 2. Generally the second cover is placed in register with the first so the circumferential line of crossovers in Fig. 2 is exactly above its counterpart in the first cover. A limited range of tubes has been fabricated in which the winding pattern has been modified to remove the circumferential ring of cross-overs, and in which the covers have been displaced, the object being to detect any weakness associated with these features of the winding pattern.

Simulation of impact damage

At the inception of this work the 50mm spherical drop-weight impact test was being used to simulate rough handling, but although this is a satisfactory method of introducing damage prior to burst testing, the rather high and variable strain rate complicates the study of failure micro-mechanics. A slow indentation test has been developed to reproduce impact type damage at slower speeds. In this test a tube is supported on a plane and indented perpendicularly to its axis by a 50mm dia. hemisphere mounted on the crosshead of an Instron machine, and differs from the drop-weight test only in *rate* of loading. Load/displacement curves are obtained at a crosshead speed of 10mm min^{-1}, displacement being the maximum deflection at the indentation point. High-speed flash photography has shown that in all essential features the damage and indentation sequence produced is similar in each test. The loading geometry is shown in Fig. 3. Both tube length and the form of the support are major factors affecting the deformations. Semi-cylindrical and vee-block supports have been tried, but found to modify the bulging of the tube in the support plane, so the plane support has been adopted to avoid this, and to simplify analysis. The load deflection curves in Fig. 4 illustrate the very marked decrease of indentation stiffness (spring constant) with decreasing tube length, and Fig. 5 shows the depression of the upper tube surface in the indentation plane, and the bulging in the support plane, for a 1200mm, $\pm 65^{\circ}$ tube. Tubes with other helix angles also show significant deformations far from the point of indentation, and to reduce errors from these end effects to below the inherent variation in properties no indentations were made closer than 300mm to a free end. Eight to ten indentations could be made at 100mm intervals on opposite sides of a nominally 1200mm long tube. This initial programme has largely been concerned with mechanisms of damage and comparisons between winding angles at a fixed indentation rate, and although strain rate does not markedly affect behaviour in the range $1 ms^{-1}$ - $2.10^{-4} ms^{-1}$, there can be significant stress relaxation and continued crack propagation if the test-piece is held statically in a deformed condition.

Microstructural studies

After impact, or indentation, damage was assessed visually and recorded as a shadowgraph on photographic paper held in contact with the inner tube surface. Optical microscopy in transmitted light did not usually show the condition of fibres in the damaged areas which appear opaque and, therefore, samples were taken from regions of interest, mounted in resin and sectioned on the desired plane. Careful lapping with 1000 grit silicon carbide, followed by 6μm and 1μm diamond polishing gives relief and artefact free surfaces for reflected-light microscopy. The most successful technique was Nomarski differential interference microscopy (described by Lang 1968, 1969) which converts phase changes on

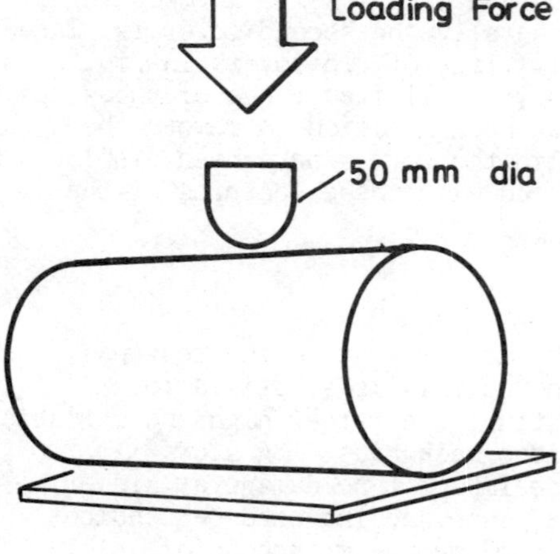

Fig. 3 Loading geometry for impact and indentation tests.

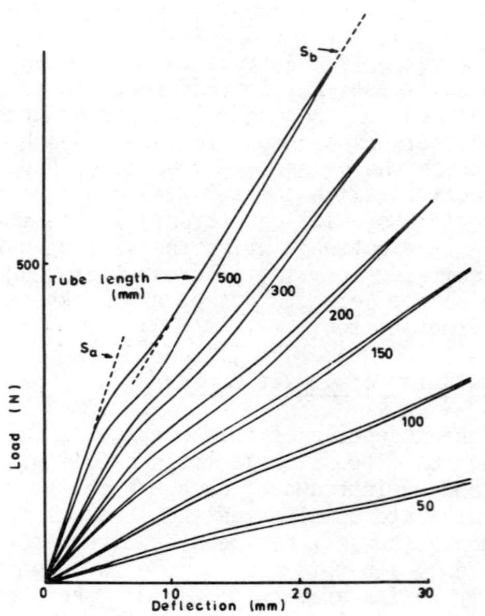

Fig. 4 Load/deflection curves for different tube lengths.

reflection into intensity contrast, overcoming the lack of contrast in straight reflection. This technique gives a false bas-relief effect which enhances the visibility of fine detail.

RESULTS

Indentation behaviour

In the initial stages of indentation, tubes of all helix angles follow a generally similar dual slope load/deflection curve, and in all cases there is considerable micro-cracking. These microcracks occur in the resin, or at the interface between fibre and resin, they run parallel to the fibres and extend through the thickness of the tape normal to the surface of the tube (Fig. 13). These cracks may occur in all four layers of the wall to give a criss-crossed effect (Fig.11). In this way a path for weepage could be established. In some cases *stress-whitening* has been observed. This consists of a concentrated region of *debonds* at fibre-matrix interfaces. The individual debonds are small (up to 3-4 fibre diameter) as shown in Fig. 12. The ultimate failure event at which the test is terminated involves a sudden loss of load bearing ability and the appearance of a localised damage zone accompanied by a sharp drop in the loading curve. The damage zone takes different forms depending on θ, and is the main subject of our investigation. The ultimate failure generally results in a large through-wall crack or considerable local fibre breakage, and represents a damage threshold above which burst strength would inevitably be lost.

The load/deflection behaviour at helix angles from $\pm 85^{\circ}$ to $\pm 45^{\circ}$ is presented in Fig. 6. These curves were produced by returning the indenter to its starting position after each 5mm increment of deflection to give a series of hysteresis loops which can be used to supplement visual observation of the development of damage. A single cycle test simply follows the upper curve and returns by the lower curve. Limited cycling in this way does not affect the type or extent of damage as cycling to a fixed deflection produces a set of hysteresis loops very closely in register.

All helix angles show a knee in the load/deflection curve at roughly 4mm deflection which corresponds to the reversal of curvature of the area immediately below the indenter, thereafter the secondary stiffness S_b is roughly half the initial stiffness S_a. (Stiffness is defined as the gradient of the load/deflection curve). Both stiffnesses pass through a maximum at $\pm 75^{\circ}$, Fig. 7, and the deflection to failure (to the load drop) is a minimum at this angle, Fig. 8.

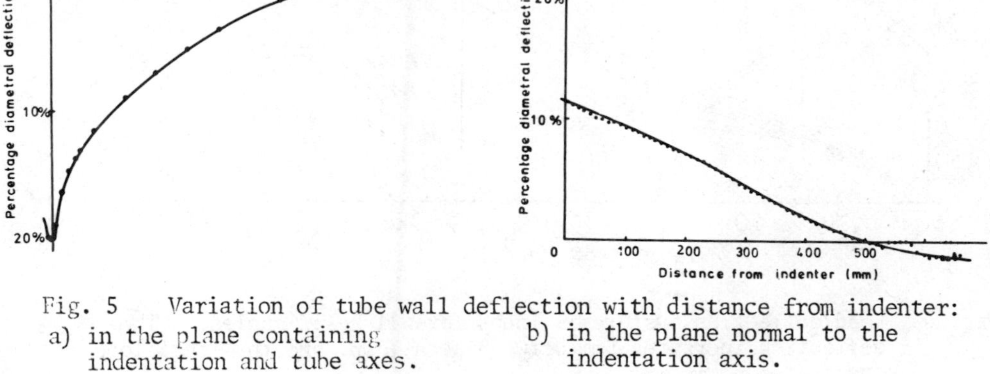

Fig. 5 Variation of tube wall deflection with distance from indenter:
a) in the plane containing indentation and tube axes.
b) in the plane normal to the indentation axis.

All curves show increasing hysteresis with deflection, Fig. 6, which is associated with the extensive multiple micro-cracking evident in the shadowgraphs, Fig. 11. The hysteresis in Fig. 6 is composed of two components, one due to formation of cracking, and the second due to other dissipative mechanisms (such as attrition at crack surfaces and viscoelastic effects in the resin) which remain, even after the multiple cracks have formed. A typical load-unload sequence is shown in Fig. 10. The second loading curve (3) lies between the first loading and unloading curves (1 and 2) but when the deflection reaches the maximum of the previous cycle (A) there is an abrupt change in slope which is followed to the next reversal (or failure). This is in agreement with the observation that further cracking occurs only when the maximum deflection is increased. The hysteresis which occurs between *unloading* and *reloading* to the same point, (the shaded area in Fig. 10 between curves 2 and 3) can only be due to the closing and opening of cracks or viscoelastic deformations and *does not* originate from crack propagation. The area of the larger loop (bounded by curves 1 and 2) followed upon *loading and then unloading* represents the energy to propagate cracks (in a

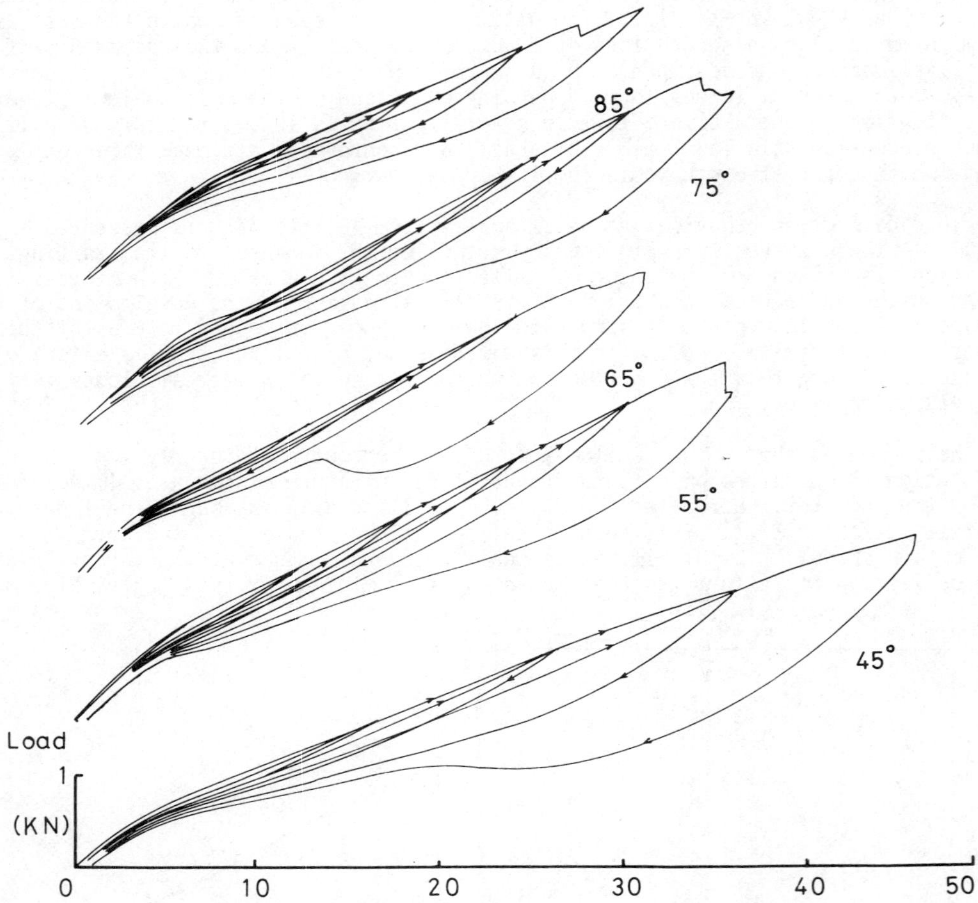

Fig. 6 Load/deflection hysteresis loops for all helix angles. The deflection increases 5mm between each loop, and the final loop terminates at failure.

5mm deflection increment) in addition to the hysteresis above. The ±45° tube shows a large hysteresis due to crack formation, and the micro-cracking is seen to be extensive in Fig. 11, the other helix angles have similar, but smaller areas of cracking and correspondingly lower hysteresis.

Tube failure modes

Three distinct failure modes have been identified. These are, however, preceded by micro-cracking which is observed to a greater or lesser extent as the initial damage mechanism in all tubes and in some cases by local stress whitening. Through-wall cracking is observed in all 85° and some 75° tubes. The second mode is local macroscopic *creasing* of the tube wall, normal to the tube axis. This is initiated by *micro-buckling* of the fibres in the critical zone. This failure mode is observed in all 75°, 65° and 55° tubes. The third mode is *delamination* between tapes parallel to the tube wall. This initiates failure by the micro-buckling and creasing mode in the 45° tubes but appears to be a consequence of failure at other angles where it often occurs as the tube is unloaded after failure.

In the 85° tubes (Fig. 11a) a small zone of widely-spaced micro-cracks forms on the inside surface of the tube, and on further deflection large through-wall cracks are initiated outside this original cracking zone; these cracks propagate rapidly right around the tube, which is then considered to have failed. In the 75° tubes the initial micro-cracks are more closely spaced, failure proceeds by stress-whitening, fibre micro-buckling and creasing (Fig. 11b). In some cases micro-cracking and delamination is also observed. The 65° and 55° tubes fail in similar modes with fibre-buckling and creasing with subsequent delamination preceeded by fine micro-cracking, (Fig. 11c). Failure in the 45° tubes is

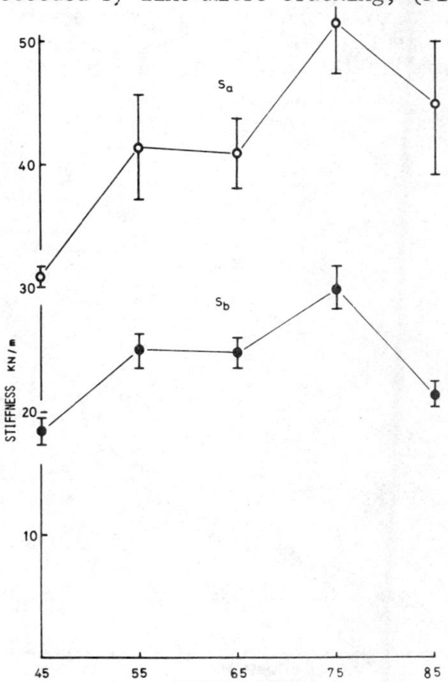

Fig. 7 Variation of initial and secondary indentation stiffnesses (S_a and S_b respectively)with helix angle.

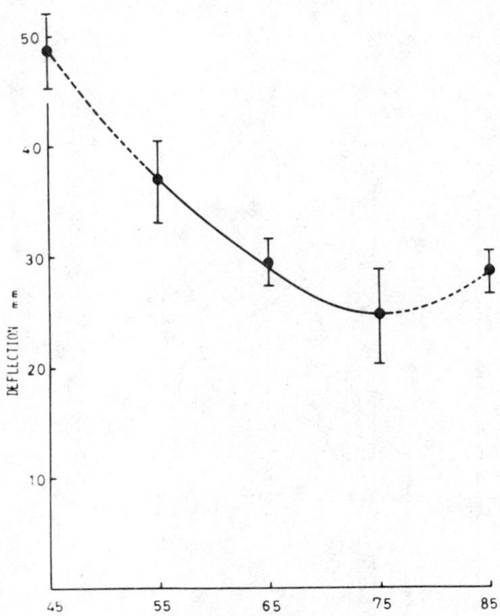

Fig. 8 Variation of deflection to failure with helix angle. Failure mode changes between 45° & 55°,and 75° & 85°.

markedly different with an extended zone of micro-cracking and delamination preceeding creasing (Fig. 11d). In all cases, except 85°, fibre fracture is associated with the creasing phenomenon and thus constitutes the effective point of failure.

DISCUSSION

Evaluation of impact and indentation resistance

The parameter which best expresses impact resistance is the maximum energy absorbed before the onset of critical damage. Ideally this energy would be stored elastically and returned to the impacter on rebound, but in practice some energy is dissipated in creating damage and in hysteresis associated with that damage. The design objective should be to maximise stored energy, and ensure that any damage is sub-critical and well distributed over a large area without degrading primary performance. The slow indentation test allows energies to failure to be measured directly from the area under the load/deflection curve. These are shown in Fig. 9, together with the hysteresis for loading to failure and then unloading (see Fig. 10).

The energy variation with angle can be empirically explained in terms of the stiffness and deflection curves of Figs. 7, 8. To a first approximation the energy absorbed to failure is $\frac{1}{2} S_b D^2$ where S_b is the stiffness prevailing for most of the test, and D the deflection to ultimate failure. The relative magnitude of the peak at ±75° in the stiffness curve Fig. 7, roughly equals the relative minimum in the deflection to failure, Fig. 8. Because the deflection appears squared in the energy expression, the energy to failure is expected to be a minimum at ±75°, and the curve should take a form similar to the deflection vs winding angle curve. This is seen to be so in Fig. 9. For the 75° tube the energy is 10J and this is in good agreement with the value obtained in the original impact tests (Fig. 1).

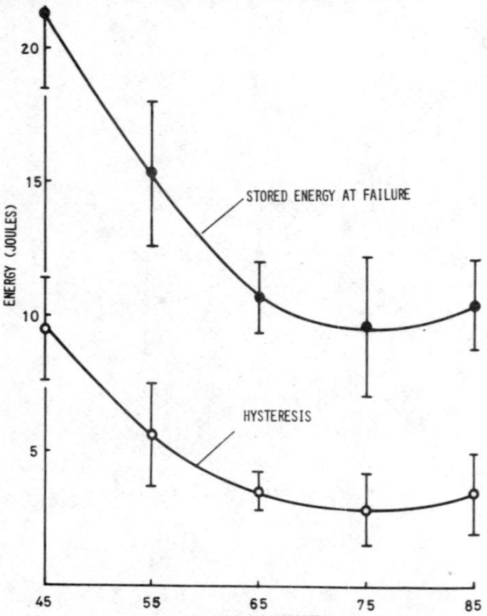

Fig. 9 Stored energy at failure, and hysteresis in loading to failure and unloading.

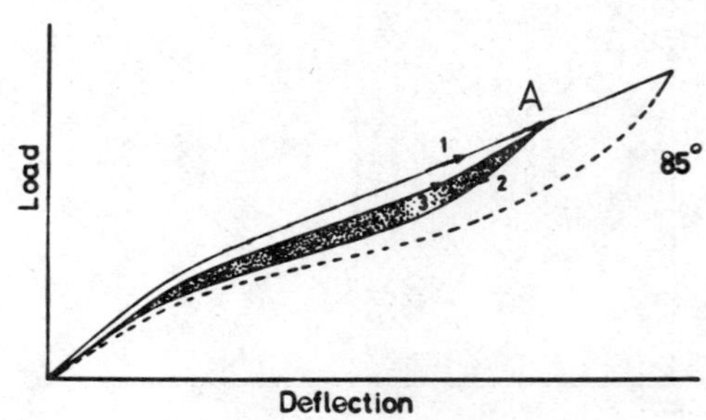

Fig. 10 Fourth and final hysteresis loops for 85° tube showing order of loading and unloading.

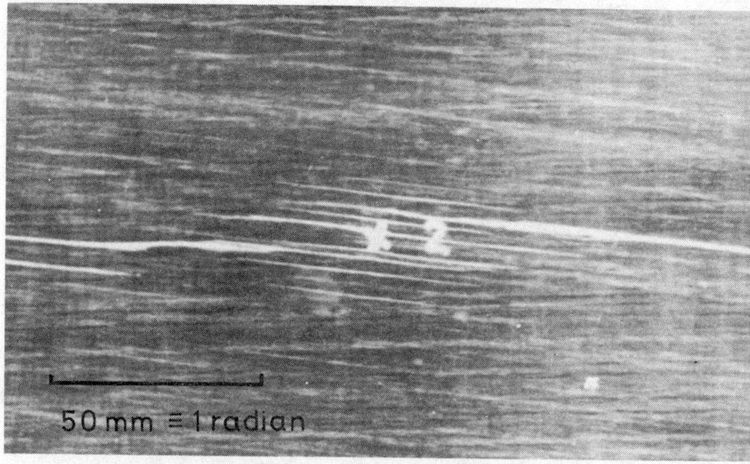

Fig. 11 Shadowgraphs of failure zones
(a) ±85° Circumferential cracking failure mode.

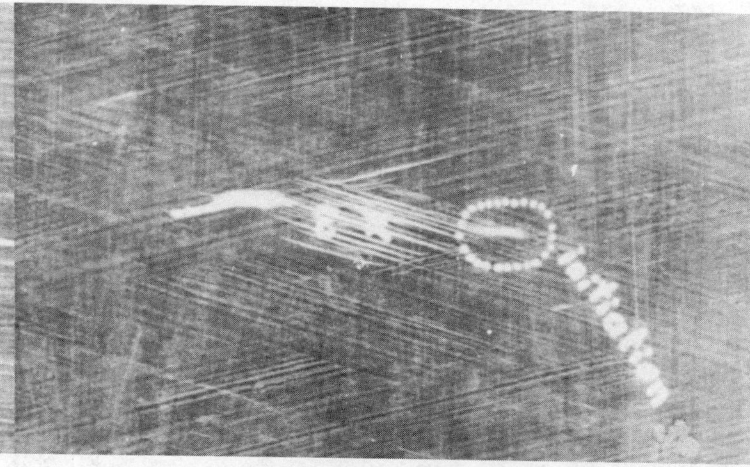

(b) ± 75° Failure by fibre buckling and creasing, initiated in stress-whitened zone.

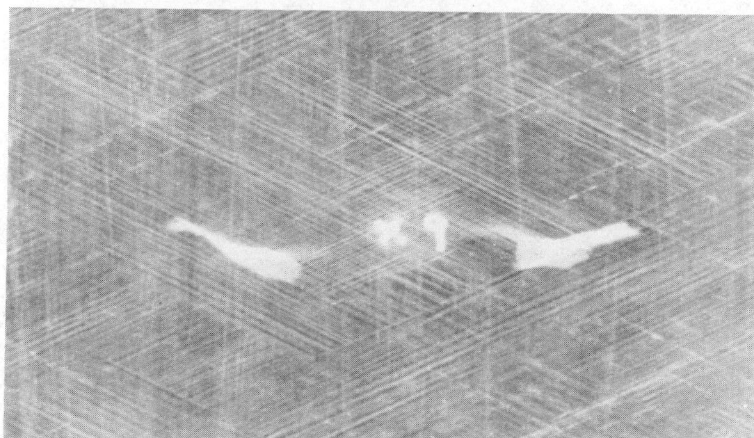

(c) ±65° Buckling and delamination failure with stress-whitening along line of crease.

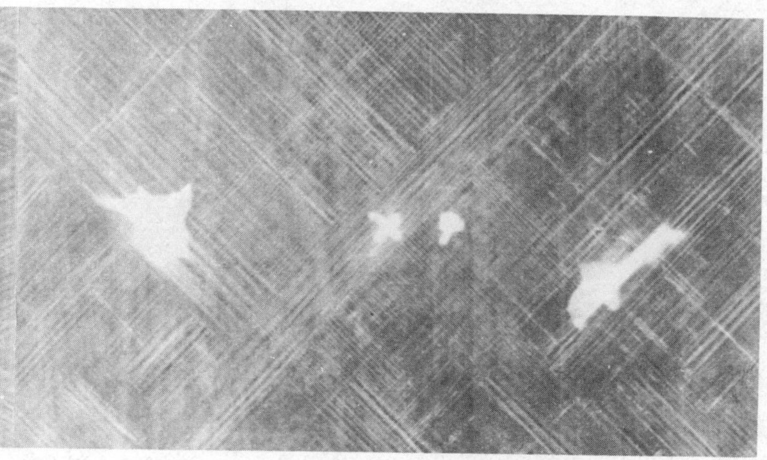

(d) ± 45° Delamination initiated failure

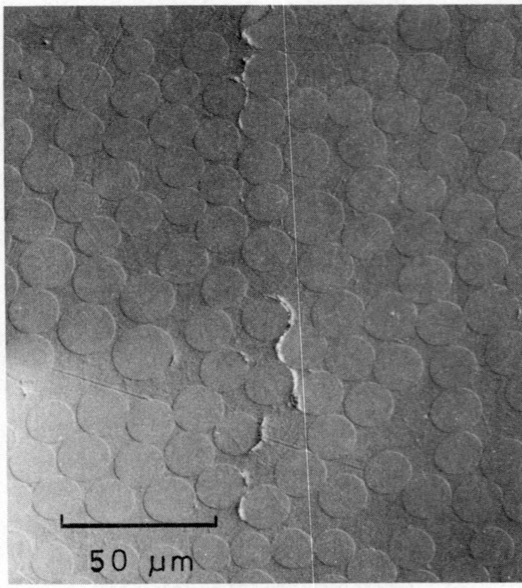

Fig. 12 Fibre debonding in stress-whitened region.

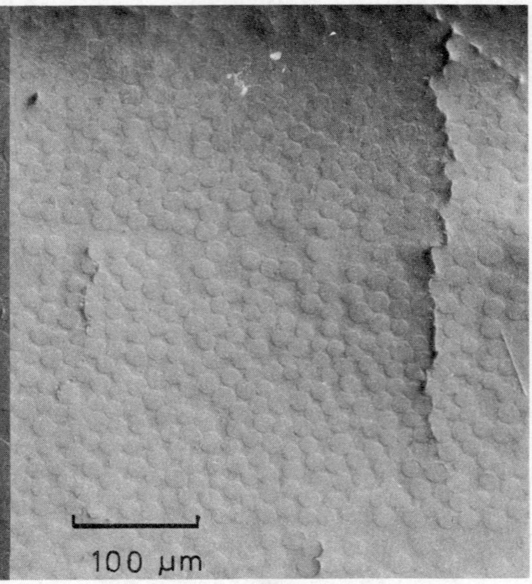

Fig. 13 Section through micro-cracked zone.

The project is continuing on two fronts. Experimentally we are building up a detailed map of the strain field for all tubes to determine the exact stress state in the failure zone, and to give some insight into why the indentation stiffness varies as it does with angle. Concurrently we are developing a finite element model of the loading and tube geometry for use as a design tool in optimizing impact resistance.

ACKNOWLEDGEMENTS

We would like to acknowledge the assistance rendered by our colleagues both at the University of Surrey and at RARDE. The project was financed by the Ministry of Defence and we would especially wish to express our appreciation for the advice and support rendered by Mr. R. Loneragan at RARDE.

REFERENCES

Cawthorne, D., Unpublished R.A.R.D.E. Document, 1977.

Hull, D., Legg, M.J. and Spencer, B. Failure of glass/polyester filament wound pipe. Composites 9, Vol. 1, p. 17-24, 1978.

Lang, W., Zeis Information 16, p. 114, No. 70, 1968,
17, p. 12, No. 71, 1969.

Spencer B., and Hull, D. Effect of winding angle on the failure of filament wound pipe. Composites 9, No. 4, p. 263-271, 1978.

Volume 3, ICM 3, Cambridge, England, August 1979

STRUCTURE VARIATIONS AFFECTING INTERFACIAL FAILURE IN STEEL FIBRE REINFORCED CEMENT

A. D. Morris and G. G. Garrett*

Ove Arup and Partners, Saxonwold, Johannesburg 2132, South Africa
***Department of Metallurgy, University of the Witwatersrand, Johannesburg 2001, South Africa**

INTRODUCTION

The poor tensile properties and ostensibly brittle characteristics of the cheap inorganic materials used principally by the civil engineer have represented a serious limitation to their more widespread application as viable engineering materials. However, their low cost of manufacture, particularly on an energy basis, presents distinct attractions and a number of schemes involving microstructural manipulation have been suggested to improve both strength and toughness. Prime amongst these has been fibre-reinforcement, and this paper examines the influence of matrix compositional variations adjacent to the all-important fibre interface on the mechanism of debonding and damage accumulation leading to failure.

The mechanical properties of any fibre reinforced cement composite will be dependent upon stress transfer between the fibre and the matrix; this in turn depends on the mechanical properties of the interface. Failure in many composites initiates through debonding between the fibre and matrix resulting in fibre pull-out and as a consequence, therefore, the strength of the fibre is generally not fully utilised. However, it is widely considered that the measurement of bond strength will provide some direct indication of bulk composite properties, although in steel-fibre reinforced cement-based materials specifically, various treatments designed to improve the bond strength have produced only marginal improvements in bulk strength properties (Mayfield and Zelly, 1973; Tattersall & Urbanowicz, 1974).

Unlike the interfacial contact zone in glass fibre reinforced cements, which has been shown to be quite complex in nature (Majumdar, 1974), the chemical stability of steel fibre in alkaline environments tends to suggest that the bond in steel fibre cement is almost entirely mechanical, i.e. of a frictional nature (Mayfield and Zelly, 1973), despite some evidence to the contrary (Majumdar, 1975). Failure in this material has therefore generally been reported to occur at the interface itself (Hughes and Fattuhi, 1975; Majumdar, 1975; Tattersall and Urbanowicz, 1974), although recently Pinchin and Tabor (1975) have observed the initiation of failure adjacent to the fibre interface which they attribute to local compositional fluctuations possibly produced by conventional manufacturing procedures. On the whole, however, little work has been carried out specific to the interfacial zone, and the mechanism of failure in the region of the fibre-matrix interface remains to be clarified. The object of this paper, therefore, is to report on a study designed to examine further the nature of the fibre-matrix interface and its influence on the mode and mechanism of failure.

MS292

EXPERIMENTAL

In order to examine the initial debonding process, single fibre pull-out tests were carried out during which load-extension curves were monitored. Experiments were conducted at a low pull-out rate (0.4mm/min) so that, following the load drop-off corresponding to initial debonding, the test could be stopped almost instantaneously without further pull-out occurring, and therefore without disrupting the interfacial zone. The interface was subsequently exposed by a technique involving simple tensile fracture across the interfacial plane (Morris, 1978) which leaves the material at, and adjacent to, the interface of control samples relatively undisturbed. (See following & Fig.3).

Pull-out specimens were horizontally cast 15 x 15 x 60mm prisms, having a single 'Duoform' steel fibre (38mm long, 0.5mm diameter) embedded 15mm into the longitudinal face of the prism; after casting, the moulds were externally vibrated for 2 minutes. Rapid hardening Portland cement was used throughout, with a constant water : cement : sand ratio of 1 : 2 : 4, by weight. Curing was carried out under water, and all samples were tested at age 7 days.

SEM fractographic studies were carried out using a Cambridge Stereoscan S180, coupled with a Link energy dispersive X-ray analyser. Previous authors (Stucke and Majumdar, 1976a ; Diamond, Young and Lawrence, 1974) have noted the difficulties involved in obtaining quantitative information by EDXA methods in these materials, although semi-quantitative results have been obtained with good reproducibility on a comparative basis (Diamond, 1972 ; Pinchin and Tabor, 1978 ; Stucke and Majumdar, 1976a). In the present study, in order to obtain some estimation of the relative proportions of calcium silicate hydrate and calcium hydroxide (CSH and $Ca(OH)_2$ in abbreviated form) existing in the regions adjacent to the interface, line scan analyses were undertaken parallel to the fibre-matrix interface. Average Ca/Si integral area ratios were subsequently compared between the selected areas at, and adjacent to, the interface and the corresponding standard bulk matrix Ca/Si ratio.

Fig.1.

Bond failure and matrix failure adjacent to the interface (fibre at left). SEM specimen opened by transverse tensile splitting immediately after initial debonding load has been reached.

RESULTS AND DISCUSSION

Fractographic examination of the interfacial zone and matrix adjacent to the interface indicates that, at initial debonding, composite failure appears to be due both to extensive matrix cracking adjacent to the interface as well as to interfacial debonding (Fig.1). Furthermore, although it might be argued that bond failure is the first to occur, resulting in subsequent matrix failure, it was observed that bond failure never occurred discreetly but was always associated with matrix failure; on the other hand, matrix failure was found to occur without corresponding bond failure (Fig.2). Thus, Fig.2(a) shows a fibre embedded within the mortar matrix after having been stressed to the initial debonding load. In addition to the thin layer of adhering cement evident on the fibre surface, the interfacial bond is obviously still intact along the length shown. The central area indicated, enlarged in Fig.2(b), shows extensive matrix cracking with an intact fibre-matrix bond. It is relevant to point out, as noted previously, that control tensile fracture surfaces of embedded fibres which have not been subjected to any form of applied pull-out stress do indeed show relatively undisturbed material at, and adjacent to the fibre interface (Fig.3), in comparison with pull-out specimens.

Fig.2(a) Thin cement layer adhering to steel fibre surface. The interfacial bond is still intact along this length of fibre.

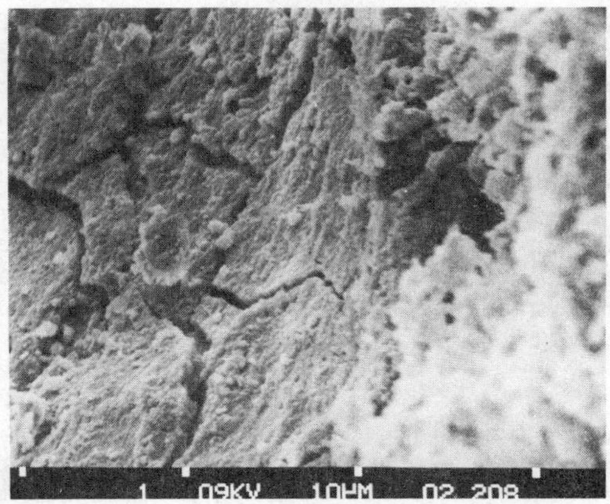

Fig.2(b) Central area in (a) enlarged, showing extensive matrix cracking with interfacial bond intact.

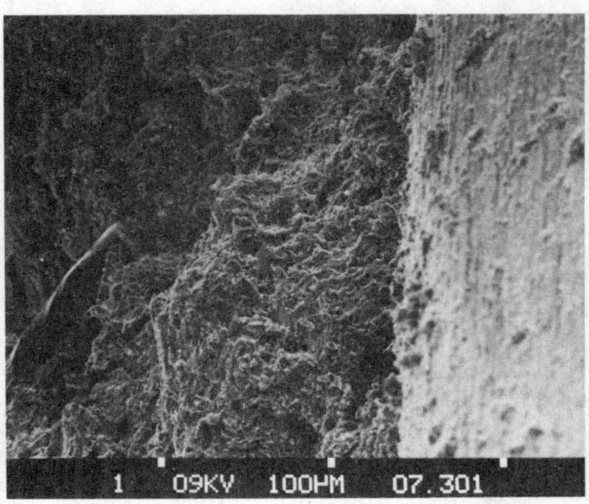

Fig.3. Transverse-split fracture surface of control specimens - no load has been applied to the steel fibre - showing no evidence of bond or matrix cracking.

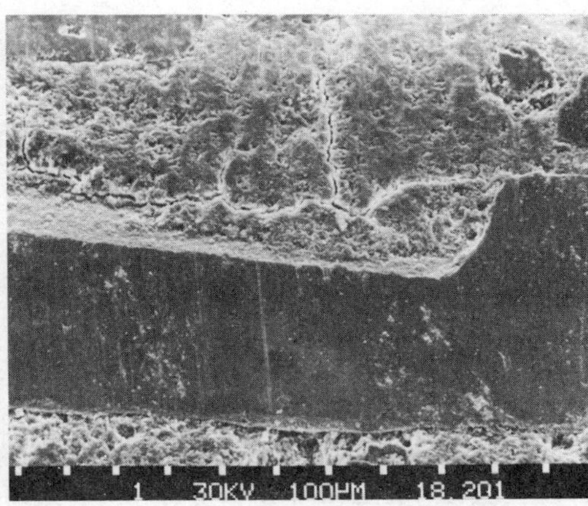

Fig.4. Preferential matrix cracking adjacent to the fibre interface in a polished cross-section rather than at the interface itself.

Further evidence of matrix cracking occurring in preference to bond cracking is shown in Fig.4. which is actually taken from a polished sample prior to the application of any load to the fibre. The cracking seems likely to have arisen either due to the polishing process itself or to the restraints imposed by the stiffer fibre upon volume changes that occur within the matrix during curing. In any event, it is evident that failure has taken place within a zone adjacent to the interface, which must be weaker than the interface itself.

Many of the regions examined close to the interface showed matrix morphologies with marked similarity to the 'platey' Portlandite ($Ca(OH)_2$)structure reported by other authors (Pinchin and Tabor, 1978; Majumdar, 1974; Williamson, 1973), for example areas marked C in Fig. 5. It is generally acknowledged (e.g. Pinchin and Tabor, 1978; Williamson, 1973) that an important feature of the formation of $Ca(OH)_2$ (a hydrolytic product resulting from the hydration of the calcium silicate phases) is that it grows within, and subsequently fills, much of the available void space (either water-filled voids or entrapped air voids). Now, it has been suggested (de Vekey, 1975) that air entrapment is higher in the vicinity of the interface due to inertia effects of a steel-fibre vibrating the small surrounding volume of cement matrix during vibration compaction. It would seem quite likely, therefore, that the areas marked C in Fig.5 are indeed $Ca(OH)_2$ which fill the initial concentration of voids in the vicinity of the steel wire. Pinchin and Tabor (1978) have also observed an enrichment of $Ca(OH)_2$ at the fibre-matrix interface, and have suggested that the fibre provides a centre for the nucleation of $Ca(OH)_2$.

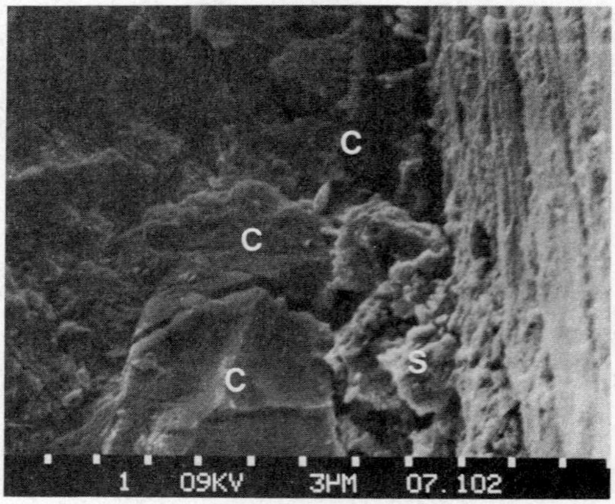

Fig.5.

The occurrence of $Ca(OH)_2$, marked C, near the interface, with calcium silicate hydrate (S) adhering to the fibre.

The first objective of the line scan elemental distribution analysis was to attempt to evaluate whether there was any merit in the suggestion (Majumdar, 1975) that the steel/cement bond in this composite may be partly chemical in nature, arising from Fe^{++} diffusion across the interface. Certainly the Fe-K distribution shown in Fig. 6(a) (taken from the interfacial region of Fig. 7, to follow) would initially seem to indicate that quite considerable diffusion has occurred. However, this result is not considered to be a true representation of Fe^{++} diffusion into the matrix since it has been shown (Diamond, Young and Lawrence, 1974) that the effective diameter/depth of X-ray generation can be of the order of 2 - 3μm so that the wire itself would be detected in probe locations in the immediate vicinity of the interface. Similar line scan analyses carried out on cement trough specimens (i.e. the matching half of a transverse-split single fibre specimen), Fig.6(b), indicate that the Fe^{++} concentration illustrated in Fig.6(a) is erroneous. The Fe-K integral count expressed as a percentage of the $Ca\text{-}K_{\alpha}$ and Si-K integral counts was in fact $\sim$3%, and therefore consistent with the average

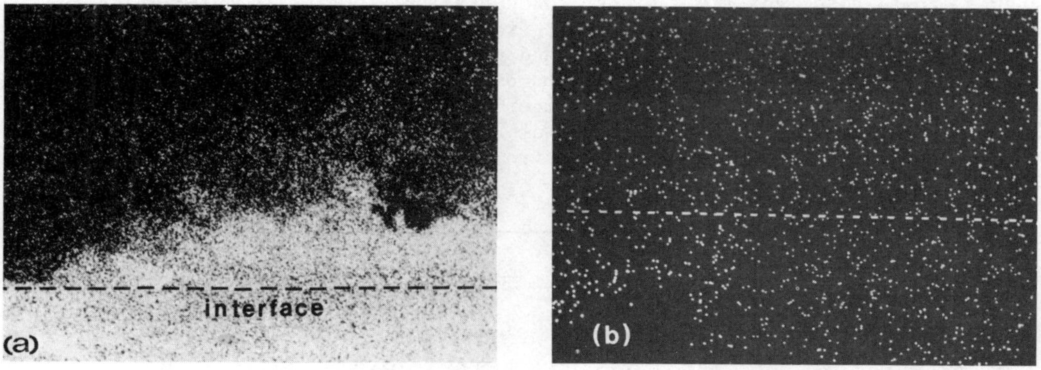

Fig.6. Fe-K elemental distribution in the interface region: (a) from an embedded fibre specimen (Fig.7); (b) from a corresponding trough section (no fibre present).

chemical composition of the cement (Morris, 1978). It must be concluded, therefore, that any chemical bonding that may be present is insignificant compared to mechanical bonding, at least in these relatively young samples, and that observed Fe^{++} concentrations in the cement matrix adjacent to the interface result from X-ray penetration effects.

The results for the variation of the integral area ratio were more positive and Fig. 7 shows a typical area over which such an analysis was carried out. Figure 8 presents the cumulative data for the integral area ratios for Ca-K_{α} and Si-K as a function of distance from the interface, taken from a number of specimens both with the steel fibre embedded and the corresponding trough samples. It is immediately apparent that there is a Ca^{++} - enriched zone within a region ~5-60 μm from the interface, with a concentration ~30μm away from the interface; this indicates an abundance of $Ca(OH)_2$ relative to CSH, which is certainly consistent with the observations of $Ca(OH)_2$ crystals close to the interface (Fig. 5). In addition, the occurrence of matrix cracking adjacent to the interface in preference to interfacial debonding (Figs. 2 and 4) is expected, in view of the weakness of $Ca(OH)_2$ platelets relative to CSH (Walsh and co-workers, 1974).

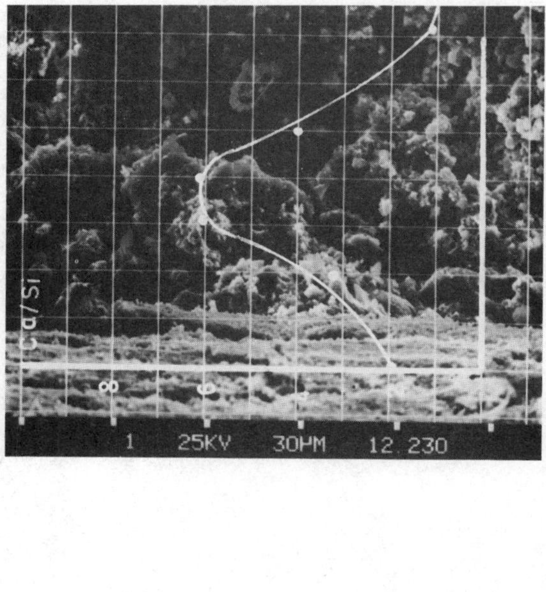

Fig.7.

Interfacial zone fracture surface area of line scan analysis, with corresponding Ca/Si concentration ratio superimposed (fibre at bottom).

The measurements taken within the trough region of cement-trough specimens (Fig. 8(a) open symbols) also indicate a higher concentration of $Ca(OH)_2$, which is in turn consistent with the occurrence of first fracture in a $Ca(OH)_2$ - enriched zone adjacent to the interface. As mentioned previously, Pinchin and Tabor (1978) observed $Ca(OH)_2$ enrichment within ~10μm from the interface, using X-ray diffraction, of material removed by a drill 20μm larger than the diameter of the hole left by a pulled-out fibre. From the results of Fig. 8(b), it would seem likely that had these authors analysed material > 10μm from the interface, they would have observed even greater enrichment.

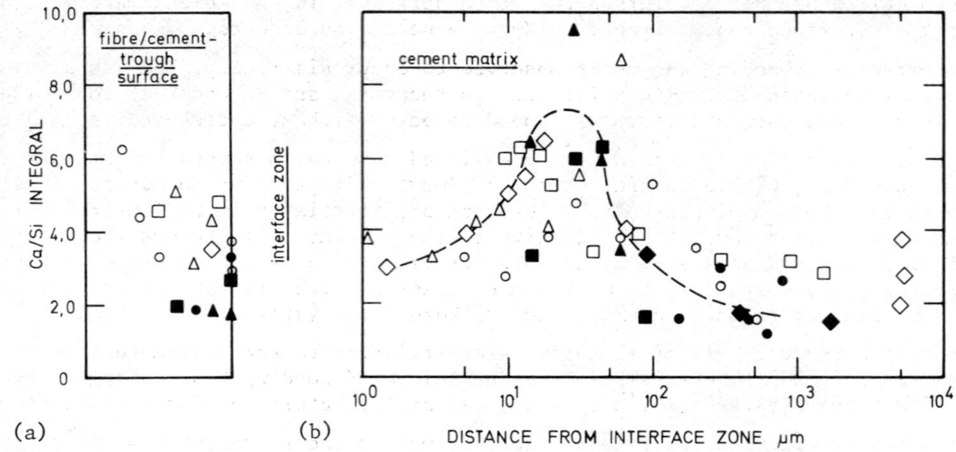

Fig.8. Variation of Ca/Si integral ratio (a) on the steel fibre surface (filled symbols) or within the cement trough (open symbols); (b) at, and adjacent to, the fibre/matrix interface: embedded steel fibre specimens (filled symbols); trough specimens - no steel fibre (open symbols). Each symbol represents a different specimen.

These observations indicate that, during the initial debonding process, the 'normal' composition hydrate (i.e. corresponding to the bulk material) in the immediate vicinity of the interface must be pulled away with the fibre, a fact which is also supported by the experimental measurements of Fig. 8(a) (filled symbols). Pinchin and Tabor (1978) on the other hand, only measured the relative ratios in the trough regions which they assumed to be in direct contact with the fibre, and they therefore suggested a progressively increasing Ca/Si ratio right up to the interface. However, since hydration of CSH results in the formation of $Ca(OH)_2$, CSH must occur before $Ca(OH)_2$ can be nucleated with the resulting $Ca(OH)_2$ occupying void spaces not already occupied by CSH. Hence, it would seem more reasonable to suggest that nucleation of the CSH phase takes place at the fibre surface, rather than the $Ca(OH)_2$ as suggested by Pinchin and Tabor (1978). Since CSH has the finest particle size of the solid phases present in hydrated cement (for example, see Stucke and Majumdar,1976b), and therefore is able to compact more readily around the fibre, the formation of $Ca(OH)_2$ at the actual interface might be expected to be inhibited, at least in the regions where the initial cement particles are themselves close to the interface. SEM observations add credence to this line of argument; for example, in Fig. 5, the area marked S which can be seen adhering to the fibre is of substantially different morphology from the surrounding areas (identified as $Ca(OH)_2$)and is probably composed of the stronger CSH phase. During subsequent pull-out, mechanical disruption leads to further extensive microcracking close to the interface and may lead to specific interfacial debonding.

CONCLUSIONS

SEM fractographic studies, in combination with energy dispersive X-ray analysis, have been carried out on the interfacial region of single steel fibre cement-based composite specimens. This investigation has identified the following features concerning the interface and the factors controlling the onset of primary debonding.

1. There is no basis for the hypothesis that the bond in steel fibre cement is partly chemical in nature. Observed Fe^{++} concentrations in the cement matrix adjacent to the interface can be traced to X-ray penetration effects.
2. Interfacial debonding was never observed to occur discreetly, but was always found to be associated with matrix failure; furthermore, and in contrast to a widely held opinion, matrix failure was found to occur without corresponding bond failure.
3. In the interfacial region, a Ca^{++} - enriched zone was observed 5-60μm away from the interface, with a concentration at $\sim 30\mu m$, indicating an abundance of calcium hydroxide (formed within voids which form preferentially in the interfacial region during vibration compaction) relative to the calcium silicate hydrate. The location of this weak zone is in good agreement with fractographic observations regarding the morphology of matrix microcracking, i.e. in many areas cracking in the matrix was associated with calcium hydroxide crystals.
4. Hydration products of normal composition (relative to the bulk material) still adhere to much of the steel fibre after initial debonding, indicating that calcium silicate hydrates are initially nucleated at the interface, forming a strong core.
5. These observations suggest that efforts aimed at incorporating new fibre types or shapes, or having particular surface treatments, may be limited in their scope for improvement by the fact that the failure - controlling characteristics lie with the cement matrix itself, and with the manufacturing operation, and it is in these areas where the challenge for significant improvement must be found.

ACKNOWLEDGEMENTS

The authors wish to acknowledge with thanks the financial support of the S.A. Council for Scientific and Industrial Research and the Portland Cement Institute. The experimental work was carried out at the University of Cape Town, and the authors are grateful to the University Research Administration and Professor A. Ball for the provision of laboratory facilities.

REFERENCES

Diamond, S. (1972). Identification of hydrated cement constituents using a scanning electron microscope - energy dispersive X-ray spectrometer combination. *Cement and Conc. Res.*, 2, 617 - 632.

Diamond, S., J.F. Young, and F.V. Lawrence (1974). Scanning electron microscopy energy dispersive X-ray analysis of cement constituents - some cautions. *Cement and Conc. Res.*, 4, 899 - 914.

Hughes, B.P., and N.I. Fattuhi (1975). Fibre bond strength in cement and concrete. *Mag. of Conc. Res.*, 27, 161 - 166.

Majumdar, A.J. (1974). The role of the interface in glass fibre reinforced cement. *Cement and Conc. Res.*, 4, 247 - 266.

Majumdar, A.J. (1975). Properties of fibre cement composites. In A.M. Neville (Ed.), RILEM Symposium on *Fibre Reinforced Cement and Concrete*. Construction Press, London. pp. 279 - 313.

Mayfield, B., and B. Zelly (1973). Steel fibre treatment to improve bonds. *Concrete*, 7, 35-37.

Morris, A.D. (1978). *Mechanical and Microstructural Properties of Steel Fibre Reinforced Cement-based Materials.* M.Sc. Thesis, University of Cape Town, South Africa.

Pinchin, D.J., and D. Tabor (1975). Mechanical properties of the steel/cement interface : some experimental results. In A.M. Neville (Ed.) RILEM Symposium on *Fibre Reinforced Cement and Concrete.* Construction Press, London. pp. 521 - 526.

Pinchin, D.J. and D. Tabor (1978). Interfacial phenomena in steel fibre reinforced cement. I : Structure and strength of the interfacial region. *Cement and Conc. Res.*, 8, 15 - 24.

Stucke, M.S., and A.J. Majumdar (1976a). Microstructure of glass fibre-reinforced cement composites. *J. Mat. Sci.*, 11, 1019 - 1030.

Stucke, M.S., and A.J. Majumdar (1976b). The composition of the gel phase in Portland cement paste. In proceedings, Conference on *Hydraulic Cement Pastes - their Structure and Properties.* Cement and Concrete Association, Slough, U.K. pp. 31 - 51.

Tattersall, G.H., and C.R. Urbanowicz (1974). Bond strength in steel fibre reinforced concrete. *Mag. Conc. Res.*, 26, 105 - 113.

Walsh, D., M.A. Otooni, M.E. Taylor, and M.J. Marcinkowski (1974). Study of Portland cement fracture surfaces by SEM techniques. *J.Mat. Sci.*9, 423 - 429.

Williamson, R.B. (1973). Solidification of Portland cement. *Prog. Mat. Sci.*, 15, 189 - 286.

de Vekey, R.C. (1975). In Discussion, Proc. 1st. Int. Cong. on Polymer Concrete, *Polymers in Concrete*, Construction Press, London. p. 107.

Volume 3, ICM 3, Cambridge, England, August 1979

SLOW CRACK GROWTH IN SOFTWOODS

P. D. Ewing and J. G. Williams

Department of Mechanical Engineering, Imperial College of Science & Technology, London SW7 2BX, U.K.

INTRODUCTION

Failures in the natural cleavage planes of wood can be described in terms of linear elastic fracture mechanics [1,2,3]. However, the majority of data published has been for toughness values at crack initiation. There is, however, considerable interest in the effect of crack speed on the fracture toughness, K_c, since this slow growth is the likely mechanism of long term failures. Several experimental methods have been used to measure crack speed. The load relaxation method used on double torsion specimens by Mindness, Nadeau & Barrett [4] showed that K_c for Douglas Fir increased as the crack speed increased. This effect has also been noted by Schniewind & Pozniak [3] using double cantilever bend (DCB) specimens, but Mai [5], in tests on Spruce using tapered cleavage specimens over a wide range of moisture contents, found that fracture toughness decreased with increasing crack speed.

This paper uses Scots Pine tapered cleavage specimens to try to resolve this apparent contradiction, and measurements were made of crack speed at various K_{Ic} values for kiln dried samples all cut from the same tree.

EXPERIMENTAL

The tapered cleavage specimens, as shown in Fig. 1, were cut from Scots Pine (*Pinus sylvestris*) kiln dried clear boards, having an approximate moisture content (MC) of 12%. These boards were selected from the tree already described by Ewing & Williams [6]. The specimens were prepared to give fractures in the radial longitudinal (RL) system. The convention used here for defining crack growth is that the first symbol defines the normal to the plane of propagation and the second the direction of the crack growth within that plane. From preliminary tests, it was found necessary to control the crack, and so side grooves were machined along each side of the specimen to a depth of approximately one quarter the thickness of the specimen.

The initial crack was formed in the specimen with a saw cut and then a razor blade was used to form a sharp crack. The specimens were loaded via pin joints in an Instron testing machine (TT-DM) at a temperature of 20°C and relative humidity of 60%. The rates of displacement used were in the range 0.05 - 5 mm/min. The crack growth was monitored with a travelling microscope and the load time

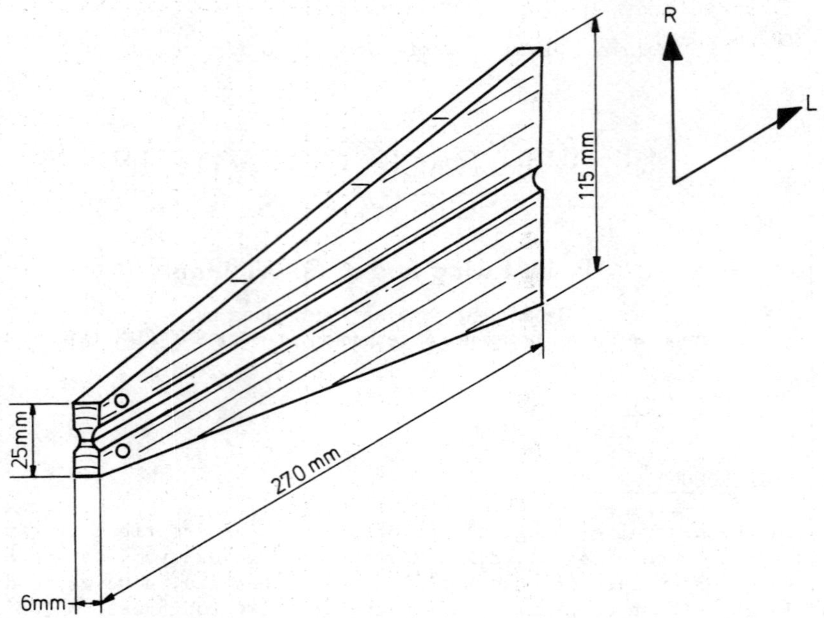

Fig. 1 Tapered cleavage specimen

Instron chart was marked at intervals as the crack tip passed gauge marks on the specimen. A typical set of data for time versus load and crack length is shown in Figure 2. For displacement rates lower than 0.05 mm/min, a single lever constant load apparatus, as described in reference [7], was used.

K_c was computed from the Irwin-Kies [8] relationship:

$$K_c^2 = \frac{P^2 E_r dC}{2 B_c da} \quad (1)$$

where C is the specimen compliance, and B_c is the crack width at the base of the groove.

From preliminary calibration tests, the compliance was determined for a number of crack lengths, so that dC/da could be found directly. The constant value (see Fig. 3) was found for just over half the specimen length as opposed to isotropic materials which give a constancy for nearly the entire length. As considerable variation in the dC/da value was found from specimen to specimen, this value was measured for each specimen prior to each test.

RESULTS

From the data obtained for each particular specimen (see Fig. 2), K_c and crack speed values were computed, and the results for all the specimens tested are shown in Fig. 4, together with their average values.

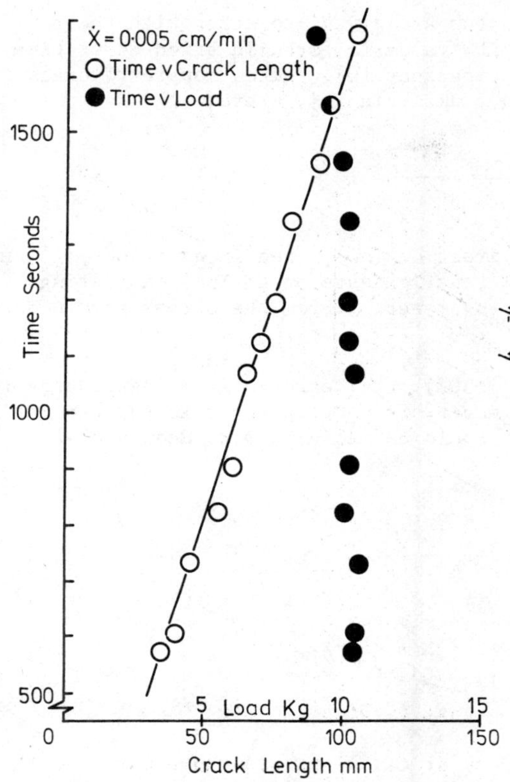

Fig. 2 Typical time versus load and crack length data

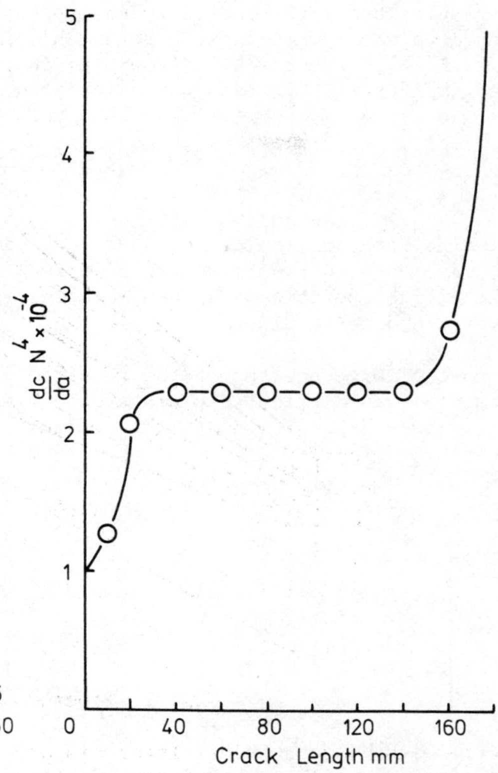

Fig. 3 Compliance/crack length curve for tapered cleavage

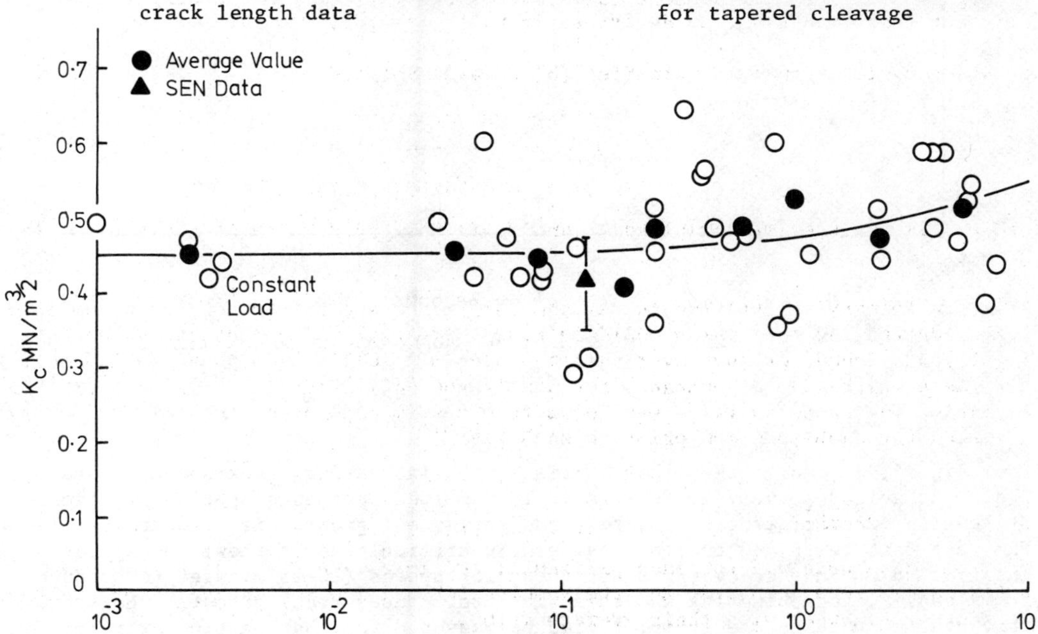

Fig. 4 Fracture toughness as a function of crack speed

The results using the constant load apparatus are given, together with their average value. The single edge notch (SEN) value is for kiln dried Scots Pine (MC = 11.2%), 6 mm thick, as reported in reference [6]. The 'apparent' crack speed ($\dot{a}$) for the SEN data was calculated from Irwin [9], where:

$$\dot{a} = \left(\frac{K_c}{E_r}\right)^2 \frac{1}{\sqrt{2}\,\pi\,t} \tag{2}$$

The value for the modulus in the radial direction (E_r), used in equations (1) and (2), was measured from a series of normal tensile tests on an Instron testing machine, incorporating a strain gauge extensometer. From the stress strain data, E_r was found to be 1100 MN/m^2.

Due to the large scatter of the results (± 30%), the degree of time dependence for Scots Pine was difficult to measure. However, from creep tests in three-point bending (see Fig. 5), the time dependence could be measured more accurately.

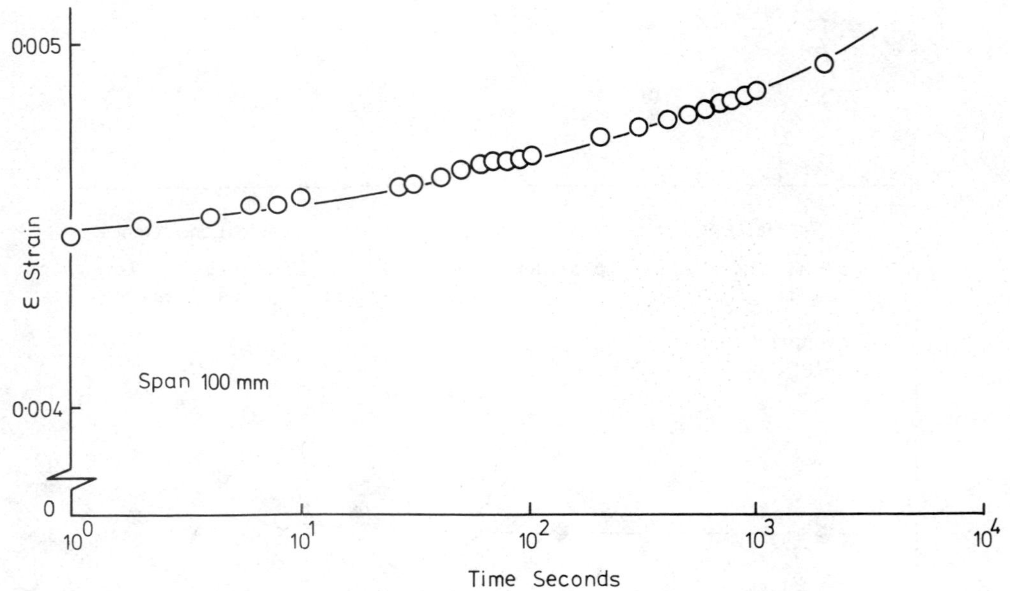

Fig. 5 Creep for Scots Pine (RL) - three-point bend

DISCUSSION

From Fig. 4, it can be seen that K_c rises with increasing crack speed. Crack growth behaviour at very low speeds (< 10^{-2} mm/sec) presents special problems. Stick-slip (i.e. crack growth arrest and subsequent growth) is evident in all but the very fast tests and caused considerable difficulties at these low speeds. The time to produce crack growth at these low speeds (10-15 minutes to produce 1 mm) causes crack blunting and this, in turn, causes crack arrest. However, providing a crack can be grown, K_c can be measured for that particular rate.

For the tests performed in the Instron testing machine ($\dot{a}$ > 10^{-2} mm/sec), the results form a set pattern. In all cases, the higher the K_c value, the lower the crack speed for any given displacement rate. (It is easily shown that $\dot{a} \propto \text{load}^{-1}$ for a given displacement rate.) In the majority of cases, the crack runs in the early wood, but no difference has been noted when the crack is growing in the late, and hence more dense, wood.

From Fig. 5, it is shown that Scots Pine is time dependent as the material creeps. Marshall, Coutts & Williams [10] and Williams [11] have shown that, if time dependence, in creep, is expressed as:

$$\varepsilon(t) = C_o t^n \sigma \tag{3}$$

then the $K_c \propto \dot{a}^n$ during crack growth. For most polymers, $n < 0.1$, and frequently $n < 0.05$, and remains constant over a substantial range of rate. For Scots Pine, $n \simeq 0.02$, as determined from the creep data in Fig. 6, and the K_c data (average values only) is also shown and agrees reasonably well, confirming that the crack growth is basically relaxation controlled [10].

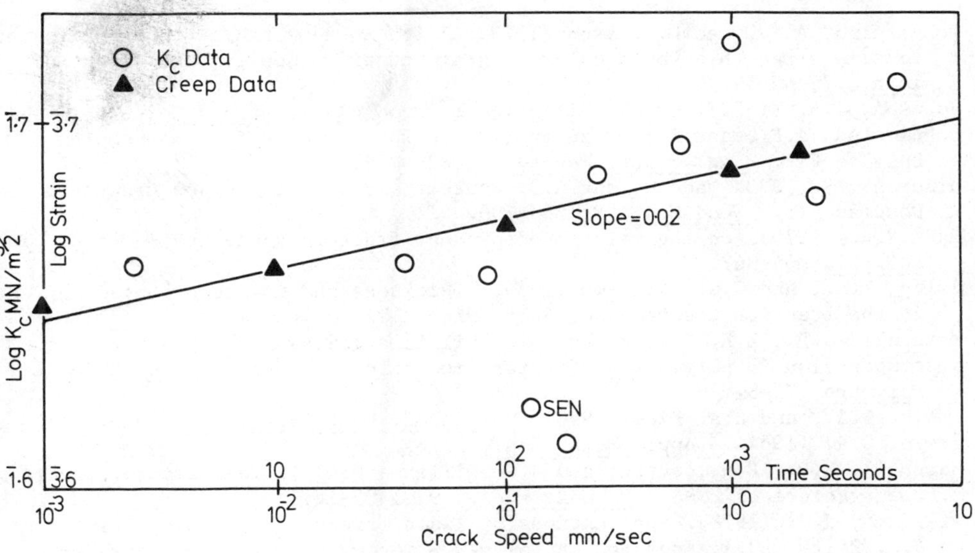

Fig. 6 Log-log plot for K_c and creep data

The test method used for crack growth in wood has considerable advantages over previously published work as the crack tip can be seen when growing. With the stress relaxation method (which utilises double torsion specimens), the crack cannot be seen and doubt must be raised if actual crack speed can be measured accurately. Schniewind & Pozniak [3], using DCB specimens, found difficulties in locating the crack tip and put an accuracy of ± 2.5 mm in their crack growth readings. It was found that an accuracy of < ± 0.5 mm was easily obtainable with tapered cleavage specimens, even at the high crack growth rates. Once the crack starts to grow from the damage zone around the initial crack tip, the crack is easily followed up to the crack speeds reported here. Above 10 mm/sec, it is not possible to follow the crack with the method used in this paper.

The crack growth data reported here are in agreement with the investigations of Mindness, Nadeau & Barrett [4] and Schniewind & Pozniak [3]. However, Mai's [5] K_c data fall with increasing crack speed and he has reported that, by varying the crosshead rates of the testing machine, a wide range of velocities could be obtained from a single specimen. This we found impossible to do in attempts to repeat the effects reported by Mai, for the following reasons.

With the crack growing under stable conditions, the crosshead rate was increased and the crack would immediately accelerate along the specimen and become unstable. The crack was now not growing at a constant rate, and the load would fall rapidly (and therefore K_c), thus giving the same effect as reported by Mai. Furthermore, if the process were reversed, on decreasing the crosshead rate when the crack is growing at a constant rate, one tends to get stick-slip conditions, and the load tends to increase, giving a false high K_c value.

With care, these two conditions can be avoided if one specimen is used for each particular rate and the crack growth is monitored over the complete section of the specimen, where the compliance is constant. All the tests reported in this paper were performed under these conditions.

REFERENCES

1. Schniewind, A.P., and D.E. Lyon (1973). A fracture mechanics approach to the tensile strength perpendicular to grain of dimension lumber. Wood Sci. & Tech., 7, 45-59.
2. Johnson, J.A. (1972). Crack initiation in wood plates. Wood Sci., 6, 151-158.
3. Schniewind, A.P., and R.A. Pozniak (1971). On the fracture toughness of Douglas Fir. Eng. Fract. Mechs., 2, 223-233.
4. Mindness, S., J.S. Nadeau, and J.D. Barrett (1975). Slow crack growth in Douglas Fir. Wood Sci., 8, 389-396.
5. Mai, Y.W. (1975). On the velocity-dependent fracture toughness of wood. Wood Sci., 8, 364-367.
6. Ewing, P.D., and J.G. Williams (1979). Thickness and moisture content effects in the fracture toughness of Scots Pine. To be published.
7. Marshall, G.P., L.E. Culver, and J.G. Williams (1969). Crack and craze propagation in polymers: A fracture mechanics approach. Plastics & Polymers, 75-81.
8. Irwin, G.R., and J.A. Kies (1954). Welding J. Res. Suppl., 33, 1935.
9. Irwin, G.R. (1964). Appl. Mats. Res., 3, 65.
10. Marshall, G.P., L.H. Coutts, and J.G. Williams (1974). Temperature effects in the fracture of PMMA. J. Mats. Sci., 9, 1409-1419.
11. Williams, J.G. (1978). Applications of linear fracture mechanics. In J.D. Ferry (Ed.), Advances in Polymer Science, Vol. 27, Springer-Verlag, Heidelberg. pp. 67-120.

Volume 3, ICM 3, Cambridge, England, August 1979

A PRELIMINARY STUDY OF TENSILE CREEP OF BEECH WITH CONCURRENT MOISTURE CHANGES

D. Hunt

Polytechnic of the South Bank, London, U.K.

INTRODUCTION

In common with other hydrophilic materials, wood creeps more under stress at a high moisture content than at a low moisture content. However,many observers in the last fifteen years (for example, Armstrong and Christensen, 1961; Armstrong and Kingston, 1962; Christensen, 1962; Hearmon and Paton, 1964; and many others) have noticed that under cyclic or changing moisture conditions the zero-load-corrected creep deflection is greatly accelerated. The essence of this behaviour, which has been called 'mechano-sorptive' (Grossman, 1976), is that with moisture cycling the creep deflection increases during drying but becomes nearly steady or even decreases during wetting. If a single moisture change takes place during creep, whether it involves wetting or drying, it always causes an increase in creep; as also does any moisture increase above the highest level previously attained during the test. A very comprehensive review of the subject was made by Schniewind (1968) and again by Grossman (1976).

Whilst much information has been published on mechano-sorptive behaviour in wood, there are two major gaps in our knowledge. Firstly there has been no confirmed explanation of the phenomenon and secondly there is still insufficient quantitative data for use in design calculations. In other words, most of the past work (with a few exceptions, e.g. Ranta-Maunus, 1975) has contributed to the qualitative but not the quantitative understanding of the problem. It is felt that any further progress requires a quantitative approach.

Most of the results of previous workers were obtained with bending measurements (although Armstrong and Kingston (1962) found similar effects with tension and compression). Such measurements are important since bending is the main mode of loading of wood. Bending measurements are also simple and inexpensive to make. However, they are not easy to analyse since they involve a combination of tensile, compressive and shear strains. It was therefore felt that a useful approach to a quantitative study would be the measurement of creep in simple tension or compression. This paper therefore describes the measurement of creep in tension along the grain of beech under constant and changing humidity conditions; using accurate creep machines that had been specially constructed for a similar study of another material.

One of the most difficult problems in obtaining quantitative measurements on wood

MS144

is its large natural variation. With bending measurements, which can be relatively simple and inexpensive, the usual solution is to use enough replicates. Such a solution is not feasible with the more complicated and expensive tensile measurements. Therefore, one of the essential requirements of making quantitative measurements of tensile creep is to develop an improved sample-matching technique. A proposed sample-matching technique is discussed in this paper.

The scope of this paper is as follows:
1) to discuss and confirm a suitable sample-matching technique.
2) to confirm that the main characteristics of mechano-sorptive behaviour, as outlined above, apply to tensile creep of beech.
3) to compare the mechano-sorptive behaviour of beech in tension with the theory of Ranta-Maunus as developed for birch in bending; at this stage, a qualitative comparison only.
4) to present the results of creep at various constant stresses and humidities, as a necessary first step towards the quantitative prediction of creep in changing humidities.

It must be noted that since items 2 and 3 were done before 1 and 4 their results are not directly comparable.

PREVIOUS WORK

Measurements of tensile and compressive creep of pine during moisture content changes were made by Armstrong and Kingston (1962). Their results give a good qualitative understanding of mechano-sorptive creep; but because accurate measuring apparatus was not available at that time, the results can not be considered as quantitative.

Important work towards quantifying mechano-sorptive creep has been done by Ranta-Maunus (1972, 1973, 1975) in developing his theory of 'hydroviscoelasticity'. He proposed that woods should fall into two classes: 'birch-like' materials, including birch, beech and pine, which show a continually-increasing strain with periodic moisture-content variations; and 'spruce-like' materials which show no cumulative effect, but have a strain that depends mainly on the maximum moisture content attained during the test; as has also been observed with a hygroscopic polymer (Hunt and Darlington, 1979). Equations were derived to fit both creep and recovery of wood. The one of most interest to this work, applied to a constant-load creep test of a birch-like material with changing moisture, is:

$$\varepsilon = \sigma\left[J + \sum_{i=1}^{n} K(u_i - u_{i-1})\right] \qquad (1)$$

where
- ε is the strain at any time t with stress σ
- J is the creep compliance at time t at the initial moisture content
- n is the number of moisture changes, u_{i-1} to u_i
- K is a material constant, denoted by:

$$\frac{K}{J_0} = \begin{cases} a^{-} & \text{for drying} \\ a^{+} & \text{for wetting while u remains below the previous maximum} \\ a^{++} & \text{for the first moisture increase above the previous maximum} \end{cases}$$

where J_0 is the instantaneous elastic compliance.

PRECISION OF RESULTS AND SAMPLE MATCHING

Importance of Accurate Matching of Samples

In order to study the mechano-sorptive effect, it is essential to be able to compare accurately the creep compliances in constant humidity conditions with those in changing conditions. The movements of the zero-load control test pieces must also be carefully matched to those of the loaded test pieces since the difference between the two movements is required as a measure of mechano-sorptive creep. Whilst these requirements may not present any problem in a uniform man-made material, they present many in a non-uniform natural material such as wood, as described below.

Variability of the material. Wood, being a natural material, is very variable. For example, in a series of preliminary tensile creep measurements on beech samples cut from the same piece and prepared for use in this project, the following coefficients of variation were obtained at a stress of 10 MNm^{-2}: strain at 40s after application of load, ± 11 percent; strain at 24h after application of load, ± 14 percent; axial dimensional change during desorption from 60% r.h. to 30% r.h., ± 19 percent. These figures may be compared with a chosen target for accumulated errors of between 2 percent and 5 percent.

Small strains involved. The strains measured at approximately 10 percent of the breaking stress were about 0.1% in constant-humidity conditions, increasing to around 0.4% in varying conditions. Strains at 30% of the breaking stress were about 0.4% and 0.9% respectively. These figures may be compared with movements due to moisture changes alone of over 0.1% (see Fig. 3). It should also be stressed that the evident viscoelastic non-linearity of the results at the higher stresses (see Fig. 1) means that it is particularly important to obtain accurate results at the lower stress levels, at which the simpler linear viscoelastic behaviour is observed.

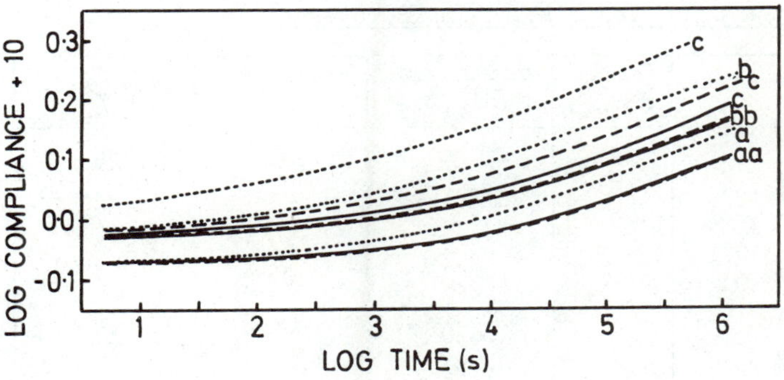

Fig. 1. Creep at constant humidities and stresses. Humidities: a, 31%; b, 61%; c, 74%. Stresses: solid, 30; dashed, 20; dotted, 10 MNm^{-2}. Compliance in SI units.

Effects of sloping grain. Sloping grain has two effects on the results: firstly because the relatively very low elastic modulus perpendicular to the grain allows the strain components in these directions to make a significant contribution to the total axial strain; and secondly because the relatively very large moisture-induced dimensional changes perpendicular to the grain can make a significant contribution to the total dimensional change along the test piece axis. Table 1 gives estimates of the errors caused by sloping grain, based on elastic modulus figures given in Wood Handbook (1955) and on dimensional changes measured in this project. The error values may be compared with the error target of 2 percent to 5 percent.

TABLE 1 Estimated Errors Caused by Sloping Grain

Angle (degrees)	Error in tensile compliance (%)	Error in shrinkage and swelling (%)
1	0.2	0.6
2	0.7	2.3
3	1.5	5.2
4	2.7	9.2
5	4.2	14.5
10	16.8	58.0

Matching of Test Pieces

Since the use of large numbers of replicates was not feasible, owing to the cost and limited number of tensile-creep machines available; it was felt that a less conventional approach was required. There were two possible types of solution: (a) an improved matching method, based on factors other than location in the tree, or (b) the maximisation of the information obtained from each test piece, possibly by re-use for later tests. These two solutions may be inter-related, since any attempt at matching by means of a preliminary creep test automatically involves re-testing, and it must be shown that the straining of the test piece in the preliminary test does not later affect the results of the real test. A number of trials were made, with the results given in Table 2. These results

TABLE 2 Repeated Creep Tests Following Recovery*

	Stress MNm^{-2}	Unrecovered strain,	Strain at 40s,%	Strain at 24h,%
Initial test	10	-----	0.100	0.128
60% rh, 10.83% m.c.	20	-----	0.168	0.221
First repeat	10	0.062	0.106	0.150
63% rh, 11.89% m.c.	20	0.053	0.171	0.240
Second repeat	10	0.050	0.105	0.140
61% rh, 11.81% m.c.	20	0.037	0.169	0.231
Third repeat	10	0.033	0.105	0.148
61% rh, 11.57% m.c.	20	0.010	0.168	0.239

* Recovery: 24h at 82% rh.
Maximum strain in initial test: 10 MNm^{-2}, 0.262%.
20 MNm^{-2}, 0.394%.

appear promising, especially since part of the strain increase in the repeat tests can be attributed to the higher moisture contents.

Based on these results, the following sample-matching procedure was adopted. Test pieces were taken to equilibrium at 61% r.h. and 23.5°C and then loaded at 10 MNm^{-2} for 24 hours. They were then unloaded and the relative humidity was increased to 83% for 8 hours to accelerate the recovery. This was followed by 61% r.h. for 40 hours, and then to 30% r.h. until the equilibrium moisture content was reached. The values used for test-piece matching were then:

(i) strain at 40s after loading.
(ii) strain at 24h after loading.
(iii) dimensional change during desorption from 61% r.h. to 30% r.h.

Re-Use of Test Pieces

Since it had been shown that an elevated-humidity recovery procedure allowed a test piece to be tested again with similar results to those of the previous test, it was felt that it would be worth considering repeated tests on the same matched test pieces. Since this would be a radical departure from normal practice, it would be important to proceed with care and to make frequent estimates of any errors caused by property changes within the test pieces. This could be done by occasional repetition of the test-piece matching procedure described above. Whilst the target error of from 2 percent to 5 percent was still to be aimed for, this figure might need to be adjusted, in the light of the results actually obtained. These errors could then be compared with those that were caused by differences between the matched test pieces.

EXPERIMENTAL APPARATUS AND TEST PIECE PREPARATION

For this type of study there was constructed a set of accurate tensile-creep machines housed in an accurately-controlled environmental chamber (Hunt and Darlington, 1978). The precision of strain measurement is such that an error in the modulus of only 0.4% was achieved at a strain level of 0.1% (i.e. a strain error of 0.0004%) when testing a PVC whose characteristics were well known.

The environmental chamber controlled the temperature to within less than ± 0.1°C and the relative humidity to within ± 1% r.h. It was also designed to give 'step changes' of relative humidity that were 90% complete within 10 minutes or less, and complete and accurately controlled within 45 minutes (these times may be compared with the time for 50% of the equilibrium moisture change in the test pieces, of about 2.3h).

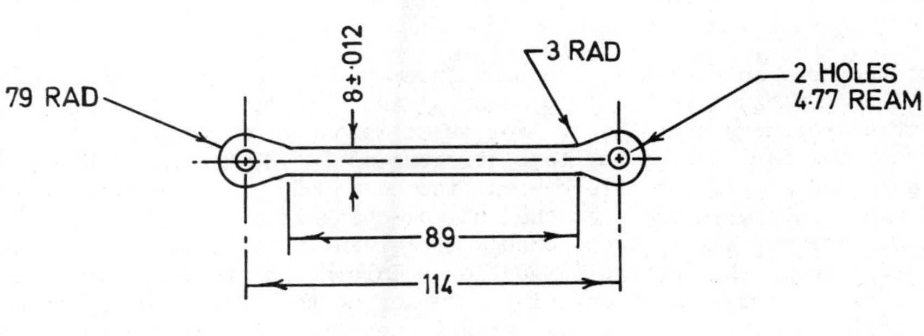

Fig. 2. Shape of test piece. Dimensions in mm.

Test pieces were taken from a single plank of wood, and were of the shape shown in Fig. 2. Care was taken that the grain direction was aligned on the axis of the test piece, the tangential direction was across its width, and the radial direction was through its 3.17 mm thickness. After rough cutting, the thickness was obtained by sanding both sides (finishing on a 100-grit belt parallel to the grain). The profile was cut on a router, using a tipped cutter. The ends were reinforced with a 0.4 mm thick vulcanised-fibre sheet.

RESULTS

The results of the first two exploratory tests on unmatched samples are shown in Figs. 3 to 6. The results of Figs. 3 and 4 are for tests run for 500 h at 30% and 60% respectively, followed by humidity cycling as shown. Some features to note are:
(a) due to lack of sample matching, the levels of the curves are spaced in an irregular manner.
(b) by plotting the constant-humidity results on a logarithmic-time base (not shown here) the level of the projected constant-humidity curves can be estimated quite accurately.
(c) the horizontal scale is the square root of time for each humidity change. Since the moisture-content change for Fickian diffusion, when plotted against the square root of time, gives a characteristic curve consisting of a straight line up to about 60% of the equilibrium change, it was hoped that plotting on this base would emphasise any anomalies either in the diffusion or the creep behaviour.
(d) anomalous dimensional changes can be observed during desorption in Figs. 3 and 4; the curve shapes approximating to that of the moisture-content curve at zero stress, but with greater deviations at higher stresses.

The results of Figs. 3 and 4 are reproduced in Figs. 5 and 6 respectively, after zero-load correction. It should be noted here that although the test pieces had not been completely matched before the tests, the zero-load dimensional changes of all test pieces had been measured over a range of relative humidity in order to adjust the zero-load corrections for each test piece individually. Features to note in Fig. 5 are that the main departures from the projected curve occur during the first desorption and during each sorption above the basic 30% r.h. There appears to be a slight recovery at the start of each wetting. The features to note in Fig. 6 are the increase in creep during desorption followed by some recovery during re-sorption. There also appears to be very little resultant creep after the first, small humidity cycle, with progressively more pronounced shape differences with the larger humidity intervals, or desorbing to lower humidity levels.

The results of the constant-humidity creep measurements on matched test pieces are shown in Fig. 1, plotted as log compliance against log time, for three stresses and three relative humidities. For these tests, four matched test pieces were used, one for each of the four stresses of 0, 10, 20 and 30 MNm^{-2}. They were then tested at 30% r.h., recovered, tested at 60% r.h., recovered, tested at 74% r.h., recovered and then checked for comparison with their pre-test characteristics. The results of these checks are given in Table 3, and an interpretation is proposed in the Discussion section. Interesting features to note in Fig. 1 are (i) the confirmation of matching accuracy at the lower humidities and stresses and shorter times, (ii) the viscoelastic non-linearity at higher stresses and humidities and longer times, as shown for example by the higher levels of the curves for 30 MNm^{-2} and (iii) the greater compliances at higher relative humidities.

DISCUSSION AND CONCLUSIONS

Reference to Figs. 5 and 6 shows that the main characteristics of mechano-sorptive behaviour, as outlined in the Introduction and Previous Work sections are confirmed. Approximate ranges of values for Ranta-Maunus' 'hydroviscoelastic constants'

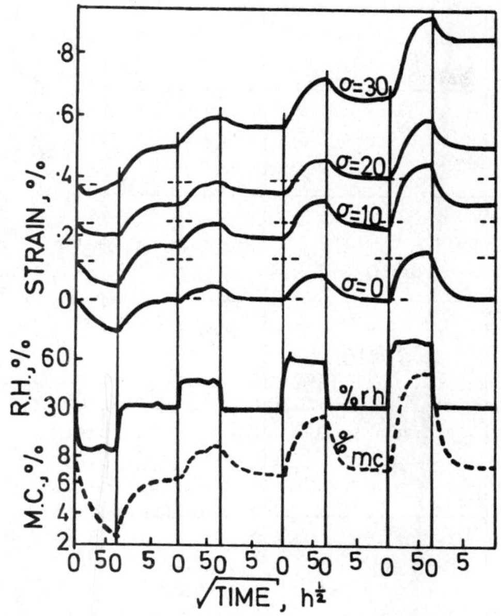

Fig. 3. Uncorrected axial creep: humidity cycling follows 500h @ 30% r.h. Projected strain curve @ 30% r.h. ------ .

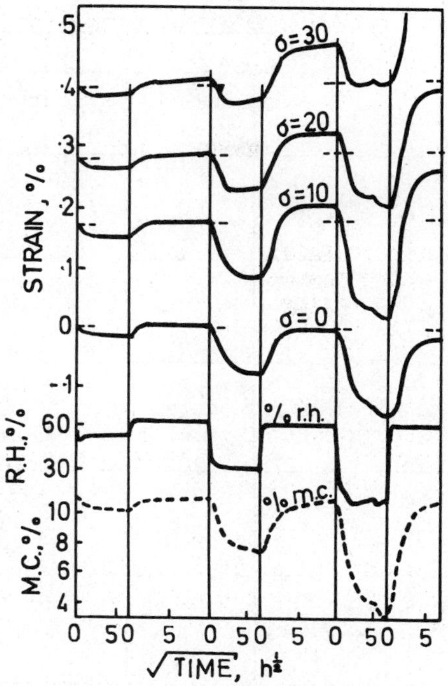

Fig. 4. Uncorrected axial creep: humidity cycling follows 500h @ 60% r.h. Projected strain curve @ 60% r.h. ------.

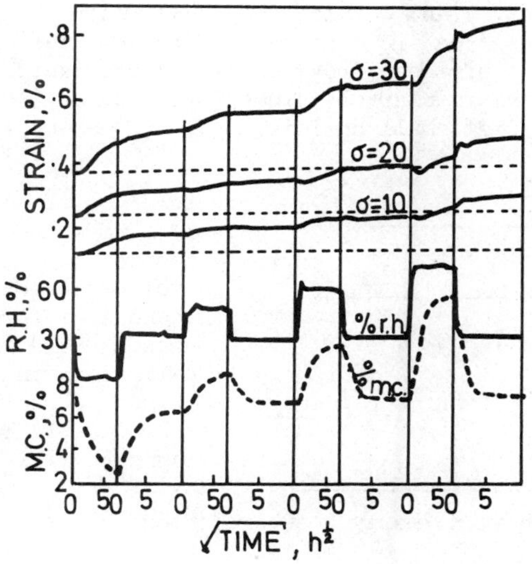

Fig. 5. Zero-load corrected axial creep: humidity cycling follows 500h @ 30% r.h. Projected creep curve @ 30% r.h. ------ .

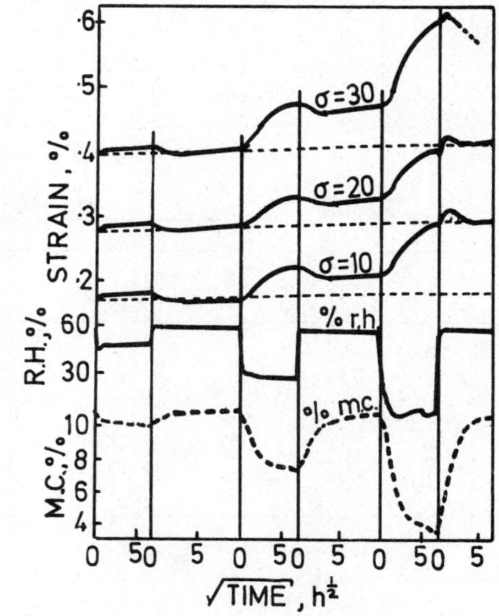

Fig. 6. Zero-load corrected axial creep: humidity cycling follows 500h @60% r.h. Projected creep curve @ 60% r.h. ------ .

TABLE 3 Checks on Effects of Re-Use of Test Pieces

	Stress ($10^6 Nm^{-2}$)	Compliance at 40s ($10^{-9}N^{-1}m^2$)	Compliance at 24h ($10^{-9}N^{-1}m^2$)	Residual* strain (%)
Matching test at 60% rh, 11.5% m.c.	10 10 10	.0906 .0924 .0912	.1203 .1241 .1239	+.015 +.019 +.023
Test at 30% rh, 7.5% m.c.	10 20 30	.0864 .0854 .0866	.1058 .1048 .1142	-.025 -.003 +.003
Test at 61% rh, 11.8% m.c.	10 20 30	.0950 .0922 .0995	.1201 .1215 .1450	-.016 +.011 +.040
Test at 74% rh, 14.4% m.c.	10 20 30	.0963 .0994 .1118	.1255 .1344 .1696	-.025 +.026 ***
Check test at ** 61% rh, 12.3% m.c.	10 10 10	.0966 .0899 .0913	.1227 .1100 .1140	----- ----- -----

*Estimated total residual zero-load-corrected strain after test.
**Transducer bridge failed after 7h. 24h values estimated.
***Fracture at grips.

are given in Table 4, with those of Ranta-Maunus for birch as a comparison. It can be seen that the values for beech and for birch are similar, but the range of values obtained in this project suggest that they may not be constants. Such a conclusion is supported by the results of Figs. 3 and 4 which show that, for example, the shapes of the curves during desorption are affected by stress. The results of Fig. 6 suggest also that the values of the hydroviscoelastic constants are affected by the size of the relative-humidity interval, since little or no permanent strain increase resulted from the first small humidity cycle (60-50-60).

TABLE 4 Hydroviscoelastic Constants

	a^-	a^+	a^{++}
Birch veneer in bending (Ranta-Maunus)	-7.4	-4.37	+12.6
Beech in tension (this project)	-6.1 to -8.6	-0.7 to +1.2	+9.8 to +14.7

In the past, mechano-sorptive creep results have normally been presented in a zero-load corrected form such as those of Figs. 5 and 6. It is of interest to note that presentation in an uncorrected form, as Figs. 3 and 4, clearly shows the differences in curve shape as a function of stress during desorption, whilst that of Figs. 5 and 6 does not. This suggests that the mechano-sorptive effect during desorption could more usefully be considered as the effect of applied loads on the moisture-induced swelling and shrinkage of wood, rather than as the effect of moisture changes on its creep. On the other hand the increased creep when the test piece is taken to a higher moisture content than had previously been reached during the test could be simply the result of larger compliances at higher moisture contents.

The results to date of creep measurements at constant stress and humidity fit the generally accepted pattern of creep in hygroscopic materials; namely, greater compliances at higher humidities (e.g. Cave, 1978) and viscoelastic non-linearity at higher stresses and humidities and longer times (e.g. Ranta-Maunus, 1972). It is hoped to use these curves as a basis of comparison with creep curves obtained with monotonically-increasing moisture contents.

The results, to date, of the proposed sample-matching technique appear promising, as shown by the relative consistency of the results of Fig. 1 and Table 2. Results given in Table 3, of repeated tests on the same test pieces suggest that this approach can be acceptable provided that caution is exercised. When the test pieces have been strained in a viscoelastically non-linear manner (e.g. at higher stresses and humidities) the residual strain appears to be greater and the creep strain (24h) in the check test appears to be less. The elastic (40s) strains are little affected. It is not known at present whether the non-linear creep has permanently affected the test pieces or whether the recovery is not sufficiently complete. It is felt that because the present tests were made at progressively higher humidities, the error caused by such effects has been minimised and the results of Fig. 1 are therefore useful. However, in future work caution should be exercised about re-using test pieces after viscoelastically non-linear creep.

REFERENCES

Armstrong, L.D., and G.N. Christensen (1961). Influence of Moisture Changes on Deformation of Wood under Stress. Nature, 191, 869.

Armstrong, L.D., and R.S.T. Kingston (1962). The Effect of Moisture Content Changes on the Deformation of Wood under Stress. Austr. J. Appl. Sci., 13, 257.

Cave, I.D. (1978). Modelling Moisture-Related Mechanical Properties of Wood. Wood Sci. Tech., 12, 127-139.

Christensen, G.N. (1962). The Use of Small Specimens for Studying the Effect of Moisture Content Changes on the Deformation of Wood under Load. Austr. J. Appl. Sci., 13, 242.

Grossman, P.U.A. (1976). Requirements for a Model that Exhibits Mechano-sorptive Behaviour. Wood Sci. Tech., 10, 163-168.

Hearmon, R.F.S., and J.M. Paton (1964). Moisture Content Changes and Creep of Wood. For. Prod. J., 14, 357.

Hunt, D.G., and M.W. Darlington (1978). Accurate Measurement of Creep of Nylon 66 at Constant Temperature and Humidity. Polymer, 19, 977-984.

Hunt, D.G., and M.W.Darlington (1979). To be published.

Ranta-Maunus, A. (1972). Viscoelasticity of Plywood under Constant Climatic Conditions. Building Technology and Community Development, Technical Research Centre of Finland. Publication 3.

Ranta-Maunus, A. (1973). A Theory for the Creep of Wood with Application to Birch and Spruce Plywood. Building Technology and Community Development, Technical Research Centre of Finland. Publication 4.

Ranta-Maunus, A. (1975). The Viscoelasticity of Wood at Varying Moisture Content. Wood Sci. Tech., 9, 189-205.

Schniewind, A.P. (1968). Recent Progress in the Study of the Rheology of Wood. Wood Sci. Tech., 2, 188.

Sliker, A. (1978). Strain as a Function of Stress, Stress Rate and Time at 90° to the Grain in Sugar Pine. Wood Sci., 10, 208-219.

Wood Handbook (1955). U.S. Dept. of Agriculture. Handbook No. 72.

Volume 3, ICM 3, Cambridge, England, August 1979

THE RELATIONSHIP BETWEEN TOUGHNESS AND FRACTURE SURFACE TOPOGRAPHY IN WOOD AND COMPOSITES

M. P. Ansell and B. Harris

School of Materials Science, University of Bath, Bath BA2 7AY, U.K.

INTRODUCTION

Synthetic fibre-reinforced composites and clear organic wood are defined as tough materials, with works of fracture greater than 10^4 J/m^2. Both classes of material derive their toughness from the structural arrangement of their component parts.

Wood, like composites, contains stiff fibres bonded into a matrix phase. In the case of practical modern composites the fibres can be glass, carbon, boron, or organic polymers, and the matrix might be a resin, metal or rubber. In wood, the fibres are large clumps of crystalline cellulose, about 15 nm in diameter, consisting of linearly-oriented molecules. These are called microfibrils. The matrix contains lignin, which is an amorphous macromolecular phase, and hemi-cellulose, composed of small molecules in semi-crystalline arrangements.

Wood and composites differ principally in the arrangement of the fibre and matrix structural elements. In softwoods the main structural units are hollow cells known as tracheids which are typically 1 to 3 mm in length and aligned parallel to the direction of growth. Tracheids are arranged in a regular annual ring structure consisting of concentric rings of thin-walled earlywood cells, alternating with rings of thick-walled latewood cells. Each tracheid has a composite structure comprising the matrix phase reinforced by the helically arranged microfibrils. The matrix also serves to bond each wood cell to its cell wall layers (Dinwoodie, 1975), the S_2 layer being the thickest with the steepest winding angle and the highest degree of order. The stiffness of each cell is principally attributed to this layer.

Man-made composites, such as GRP and CFRP, resemble wood inasmuch as they contain a low stiffness matrix reinforced with high stiffness fibres. However, the diameter of the reinforcing element in such composites (5μm upwards) is far higher than that of the wood microfibril (10 to 30 nm). The stiff aligned continuous fibres or cross-woven fibre mat in a composite take up the applied load directly whereas in wood the reinforcing phase is effectively in the form of closely packed, concentric springs some 30μm in diameter which are constrained not only by the matrix phase but by adjacent cellulose molecules. Extension of the wood cell therefore involves a multiplicity of intricate microshear interactions between the cellulose microfibrils and the matrix, whereas in the fracturing of a composite, aligned fibres are often able to pull out of their matrix sheaths in a relatively simple fashion. Furthermore, the softwood cell is hollow and in longitudinal tension this endows the wood with values of specific stiffness, failure stress and work of fracture that are very favourable by comparison with other composites and conventional engineering materials.

Wood and composites both contain fibre-matrix interfaces which can check the progress of cracks during failure, thereby increasing the material toughness. At a higher structural level, however, wood is in addition composed of concentric rings of earlywood and latewood tracheids which have substantially different elastic modulus and strength and therefore form an additional macro-interface at their boundaries, seen in cross-section as the grain of the wood. The material is thus a two-phase composite 'laminate' at a macroscopic level and a sophisticated, helically-wound cellular composite at a microstructural level. Practical reinforced plastics also often possess macro-structural interfaces since they are pressed up from separate, pre-impregnated laminae. Interfaces between these separate laminations in the finished composite frequently lead to interlaminar shear failure at low stress levels in bending. But the ability of these interfaces to divert translaminar cracks, thereby increasing the composite toughness, has been deliberately exploited on occasion.

In this paper the topography of fracture in a softwood is compared with those of a 0/90° woven GRP and a unidirectionally reinforced CFRP. These materials were tested in tension to failure and acoustic emissions (AE's), generated as a result of the growth of flaws, were simultaneously recorded. Characteristic stages in the development of structural damage can thus be detected and related to the level of tensile strain. The topography of the fractured surfaces were studied by SEM and the effect of the arrangement of fibres and matrix on the path of failure in wood and composites is discussed. Finally, fracture energies are compared in the two classes of tough materials.

EXPERIMENTAL DETAILS

Clear Scots Pine sapwood (Pinus sylvestris) was kiln-dried and seasoned to 65% RH at 20°C. Samples were cut with the dimensions shown in Fig.1, by rough cutting

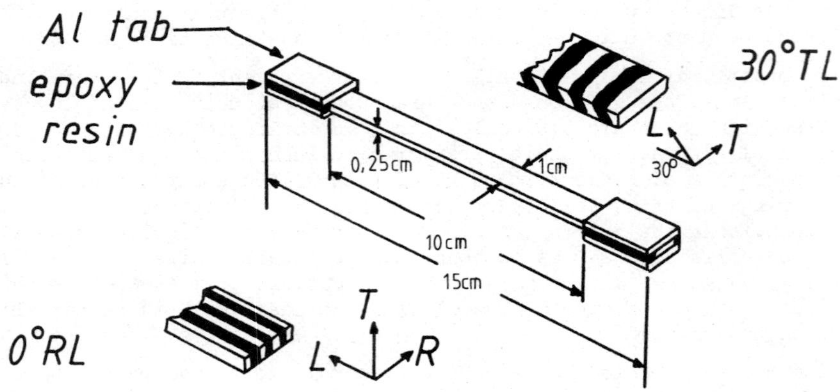

Fig. 1. Dimensions and grain orientation of wood samples

followed by sanding in metal jigs. Specimens with a 0° grain angle were radial - longitudinal (RL) in orientation, and those with a 30° grain angle were tangential - longitudinal, (TL).

GRP samples were cut from sheets of 9-ply glass-epoxy composite manufactured by Formica Ltd. This was a 60 vol.% laminate 1.4 mm thick, and the plain weave glass cloth, which had a thread count of 16.5 per cm (warp) and 12.6 per cm (weft) was silane coated. Glass fibres, cross-woven at right angles in an epoxy resin matrix, were aligned at 0° and 90° to the sample length.

Abraded aluminium tabs of 1.5 × 2.5 cm in surface dimension were glued to the ends of the samples with epoxy resin, a clear 10 cm length of sample being exposed between the tabs. The space between the overlapping edges of the tabs was filled with epoxy resin to prevent crushing of the samples before testing.

Specimens were tested in tension at a constant cross-head speed of 0.1 mm/min. The serrated jaws of the Instron grips bit into the soft aluminium preventing shear between the tabs and the jaws. This procedure is necessary to eliminate the risk of spurious emission of noise from the grip region.

CFRP samples (Fuwa, 1975) were also un-notched with the dimensions of Fig.2.

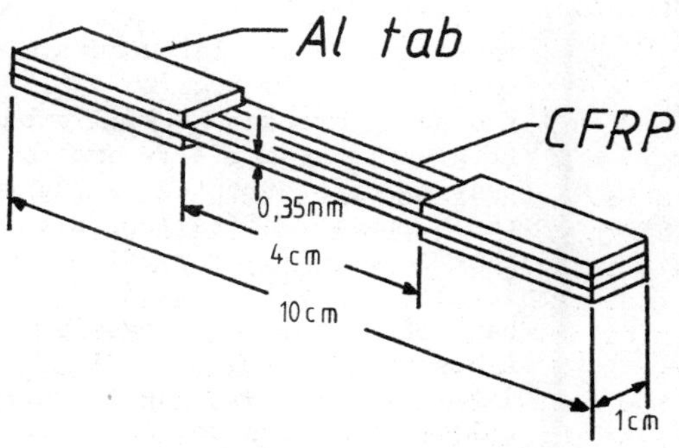

Fig.2. CFRP specimen dimensions

Single layers of Ciba-Geigy epoxy resin pre-impregnated sheet were hot pressed and cured. Samples were uniformly 0.35 mm in thickness and contained 35 vol. % of type I, untreated carbon fibre. CFRP specimens were tensile-tested with a cross-head speed of 0.05 mm/min. Strain measurement for all samples was made with an Instron extensometer.

AEs were detected with a PZT-5 (lead-zirconate-titanate) piezo-electric transducer coupled firmly to the specimen with silicone grease. Dunegan 3000 series apparatus amplified and processed the signals. Overall amplification was 10,000x (80 dB) and a band pass filter passed signals between 100 and 300 kHz. Each rising positive signal of amplitude 100 μV or greater contributes to the total ring-down count recorded by a digital counter which, after amplification, accepted signals of greater than one volt. Accumulated counts were recorded at approximately 0.01% strain intervals, providing detailed information of structural change throughout the duration of the test.

SCOTS PINE RESULTS

Stress-strain and AE-strain curves for 0° radial longitudinal wood are co-plotted in Fig. 3. Each AE characteristic corresponds to one of the stress-strain curves. The rate of generation of stress waves at small strains is low, whereas near failure the rate is high. By plotting AE on a logarithmic scale a more detailed examination of stages in flaw development can be achieved by amplifying the low strain end of the characteristic and compressing the rapid emission rate at higher strains. In all wood data the slopes of the stress-strain curves vary, but individually they are close to linear right up until failure. This spread is a result of variations in the proportion of latewood and earlywood which determine the elastic modulus of each specimen. It is impossible to keep this proportion constant in practical samples.

The mean failure strain of the samples represented in Fig. 3 is 0.8 ± 0.1%. The total AEs to failure are also fairly close, although the shape of the individual characteristics is variable, indicating a non-uniform rate of flaw growth. The AE curves are a combination of sharp vertical increases interspersed with more gradual accumulations of counts. Slow count rates are attributed to gradual opening of microflaws as the helically-wound cell wall reinforcement extends elastically within the matrix of hemicellulose and lignin. The rapid increases in AE are attributed to either interlaminar shear in planes of weakness, for example at ray cell-tracheid interfaces and earlywood-latewood interfaces, or to brittle failure of tracheids, both of which are fast failure events.

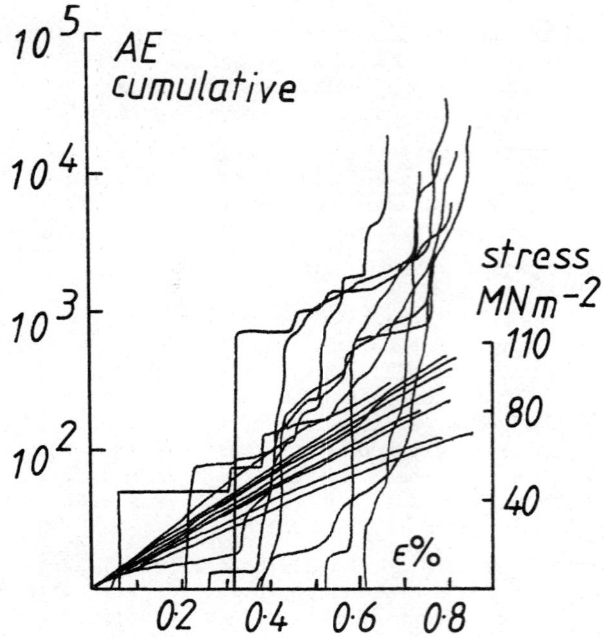

Fig.3. Acoustic emission and stress vs. strain for Scots Pine, 0°, RL.

The appearance of the fractured wood can be described at two levels. The sketch of failed 0° R.L. wood in Fig.4 demonstrates typical paths of crack propagation, either along the winter interface between latewood and earlywood in the direction of cell growth or perpendicular to the cell axis.

At a microscopic level the essential differences between transverse crack propagation in earlywood and latewood are demonstrated in Figs. 5 & 6. Fracture through the earlywood (Fig.5) is essentially smooth, passing uninterrupted through the tracheid and ray cell interfaces. The ray cells are identified by the diagonal band of pitted cells passing across the micrograph. In latewood, however, the thicker-walled tracheids act as barriers to crack propagation and fracture occurs in steps. Vertical steps in Fig.6. occur in the radial-longitudinal plane, particularly where ray cells exist. Intercellular shear at such an interface is illustrated in Fig.7. On fractured wood surfaces there is little evidence of the fibre pull-out that is so common a feature of the failure of reinforced plastics.

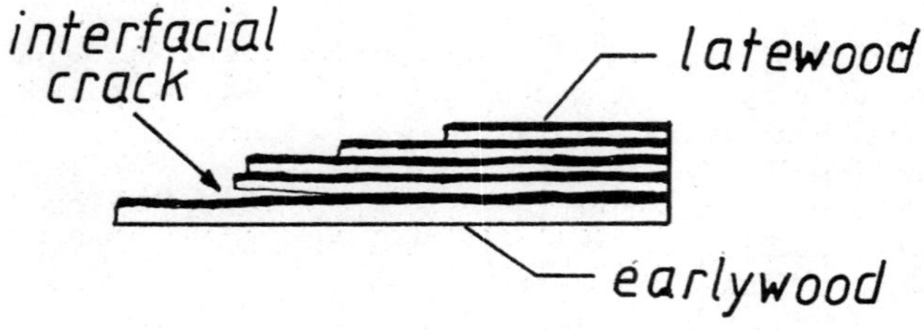

Fig.4. Typical fracture in Scots Pine (0°RL).

As wood is deformed in tension the S_2 layer of longitudinally-stressed tracheids will tend to unwind and this can occasionally be observed (Fig.8). Here the right-hand spiral of the S_2 microfibrils is clearly seen, indicating that micro-flaws were developing along the helical windings prior to cell fracture.

From Fig.3. it is seen that the multiplicity of available fracture paths results in substantial variations in the shape of AE curves. On the other hand, failure strains and total counts to failure do not vary much from sample to sample, which suggests that fracture energies would also be uniform.

The many cellular interfaces and the annual ring macro-structure in wood retard the growth of microcracks to critical dimensions, thereby inhibiting catastrophic crack propagation until a sufficiently large stress concentration has built up. When this point is reached a rapidly accelerating crack passes through the specimen, instantaneously, along the path of greatest weakness, and no deviation from linearity is seen in the stress-strain curve.

The AE characteristic for the 30° grain angle TL Scots Pine is seen in Fig. 9.

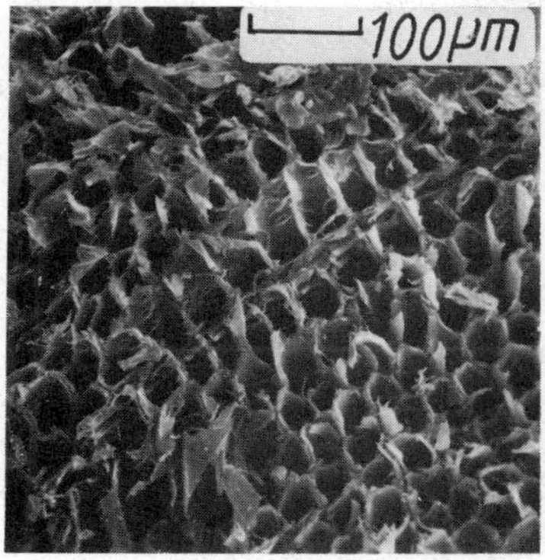

Fig.5. Fracture plane in Scots Pine earlywood

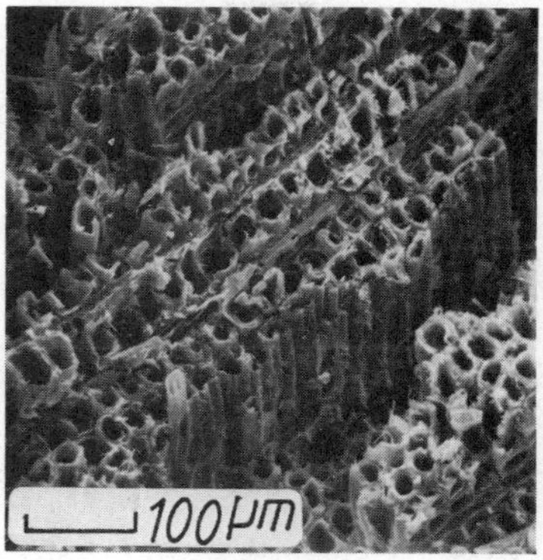

Fig.6. Stepped fracture in Scots Pine latewood

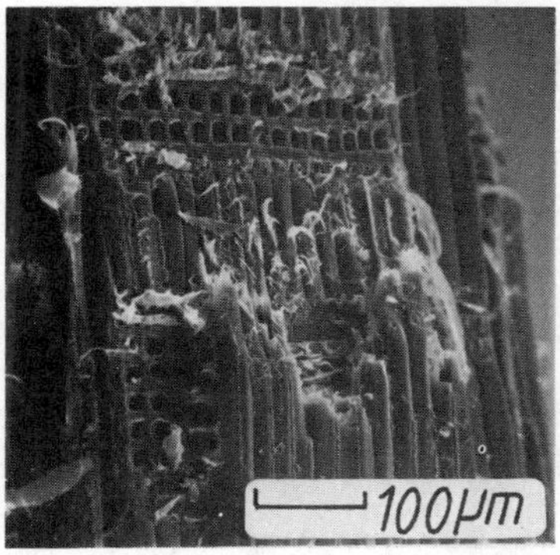

Fig.7. Shear failure at a ray cell interface

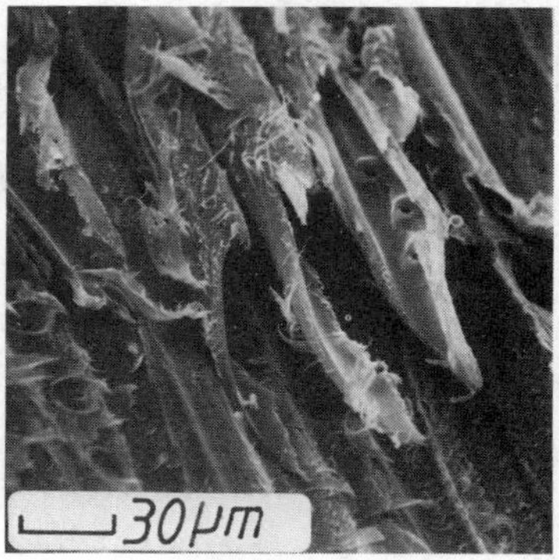

Fig.8. Helical fracture in latewood

The AE curves are now much closer at intermediate strains but there is greater scatter in the failure strains. The almost straight log-linear AE characteristics are a result of fracture exclusively in planes of earlywood, the crack travelling through the TL plane, (Fig.10). The constraint imposed on the fracture path by the test geometry limits the type of failure event to shear and tracheid fracture in the earlywood and thus the AE output is more uniform and reproducible.

GRP RESULTS

The stress-strain and AE plots for $0/90^{\circ}$ woven GRP are seen in Fig.11. Although the average modulus of elasticity 20GPa and fracture strain (1.2%) are somewhat higher than those for wood, these values are similar enough for the AE characteristics to be compared. As might be expected for specimens cut from a uniform sheet of synthetic material the stress-strain curves are very close together and smooth in shape. Furthermore, the AE-strain curves all have the same general shape. AEs begin to accumulate at about 0.12% strain, building up smoothly until failure when the usual rapid increase in AEs is noted.

GRP deforms differently from wood. Initially the load is carried mainly by the fibres but, as the weakest brittle fibres fail, work is done in debonding these fibres and subsequently in pulling their broken ends out of the matrix. As a result of the regular arrangement of fibres in the resin matrix, the AE curves

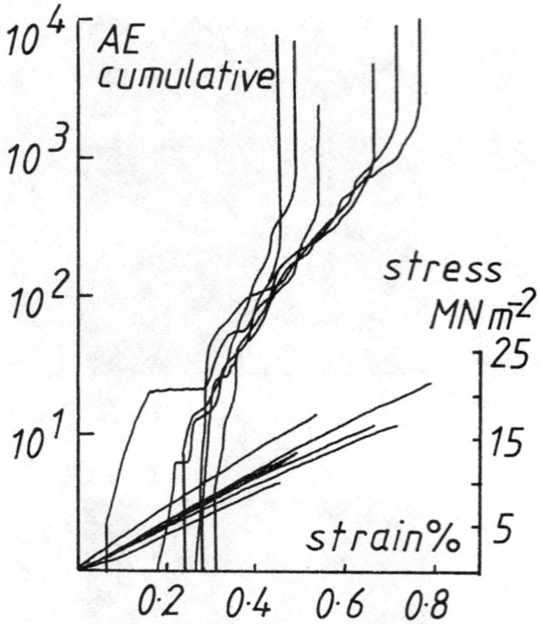

Fig.9. Acoustic emission and stress vs. strain for Scots Pine, 30°, TL.

are smooth and progressive in shape. Failure of the GRP, a fast, brittle event, is accompanied by a rapid burst of emissions as in wood. Pulled-out glass fibres are seen in Fig.12. Particles of the resin matrix remain adhered to the fibres and surface debris is evident. Macroscopically all cracks in these specimens propagated linearly at about 90° to the long axis. At the microscopic level cracks passed cleanly through the brittle matrix but were at intervals diverted vertically along planes of fibres (Fig.13). Three of the 90° fibres are located in the lower left hand corner of the micrograph.

CFRP RESULTS

Typical AE-strain and stress-strain curves for 0° uniaxially reinforced CFRP have been replotted (Fig.14) from the results of Fuwa (1974). The stress/strain characteristic for the 35 vol.% composite is close to linear but the failure strain is much lower than that of GRP. In all specimens tested by Fuwa many successions of fibre breaks of short length were observed. In stronger samples the amount of such damage is greater as is the length of broken fibre ends.

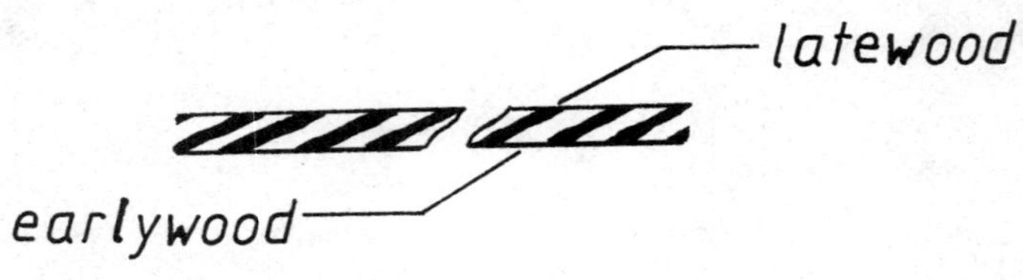

Fig. 10. Typical fracture surface in Scots Pine 30° TL wood.

Fibre breaks occur both perpendicular to the direction of fibre alignment and at angles to this direction. Accumulation of AE (Fig.14) is close to log-linear and similar to the behaviour of fibre bundles, denoting a progressively smooth increase in the generation of stress waves as a result of random fibre failures. There was little evidence that acoustic signals in these unidirectional CFRP were generated by any other micro-failure event.

The table compares the properties and acoustic behaviour of the wood and reinforced plastics which we have studied. No obvious conclusions can be drawn about the differences between total acoustic response to failure of these different materials, for despite the similarity of their fracture toughnesses, the total count varies between 15 and 500 thousand counts. These differences are clearly related, however, to much higher brittleness (and accompanying high stored elastic energy at failure) of the reinforcing elements. Thus, a substantial part of the AE from GRP and CFRP derive from single fibre and bundle failures, whereas AE from wood is unlikely to be generated by tracheid failures.

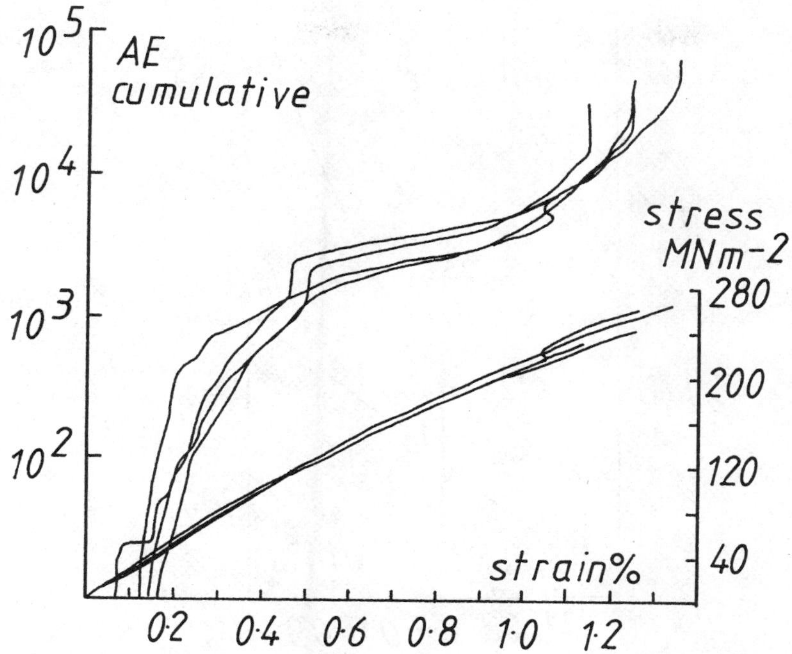

Fig.11. Acoustic emission and stress vs. strain for glass/epoxy 0/90°laminate.

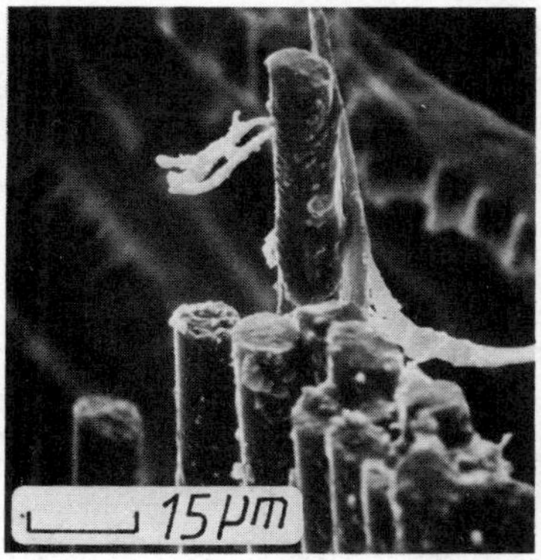

Fig.12. Pull-out of glass fibres

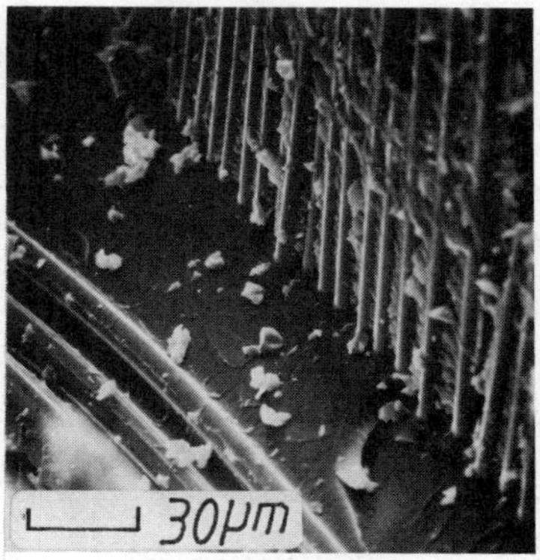

Fig.13. Crack-stopping at matrix/ fibre interface

TOUGHNESS OF WOOD AND COMPOSITES

Wood and most reinforced plastics (RP) are composites consisting of a brittle matrix reinforced by a strong fibrous phase which is also brittle in character, yet the fracture energies of all of the materials discussed here are high, being of the order of 20 to 30 kJm^{-2}. In neither case can this level of toughness derive from the properties of the separate constituents, and must therefore be a result of composite action and the presence of fibre/matrix interfaces. Another similarity between wood and composites is that both are affected by a substantial level of residual stress in the condition in which they are normally used. In wood these stresses arise during drying out from the 'green' state, a process in which some 6% of the volume is lost, the losses being non-uniform through the cellular structure. In consequence, seasoned clear wood contains a distribution

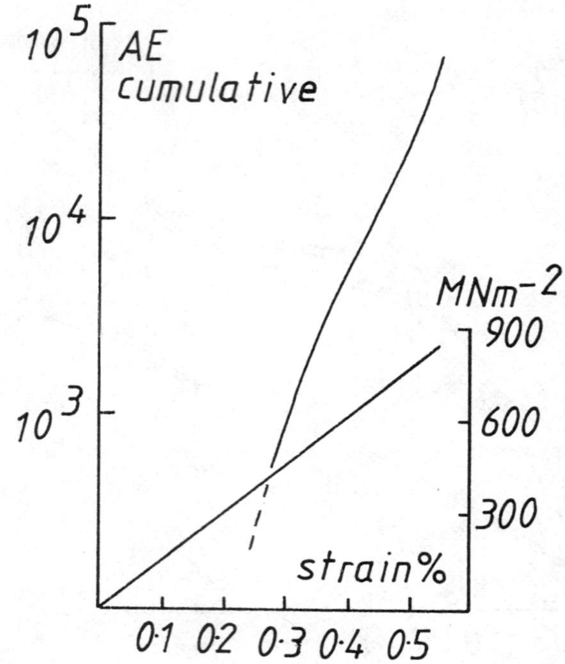

Fig. 14. Acoustic emission and stress vs. strain for unidirectional carbon fibre/epoxy composite.

Material	σ_f,MP$_a$	ε_f, %	E, GP$_a$	ρkgm^{-3}	ΣAE 10^3counts	γ_f,kJm^{-2}	$\frac{1}{2}\sigma_f\varepsilon_f$ J/m^3
Scots Pine 0°RL (12.5% emc)	85.4	0.76	11.3	505	14.6	23	32
Scots Pine 30°TL (12.5% emc)	17.7	0.64	2.7	505	7.9	-	5.7
GRP (0/90°)	240	1.2	20	2000	97.5*	20	144
CFRP Unidirectional	850	0.55	155	1500	486*	30	234

*normalised to same cross-sectional area as wood results

of microscopic defects such as ruptured cellwalls, torn pit membranes and minute checks or fissures between the phases. R.P. develop residual stresses due to differential thermal contraction during cooling from the curing temperature (often in excess of $100^{\circ}C$), and even cold-cured composites contain stresses induced by cure-contraction of the matrix resin. These residual stresses may be sufficiently high to cause resin cracking and interfacial debonding in the transverse plies of a laminate. Furthermore, the best of commercial laminates will contain a proportion of broken, damaged, or misaligned fibres, areas of poor wetting-out of fibres, or resin-rich regions. Both wood and composites must therefore be considered as highly defective structures on a microscopic scale.

In attempting to account for the comparable toughnesses of wood and some composites reference has sometimes been made to the familiar 'brushy' failure surfaces so often observed in broken GRP and CFRP and apparently similar features of wood failure surfaces. As we have shown, however, this similarity is purely superficial and on a microstructural scale there appear to be more points of difference than similarity in their behaviour.

In discussing the toughness of GRP and CFRP many separate mechanisms have been invoked, (Phillips and Harris, 1977). Cracks passing through laminates encounter frequent planes of weakness which are able to divert the main crack. Frequent diversion makes for a tortuous crack path and requires increased work of fracture. On the microstructural level fibre/resin interfaces also serves as crack stoppers. Fibre/resin debonding will often occur at the crack tip, requiring further energy expenditure, and again there may be small-scale diversion of the crack. Fibres often fracture randomly throughout a stressed region as a result of the statistical distribution of flaws in brittle solids. During the subsequent pulling-apart of the broken halves of the composite the ends of fractured fibres must be withdrawn from their matrix holes. Because of the residual thermal stress system the resin usually grips the fibres tightly, and pulling out therefore calls for more energy of fracture. Under some circumstances single cracking is suppressed and multiple cracking occurs instead. (Aveston and Kelly, 1971). The total work of fracture will therefore be increased because of the multiplication of crack tip energy absorbing events. Finally, when a composite structure containing components of differing degrees of anisotropy is elastically deformed elastic constraints are set up within the structure which are capable of modifying crack behaviour so as to increase or decrease the apparent fracture toughness, depending upon particular properties and geometry of the phases (Guild et al, 1978). All of these mechanisms may contribute to the toughness of a composite but approximate calculations relating to GRP and CFRP suggest that the mechanisms of debonding and fibre pull-out largely determine their work of fracture. Where delamination is deliberately fostered, this also contributes significantly to the toughness of laminates.

Softwoods often exhibit brushy fractures and the work of Mark (1967) implied that pull-out could be contributing to toughness. We find, however, that the brushiness is defined at a structural level much higher than that of the individual tracheid. Crack diversion and the production of indented fracture surfaces in Scots Pine results from the tendency of the crack to follow the well-defined planes of weakness between earlywood and latewood at the winter-spring interface. We have also shown that in softwoods where there is little early and latewood cell differentiation, such as Parana Pine, both the toughness and the total AE count to failure are much lower than in Scots Pine or Douglas Fir which have a more distinctly marked ring structure (Ansell, 1979). Cracks in wood propagate in a markedly intermittent fashion, as indicated by the stepped AE vs. strain curves. Such curves rarely occur in RP where fracture appears to develop more uniformly except near final failure.

Although fibre pull-out does not constitute a toughening mechanism in wood, there is evidence that extensive fibre deformation, effectively by the unravelling of the helically-wound tracheids, does contribute to the post-cracking fracture energy. This mechanism, which has been beautifully modelled by Gordon and Jeronimidis (1978), is similar in effect to the behaviour of the coiled, duplex element model reinforcements discussed by Morley (1971).

ACKNOWLEDGEMENTS

We are happy to record our gratitude to S.R.C. for support of our work and to Dr.J.M.Dinwoodie for valuable advice and discussions.

REFERENCES

Ansell, M.P. (1979), Acoustic emission from softwoods in tension Wood Science and Technology, in press

Aveston, J. Cooper, G.A. & Kelly, A., (1971), N.P.L. Conference on "The Properties of Fibre Composites", IPC Press (London)15-24

Dinwoodie, J.M., (1975), Journal of Microscopy, 104, 3 - 32

Fuwa, M., (1974), D.Phil. Thesis, University of Sussex

Gordon, J.E. & Jeronimidis, G., (1978), Royal Society Discussion Meeting "Newer Fibres and their composites, in press

Guild, F.J., Harris, B. & Atkins, A.G., (1978), J.Materials Science, 13, 2295-2299

Mark, R.E., 1967, "Cell Wall Mechanics of Tracheids", Yale University Press (New Haven, Conn)

Morley, J.G., (1971), N.P.L. Conference on "The Properties of Fibre Composites", IPC Press (London), 33 - 35.

Phillips, D.C. & Harris, B., (1977), The Strength, Toughness and Fatigue Properties of Polymer Composites,in "Polymer Engineering Composites", ed. M.L. Richardson, Applied Science/Elsevier, (London), 45-154.

Volume 3, ICM 3, Cambridge, England, August 1979

TIME - DEPENDENT FAILURE OF WOOD

S. Mindess, J. D. Barrett* and R. A. Spencer

Department of Civil Engineering, University of British Columbia, Vancouver, British Columbia, Canada
***Department of the Environment, Canadian Forest Service, Western Forest Products Laboratory, Vancouver, British Columbia, Canada**

INTRODUCTION

It has long been known that the strength of wood depends upon the load history. Experimental studies had shown that the strength of wood increased as the stress rate increased (Liska), and that the time to failure at constant stress decreased with increasing stress (Wood); these effects were recognized by the building codes. However, it was only in 1961 that fracture mechanics was applied to the study of wood (Atack and co-workers), when it was shown that a Griffith-type relationship existed between the stress at fracture and crack length in precracked specimens. Since that time, a substantial literature on fracture mechanics studies of wood has appeared.

It has recently been suggested (Mindess, Nadeau and Barrett, 1976) that the dependence of the strength of wood upon the loading history is a consequence of subcritical crack growth, that is, the slow growth of cracks that are too small to cause failure initially under the applied stress. Such a process would explain not only delayed failure, but also the dependence of the failure stress on the loading rate, since in specimens loaded slowly more time is available for slow crack growth than in specimens loaded rapidly.

For many brittle materials, the relationship between the velocity of cracks and the stress intensity (at least for low velocities) can be expressed as

$$\log V = n \log K_I + \log A \tag{1}$$

where V is the crack velocity, and n and $\log A$ are constants. On a log-log graph, n is then the slope of a V-K_I plot. In addition, it has been shown (Evans; Wiederhorn) that equation (1) can be integrated to give an expression for the effect of strain rate, $\dot{\varepsilon}$, upon strength when subcritical crack growth occurs, i.e.

$$\sigma_f^{n+1} = 2(n+1) E \varepsilon \sigma_{IC}^{(n-2)} AY^2 K_{IC}^{(n-2)}(n-2) \tag{2}$$

where σ_{IC} is the value of σ_f in the complete absence of slow crack growth, and Y is a parameter that depends upon the specimen geometry and mode of loading. The logarithmic ratio of failure stresses at two different strain rates is then

$$\log \frac{\sigma_2}{\sigma_1} = [1/(n+1)] \log (\dot{\varepsilon}_2/\dot{\varepsilon}_1) \tag{3}$$

Thus, as shown in Fig. 1, there are potentially three independent methods available of evaluating the slope of a log V - log K_I plot:

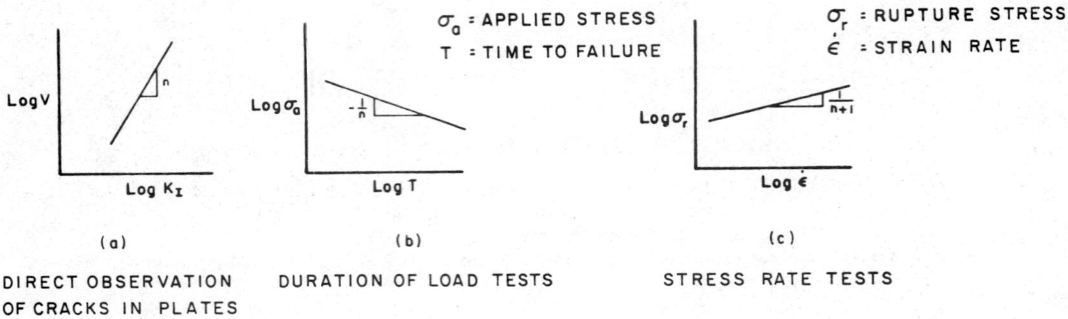

Fig. 1. Three ways of determining n, the slope of the log crack velocity V versus log stress intensity K_I plot.

1. direct observation of controlled crack growth, with the stress intensity obtained analytically;

2. in constant load tests, by plotting log T versus log σ_a, where T and σ_a are the time to failure and the applied stress, respectively; the slope of this plot is - 1/n; and

3. in specimens loaded at a constant rate of strain to failure, the logarithm of fracture strength plotted against $\dot{\varepsilon}$ is a straight line with a slope of 1/(n+1).

The agreement between values of n determined by the three methods is a test of the assumption that delayed failure and the stress rate dependence of strength are governed by the kinetics of subcritical crack growth. The work reported here is an attempt to compare these three methods. In order to do so, five different studies were carried out on Douglas fir, as indicated in Table 1.

TABLE 1 SCHEDULE OF TESTS

Type of Test	Type of Specimen
1. Direct determination of V-K_I plot, using the double torsion test.	small, clear; RL mode of crack propagation (i.e. specimen stressed perpendicular to the grain).
2. Rate of loading tests in flexure.	small, clear; RL mode of crack propagation (failure in tension perpendicular to the grain.
3. Rate of loading tests in direct tension and in flexure.	larger, not clear; failure in tension perpendicular to the grain.
4. Time to failure for constant load tests, in direct tension and in flexure.	specimens matched with those in (3).
5. Rate of loading tests in flexure	structural lumber, failure in tension parallel to the grain.

EXPERIMENTAL PROCEDURES

The experimental procedures have been described in detail elsewhere. They may be summarized briefly as follows:

1. Double torsion tests (Mindess, Nadeau and Barrett, 1975). The load relaxation method (Williams and Evans) was used on double torsion specimens to determine the relationship between crack velocity and stress intensity. The type of loading and specimen dimensions are shown in Fig. 2. As may be seen, the crack propagates by an opening mode; for this load arrangement, the stress intensity at the crack tip is independent of the crack length, and depends only on the load and the fixed dimensions of the specimen and loading fixture. Since wood is orthotropic, the stress intensity factor was computed from the strain energy release rate (Sih and Liebowitz) according to

$$G_I = \left(\frac{A_{11}\ A_{22}}{Z}\right)\left[\frac{A_{22}}{A_{11}}^{2} + \frac{2A_{12}+A_{66}}{2A_{11}}\right]^{\frac{1}{2}} K_I^2 \tag{4}$$

where the 1 direction corresponds to the direction of crack propagation; the 2 direction is normal to the crack surface; and A_{ij} are the appropriate elastic compliances. For these specimens, 1 and 2 correspond to the longitudinal and radial direction, respectively. Evaluating equation (4) for

$E_L = 15.65\text{x}10^3$ MPa, $E_R = 0.985\text{x}10^3$ MPa, $G_{LR} = 0.882\text{x}10^3$ MPa, $\nu_{LR} = 0.30$, the stress intensity can be calculated from

$$K_I = 6.44\ PW_m\ t_n^{-\frac{1}{2}}\ \text{kNm}^{-3/2} \tag{5}$$

where W_m and t_n are defined in Fig. 2. Crack velocity was determined by loading the specimen till a sharp crack began to grow. The cross-head was then stopped, and a load-relaxation curve obtained. Each point on the resulting load-time curve corresponded to a different crack velocity and stress intensity; thus, a complete V-K_I plot could be obtained from a single load-relaxation curve. The velocity

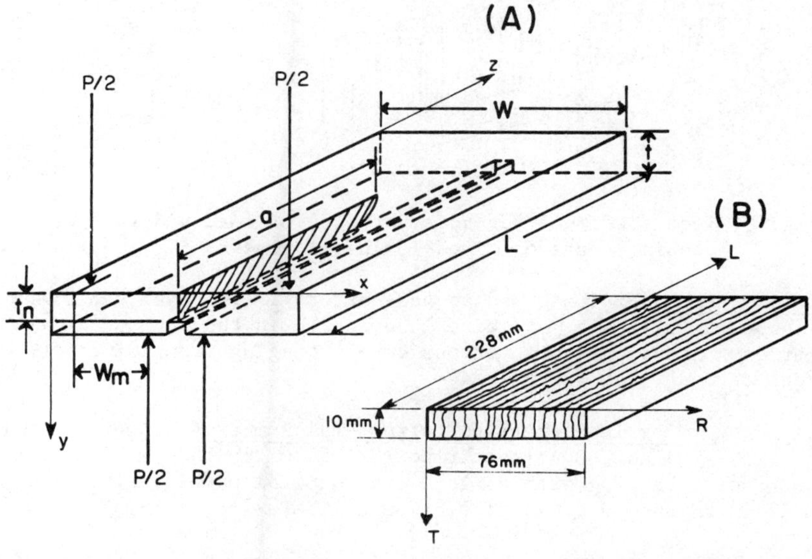

Fig. 2. Double torsion test specimens. (a) Orientation of loading points. (b) Orientation of grain.

depends only on the slope of the relaxation curve and the load,

$$V = -\left[(A_i P_i)/P^2\right](dP/dt) \tag{6}$$

where A_i and P_i are the initial crack length and load, respectively, and P is the load at any given point along the relaxation curve.

2. Rate of loading tests in flexure for small, clear specimens (Mindess, Nadeau and Barrett, 1976). Specimens with cross-sectional dimensions 25.4 x 25.4 mm were tested in three point loading on a span length of 254 mm, so that failure occurred essentially in tension perpendicular to the grain. Seven different displacement rates were used, varying from 5.46×10^{-8} m/s to 8.47×10^{-3} m/s, corresponding to stress rates of 89.6 Pa/sec to 1.73 MPa/sec.

3. Rate of loading tests in tension and flexure for larger, non-clear specimens (Mindess, Madsen and Barrett). For these tests, the specimens consisted of glue-laminated blocks, as shown in Fig. 3. The bending tests were again carried out in third-point loading. For both types of test, failure occurred in tension perpendicular to the grain. In direction tension, displacement rates ranged from 6.0×10^{-6} m/sec to 2.20×10^{-3} m/sec; in flexure, they ranged from 1.18×10^{-5} m/sec to 5.49×10^{-2} m/sec.

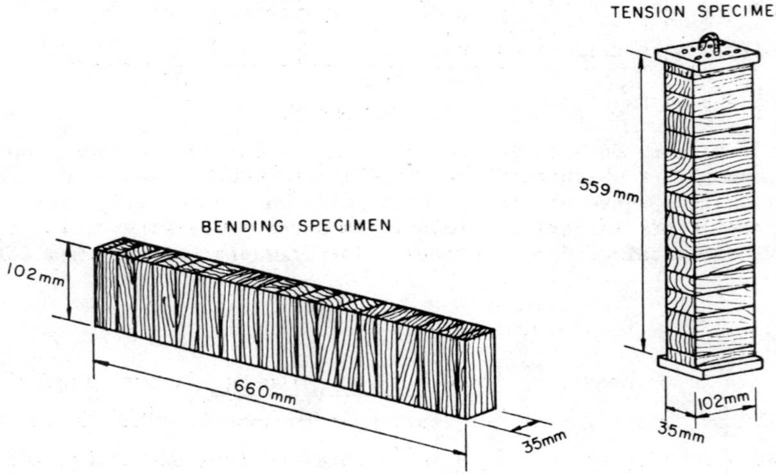

Fig. 3. Test specimens for both duration of load and rate of loading tests on glued-laminated blocks.

4. Constant load tests (Mindess, Madsen and Barrett). The test specimens were matched to those in (3) above, and were loaded both in tension and in three point loading. For both types of tests, stress levels ranging from 10% to 50% of the ultimate short-term strength were used.

5. Rate of loading tests in flexure for structural lumber (Spencer). Specimens were commercially available Douglas fir boards, 51 mm by 152 mm by 3.66 m long, tested in four point bending with a clear span of 2.51 meters. Rates of loading ranged from 222.4 kN/sec to 0.045 N/sec (752×10^6 MPa/sec to 151.7 Pa/sec).

EXPERIMENTAL RESULTS

The double torsion tests gave $V\text{-}K_I$ plots as shown in Fig. 4. The curves labelled A, B, and C represent average $V\text{-}K_I$ plots for specimens cut from three different boards, indicating the variability to be expected from one sample to another. Typically, even the within-board coefficient of variation in n-values was about 33%. The average n-value for all of the samples tested (49) was 61.9 ms^{-1}, with a coefficient of variation of 27.5%.

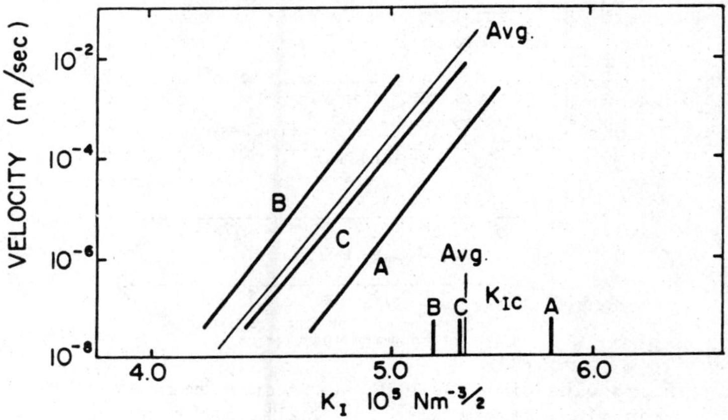

Fig. 4. $V\text{-}K_I$ plots and K_{IC} values for Douglas fir.

The results of the rate of loading tests on small clear specimens are shown in Fig. 5, based on a total of 124 specimens. The equation of the least squares line shown in Fig. 5 is

$$\log \sigma_r = 6.99 + 0.015 \log \dot{u} \qquad (7)$$

where $\dot{u}$ is the cross-head displacement rate. From Eq. 3, the slope of this line should be $1/(n+1)$, which gives the result $n = 65.4\ ms^{-1}$. This is in excellent agreement with the value of $n = 61.9\ ms^{-1}$ obtained from the double torsion test. However, as will be shown below, it is not in agreement with other results in the literature.

The rate of loading tests on the larger, non-clear, glued-laminated specimens were, on the other hand, quite inconclusive. As may be seen from Fig. 6, there is little apparent effect of loading rate on strength in either tension or flexure. Considering the scatter in the tests, the very small negative slopes of the regression lines (and consequently the n-values) are not meaningful. The ranked data for these tests (Figs. 7 and 8) also indicate that loading rate had essentially no effect on the strength distribution of the specimens.

The constant load test results carried out on specimens matched to the rate of loading tests, however, did show that a constant applied load affected the strength as shown in Fig. 9. In flexure, the n-value calculated from the early portion of the curve (where the time to failure was less than 100 hours) is about 47 ms^{-1}. The curve for tension, however, shows very little duration of load effect at short life, which is consistent with the rate of loading results described above.

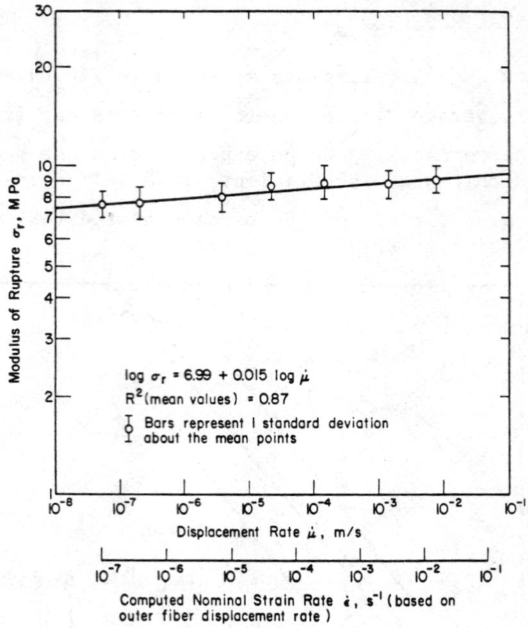

Fig. 5. Modulus of rupture versus displacement rate and nominal strain rate for small, clear specimens in bending.

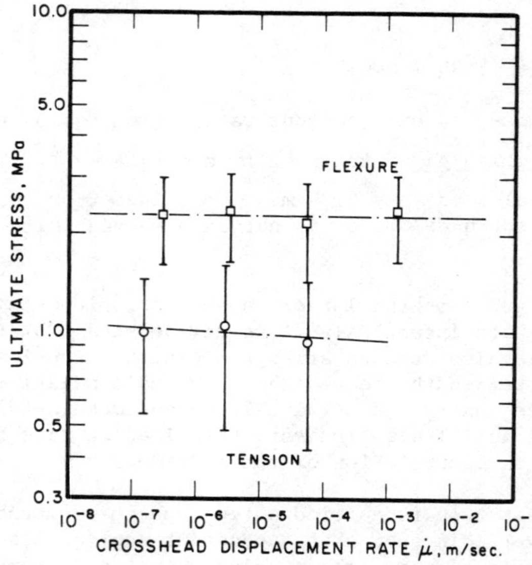

Fig. 6. Modulus of rupture versus crosshead displacement rate in flexure and tension for glued-laminated blocks.

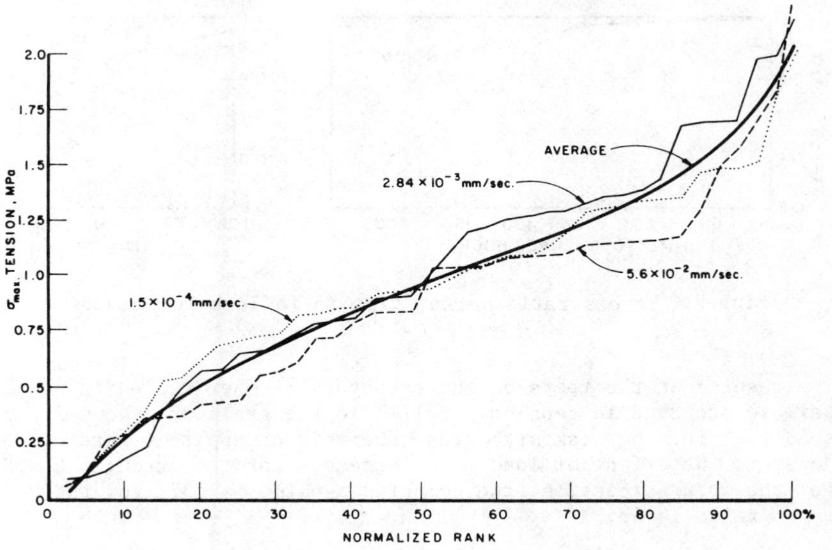

Fig. 7. Short term tensile strength for glued-laminated blocks.

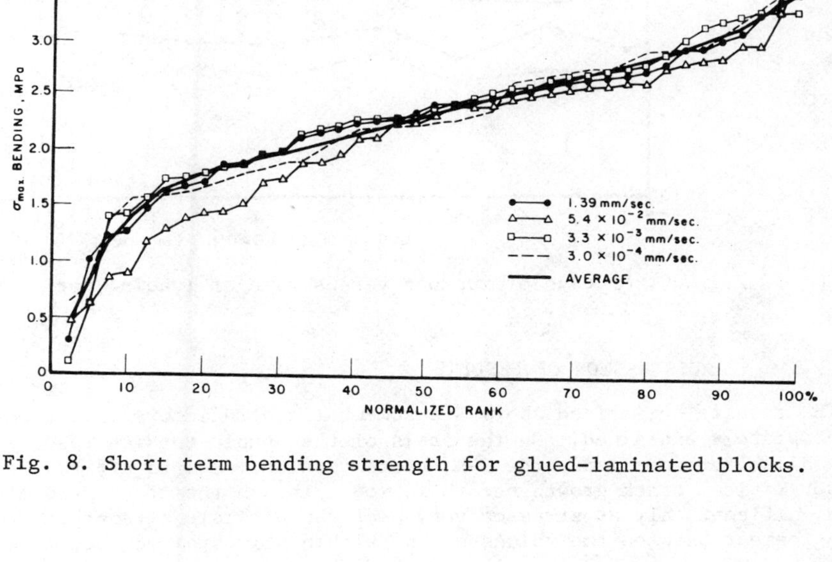

Fig. 8. Short term bending strength for glued-laminated blocks.

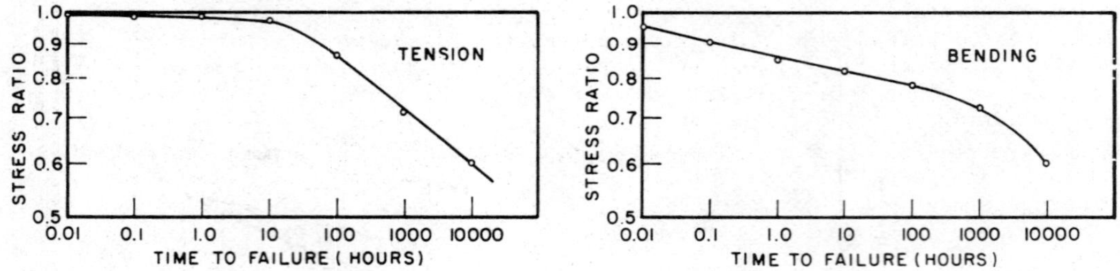

Fig. 9. Stress ratio versus time to failure for glued-laminated blocks.

The results of the tests on the structural lumber tested in bending so that failure occurred in tension parallel to the grain are shown in Fig. 10. It may be seen that for the weakest boards, there is no apparent rate of loading effect. However, the effect of load rate increases for the stronger boards in the sample. For the 50th percentile, the computed n-value is 137; for the 95th percentile, the n-value is 53.

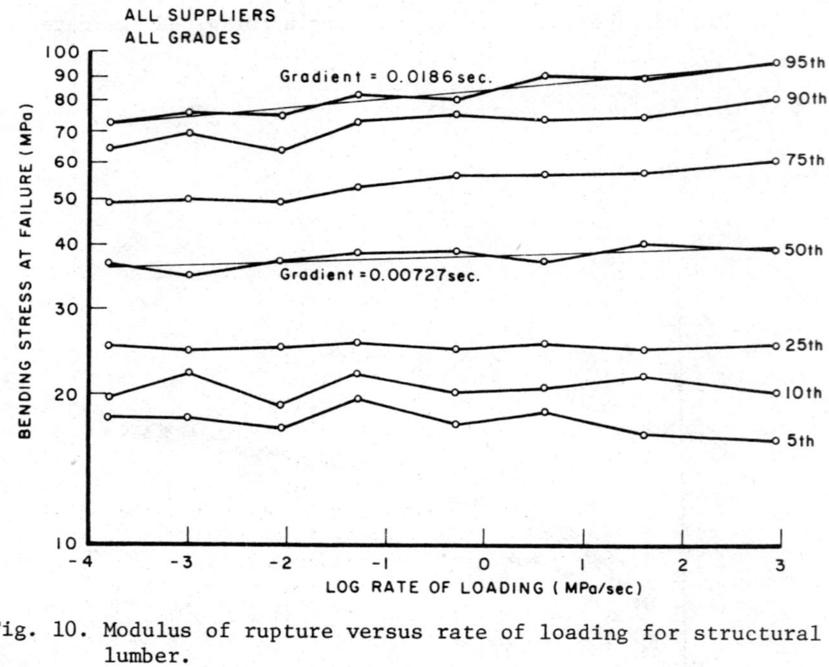

Fig. 10. Modulus of rupture versus rate of loading for structural lumber.

DISCUSSION OF RESULTS

The results summarized above are somewhat contradictory in nature, and cannot be completely explained. On the basis of the double torsion tests carried out on small clear specimens, the V-K_I plots obtained seem to affirm the view that subcritical crack growth occurs in wood, though the crack velocity becomes significant only at stresses very near the ultimate strength. Furthermore, the agreement between the values of n (within 6%) obtained from this test and the rate of loading tests on small clear specimens suggests that time dependent failure is indeed governed by crack growth kinetics. Unfortunately, there appear to be significant discrepancies between these results and the n-values which can be deduced from other reports in the literature, which range from about 25 to 39. The lower n-values reflect a shorter life expectancy under constant load than would be predicted from the present data. The reasons for this discrepancy are not clear. There may be experimental difficulties in comparing short term tests and long term tests, because cyclic changes in either relative humidity or temperature may weaken wood, as indicated by Schiewind. Also, because wood is viscoelastic, the use of a linear elastic fracture mechanics model may simply be inappropriate.

Turning to the tests on the larger, non-clear glued laminated specimens, the results become even more difficult to interpret. Because there was no effect of loading rate on strength, on the basis of the slow crack growth model one would expect that a constant load would also have no effect on strength, but this is not the case. Again, the reasons for this apparent lack of agreement are not clear. Possible reasons are:

1. As suggested above, ramp loading and constant load tests may not be truly comparable because of the different time scales involved. For the rate of loading tests, failure times were less than about 6 hours, while some of the constant load tests continued for over one year. At very early times, the slopes of the constant load versus time to failure curves (Fig. 10) are very small, which agrees at least qualitatively with the lack of rate of loading effects (Fig. 7). At long times, the slopes seem to increase very considerably, an effect not predicted by the simple linear elastic fracture mechanics models assumed by Evans and Wiederhorn. This may suggest that different fracture mechanisms exist for short and long time tests, or it may reflect the viscoelastic nature of wood.

2. Small clear specimens cannot be compared to larger specimens that are not clear. The flaws that exist in non-clear specimens may be of a sufficient magnitude to mask the rate of loading effects that are noted with small clear specimens.

Turning finally to the tests on structural lumber, it was noted that while the stronger boards show an increasing strength as the rate of stressing increases, as would be predicted by the slow crack growth model, this effect becomes much less pronounced for the weaker boards. Similar results have been reported by Madsen and Barrett for Hem-fir. These results suggest that the best select structural grade of lumber is not much different in its behaviour from small clear specimens. However, the behaviour of normal structural lumber containing significant flaws is not comparable to that of small clear specimens. In this regard it should be noted that current design codes and specifications allow increases in allowable stresses for loads of short duration. However, these results indicate that this increase is not justified, at least for specimens loaded to failure in less than 100 hours. This is particularly true for the weaker boards, on which design criteria are based.

It may be possible to explain these results using the simple relationship between crack size and K_{IC} for a beam in bending, which for the loading system used may be written as

$$K_{IC} = 1.12\,\sigma\sqrt{a} \qquad (8)$$

Thus, for $K_{IC} \cong 2.2\ \mathrm{MNm}^{-3/2}$, the crack size can be related to the applied stress. For the results at the 95th percentile, this gives a value for the predicted crack size of about 0.4 mm at the highest rate of loading, and about 0.66 mm at the lowest rate of loading. That is, the crack length increases by about 60 percent. This is consistent with the theory that at low loading rates, the crack has time to grow significantly, corresponding to a lower fracture stress. On the other hand, for the lower stresses at the 5th percentile, the predicted crack size is about 13 mm. Here, the small amount of crack growth that might take place at the lowest rate of loading is only a small fraction of the apparent crack size, and no significant rate of loading effects on fracture strength would be expected.

The tests on the structural lumber also revealed that a larger number of splintering tension failures, with the boards breaking into completely separate pieces, occurred at the highest stress rates. This may also imply different failure mechanisms between high and low rates of loading. It also suggests that knots and other defects may have a significant influence on rate of loading effects.

CONCLUSIONS

Based on these results, it may be concluded that some form of slow crack growth occurs in wood under stress. However, in developing a model for time-dependent failure based on linear elastic fracture mechanics, it is assumed that if n and A are known in Equation 1, then we can predict the effects of an arbitrary load history on the strength of wood. The results presented above, unfortunately, do not appear to confirm the assumed model, except perhaps for the results on the small, clear specimens. Thus, we still cannot reliably predict the long-term behaviour of wood from short time tests, or from tests on small clear specimens. Investigations along the lines of the simple fracture mechanics models described above have not been very fruitful. However, work in this area is still necessary, because of the need for being able to predict duration of load effects reliably. This is particularly important when limit state design or probabilistic assessments of strength are being considered. A general model to describe the time-dependent failure of wood still remains to be found. In addition, it would appear the increase in allowable stress permitted for short-term loading is not justified.

REFERENCES

Atack, D., W.D. May, E.L. Morris and R.M. Sproule (1961). The energy of tensile and cleavage fracture of black spruce. Tappi, 44, 555-567.

Evans, A.G. (1974). Slow crack growth in brittle materials under dynamic conditions. Int. J. Fracture, 10, 251-279.

Liska, J.A. (1950). Effect of rapid loading on the compressive and flexural strength of wood. Report No. R1767, USDA Forest Service, Forest Prod. Lab, Wis.

Madsen, B. and J.D. Barrett (1976). Time-strength relationship for lumber. Civil Eng. Structural Research Series No. 13, Univ. of British Columbia, Vancouver, B.C.

Mindess, S., B. Madsen and J.D. Barrett (1978). Rate of loading and duration of load tests on Douglas fir in tension perpendicular to the grain. Presented at the First International Conference on Wood Fracture, Banff, Alberta.

Mindess, S., J.S. Nadeau and J.D. Barrett (1975). Slow crack growth in Douglas fir. Wood Sci., 8, 389-396.

Mindess, S., J.S. Nadeau and J.D. Barrett (1976). Effect of constant deformation rate on the strength perpendicular to the grain of Douglas fir. Wood Sci.8,262-266.

Schniewind, A.P. (1967). Creep-rupture life of Douglas fir under cyclic environmental conditions. Wood Sci. and Tech., 1, 278.

Sih, G.C. and H. Liebowitz (1968). Mathematical theories of brittle fracture. In H. Liebowitz (Ed.), Fracture, Vol.2. Mathematical Fundamentals, Academic Press, N.Y., pp.68-190.

Spencer, R.A. (1978). Rate of loading effects in bending for Douglas fir lumber. Presented at the First International Conference on Wood Fracture, Banff, Alberta.

Wiederhorn, S.M. (1974). Subcritical crack growth in ceramics. In R.C. Bradt, D.P. H. Hasselman and F.F. Lange (Eds.), Fracture Mechanics of Ceramics, Vol.II, Plenum Press, New York, pp.613-646.

Williams, D.P. and A.G. Evans (1973). A simple method for studying slow crack growth. J. of Testing and Evaluation, 1, 264-270.

Wood, L.W. (1951). Relation of strength of wood to duration of load. Report No. 1916, USDA Forest Service, Forest Prod. Lab., Madison, Wis.

Volume 3, ICM 3, Cambridge, England, August 1979

MORPHOLOGICAL ASPECTS OF WOOD FRACTURE

G. Jeronimidis

Department of Engineering, University of Reading, Whiteknights, Reading RG6 2AY, U.K.

INTRODUCTION

An important aspect of modern technology is the search for stiffer, stronger and lighter materials. This trend is reflected by the considerable financial effort which goes into the production and development of composites based on a variety of organic and inorganic fibres. These new materials are in competition with the more traditional ones for structural applications but their high stiffnesses and strengths are often obtained at the expense of other desirable qualities, toughness in particular. In many practical applications, the lack of adequate fracture energy puts a limit to the full utilisation of the other mechanical properties (Berg, 1970).

It is therefore debatable whether it would not be more realistic to design composite materials with better all round performance. This is where materials scientists and engineers may have something to learn from biological materials. Wood is a good example of a structure where compromise between different mechanical requirements has led to a remarkable combination of lightness, stiffness, strength and toughness.

WOOD STRUCTURE AND MECHANICAL PROPERTIES

The success of wood as a structural material, both in biology and in technology, is the result of various factors, among which the strength of the cellulose molecule and the complex arrangement of microfibrils in the cell walls are of particular importance.

In Table 1 an attempt has been made to compare the mechanical properties of some engineering materials with wood. The table has been obtained considering the following criteria: specific Young's modulus (E/ρ), efficiency of columns (E/ρ^2) and panels (E/ρ^3) in compression, specific strength (S/ρ), specific flexural strength (S/ρ^2) and specific work of fracture[1].

A similar comparison which also includes production costs can be found in Gordon (1978). From the table, it can be seen that wood compares very favourably with other materials on a weight to weight basis and it is perhaps worth recalling that,

---

1 The work of fracture or effective surface energy γ is related to the strain energy release rate G and the stress intensity factor K by equations (2)

even today, the annual world consumption of timber is comparable to that of steel (Dinwoodie, 1975).

TABLE 1 Order of Merit of Typical Engineering Materials based on Mild Steel = 1

Material	E/ρ	E/ρ^2	E/ρ^3	S/ρ	S/ρ^2	γ/ρ
Mild steel	1.0	1.0	1.0	1.0	1.0	1.00
High tens. steel	1.0	1.0	1.0	3.0	3.0	0.02
Aluminium	1.0	2.8	8.2	2.9	8.4	1.00
GRP	0.4	1.7	7.8	6.4	29.0	0.05
Wood (average)	0.9	13.0	194.0	3.7	53.7	0.27

Considering wood as a composite material, the study of the relationships between structure and mechanical behaviour can be done at different levels. On the one hand, the cells can be considered as a system of parallel discontinuous fibres embedded in a matrix system and some of the properties of the composite can be obtained from the knowledge of the corresponding 'fibre' properties. In doing so, the fine morphology of the cells is not explicitly taken into account but the results are satisfactory.

For sitka spruce (Picea Sitchensis) for example, the mean axial stiffness and tensile strength of single cells are 12.5 GN/m^2 and 170 MN/m^2, respectively (both quantities referred to the gross cross sectional area). If the matrix material has negligible stiffness and strength, compared to the fibres, then by the law of mixtures:

$$E_c \simeq E_f\, V_f \qquad (1a)$$

$$\sigma_c \simeq \sigma_f\, V_f \qquad (1b)$$

where E is the Young's modulus, σ the tensile strength, V_f the volume fraction and the subscripts c and f refer to composite and fibre.

The volume fraction of cells in sitka spruce is about 80% (Scallan and Green, 1974) and from equations (1a) and (1b), the calculated values of E_c and σ_c for wood are 10 GN/m^2 and 136 MN/m^2 respectively, in good agreement with the mean experimental values of 10 GN/m^2 and 100 GN/m^2.

Alternatively, one could consider the cellulose microfibril (20 nm in diameter) as the basic unit of the composite and derive the mechanical properties from the information available on the laminated structure of the cell wall (Fig. 1). A comprehensive account of this type of approach can be found in Mark (1967).

For the purpose of this paper, two features of the cell morphology are of particular interest: the asymmetric arrangement of cellulose microfibrils in the secondary wall, S_2, and the fact that it accounts for up to 80% of the nett cross sectional area of material and, as such, it can be considered the major load bearing component. As with other fibrous composites, the strength of wood in compression is lower than in tension and the compressive failure has been related to typical 'slip lines' occurring in the cell walls (Dinwoodie, 1968, 1974).

FRACTURE OF WOOD

The anisotropic behaviour of wood is reflected in its fracture properties and it is an everyday experience that it requires less energy to split wood along the grain than to break it across the grain. However, the application of fracture mechanics concepts to this material is comparatively new. Early work (Atack and co-workers, 1961; Debaise, Porter and Pentoney, 1966; Porter, 1964) has been mainly concerned with crack propagation along the grain which shows a brittle type behaviour similar to that of some artificial polymers. More recently, various experiments have been done on fracture across the grain (Boatright, 1977; Jeronimidis, 1976; Pearson, 1974; Schniewind and Centeno, 1973) and the growth of cracks at an angle to the grain direction (Williams and Birch, 1976). The different parameters which have been determined, i.e. stress intensity factor K, strain energy release rate G and work of fracture (effective surface energy or fracture energy) γ, are related by the following equations (Hayes, 1975a, 1975b):

$$G = 2\gamma = K^2 f(Aij) \tag{2}$$

where f(Aij) is a function of the elastic constants of the material and takes into account its anisotropy (Sih, Paris and Irwin, 1965).

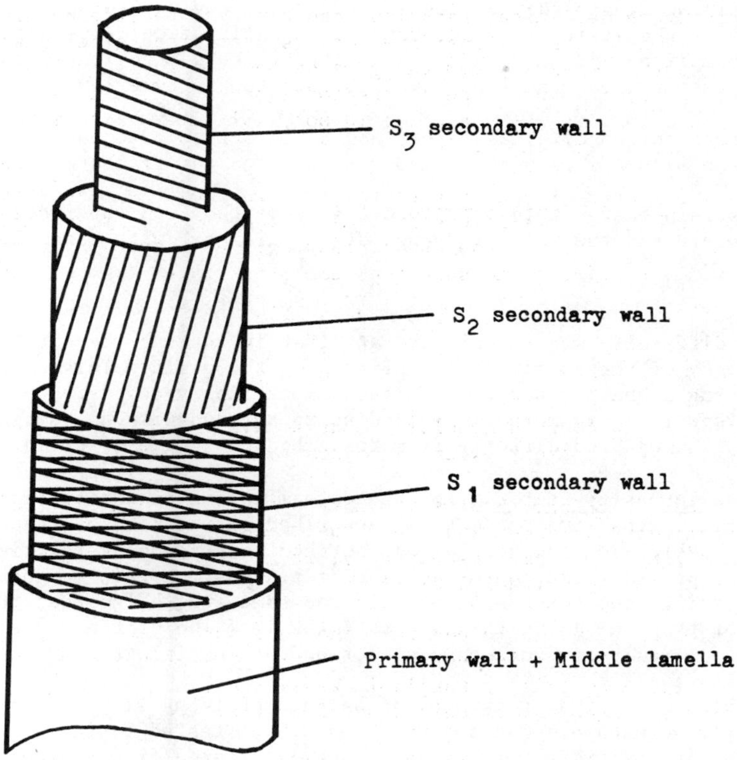

Fig. 1. Simplified structure of the wood cell

Experimental Determination of Fracture Parameters in Wood

In order to verify experimentally the validity of equations (2) for the fracture of wood across the grain, the three quantities γ, G and K have been measured independently. For all the tests, the LT system of crack propagation was chosen (crack front advancing in the tangential direction due to a tensile stress in the longitudinal direction, Schniewind and Pozniak, 1971). The results obtained by Schniewind and Centeno (1973) show that there is little difference between the LR and LT systems. Although various species of wood have been used, the results reported in this paper are those obtained on sitka spruce and balsa (Ochroma sp.)

Work of fracture. This has been measured by 'three point bending' fracture tests on notched beams. The ratio of span length to depth was large enough to avoid crushing at the loading points.

The beams were of square cross section (15 x 15 x 120 mm) with an 'inverted roof' notch cut across the grain (Tattersall and Tappin, 1966). The energy required to propagate the cracks quasistatically across the specimen is computed from the area under the load-deflection curve and the work of fracture γ is calculated directly by dividing this figure by twice the cross sectional area of the sample. A few experiments have been done using a straight edge notch but no significant differences have been found between the results.

Strain energy release rate. Strips of sitka spruce and balsa ranging from 0.25 mm to 1 mm in thickness and 100 mm long have been tested following the procedure described by Srawley, Jones and Gross (1964). From a first series of test pieces containing single edge notches of length a ranging from 0 to w/2 (where w is the width of the strip, 10 mm) a compliance calibration curve was obtained by plotting the compliance of each specimen ($c = P/\delta\ell$) as a function of the ratio a/w. Care was taken to keep the load P well below the fracture level.

A second series of sitka spruce and balsa strips of equal dimensions was tested to failure with a/w ranging from 0.0 to 0.5 and the fracture loads recorded.

The strain energy release rate G_{IC} is computed from the fracture load P_f, the ratio a/w and the rate of change of compliance C with crack length $\partial C/\partial(a/w)$:

$$G_{IC} = \frac{P_f^{\,2}}{2w} \quad \frac{\partial C}{\partial(a/w)} \tag{3}$$

Some difficulty was experienced at first in cutting the notches because of the tendency of the specimens to split along the grain. This was eventually overcome by using a sharp razor blade heated electrically and driven into the wood. For both species this method provided sharp notches with no damage and the notch size could be measured directly from the embedded length of blade.

Stress intensity factor. For the determination of the stress intensity factor K, tensile strips similar to those described in the previous experiment have been used. Both double edge and single edge notched test pieces were tried at first but the former proved rather unreliable. The reasons for this may have been the difficulty of cutting two notches of exactly the same length (so that the crack spreads from one of them only) and the fact that the location of the crack tip, between cells or within a cell, may have some effect on the initial stages of propagation.

A series of strips of spruce and balsa containing single edge notches of length varying between a = 0 and a = w/2 have been tested to failure in tension. The stress intensity factor K_{IC} has been calculated from the equation:

$$K_{IC} = \sigma_f \, (\pi a)^{\frac{1}{2}} \, k(a/w) \tag{4}$$

where σ_f is the remote fracture stress, a the crack length and k(a/w) a correction factor which takes into account the finite width of the specimens (ASTM, 1967).

The results obtained from the three types of experiments are reported in Table 2.

TABLE 2 Fracture Parameters for Spruce and Balsa

Species	γ $J/m^2 x10^4$	G_{IC} $N/m x10^4$	K_{IC} $N/m^{3/2} x10^7$
Sitka spruce	1.20(0.20)	1.80(0.20)	0.70(0.12)
Balsa	0.28(0.10)	0.50(0.15)	0.25(0.10)

All the results are the average of ten or more tests. The standard deviations are reported in brackets.

In order to have a more direct comparison between the three types of tests, the G_{IC} and K_{IC} values have been converted into fracture energy γ using equations (2). The functions f(Aij) for spruce and balsa have been calculated from published data on the elastic constants of these materials (Bodig and Goodman, 1973; Dinwoodie, 1975). This comparison is shown in Table 3.

TABLE 3 Comparison between Fracture Energy Values obtained by the three methods

Species	γ(measured) $J/m^2 x10^4$	γ(from G_{IC}) $J/m^2 x10^4$	γ(from K_{IC}) $J/m^2 x10^4$	f(Aij) $m^2/N x10^{-9}$
Sitka spruce	1.20	0.90	0.72	0.7
Balsa	0.28	0.25	0.22	0.3

Considering the variability of wood, the results are in good agreement and suggest that this type of approach is applicable to the study of the fracture behaviour of wood. Where comparison is possible, the values for γ and K_{IC} reported in this paper are consistent with results obtained by other workers (Boatright, 1977; Chappell and Morley, 1976; Pearson, 1974; Schniewind and Centeno, 1974; Tattersall and Tappin, 1966). It is interesting to note that although the work of fracture of balsa is four times lower that that of spruce in absolute, the specific work of fracture for the two species is comparable (2×10^4 J/m^2 for balsa and 3×10^4 J/m^2 for spruce).

The Effect of Temperature and Moisture Content on the Fracture of Wood

Comparatively little work has been done on the effect of temperature and humidity on the fracture of wood. Porter (1964) and Debaise and co-workers (1966) have reported the results for crack propagation along the grain at various moisture contents and temperatures. They found that for constant moisture content the fracture energy decreases with temperature (Fig. 2). This result is consistent with the view that the fracture energy of wood for cracks parallel to the grain is due to the matrix component which behaves like a semibrittle polymer.

In order to assess the influence of temperature on the fracture energy of wood across the grain, a series of work of fracture measurements on sitka spruce and teak (Tectona Grandis) have been made at different temperatures, ranging from +100°C to - 150°C. The results obtained are shown in Fig. 3.

It has been found that a decrease in temperature increases the work of fracture. Similar results have been reported by Bennett (1977). This effect of temperature

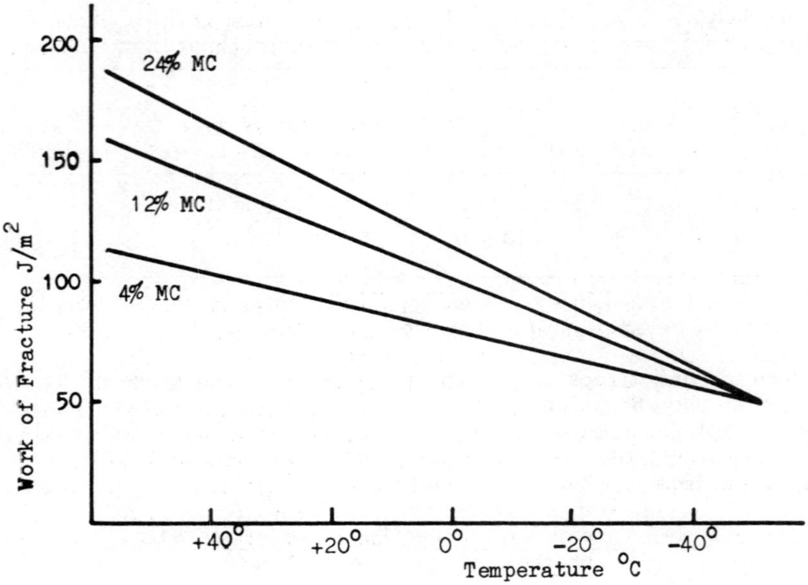

Fig. 2 The effect of temperature on the fracture energy of wood (crack propagation along the grain).

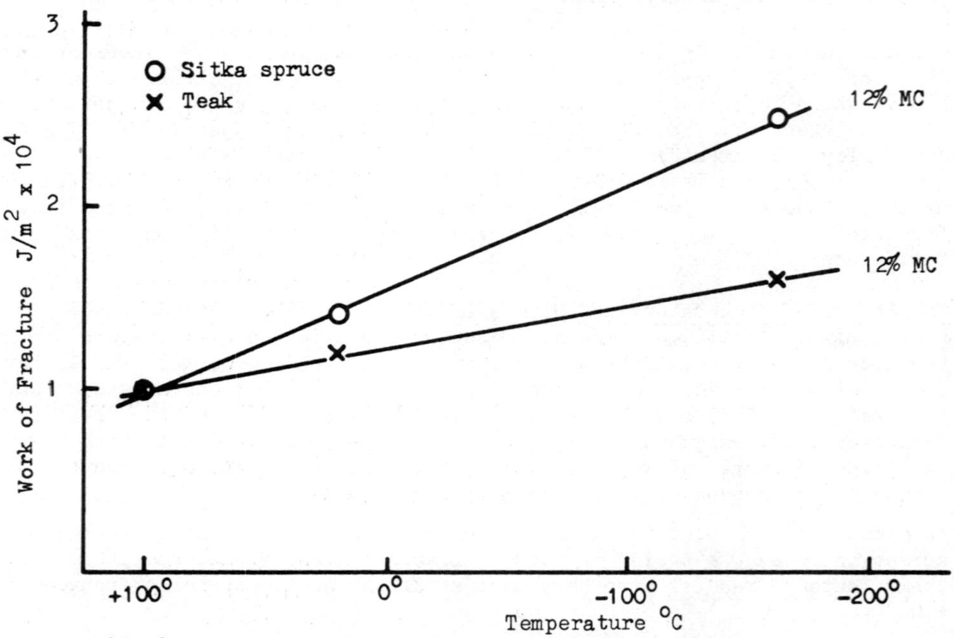

Fig. 3 The effect of temperature on the fracture energy of wood (crack propagation across the grain).

on the fracture behaviour of wood is quite different from what has been observed in other materials which, as a rule, show a decrease in toughness with decreasing temperature.

The reasons for this are not clear but this behaviour is important in relation to the physical mechanisms for energy absorption during fracture. Also, since a great proportion of the world's forests grow at latitudes where temperatures of -30°C are not exceptional, a work of fracture mechanism which is little affected by low temperatures is an advantage for the trees.

FRACTURE ENERGY AND MORPHOLOGY OF WOOD

It has been suggested (Tattersall and Tappin, 1966) that the high value of the work of fracture of wood is associated with some kind of fibre pull out similar to that observed in artificial composites. The observation of the fracture surface of wood does not support this view, at least insofar as the cells as a whole are considered as being the reinforcing fibres. If fibre pull out were the main contributing factor then, on the assumption that the material is optimised in this respect (i.e. fibre length = critical length, Kelly, 1970), one would expect to see a significant proportion of fibre ends. The evidence is that this is not so and, furthermore, it can be shown that, under the conditions of optimum fibre length, the work of fracture which is calculated is one order of magnitude less than the experimental value.

For a system of parallel discontinuous fibres of critical length, the work of fracture due to pull out is given by (Cottrell, 1964):

$$\gamma_{(\text{pull out})} = \frac{1}{48}\,\frac{V_f\,\sigma_f^{\,2}\,d}{\tau} \qquad (5)$$

where V_f is the volume fraction of fibres, σ_f their tensile strength, d their diameter and τ the interfacial shear strength between fibre and matrix. Table 4 shows the comparison between fracture energy values calculated with equation (5) and measured experimentally for softwoods and hardwoods.

TABLE 4 Comparison between Calculated and Experimental values of the fracture energy of woods

Wood type	σ_f (1) N/m^{2}x10^8	V_f (2) %	d (2) m x10^{-5}	τ (3) N/m^{2}x10^7	γ_f J/m^{2}x10^4 (Calculated)	γ_f J/m^{2}x10^4 (Measured)
Softwoods	1.8	0.96	3.0	1.2	0.16	1.2
Hardwoods	2.4	0.73	1.5	1.4	0.094	1.1

(1) Based on gross cross sectional area (Mark, 1967).
(2) Average values obtained from Scallon and Green (1974).
(3) Average values obtained from Lavers (1969).

In theory it would be possible to carry out a similar calculation based on the cellulose microfibre but, apart from the great complexity involved in taking into account their different orientations and the laminations in the cell wall, there is not at present sufficient experimental information about the various quantities. The problem can, however, be analysed by considering in more detail the deformation properties of single wood cells.

It has been shown that single wood tracheids tested in direct tension can either undergo a 'tension buckling' type of deformation (Hardacker and Brezinski, 1973; Page and co-workers, 1971) when the ends are prevented from rotating, or unwind under load if one end is allowed to rotate with respect to the other (Mark and co-workers, 1971). Both kinds of behaviour are associated with the helical pattern of cellulose microfibrils in the S_2 wall. If the matrix material in which the microfibrils are embedded has a brittle behaviour, then the high shear stresses induced by the first type of deformation will cause multiple failure between the microfibrils, the production of cracks in the S_2 parallel to the microfibrils and ultimately the inward collapse of this cell wall (tension buckling). This type of deformation (as well as the untwisting) is to a great extent irreversible and considerable amounts of energy are associated with it.

A model has been proposed which takes these effects into account and can predict a work of fracture in reasonable agreement with the experimental values (Gordon and Jeronimidis, 1974; Jeronimidis, 1976).

The stress-strain behaviour of cells which undergo 'tension buckling' is qualitatively similar to that of a ductile material and the energy absorbed during the irreversible phase of the deformation can be estimated from the area under the stress-strain curve. From published results (Hardacker and Brezinski, 1973; Page and co-workers, 1971), the energy absorbed per unit volume is of the order of 6×10^7 J/m^3. If this figure is multiplied by the fibre length (typically between 2 and 4 mm.) the work of fracture which is obtained for the fibres themselves has a value of the order of $1 - 2 \times 10^5$ J/m^2.

In a composite reinforced with short ductile fibres with work of fracture γ_f, the fracture energy of the composite can be calculated from (Piggott, 1974):

$$\gamma_c = \gamma_f V_f \left(1 - \frac{s_c}{s}\right) \tag{6}$$

where s_c is the critical aspect ratio and s the actual aspect ratio (length/diameter) of the fibres.

Substituting into equation (6) typical values for s_c and s for wood cells, the maximum calculated value for the work of fracture of wood is about 10^5 J/m^2. This figure is one order of magnitude higher than the experimental one but the calculation assumes that all the cells on the crack plane behave in a ductile manner and also that the extent of 'plastic' strain is, for all cells, the maximum value obtained experimentally ($\simeq$ 18%). Neither of these conditions is likely to be true in practice.

In order to confirm that the tension buckling mechanism can take place in wood, the fracture surfaces of the tensile specimens used in the experiments have been observed in a scanning microscope.

The features which characterise the cell deformation during fracture are the following:

- the cracks which develop because of the high induced shear stresses should run in a direction parallel to that of the microfibrils in the S_2 wall (Fig.4)
- the S_2 wall should decrease locally in diameter at the early stages of buckling (an effect similar to the local necking in ductile metals) and separate from the outer primary wall (Fig. 4). This effect has been postulated also by Mark (1967);
- later stages of fracture will produce folding inwards of the S_2 (Fig. 5).

COMPRESSION FAILURE, FATIGUE AND REACTION WOOD

The mechanism for absorbing energy during crack propagation in tension is related to the morphology of the wood cells and circumstances which affect the structure of the S_2 wall in particular, will have a great effect on the capacity of the system of absorbing work.

It has been shown that the failure of wood in compression is associated microscopically with extensive damage in the S_2 wall; macroscopically, the failure is characterised by a compression crease which runs at an angle to the grain direction. If the load is reversed, the damaged zone acts as a crack in tension and because a substantial amount of energy has already been dissipated in the cell wall, there is little resistance left to resist the propagation of the tensile crack and this leads to brittle failure (Dinwoodie, 1971).

Not much work has been done on the fatigue properties of wood but, as a rule, the resistance to fatigue failure is good (Boatright, 1977) provided that the stress reversals occur between tensile limits. If the lower stress level is compressive it can reduce considerably the fatigue life of wood.

Under certain circumstances related to growth, the living tree can produce a type of supporting tissue known as reaction wood which, depending on the species, is either compression wood or tension wood. In the former the cell walls are thicker, often containing cracks, and the helix pattern of microfibrils in the S_2 is much flatter, whereas in the latter the cell cavity of lumen is almost entirely filled by an extra layer and it is thought that the cellulose microfibrils are axially orientated. Both kinds of reaction wood show a lower toughness than normal and this is presumably associated with the loss of ductile behaviour which results from the alteration of the structure.

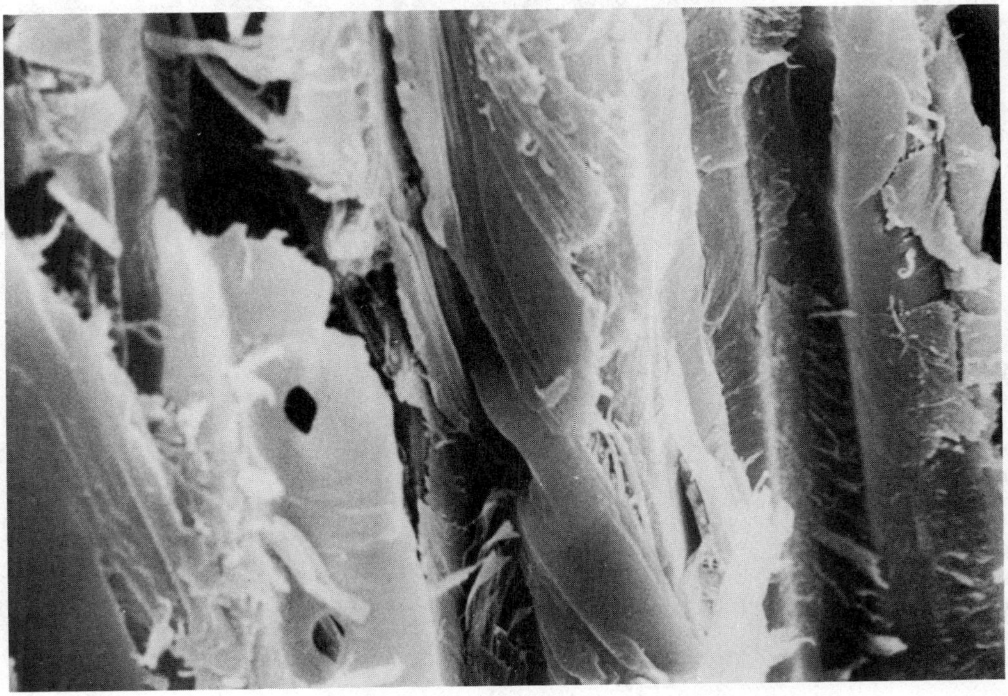

Fig. 4 Tensile failure in sitka spruce. Local necking in the S_2 wall due to 'tension buckling'. Magn. X 1000.

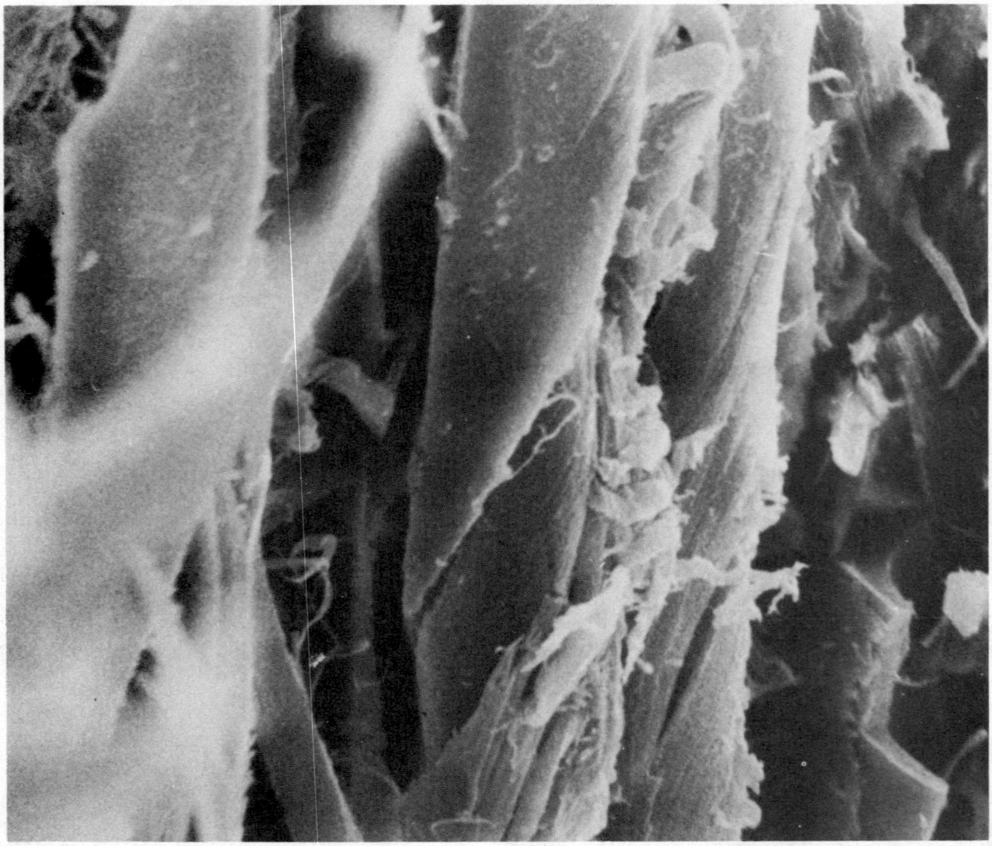

Fig. 5 Tensile failure in sitka spruce. Folding inwards of the S_2 wall (tension buckling) as a result of crack propagation parallel to the cellulose microfibrils. Magn. X 1000.

CONCLUSIONS

For cracks propagating across the grain, the effective fracture energy of wood is related to the fine scale morphology of the plant cell wall. It is interesting to speculate that the preferred orientation of the cellulose microfibrils in the S_2 wall can be considered as a sacrificial mechanism intended to increase the toughness of the material. In some instances, the direction of microfibrils is nearly parallel to the cell axis and this leads to an increase in stiffness and tensile strength but this type of structure is found commonly in grasses (hemp, ramie, flax) which are not exposed to external loads in the same way as trees are. Some of the features described in this paper have been applied to the design of artificial composites and the results which have been obtained (Gordon and Jeronimidis, 1978) show that the toughness of such materials can be increased by a factor of more than ten at the expense of a moderate reduction in axial stiffness.

ACKNOWLEDGEMENTS

The author wishes to thank the Science Research Council for financial support and Professor J. E. Gordon and Dr. C. R. Chaplin for many helpful discussions and suggestions.

REFERENCES

A.S.T.M. (1965). Fracture toughness testing and its applications. STP 381. Philadelphia.

Atack, D., W. D. May, E. L. Morris and R. N. Sproule (1961). The energy of tensile and cleavage fracture in black spruce. Tappi, 44, 555-567.

Bennett, C. P. (1977). Wooden pipelines. Chart. Mech. Eng., June 1977, 70-73.

Berg, K. A. (1970). Problems in the design of joints and attachments. In Wendt, F. W., H. Liebowitz and N. Perrone (eds.), Mechanics of Composite Materials, Pergamon Press, Oxford. pp 467-479.

Boatright, S. W. J. (1977). Studies in the fracture and fatigue of wood. Dissertation for the degree of Master of Science, University of Cape Town,S.A.

Bodig, J. and J. R. Goodman (1973). Prediction of elastic parameters for wood. Wood Sci. , 5, 249-264.

Chapell, M. J. and M. J. Morley (1976). The fracture toughness of conventional materials and composite systems containing non-fracturing reinforcing elements J. Mat. Sci., 11, 57-70.

Cottrell, A. H. (1964). Royal Society Discussion on New Materials, Proc. Roy. Soc., London, A282, 2.

Debaise, G. R., A. W. Porter and R. E. Pentoney (1966). Morphology and mechanics of wood fracture. Mat. Res. Stand., 6, 493-499.

Dinwoodie, J. M. (1968). Failure in timber. Part 1. Microscopic changes in cell wall structure associated with compression failure. J. Inst. Wood Sci., 21, 37-53.

Dinwoodie, J. M. (1971). Brashness in timber and its significance. J.Inst. Wood. Sci., 5, 3-11.

Dinwoodie, J. M. (1974). Failure in wood. Part 2. The angle of shear through the cell wall during longitudinal compression stressing. Wood Sci. Technol., 8, 56-57.

Dinwoodie, J. M. (1975). Timber. A review of the structure-mechanical property relationship. J. Microsc., 104, 3-32.

Gordon, J. E. and G. Jeronimidis (1974). Work of fracture of natural cellulose. Nature Lond., 252, 116.

Gordon, J. E. (1978). Structures, Pelican, Chap. 11.

Gordon, J. E. and G. Jeronimidis (1978). Composites with high work of fracture. Phil. Trans. Roy. Soc. Lond. (in press).

Hardacker, K. W. and J. P. Brezinski (1973). The individual fiber properties of commercial wood pulps. Tappi, 56, 154-157.

Hayes, D. J. (1975a). Origins of the energy balance approach to fracture. J. Strain Analysis, 10, 195-197.

Hayes, D. J. (1975b). Origins of the stress intensity factor approach to fracture. J. Strain Analysis, 10, 198-200.

Jeronimidis, G. (1976). The fracture of wood in relation to its structure. In Baas, P., A. J. Bolton and D. M. Catling (Eds.), Wood structure in biological and technological research, Leiden University Press (Leiden Botanical Series No.3). pp 253-265.

Kelly, A. (1970). Interface effects and the work of fracture of a fibrous composite. Proc. Roy. Soc. Lond. , A319, 95-116.

Lavas, G. (1969). The strength properties of timbers. Bull. Forest Prod. Res.Lab. 50, 2nd ed.

Mark, R. E. (1967). Cell wall mechanics of tracheids. Yale University Press.

Mark, R. E., J. L. Thorpe and A. J. Angello (1971). Twisting energy of holocellulose fibers. J. Polym. Sci., C36, 177-195.

Page, D. H., F. El-Hosseiny and K. Winkler (1971). Behaviour of single wood fibres under axial tensile strain. Nature Lond. 229, 252-253.

Pearson, R. G. (1974). Application of fracture mechanics to the study of the tensile strength of structural lumber. Holzforschung, 28, 11.
Pigott, M. R. (1974). The effect of aspect ratio on toughness in composites. J. Mat. Sci., 9, 494-502.
Porter, A. W. (1964). On the mechanics of fracture in wood. For. Prod. J., 14, 325.331.
Scallan, A. K. and H. V. Green (1974). A technique for determining the transverse dimensions of the fibres in wood. Wood and Fiber, 5, 323-333.
Schniewind, A. P. and R. A. Pozniak (1971). On the fracture toughness of douglas fir. Eng. Fract. Mech., 2, 223-233.
Schniewind, A. P. and J. C. Centeno (1973). Fracture toughness and duration of load factor. I. Six principal systems of crack propagation and the duration factor for cracks propagating parallel to the grain. Wood and Fiber, 5,152-159.
Sih, G. C., P. C. Paris and G. R. Irwin (1965). On cracks in rectilinearly anisoptropic bodies. Int. J. Fract. Mech., 1, 189-203.
Srawley, J. E., M. H. Jones and B. Gross (1964). Experimental determination of the dependence of crack extension force on crack length for a single edge-notched tension specimen. NASA TN-D 2396.
Tattersall, H. G. and G. Tappin (1966). The work of fracture and its measurement in metals, ceramics and other materials. J. Mat. Sci., 1, 296-301.
Williams, J. G. and M. W. Birch (1976). Mixed mode fracture in anisotropic media. In Cracks and fracture, A.S.T.M., Philadelphia, pp 125-237.

Volume 3, ICM 3, Cambridge, England, August 1979

CRACKING OF LAYERED STRUCTURES: THE SUPPRESSION OF YIELDING BY CASE HARDENING

A. G. Atkins

Delta Materials Research, Box 22, Hadleigh Road, Ipswich, Suffolk IP2 0EG, U.K.

INTRODUCTION

The loads at which layers in a laminate crack are different from the loads at which they crack when free, owing to interlaminar tractions set up in the adhesive layers during deformation. In broad terms, less tough layers crack at greater displacements and loads than they would if free and more tough layers crack at smaller displacements and loads : the action of the interfacial tractions is as if to put self-equilibrating crack tip closing moments in the less tough layers and opening moments in the more tough layers (Atkins & Mai, 1976).

Two extremes of behaviour may be identified for a series of cracked plates loaded in parallel through common loading pins (Fig. 1), viz :

(i) where there is no interlaminar adhesive, so that the layers crack individually with different propagated lengths in each layer. The total load X_T is then simply the sum of the separate n loads at all common boundary displacements u, i.e.

$$X_T\Big)_{free} = \sum_n X_i \qquad (1)$$

and

(ii) where the interlaminar glue tractions are capable of forcing a common composite flat crack front with the same crack length in every layer (different from the individual unglued lengths). Atkins & Mai (1976) showed that, for the special case of all layers having equal stiffness, the total load for n layers is given by

$$X_T\Big)_{full\ restraint} = \sqrt{n \sum X_i^2} \qquad (2)$$

In between these extremes are those cases where the interlaminar glue action is only partially capable of restraining the free behaviour of the layers, so that a stepped crack front is produced, with crack lengths and crack loads lying between the free and fully restrained values.

The analysis which gives equation (2) is based on the physically reasonable concept of adding up the individual works of fracture required in every layer. This leads to a rule-of-mixtures for the overall fracture toughness (R) of the layered structure. A rule of mixtures approach — well known for modulus and strength in fibre-reinforced composites — seems first to have been applied to toughness (in terms of strain energy release rates) by Bluhm (1961 a,b; 1962) in a series of papers dealing with work of fracture in both laminated structures and in thick solid plates possessing low centreline toughness coupled with high edge toughness. Gurney & Mai (1972) also employed the rule of mixtures in their crack stabilising sandwich experiments. Neither Bluhm nor Gurney & Mai considered interlaminar tractions and were concerned with the total energy of fracture rather than the separate loads and displacements at which cracking takes place which is the feature of the present line of attack. It may be noted that a rule of mixtures approach has been suggested for the stress intensity factor K as opposed to toughness R (e.g. Parvin & Williams, 1975). Such an approach is incompatible with a rule of mixtures on R since, in general $R \propto X^2 \propto K^2$ so a linear addition of R cannot be the same as a linear addition of K; as explained by Guild, Harris & Atkins (1978) a rule of mixtures for K ought only to be valid in the absence of interfacial tractions.

It is noted that X_T from equation (2) is greater than X_T from equation (1) owing to the addition of terms in X^2 rather than terms in X. This interesting and perhaps unexpected synergistic effect has been looked for experimentally and indeed observed in PMMA/PVC DCB sandwich testpieces glued together with cyanoacrylate adhesive (Guild, Harris & Atkins, 1978); clearly such behaviour has implications for hybrid composites.

Perhaps of more interest for the present paper are the magnitudes of the loads at which the layers crack in the presence of interfacial tractions in comparison with their free cracking loads. For the case of layers of equal stiffness, all layers crack at the same load given by $\sqrt{\sum X_i^2/n}$ when the interfacial tractions are capable of forcing a common flat crackfront (Atkins and Mai, 1976); when the layers have different stiffnesses, they each crack at a different traction-modified load, expressions for which are derived later. Now pre-cracked solids with large toughness-to-strength (K/σ_y) ratios very often flow plastically rather than crack, as the load to cause general yield is smaller than the load to cause cracking. However if such a solid be sandwiched with a brittle layer, there is clearly the possibility that the traction-modified cracking load could be brought down below the yielding load, which would result in the ductile solid fracturing in a brittle fashion instead of yielding. This proves to be so, experiments in tension with PVC sheet on which epoxy had been smeared demonstrating that the PVC will crack brittly instead of drawing down to a stable neck (Williams, 1975; Guild and co-workers, 1978). This presents a method of determining the toughness of ductile solids in the laboratory, given appropriate materials to act as the brittle layers along with appropriate interlaminar adhesives, and given a general analysis for converting the testing machine total cracking load down into the cracking loads for individual layers.

In place of laminates, we may think of solid plates which possess toughness varying through the thickness. The cracking theory for discrete laminae may be applied to such solid plates, where glue actions are replaced by a distribution of through-thickness shear stresses. Depending on circumstances, uniform flat crack fronts may be produced; in other cases shear yielding local to the crack tip will limit the magnitude of through-thickness tractions and the well-known thumbnail crack front profile will be produced, for example, as in plates of low (K/σ_y) metals having plane strain/plane stress transitions in toughness. This is the equivalent of a stepped crack front in laminates when the adhesive is incapable of transmitting the interfacial traction required for a flat crack front. A quasi-empirical adaptation of the discrete layer analysis to the problem of cracking in ductile solids can successfully predict thumbnail tunnelling, pop-in and slow crack growth (Atkins and Mai, 1977).

Instead of solids with varying toughness that are usually ductile in their overall behaviour, we may consider solids which have some regions of very low toughness (such as case-hardened soft steels). Thinking of the parallel with PVC/epoxy laminates, the composite behaviour of pre-cracked testpieces made from such solids may be brittle, with generalised plastic flow in the soft core being suppressed by the actions of the through-thickness shear stresses. In this way, it may be possible to determine the toughness of the soft core (which, in the absence of the case hardened layer, would yield generally before cracking in laboratory-sized specimens). The present paper explores these ideas by (a) generalising the flat-front cracking analysis to include layers of different thickness and different moduli, (b) describing tentative experiments on soft steel testpieces having different depths of case hardening in order to see if, and under what conditions, yielding is inhibited, and (c) using the analysis to determine values of R and K_I for the soft steel core.

THEORY FOR CRACK PROPAGATION IN LAYERED STRUCTURES

Consider a series of laminae with the same starting crack length L_s as shown in Fig. 1. Each may have different thicknesses and different mechanical properties, so that their individual stiffnesses will be different, and the loads and displacements at which cracking would occur were each layer free will be different. In Fig. 2, the load-displacement Xu behaviours for two laminae are shown where, anticipating the application of the analysis to brittle case-hardened layers on soft cores, the Fig. shows a thin brittle component and a thick tough component; other combinations are of course possible. The Xu trace of the brittle lamina is Oab; part of the R locus during propagation is given by ab.., which for purposes of illustration is taken to correspond with stable propagation, i.e. displays a gradually decreasing X with increasing u, (Gurney & Hunt, 1967; Gurney & Mai, 1972).

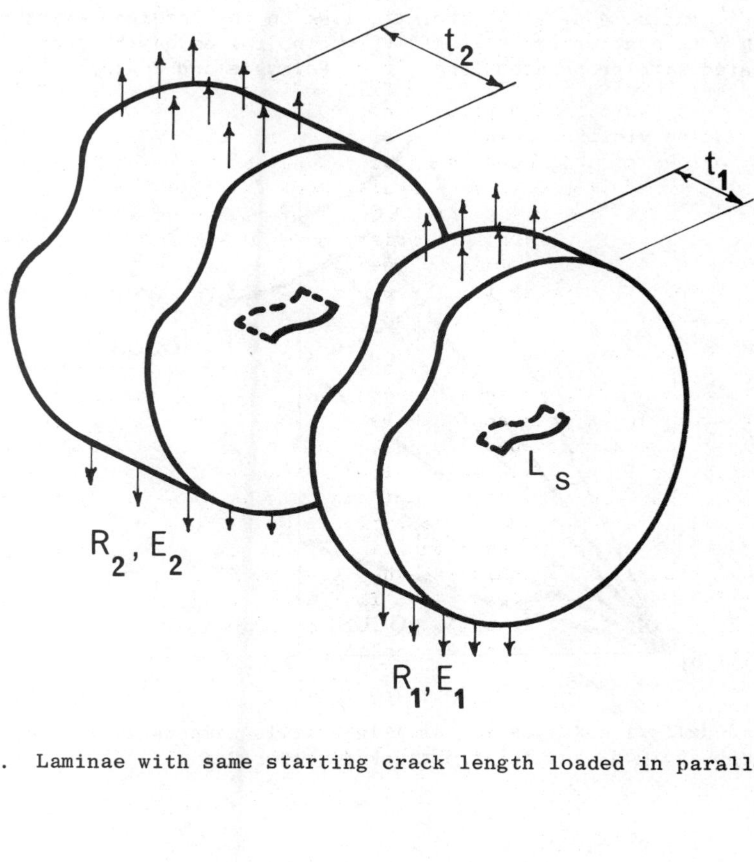

Fig. 1. Laminae with same starting crack length loaded in parallel.

Providing that there is no remote yielding in the lamina and that all irreversibilities are contiguous with the crack faces, the cracked body behaves in a globally elastic manner with an unloading line bO returning to the origin of displacement; these are the conditions required for the application of fracture mechanics methods. The segmental work area OabO thereby generated represents the work of fracture required to propagate the crack from the starting length L_s (which gives a stiffness Oa) to the unloaded propagated length L_f (which gives a stiffness Ob) i.e.

$$\text{work area OabO} = R_1 t_1 (L_f - L_s) \qquad (3)$$

where R_1 is the fracture toughness and the thickness of the layer is t_1 (Gurney & Hunt, 1967). Similarly, the Xu trace for the thick tough component is Ocd and, as before,

$$\text{work area OcdO} = R_2 t_2 (L_f - L_s) \qquad (4)$$

with stiffness Od being that of the tough layer after the same amount of crack propagation as in the brittle layer. In some cases it could be that $X_c > X_{yield}$ where X_{yield} is the load to cause remote generalised yielding of the testpiece in which case X_c would never be attained in the free tough lamina. The purpose of the analysis is to determine what happens to X_c when the laminae are stuck to one another and to see whether and under what conditions $X_c < X_{yield}$.

When free, crack propagation from L_s to L_f in each layer takes place over different displacements, viz : early on from u_a to u_b in the brittle layer, and later from u_c to u_d in the tough layer. If the components are glued together, and if the adhesive is capable of transmitting the necessary tractions to force a common crack front in both layers, it is clear that simultaneous crack propagation from L_s to L_f must take place over a compromise displacement u_e to u_f where $u_a < u_e < u_c$ and $u_b < u_f < u_d$. Point e lies on the combined stiffness line of the sandwich with starter crack length L_s; point f lies on the combined stiffness line associated with propagated length L_f. Points e and f also lie on the toughness

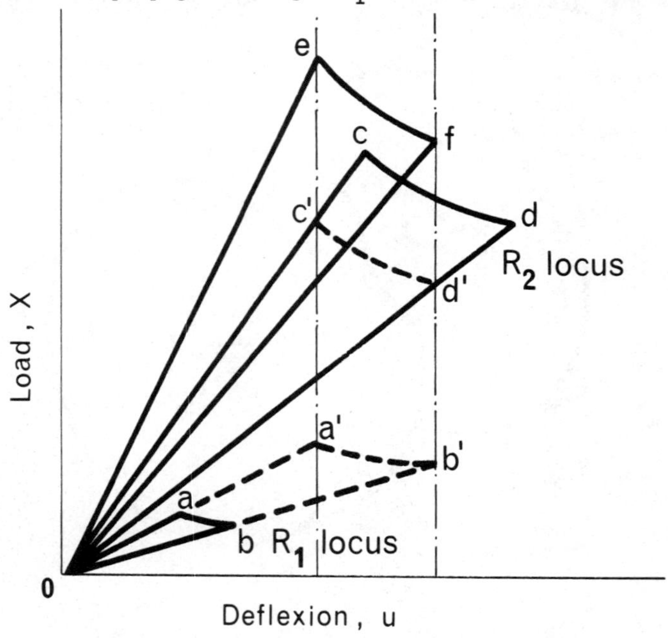

Fig. 2. Load-deflexion Xu traces for thin brittle component Oab, thick tough component Ocd, and the two stuck together Oef.

locus for the combined sandwich, and their location is controlled by the work area requirement

i.e. area OabO + area OcdO = area OefO

$$R_1 t_1 \ (L_f - L_s) + R_2 t_2 \ (L_f - L_s) = R^* \ (t_1 + t_2) \ (L_f - L_s) \quad (5)$$

where R* is the average toughness for the two-layer sandwich. R* is thus given by

$$R^* = (R_1 t_1 + R_2 t_2)/(t_1 + t_2) \quad (6)$$

which is, of course, a rule-of-mixtures on R. As pointed out by Guild and co-workers (1978), these R must be at comparable crack velocities in rate sensitive solids.

With A as crack area, we may follow Gurney & Ngan (1971) and write for each layer

$$R = -\partial\Lambda/\partial A)_u = (1/2t)X^2 \, d/dL(u/X) \quad (7)$$

in a linear elastic system where the strain energy $\Lambda = \frac{1}{2}Xu$. Equation (5) then becomes

$$X_1^2 t_1 d/dL(u/X)_1 + X_2^2 t_2 d/dL(u/X)_2 = X_T^2 (t_1 + t_2) d/dL(u/X)_T \quad (8)$$

where X_T is the total load in the sandwich (the subscript 'full restraint' is omitted here) and $(u/X)_T$ is the compliance of the two-layer sandwich.

Fracture mechanics formulae are of the general form

$$K = \sigma\sqrt{L} \, Y(L) \quad (9)$$

(e.g. Tada, Paris & Irwin, 1973) and as K^2 = ER (omitting for simplicity any term in Poisson's ratio) we have

$$R = \sigma^2 L Y^2/E = (1/2t)X^2 d/dL(u/X) \quad (10)$$

from which

$$d/dL(u/X) = 2LY^2/EW^2 t \quad (11)$$

where W is some characteristic width dimension. This may be substituted in equation (8) to give

$$X_1^2/E_1 t_1 + X_2^2/E_2 t_2 = X_T^2/E_T(t_1 + t_2) \quad (12)$$

where E_T is the Young's Modulus for the (crack free) sandwich combination. We are assuming that there is no interfacial slip, so the well-known Voigt rule-of-mixtures expression may be used for E_T, i.e. $E_T = (E_1 t_1 + E_2 t_2)/(t_1 + t_2)$. We then obtain

$$X_T^2 = X_2^2 (1 + \alpha)\{X_1^2/\alpha X_2^2 + 1\} \quad (13)$$

where $\alpha = E_1 t_1/E_2 t_2$. Furthermore, from the general K formula,

$$X_1^2/X_2^2 = E_1 R_1 t_1^2/E_2 R_2 t_2^2 = \alpha\beta \quad (14)$$

where $\beta = R_1 t_1/R_2 t_2$

Therefore

$$X_T)_{\text{full restraint}} \left\{ \begin{aligned} &= X_2\sqrt{(1+\alpha)(1+\beta)} \\ &= X_1\sqrt{(1+\alpha)(1+\beta)/\alpha\beta} \end{aligned} \right\} \quad (15)$$

which is the expression for X_T along ef.. in terms of X_2 along cd .. or X_1 along ab .. at the same L; if free the separate layers would see X_2 and X_1 at that same L at different u's of course. Equations (12) and (15) may be generalised to many layers, and when α = 1, equation (13) reduces to equation (2).

The actual loads in each layer when cracking takes place under conditions of full restraint are given by the loci a'b'.. and c'd'.. where a' and b' lie on Oa and Ob extrapolated to encompass the displacement over which the sandwich cracks (i.e. u_e to u_f); similarly for c' and d'. To arrive at X_a' and X_c' we require the compliances of the cracked laminae, which, using equation (11) are

$$(u/X) = (2/EW^2t)\int LY^2 dL + C \quad (16)$$

where C is the compliance of the crack-free lamina. C varies inversely with Et, and we may write

$$u/X = D(L)/Et \quad (17)$$

for all laminae with congruent boundaries where D(L) is the same for all laminae.

Thus from the geometry of similar triangles in Fig. 2,

$$\left. \begin{aligned} X_{a'} &= \sqrt{\alpha(1+\beta)/\beta(1+\alpha)}\,.\,X_a = [\alpha/(1+\alpha)]\,X_e \\ X_{c'} &= \sqrt{(1+\beta)/(1+\alpha)}\,.\,X_c = [1/(1+\alpha)]\,X_e \end{aligned} \right\} \quad (18)$$

where X_e is (X_T) at the start of cracking. When $\alpha = 1$, equations (18) give the same full restraint load for each layer as mentioned in the Introduction. The loci a'b'.., c'd'.. describe the apparent toughnesses with which the layers crack in the presence of the interfacial tractions. A measure of the tractions that have to be withstood by the interface between a two component laminate is given by

$(X_a - X_{a'})$ which produces a crack tip closing effect in the less tough layer,
and $(X_c - X_{c'})$ which produces a crack tip opening effect in the tougher component.

This pattern may be generalised for more than two layers. Cracks propagating in residual stress fields experience similar imposed opening or closing stress fields (Mai, 1977). The distribution of such tractions at interfaces in laminates or in solid continua is not clear, and the manner in which interfacial breakdown might occur in practice warrants further study.

Fig. 2 happens to show the combination of low stiffness/low toughness in one layer with high stiffness/high toughness in the other but diagrams may be constructed for any two combinations from the four cases of high or low stiffness and high or low toughness. As far as the use to which the analysis is put in this paper, it is clear from the second of equation (18) that reduction in the cracking load for the tougher component comes about when $\alpha > \beta$, which is the same as saying that $u_a < u_c$ in the diagram.

EXPERIMENTS AND RESULTS

Three point bending precracked testpieces with 10mm x 10mm cross-section were manufactured from mild steel barstock of specification EN3B (nominal % composition 0.15 C, 0.22 Si, 0.04 S, 0.01 P, 0.80 Mn). The specimens, with 50.4 mm span, did not conform to any particular recommended geometry in this series of exploratory tests, and the starter cracks were not fatigued for simplicity, it being considered that the attainment of a brittle fracture at a non-fatigued crack tip should be a good test of whether sandwiching would be able suppress yielding. All specimens were annealed to a hardness of about 130 VPN before some were carburised and stress relieved; the depths of case-hardening were 125, 250 and 500 μm and 1 mm. The carburising heat treatment increased the Vickers hardness of the soft core to 270, (the specimens were not tempered on this occasion); the hardness of the case was greater than 700 VPN. Conditions along the starter crack front thus consisted of hard cases flanking the softer central region on either side, the crack front being ground out after carburising. Notch depth (a) to beam depth (b) ratios varied between 0.2 and 0.7.

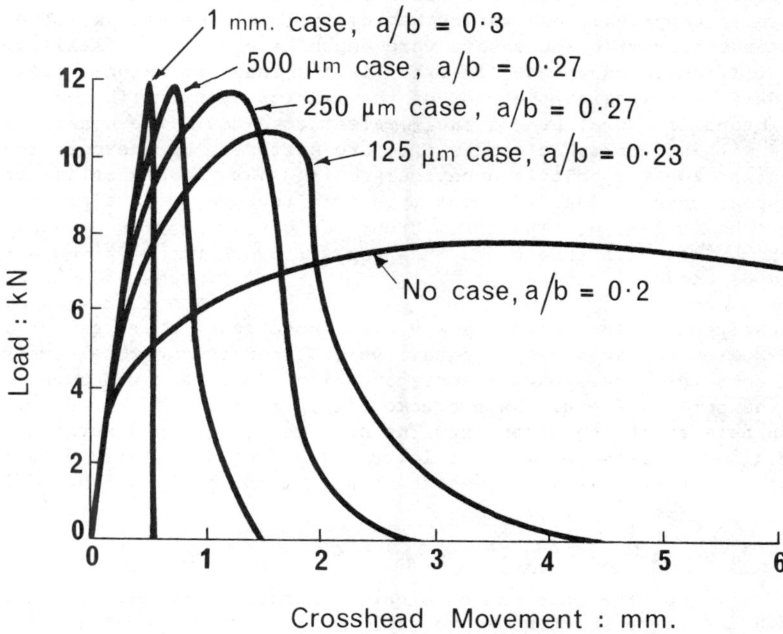

Fig. 3 Load-deflexion on traces of 3-point bending beams with various case depths.

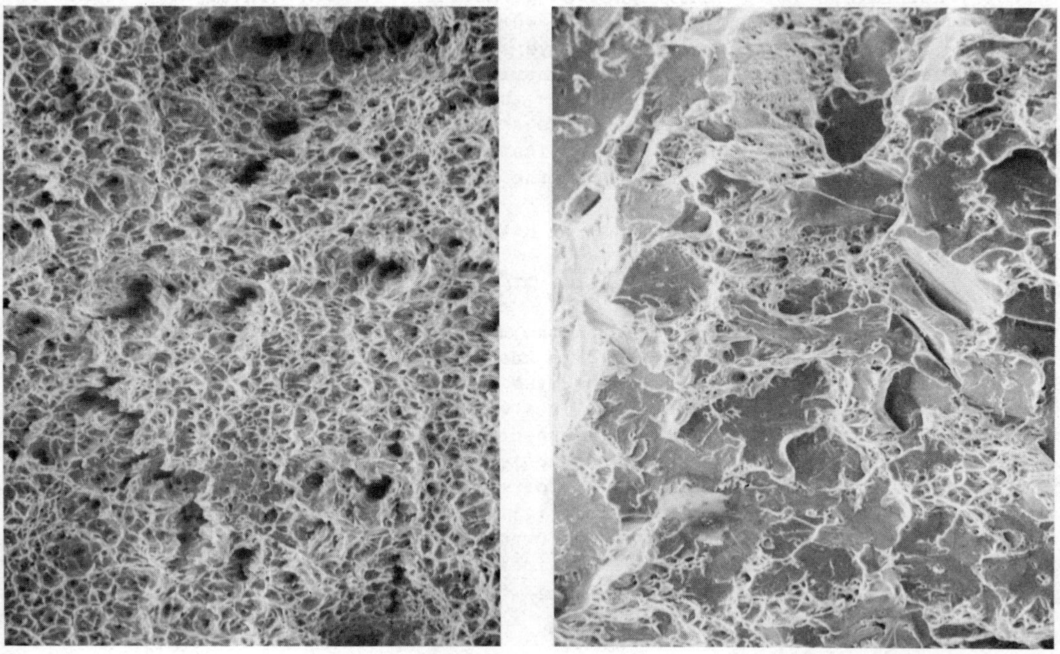

Fig.4 (left) Fibrous fracture surface typical of thin case depths (x 320); (right) cleavage fracture surface of 1 mm case depth (x 320). Crack propagation is from left to right in both pictures.

Fig. 3 shows the room temperature load-deflexion response at a crosshead speed of 85 μm/s of beams with comparable a/b ratios having various case depths. The gradual change in behaviour with increasing case depth is clear : at small case depths there is still considerable non-linearity in the load-deflexion trace, accompanied by sideways necking at the root of the starter crack, with general yielding before fibrous fracture, but at the greatest case depth the behaviour is completely brittle with a facetted cleavage fracture surface. Furthermore the total load is greatest for the brittle behaviour(noting that the a/b ratios vary somewhat). The appearances of the different types of fracture surface are shown by the SEM micrographs in Fig. 4. The crack fronts were flat, as far as could be ascertained, in the brittle fracture cases thus suggesting that the laminate cracking analysis could be used.

The results were analysed in the following way : in those cases where generalised yielding before cracking was suppressed, equations (15) permit the total cracking load to be broken down into the separate loads at which the case would have cracked (X_1) and the soft core would have cracked (X_2) if free; of course, without the shear traction between the hard case and the soft core, X_2 would never be attained as yielding would have set in at a lower load. For simplicity we assume that the toughness of the case is much less than that of the core, so that $\beta \approx 0$ and therefore

$$X_T\Big)_{\text{full restraint}} = X_2\sqrt{1+\alpha}$$

Furthermore, $\alpha = t_1/t_2$ as the case and core have comparable moduli. In Fig. 3 the 1 mm case depth testpiece which shows no yielding, cracks at about 12 kN. With a 1 mm case on both flanks of the core $\alpha = 2/8 = 0.25$, so $12 = X_2\sqrt{1.25}$ whence $X_2 = 10.7$ kN. This load may be used in a standard fracture mechanics formula to determine K_I for the core; instead of K_I, R for the core could be directly determined using other specimen geometries with more favourable geometric stability factors, by Gurney's segmental work area method (Gurney & Mai, 1972). It is realised that the case hardened layers are probably already in a state of compressive residual stress from the carburising process; this aids the shear traction required to suppress yielding in the core, the loci ab.. and cd.. in Fig. 2 already 'having moved' towards the a'b'.., c'd'.. positions. Such residual stresses do not affect the application of the analysis for the determination of the free X_2 load level from the testing machine total load.

For the present series of experiments we have, for example from Tada and coworkers (1973)

$$K = 3Xs\sqrt{\pi a}\,.\,F(a/b)/2tb^2$$

where s is the 3-point bending span, a the notch depth, b the beam depth and t the thickness of the core. For a = 3 mm and b = 10 mm, F = 1.06 which, with X = 10.7 kN, t = 8 mm and s = 50.4 mm gives $K_I = 104$ MPa$\sqrt{\text{m}}$. Other estimates for K_I from 1 mm case hardening testpieces with different a/b ratios were 112 and 95 MPa$\sqrt{\text{m}}$ for a/b = 0.4 and 0.7 respectively. The R values converted from the three K_I values, using $K^2 = ER/(1-\nu^2)$, are 46, 53 and 38 kJ/m^2 respectively. Although testpieces with 500 μm cases (for which $\alpha = 1/9$ and t = 9 mm) showed some non-linearity before crack propagation, it is of interest to calculate their K_I and R by a similar route using the peak cracking load, remembering 'offset procedures' that are sometimes employed in the determination of K_{IC} (e.g. Knott, 1973). Results gave 100 MPa$\sqrt{\text{m}}$ and 43 kJ/m^2 for both a/b = 0.2 and 0.6; with 118 MPa$\sqrt{\text{m}}$ and 60 kJ/m^2 for a/b = 0.3.

CONCLUSIONS

Suppression of general yielding, by case hardening the flanks of precracked testpieces, permits K_I to be determined using the laminate analysis for the distribution of loads across different regions of the crackfront. Sensible values for K_I (circa 100 $MPa\sqrt{m}$) and R (40-50 kJ/m^2) seem to be given for mild steel testpieces that would normally flow plastically before cracking. This method of toughness determination could be a valuable tool for generating data otherwise difficult to obtain in solids with large K/σ_y ratios. Further work has to be done to put these ideas on a firm footing.

ACKNOWLEDGEMENTS

Thanks are owed to A. Hubbard and H.R. Pettit for assistance with the experiments and microscopy.

REFERENCES

Atkins, A.G., and Y.W. Mai (1976). Int. J. Fract., 12, 923
Atkins, A.G., and Y.W. Mai (1977). Submitted to Int. J. Fract.
Bluhm, J.I. (1961 a). Proc. Am. Soc. Test. Mater., 61, 1324.
Bluhm, J.I. (1961 b) Proc. Am. Soc. Test. Mater., 61, 1332.
Bluhm, J.I. (1962) Proc. Am. Soc. Test. Mater., 62, 914.
Guild, F.J., B. Harris and A.G. Atkins. (1978). J. Mater. Sci., 13, 2295.
Gurney, C., and J. Hunt. (1967). Proc. R. Soc., A299, 508.
Gurney, C., and K.M. Ngan. (1971). Proc. R. Soc., A325, 207.
Gurney, C., and Y.W. Mai. (1972). Engr. Fract. Mech., 4, 853.
Knott, J.F. (1973). Fundamentals of Fracture Mechanics. Butterworths, London.
Mai, Y.W. (1977). Int. J. Mech. Sci., 19, 325
Parvin, M., and J.G. Williams (1975). Int. J. Fract., 11, 963.
Tada, H., P.C. Paris, and G.R. Irwin (1973). The Stress Analysis of Cracks Handbook. Del Research Corp., Hellertown, Penna.
Williams, J.G. (1975) Private Communication

Volume 3, ICM 3, Cambridge, England, August 1979

MECHANICAL PROPERTIES OF SILICON CARBIDE REINFORCED ALUMINIUM

A. P. Divecha and S. G. Fishman

Naval Surface Weapons Center, White Oak, Silver Spring, Maryland 20910, U.S.A.

INTRODUCTION

A metal matrix composite system using potentially low cost SiC whiskers in an aluminum matrix is described. Unlike most metal matrix composites, conventional powder metallurgy and metal forming techniques are used in processing. Although most work to date has been with 2024 Al matrix, composites with 5456 Al and 6061 Al have also been fabricated. Typical tensile strength values of 2024 Al reinforced with 25 volume percent SiC whiskers are between 530 and 640 MPa. In most composites ductility is low, on the order of one percent elongation to failure. Fracture toughness data indicate that the inherent characteristics of the matrix have a strong influence on the toughness of the composite. Metallographic evidence indicates that whisker damage is proportional to the amount of mechanical deformation. Methods of composite fabrication and typical properties are presented and discussed in detail.

Advanced composites offer the designer a high performance and versatile class of engineering materials. Use of filament reinforced organics began in 1965 with boron fibers. Since that time much progress has been made toward application of advanced composites. Metal matrix composites are currently at approximately the same stage of development as the organics were in the 1960 s and now appear to be on the verge of a period of accelerated progress. A primary reason is the requirement for strong, lightweight, and corrosion resistant materials. Aluminum is an ideal matrix to meet these requirements. Incorporation of a reinforcing agent into aluminum yields a high performance material with modulus and strength increases that lead to structural weight savings and enhanced elevated temperature capability. This paper describes a particular type of composite in which a single crystal, or a whisker, is used as reinforcement in aluminum alloy matrices.

Whisker-reinforced metal matrix composites are not new. In the 60s and 70s (Divecha et al, 1967; Parrott, 1972) conducted a number of investigations using single crystal whiskers in composite fabrication. Properties of whisker-reinforced composites were quite impressive and it was demonstrated that a large percentage of the room temperature composite strength could be retained at elevated temperatures. A decline in interest in these composites occurred, however, because of the high cost of whiskers. Attempts to streamline the whisker growth processes met with limited success. Development of boron filaments by Christian (1975) and Jones (1975), having 3450 N/mm^2 ultimate tensile strength and a relatively easy production technology based on diffusion bonding foil/continuous filament layups, made boron fiber composite more attractive. Recent advances in processing kinetics, however, have led to the development of a high-performance whisker for which the potential cost is no more than \$22/kg.

Whisker-reinforced or short-fiber-reinforced metals offer an intermediate alternative between homogeneous metals and continuous-filament-reinforced composites. That is, while the continuous-filament reinforced metals cannot be processed via deformation in the classic sense, such processes as extrusion, rolling, swaging, and press-brake operations can be performed relatively easily on short-fiber composites. Furthermore, because the fibers or whiskers are extremely small, the damage from deformation processing is likely to be accommodated without serious degradation of mechanical properties. Indeed, it is seen in the present study that even after large reductions the fiber lengths remain of adequate size to enhance elastic properties.

EXPERIMENTAL METHODS

Commercially available atomized aluminum alloyed powders of -325 mesh have been used in the present work. Major emphasis was placed on 2024 aluminum alloy initially because it is well-known and is a high strength alloy used in aerospace structures. Upon achievement of significant reinforcement of the 2024 Al matrix, other aluminum alloys including 5456 and 6061 were included in the study. The nominal chemical compositions of these matrices or alloys are presented in Table 1.

TABLE 1 Typical Compositions of Aluminum Alloy Matrices

Alloy	Percent by Weight
2024	4.5 Cu, 1.5 Mg, 0.6 Mn, balance Al
5456	5.0 Mg, 0.7 Mn, 0.15 Cr, 0.15 Cu, balance Al
6061	1.0 Mg, 0.6 Si, 0.25 Cu, 0.25 Cr, balance Al

The whiskers or the reinforcing agents utilized in this work are manufactured by Silag, Inc., a subsidiary of Exxon Enterprises, Inc., in New York City. The whiskers are extremely fine and their diameters range from 0.2 to 0.5 microns. There is a large amount of fine particulate, nonwhisker, debris which amounts for as much as 85 percent of the total whisker product. Selected view of the whiskers at two magnifications is presented in Fig. 1. Nonwhisker material is readily seen at lower magnifications (left). At higher magnification individual fibers are more clearly resolved. Most of the whiskers are of hexagonal cross-section although triangular ones have been occasionally observed.

Billet Synthesis

In a typical case, the composite billet synthesis comprises of the following: (1) mixing of the components, (2) cold compaction, (3) hot compaction, (4) ejection from the mold.

For the mixing operation, the matrix powder (-325 mesh) and the SiC whiskers are placed in a container and mechanically agitated. The mixture is then cold compacted at ~ 13.78 MPa in a hydraulic press (American Steel Foundries, 609.6 x 914.4 mm platens, upacting model.) The mold utilized for compaction consists of

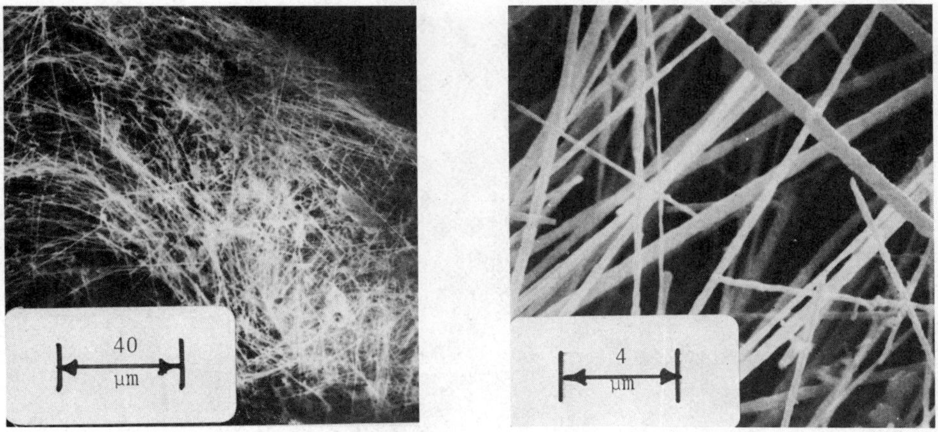

Fig. 1. Selected views of Silicon Carbide Whiskers via SEM at two magnifications.

a graphite liner in a steel retainer. The former is preferably of a high density type (e.g., HPD-1 grade from Poco graphite, Decatur, Texas). The steel retainer is necessary to prevent graphite die rupture during hot and cold compaction. The die cavity punch and plug are cylindrical. Most of the composite billets are 31.75 mm in diameter although a number of billets measuring 63.5 mm in diameter have been prepared without difficulty.

Hot Extrusion

Composite billets were processed via extrusion in a die assembly of conventional design. A sophisticated extrusion facility was not available and hence an upacting, semi-automatic, 4 poster, 368 mm travel, 864 x 610 mm platen transfer molding press, 182,000 kg capacity, manufactured by Baldwin Lima Hamilton, Philadelphia, PA, was utilized. The billets were heated inside the container. The ram speed was always slow (under 50 mm/in.) to reduce friction. As more experience was gained an Inco 718 ram and the extrusion die were utilized in place of tool steel. It is now established that Inco 718 can be utilized almost indefinitely without encountering softening associated with tool steels under these conditions.

Composite billets containing up to 25 v/o SiC are readily extruded into shapes such as rectangles, rounds, tubes, and crosses. Examples are shown in Fig. 2.

Hot Rolling

A two high rolling mill* with 152.4 mm diameter rolls, with a single speed was utilized for (cross) rolling experiments.

Mechanical Testing Composite Rod and Sheet

Due to its brittleness and anisotropic nature, composite material testing requires greater attention than homogeneous metallic materials. A special grip designed

*Manufactured by Standard Machinery Corp., Providence, RI

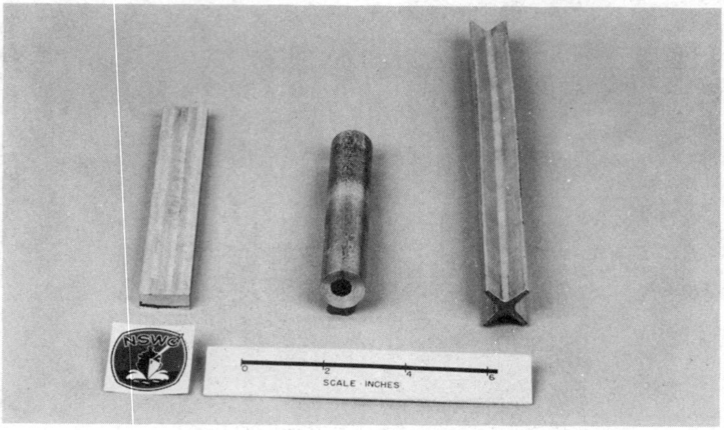

Fig. 2. Examples of flat, tube and a cross obtained from SiC/Al billets via hot extrusion.

by Christ and Swanson (1976) was utilized in tensile tests conducted on extruded rods. The most important feature of this method is the spherical ball and seat grip assembly which assures good specimen alignment.

Room temperature tensile testing of composite sheet was conducted on straight-sided coupons machined to provide square edges. The maximum allowable width variation was ± 0.076 mm. The ends of the test specimen were tabbed with thick aluminum over a length of 25.4 mm from each end. The aluminum tabs were bonded to the composite sheet with an epoxy (epibond 1210) adhesive. The specimen gage length was 76.2 mm.

All testing was performed on standard Instron Tester Model 1115, 90 kN capacity, at a strain rate of 0.127 min^{-1}. Extensometers and strain gages were utilized as necessary to obtain moduli values.

Fracture Toughness, K_{Ic} Testing

The fracture toughness of extruded rods of 15 vol % SiC composites of three different aluminum matrices was determined using the Fractometer II system developed by Terra Tek, Salt Lake City, Utah. This technique uses a short rod with a diameter two-thirds its length with a V-shaped region formed by cutting longitudinal slots along the axis of the test rod, and hence is convenient for measuring fracture toughness in extruded composite rods. The Fractometer system has a thin inflatable stainless steel bladder which, when inserted into the slot and expanded by a fluid, forces the crack open in a Type I mode and causes a crack to initiate from the point of the vee and propagate into the sample. Data from a load-load point opening curve and the maximum applied load are used to determine values of K_{Ic}. The method assumes that the recoverable energy put into the specimen during the test is absorbed by plastic deformation of the specimen and in the formation of fracture surfaces. It also is assumed that the fracture surface formation energy per unit area remains constant. Detailed descriptions of the Fractometer System and method are provided in the literature by Barker (1978a, 1978b, 1978c).

RESULTS

Structure and Mechanical Behavior of As Cast Billets

The initial investigation was directed at examining the deformation characteristics of the composite billet. Metallographically, it was evident that billets were nearly fully dense but lacked homogeneity (Fig. 3). That is, the distribution of SiC in the aluminum matrix was not uniform. When hot rolling at above 400°C ,

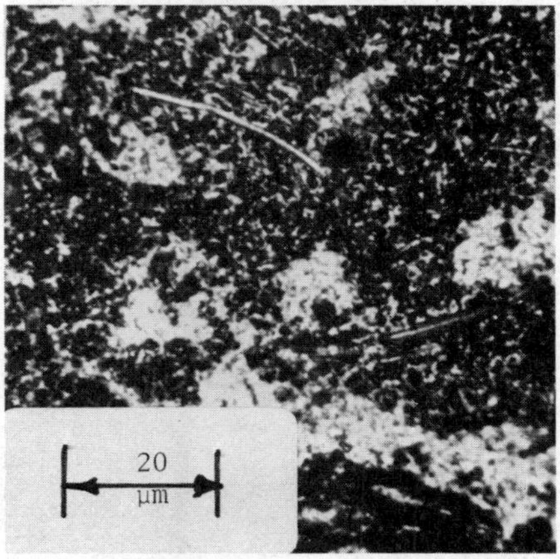

Fig. 3. An optical micrograph of an as fabricated billet exhibiting poor fiber-matrix distribution As Polished.

severe cracking was encountered--even when reduction per pass was kept as low as 1 percent. It was reasoned that this apparent lack of ductility in the composite was analogous to that of as cast billets containing a large percentage of alloying elements, as in alloys 7075 or 2024 Al. Conventionally cast ingots of these alloys require homogenization treatment prior to extrusion and/or rolling. During hot extrusion, the brittle intermetallic compounds are dispersed very uniformly in the entire structure. Consequently, the composite billets were extruded at elevated temperature to eliminate or minimize large matrix-rich and whisker-rich areas. It was experimentally determined that 1:10 extrusion ratio is adequate while 1:20 ratio is impractical, at least under laboratory conditions. Attempts to extrude at 1:20, slow speeds, high temperature (only 10°C below that of the matrix solidus) failed because of severe surface tearing of the extrusion. Since all preliminary deformation studies were conducted @ 25 v/o SiC whisker concentration in a 2024 Al matrix, most data presented herein concern the 2024 Al composites.

Effect of Hot Extrusion

As expected, the extruded composites exhibited much superior fiber/matrix distribution. It was also observed that the degree of homogenization induced in the composite was comparable at low (1:6.5) and high (1:10) extrusion ratios as shown in Fig. 4. There is some evidence that the aspect ratios of the whiskers in mildly deformed extrusion are higher (Fig. 5) than those in composite extruded at 1:10 extrusion ratio. At the time of this writing, the effect of extrusion

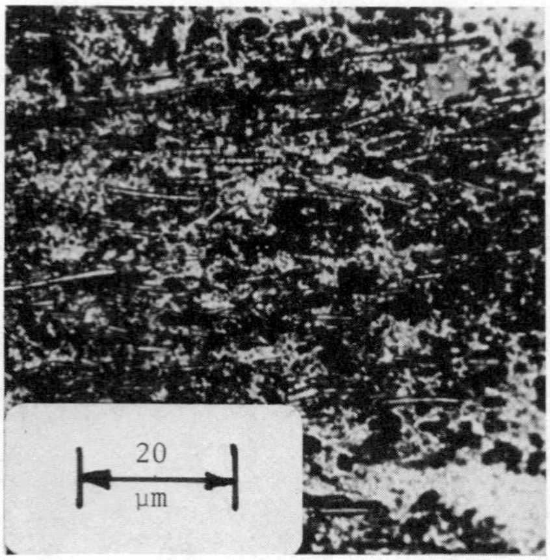

Fig. 4. Longitudinal cross-section of an SiC/Al billet extruded at 1:6.5 extrusion ratio (As Polished).

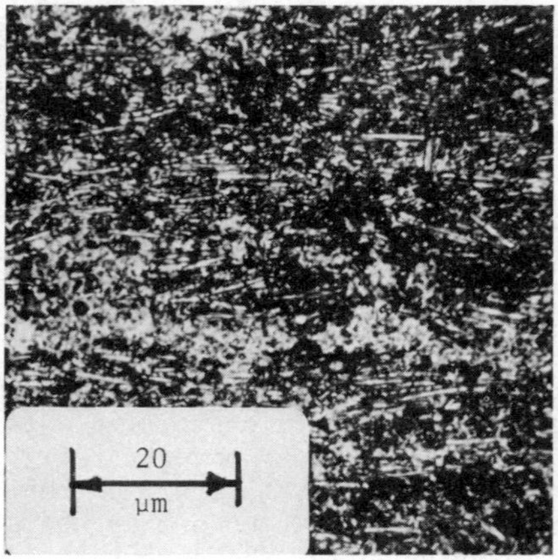

Fig. 5. Same as Fig. 4 except that the billet was extruded at a higher (1:10) extrusion ratio. Whisker aspect ratio is lower.

parameters on mechanical properties is not available. It is, however, listed for a future study.

Mechanical properties of 2024-25 v/o SiC, in extruded and heat-treated condition are presented in Table 2. There is a significant scatter in the properties, particularly the elastic modulus values. The reasons for this scatter are not known. From microstructures of the "as cast" billets and extrusions the following factors appear to be responsible singly or simultaneously: (a) variability in the whisker product, (b) billet synthesis procedure, and fiber/matrix distribution, (c) condition of the matrix in the as-extruded condition, (d) matrix response to heat treatment, and (e) variability in the extrusion process from one

TABLE 2 Properties of Extruded 2024 Al - 25 v/o SiC Rods

	As-Extruded	In T4 Condition
Ultimate Tensile Strength, MPa	400 - 413	530 - 640
Elastic Modulus, GPa	124 - 138	124 - 138
Elongation to Fracture, percent	1.0 max.	1.0 max.

to the next run.

Despite these variables which are difficult to control at the present time, the properties of the composite system are impressive. The elastic modulii of 124.02 to 137.80 GPa are competitive with titanium. The degree of difficulty encountered in deformation processing of titanium is believed to be greater than that of the whisker reinforced composite primarily because the former requires much higher working temperatures (700°C). It is admittedly true that the whisker composite lacks extensive ductility which characterizes most metals and alloys. Enhancing the composite ductility to match homogeneous metals may be possible by lowering the whisker volume fraction which in turn would lower the elastic modulus proportionately. As discussed later, this indeed is the case for composites prepared using 15 v/o SiC reinforcement in 2024, 5456 and 6061 alloys.

Effect of Hot Rolling

Ability to roll a metallic material into a sheet is an important aspect of its usefulness. Subsequently, thin sheet can be roll-formed into channels, hot sections, Zees, etc. SiC/Al billets, as mentioned briefly, earlier, were not rollable. It was reasoned that the rollability of the composite after extrusion may be improved. In fact, if rolled sheets as thin as 0.5 mm could be obtained, it could be considered for a number of applications for the U. S. Navy. Rectangular composite bars measuring 38.1 mm wide x 15.2 mm thick obtained by extruding 63.5 mm diameter billets containing 15 and 25 v/o SiC whiskers were prepared for the study.

It became immediately apparent that hot rolling of these composites in the direction of extrusion was not feasible. For the composite containing 25 v/o SiC, extensive cracking began after less than 5 percent reduction in thickness despite the high temperature (450°C). Although the composite containing 15 v/o SiC exhibited superior rollability, cracking appeared after 10 percent reduction. It was reasoned that this behavior was largely attributable to the fibers or whiskers predominantly aligned in the extrusion direction. During rolling, the whiskers restricted the matrix flow in the axial direction so efficiently that the composite deformed laterally rather than in the rolling direction. Based on this observation, cross-rolling was attempted successfully.

Typical properties of the composite sheets containing 15 v/o SiC thus prepared are presented in Table 3. It is seen that the properties of the composites are quite isotropic, that is, the extrusion and rolling effects are not readily discernible from the values of strength and moduli . Once again, the ductilities are quite low, regardless of the matrix composition. Equally important is the fact that improved whisker product utilized in one case did not have any beneficial effect on the composite properties.

TABLE 3 Properties of Three Aluminum Alloys Reinforced with 15 v/o SiC

Matrix	Whisker Quality	Orientation	Ultimate Tensile Strength MPa	Elastic Modulus GPa	Elongation %	No. of Specimens
2024 Al	High	extrusion	407.8	93.0	0.006	8
2024 Al	High	rolling	394.1	100.6	0.006	2
2024 Al	Low	extrusion	404.4	90.3	0.011	5
2024 Al	Low	rolling	442.3	95.8	0.006	7
6061 Al	Low	extrusion	305.9	91.0	0.006	5
6061 Al	Low	rolling	265.9	91.0	0.005	5
5456 Al	Low	extrusion	378.9	95.8	≤ 1	2
5456 Al	Low	rolling	395.5	108.9	≤ 1	2

Enhancement of Composite Ductility

For any metallic material to be considered for use in structural applications, the designers customarily expect ductilities that exceed 5 to 7%. In continuous filament reinforced composite (with a metallic or a nonmetallic matrix) the elongation to failure is necessarily restricted to that of the reinforcing filament. Since most high modulus filaments are ceramics (e.g., boron, silicon carbide, alumina, graphite) or ceramic-like, their "elongation" to failure is completely elastic and never exceeds 0.5 to 1%. Fortunately, with short fiber or whisker reinforcement, this constraint is not as rigid. Plastic flow in the matrix is not only possible but occurs extensively via yielding in shear at the fiber or whisker ends. Given proper conditions, useful gain in ductility can be achieved at the same time retaining to a large degree the stiffness induced by the fiber.

Table 4 presents data on 15 v/o SiC whiskers in three aluminum alloys in as extruded form. The ductilities are well above 1% in each case including the 5456 alloy which contains magnesium. It is seen that when the degree of deformation is not severe (as in rolling), and when the whisker concentration is only 15 percent by volume, ductilities are higher in all composites, including the one containing improved whisker product.

TABLE 4 Very Recent Data on As-Extruded* Composites Contain 15 v/o SiC Whiskers

Matrix Alloy	Ultimate Tensile Strength MPa	Elastic Modulus GPa	Elongation %	No. of Specimens	Remarks
2024 Al	378.9	97.8	1.3 to 1.7	2	Standard Whiskers
2024 Al	391.3	111.6	1.3 to 1.6	2	High Quality Whiskers
5456 Al	369.3	90.2	1.1 to 1.7	2	Standard Whiskers
5456 Al	352.0	121.9	Not Measured	2	Standard Whiskers

*Extruded at 480°C at an extrusion ratio of 1:10 from a round billet into a rectangle.

Fracture Toughness (K_{Ic}) of Extruded Composites

Since a metal matrix composite is inherently heterogeneous, it is possible that its "low" ductility may not be a true measure of its resistance deformation in the classic sense. A more meaningful measurement may be the fracture toughness of a composite. Unfortunately, the size specimens needed for K_{Ic} tests are quite large and would have involved making special material with different mechanical history than that used for mechanical property determination. It was decided to use the TerraTek method described above. Although this method has the disadvantage of not being a standard technique, the same type extruded rods could be used for tensile specimens and fracture toughness determination.

Measured values of K_{Ic} for 15 vol % SiC/Al composites with 2024 Al, 5456 Al and 6061 Al matrices are presented in Table 5. Values of K_{Ic} for the SiC/6061 Al for the six samples of as-extruded rods average 30.0 MPa$\sqrt{m}$, as compared to 34.0 MPa$\sqrt{m}$, reported by Nelson and Kaufman (1973), for 6061 Al-T6. The K_{Ic} values for 2024 Al and 5456 Al SiC/Al composites were substantially lower, averaging 15.0 and 17.0 MPa$\sqrt{m}$, respectively, for two measurements. As mentioned previously, there are inhomogeneous regions in the as-fabricated billet in which localized volumes of SiC-rich and SiC-poor composite exist. During the extrusion process, these inhomogeneous regions are elongated in the extrusion direction and smooth, "ductile-appearing" regions and "granular" appearing regions, such as shown in

TABLE 5 K_{Ic} Values for 15 Vol % SiC/Al of Different Matrices

K_{Ic}, MPa$\sqrt{m}$	2024 Al	6061 Al	5456 Al
	14.4	25.9	16.0
	15.5	28.1	18.0
		37.5	
		32.1	
		28.8	
		28.0	
Average	15.0	30.0	17.0

Figure 6 can be observed. In the smooth "Ductile-appearing" region to the right in Fig. 6, it is thought that the SiC particles and whiskers are well-bonded to the matrix. The dimples are probably caused by nonaligned fibers pulling out of the matrix and/or by microvoid coalescence as the main crack propagates through the composite. With well-bonded material, the crack is expected to propagate around fibers aligned in the extrusion direction which is also the crack propagation direction. Indeed, in these regions no fibers were observed on the fracture surface. In the "granular" regions, to the left in Fig. 6, secondary cracks are often visible. Matrix-whisker bonding is probably not good in the granular regions, and as a result, it is reasonable that cracks propagate preferentially through them. Evidence of poor whisker-matrix bonding, is shown by SiC debris on the fracture surface. Regions of well-bonded reinforcement and clumps of poorly bonded reinforcement were common to all three of the Al matrix composites. The large differences in the measured values of K_{Ic} for the 6061 composites and the 2024 and 5456 composites are probably attributable to the matrices themselves. Although all three types of composites had large regions of apparently well-bonded Al-SiC, the localized ductility for the 6061 composite, as demonstrated by large "pull-out" dimples, was much higher than for the other two composites. Figures 7(a) and 7(b) for 6061 Al and 5456 Al matrix composites compare similar regions

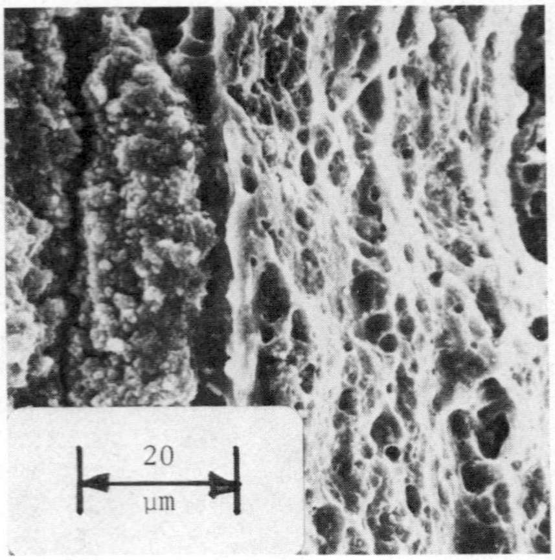

Fig. 6. Fracture surface of composite toughness specimen illustrating "ductile-appearing" and "grandular" regions (6061 Al matrix).

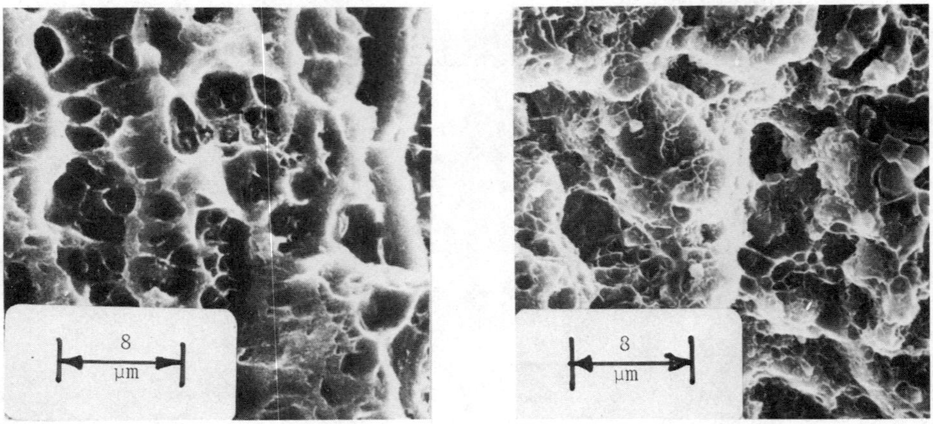

Fig. 7(a). SiC/6061 Al toughness specimen illustrating high matrix ductility.
7(b). SiC/5456 Al toughness specimen illustrating low matrix ductility.

on the composite fracture surfaces. Not only are the dimples much smaller in the 5456 matrix composite, but brittle cracked areas, possibly due to the presence of oxide inclusions in the high Mg alloy, are also apparent.

CONCLUSIONS

1. Silicon carbide whiskers are an effective reinforcing agent in the systems studied. Modulus and strength of three different matrix materials representing most aluminum alloy categories can be improved significantly.

2. Composite ductilities increase as the volume fraction of the whiskers decreases.

3. Composite rollability improves after extrusion. It is necessary to cross-roll the extruded composite due to matrix flow restriction by the aligned reinforcing whiskers.

REFERENCES

Barker, L. M. (1978). Theory for determining K_{Ic} from small non-LEFM specimens, supported by experiments on aluminum, TerraTek Report TR 78-6, TerraTek, Inc., 400 Wakara Way, Salt Lake City, Utah.

Barker, L. M. (1978). Short rod K_{Ic} measurements of Al_2O_3, from Fracture Mechanics of Ceramics, 3, Ed. by R. C. Brandt, D. P. H. Hasselman, and F. F. Lange, Plenum Publishing Corp.

Barker, L. M. (1978). Low cost test method determines K_{Ic}'s of brittle and ductile materials, Metal Progress, pp. 15-20.

Christ, B. W., and S. R. Swanson (1976). Alignment problems in the tensile test, Journal of Testing and Evaluation, pp. 405-417.

Christian, J. L., Fabrication methods and evaluation of boron/aluminum composites, Proceedings of the 1975 International Conferences on Composite Materials, 2 p. 706, AIME, NY 10017.

Divecha, A. P., and P. Lare, Research on whisker reinforced metal composites, Final Report Contract No. NOw-66-0213-d, Naval Air Systems Command, May 1967.

Jones, B. H. The application of advanced automated fabrication methods to manufacture products from fialmentary composites, Proceedings of the 1975 International Conference on Composite Materials, 2, p. 789, AIME, NY 10017.

Nelson, F. G., and J. G. Kaufman (1973). Fracture toughness of plain and welded 3-in.-thick aluminum alloy plate, in Progress in Flow Growth and Fracture Toughness Testing, ASTM STP 536, pp. 350-376.

Parratt, N. J. (1972). Fiber reinforced materials technology, pp. 146-151, Van Nostrand Reinhold.

Volume 3, ICM 3, Cambridge, England, August 1979

CRAZING AND SHEAR LIP FORMATION IN THE NOTCHED CRACK PROPAGATION OF A HIGH DENSITY POLYETHYLENE IN AIR

S. Bandyopadhyay* and H. R. Brown

Department of Materials Engineering, Monash University, Clayton, Victoria, Australia
***Present address: Australian Dental Standards Laboratory, 240 Langridge Street, Abbotsford 3067, Australia**

INTRODUCTION

The presence of a sharp notch can change the fracture of a material from a ductile to a brittle mode. Under uniaxial tensile loading a local state of triaxial tension can be created (Parker, 1957) in notched specimens. Consequently, for a material which is capable of plane stress yielding, notched tensile loading of a thick specimen may cause plane strain crack propagation at the specimen interior and simultaneously plane stress deformation near the edges thus forming shear lips. Shear lip formation in thermoplastics has been reported in the literature for polycarbonates (Fraser and Ward, 1978; Kambour, Holic and Miller, 1978; Kambour and Miller, 1976; Key and Katz, 1969; Mason and Hertzberg, 1973; Parvin and Williams, 1975) and vinyl urethane polymers (Brown and Ward, 1973; Harris and Ward, 1973).

High density polyethylenes (HDPE) are normally ductile materials and in unnotched tensile tests in air fail by necking and cold drawing, although in service they frequently fail in a brittle manner. Naturally, brittle failure of such a normally ductile material is of concern from a design point of view. In the presence of certain environments, polyethylenes fail in an apparently brittle manner, and the phenomenon, known as environmental stress cracking (esc) has been widely investigated. There does not however, seem to exist much information in the literature about the brittle failure of polyethylenes in air.

In this paper are presented the results of an investigation of the crack propagation characteristics of notched thin sheets of a HDPE in air.

EXPERIMENTAL

Materials

The material used was a low molecular weight linear HDPE, GA 7260 (melt flow index = 18) supplied by Hoechst Australia Pty. Ltd. Compression moulded sheets of thicknesses 0.9 mm and 1.5 mm were made by heating the polymer at 180° C for 15 minutes then fast cooling to room temperature - a cooling rate of 1° C per second was applied between 180° C and 100° C. The sheets were subsequently annealed at 85° C for 20 hours followed by furnace cooling. The average size of the spherulites was found to be about 30 μm. Tensile tests using unnotched specimens showed that the material failed by necking and cold drawing with little work hardening (this has commonly been observed for high melt flow index polyethylenes; Haward and Owen, 1977; Vincent, 1966). E and σ_y were found to be 118 and 23 MPa respectively.

Crack Propagation Tests

Single edge notched (SEN) specimens were cut from the compression moulded sheets and a sharp razor cut was provided at the notch tips. Crack propagation tests in air were subsequently carried out by loading these specimens in tension in a dead weight load apparatus at 24.5 ± 0.5° C. During crack propagation, the crack lengths were monitored, with a light source behind the specimen, by an inbuilt travelling microscope which gave an accuracy of 0.02 mm. Also when required, the crack tips were photographed by an Olympus PM-6 camera that could be fitted on to the microscope tube. K_I values were obtained by using Brown and Srawley equation (1966). The gross stress applied ranged between 1 and 3.5 MPa.

RESULTS AND DISCUSSION

Crack Tip Observations

It was observed that under load, a dark zone gradually appeared ahead of the sharp razor cut and the razor cut eventually opened, after which the crack propagation started. Figure 1 shows such a typical crack tip situation when the razor cut had opened and subsequent crack propagation was just about to begin.

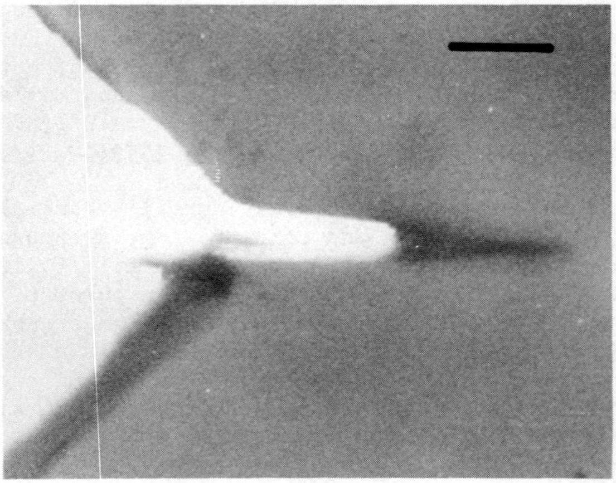

Fig. 1 Appearance of a dark region at the tip of the sharp razor cut at the onset of crack propagation. Scale: 0.5 mm.

Once the crack propagation began, the crack speed subsequently increased with time until final failure. A typical crack propagation curve has been shown in Fig. 2. Similar crack propagation curves have been observed in the case of brittle plastics like polymethylmethacrylate (PMMA) in air (Marshall et al, 1969) and also in brittle esc failure of high melt index polyethylenes (Bandyopadhyay and Brown, to be published; Marshall et al, 1970).

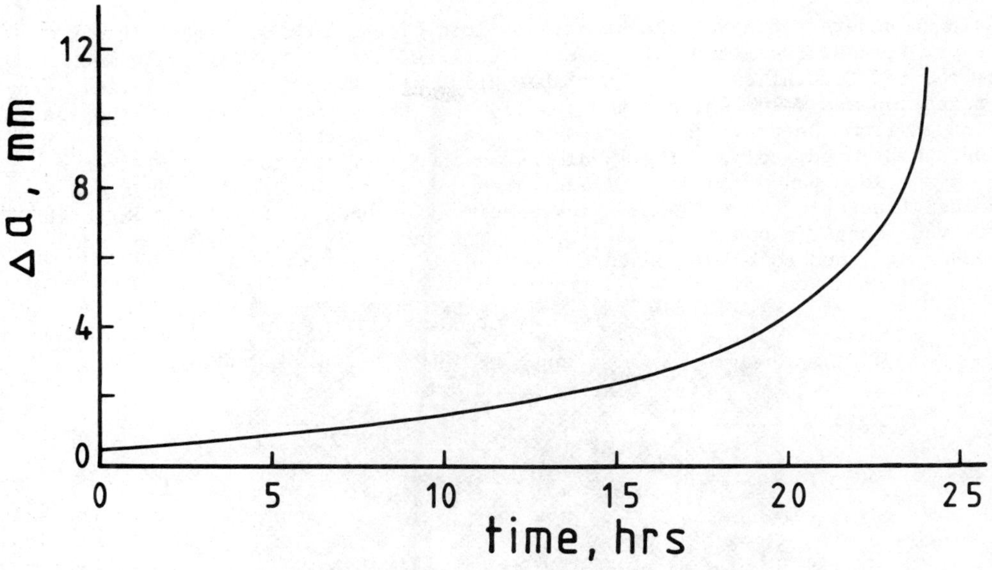

Fig. 2 A typical crack propagation curve showing brittle nature of failure.

During the crack propagation there always existed a dark zone at the crack tip. A close observation however, revealed that the dark zone actually consisted of two regions: a dense dark region and a less dark region. The sizes of these two dark regions changed during the crack propagation. One such crack tip situation has been presented in Fig. 3(a) in which both the dense dark and less dark regions are quite large. A schematic diagram of such crack tip situations has been shown in Fig. 3(b) where point A denotes the end of the dense dark region and the distance AB is the length of the less dark region.

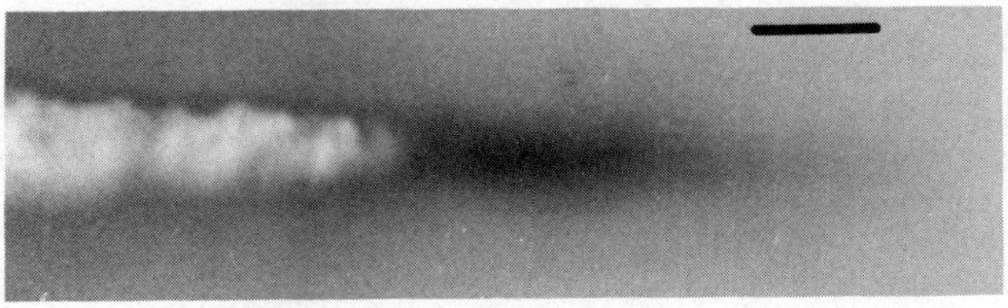

a.

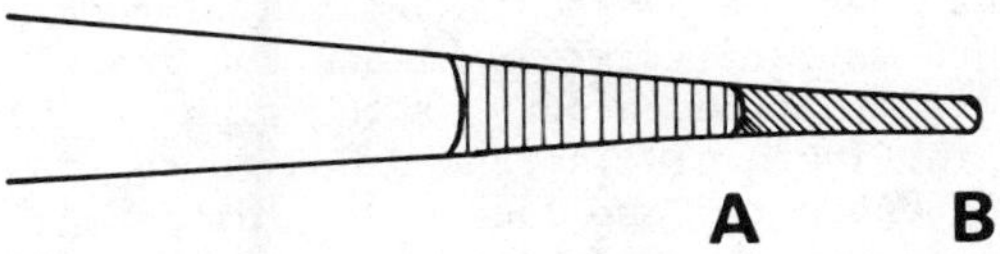

b.

Fig. 3 Crack tip showing dense dark and less dark regions. Scale = 0.5 mm. (a); schematically shown in (b).

To gain an understanding of the brittle failure of the HDPE specimens it was necessary to know the nature of these two dark regions. The following method was first adopted for this. In the stable crack propagation region a certain crack tip situation was selected, and the positions of the points A and B on the specimen were noted from the travelling microscope scale. The crack tip was then photographed immediately. The specimen was then fast fractured by placing a high load on the load pan. An optical micrograph of a fracture surface obtained by such fast fracturing of a 1.5 mm thick specimen is shown in Fig. 4(a) which shows two curved fronts (shown schematically as 1 - 1 and 2 - 2 in Fig. 4(b)), although the curved fronts tend to become obscured near the edges by big fibrils.

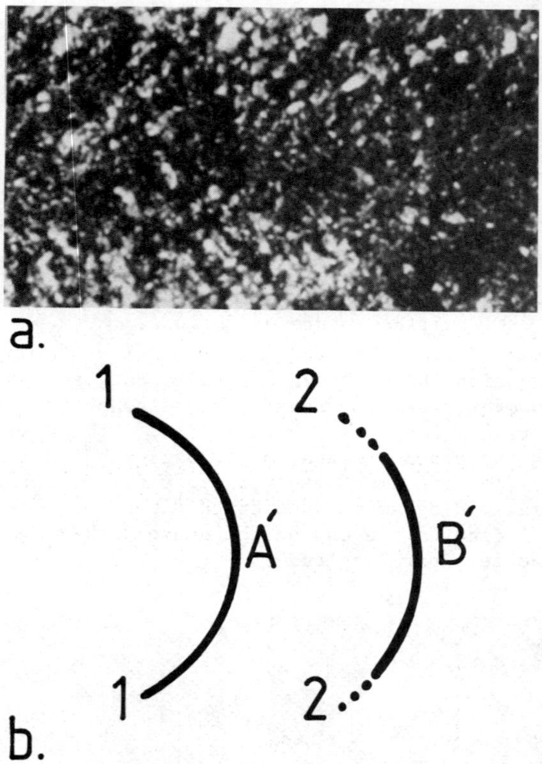

Fig. 4 Optical micrograph of a fast fracture surface showing two parallel curved fronts (a); shown schematically in (b).

A comparison of distances measured by the travelling microscope scale, the optical micrograph of the fracture surface (Fig. 4(a)) and the crack tip photograph taken prior to the fast fracture revealed that the tips of the two curved fronts, namely points A' and B' in Figure 4(b) corresponded to the points A and B of Figure 3(b). This suggested the presence of two different regions at the crack tip for a stable crack propagation. The distance A'B' of Figure 4(b) corresponded to the less dark region in the crack tip photograph prior to fast fracture. It is also seen in Figure 4(a) that additional fracture markings originated from the curved front 1,1 then propagated through up to the front 2,2. It thus appears that the front 1, 1 was the crack front prior to the fast fracture whose tip was A'. Again, the region bounded by the two fronts (this region corresponds to the

less dark region) shows a difference in contrast from the surfaces both on its left and on its right. The crack region would not be affected by the fast fracture. Further, the region on the right of the front 2-2 shows a fracture surface distinctly different from that of the crack. All these tend to suggest that the material in the zone between these two curved fronts was different from that on the left (the crack) and also that on the right (possibly unyielded material). This material could then be partially deformed yet load bearing. Consequently this less dark region ahead of the crack tip can be termed as a 'craze'. The difference in the intensity of darkness could be due to a different extent of scattering from the two different regions. The existence of such craze regions was confirmed by repeated experiments.

Application of the Dugdale Model

The Dugdale model originally developed for mild steel (Dugdale, 1960), then modified for small scale yielding at the crack tip (Burdekin and Stone, 1966) has successfully been applied to predict the craze zone and displacements at the crack tip in glassy polymers (Brown and Ward, 1973; Doyle and Wagner, 1976; Ferguson et al 1973; Kambour, 1973; Marshall et al. 1970; Skibo et al, 1977. Attempts were made to see if the model could be used to predict the crack tip conditions in the present case.

Following the model, the constant craze stress for a particular crack tip position is given by

$$\sigma_c = \frac{\pi}{8} \cdot \frac{\delta}{\Delta} \cdot E \qquad \Delta << a \qquad \text{(where } \Delta = \text{craze length)}$$

and stress intensity factor $K_D = \sqrt{\frac{8\Delta}{\pi}}\ \sigma_c$

It is already known that the less dark region corresponds to Δ. For a crack tip the δ value can be obtained directly from the crack tip photograph at the point A (Fig. 3). Consequently K_D values can be obtained from the crack tip photographs using the measured value of E. A plot of K_D vs K_I has been shown in Fig. 5, for five crack tip positions for two of which (marked x) Δ was

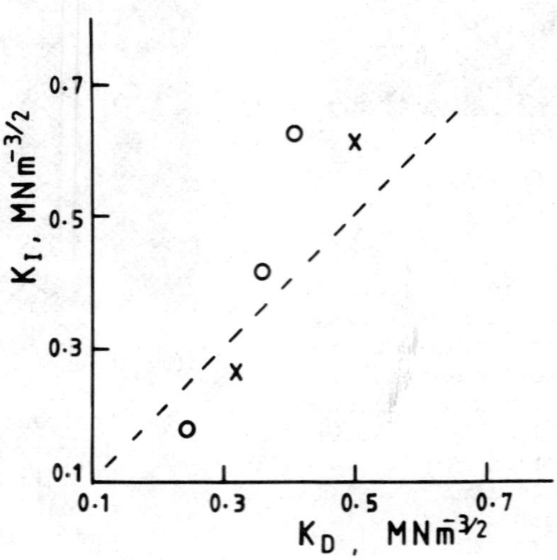

Fig. 5 Comparison of K_I and K_D for a few crack tip positions. The dotted line represents K_I=K_D.

measured from actual fracture surfaces. The agreement between K_D and K_I is fairly good considering the difficulty of the measurements although at high K_I values the agreement tends to decrease.

Other Methods of Observing the Craze

Although the fast fracture method satisfactorily demonstrates the presence of a craze at the crack tip, during the fast fracture the craze region also becomes highly fibrillated. (The presence of such big fibrils caused difficulty in scanning electron microscopic studies because of charging.) For this reason, it was felt necessary to carry out experiments in which the craze regions would fail in a brittle manner. This would also help to understand the nature of the material in the craze. We have recently reported (Bandyopadhyay and Brown, 1978) that when esc of polyethylene is carried out at extremely low crack speeds, truly brittle failure can take place in an interlamellar manner within the spherulites without any deformation. To use this technique in the present study, when the crack in air had propagated through a certain distance with a craze ahead of it, the crack tip was photographed and the load reduced to about one fifth of its previous value to give very low esc crack speed and subsequent crack propagation was carried out at this low load in a 10% solution of the detergent Igepal Co 630.

Figure 6(a) shows the scanning electron micrograph of such a fracture surface in which the brittle esc failure (on the right) is separated from fibrillar failure in air (left) by a transition region in the centre. This transition region matches with the craze length obtained from the crack tip photograph prior to inducing the brittle esc failure. Figure 6(a) also shows the formation of shear lip during the crack propagation in air, which has been shown at a high magnification in Figure 6(b). In the shear lip the material failed by drawing and

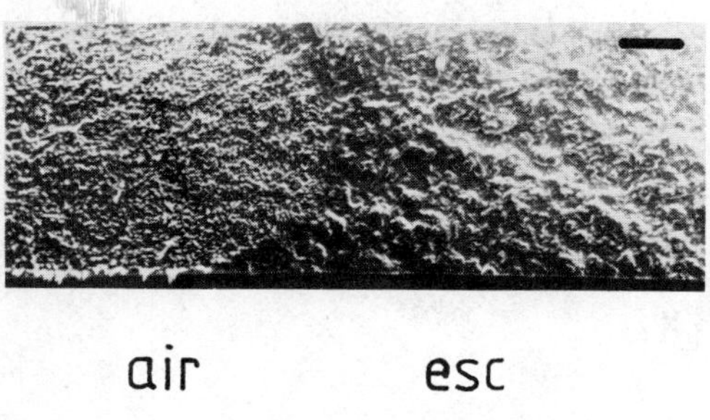

Fig. 6(a) Scanning electron micrograph of a fracture surface showing crack propagation in air, a transition region (craze) and brittle esc crack propagation. Also shear lip is visible in the region of crack propagation in air. Scale: 200μm.

Fig. 6(b) Higher magnification of the shear lip of Fig. 6 (a). Contraction in thickness decreases in the craze region and disappears at the onset of esc failure. Scale: 100μm.

(In these figures, K_I was 0.62 $MNm^{-3/2}$ at the crack tip in air and reduced to 0.12 $MNm^{-3/2}$ at the onset of esc failure.)

necking, in a manner reminiscent of plane stress failure of semicrystalline polymers (e.g. Peterlin, 1977). On the other hand, when the material in the specimen interior failed under plane strain because of the constraint, it deformed by void formation and growth. The shear lip and the associated thickness contraction are distinctly visible in Fig. 6(b). The shear lip size is about 0.03 mm on one side, compared with the sheet thickness of 0.9 mm. This indicated that the fracture was predominantly in plane strain. It is interesting to note in Fig. 6(b) that the contraction in thickness decreases in the craze region then ends at the onset of the brittle esc failure, a fact that demonstrates the existence of a craze at the crack tip. The shear lips are also visible near the specimen edges in the optical micrograph shown in Fig. 4(a). But the scanning electron micrograph of Fig. 6 has the advantage over the optical micrograph in that the thickness contraction is revealed much better in the former.

The minimum thickness to induce plane strain failure is given by

$$B_o > \frac{1}{\pi}\left(\frac{K_I}{\sigma_y}\right)^2$$

σ_y for this material $\simeq$ 23 MPa

Using a lower value of K_I for crack propagation e.g. 0.2 $MNm^{-3/2}$

$$B_o = 2.4 \times 10^{-5} \text{ metre}$$

$$= 0.024 \text{ mm}$$

for a higher value of K_I, say 0.8 $MNm^{-3/2}$ and assuming σ_y remains unchanged,

$$B_o = 0.39 \text{ mm}$$

It can thus be expected that the fracture would give evidence of a combined mode of failure and hence shear lips would be seen.

The other technique used to observe the presence of craze was the liquid nitrogen fracture (LNF). This technique was very similar to the impact failure described earlier, but here the specimen was cooled to liquid nitrogen temperature before fracture.

Observation of the fracture surfaces under the optical microscope confirmed the presence of a parallel curved zone, the craze, ahead of the crack. The fracture surface was then sputter coated with gold and observed in the scanning electron microscope. Such a crack/craze region obtained by the LNF has been shown in Fig. 7(a) which shows: (i) the crack region (on the left) that has a fibrillated and voided appearance; and (ii) the craze region which is partly voided but where the material has not yet fibrillated. The crack propagation takes place by the growth of the voids (in the craze) and subsequent failure of the supporting material by necking and drawing.

In the top right corner of Fig. 7(a) the LNF has gone off plane after travelling through some distance in the craze. The LNF technique has been applied by the authors to obtain direct evidence of craze at the crack tip in esc failure of polyethylene and the off-plane propagation was observed there also.

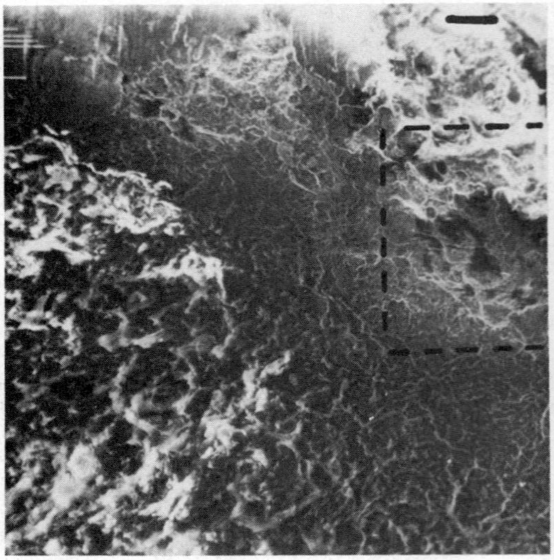

Fig. 7(a) Scanning electron micrograph of the crack/craze region obtained by liquid nitrogen fracture. Scale: 20μm.

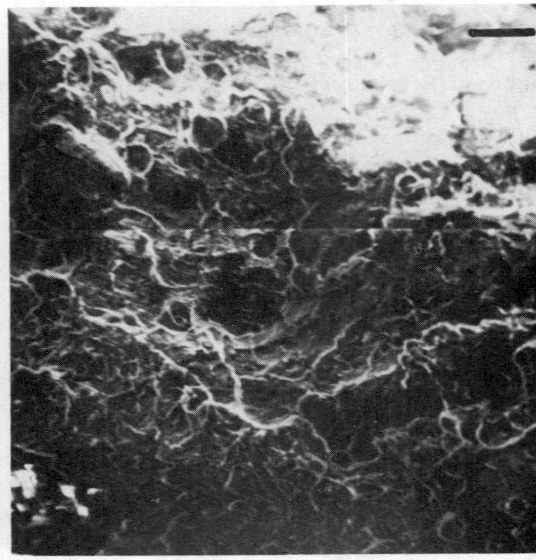

Fig. 7(b) A small region of Fig. 7(a) at higher magnification showing voids in the craze more clearly. Scale: 10μm.

This happens because the LNF fracture was carried out in an impact manner which provides a large amount of energy to the fracture process and as a result, the LNF need not always follow the weakest plane, i.e. the plane of the craze. As long as the LNF propagates through some distance in the craze before going off plane, the presence of a partly voided but unfibrillated region (craze) can always be observed at the crack tip.

A small portion of the craze region of Fig. 7(a), shown by the dotted square, has been presented at a higher magnification in Fig. 7(b) which shows the voids in the craze very clearly.

SUMMARY

Notched crack propagation tests of an HDPE at low stress levels have shown that even with very thin sheets, the material can fail in a brittle manner. The fracture surface shows evidence of predominantly plane strain failure associated with shear lips.

The mechanism of the brittle failure is the formation of a craze at the crack tip, and subsequent breakdown of the craze structure by coalescence of the voids. The Dugdale model for small scale yielding can be satisfactorily used to predict the craze stress.

Evidence of the voids in the craze region has been established by liquid nitrogen fracture methods.

REFERENCES

Bandyopadhyay, S., H.R. Brown (1978). Environmental stress cracking and morphology of polyethylene. Polymer, 19, 589-592.

Bandyopadhyay, S., H.R. Brown. Direct evidence of void formation (craze) ahead of the crack in the environmental stress cracking of polyethylene. Submitted for publication.

Brown, H.R., I.M. Ward (1973). Craze shape and fracture in polymethylmethacrylate. Polymer, 14, 569-575.

Brown, W.F., J.E. Srawley (1966). Plane strain crack toughness testing of high strength materials. A.S. .M. S.T.P. 410.

Burdekin, F.M., D.E.W. Stone (1966). The crack opening displacement approach to fracture mechanics in yielding materials. J.Strain Analysis, 1, 145-153.

Doyle, M.J., J.G. Wagner (1976). Strain in a polystyrene craze layer. A.C.S. Adv.in Chem.Ser., 154, 63-71.

Dugdale, D.S. (1960). Yielding of steel sheets containing slits. J.Mech.Phys. Solids, 8, 100-104.

Ferguson, R.J., G.P. Marshall, J.G. Williams (1973). The fracture of rubber-filled polystyrenes. Polymer, 14, 451-459.

Fraser, R.W., I.M. Ward (1978). Temperature dependence of craze shape and fracture in polycarbonate. Polymer, 19, 220-224.

Harris, J.S., I.M. Ward (1970). Craze of oriented polyethylene tetrapthalate. J.Mater.Sci., 5, 573-81.

Haward, R.N., D.R.J. Owen (1977). The detergent stress cracking of polyethylene. Proc.Roy.Soc.(London), A352, 505-521.

Kambour, R.P. (1973). A review of crazing and fracture in thermoplastics. J.Polym.Sci. D., 7, 1-152.

Kambour, R.P., S. Miller (1976). G_c vs crack velocity in single groove double cantilever beam specimens of polycarbonate. J.Mater.Sci., 11, 1220-1226.

Kambour, R.P., A.S. Holic, S. Miller (1978). Plane strain fracture energy instability and mixed mode crack propagation in polycarbonate at room temperature. J.Polymer Sci. Polym.Phys.Ed., 16, 91-104.

Key, P.L., Y. Katz (1969). On the popin mode of fracture. Inf.J.Fract.Mech., 5, 63-68.

Marshall, G.P., L.E. Culver, J.G. Williams (1969). Crack growth in PMMA in air. Plastics and Polymers, February, 75-81.

Marshall, G.P., L.E. Culver, J.G. Williams (1970). Craze growth in polymethyl-methacrylate: a fracture mechanics approach. Proc.Roy.Soc.(London), A319, 165-187.

Mason, J.A., R.W. Hertzberg (1973). Fatigue failure in polymers. Crit.Rev.Macro. Sci., 7, 433-467.

Parker, E.R. (1957). Brittle Failure of Engineering Structures. Wiley, New York.

Parvin, M., J.G. Williams (1975). Ductile brittle fracture transition in polycarbonates. Int.J.Fracture, 11, 963-71.

Peterlin, A. (1977). Plastic deformation of crystalline polymers. Polymer Eng.Sci., 17, No.3, 183-193.

Skibo, M.D., R.W. Hertzberg, J.A. Mason, S.L. Kim (1977). On the generality of discontinuous fatigue crack growth in glassy polymers. J.Mater.Sci., 12, 531-542.

Vincent, P.I. (1966). A working hypothesis of polymer fracture. Physical basis of yield and fracture. Institute of Physics and Physical Society, Oxford.

Volume 3, ICM 3, Cambridge, England, August 1979

FATIGUE AND FRACTURE RESPONSE OF METAL-INFILTRATED SINTERED POWDER METALS

R. A. Queeney

Department of Engineering Science and Mechanics, The Pennsylvania State University, University Park, PA 16802, U.S.A.

INTRODUCTION

The present study was addressed to characterizing the influence of metal infiltration on the resistance of a sintered metal powder to fatigue crack propagation and fast fracture. Experimental data on crack propagation and instability were gathered for sintered 4600 alloy steel powder infiltrated with pure copper. Although infiltration is known generally to improve conventional strength, ductility, and toughness measures for sintered powders, the growth rates and stability of flaws, always present in the form of residual porosity, can be adversely affected by the presence of the weaker infiltrant in the sintered powder, infiltrant composite. Further, sintered forms with the least porosity before infiltration, usually considered to have the best mechanical strength and ductility levels, were found to have higher crack growth rates and potentially lower fracture resistance. Designs using infiltrated composites, although analyzed to be yield-safe, may be highly prone to quasi-brittle failure, as critical flaw sizes approach that of the residual pore diameter.

The utilization of metal powders to manufacture machine elements and structural members is a direct result of the total economics of the manufacturing operation. Although the starting powders themselves are more expensive, by weight, than cast or simple wrought forms of the same composition, the finished item may cost less. This is so, due to the fact that the powder may be pressed and sintered to a shape that requires no, or little, finish machining. Further, the extreme composition gradients that can be encountered in alloy casting are restricted in atomized powders to the slight inhomogeneities of the powder grain itself, usually only tens or hundreds of a micrometer in diameter. The designer gains these advantages, however, at the expense of degraded mechanical response when the response of the sintered powder is compared to that of the wrought form of the same material. Weakened and embrittled material is caused by the porosity contained in the sintered material. Although the porosity can be minimized by designing powder grains of certain shapes and distribution in size, porosity fractions of from 0.15 to 0.05 are routinely encountered.

Amelioration of strength and dutility response levels in sintered powders can be achieved by secondary thermo-mechanical processing, which might also produce more closely defined geometries. Cold processing, such as coining, will reduce, but not eliminate, porosity. Hot processing, using a conventional metalworking technique, can reduce or eliminate porosity, depending on the extent of deformation. All such operations, however, adversely affect the economic advantage inherent in the powder option. A compromise technique for porosity reduction is to infiltrate the interconnected pore structure with a metal or alloy of a lower melting point than that of the sintered material. The molten infiltrant can be drawn into the pores by capillary action alone, or the process can be pressure-aided. Since the infiltrant has a lower melting point than the sintered matrix, it will likely be weaker than fully dense forms of the matrix composition, and residual porosity will be retained to some degree, as not all porosity is interconnected. Excessive infiltration must be avoided to prevent powder grain separation and, through nonuniform swelling, warpage.

The conventional wisdom in powder metal technology is that infiltration leads to ductility and strength enhancement, and extensive studies support this view; however, excessive infiltration, leading to dil tational strains and stresses can be degrading to mechanical response. Given a state of successful infiltration, there are two aspects of the material that are of concern mechanically. One is the existence of residual porosity that might serve as starter cracks if cyclic or fatigue loading is to be encountered in application. The other aspect is the still-inferior ductility, or potential notch sensitivity that many such composites will evidence. The residual porosity itself, or fatigue cracks developing from the same, would be the notches in question. Contemporary linear elastic fracture mechanics is widely accepted as one method for assessing the mechanical integrity of load-bearing machine elements and structural members containing sharp flaws. For flaws subjected primarily to displacement tending to separate their faces, or Mode I loadings, the pertinent stress analysis parameter in LEFM is the stress intensity factor K_I, which, as reviewed by Paris and Sih (1965) is of the form:

$$K_I = \alpha\sigma\sqrt{\pi a} \tag{1}$$

Here, σ refers to section stresses away from the flaw, a is the flaw size or length, and α is a general term uniquely defined for the structural and flaw geometry and for the load application particulars.

Two applications of the stress intensity factor will be exploited in the present work. One is the existence of an upper limit to K_I for a given material microstate. If the loaded crack, or flaw, is constrained by sufficient material in the direction of the crack front, this upper limit is a minimum, the plane strain fracture toughness K_{IC}. Methods for the determination of K_{IC} were proposed by Brown and Srawley, (1966) and such determinations have been performed on various sintered steels by Clarke and Queeney (1972) and Barnby, Ghosh, and Dinsdale (1973). The second application is to the correlation of fatigue crack propagation (FCP) data, as first demonstrated by Paris and Erdogan (1963). In their analysis, if the crack is subjected to load excursions that lead to a stress intensity range ΔK_I ($= K_{I,max} - K_{I,min}$) then the rate of crack extension per cycle of load excursion N is given by:

$$da/dN = A(\Delta K_I)^m \tag{2}$$

Here, A and m are dependent on material variables, frequency of cycling, temperature, environment, and stress state, such as the ratio of mean to cyclic stress amplitude. Only the materially dependent contribution to A and m will be addressed herein. Although Barnby, Ghosh, and Dinsdale (1973) recognized the application of equation (2) to FCP in sintered metals, they used the concepts only to provide starter cracks for K_{IC} measurements.

EXPERIMENTAL STUDIES

A heat-treatable commercial steel powder was chosen for this study with an intended final composition close to that of AISI/SAE 4600 alloy. The 100 mesh powder was blended with one volume percent of both graphite and zinc stearate in a tumbling can mixer prior to pressing. The blended powder was then pressed in a floating, double-acting ram cylindrical die to pressures of 550, 690, and 830 MN/m^2. Lubricants were then burned off in air at 500°C for two hours, after which the billets were sintered in cleaned and dried (dew point below -40°C) flowing hydrogen at 1150°C for one hour. The billets were then slowly furnace cooled to 500° to facilitate machining. Two types of test specimen configurations were cut from the cylindrical sintered steel billets--two plates or two beams per billet--shown in Fig. 1 below.

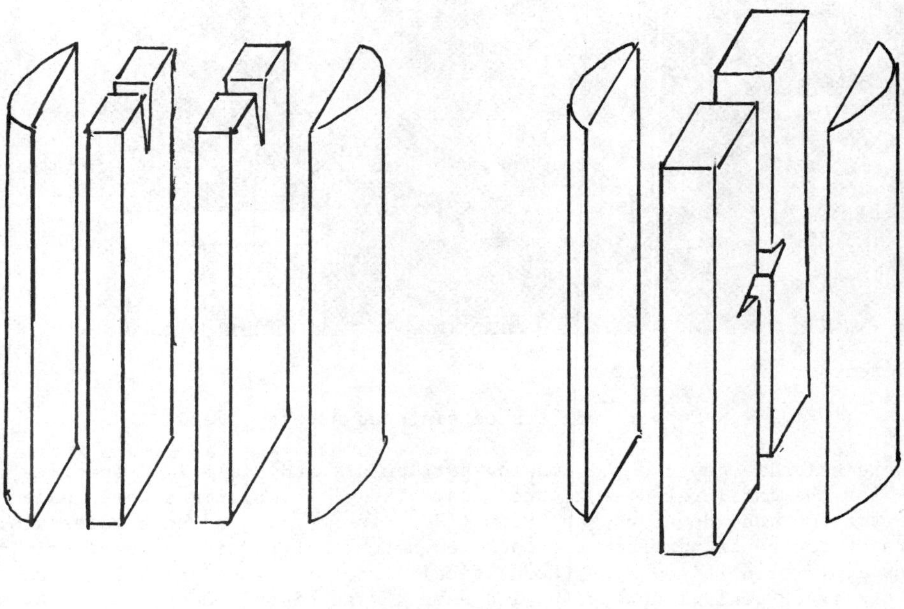

Fig. 1. Specimen positions in sintered steel billet.

Density and composition were then characterized in the billet scraps in regions immediately adjacent to the specimen positions. The final sintered steel composition was found to be (in volume percent): 1.9 Ni, 0.05 C, 0.6 Mn, 0.3 Mo, 0.001 P, 0.3 Si, 0.02 S, remainder Fe. The porosity fraction, as it depended on pressing pressure, is given in Table 1 below.

TABLE 1 Porosity Fraction in Sintered 4600 Steel

Pressing Pressure, MN/m^2	Porosity Fraction
550	0.145
690	0.115
830	0.093

After separation from the billets, the plates--25 mm wide, 70 mm long, 3.2 mm thick--and the beams--63 mm wide, 12.5 mm deep, and 70 mm long--were ultrasonically cleaned in acetone, vacuum dried, and infiltrated with 99.999% pure copper. Infiltration was accomplished at 1200°C in one hour in the same atmosphere as for sintering; subsequently the specimens were furnace cooled to 880°C and oil quenched. A typical infiltrated microstructure is shown in Fig. 2, with powder grains, infiltrant, and residual porosity evident.

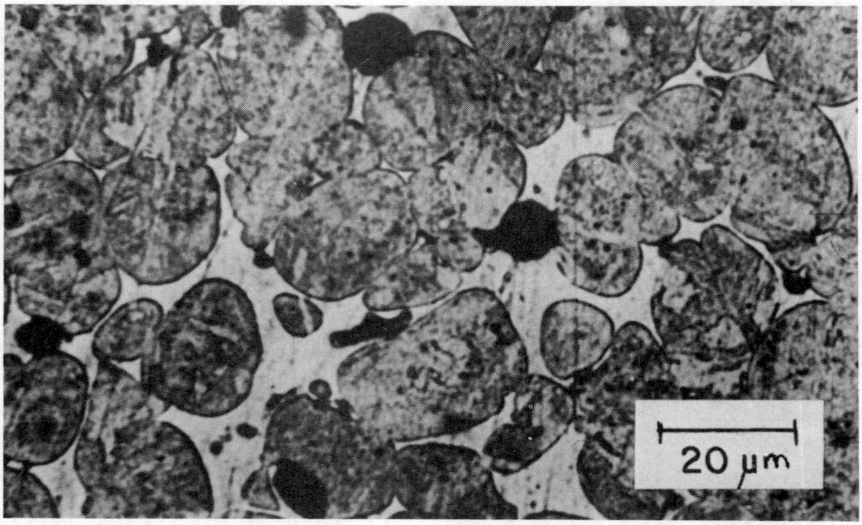

Fig. 2. Copper-infiltrated sintered steel powder.

Figure 3 schematically depicts the loading particulars and analytical specimen dimensions for the infiltrated test specimens. The plate specimens were loaded in the double torsion mode developed by Evans (1973) for testing ceramic materials. Starter cracks can be driven from the machined notch, after which several critical load values could be obtained. As the critical load for crack instability was achieved, the crack would lengthen and increase the specimen's compliance. As the load thereby decreased, the cracks became stabilized until the load was again increased, and so forth. In this manner, several critical load experiments could be completed on a given specimen with fixed microstate. All experiments were conducted in air, at room temperature, on an Instron TTD universal testing machine with a crosshead speed of 1.0 mm/minute. The beam specimens were tested under similar environmental conditions on an Instron Cycler operated at 30 Hz. The applied loads never changed sign, varying from a maximum value to 1/10 of the maximum value; thus, the propagating crack was subject to tension-tension loading. Finally, indentation hardness tests were carried out to approximately determine static yield strengths-Rockwell B-scale to measure overall composite behavior, and Vickers microhardness to ascertain individual constituent phase strengths in situ.

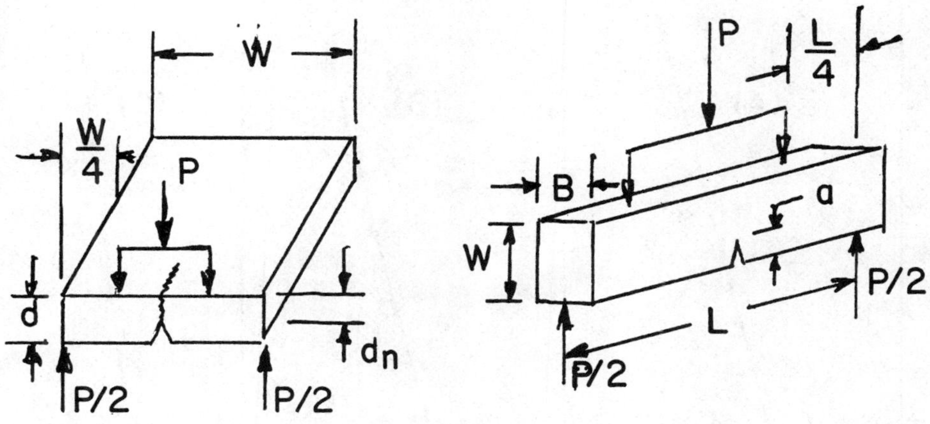

Fig. 3. Specimen configurations for mechanical testing.

RESULTS OF EXPERIMENTS

Table 2 summarizes the result of indentation hardness tests averaged over the composite microstructure with a Rockwell B-scale indenter. For a given pressing pressure, or sintered billet density (or, conversely, porosity fraction) there was no systematic variation in hardness for slight differences (one volume percent) in infiltrant content; thus, only one average B-scale hardness is indicated.

TABLE 2 Indentation Hardness for Cu-Infiltrated Sinter

Pressing Pressure, MN/m^2	Rockwell B Hardness
550	65
690	69
830	71

Vickers microhardness numbers of 185 and 255 were measured for the copper and steels phases, respectively, in situ in the microstructure. Using Hetényi's (1950) correlation tables between strength measures and hardness numbers, the estimated yield strengths were 515 MN/m^2 for the copper infiltrant and 690 MN/m^2 for the steel.

The results of the FCP experimentation are shown in Fig. 4, where the rate of crack growth per cycle of loading, da/dN, is plotted against stress intensity range ΔK_I.

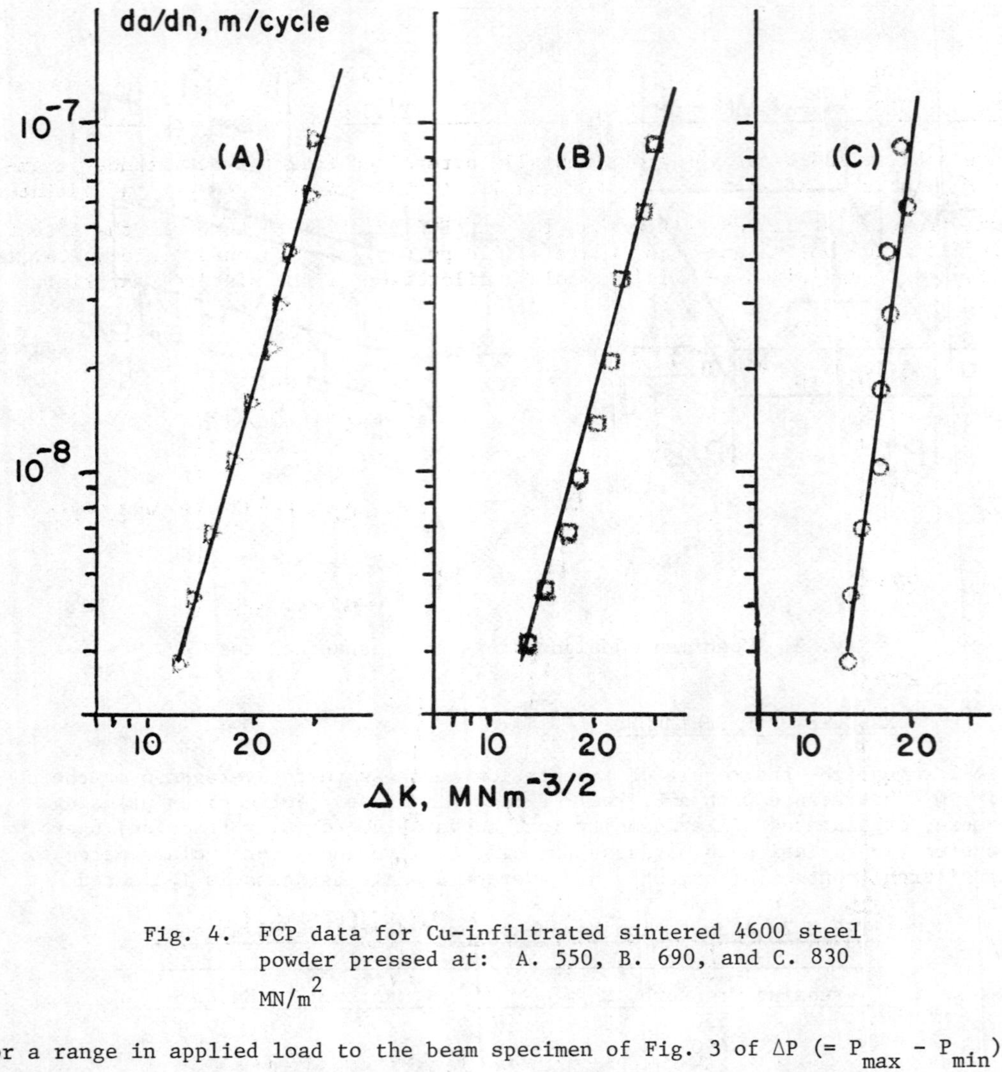

Fig. 4. FCP data for Cu-infiltrated sintered 4600 steel powder pressed at: A. 550, B. 690, and C. 830 MN/m^2

For a range in applied load to the beam specimen of Fig. 3 of ΔP (= $P_{max} - P_{min}$), Brown and Srawley (1966) found the stress intensity factor range to be:

$$K_I = \frac{3PL}{4BW^2}\sqrt{a}\ [1.99 - 2.47\ (a/w) + 12.47\ (a/w)^2 - 23.17\ (a/w)^3 + \ldots \ldots + 24.8\ (a/w)^4] \quad (3)$$

The lines drawn through the data points are best-fit lines determined by the least squares method. Table 3 summarizes the line slopes and intercepts--exponent m and coefficient A, respectively, in equation (2) and material variables for the three cases.

TABLE 3 Fatigue Crack Growth Parameters and Material Variables

Pressing Pressure MN/m^2	Volume Fraction Copper	FCP Coefficient A	FCP Exponent m
550	0.12	7.89×10^{-22}	3.65
690	0.10	4.93×10^{-23}	3.9
830	0.07	$1{:}00 \times 10^{-39}$	8.1

Figure 5 below indicates the experimentally determined fracture resistance parameter K_{IC} for the infiltrated sintered steel. A wider range of copper infiltrant volume fraction was investigated than for the FCP experiments to probe the effect of infiltration beyond the point where simple pore filling is encountered; i.e., infiltration sufficient to initiate volume dilatation of the sintered material.

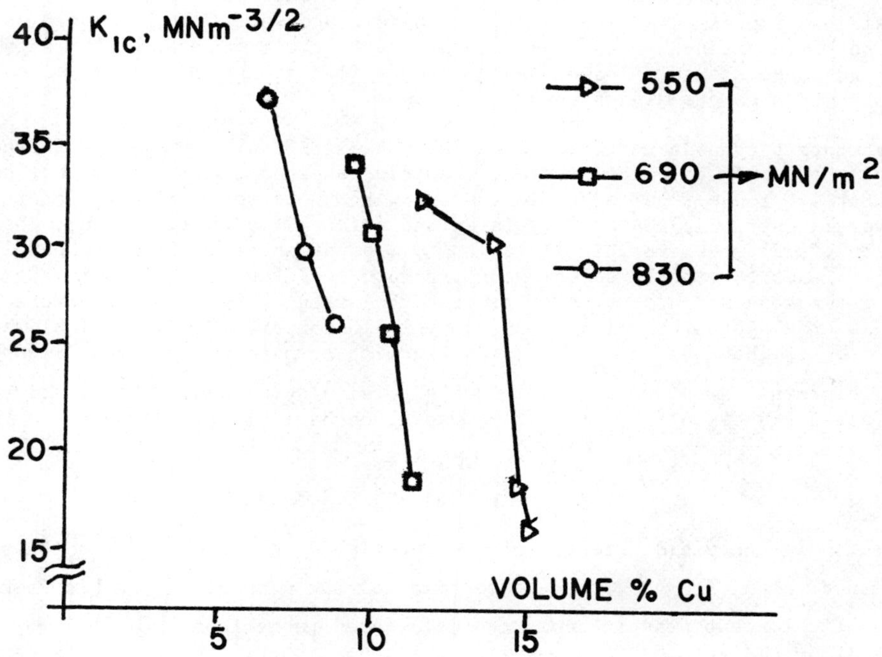

Fig. 5. Fracture toughness K_{IC} in Cu-infiltrated sintered 4600 steel and pressing pressure.

Referring to the specimen loading schematic of Fig. 3, the stress intensity factor, due to Evans (1973), is found to be:

$$K_I = \frac{PW}{4} \left[\frac{3(1 + \nu)}{W\, d^3 d_n}\right]^{1/2} \tag{4}$$

Here, ν is the Poisson ratio taken as 0.30, and the fracture toughness K_{IC} results when the load P is the critical value.

DISCUSSION OF RESULTS

Many investigators (e.g., Rennhack, 1961, Taubenblatt, Smith, and Lewis, 1974) have demonstrated the beneficial effects of infiltration on strength and ductility measures. The indentation hardness values of Table 2, an indirect strength measure, demonstrate the same effect, with the highest hardness values being associated with the densest sintered steel. The hardness values were consistent in both beam and plate specimens, a variation in infiltrant fraction in the latter did not systematically effect the results.

The fatigue strength of sintered powder metals has been studied (Haynes, 1970 and Williams and Haynes, 1971, to cite but two) and the results of the studies indicate that, the lower the porosity, the better the fatigue strength. However, such studies encompass three separate, and perhaps unrelated, physical phenomena--crack initiation, crack growth, and unstable crack propagation. Fatigue crack growth rates, as determined in this study, seem to follow a different pattern--the denser the composite, the faster the growth rate. There are no simple extant analytical models to predict why this composite microstructure, with sharp response delineations between phases, should have a lowered resistance to crack growth. Recent analysis and experimental studies by Reifsnider and Kahl (1973) strongly suggest that the yield strength gradients encountered by the crack, moving through the steel and copper, should indeed cause more rapid crack growth in the more compact microstructure as pressing pressure increases.

The fracture toughness tests indicate that the infiltrated powders with the highest sintered densities have the greatest fracture resistance when fully, but not over infiltrated. These results are consistent with those reported for sintered metals with varying porosity content by Clarke and Queeney (1972) and Barnby, Ghosh, and Dinsdale (1973). However, infiltration to a point where swelling is barely discernible leads to rapidly diminishing fracture resistance, as seen in Fig. 5. Here, grain separation apparently reduces the strength aspect of fracture resistance as crack paths travel more extensively in copper. The analytical validity of the K_{IC} determinations must be established with respect to specimen dimensions, the plate thickness in the case of the double torsion specimens. The elastic-plastic interface r_p, ahead of the crack tip, was given by Irwin (1960) to be

$$r_p = \frac{1}{6\pi} \left(\frac{K_{IC}}{\sigma_o}\right)^2 \tag{5}$$

wherein σ_o is the yield strength of the material in question. Using the previously determined value of σ_o for the weakest part of the composite, the highly-constrained copper, the largest plastic zones are calculated to be less than 0.5 mm, an order of magnitude smaller than the specimen thickness; therefore, the tests may be judged valid.

Finally, the question of fracture safe design practices may now be addressed. If design stresses, based on the Rockwell B-scale hardness values and their correlation to yield strength (Hetényi, 1950), were to be in the order of 500 MN/m^2, and the material in question had the K_{IC} response of 15 $MNm^{-3/2}$, one of the lowest values associated with over-infiltration, a potentially dangerous situation would be at hand. The critical flaw size, of a small flaw in a tensile field, can be estimated from:

$$K_I = \sigma_d \sqrt{\pi a} \tag{6}$$

Using $K_I = K_{IC} = 15\ MNm^{-3/2}$, $\sigma_d = 500\ MN/m^2$, the critical flaw size is about 0.1 mm. Residual porosity, or such porosity extended by cyclic stress, may well be of this size, rendering the machine part or structural element unsafe. While these levels of infiltration must obviously be avoided, their existence, particularly on a local basis, demands fracture-safe analysis in critical applications.

CONCLUSIONS

The present studies indicate that the infiltration of a sintered metal powder by another metal, or alloy, of lower melting point may lead to some property improvement but, simultaneously, degrade other mechanical response measures. General strength measures may be improved, as may fracture resistance, but the latter response may be degraded through poor control of the infiltration process with no impairment of the former. Overall fatigue strength may be improved by reduction of porosity, but the growth of residual pores might be vastly accelerated. Finally, poor control of infiltration may lead to mechanical designs which are yield-safe, but perilously close to being fracture unsafe.

REFERENCES

Barnby, J. T., D. C. Ghosh, and K. Dinsdale (1973). The fracture resistance of a range of sintered steels. Powder Met., 16, 55-71.

Brown, W. F., Jr., and J. E. Srawley (1966). Plane Strain Crack Toughness Testing of High Strength Metallic Materials. American Society for Testing and Materials, Philadelphia.

Clarke, G. A., and R. A. Queeney (1972). Fracture toughness and density in sintered 316 stainless steel. Internat. J. Powder Met., 8, 81-87.

Evans, A. G. (1974). Fracture Mechanics Determinations. In R. C. Bradt (Ed.), Fracture Mechanics of Ceramics, Vol. 1, Plenum Press, New York. pp. 17-48.

Haynes, R. (1970). Fatigue behaviour of sintered metals and alloys. Powder Met. 13, 465-510.

Haynes, R., and S. H. Williams (1973). Effect of porosity on the fatigue behaviour of sintered precipitated nickel powder. Powder Met., 16, 387-404.

Hetényi, M. (1950). Handbook of Experimental Stress Analysis, Wiley and Sons, New York. pp. 22-23.

Irwin, G. R. (1960). Plastic Zone Near a Crack and Fracture Toughness. In Proceedings of Seventh Sagamore Ordinance Material Research Conference. Syracuse University Press, Syracuse, New York. pp. IV-63.

Paris, P. C., and G. C. Sih (1970). Stress Analysis of Cracks. In Fracture Toughness and Its Applications. American Society for Testing and Materials STP381, Philadelphia. pp. 30-81.

Reifsnider, K., and M. Kahl (1974). Effect of local yield strength gradients on fatigue crack propagation. Int. J. Mech. Sci., 16, 105-119.

Rennhack, E. H. (1961). Strengthening effect of copper in infiltrated sintered iron parts. Prog. in Powder Met., 17, 12-18.

Taubenblat, P. W., W. E. Smith, and R. Lewis (1974). An infiltration system for achieving high strength in P/M parts. Modern Dev. in Powder Met., 8, 149-162.

Volume 3, ICM 3, Cambridge, England, August 1979

ELLIPTICAL PLANAR CRACK IN A THREE DIMENSIONAL BODY FRACTURE CRITERIA

Y. Katz, A. Busiba and H. Mathias

Nuclear Research Center Negev, Beer-Sheva, P.O.B. 9001, Israel

INTRODUCTION

Plane discontinuities in a three-dimensional (3D) elastic body, bounded by closed curves, have been treated theoretically in the literature (Green and Sneddon, 1950; Irwin, 1962; Kassir and Sih, 1966; Westmann, 1966). The main effort has been expended in establishing fracture mechanics parameters, in terms of the local environment of the elastic crack border. On the other hand this theoretical investigation has been followed by relatively limited experimental studies. The lack of such complementary information in 3D, is mainly connected with difficulties in introducing controlled flaws in the bulk, particularly, such that could be observed during the loading cycle.

Two of the basic aspects involved in the growth nature of a more general planar crack, are directly relevant to the present investigation. Firstly, the mechanism of the crack growth stage, is an important factor in evaluating fracture criteria, particularly in the current case, demonstrated by an elliptical planar crack. Since the stress intensity factor is non-uniform around the crack border, the physical meaning of the critical stress intensity factor and the critical elastic strain energy release rate concepts are no more equivalent.

Secondly, the expected improvement of the crack growth resistance is generally attributed to a more desirable and effective microstructure, in terms of fundamental plastic-flaw behaviour. The role of the crack front is seldom considered, although it probably affects the character of crack instability. On the other hand, the crack front might be influenced by a particular microstructure, as in duplex structure, or composite systems. Therefore, certain gain in fracture toughness might include a cooperative influence of basic plastic properties and geometrical features.

Contributions concerning the penny shaped crack with various boundary conditions, provided the foundations for the 3D elastic solution of an elliptical crack. Irwin (1962) treated the stress intensity factor and the elastic strain energy release rate for an elliptical crack by applying Green and Sneddon (1950) elastic displacements for a flat, sharp elliptical crack front. Kassir and Sih (1966) proposed the constant shape mode to be the typical growth mode for the elliptical crack. Key (1969) formulated the theoretical strain energy release rate, indicating the effects of the crack extension modes on the calculated strain energy release rate. Smith (1971) discussed the problem of a disc shaped crack under open mode and general shear conditions, emphasizing growth characterizations as well. Following the mentioned studies one could conclude that basic assumptions regarding the crack growth

MS213

mode influence the strain energy release rate function. Thus, direct reflection on fracture criteria remains obvious.

The objective of the present study has been to search for experimental verification of the cracking mode and critical loading conditions, in case of an elliptical planar crack. Actually combined information of this kind is essential in order to make fracture criteria evaluation possible. With this goal in mind, an experimental program was scheduled taking into account the quantitative background gathered so far for local and global critical condition. Controlled bulk flaws were embedded in a polyester resin to produce transparent tensile specimens, containing flaw in a plane perpendicular to the tensile axis. Open mode tests were performed followed by crack growth observations and critical load measurements. Consequently, experimental and theoretical aspects regarding growth mechanisms and proper fracture criteria are investigated. In the current study the elliptical planar elastic crack is representative in class of planar 3D cracks, where non-uniform stress environment around the crack periphery controls the early stages of crack instability.

EXPERIMENTAL PROCEDURE

Open mode tests were performed using single edge notched specimens (SEN) and cylindrical cracked specimens. Polyester resin was selected and tensile segments were produced by molding technique. Cracking was carried out by utilizing fine Teflon (P.T.F.E.) foils of 0.075 mm in thickness. The foil was cut to the desired crack geometry and then embedded into the bulk. Preliminary work indicated no adhesion between the two materials. Thus, well controlled discontinuity could be introduced in terms of its shape and location.

Two types of polyester were used, glassy material designated PO-1 and modified material PO-2 which corresponded to higher degree of viscoelastic behaviour. The latter was produced by minor addition of plasticizers in order to reduce the extreme notch sensitivity of the material. The PO-2 material still remained sensitive to existing crack but allowed better observation of the crack growth process. Uniaxial tensile tests of both materials showed typical linear elastic region for a crosshead speed of 1 mm min^{-1}.

Although the material was not suited for photoelastic studies the loading cycle was observed by polarized light, with quarter-wave plates in opposition. This procedure enabled better observation of the local events which occurred at the crack periphery.

Generally critical load measurements were performed by using the PO-1 polyester, while the PO-2 material was mainly used for relatively slow crack growth observations at different crack extension stages. Testing procedure included the following series; firstly SEN specimens (Fig. 1) with a straight crack front, center loaded and geometry of $a'/W=0.3$ were tested (a' and W are the crack length and specimen width, respectively). Secondly, cylindrical specimens with two types of flat geometries, namely symmetrical penny shaped and elliptical cracks were used (Fig. 1 and Fig. 2).

The SEN specimens and the penny shaped crack specimens provided K_{IC} and G_{IC} values for the resin. Prior tests were performed for comparison between SEN specimens, with embedded Teflon foils and SEN specimens pre-cracked, by means of razor blade. The crack radii for the penny shaped crack specimen varied between 3-6 mm with r/R ratio between 0.2-0.4 (r is the crack radius and R the specimen radius). The cylindrical specimens were about 115mm in length and 25.4mm in diameter. The already mentioned crack geometries, i.e. the SEN specimen with straight crack front and penny shaped crack, gave a uniform stress environment and were used for calibration.

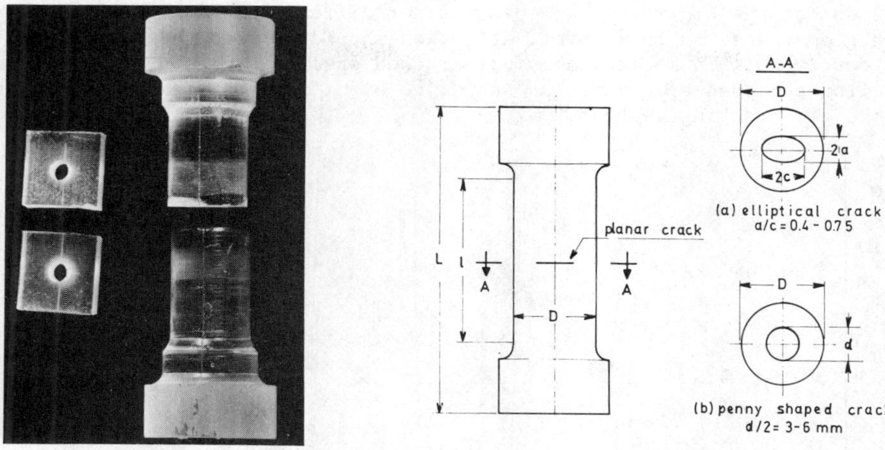

Fig. 1. Polyester molded SEN and cylindrical specimens. Scheme illustrating specimen geometry and dimensions for open mode testing.

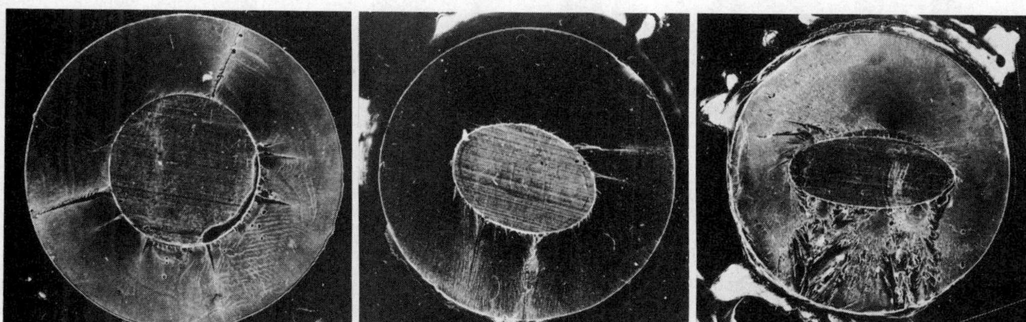

Fig. 2. Typical fracture surface in PO-1 polyester for diferent planar crack geometries.

In additon another serieswas tested with elliptical planar crack, produced by similar specimen preparation and cracking method.Here again, the cylindrical specimens were 25.4mm in diamter, while the crack geometry varied in the range of a/c ratio between 0.4-0.75 (a and c are the minor and the major axis of the elliptical crack, respectively).

All tests were carried out at room temperature with identical conditioning. The specimens were tested using standard tensile test procedure with cross head speed of 1mm min^{-1}. The tested series included 20 specimens for the SEN, while the penny shaped and the elliptical crack series were based on measurement of 4 specimens each.

Direct crack observations and examination were carried out in the PO-2 material. Due to the glassy behaviour of the PO-1 material, crack extension was further studied by means of scanning electron fractographs.

EXPERIMENTAL RESULTS

Figure 3 summarizes K_{IC} values as obtained for different PO-1 series with various specimen geometries. At this point, attention should be paid to the SEN specimens. Series1 corresponds to mechanically pre-cracked specimens, while series 2 corresponds to Teflon embedded specimens. As shown, no significant differences were obtained due to the two cracking methods.

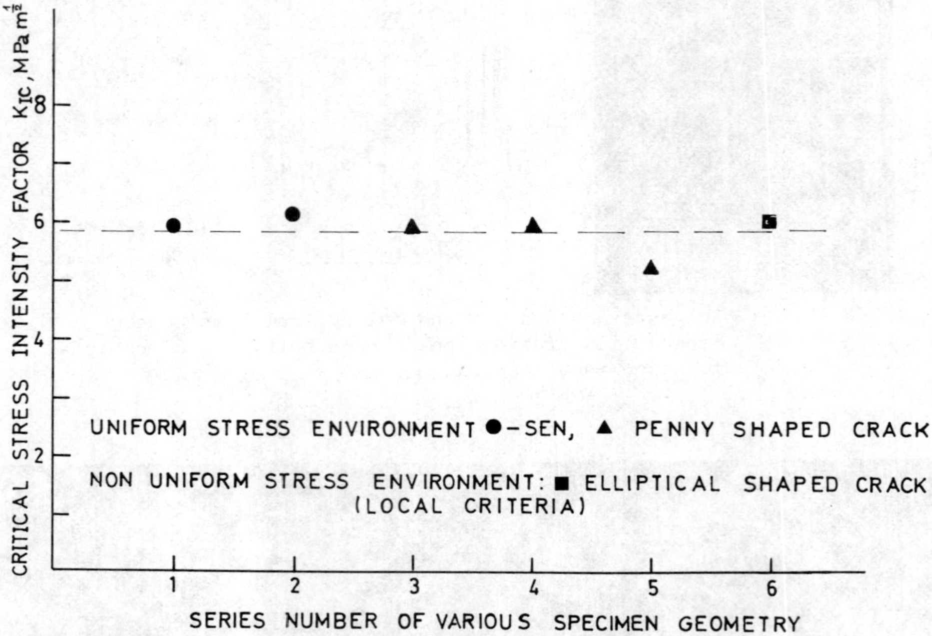

Fig. 3. K_{IC} values for PO-1 material obtained from series with various specimen geometries.

In case of the glassy PO-1 material, catastrophic fracture happened instantaneously, with no previous indication, neither by load relaxation, nor by slow crack front extention. In case of the PO-2 material, crack propagation could by arrested after instability by load removal and the relatively slow crack growth region was observed at different crack extension stages.

Figure 4 illustrates the PO-2 specimens sliced parallel to the crack plane and later examined by light microscopy. These macrographs show the geometrical features of the crack front at the early stages of instability. The crack extension in all specimens started along the minor axis front, namely by a local burst, similar to the so called "pop-in" phenomena. Following this typical crack extension, shown in Fig. 4a, an intermediate stage developed with additional area swept out (Fig. 4b) until a more stable almost penny shaped crack geometry developed, as illustrated in Fig. 4c.

Fractographic studies regarding the PO-1 glassy material indicated, that crack initiation process and extension were similar to the PO-2 polyester, although high crack front velocities resulted in the PO-1. For example, Fig. 5 shows a scanning electron fractograph taken from the PO-1 fracture surface. Gold coating served as a conducting layer of about 300Å in thickness in this case. As shown the crack extension burst along the minor axis front can readily be observed. The critical location along the crack periphery happened to be consistent and could be visualized

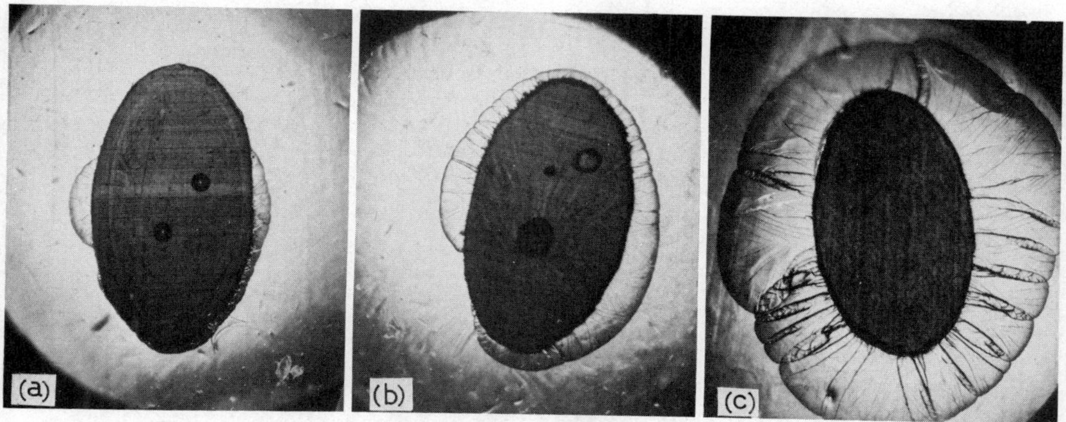

Fig. 4. Crack growth nature in PO-2 polyester. Optical macrographs. (a) "pop-in" at the minor axis front. (b) Intermediate crack extension stage. (c) Crack geometry transition toward a uniform stress intensity environment.

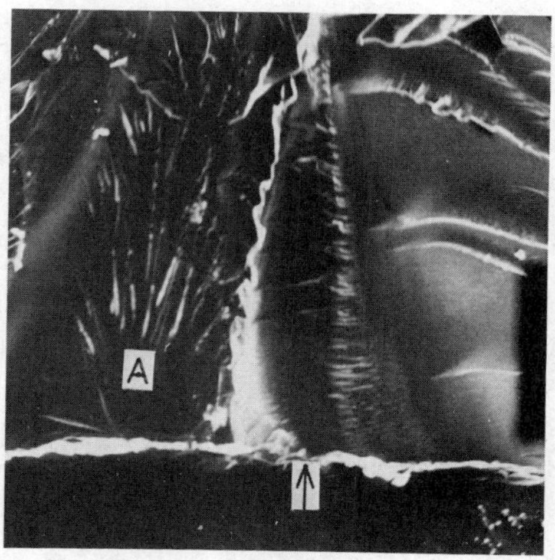

Fig. 5. Scanning electron fractograph in PO-1 polyester indicating burst (region A) at the minor axis front. Arrow indicate the crack periphery of the elliptical crack, 500X.

clearly on the mirror plane. More information supporting this argument was obtained by a macro-replication technique taken from the PO-1 fracture surface. Typical Wallner lines indicated the origin of crack extension. Localized crack extension could be observed along a limited segment of the elliptical border which corresponds to high values of the stress intensity factor.

Quantitative results regarding the critical loading conditions are shown in Fig. 6 and demands some additional explanation. Following the analysis by Key (1969), three different cases were compared in terms of critical loading conditions. Firstly, two global cases namely, the constant shape mode a/c = const. and the fixed major axis mode c = const. Secondly, the local G_{Imax} criteria as obtained from K_{Imax} at the minor axis, by the use of Irwin's relationship.

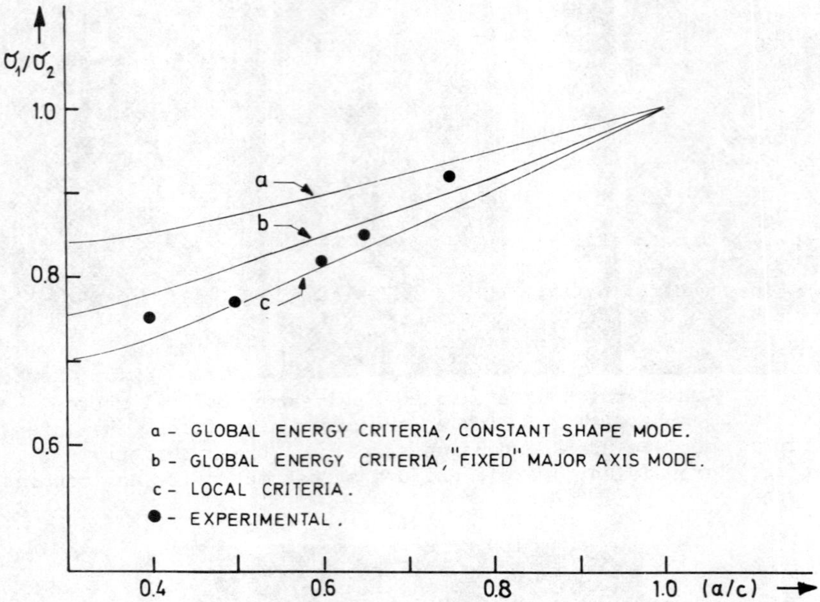

Fig. 6. The relative critical stress σ_1/σ_2 vs. a/c ratio range. σ_1 and σ_2 are the critical stress for penny and elliptical crack respectively.

Key extended the more general expression for the initially planar flaw in order to study the mentioned different modes.

Since for Z=0 (planar flaw) the elastic strain energy release rate is given by;

$$G = \frac{\sigma}{2}\frac{\partial V}{\partial A_c} \tag{1}$$

where; G is the nominal applied stress, V is the volume of the flaw under load, A_c is the area of the flaw in a plane perpendicular to the tensile axis.

Thus for a/c = const.

$$G = \frac{2\sigma^2(1-\nu^2)a}{E\cdot E(k)} \tag{2}$$

and for c = const.

$$G = \frac{4\sigma^2(1-\nu^2)a}{E}\frac{\pi}{3E^2(k)}\{E(k) + K(k) + \frac{E(k) - K(k)}{k^2}\} \tag{3}$$

where K(k) and E(k) are the complete elliptic integrals of the first and second kind, respectively, and $k = [1-(a/c)^2]^{\frac{1}{2}}$.

On the other hand, the local value of the maximum G for the elliptical planar crack, has been determined by Irwin (1962) at the end of the minor axis, namely;

$$G_{max} = \frac{\sigma^2(1-\nu^2)\pi a}{E^2(k)\cdot E} \tag{4}$$

Actually Fig. 6 is based on elastic strain energy release rate terms, as indicated in equations (2) to (4).

For the purpose of non-dimensional presentation consider the ratio G_p/G_{el}^m, where G_p is the value of G for the planar penny shaped crack, according to Sneddon (1946). G_{el} is the value of G for the particular local or global mode m, for the elliptical planar crack (Eq. (2)-(4)).

Alternatively, consider the ratio σ_1/σ_2 for $G_p = G_{el}^m$, where σ_1 and σ_2 correspond to the critical nominal stresses for the penny shaped and the elliptical planar crack, respectively. In fact, this relationship is applied in Fig. 6 consequently, the dependence of σ_1/σ_2 vs. a/c is shown in Fig. 6, with experimental results as obtained from critical load measurements, also shown in Fig. 3. Such approach, combined with experimental results presentation allows improved fracture criteria evaluation.

Accordingly, as observed during the present investigation, the "fixed" major axis mode appeared to be the most favourable mode, suporting to concept of crack shape transition from elliptical to a more symmetrical almost disc shaped crack. This behaviour is consistent and actually expected, from both, local as well as global theoretical predictions shown in Fig. 6. In terms of fracture criteria, experimental results indicated the domination of the local criteria, which is the most conservative criterion for the elliptical shaped crack. However, at high value of a/c the theoretical gap between the different criteria narrows, as was found experimentally.

DISCUSSION

Referring to the crack growth process, it seems that the crack extension which starts at the critical location, extends similar to the typical "pop-in" crack propagation. Obviously "pop-in" which occurs in a straight front, results from the local variations of the plastic constraint, due to the triaxial state of stress. This issue has been investigated and demonstrated in polycarbonate by Key and Katz (1969), where a general plane stress condition prevailed. The kind of "pop-in" extension illustrated in Fig. 4, results from non-uniform K_I along the crack border, which is a different origin for a local crack extension.

Dobbs, Field and Maitland (1973) used laser techniques to induce disc shaped cracks in transparent materials,namely, glass and polymethylmetacrylate systems. They pointed out the advantage of laser-induced internal crack in 3D problems over other techniques (Ernsberger, 1969; Kolsky, 1959). Although their work is mainly concerned with the determination of fracture surface energies, some fracture mechanics topic has been included as well. In the case of the penny shaped crack, they observed uniform crack growth as expected. On the other hand, the preferential or off-center crack growth which was attributed to specimen surface effects, can not simply be extended to the elliptical crack, as proposed by the mentioned investigators. In finite specimen which include an elliptical planar crack, the local event is influenced by combined factors, the K variation and the external specimen surface correction factor. Only for high ratio of a/c and close surface cracks, preferential growth along the major axis front is possible, if homogeneous material is considered.

In the current study the specimen boundary never disturbed the crack initiation at the minor axis front. When the distance of the crack border from the external surface is sufficient, the correction function is a weak increasing function.

Westmann (1966) looked at the general growing crack front, considering the tendency to shape a contour in a way that the stress intensity becomes nearly uniform. The transition from the elliptical toward a circular one, follows this trend as verified and demonstrated in the current investigation.

However, such local behaviour results a much lower critical stress for instability. The local "pop-in" alters the crack geometry and local boundary dragging becomes the dominating factor in the growth mechanism. In the present study brittle material was used and the extension stage followed without an increase of the applied load.

The present proposed technique provides an experimental procedure which seems capable to clarify the mentioned aspects. Generally, fracture criteria evaluation should be supported by direct crack growth initiation and propagation. A combined approach of this kind is needed in order to deal with the complex problem of fracture criteria in non-uniform stress environment flaw. For example, the constant shape mode suggested by Kassir and Sih (1967) has not been confirmed, according to the experimental results obtained during the present study. Acception of such proposed mode, yields a less conservative fracture criterion, but as shown in the current study this mode is not favourable for the elliptical planar crack in 3D.

In addition, the proposed approach as described in the present work, can be extended to other related problems. For example, the finite specimen correction factor in case of elliptical crack, has been studied by Murakami (1974) and needs experimental support. Studies concerning the possible gain due to preferential crack front propagation, like the work by Evans (1973) might promise a wide range of application in material design. However, sound information regarding general crack front behaviour, growth nature and fracture criteria is needed. Fundamental approach along this line has been attempted in the present study, demonstrated by the elliptical planar elastic crack.

CONCLUSIONS

The proposed experimental technique provides an effective method to investigate fracture mechanics concepts in 3D problems. Particularly, problems of the general crack behaviour.

The exact growth mode is an important factor whenever fracture criteria is considered.

In non-uniform local elastic stress intensification along planar crack periphery, local fracture criterion is no more equivalent to integrated energy and global fracture criteria.

In the case of elliptical planar crack, the constant major axis mode extension is favourable. Early stage of crack extension consist of geometrical crack transition, toward more stable penny shaped geometry.

Both, local and global concepts predict the geometrical transition, while quantitative critical conditions indicate that the conservative local fracture criterion dominates in case of the PO-1 material.

The constant shape crack extension mode is not favorable in the case of elliptical planar crack

ACKNOWLEDGEMENT

The autors wish to express their appreciation to Mr. M. Kupiec and Mr. Y. Fachima of the Nuclear Research Center Negev, for experimental assistance.

REFERENCES

Dobbs, H.S., J.E. Field, and R.H. Maitland (1973). Phil. Mag., 28, 33.

Ernsberger, F.M. (1969). Phys. Chem. Glasses, 10, 240.

Evans, A.G. (1973). Crack propagation between blocking obstacles. Proc. of the Third Int. Congress on Fracture, Munchen, Vol. 10.

Green, A.E., and I.N. Sneddon (1950). Proc. Cambridge Phil. Soc., 46, 159.

Irwin, G.R. (1962). J. Applied Mechanics, 29, 651.

Kassir, M.K., and G.C. Sih (1966). J. Applied Mechanics, 33, 601.

Kassir, M.K., and G.C. Sih (1967). Int. J. of Eng. Sci., 5, 899.

Key, P.L. (1969). Int. J. Fracture Mechanics, 5, 287.

Key, P.L., and Y. Katz (1969). Int. J. Fracture Mechanics, 5, 63.

Kolsky, H. (1959). In B.L. Averbach, D.K. Fehlbeck, G.T. Hahn and D.A. Thomas (Ed.), Proc. of the Swampscott Conf. on Fracture, Willey, New-York.

Nisitani, H., and Y. Murakami (1974). Int. J. of Fracture, 10, 353.

Shah, R.C., and A.S. Kobayashi (1971). Eng. Fracture Mechanics, 3, 71.

Shah, R.C., and A.S. Kobayashi (1973). Int. J. Fracture, 9, 133.

Sneddon, I.N. (1946). Proc. Roy Soc., Ser. A, 187, 229.

Westmann, R.A. (1966). Int. J. Fracture Mechanics, 2, 561.

Volume 3, ICM 3, Cambridge, England, August 1979

A DUGDALE-BARENBLATT EQUILIBRIUM MODEL FOR CRAZES IN GLASSY POLYMERS

S. J. Israel, C. S. Kantamneni and W. W. Gerberich

University of Minnesota, Department of Chemical Engineering and Materials Science, Minneapolis, Minnesota 55455, U.S.A.

INTRODUCTION

The Modified Dugdale-Barenblatt Model

In the last decade many investigators (Brown and Ward, 1973; Cotterell, 1968; Israel, Thomas and Gerberich, 1979; Mills and Walker, 1976; Morgan and Ward, 1977; Weidmann and Döll, 1976) have attempted to make a quantitative correlation between the craze profile and the plastic zone profile described by the Dugdale-Barenblatt model (Barenblatt, 1962; Dugdale, 1960). The results of these investigations seem to indicate that this model gives a reasonable but not exact description of the craze profile. The results reported in a recent paper (Israel, Thomas and Gerberich 1979) indicate that for craze and plastic zone profiles determined for equivalent stress intensities, the general shape of the plastic zone profile was very close to that of the craze; however, the plastic zone profile was found to be much larger. It was concluded that the Dugdale plastic zone model was not a fully adequate model to describe air craze geometries in PMMA.

With consideration of these results a modification of the Dugdale-Barenblatt model has been proposed to include two distinctive non-linear zones at the crack tip, Fig. 1 (Gerberich, 1977). The zone immediately next to the crack opening is "semi-cohesive" in that it is filled with microcracks and voids while the furthest cohesive zone is the standard plastic zone. This model is not too different from the concept proposed independently by Williams (1977) and the physical picture of a combined craze/plastic zone suggested by Argon, Hannoosh and Salama (1977). It was demonstrated that with this modification the Dugdale approach can be used to describe craze arrest or equilibrium in terms of the individual properties of the plastic zone and craze.

In the application of this model to fracture in glassy polymers it is assumed that the craze (semi-cohesive zone) is preceded by a plastic zone. Considering that the material in the craze is drawn well beyond its yield point, it is intuitive that there exists such a plastic zone beyond the craze. The equilibrium crack/craze/plastic zone shown in Fig. 1 is expressed in terms of a modified Barenblatt (1962) model as

$$K_{2a} + K_{2(c \to b)} + K_{2(b \to a)} = 0 \tag{1}$$

where the elastic stress intensity term corresponds to the fictitious crack length, the cohesive term corresponds to the plastic zone contribution, and the semi-cohesive term to the void filled craze region. This modified Dugdale-Barenblatt model was then treated as Eshelby (1968) has done by considering the local stress intensity solution applicable to each region. This gives equation 1 as

$$\underbrace{\sigma\sqrt{\frac{a}{\pi}}\int_{-a}^{a}\frac{dx}{\sqrt{a^2-x^2}}}_{\text{elastic}} - \underbrace{2\sigma_c\sqrt{\frac{a}{\pi}}\int_{c}^{b}\frac{dx}{\sqrt{a^2-x^2}}}_{\text{semi-cohesive}} - \underbrace{2\sigma_{ys}\sqrt{\frac{a}{\pi}}\int_{b}^{a}\frac{dx}{\sqrt{a^2-x^2}}}_{\text{cohesive}} = 0 \tag{2}$$

where σ is the applied stress, σ_c the stress within the craze, and σ_{ys} the yield strength within the plastic zone. The only assumption here is that σ_c is constant, which is reasonably well justified for much of the craze length (Doyle and Wagner, 1976; Knight, 1965). Equation 2 reduces to

$$\pi\sigma - 2(\sigma_{ys} - \sigma_c)\cos^{-1} b/a - 2\sigma_c \cos^{-1} c/a = 0 \tag{3}$$

The physical picture of this model includes a threshold applied stress intensity and hence a plastic zone below which a craze will not form. Above this, a craze will start growing with an individual plastic zone larger than the threshold value. As the craze becomes longer, the fibrils within the craze pull back on the opening surfaces gradually reducing the stress intensity at the craze tip. Finally, at equilibrium the plastic zone at the tip of the craze has returned to the threshold value.

To see if the equilibrium model could predict craze lengths at arrest it was applied to data obtained by Andrews and Levy (1974) for ethanol crazing in PMMA. An excellent agreement between the calculated semi-cohesive zone lengths and the equilibrium craze lengths was found by taking the plastic zone length $R_p = 0.72$ µm and $\sigma_c = 1.4$ MPa (Gerberich, 1977). This value of σ_c correlates well with the range of 1-2 MPa which had been estimated by Andrews and Levy (1974).

Activation Energy for Crack (Craze) Growth

Charles and Hillig (1962, 1965) have proposed a theoretical model for thermally activated crack growth mechanisms in glasses. In this model they assumed that the velocity of crack growth was determined by the rate of chemical dissolution of the glass, as affected by the environment. They described the process as being thermally activated where the apparent activation energy is a function of the applied stress. Other approaches to this problem have been physio-chemical or mechano-chemical in nature. For example, the rate of crack propagation has been considered in terms of bond breakage (Bartenev, 1955; Bartenev, Yudina and Rehbinder, 1958; Bartenev and Razumovskaya, 1966), or from a fracture mechanics approach (Irwin, 1966; Schonert, Umhauer and Klemm, 1966; Kies and Clark, 1969; Wiederhorn and Bolz, 1970). In one such investigation a fracture mechanics modification of the Charles and Hillig model has been proposed (Wiederhorn and Bolz, 1970; Wiederhorn and Johnson, 1973a, 1973b; Wiederhorn, 1974; Wiederhorn and co-workers, 1974). Here the velocity of crack growth is given by

$$v = v_o \exp -\left(\frac{E_{app}}{RT}\right) \tag{4}$$

$$E_{app} = E^* - \frac{2V^*K_I}{(\pi\rho)^{\frac{1}{2}}} \tag{5}$$

where v_o is a constant, E^* the activation energy, V^* the activation volume and ρ the crack tip radius.

If crack growth follows this model the results from an appropriate test program will yield E_{app} from the slope of an Arrhenius type plot (ℓn v versus 1/T). Furthermore, as typically done in thermal activation analysis, a plot of E_{app} versus K_I will give the value of the activation energy E^* when extrapolated to zero stress intensity.

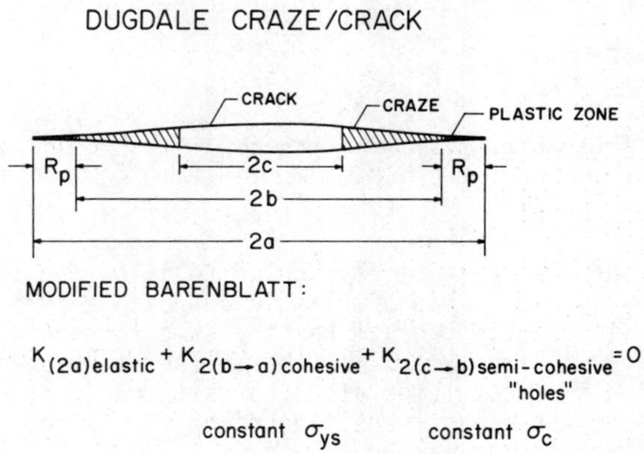

Fig. 1. Concept of a three region craze/crack/plastic zone at equilibrium.

EXPERIMENTAL METHODS

Testing Apparatus

A system was constructed to determine the craze tip velocity (v) and equilibrium craze length (ℓ_{eq}) as a function of time for a prescribed initial stress intensity. With this apparatus the sample was subjected to a constant (dead-weight) load while being submerged in a methanol bath. The temperature of the methanol was maintained within 0.5°C. After applying the load the craze length was measured to ± 0.01 mm at regular intervals with the aid of a travelling microscope.

Specimen Preparation

Rohm and Haas general purpose (grade G) plexiglass sheet was used in this investigation. The molecular weight of this material was measured with gel permeation chromatography and found to be $\overline{M}_n$ = 240,000, $\overline{M}_w$ = 940,000 and $\overline{M}_w/\overline{M}_n$ = 3.9 (Rohm and Haas). Birefringence studies on the as received material indicated no detectable molecular orientation.

Compact tension specimens (CTS) were milled to size in accordance with ASTM specifications from three thicknesses of material; 12.75mm, 6.4mm and 1.6mm. The initial crack was introduced by sharpening the machined slot with a jewellers saw. A razor blade was then slowly driven into the tip of the notch to produce a crack preceded by a single craze. With this technique sharp and uniform craze-crack fronts were produced thus minimizing the possibility of craze bifurcation.

RESULTS AND DISCUSSION

Calculation of the Semi-Cohesive Zone Length

The experimental investigation involved a series of craze stability experiments run at 298,293,278 and 270°K for PMMA in methanol. At each temperature, runs were made over a range of initial applied stress intensity and craze lengths were recorded as a function of time. The equilibrium craze lengths could then be compared with the semi-cohesive zone lengths predicted by the three-zone model. This was accomplished by first replacing the ficticious crack length a by $b + Rp$ in equation 3. The equivalent applied stress σ was found by determining the stress intensity appropriate to the CTS specimen, equation 5 (Knott, 1973). This value was then used in the infinite plate solution, equation 6 (Knott, 1973) to back calculate σ.

$$K_I = \frac{P}{BW^{\frac{1}{2}}} f(a/w) \qquad (5)$$

where $f(a/w) = 29.6(a/w)^{\frac{1}{2}} - 185.5(a/w)^{3/2} + 655.7(a/w)^{5/2} - 1017(a/w)^{7/2} + 638.9(a/w)^{9/2}$

$$K_I = \sigma\sqrt{\pi a} \qquad (6)$$

The parameters σ_{ys} and σ_c were simply scaled to allow for the temperature dependence of the flow stress. This simplistic approach assumes that the flow stress of a microfibril in the wet craze had the same temperature dependence as the yield stress for the bulk polymer. If the solvent pick-up does not lower the T_g of the fibrils below the test temperature, or alternatively, if the rate equation holds right through the glass transition range (Roetling, 1965), then this may be a reasonable first approximation.

The plastic zone length R_p is found by evaluating the two zone Dugdale-Barenblatt model at the threshold stress for craze formation. This is given by,[1]

$$Rp = a\left[\sec\left(\frac{\pi\sigma}{2\sigma_{ys}}\right) - 1\right] \approx \frac{\pi K_{th}^2}{8\sigma_{ys}^2} \qquad (7)$$

To obtain the best fit of equation 3 to the data, a full series of computerized curves was generated with σ_c and K_{th} as the variables. One such set of curves for $K_{th} = 0.11$ MPa-m$^{\frac{1}{2}}$ and T = 293°K is shown in Fig. 2. In comparing this figure

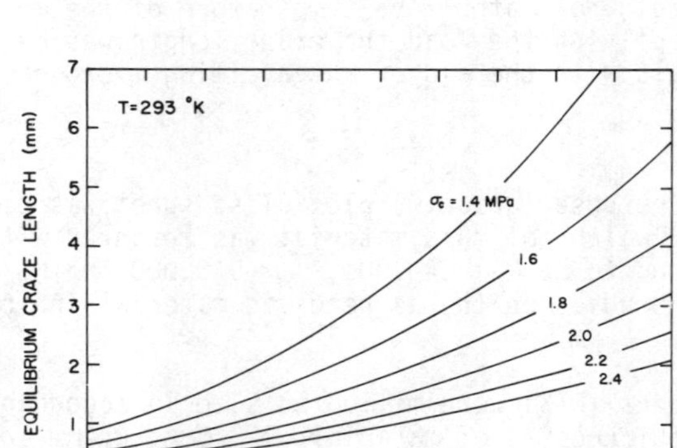

Fig. 2 Family of curves generated by equation 3 using $K_{th} = 0.11$ MPa-m$^{\frac{1}{2}}$, T = 293 K.

[1] Over the range of K_{th} used in this investigation the approximation differs from the exact solution by no more than 0.3%.

and other computer printouts at different K_{th}'s to the data, it was found that the best fit occurred with σ_c = 1.5 MPa and K_{th} = 0.11 MPa-$m^{1/2}$ (Rp = 0.83 μm). These parameters are very consistent with those which were found to give the best fit to the Andrews and Levy (1974) data.

Correlation of the 3-zone Model with Experimental Data

Representative results from the test program outlined above are shown in Fig. 3. It is immediately seen that the higher the initial applied stress intensity the longer the craze grows before equilibrium is reached. This behavior is consistent with the model and can be explained using the Barenblatt equilibrium expression appropriate to the 3-zone geometry, equation 2. In the cohesive term the range over which the integral is taken is equal to the plastic zone length, R_p. This value is independent of K_I and determined solely by the critical stress intensity for craze formation; consequently, the cohesive term remains constant. Through equation 6 the elastic term will increase in proportion to the increase in K_I. Assuming that the retractive stress exerted by the craze is constant over its length, an increase in the applied stress intensity must be offset by a corresponding increase in the semi-cohesive zone length. Another observation from Fig. 4 which lends support to this model is the increase in equilibrium craze length with increasing test temperature. Inspection of equation 3 indicates that σ_{ys}, σ_c and R_p are the only parameters that show a temperature dependence.[2] Using the temperature coefficient of the yield strength for dry PMMA reported by Vincent (1967), σ_{ys} was found to vary from 97.5 MPa (270°K) to 75.6 MPa (293°K). Assuming that σ_c changes in a proportional fashion, σ_c was found to vary from 1.935(270°K) to 1.5 MPa (293°K) over this temperature range. Through equation 3 it can be seen that to achieve an equilibrium craze at an increased temperature both the plastic zone and the semi-cohesive zone must increase in length.

Fig. 5 shows the fit of equation 3 using K_{th} = 0.11 MPa $m^{1/2}$ and σ_c = 1.5 MPa to the experimental data for all four test temperatures. It was surprising that the agreement was found to be quite good by simply using the temperature coefficient of σ_{ys} for dry PMMA and assuming that σ_c changes proportionately. One might think that over this range the temperature dependence of the diffusivity and solubility would appreciably change σ_{ys} and σ_c and thus be a factor in controlling

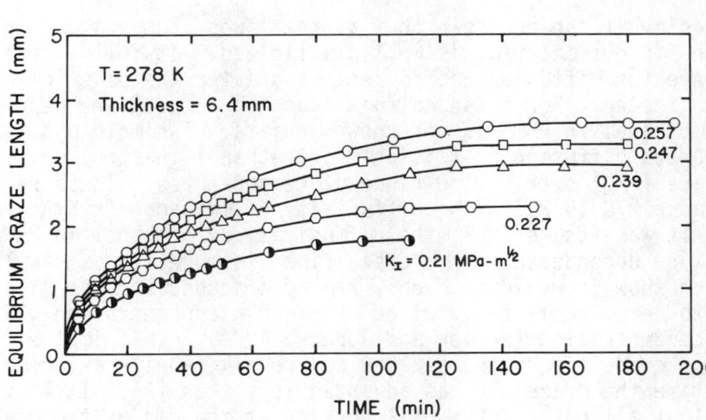

Fig. 3 Representative craze growth data showing the effects of initial applied stress intensity.

---

[2]The change in Rp occurs through the temperature dependence of σ_{ys}, equation 7.

equilibrium craze lengths. By simple diffusion calculations it can be shown that the conditions in the craze and plastic zone material are such that the temperature dependence of the diffusivity will have little effect on the average properties of these regions. In these calculations the methanol/PMMA system is assumed to show classical diffusion or Fickian behavior (Andrews, Levy and Willis, 1973). The diffusion coefficient of methanol in PMMA was taken as $5 \times 10^{-11} cm^2/sec$ (293°K), this lies between the value $10^{-11} cm^2/sec$ reported by Kramer and Bubeck (1978) and, a value of $1.8 \times 10^{-10} cm^2/sec$ extracted from methanol absorption isotherms presented by Andrews, Levy and Willis (1973). We are currently conducting research which intends quantitatively to determine the effects of temperature and stress on the diffusivity and solubility of methanol in PMMA. Another factor which must be dealt with in making these calculations is the fact that a complete state of static equilibrium was never obtained in these experiments. One can see from inspection of Fig. 3 that after approximately 100 minutes a quasi-equilibrium is reached where the craze tip velocity reaches a constant value of approximately $10^{-6} cm/sec$.

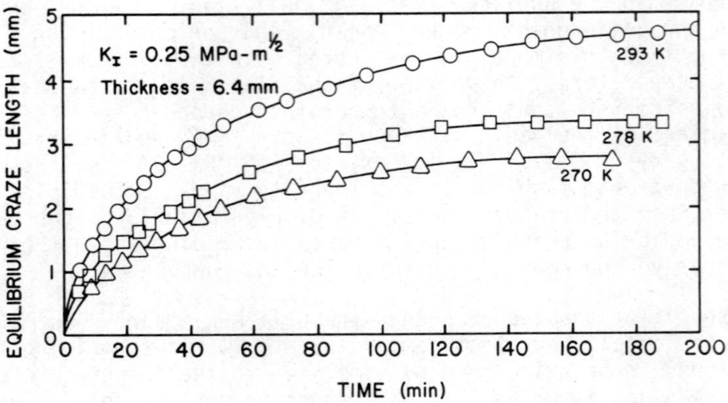

Fig. 4. Representative craze growth data showing the effects of test temperature.

For the craze region it can be shown that at the lowest temperature in the experimental range the fibrous material is well plasticized. It follows that at higher temperatures where the diffusion coefficient is greater the plasticization process would occur even faster. For these calculations the temperature dependence of the diffusivity of methanol in PMMA is not known. Fujita, Kishimoto and Matsumoto (1960) have reported diffusion coefficients of methanol in structurally similar polymethyl acrylate (PMA) over a wide range of temperatures. These results indicate that over the range 270 to 293°K the diffusivity of methanol in PMA undergoes a 30 fold increase. It was assumed that the diffusivity of methanol in PMMA shows a similar temperature dependence. If a craze fiber is modelled as a solid cylinder, calculations then show it would take only 3/4 of a second for the axis of a 500Å diameter fiber to become well plasticized or reach a concentration which is 50% of the surface concentration (Kramer and Bubeck, 1978). This implies that for a quasi-equilibrium craze tip moving at 10^{-6} cm/sec all fibrils are completely plasticized by the time the craze tip has advanced less than 75Å. It is therefore safe to assume that all but a very small portion at the tip of the craze is fully plasticized at 270°K. Increasing the temperature will only increase the rate of this process and consequently the temperature dependence of the diffusivity will not produce an additional decrease in σ_c.

The methanol concentration profile in the plastic zone preceding the craze can be estimated from the solution of a constant concentration moving boundary problem.

The solution to this problem is given by

$$c/c_o = 1/2 \left\{ 1 - \text{erf}\left(\frac{x}{2\sqrt{Dt}}\right) + e^{\frac{-v(x-vt)}{D}} \left[1 - \text{erf}\ \frac{x-2vt}{2\sqrt{Dt}} \right] \right\} \quad (8)$$

where the boundary (craze tip) is assumed to move at a constant velocity v and initially the boundary is located at $x = 0$ (Carslaw and Jaeger, 1959). Taking the growth rate v as 10^{-6} cm/sec and a time 150 minutes after the onset of craze growth a sufficient concentration for plasticization ($c/c_o \geq 0.5$) will extend 0.3 μm into the plastic zone at 293°K. As this is a small portion of the total plastic zone; it is concluded that in the temperature range of these tests the temperature dependence of methanol diffusivity has little effect on the properties of the plastic zone. The results of these calculations are also consistent with the values of σ_c and σ_{ys} used in evaluating the three-zone model in making the correlation with the experimental data. The material within the craze is well plasticized and consequently has a very low cohesive stress (σ_c = 1.5 MPa). Conversely, the concentration of methanol over most of the plastic zone is small and this region is assumed to have the properties of yielded bulk polymer.

Thickness Effects on the Quasi-Equilibrium Craze Lengths

At constant temperature and stress intensity it was noted that the quasi-equilibrium craze length increased with specimen thickness, Fig. 6. As sample thickness increases the state of stress at the craze tip approaches more closely plane strain behavior. The additional lateral constraint accompanying this transition has the effect of decreasing the plastic zone size. Under these conditions equation 3 predicts that a state of quasi-equilibrium can be reached only if the increase in specimen thickness is accompanied by an increase in craze length. Alternatively, other factors such as the state of stress on craze formation or solvent solubility may be involved.

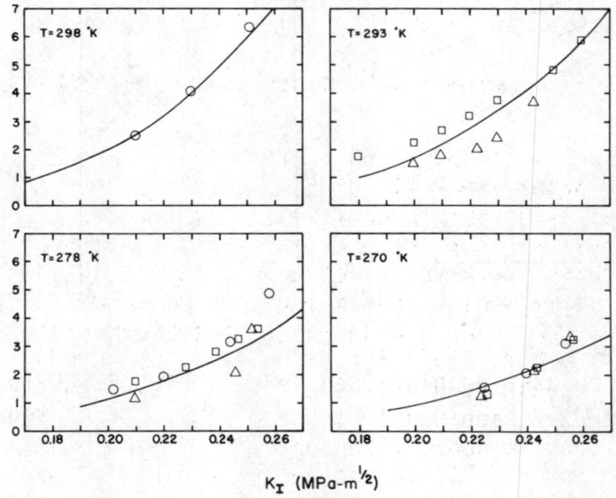

Fig. 5. Best fit of equation 3 to experimental data (K_{th} = 0.11 MPa-m$^{1/2}$), σ_c = 0.15 MPa); specimen thickness, ○ = 1.6mm, □ = 6.4mm, △ = 12.75mm.

Activation Energy Determination

Craze growth data were analyzed using the procedure outlined in a previous section. For constant specimen thickness a plot of $\ell n\ v$ versus $1/T$ yielded a family of straight lines for different stress intensities, Fig. 7. From the slopes of these lines it was found that E_{app} decreases with increasing stress intensity which is

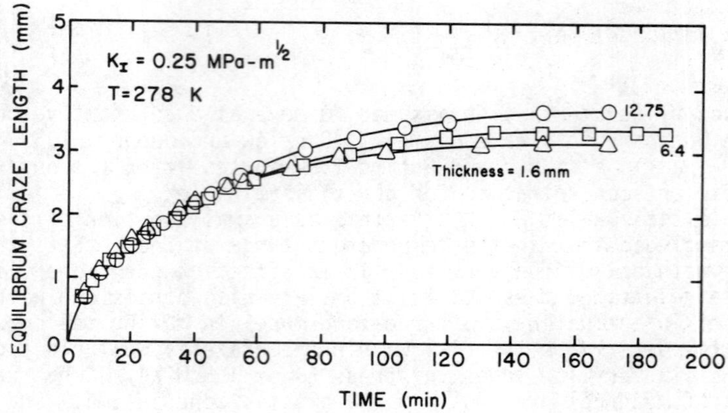

Fig. 6. Representative craze growth data showing the effects of specimen thickness

consistent with equation 4. For each specimen thickness the activation energy, E*, was found from a plot of E_{app} versus K_I as seen in Figure 8. Using a least squares technique to extrapolate this data the results indicate that the activation energy increases from 47.7 kJ/mol for the 1.6 mm thick samples to a constant value ≈ 75.3 kJ/mol for the thicker specimens.

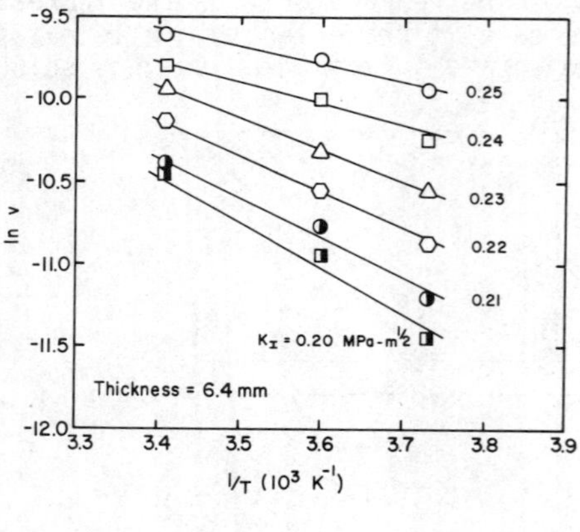

Fig. 7. Arrhenius type plot used to determine E_{app}.

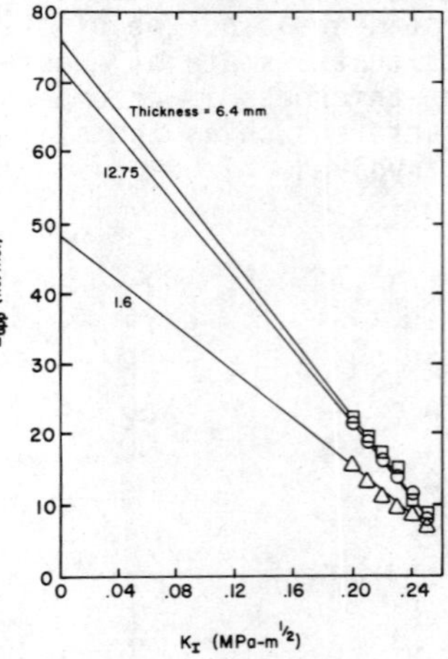

Fig. 8. Extrapolation technique used to determine E*.

The processes thought to control the rate of craze propagation are the diffusion of methanol into the newly formed fibers at the craze tip and, the diffusion controlled penetration of methanol into the plastic zone. Diffusion calculations similar to those used in the quasi-equilibrium craze length discussion indicate that for a craze velocity of 10^{-4} cm/sec there is a region approaching 0.8 μm in length of unplasticized fibrils behind the moving craze tip. Although this is generally less than 1% of the total craze length, if the yield stress of the material in this region averaged 50% of that for dry PMMA it could contribute up to 20% of the total retractive stress exerted by the craze. Consequently, the rate at which methanol diffuses into these fibrils now becomes a contributing factor in controlling craze growth. If we again assume that the methanol/PMMA system behaves like the methanol/PMA system investigated by Fujita, Kishimoto and Matsumoto (1960) we can obtain additional support for this hypothesis. Calculations show that the activation energy for methanol diffusion in PMMA is in the order of and even slightly higher than the activation energy for craze growth determined experimentally. The activation energy for methanol diffusion in PMA was calculated to be 102.1 kJ/mol, whereas for craze growth in the thicker PMMA samples, it was 75.3 kJ/mol.

Thickness Effects on Craze Propagation Rate

The increase in craze propagation rate with specimen thickness can be rationalized by considering the change in state of stress. Deformation in the thicker samples approaches more closely plane strain conditions. With consideration of the amount of energy dissipated plastically during crack propagation, under equivalent conditions the thicker sample with the smaller plastic zone would be expected to exhibit a larger craze velocity.

CONCLUSIONS

Quasi-Equilibrium Craze

1. An application of the modified Dugdale-Barenblatt model to quasi-equilibrium of a craze/plastic zone configuration is able to predict craze lengths for any given initial condition of applied stress intensity and crack length.
2. Scaling of the flow properties to the temperature dependence of the yield strength has allowed prediction of quasi-equilibrium craze lengths over the temperature range 270-298°K.
3. The increase in quasi-equilibrium craze length with increasing specimen thickness can be attributed to the state of stress approaching more closely plane strain conditions.

Propagating Craze

1. An analysis of craze growth data with a thermally activated crack growth model indicated that the activation energy for craze propagation ranges from 47.7 to 75.3 kJ/mol, increasing with specimen thickness.
2. Diffusion calculations and the correlation between the activation energies for diffusion and craze propagation both support the hypothesis that diffusion of the plasticizing solvent into the fibrils at the tip of the craze controls the rate of craze growth.
3. The increase in craze propagation rate with specimen thickness can be attributed to the decrease in the energy dissipated plastically in the smaller plastic zone produced under more nearly plane strain conditions.

REFERENCES

Andrews E. H. and G. M. Levy (1974). Solvent stress crazing in PMMA: 1. Geometrical effects. Polymer, 15, 599.

Argon A. S., J. G. Hannoosh and M. M. Salama (1977). In D.M.R. Taplin (Ed.), Fracture 1977, Vol. 1, University of Waterloo Press, Waterloo, Canada, p. 445.

Barenblatt G. I. (1962). The mathematical theory of equilibrium cracks in brittle fracture. In R. von Mises and T. von Karman (Eds.), Advances in Applied Mechanics, Vol. 7, Academic Press, New York, p. 30.

Bartenev G. M. (1955). Izv. Akad. Nauk S.S.S.R., Otdel. Tekh. Nauk, 9 53.

Bartenev G. M. and I. V. Razumovskaya (1966). The theory of time dependent strength of solids in surface-active media. In T. Yokobori, T. Kawasaki and J. L. Swedlow (Eds.), Proc. First International Conference on Fracture, Vol. 3, Japanese Society for Strength and Fracture of Materials, Sendai, p. 1871.

Bartenev G. M., I. V. Yudina and P. A. Rehbinder (1958). Theory of spontaneous dispersions of solids. Kolloid. Zh., 20, 655.
Brown H. R. and I. M. Ward (1973). Craze shape and fracture in poly(methyl methacrylate). Polymer, 14, 469.
Carslaw H. S. and J. C. Jaeger. In Conduction of Heat in Solids, Second edition, Oxford, London.
Charles R. J. and W. B. Hillig (1962). In Symposium on Mechanical Strength of Glass and Ways of Improving It, Florence Italy, Sept. 25-29, 1961, Union Scientific Continentale du Verre, Charleroi, Belgium.

Cotterell B. (1968). Fracture propagation in organic glasses. Int.J.Fract., 4, 209.
Doyle M. J. and J. G. Wagner (1976). Strain in a polystyrene craze layer. In R. D. Deanin and A. M. Crugnola (Eds.), Toughness and Brittleness in Plastics, American Chemical Society, Washington, p. 63.
Dugdale D. S. (1960). Yielding of steel sheets containing slits. J. Mech. Phys. Sol, 8, 100
Eshelby J. D. (1968). In Fracture Toughness, The Iron and Steel Institute, ISI Publication 121, London England, 55-129.
Fujita H., A. Kishimoto and K. Matsumoto (1960). Concentration and temperature dependence of diffusion coefficients for systems polymethyl acrylate and n-alkyl acetates. Trans. Faraday Soc., 56, 424.
Gerberich W. W. (1977). A Dugdale Barenblatt model for an equilibrium craze. Int. J. of Fracture, 13, 535.
Hillig W. B. and K. J. Charles (1965). In V. F. Zackay (Ed.), High Strength Materials, John Wiley and Sons, Inc., New York.
Irwin G. R. (1966). In NRL Memorandum Report 1678.
Israel S. J., E. L. Thomas, W. W. Gerberich (1979). Correlations of the craze profile in PMMA with Dugdale's plastic zone profile. in press J. Mat. Sci.
Knight A. C. (1965). Stress crazing of transparent plastics. Computed stresses at a nonvoid craze mark. J. Poly. Sci., A3, 1845.
Knott J. F. (1973) Fundamentals of Fracture Mechanics. Butterworth and Co., London.
Kramer E. J. and R. A. Bubeck (1978). Growth kinetics of solvent crazes in glassy polymers. Dept. of Mat. Sci. and Eng., Cornell Univ. report number 2933.
Mills N. J., N. Walker (1976). Craze growth and crack growth in poly(vinyl chloride) under monotonic and fatigue loading. Polymer, 17, 335.
Morgan G. P., I. M. Ward (1977). Temperature dependence of craze shape and fracture in poly(methyl methacrylate). Polymer, 18, 87.
Roetling J. A. (1965). Yield stress behavior of poly(ethyl methacrylate) in the glass transition region. Polymer, 6, 615.
Rohm and Haas Plexiglass G Product Literature, Rohm and Haas Company, Philadelphia
Schonert K., H. Umhauer and W. Klemm (1969). The influence of temperature and environment on slow crack propagation in glass. In P. L. Pratt (Ed.), Fracture 1969), Chapman and Hall Ltd., London, p. 474.
Vincent, P. I. (1967), In Encyclopedia of Polymer Science and Technology, Vol 7 Interscience, New York, p. 321.
Weidman G. W., W. Döll (1976). Abstracts from Third International Conference on Deformation, Yield and Fracture of Polymers, p. 12.1.
Wiederhorn S. M. (1974) In R. C. Brandt, D.P.H. Hasselman and F.F. Lange (Eds.), Fracture Mechanics of Ceramics, Vol. 2, Plenum Press, NY, p. 613.
Wiederhorn S. M. and L. H. Bolz (1970). J. Am. Ceram. Soc., 53(10), 543.
Wiederhorn S. M. and H. Johnson (1973a). J. Am. Ceram. Soc., 56(2), 108.
Wiederhorn S. M. and H. Johnson (1973b). Effect of electrolyte pH on crack propagation in glass. J. Am. Ceram. Soc., 56(4), 192.
Wiederhorn S. M., H. Johnson, A. M. Diness and A. H. Heuer (1974). J. Am. Ceram. Soc., 57, 336.
Wiederhorn S. M., H. Johnson, A. M. Diness and A. H. Heuer (1974). Fracture of glass in vacuum. J. Am. Ceram. Soc., 57, 336.
Williams J. G. (1977). A model of fatigue crack growth in olymers. Private communications.

Volume 3, ICM 3, Cambridge, England, August 1979

DESIGN OF CORONA 5 (Ti-4.5Al-5Mo-1.5Cr) TO MEET APPLICATION

F. H. Froes, C. F. Yolton*, J. C. Chesnutt, C. H. Hamilton** and W. T. Highberger*****

Air Force Materials Laboratory, Dayton, Ohio, U.S.A.
***Crucible Research Center, Pittsburgh, PA, U.S.A.**
****Science Center, Rockwell International, Thousand Oaks, CA, U.S.A.**
*****Naval Air Systems Command, Washington, D.C., U.S.A.**

INTRODUCTION

During the past few years the strong drive for increased performance in advanced aircraft, military aircraft in particular, has led to a necessity for materials with improved mechanical property combinations. The high strength to density ratio of titanium, in conjunction with excellent corrosion resistance and fracture behavior, has resulted in extensive use of this material in advanced systems. The Ti-6Al-4V alloy, developed twenty-five years ago and presently the workhorse of the USA industry, exhibits attractive mechanical property combinations in varied applications and is the bench mark against which other titanium alloys are judged.

A recently introduced titanium alloy, CORONA 5 (Ti-4.5Al-5Mo-1.5Cr), was originally developed for application in fracture critical components (Berryman, 1974, 1975, 1976, 1979). As such, chemistry/microstructure were designed to produce moderate strength levels (900-950 MPa ultimate tensile strength) with concurrent high fracture toughness (>110MPa$\sqrt{m}$). The attractive property combinations exhibited by the alloy have resulted in attempts to fabricate the alloy by advanced techniques such as superplastic forming and diffusion bonding. For optimum use of these forming methods a different type of microstructure is required from that which produces high fracture toughness. The present paper will discuss the basics of microstructure-property interrelationships in this alloy and will demonstrate how a knowledge of this understanding is allowing optimum utilization of the alloy in various advanced design concepts. To put the characteristics of the alloy in perspective, comparisons will be made with the workhorse Ti-6Al-4V alloy as used in equivalent design concepts.

ALLOY DESIGN

Chemistry

As discussed in the introduction, the CORONA 5 alloy was originally developed to produce, moderate strength combined with high fracture toughness (930MPa-110 MPa $\sqrt{m}$). However a secondary consideration was that the alloy should be capable of achieving higher strength levels (albeit at lower fracture toughness values). A spectrum of alloy chemistries ranging from lean alpha-beta to rich metastable beta compositions, was evaluated (Berryman, 1974). This range being chosen so that the required strength goals could be achieved. The alloys were further selected so that strength level was attained by dispersion strengthening (by the alpha phase

MS283

in the beta-matrix) rather than by solid solution strengthening (due to alloy additions in solid solution in the beta matrix). This criterion was used since it was recognized that dispersion strengthening should lead to a strength increase with a reduced fracture toughness debit compared to that achieved with solid solution strengthening. This a result of the tortuosity factor introduced by the alpha phase which increases the effective fracture toughness (Hirth, 1977).

As a result of this alloy selection facet of the program, the composition Ti-4.5Al-5Mo-1.5Cr was selected for scale-up to larger sized ingots and thermomechanical processing (TMP) optimization (including final heat-treatment). This material is an alpha-beta alloy with beta transus of approximately 925°C.

Thermomechanical Processing (TMP)

Based on an evaluation of a variety of TMP sequences, it was determined that the alpha phase within the beta matrix strongly influenced the microstructure and hence the mechanical properties. Two basic processing options are available: (a) to beta process (where processing is carried out wholly above the beta transus or in which finish processing is completed below the beta transus, but at a temperature high enough that very little alpha phase is present) or (b) to alpha-beta process (in which case processing is carried out below the beta transus temperature in the presence of alpha phase). After either processing route, annealing below the beta transus temperature (within about 175°C of the transus temperature) results in a distribution of primary alpha which is related to the processing sequence and annealing temperature. With beta processed material, a lenticular alpha morphology is achieved and maintained. With alpha-beta processing the primary alpha become globular during subsequent heat treatment. The change in morphology of the alpha phase from lenticular to globular is a direct result of the prior working (deformation) of the alpha. The strain energy in the alpha is thought to cause the alpha to recrystallize and in the course of recrystallization to relax to the lower surface energy globular configuration (Froes, 1979). This is demonstrated in Fig. 1. The rate at which the lenticular alpha transforms to globular alpha is a function of (a) annealing temperature, (b) annealing time and (c) the amount of working the alpha has received. Hence lightly worked alpha remains lenticular longer than heavily worked alpha. Figure 2 illustrates this effect for three levels of deformation. Time-temperature combinations to the bottom right of the appropriate deformation line produce lenticular alpha, while to the top left, globular alpha results.

It was found that while strength was virtually unaffected by the primary alpha, other properties such as fracture toughness and elevated temperature flow characteristics (particularly as applied to superplastic forming and diffusion bonding) were strongly influenced by the morphology of this type of alpha. High fracture toughness was associated with a high aspect ratio (lenticular) alpha, while lower fracture toughness values, at the same strength level, corresponded to a low aspect ratio (globular) alpha. This effect has been related to the tortuosity of the fracture path (Hirth, 1977). In direct contrast, optimum superplastic forming and diffusion bonding were found in material with a globular microstructure (Chesnutt, 1979). This latter effect can be related to the presence of a fine equi-axed two phase microstructure (Edington, 1976).

The final facet of the TMP is to age the alloy at temperatures 200-400°C below the beta transus to precipitate fine alpha between the coarse primary alpha regions. This treatment results in an increase in strength level which is proportional to the interparticle spacing of the fine alpha.

MECHANICAL PROPERTIES

During a series of contract (Berryman, 1974, 1975, 1976, 1979) and company (Froes, 1978a; Chesnutt 1979) sponsored programs, most of the significant mechanical properties of CORONA 5 have been developed. These include strength (yield and ultimate tensile strength), ductility (elongation and reduction in area from a tensile test), elastic modulus, fracture toughness, stress corrosion cracking susceptibility (K_{Iscc}), fatigue crack growth data (da/dN), smooth and notched fatigue (S-N), creep behavior (to 370°C) and elevated temperature flow characteristics (in the temperature range 825-925°C). Thus the potential of the alloy has been defined and appears to have many applications in the aerospace industry (Froes, 1978a). For the present paper, two of the mechanical properties of the alloy requiring quite different contrasting microstructures have been selected for consideration --high fracture toughness and superplastic forming/diffusion bonding behavior. Only these properties will be considered here, other properties can be found in the referenced work.

Fracture Toughness

Fracture toughness generally shows an inverse relationship with strength level. Hence maximum fracture toughness can be achieved by minimizing strength. Thus, by the appropriate selection of thermomechanical processing and final heat treatment parameters, CORONA 5 may be tailored to optimize fracture toughness, while meeting minimum strength levels which are dictated by design criterion. Conversely, the alloy may be tailored to provide a minimum fracture toughness level, as determined by inspectability imposed limitations on critical crack length, and the strength level may be maximized.

The knowledge of microstructure development described in the previous section has resulted in three recommended thermomechanical processing routes for high fracture toughness. The first (I) is either beta of alpha-beta processing followed by a beta anneal and subsequent high alpha-beta heat-treatment (to produce the required lenticular primary alpha) and final aging (to adjust the strength level). The second (II) is beta processing again followed by a high alpha-beta heat-treatment and age. And finally, the alloy may be alpha-beta processed (III). In this case a duplex alpha-beta anneal is required: the first anneal being close to the beta transus to globularize the alpha which has been worked and to minimize the volume fraction of this microstructural constituent. This is followed by a second lower temperature anneal to precipitate new lenticular alpha between the globular alpha particles. This duplex treatment is then followed by an aging cycle in a similar manner to the other two processes.

The three recommended processing sequences are shown schematically in Fig. 3 along with the microstructures resulting from each of these routes. Corresponding typical fracture toughness values are given in TABLE 1 (Berryman 1974, 1975, 1976). The fracture toughness levels after processing route III (alpha-beta) are somewhat lower than those resulting from the other two routes. This is a reflection of the proportion of primary alpha which is globular and thus does not contribute most effectively to producing a tortuous crack path. A further refinement of processing route III would be to anneal even closer (below) to the beta transus to minimize the proportion of globular alpha.

By adjusting the final aging treatment the strength level attained can be varied between 800 MPa UTS to a maximum in excess of 1400 MPa UTS. Throughout this strength range the same criterion for fracture toughness is found to apply -- lenticular primary alpha for high values, globular primary alpha for low values. The further effect of local yield strength (as dictated by the fine alpha dispersion) also influences fracture toughness in an inverse relationship (fracture toughness decreasing in direct proportion to strength level for a constant microstructure type (Hirth, 1977)). These effects are shown in Fig. 4 over the strength range 800-1400 MPa UTS and for primary alpha morphologies varying from lenticular to globular.

A comparison of CORONA 5 with Ti-6Al-4V is shown in TABLE 2. Here design allowables are compared (which generally are towards the low end of the "typical values" range) for the two alloys in a moderate strength condition. It can readily be seen that the CORONA 5 exhibits a significantly superior property combination to the Ti-6Al-4V. Of particular practical importance in design is the critical crack length parameter, $2a_c$, which for the CORONA 5 is twice that of the Ti-6Al-4V alloy at the same UTS level. For various reasons, including the lack of deep-hardening (Froes, 1978b), Ti-6Al-4V is not used in higher strength conditions. However, as discussed previously CORONA 5 may be readily thermomechanically processed to strengths >900 MPa UTS with acceptable fracture toughness. Thus the newer alloy should receive increased consideration for use in fracture critical applications where the general properties of titanium make this alloy type the designers choice. In fact, the superior property combination of CORONA 5 should result in some replacement of competitive materials such as steel in applications in which the characteristics of Ti-6Al-4V have not resulted in the selection of titanium.

Superplastic Forming/Diffusion Bonding (SPF/DB)

Because of extraction techniques and difficulties with both processing and fabrication, titanium is an inherently expensive material as currently used. Thus only when the superior mechanical characteristics significantly outweigh the cost disadvantages which exist in comparison with some competitive materials is titanium selected for use. Recently widespread results have shown that SPF with or without concurrent DB can result in substantial cost-savings (>50% of conventionally fabricated components) creating great interest in this fabrication method (Hamilton, 1976). To date, most work has concentrated on the Ti-6Al-4V in the temperature range 900-925°C (Hamilton, 1979). While the behavior of this alloy has been more than adequate to meet current design concepts, the imaginable innovations possible with SPF/DB have led to desires for improved performance. The two major directions for advance are (i) to lower temperature/lower pressure/shorter time fabrication and (ii) to a potential for hardening to higher strength levels subsequent to the SPF/DB operation. Consideration of the microstructural requirements for optimum SPF (and DB) behavior (Edington, 1976) indicates that the first advancement can be achieved by using a fine equiaxed microstructure with approximately equal properties of the two phases at the forming temperature. Because of the phase proportions present at various temperatures in the Ti-6Al-4V alloy and due to the inherent difficulty in imparting substantial amounts of strain to this material (which would lead to a very fine equi-axed microstructure on subsequent heat-treatment) advances of the required type are difficult in this alloy (though possible with imaginative processing procedures (Froes, 1978c) and agressive implementation (Hamilton, 1979)). The second advance is also difficult with Ti-6Al-4V for the reasons alluded to earlier. In contrast, CORONA 5 can exhibit the required phase proportions at lower temperature (~840°C) (Rosenblum, 1979), can be heavily worked to give a fine equi-axed microstructure (Chesnutt, 1979), and does have the potential to achieve high strength levels after forming (Chesnutt, 1979; Rosenblum, 1979).

The excellent chance for advancement in material performance using the CORONA 5 alloy led to a program to investigate the SPF/DB characteristics of the material (Chesnutt, 1979). Since the requirement for this application is a fine equi-axed microstructure, the required thermomechanical processing sequence is in direct contrast to that recommended for high fracture toughness (lenticular primary alpha). Here the requirement is substantial working in the alpha-beta region followed by annealing approximately 85°C below the beta transus temperature so that a fine globular alpha results.

Initial efforts were directed towards producing sheet by a warm rolling process. However, the material produced by this method was highly directional (Fig. 5) with superplastic behavior acceptable in the longitudinal direction but inadequate in the sheet transverse direction.

In order to eliminate this directionality it was decided to impart sufficient strain to the material that a subsequent anneal would result in complete recrystallization and uniform equi-axed fine microstructure. This was achieved by modifying the processing sequence to include a final cold rolling campaign. First attempts were unsuccessful due to the formation of strain induced orthorhombic martensite during the rolling operation which severely curtailed the amount of cold rolling possible (<10%). However, by annealing at 845°C for 4 hours prior to cold rolling, the beta matrix was sufficiently enriched in beta stabilizer (Mo and Cr) to prevent subsequent formation of the orthorhombic martensite during cold rolling.

After cold rolling approximately 40% and annealing at 845°C for 4 hours, a fine (~2 μm) equi-axed microstructure was produced (Fig. 6). A total of eleven tensile specimens machined from the cold-rolled fine-grained sheet, were tested to determine the superplastic properties (Hamilton, 1976) of the CORONA 5 sheet. The tests included the step/strain-rate test as well as total elongation test under constant strain-rate conditions (Hamilton, 1976). During the total elongation tests, the strain-rate was periodically and temporarily changed by 40% to permit a determination of the strain-rate sensitivity exponent, m, as a function of strain.

The m vs. $\dot{\varepsilon}$ curves for these data are shown in Fig. 7. For the temperature of 870°C, the m values appear to be superior to Ti-6Al-4V especially over the higher strain rates (e.g. above $10^{-4}s^{-1}$), exhibiting somewhat higher maxima in m than the Ti-6Al-4V alloy. In comparison with the Ti-6Al-4V alloy at 925°C, the CORONA 5 m values are somewhat lower. It is, however, significant that the m for CORONA 5 at 870°C is in excess of 0.5 for strain rates below about $6X10^{-4}$ up to $10^{-3}s^{-1}$, rates for which corresponding forming times are considered to be practical.

The results of the total elongation tests showed that elongations of 280% to 500% were possible at 870°C for strain rates of 10^{-3} and $2X10^{-4}$ respectively. These values confirm that significant superplastic forming of this alloy is possible. Very recently, a small pan was fabricated from the CORONA 5 alloy at 870°C using the differential gas pressure superplastic forming process. Complete results will be reported elsewhere (Chesnutt, 1979).

Preliminary determinations of the diffusion bonding characteristics suggest that similar comparisons exist with Ti-6Al-4V as were observed in the case of superplastic forming behavior. That is, flow stress for CORONA 5 at 870°C will be somewhat higher than for Ti-6Al-4V at 925°C. This would dictate an increase in diffusion bonding pressure and/or time compared to Ti-6Al-4V (the latter alloy at the higher temperature).

As discussed previously, the Ti-6Al-4V alloy to date has only been used in the as fabricated condition after SPF or SPF/DB. This is, in part, a result of the rapid decomposition characteristics of the alloy and hence low potential for aging following cooling from the forming temperature (Froes, 1978b). In contrast, CORONA 5 exhibits slower beta to alpha decomposition and a greater inherent hardening capability (Rosenblum, 1979). As a result, work is in progress further to enhance the usefulness of CORONA 5 after SPF or SPF/DB by defining past-fabrication heat-treatments to allow the material to be used at high strength levels ($\geq$ 1060 MPa UTS) (Rosenblum, 1979). A further refinement will be to design these heat treatments to develop a lenticular microstructure for fracture critical applications. This will be done using a duplex high-low sub-beta transus duplex anneal followed by an air cool (to avoid the distortion resulting from water quenching) and final aging treatment.

SUMMARY AND CONCLUSIONS

The thermomechanical processing - microstructure - mechanical property interrelationships for the titanium alloy CORONA 5 (Ti-4.5Al-5Mo-1.5Cr) have been presented. It has been demonstrated how two quite different mechanical properties - fracture toughness and elevated temperature flow characteristics -- can be optimized by producing two basically different microstructures -- lenticular alpha in a beta matrix and a fine equi-axed alpha-beta mixture respectively. The salient mechanical properties of CORONA 5 have been compared with the workhorse Ti-6Al-4V characteristics, in each case demonstrating the attractive performance of the newer alloy. Prospects for further improvements after SPF or SPF/DB have been defined which should lead to early use of the alloy.

Finally, it can be concluded that the present work demonstrates how the metallurgist can contribute to effective material utilization not only by selection of the correct alloy system and specific chemistry, but also by designing thermomechanical processing to obtain the desired microstructure and hence mechanical properties.

ACKNOWLEDGEMENTS

The authors gratefully acknowledge helpful discussions with Professor J. C. Williams, Carnegie-Mellon University; Professor J. P. Hirth, The Ohio State University and M. E. Rosenblum, University of Cincinnati. The work was supported by Naval Air Systems Command under contracts N00019-73-C-0335, N00019-74-C-0273, N00019-75-C-0208 and N00019-76-C-0427. Thanks are due to the Naval Air Systems Command for permission to publish this paper.

REFERENCES

Berryman, R. G., F. H. Froes, J. C. Chesnutt, C. G. Rhodes, J. C. Williams and R. F. Malone (1974). High-fracture toughness titanium alloy development. Final Engineering Report, No. TFD-74-657, to Naval Air Systems Command. Contract N00019-73-C-0335.

Berryman, R. G., F. H. Froes, J. C. Chesnutt, C. G. Rhodes, J. C. Williams and R. F. Malone (1975). High-fracture toughness titanium alloy development. Final Engineering Report, No. TFD-75-640, to Naval Air Systems Command. Contract N00019-74-C-0273.

Berryman, R. G., J. C. Chesnutt, F. H. Froes and C. G. Rhodes (1976). High-fracture toughness titanium alloy development. Final Engineering Report, No. TFD-76-471, to Naval Air Systems Command. Contract N00019-75-C-0208.

Berryman, R. G., G. R. Keller, J. C. Chesnutt, F. H. Froes and C. G. Rhodes (1979). The development of a new titanium alloy with high fracture toughness. Final Engineering Report to Naval Air Systems Command. Contract N00019-76-C-0427. Under Preparation.

Chesnutt, J. C., C. H. Hamilton, C. F. Yolton and F. H. Froes (1979). Superplastic forming/Diffusion Bonding (SPF/DB) of CORONA 5 (Ti-4.5Al-5Mo-1.5Cr). To be submitted for publication in Metallurgical Transactions.

Edington, J. W., K. N. Melton and C. P. Cutler (1976). Superplasticity. Progress in Material Science, 21, 63-146.

Froes, F. H., J. C. Chesnutt, R. G. Berryman, G. R. Keller and W. T. Highberger (1978a). CORONA 5 (Ti-4.5Al-5Mo-1.5Cr): An alpha-beta titanium alloy with improved properties. Proceedings of 10th National SAMPE technical conference, SAMPE National Business Office, Azusa, Ca. pp. 522-533.

Froes, F. H., R. F. Malone, J. C. Williams, M. A. Greenfield and J. P. Hirth (1978b). Alloy design and property interrelations of metastable beta titanium alloys. Proceedings of forging and properties of aerospace materials, The Chameleon Press, London. pp. 143-163.

Froes, F. H., C. F. Yolton, J. C. Chesnutt and C. H. Hamilton (1978c). Microstructural control in titanium alloys for superplastic behavior. Proceedings of forging and properties of aerospace materials, The Chameleon Press, London. pp. 371-398.

Froes, F. H., J. C. Chesnutt and J. C. Williams (1979). Effect of thermomechanical processing on the shape of alpha titanium particles in a beta titanium matrix. To be submitted for publication in Metallurgical Transactions.

Hamilton, C. H. and G. W. Stacher (1976). Superplastic forming of Ti-6Al-4V beam frames. Metal Progress, March, pp. 34-37.

Hamilton, C. H., J. C. Chesnutt, C. F. Yolton and F. H. Froes (1979). Superplastic forming of cold rolled Ti-6Al-4V sheet. To be submitted for publication in Metallurgical Transactions.

Hirth, J. P. and F. H. Froes (1977). Interrelations between fracture toughness and other mechanical properties in titanium alloys. Metallurgical Transactions, 8A, 1165-1176.

Rosenblum, M. E., C. F. Yolton and F. H. Froes (1979). Microstructure and mechanical properties of fine-grained CORONA 5 (Ti-4.5Al-5Mo-1.5Cr). To be submitted for publication in Metallurgical Transactions.

TABLE 1

FRACTURE TOUGHNESS OF CORONA 5 AFTER OPTIMUM PROCESSING

	Processing Route	Product Form*	Fracture Toughness ($MPa\sqrt{m}$)	Tensile YS (MPa)	Strength UTS (MPa)
I.	Beta Anneal	Plate[1]	129	775	865
		Plate[2]	156	860	950
II.	Beta Processed	Plate[1]	130	920	980
		Plate[2]	145	760	855
		Plate[3]	160	890	1000
III.	Alpha-Beta Processed	Plate[4]	111	905	935

* Numbers refer to references: (1) Berryman (1974), (2) Berryman (1975), (3) Berryman (1976) and (4) Berryman (1979).

TABLE 2

COMPARISON OF CORONA 5 and Ti-6Al-4V

DESIGN ALLOWABLES FOR FRACTURE CRITICAL APPLICATIONS

Parameter	CORONA 5	Ti-6Al-4V
UTS (MPa)	900	900
YS (MPa)	825	795
K_{Ic} ($MPa\sqrt{m}$)	120	85
E ($MPa \times 10^3$)	110	110
ρ (GM/cc)	4.5	4.4
$2\,a_c^*$ (m)	0.26	0.13

$$* \; 2\,a_c = \frac{2}{\pi}\left\{\frac{K_{Ic}}{2/3 \times UTS}\right\}^2 \text{, meters.}$$

This is a critical crack length design parameter.

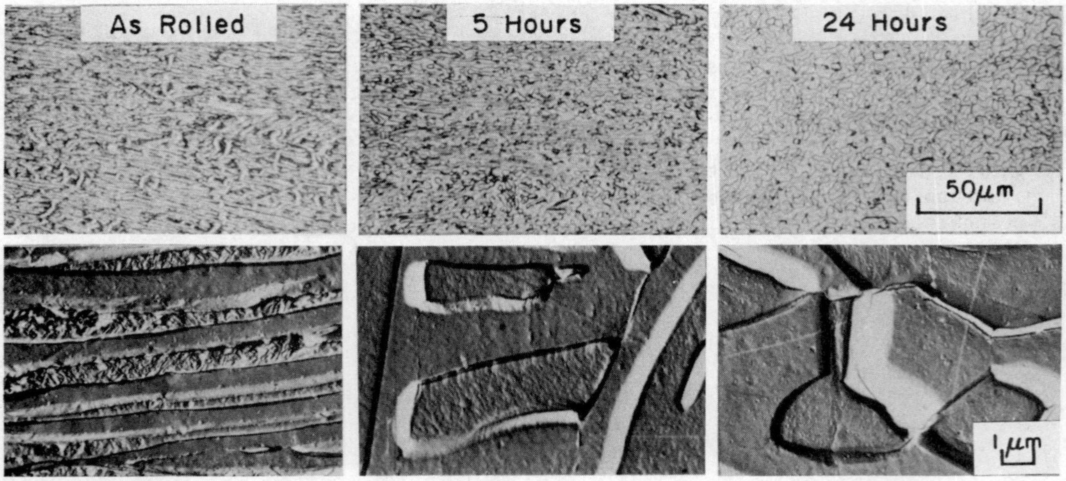

Figure 1 - Transformation of lenticular alpha to globular alpha by holding at 800°C after deforming material 70% in the alpha-beta field.

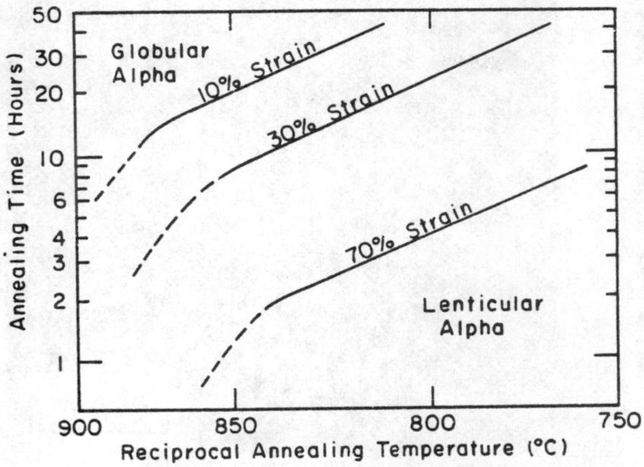

Figure 2 - Relationship of alpha morphology to annealing time and temperature after various amounts of deformation (strain) in the alpha-beta phase field.

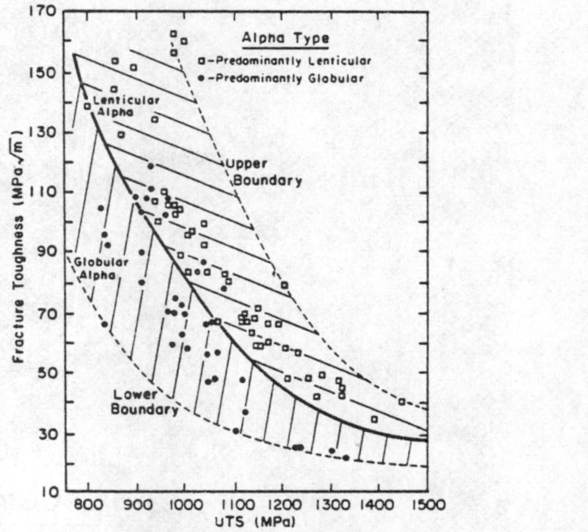

Figure 4 - Variation of fracture toughness with strength level for CORONA 5. At a given strength level, high fracture toughness results from lenticular primary alpha; low fracture toughness from globular primary alpha.

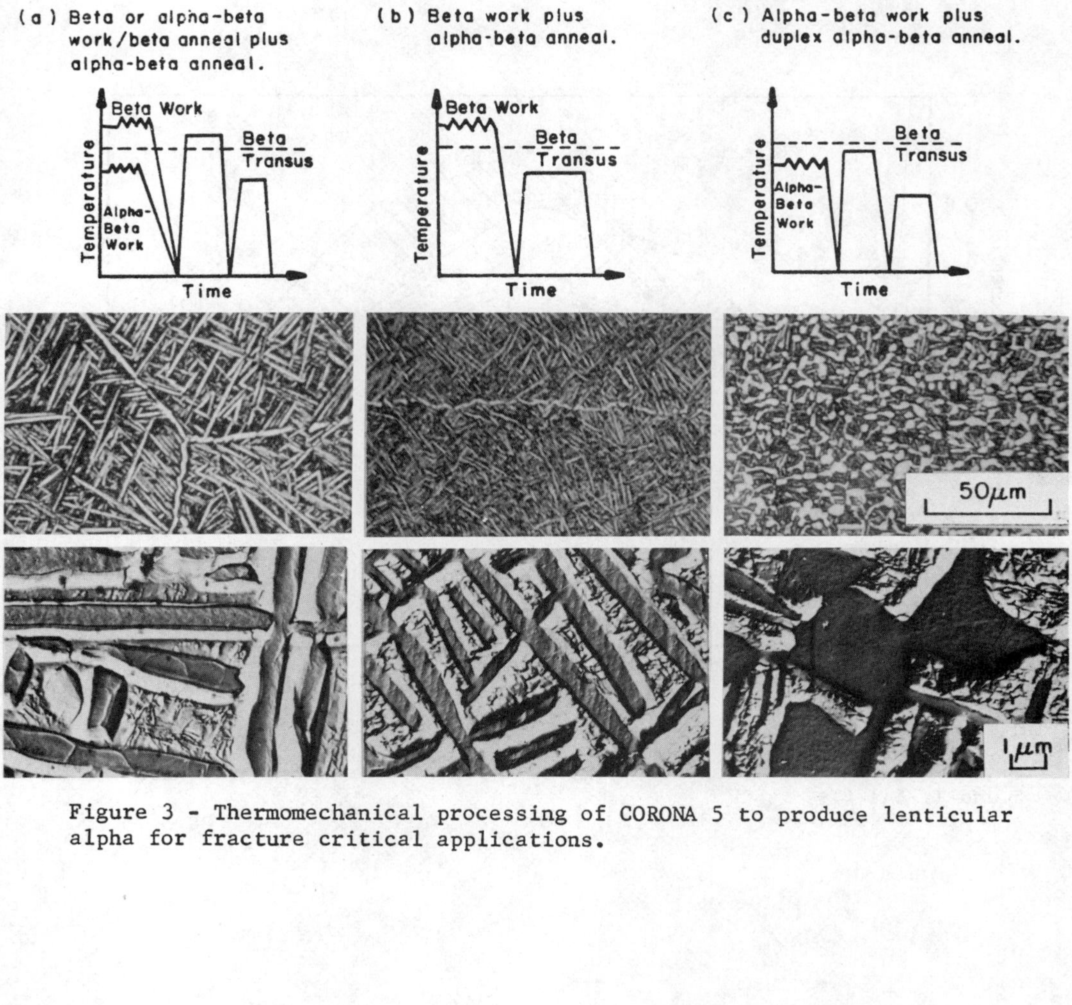

Figure 3 - Thermomechanical processing of CORONA 5 to produce lenticular alpha for fracture critical applications.

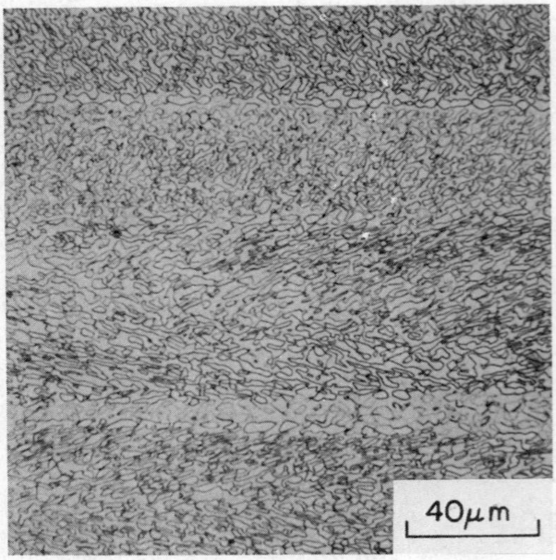

Figure 5 - CORONA 5 sheet produced by a warm rolling sequence. Note directionality of microstructure.

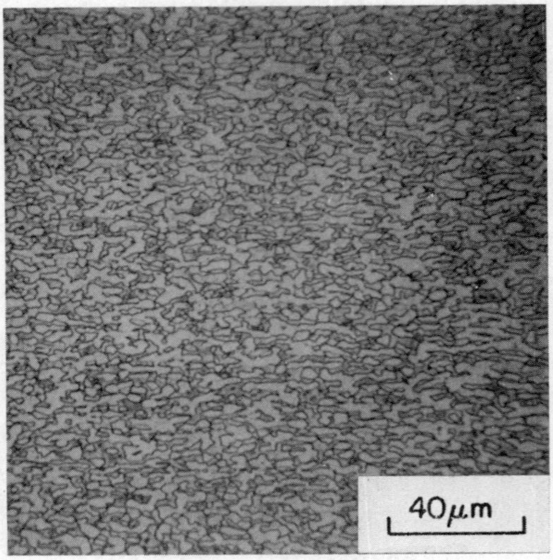

Figure 6 - CORONA 5, produced by final cold rolling sequence. Note fine equiaxed microstructure.

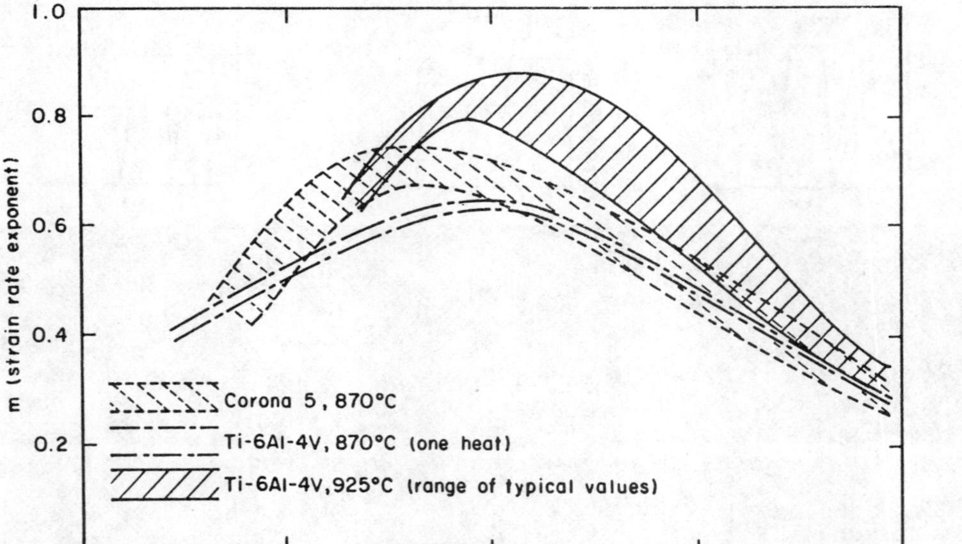

Figure 7 - Strain rate exponent (m) as a function of strain rate for CORONA 5 and Ti-6Al-4V.

THEME VI

FAST FRACTURE IN DUCTILE MATERIALS

Volume 3, ICM 3, Cambridge, England, August 1979

CHARPY ENERGY TRANSITIONS FOR WELD METAL AND HEAT AFFECTED ZONES IN SA533B CLASS 1 PLATE

I. Milne and D. A. Curry

Central Electricity Generating Board, Central Electricity Research Laboratories, Kelvin Avenue, Leatherhead, Surrey, U.K.

INTRODUCTION

In a welded structure crack-like defects are frequently associated with welds or heat affected zones, where material properties are at their most uncertain. Fracture toughness testing is costly and time consuming. For heat affected zones, K_{1C} can be measured only with difficulty and not a little good fortune since the shape of the HAZ precludes accurate location of the fatigue crack tip. Moreover at temperatures where K_{1C} is high, or where cracks propagate in a ductile mode under monotonic loading, plastic zones develop outside the HAZ. Here even with satisfactory crack tip positioning the resulting K_{1C} or J_{1C} is of questionable validity.

A programme has been designed to investigate elastic-plastic effects in fracture toughness testing of submerged and manual metal arc weld metal and HAZs located in 130mm thick pressure vessel steel plate. To characterise the material comprehensively, and provide the basis for a Charpy test - fracture toughness test correlation, Charpy tests have been performed on each weld metal and heat affected zone. In addition Charpy tests have also been performed on parent plate and simulated HAZs obtained by austenitizing 2 blocks of plate material for 30 minutes at 1250°C and water quenching. For the latter series of tests, following stress relief annealing for 3 hours at 600°C, one of these blocks of material was water quenched and the other cooled at 20°C per hour.

This paper describes these Charpy tests and with the aid of the data currently available from the fracture toughness testing programme examines the Charpy-Fracture Toughness correlations commonly used for failure assessment.

MATERIAL AND SPECIMENS

The plate was supplied to meet the ASTM specification SA533B Class 1, with a yield stress of 450 MPa and an ultimate tensile stress of 600 MPa. It was prepared for welding by grinding to form one vertical face and one face inclined at 7° to the vertical, the vertical face being designed to test the HAZ properties. Both welds were made using commercially available consumables, the root gap being 30mm for the submerged arc (SA) process and 20mm for the manual metal arc (MMA) process. Preheat was 200°C with a maximum interpass temperature at 300°C. Each weldment was finally stress relieved for 3 hours at 600°C and cooled at 20°C per hour.

RESULTS

Hardness Profiles

Hardness profiles were measured across selected heat affected zones for both types of weldments. As expected these displayed the general pattern of a hardness peak of 280-320 VPN in the HAZs; the weld metals, at 230 VPN, were about 30 VPN harder than the plate. In the weld metals the hardness was of course more variable than in the plate because of the greater variety of microstructures in the welds.

Charpy Energy Data

Figures 1a-f display the Charpy energy data; for the weldments each symbol represents a specific layer through the thickness, as indicated in Fig. 1c. There was no general layer effect although the surface level of the MMA HAZ (Fig. 1f) was consistently tougher than the remaining HAZs. From these Figs. the following points are of interest:

(a) In terms of ductile-brittle transition temperature and upper shelf level the toughest material was the plate as stress relieved.

(b) The upper bounding line to the HAZ data from both weldments was similar to the upper bounding line of the plate data.

(c) The least tough material was also from the plate, but after heat treating to produce the simulated HAZ structure, stress relief annealing and slow cooling (at 20°C per min.).

(d) The lower bounding line from either set of weld metal data was similar to the lower bounding line of (c) above, as indeed was the lower bounding line of both of the HAZ data sets.

(e) In terms of data scatter at the mid-transition point, the HAZ data exhibited much more scatter than data from any of the other microstructures (see Table 1).

The ductile brittle transition temperature can be expressed in many ways; we have chosen three, the 50% fracture appearance transition temperature, FATT, and the temperature of the lower bounding line at 40 joules and at 68 joules. In this latter case, all of the specimens at and above this energy level exhibited greater than 0.89mm lateral expansion, so this temperature could be related directly to the reference nil-ductility temperature RT_{NDT}, as defined in the ASME (1977) Boiler and Pressure Vessel Codes. The 40 joule temperature is used for defining a lower bound K_{1C} correlation in the draft B.S.I. Public Document (1978). These temperatures are all listed in Table 1, along with the upper shelf energy levels.

Fractography

Weld metals. Although the SA weld metals exhibited coarser brittle fractographic features than the MMA there was little difference in the general appearance of the fracture samples of the two welds. On the macro scale, when fracture was brittle, the underlying columnar and equiaxed structures were evident on the fracture surface. However at higher magnification the scanning electron microscope revealed that all the brittle failure was by cleavage rather than along grain boundaries. The regions of ductile fracture in mid-transition and on the upper shelf were entirely by ductile dimpling, each dimple containing a small second phase particle. An EDAX analysis of these showed that they contained a high level of sulphur and manganese.

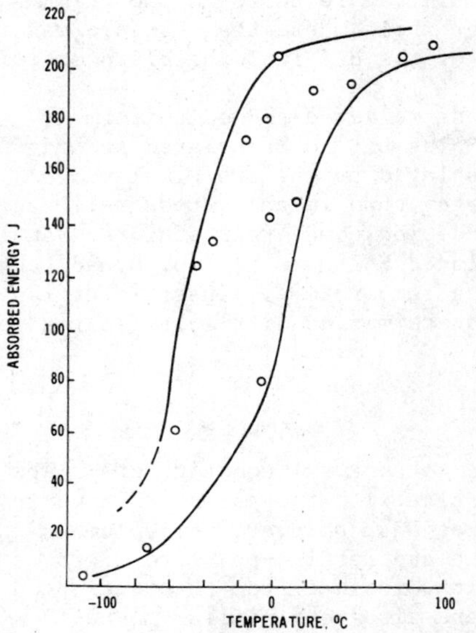

Fig. 1a. Chapry data from plate.

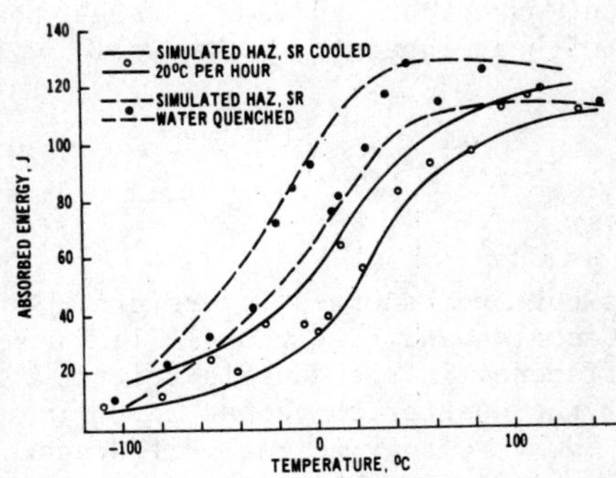

Fig. 1b. Charpy data from simulated HAZ

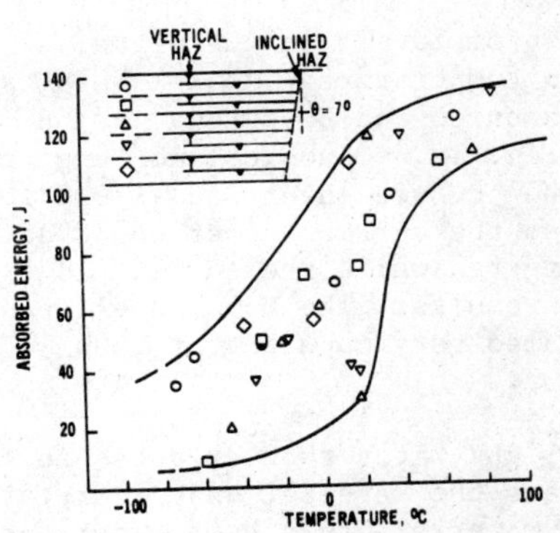

Fig. 1c. Charpy data from Sub Arc Weld Metal

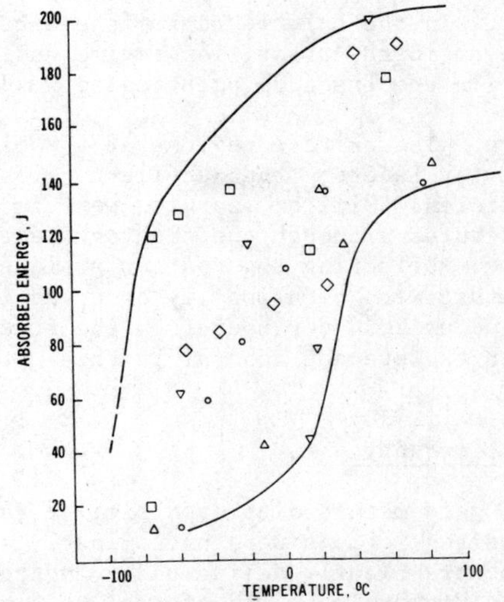

Fig. 1d. Charpy data from Sub Arc HAZ

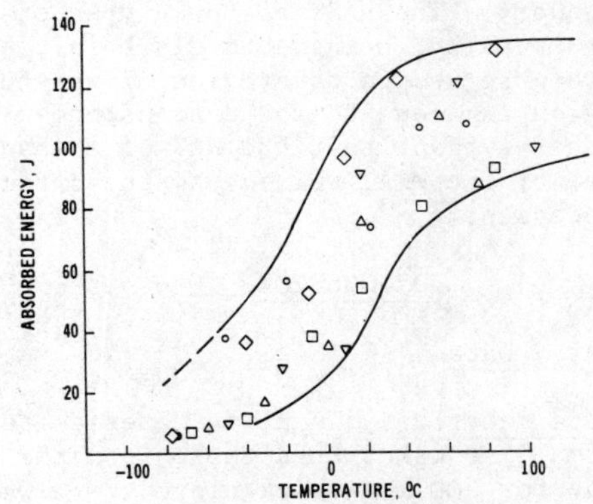

Fig. 1e. Charpy data from MMA Weld Metal

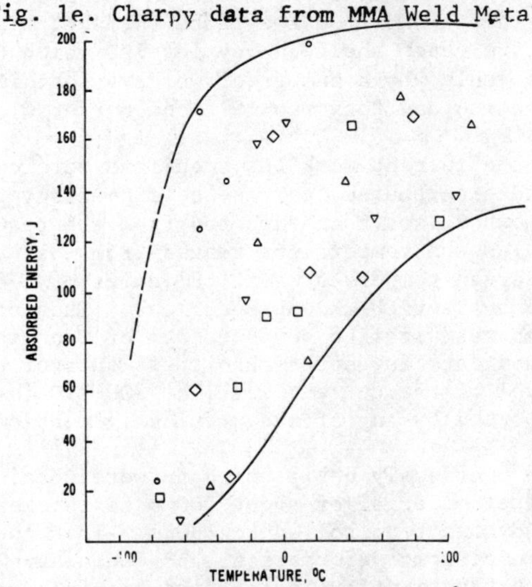

Fig. 1f. Charpy data from MMA HAZ

HAZs. In the brittle regime the HAZs displayed similar features to the weld metals. However in the transition regime and on the upper shelf where the fracture was ductile the fracture path tended to deviate out of the HAZ into the plate material.

Plate. The brittle regions of the plate as stress relieved showed uniform cleavage facets; the ductile regions showed dimples but no associated second phase particles. Similar features were in general displayed by the two simulated HAZ structures although the cleavage facets were larger than in the stress relieved plate, reflecting the coarser grain size. In addition some areas of intergranular fracture were occasionally observed in the simulated HAZ material which had been cooled at 20°C per hour from the stress relieving temperature. These intergranular features were not general in this material but there was no difficulty in finding them.

Metallography

Both weld metals displayed similar ranges of microstructural constituents. The as-deposited regions were predominatly acicular ferrite with the parent columnar grain structure clearly delineated. In the reheated regions, however, the columnar structure had been transformed to a mixed bainite and ferrite-pearlite microstructure whose coarseness varied as expected with distance from the inter-bead boundary. The HAZs of both types of weldment were also very similar microstructurally, being entirely bainitic. Again as expected, very large grain sizes were observed in the region of the fusion zone in contrast to the grain refined region adjacent to the untransformed plate microstructure. On a macroscopic scale it was evident that the HAZ of the sub arc weldment was wider and less straight than that of the MMA, reflecting the different levels of heat input of the two welding processes.

DISCUSSION

Charpy Data

Plate material. In plate material test specimens cut from surface layers are normally tougher than those from the interior. Canonico and Berggren (1971) found that for 300 mm thick plate there was a 90°C difference in the ductile-brittle transition temperature between specimens cut from the quarter thickness level in the plate and from the plate surface. There was however, only a small difference in the upper shelf energy levels, which have been repeatedly observed to be about 125 joules by a number of workers spanning the last 10 years (e.g. Wessel 1969, Canonico and Berggren 1971 and Server, Oldfield and Wullaert 1977).

In the current work the specimens were cut from random levels in the plate, producing the scatter in the mid-energy transition temperature. At 60°C this was somewhat less than expected from the results of Canonico and Berggren. At 68 joules the temperature ranged from -65° to -6°C, Fig. 1a. Canonico and Berggren observed temperatures at 68 joules of -70°C for the surface layers and +20°C for the quarter thickness specimens. Thus at 68 joules the current upper bound data compares with the surface data of Canonico and Berggren while the current lower bound data is better than their quarter thickness results. The upper shelf energy level of the current work, at 200-210 joules, was also very much better than reported by any of the previously mentioned workers.

The previously published data were obtained from light water reactor pressure vessel material of about 300mm in thickness, whereas the material under current study had been rolled to 130mm. Thus the present material would have experienced more plastic deformation, and would have cooled at a more uniform rate during the heat treatment cycles; these two factors are likely to be the cause of the observed improvements in the material properties, especially at the mid-plate thickness, and in their uniformity.

Weld metal. The data of Wessel (1969) Canonico and Berggren (1971) and Server Oldfield and Wullaert (1977) all indicate that weld metal produces superior Charpy data to the plate material. The present results contradict this, both from the viewpoint of the ductile-brittle transition temperature and the upper shelf energy levels. For example, in the current work there was an upward shift of 40°C in d/B transition between the plate and either weld metal while both Wessel (1969) and Canonico and Berggren (1971) obtained a downward shift of similar magnitude. The upper shelf level behaved similarly, reducing from about 200 to between 95 and 130 joules in the present work while all the aforementioned workers observed an increase.

Unlike the plate properties, the properties of the weld metal will not be influenced by thickness differences, so it must be concluded that the weld metals under current study have inferior Charpy properties to those referred to above. This, together with the improved properties of the thinner plate produced the above contradiction. Whether or not this produces inferior fracture toughness data remains to be seen.

Embrittlement effects. The lower bound curves of all the data derived from either of the weldments were broadly delineated by the data from the simulated HAZ material, when slowly cooled. This was displaced +30°C from the water quenched simulated HAZ data, indicating some degree of temper embrittlement. The regions of intergranular fracture on the fracture surfaces of the slowly cooled simulated HAZ specimens substantiate this belief. The lack of similar fractographic evidence in the HAZ or weld specimens taken from the weldments does not, however, mean temper embrittlement is absent in these weldments. The complex overlapping microstructures make intergranular fracture both difficult to identify and unlikely to be sustained over a large area.

The treatment chosen to simulate the HAZ structures was designed to reproduce, as far as was observable, both the microstructure and the hardness found in the HAZs of the weldments (Brear 1976). In reproducing both the ductile brittle transition temperature and the upper shelf level of the welds and lower bound HAZ data, this treatment can be considered to have been highly successful.

Metallography and Fracture Processes

The columnar features observed on the brittle fracture surfaces of both weld metals clearly correspond to the as deposited regions of acicular ferrite. The reheated region of the weld metal will have been subjected to temperatures and cooling rates closely similar to those experienced by the HAZs during subsequent bead deposition. In consequence, it is not surprising that, with similar coarse bainitic regions, the weld metals and the simulated HAZs should have displayed similar toughness over the lower half of the ductile brittle transition. This is despite the fact that there was no evidence of intergranular fracture in the weld metal. As previously noted this does not completely preclude the possibility of temper embrittlement in the welds. Certainly there were plenty of regions in the slowly cooled simulated HAZ structures that failed by cleavage suggesting that the temper embrittlement in this structure was neither universal nor entirely necessary for the development of this lower bound curve.

What is, however, surprising is the similarity of the upper shelf levels of both the weld metals and the simulated HAZ structures. A high volume fraction of inclusions in the weld metals was indicated by the presence of many inclusions on the ductile fracture surfaces, whereas inclusions were difficult to find in the simulated HAZs. Thus it might be expected that the weld metals were less ductile than the cleaner simulated HAZs. The simulated HAZs were substantially harder than the weld metals (320 as opposed to 230 VPN) reflecting the exclusively bainitic micro structure of the former and the mixed, less lenticular microstructures of the latter. It is possible that the higher ductility associated with the acicular ferrite of the weld metals compensates for the higher strength and lower inclusion content of the bainitic simulated HAZ structure, resulting in similar upper shelf energy levels.

Fracture Toughness and Impact Data Correlations

For fracture mechanics assessments, lower bound values of fracture toughness are needed. In the transition regime there are two appropriate correlations, BSI (1978) and Jones (1972), which use respectively the temperature at 40 joules and the 50% fracture appearance transition temperature, FATT. A further correlation can be inferred from the ASME (1977) codes using the requirement in article NB2331 of Section III that at 33°C above RT_{NDT} the Charpy energy should exceed 68 joules, together with the K_{1C} curve of Section XI. These correlations, when superimposed on each other, Fig. 2, imply the following equivalences:

$$T_{40J} = RT_{NDT} - 40^{\circ}C = T_{68J} - 73^{\circ}C = FATT - 70^{\circ}C \qquad (1)$$

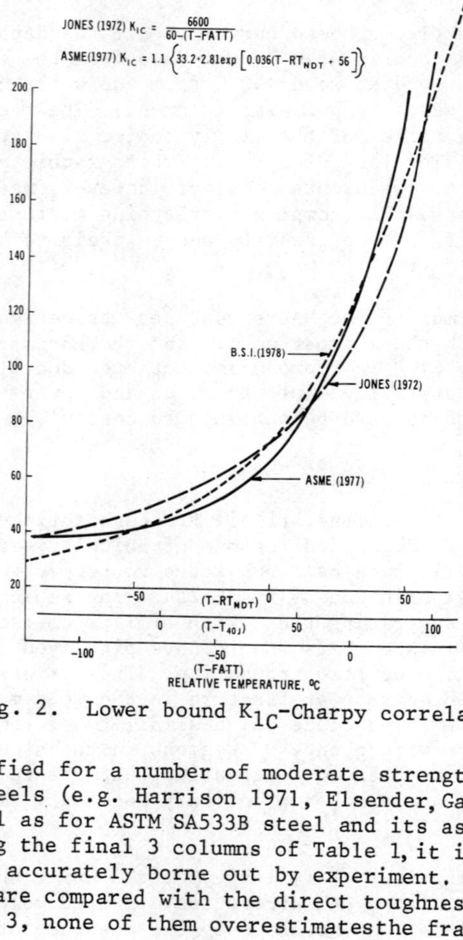

Fig. 2. Lower bound K_{1C}-Charpy correlations.

They have been justified for a number of moderate strength bainitic and fine grained pearlitic steels (e.g. Harrison 1971, Elsender, Gallimore, Murphy and Poynton 1978) as well as for ASTM SA533B steel and its associated weldments. However, by comparing the final 3 columns of Table 1, it is apparent that these equivalences are not accurately borne out by experiment. Nevertheless, when the Charpy correlations are compared with the direct toughness measurements of Milne and Curry 1979, Fig. 3, none of them overestimatesthe fracture toughness.

It is apparent from Fig. 2 that as the temperature difference expression on the abscissa increases, all of the correlations represented in equation 1 rise to infinity so that the upper level of toughness determined by the onset of ductile tearing cannot be defined by these expressions. In the ductile tearing regime, there is only one correlation claiming to predict lower bound toughness data, that in the BSI (1978) document Fig. 4. This requires Charpy energies at the temperature

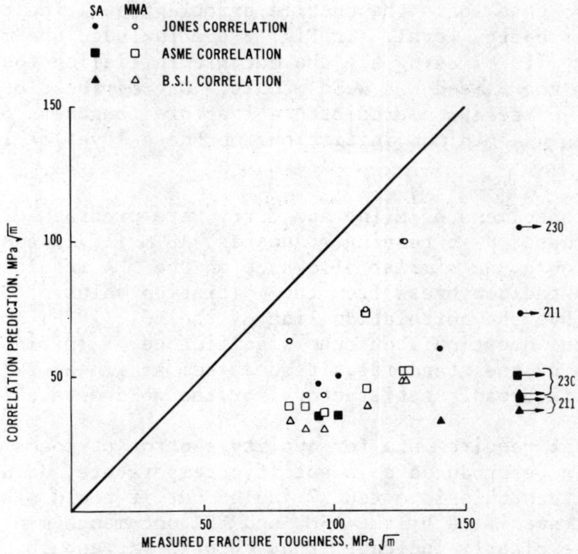

Fig. 3. Plot of correlated versus measured K_{1C}.

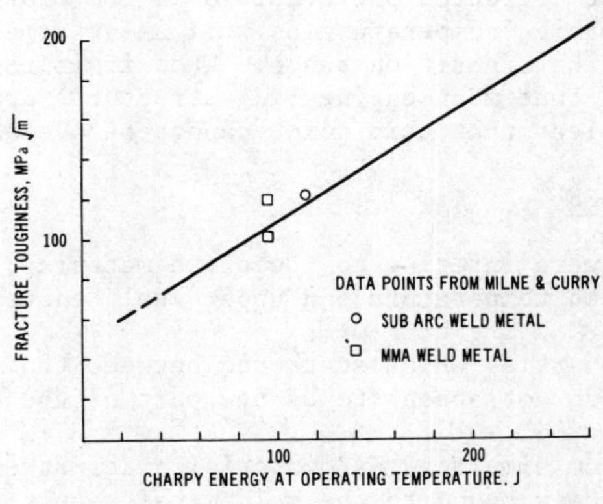

Fig. 4. B.S.I. (1978) Upper shelf energy correlation.

of interest, so that above the ductile-brittle transition temperature it refers to the upper shelf energy level. In Fig. 4 are included the minimum data points of Milne and Curry (1979) using all the ductile initiation toughness levels currently available from the SA and MMA weld metals. One of these points falls below the line. The BSI correlation predicted a fracture toughness of 108 $MPa\sqrt{m}$ while Milne and Curry measured a minimum initiation toughness level of 102 $MPa\sqrt{m}$ for the MMA weld metal.

The toughness specimens of Milne and Curry were precracked to a depth of 25mm. The fracture toughness, K_{1C}, testing standards, ASTM (1972) and BSI (1977) allow up to 2% of crack growth. A similar allowance on the MMA results of Milne and Curry would increase the toughness from the initiation value of 102 $MPa\sqrt{m}$ to about 150 $MPa\sqrt{m}$, well above the correlation line of the BSI (1978) document. Thus, apart from raising the question about the significance of initiation toughness versus K_{1C} as defined by the standards, it does seem as if the BSI correlation for upper shelf cut off is broadly satisfactory for the weld metals in question.

The conventional requirements for quality control purposes are for three Charpy test results to be produced at a specific temperature and above a certain energy level. In Britain this is often 27 joules but it could easily be the 40 joule or 68 joule level specified by the ASME and BSI documents mentioned above. The data presented above clearly indicate that this is not enough. The upper shelf level cannot be deduced from this limited number of test results, so that even though the general level of toughness in the transition regime can be determined by these measurements, no judgement can be made about the upper shelf cut-off. In the current series of tests, had the objective been merely to establish a temperature with a guaranteed energy level of 40 or 68 joules, no information would have been gained about this upper shelf-cut off and the weld metals' inferior upper shelf would have gone unnoticed. Thus it appears that, even for quality control purposes, Charpy data should be collected preferably over the whole transition range, but definitely at the service temperature as well as at some specific temperature or energy level within the transition range. When it is remembered that it is the deliberate intention that most engineering structures are operated in the ductile regime, it becomes clear that this point cannot be overemphasised.

CONCLUSIONS

1. The weld metals were inferior to the plate material, both in terms of ductile brittle transition temperature and upper shelf energy level.

2. The HAZs had properties which scattered between those of the plate and the weld metal, and do not appear to be the part of the weldment most likely to fail.

3. Data obtained from simulated HAZ material after stress relieving satisfactorily delineated the lower bound to the weld metal results. This indicated some level of temper embrittlement in both the weld metals and the HAZs.

4. In the transition regime all the current lower bound correlations with fracture toughness were successful. The least pessimistic of these were the RT_{NDT} correlation of ASME (1977) and the FATT correlation of Jones (1972).

5. In the upper shelf regime, the only lower bound correlation currently available satisfactorily predicted data for fracture toughness as defined by the relevant standards, but not necessarily for the toughness level at the initiation of ductile cracking.

6. The current engineering practice of defining a certain Charpy energy level at a certain temperature is not sufficient even for quality control purposes. The minimum requirement must include Charpy data collected at the service temperature.

ACKNOWLEDGEMENTS

This work was carried out at the Central Electricity Research Laboratories and this paper is published by permission of the Central Electricity Generating Board.

REFERENCES

ASME (1977). Boiler and Pressure Codes, Section III article NB2331 and Section XI Appendix A Article A4000.

ASTM 1972. "Standard Method of Test for Plane-strain Fracture Toughness of Metallic Materials". American Society for Testing of Metals E399-72.

Brear, J., 1976. "Stress Relief Embrittlement in Mn-Ni-Mo Pressure Vessel Steels", in "Grain Boundaries", published by Institution of Metallurgists, p.C13, April 1976.

B.S.I., 1977. "Methods of Plane Strain Fracture Toughness (K_{1C}) Testing". British Standards Institution BS.5447-1977.

B.S.I. 1978. "Draft P.D. for Guidance on Some Methods for the Derivation of Acceptance Levels for Defects in Fusion Welded Joints. B.S.I. Document 78/75394.

Canonico, D.A. and Berggren, R.G., 1971. "Tensile and Impact Properties of Thick Section Plate and Weldments". Nuclear Engineering and Design, 17, 4-15.

Elsender, A., Gallimore, R., Murphy, M.C. and Poynton, W.A. 1978. "Strength and Toughness of 3.5 Ni-Cr-Mo-V Steam Turbine and Generator Rotor Forgings". Metals Technology, 45-56.

Harrison, J.D., 1973. "The Selection of Material to Avoid Fracture in Large Offshore Structures" paper OTC 1909 presented at 5th Offshire Technology Conference, Houston, Texas.

Jones, G.T., 1972. "Charpy Correlations for Rotor and Disc Steels". Proc. Inst. Mech. Eng. 180, 122.

Milne, I. and Curry, D.A., 1979. "Fracture Toughness and Crack Growth Resistance of Submerged Arc and Manual Metal Arc Weldments in SA533B Cl.1 Plate". To be published.

Server, W.L., Oldfield, W. and Wullaert, R.A., 1977. "Experimental and Statistical Requirements for Developing a Well-defined K_{1R} Curve", E.P.R.I. doc. NP-372, Project 696-1, Final Report.

Wessel, E.T., 1969. "Linear Elastic Fracture Mechanics for Thick Walled Welded Steel Pressure Vessels. Material Property Consideration". Proc. of Symposium on Fracture Toughness Concepts for Weldable Structural Steel, Section H, Risley, April, 1969.

TABLE 1 Temperature and Energy Data from Charpy Transition Curves

MATERIAL	Upper Shelf Energy (J)	Mid Transition Temp. Scatter (°C)	Temp. at 68J (°C)	RT_{NDT} (°C)	FATT (°C)	Temp. at 40J (°C)	$RT_{NDT}-40$ $T_{68}-73$ (°C)	FATT-70 (°C)
Plate	200 - 210	60	-6	-39	-5	-25	-79	-75
Simulated HAZ, quenched	110 - 130	35	+5	-28	+18	-25	-68	-52
Simulated HAZ, slowly cooled	110 - 115	35	+35	+2	+35	+10	-38	-35
SA Weld	115 - 130	60	+35	+2	+23	+22	-38	-47
MMA Weld	95 - 135	55	+35	+2	+26	+14	-38	-44
SA HAZ	135 - 220	110	+30	-3	ND	+15	-43	ND
MMA HAZ	135 - 200	105	+15	-18	ND	-15	-58	ND

ND - Not Definable

Volume 3, ICM 3, Cambridge, England, August 1979

EVALUATION OF CONSTRAINT FACTOR AND J-INTEGRAL FOR SINGLE-EDGE NOTCHED SPECIMEN

M. Shiratori and T. Miyoshi*

Department of Mechanical Engineering, Yokohama National University, Yokohama, Japan
***Department of Precision Machinery Engineering, University of Tokyo, Tokyo, Japan**

INTRODUCTION

The single-edge notched (SEN) specimen is one of the most important specimens in fracture toughness testing. Ewing and Richards (1974) have given the plane strain general yielding load of rigid-plastic, pin-loaded specimens by means of slip line field and upper bound approaches. In this paper their approach is extended to more general cases where axial force and bending moment are applied to SEN specimens at the same time. As a result general yielding loci have been obtained for an arbitrary combination of axial force and bending moment. The constraint factor and rotational factor for pin-loaded SEN specimens, as well as compact tension (CT) specimens, have been given in explicit forms which are similar to the results of Ewing and Richards but are more practical in the analysis of fracture toughness testing data. These formulae are useful when the J-integral is analyzed in the way used by Merkle and Corten (1974) for CT specimens. Explicit formulae, which give the J-integral from a single load vs. load-point-displacement record, have been obtained and the results are discussed.

PLASTIC CONSTRAINT OF SEN SPECIMEN DUE TO COMBINED AXIAL FORCE AND BENDING

Description of the Problem and Methods of Analysis

Let us consider the case shown in Fig. 1 where an arbitrary combination of axial force T and bending moment M (both defined per unit thickness) is applied to a SEN specimen with width W, crack length a and ligament length b. M is taken about the mid-point of the ligament. We investigate yielding of the specimen by this combination of T and M whose magnitudes have to be evaluated. For simplicity, rigid-plastic material and plane strain deformation are assumed. There are three approaches considered: a lower bound approach, a slip line field approach and an upper bound approach.

Lower Bound Approach (LBA)

Stress elevation due to plastic constraint around the crack tip is disregarded and the stress distribution across the ligament is simply assumed to be as shown in Fig. 2, where k is the yield stress in shear and rb is the distance from the crack tip to the stress reversal point. The parameter r is the rotational factor. The equilibrium condition for this assumed stress distribution is

$$T=2kb(2r-1), \qquad M=\frac{1}{2}kb^2 4r(1-r) \tag{1}$$

Slip Line Field Analysis (SLFA)

When the bending moment is sufficiently large compared with the axial force, the possible slip line field is shown in Fig. 3. This is similar to the one presented by Green (1953) for pure bending. The effect of axial force is to change the radius R of the hinge and to move the point Q which moves downwards with increasing tensile force. Stresses p_θ and k along slip line OQR can be equilibrated to external traction, and the resulting equations are

$$T=2kb(C_2\frac{R}{b}-1), \qquad M=\frac{1}{2}kb^2[2\sqrt{2}C_3\frac{R}{b}+4\lambda(\frac{R}{b})^2+2C_3{}^2-4C_1\frac{T}{2kb}], \tag{2}$$

where

$$\begin{aligned} \lambda &= 1.021 = 58.5^\circ, \\ C_1 &= (R/b)\sin(2\lambda-\pi/4)-0.5=0.951(R/b)-0.5, \\ C_2 &= (2\lambda+0.5)\sin(2\lambda-\pi/4)+0.5\cos(2\lambda-\pi/4)=2.572, \\ C_3 &= 1-(R/b)\{\sin(2\lambda-\pi/4)+\sqrt{2}/2\}=1-1.658(R/b). \end{aligned} \tag{3}$$

The condition that the point Q should remain on the specimen is

$$C_3 \geqq 0, \qquad \text{i.e.} \qquad R/b \leqq 0.603 \tag{4}$$

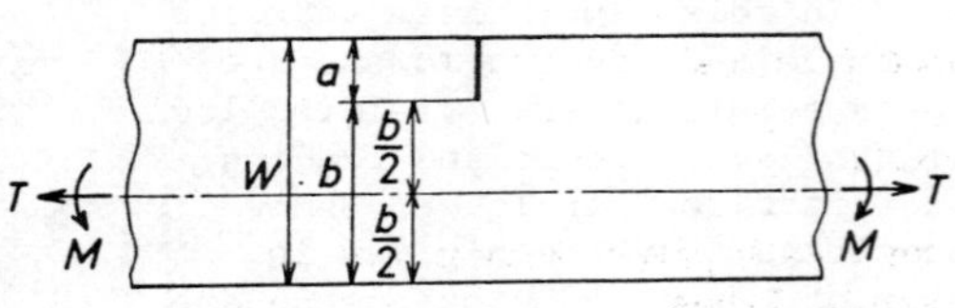

Fig. 1. SEN specimen under combined axial force and bending moment

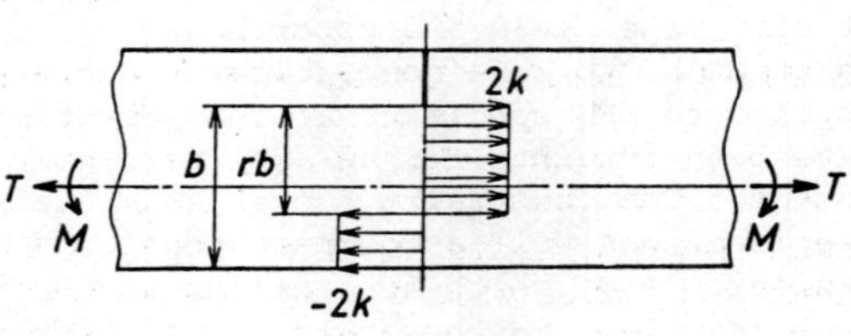

Fig. 2. Lower bound approach

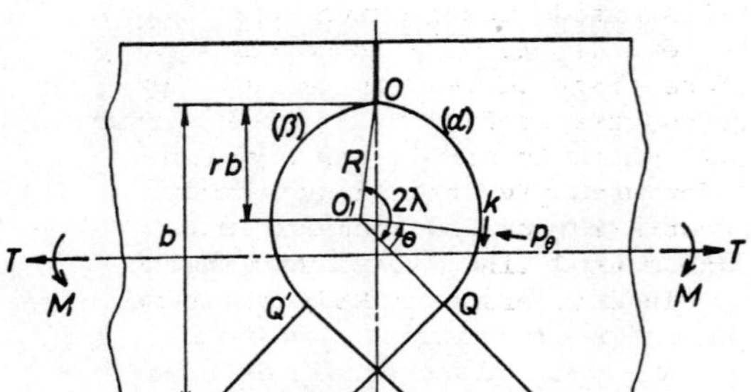

Fig. 3. Slip line field approach

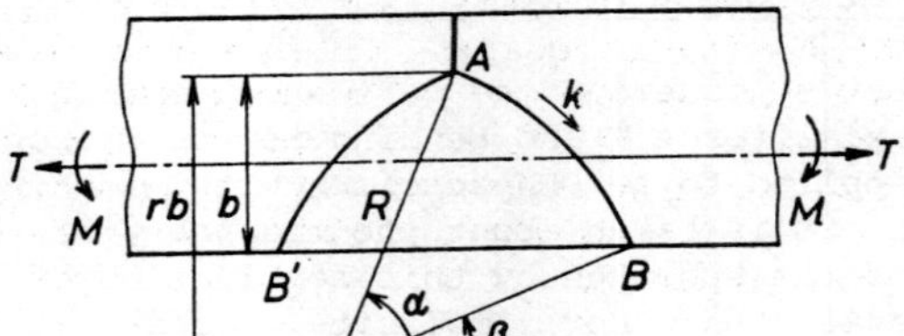

Fig. 4. Upper bound approach

Upper Bound Approach (UBA)

When (4) is not satisfied Ewing and Richards (1974) and Rice (1972) proposed a kinematically admissible slip line field shown in Fig. 4 where rigid strips on both sides rotate along circular arcs AB and AB'. This field gives an upper bound general yielding load to the true solution. From the upper bound theorem (Rice, 1972),

$$M+T(r-0.5)b \leqq kR^2(\alpha-\beta), \tag{5}$$

where

$$rb=R\sin\alpha, \qquad (r-1)b=R\sin\beta, \qquad 2(\alpha-\beta)=\tan\alpha-\tan\beta. \tag{6}$$

General Yielding Loci

When an axial force or a bending moment is applied independently to a plain strip of width b, the magnitude of the axial force T_o or the bending moment M_o at which the strip yields generally is

$$T_o=2kb \qquad \text{or} \qquad M_o=0.5kb^2. \tag{7}$$

The general yielding axial force T and bending moment M in equations (1), (2), (5) can be normalized by T_o and M_o respectively to give the following relations:

$$T_1=T/T_o=T/2kb, \quad M_1=M/M_o=M/0.5kb^2, \tag{8}$$

where T_1 and M_1 are the plastic constraint factors defined for axial force and bending respectively. Elimination of r in equations(1) gives the following relation between T_1 and M_1 for LBA,

$$M_1+T_1^2=1, \tag{9}$$

Similarly the elimination of R/b in equations(2) gives

$$M_1+0.739T_1^2-0.521T_1=1.26 \tag{10}$$

for SLFA, where the range of valid T_1 is given by $0<T_1 \leqq 0.551$ from the inequality of equation (4). The dotted and solid lines in Fig. 5 are the plots of equations (9) and (10) respectively. On the other hand a numerical analysis based upon the algorithm of equations (5) and (6) results in the chain line in Fig. 5. These lines show the relations between the plastic constraint factors T_1 and M_1 when axial force and bending moment are applied to the SEN specimen at the same time. When a point which is prescribed by an axial force T and a bending moment M, falls inside these curves only local yielding will occur around crack tip, while general yielding is possible only when the point falls on the curves. Therefore these curves can be regarded as general yielding loci or loci of plastic constraint factors.* Substitution of $T_1=0$ into equation (10) results in $M_1=1.26$ which is consistent with the solution of Green (1953) while LBA and UBA for $M_1=0$ give the value of $T_1=1.0$ which is consistent with the solution for centrally cracked specimen under pure tension (Knott, 1973). Furthermore, the dotted line, i.e., equation (9), is the same as the general yielding loci obtained by Prager (1957) for a plain strip of width b.

General Yielding Load of a Pin-Loaded SEN Specimen

Fig. 6(a) shows the standard pin-loaded specimen which is commonly used in plane strain fracture toughness testing. When the axial pin-force per unit thickness is denoted by T, the bending moment per unit thickness about the mid-point of the ligament is

* According to the analysis for a wedge-shaped notch with large flank angle, these general yielding loci have been proved to be a kind of plastic potential, and the loci have been obtained for four quadrants on T_1-M_1 plane (Shiratori and Dodd, 1978).

$$M=T(W-b)/2 \tag{11}$$

Substituting equations (8) into equation (11),

$$M_1=2T_1(W/b-1), \tag{12}$$

a straight line passing through the origin of the T_1-M_1 plane, whose slope is dependent on b/W. The intersection points of this line with the general yielding loci give the general yielding load T_1 for the SEN specimen. The explicit formula for T_1 is

$$T_1=r/r_o-1, \quad r=-(W/2b-1)+\sqrt{[(W/2b-1)^2+2r_o(W/2b-0.5)]} \tag{13}$$

where r_o is the rotational factor for pure bending given by

$$\begin{array}{llll} & r_o=0.5 & \text{for} \quad 0<b/W<1 & \text{in LBA} \\ \text{and} & r_o=0.370 & \text{for} \quad 0<b/W\leqq 0.455 & \text{in SLFA.} \end{array} \tag{14}$$

Equations (13) with r_o=0.370 are essentially the same as the result of Ewing and Richard (1974) but the expressions are a little different. Equations (13) are more convenient for practical use because T_1 and r can be calculated directly when the ratio of b/W is prescribed. The intersection point with the upper bound line gives an upper bound general yielding load T_1 and a rotational factor r, expressible as

$$\begin{aligned} T_1&=a_o+a_1(b/W)^2+a_2(b/W)^4+a_3(b/W)^6+a_4(b/W)^8, \\ r&=b_o+b_1(b/W)^2+b_2(b/W)^4+b_3(b/W)^6+b_4(b/W)^8, \end{aligned} \tag{15}$$

which are interpolation formulae obtained by a numerical analysis. The coefficients a_o-a_4 and b_o-b_4 are given in Table 1. These interpolation formulae have the maximum error of 2.6% in the range of $0.2\leqq b/W\leqq 0.8$. Figure 7 shows the general yielding load vs. a/W=1-b/W obtained by these three approaches, while Figure 8 shows the rotational factor r vs. a/W.

General Yielding Load of a Compact Tension Specimen

In the case of a CT specimen the parameter W is taken to be the distance from the load line to the bottom surface of the specimen as shown in Fig. 6(b). The general yielding load is obtained by replacing W in equations (13) with 2W to give

$$T_1=r/r_o-1, \quad r=-(W/b-1)+\sqrt{[(W/b-1)^2+2r_o(W/b-0.5)]}. \tag{13'}$$

Here again

$$\begin{array}{llll} & r_o=0.5 & \text{for} \quad 0<b/W<1 & \text{in LBA} \\ \text{and} & r_o=0.370 & \text{for} \quad 0<b/W\leqq 0.910 & \text{in SLFA.} \end{array} \tag{14'}$$

There is no need to give the formula for an upper bound general yielding load corresponding to equations (15) because equations (13)' cover almost all the range of b/W for practical purpose. The second of equations (13)' with r = 0.370 seems to be useful because it gives an explicit formula of the rotational factor which is an important parameter in COD testing.

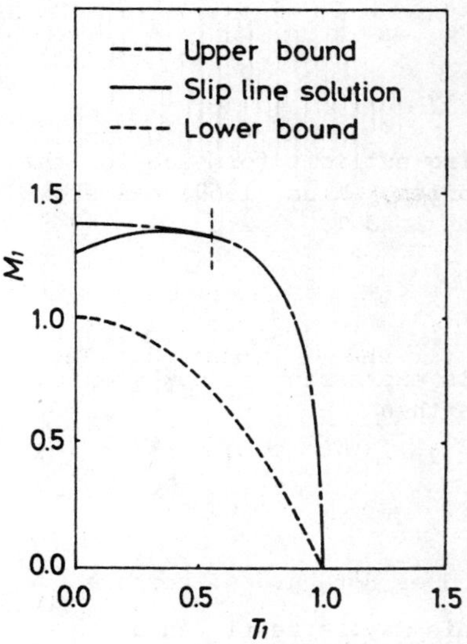

Fig. 5. General yielding loci

TABLE 1 a_0-a_4 and b_0-b_4	
a_0= 0.09879	b_0= 0.4836
a_1= 1.766	b_1= 0.9993
a_2= 4.930	b_2=-8.952
a_3=-15.76	b_3= 35.66
a_4= 11.09	b_4=-23.12

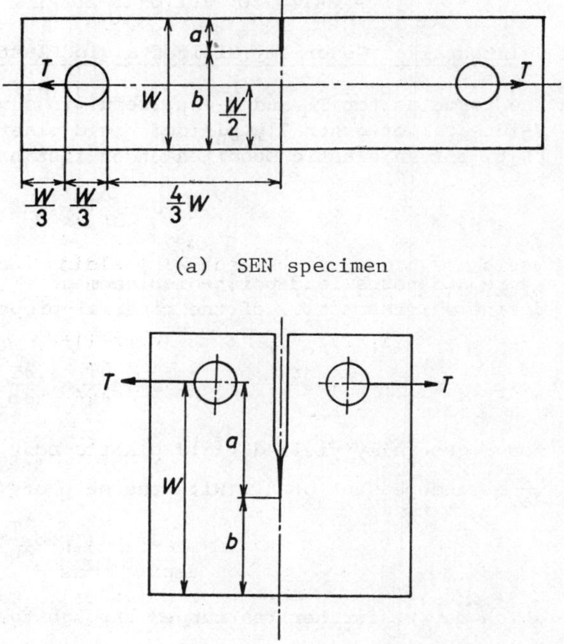

(a) SEN specimen

(b) CT specimen

Fig. 6. Shapes of the specimens

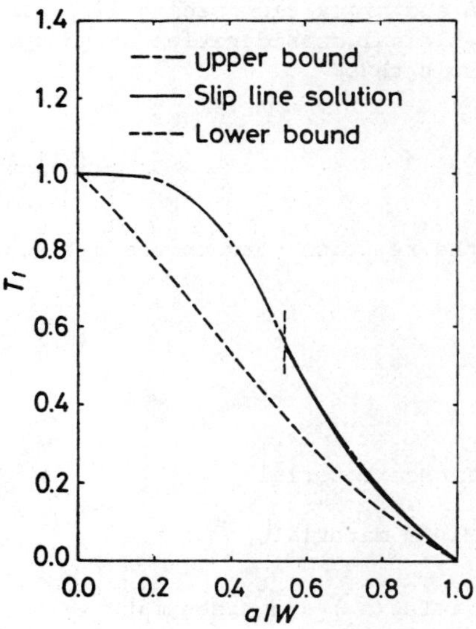

Fig. 7. Constraint factor

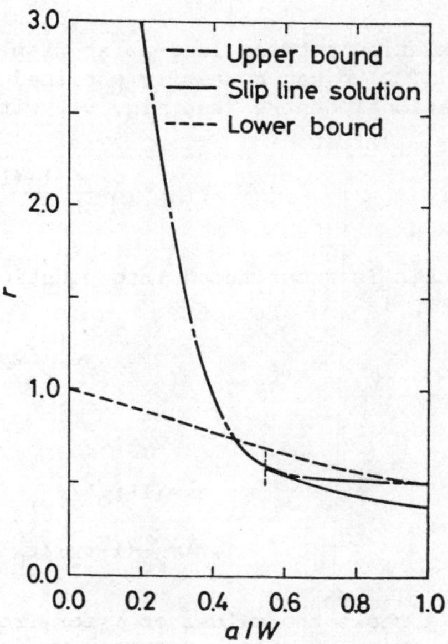

Fig. 8. Rotational factor

EVALUATION OF THE J-INTEGRAL

J-Integral of Generally Yielded Rigid Plastic Body

The formulae for T_1 and r are useful in developing explicit formulae for the J-integral of generally yielded rigid-plastic bodies. Rice (1968) has shown that, for an elastic specimen under instantaneous load T,

$$J=-\int_0^{\Delta}\left(\frac{\partial T}{\partial a}\right)_{\Delta}d\Delta \tag{16}$$

where Δ denotes load-point-displacement. If this expression is now used to define J irrespective of the material properties then

$$J=-\Delta\frac{\partial T}{\partial a}=\Delta\frac{\partial T}{\partial b} \tag{17}$$

for a generally yielded rigid plastic body that does not harden, because $\frac{\partial T}{\partial a}$ is independent of Δ. With the help of (8) this may be rewritten as

$$J=2k\Delta\left(T_1+b\frac{\partial T_1}{\partial b}\right) \tag{18}$$

which may be further rearranged through the use of (13) and (13)' to

$$J=\frac{1-r_o}{r_o}2k\Delta\left[1-\frac{1-\alpha b/W}{1-(1-r)\alpha b/W}\right] \tag{19}$$

where

$\alpha=2$ and r is given by equations (13) for SEN specimen

and $\alpha=1$ and r is given by equations (13)' for CT specimen. (20)

The relation between load-point-displacement Δ and crack tip opening displacement (CTOD) δ can be easily obtained by a kinematical consideration based upon a rotational centre (see Fig. 9), with the result that

$$\Delta=\frac{W}{\alpha b}\,\frac{1-(1-r)\alpha b/W}{r}\,\delta . \tag{21}$$

When this is substituted into equation (18), the relation between the J-integral and CTOD is

$$J=\frac{1-r_o}{r_o}2k\delta=\lambda\sigma_Y\delta \tag{22}$$

where $\lambda=(1-r_o)/r_o$ for a Tresca material

and $\lambda=\frac{2}{\sqrt{3}}(1-r_o)/r_o$ for a Mises material. (23)

Table 2 shows the values of λ for Tresca materials (λ_T) and Mises materials (λ_M) obtained by LBA and SLFA. Values of SLFA are more realistic because the plane strain plastic constraint at the notch root is taken account of, while values of LBA can be thought to correspond to those of plane stress deformation because no constraint is considered at the notch root. A similar analysis is possible for SEN specimen in the range of $0.455 \leq b/W < 1$ where the slip-line field solution is invalid. In this case equations (15), which are the results of UBA, are combined with equations (18) and (21). Figure 10 shows the results of numerical analysis for a Tresca material. It is interesting to see that the value of λ_T is changing from 1.0 to 1.70 with increasing crack length. Very short cracks correspond to nearly pure tension where there is no constraint, suggesting that λ is a parameter depending on notch constraint.

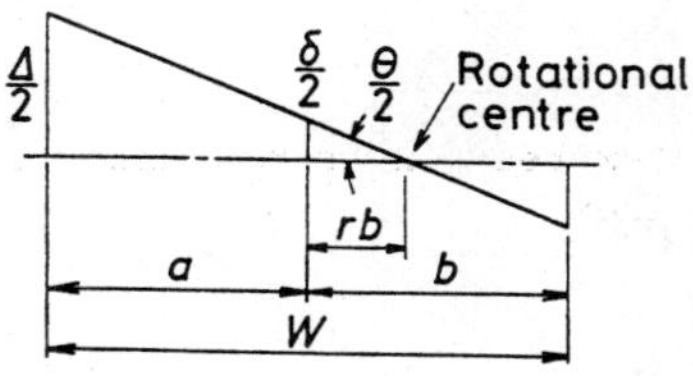

Fig. 9. Kinematical relation between COD and rotational factor

TABLE 2 λ_T and λ_M

	r_0	λ_T	λ_M
SLFA	0.370	1.70	1.97
LBA	0.500	1.00	1.15

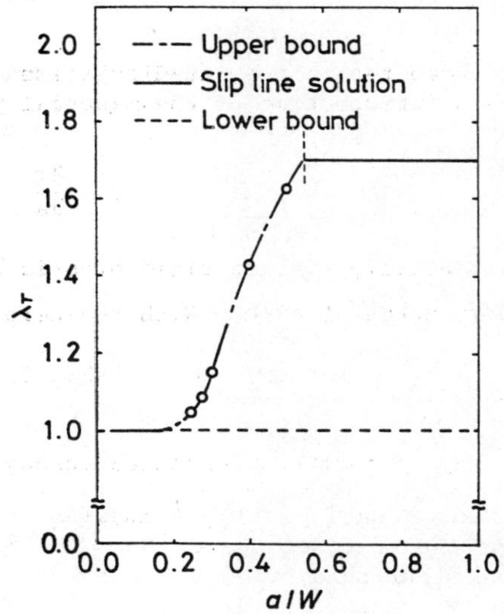

Fig. 10. λ_T vs. a/W

Simple Formulae to obtain J from a Single Load vs. Load-Point-Displacement Record

Merkle and Corten (1974) have presented a simple formula for CT specimen to obtain J from a single load vs. load-point-displacement record. The formula is shown to be applicable to real materials if the crack length and width ratio a/W is chosen properly. Their method has been extended to the more general cases treated in this paper. For the range of plastic deformation where elastic strain can be neglected but the total deformation is not so large as to be treated by large deformation theory, they give the following expressions:

$$J=-\frac{1}{B}\int_0^{\Delta}\left(\frac{\partial P}{\partial a}\right)_{\Delta} d\Delta, \tag{24}$$

where

$$-\left(\frac{\partial P}{\partial a}\right)_{\Delta}=\frac{1}{2}\left[\left(\frac{\partial P_{fp}}{\partial c}\right)_{\Delta}/P_{fp}\right]P+\frac{1}{2}\left[\left(\frac{\partial \theta}{\partial c}\right)_{\Delta}/\left(\frac{\partial \theta}{\partial \Delta}\right)_c\right]\left(\frac{\partial P}{\partial \Delta}\right)_c, \tag{25}$$

$$P_{fp}=TB, \qquad 2c=b, \qquad \Delta=\theta[W/\alpha-2(1-r)c] \tag{26}$$

In these equations P and Δ are load and load-point-displacement of the specimen of thickness B, P_{fp} is the general yielding load, and θ is the angle of rotation. The parameters T and r in equations (26) can be evaluated by the explicit formulae presented in the previous section. Generally T and r are expressible as

$$T=2kbT_1(b/W), \quad \text{and} \quad r=r(b/W), \tag{27}$$

and then (24) and (25) take the form

$$J=\frac{1}{b}\,\eta_r\int_0^{\Delta}(\frac{P}{B})d\Delta+\eta_c\int_0^{P/B}\Delta d(\frac{P}{B}) \tag{28}$$

where

$$\eta_r=1+b(\frac{\partial T_1}{\partial b})/T_1, \qquad \eta_c=\frac{\alpha b/W}{1-(1-r)\alpha b/W}\{1-r-b(\frac{\partial r}{\partial b})\}. \tag{29}$$

Further substitution of equations (13), (13)' and (15) into the above yields

$$\eta_r=1+\frac{1}{1-(1-r)\alpha b/W}, \quad \eta_c=\frac{\alpha b/W}{1-(1-r)\alpha b/W}\{(1-r)-\frac{r-r_o}{1-(1-r)\alpha b/W}\} \tag{30}$$

for LBA and SLFA, and

$$\eta_r=1+\frac{2\Sigma m a_m(b/W)^{2m}}{\Sigma a_{m-1}(b/W)^{2(m-1)}}, \quad \eta_c=\frac{2b/W}{1-(1-r)2b/W}\,(1-r)-2\Sigma m b_m(b/W)^{2m} \tag{31}$$

for UBA. Equations (30) are rather complicated to understand because many things are included there. For SEN specimens $\alpha=2$ while for CT specimens $\alpha=1$. On the other hand, $r_o=0.370$ for SLFA and $r_o=0.5$ for LBA. The parameter r is given by equations (13) and (13)'. Equations (30) with $\alpha=1$ and $r_o=0.5$ are essentially the same as those of Merkle and Corten (1974) though the expressions are a little different. The valid range of equations (30) in the case of SLFA is $0<b/W\leq 0.910$ for CT specimens but $0<b/W\leq 0.455$ for SEN specimens. Therefore the formula for UBA given by equations (31) is necessary only for SEN specimens with short cracks ($0.455<b/W<1.0$).

Table 3 shows values of J in a CT specimen with a typical P-Δ curve of A533B steel at 120°C. The P-Δ curve is normalized by P_Y and Δ_Y, respectively, which are values at the proportional limit. ϕ_r and ϕ_c are the potential energy and complementary energy obtained from the P-Δ curve. J_L and J_S denote J values obtained by LBA and SLFA, while J_R is obtained from the equation proposed by Rice, Paris and Merkle (1973), which is expressed by taking $\eta_r=2.0$ and $\eta_c=0.0$ in equation (28). The results of Table 3 show that J_S is about 5-10% higher than J_L. Finally, a finite element analysis has been carried out to compare our formulae for J evaluated for a SEN specimen. Plane strain deformation theory was used in the finite element analysis. The shape of the specimen is the same as shown in Fig. 6(a) where a/W=0.5 and 0.75 with W=50 mm. The mechanical properties were chosen to simulate A533B steel at room temperature

Figure 11 shows the results where J_U, J_S and J_L are obtained by UBA, SLFA and LBA, respectively, while J_R is obtained from Rice's equation. These are obtained from a single load vs. load-point-displacement record of the finite element analysis. But $\Delta_{crack}=\Delta-\Delta_{no\ crack}$ is used instead of Δ to remove the effect of plastic deformation around the pin-hole. J_{path} was obtained from the original definition of J as a path independent integral. Values of five paths were averaged to give the value of J_{path}. Figure 11 shows that J_S is a good approximation to J_{path} for a/W=0.5, while J_L, J_S and J_U are reasonable for a/W=0.75.

TABLE 3 Values of J_S/J_L and J_S/J_R

Δ/Δ_Y	P/P_Y	ϕ_c/ϕ_r	J_S/J_L	J_S/J_R
1.00	1.00	1.000	1.086	1.338
2.00	1.38	0.619	1.072	1.281
3.00	1.55	0.459	1.067	1.257
4.00	1.63	0.365	1.063	1.243
5.00	1.68	0.306	1.061	1.235
6.00	1.70	0.255	1.059	1.227
7.00	1.72	0.225	1.058	1.222
8.00	1.73	0.916	1.057	1.218
9.00	1.74	0.179	1.056	1.216
10.00	1.75	0.163	1.056	1.213

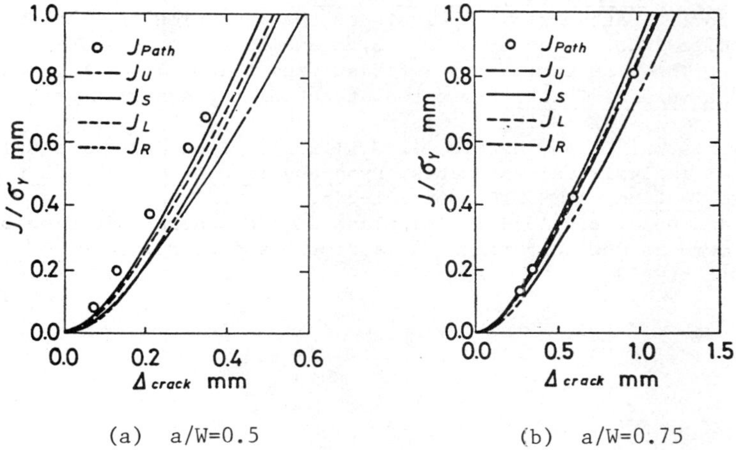

Fig. 11. J-integral vs. Δ_{crack}

SUMMARY

Lower bound, slip line field and upper bound analyses have been carried out for plane strain, rigid-plastic, single-edge notched specimens under combined axial force and bending moment. As a result general yielding loci have been obtained for an arbitrary combination of axial force and bending moment, and the results of Ewing and Richards (1974) have been found to be special cases of this general approach. Constraint factors and rotational factors for pin-loaded SEN specimens and compact tension specimens have been given in explicit forms which are essentially the same as the results of Ewing and Richards but more practical in the analysis of fracture toughness testing data. These formulae have been applied to obtain explicit formulae for J. For a rigid-plastic body the relation between J and crack tip opening displacement δ is obtained in the form of $J=\lambda\sigma_Y\delta$ where σ_Y is the tensile yield stress and λ is a parameter varying from 1.0 to about 2.0 depending on the material and the state of notch constraint. For real materials a similar analysis to that of Merkle and Corten (1974) has been carried out, and explicit formulae which give J from a single load vs. load-point-displacement record have been obtained. These formulae are shown to give good estimates of J for both of SEN and CT specimens.

Finally the authors would like to acknowledge the helpful discussions of Dr. Dodd at University of Oxford and Professor Kudo and Mr Shigeta at Yokohama National University.

REFERENCES

Ewing, D. J. F. and Richards, C. E. (1974). The yield-point loads of singly-notched pin-loaded tensile strips. J. Mechs. Phys. Solids, 22, 27-36.

Green, A. P. (1953). The plastic yielding of notched bars due to bending, Quart. J. Mech. Appl. Math., 6, 223-239.

Knott, J. F. (1973). Fundamentals of Fracture Mechanics, Butterworths, London.

Merkle, J. G. and Corten, H. T. (1974). A J-integral analysis for compact specimen, considering axial force as well as bending effects, J. Press. Vessel Tech., Trans. ASME, J96, 286-292.

Prager, W. (1959). An Introduction to Plasticity, Addison-Wesley, New York.

Rice, J. R. (1968). Mathematical analysis in the mechanics of fracture, in H. Liebowitz (Ed.), Fracture, Vol. 2, Academic Press, New York, 191-311.

Rice, J.R. (1972). The line spring model for surface flaws, In J. L. Swedlow (Ed.), The Surface Crack, Presented at the Winter Annual Meeting of ASME, New York, 171-185.

Rice, J. R., Paris, P. C. and Merkle, J. G. (1973). Some further results of J-integral analysis and estimates, Progress in Flaw Growth and Fracture Toughness Testing, ASTM STP 536, 231-245.

Shiratori, M. and Dodd, B. (1978). The plane strain general yielding of notched members due to combined axial force and bending, Int. J. Mech. Sci., 20, 451-463.

Volume 3, ICM 3, Cambridge, England, August 1979

THE STRENGTH OF WELDS IN HSLA STEEL AFTER INITIAL PLASTIC DEFORMATION

S. Sedmak, A. Radović and Lj. Nedeljković

Faculty of Technology and Metallurgy, University of Belgrade, Yugoslavia

INTRODUCTION

The underground penstock for the Bajina Bašta Hydroelectric Pumping-Up Power Plant, that is under construction in Yugoslavia, is considered to be highly loaded, HxD being as high as 2789 mxm (the diameter of the penstock D measures 4.2 m and the high water level H is 664 m).

The aim of the designer to have only one penstock in spite of the high nominal power of the turbines dictated the selection of high strength steel for the construction of the penstock. The choice fell on Japanese HSLA steel, quenched and tempered, class HT80 . This is considered to be the steel with the highest mechanical properties used in recent times for penstock construction (Ohira, Okuyoshino, Nabara, after Horikawa, 1972). However, even this high quality steel could only be used for the penstock construction with the allowable stress calculated fety factor 1.7, a value which is recommended in Europe (Deutscher Ausschuss für Stahlbau, 1974). With this safety factor the thickness of the wall could be lower than 50 mm, which is considered as an upper limit for HT80 class steel. In Japan, the recommended safety factor for HT80 class steel is 2.07 (Water Gate and Penstock Association, 1971), and practically used values are even higher (e.g. for Ohira penstock the factor was as high as 2.19). One of the main reasons for the adoption of the high safety factor in Japan is the small difference between the yield stress and ultimate tensile strength of HT80 class steel.

The safety factor of 1.7 for the Bajina Bašta penstock could be adopted only by taking into account the additional support of tunnel rocks. This mean that the stress level in the penstock would be lower than that calculated on the basis of the yield stress of HT80 steel and its welded joints ($\sigma_{0.2}$ = 687 MPa). But even so, very strict requirements for the safety of the steel penstock and for a high quality of the welded joints remained essential. Therefore, a wide program of investigation was initiated before the beginning of construction of the penstock. Special attention was paid to the problem of plasticity in the steel and welded joints of the penstock, bearing in mind that plasticity can be reduced by temporary effects of water hammer. The estimated level of water hammer at the Bajina Bašta penstock corresponds to 30% of the design load. Under these conditions circumferential stress σ_φ can be as high as 75% of the yield stress, and, due to stress concentrations and residual stresses, local plastic deformation might develop. In order to investigate this important circumstance in more detail and to gain insight into the possible effects of initial plastic deformations on the overall safety of the penstock, it was decided to include a test of a prototype construction in the program of investigations. This investigation had to determine the behavior of the prototype when loaded to the upper limit of pressure and calculated overloads, and to evaluate the effects of overloading on the plasticity. This evaluation was to be obtained on the basis of static tests and tests under conditions of fast fracture, by comparing the behaviour of specimens taken from the prototype and trial samples, welded under the same conditions and at the same time.

EXPERIMENTAL DETAILS

Materials

For the construction of the prototype, six plates, 47 mm thick, made of SUMITEN 80P (SM80P) steel, from the Japanese Sumitomo Steel Works, were used. The plates were made from two converter steel heats, treated in a vacuum and cast in ingots. The chemical composition and mechanical properties are given in Table 1.

As can be seen from the chemical composition and mechanical properties, the material is complex alloyed, low carbon steel. The SM80P plates have good impact toughness, characterized by high values of absorbed energy at $-40^{\circ}C$ (Table 1) and a low transition temperature (lower than $-60^{\circ}C$, but often even lower than $-100^{\circ}C$). But in spite of satisfactory impact properties, a high level of anisotropy was observed, that varied from plate to plate of the same steel heat. The degree of anisotropy, expressed by the ratio of absorbed energy at $-40^{\circ}C$ for the Charpy V specimens taken in a rolling direction (L) and transverse to it (C), varied between 1.5 and 2.6.

Manual arc welded joints were made by low hydrogen basic coated electrodes, and submerged arc welded joints with core wire and melted welding powder. The chemical composition, tensile and impact properties of both weld metals are given in Table 1.

Construction of the Prototype

The design of the prototype took into consideration the shape and dimensions of the most highly stressed part of the penstock (diameter D = 4.2 m, knee 5°, material thickness 47 mm) (Fig. 1). In order to reduce the influence of the lids, stiffeners were placed at both ends of the shell.

The plates were cold formed. The edges were prepared by an automatic gas cutter and ground before welding. All combinations of weld positions and welding procedures that appear on the penstock were made on the prototype. The longitudinal welds LS1, LS2, LS3 and LS4, and the circular weld CS were made by submerged arc welding. The longitudinal welds LM1 and LM2 and the circular weld CM were made by manual arc welding. Welds were repaired manually. The prototype was workshop welded. The weld CM was made under humidity conditions close to those in the tunnel.

TABLE 1 Chemical Composition and Mechanical Properties of the Plates and Weld Metals

Material SM 80 P	Chemical composition (%)											
	C	Si	Mn	P	S	Cu	Cr	Ni	Mo	V	B	C_{eq}^{*}
Plates Heat A	0.10	0.30	0.90	0.01	0.008	0.24	0.48	1.01	0.47	0.03	0.0016	0.50
Plates Heat B	0.12	0.27	0.87	0.011	0.007	0.27	0.47	0.99	0.45	0.03	0.0012	0.51
Weld Metal MAW**	0.06	0.53	1.48	0.011	0.005		0.24	1.80	0.43			
Weld Metal SAW***	0.07	0.37	1.87	0.01	0.011		0.44	0.13	0.73			

Material SM 80 P	Direction	Mechanical properties: Tensile test			Charpy impact test	
		Y. P. [MPa]	U. T. S. [MPa]	El. [%]	Ch V [J]	°C
Plates Heat A	L	755 – 794	804 – 834	24 – 29	156 – 223	–40
	C	755 – 794	795 – 834	22 – 23	60 – 147	–40
Plates Heat B	L	775 – 795	824 – 844	25 – 26	216 – 237	–40
	C	775 – 804	824 – 853	19 – 23	83 – 121	–40
Weld Metal MAW**		722	810	23	99	–5
Weld Metal SAW***		687	804	23	78	–18

* $C_{eq} = C + \frac{Si}{24} + \frac{Mn}{6} + \frac{Ni}{40} + \frac{Cr}{5} + \frac{Mo}{4} + \frac{V}{14}$ ** MAW – Manual arc welding *** SAW – Submerged arc welding

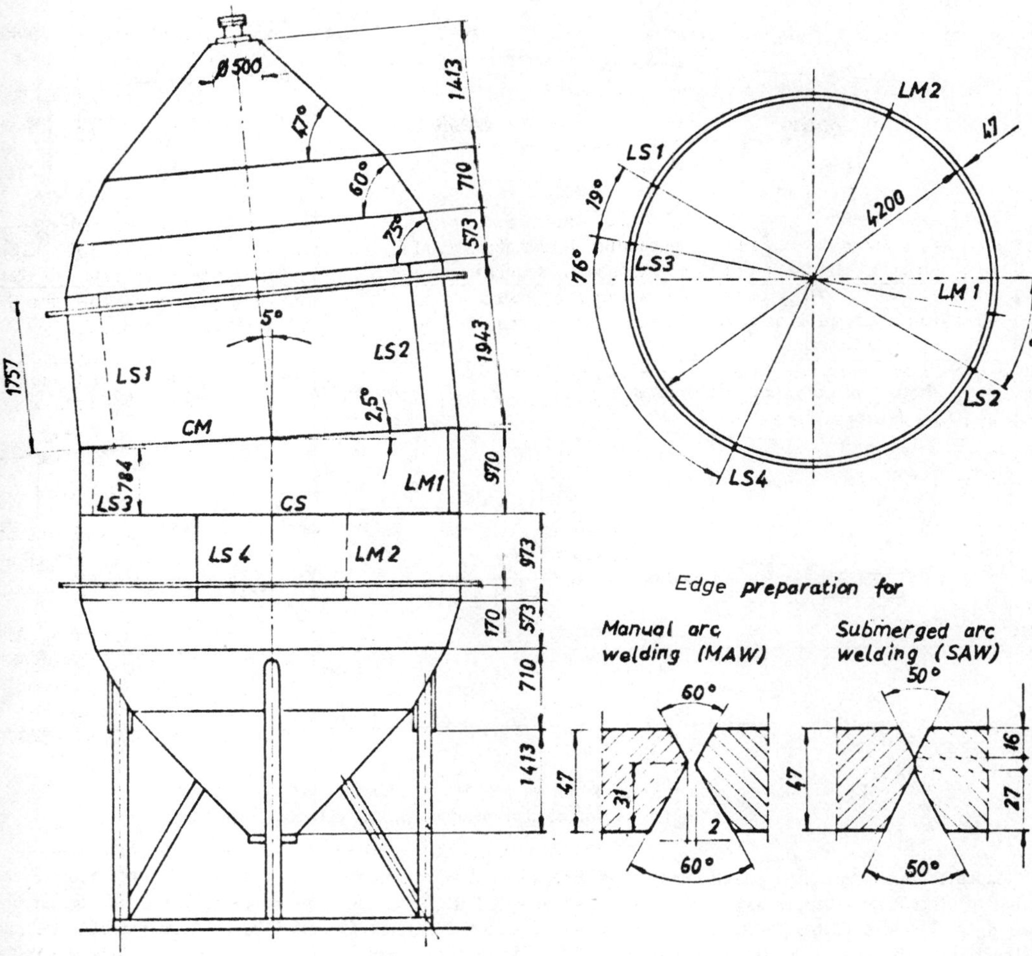

Fig. 1. Design and dimensions of the prototype

From the same cold formed plates, under the same conditions and at the same time two trial samples 1100x500 mm were also welded, so it was possible to cut out the same number of specimens as from the corresponding welds of the prototype.

Heat input by submerged arc welding was 1.5 – 4.8 MJ/m, and by manual arc welding 0.5 – 1.85 MJ/m. All the welds on the prototype and on the trial samples were inspected ultrasonically and radiographically and satisfied the "black" requirements according to IIS/IIW recommendations for radiographic inspection.

Testing of the Prototype

The prototype as a whole was pressurized at an ambient temperature which was between 6°C and −3°C. Strains on the prototype were monitored by strain gauges, Moiré grids and acoustic emission. Altogether, 28 strain gauges, placed outside, were active. Some of the outer wall and all of the inner wall strain gauges failed. Moiré grids were used to observe the overall picture of the prototype deformation. Acoustic emission sensors were positioned in such a way as to enable the location and identification of possible critical sources of noise, as an indication of the development of local large plastic deformation or crack initiation, in order to be able to prevent a catastrophic failure during pressurizing. Fig. 2. shows the disposition of measuring devices on the prototype. The prototype

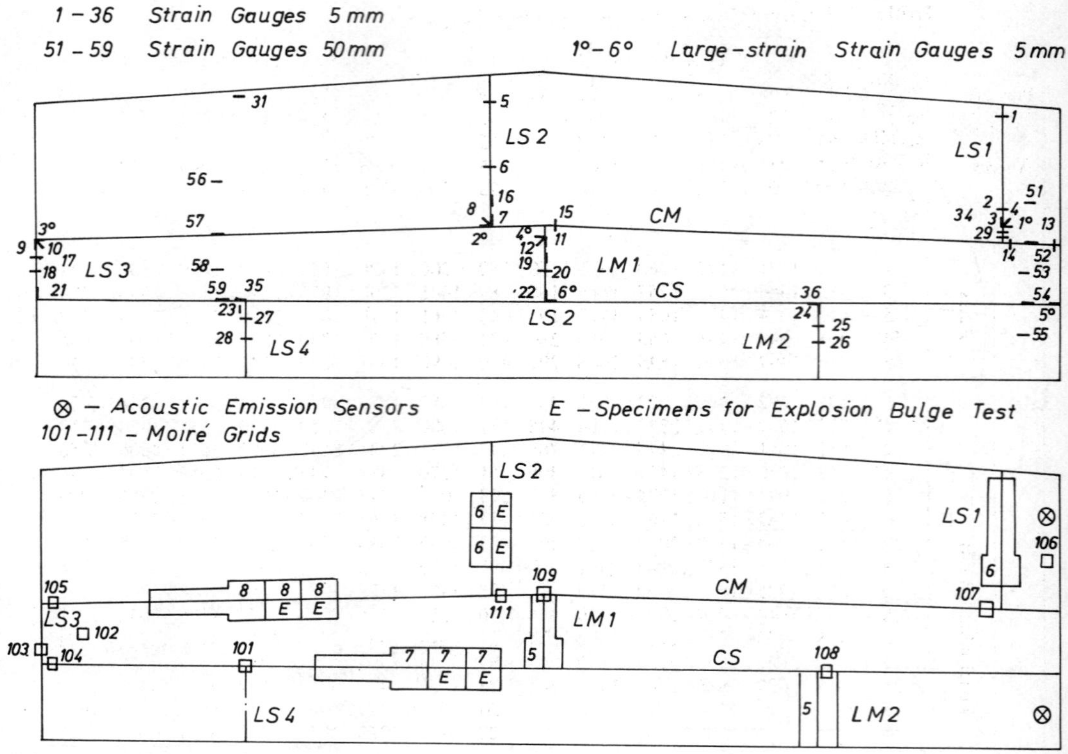

Fig. 2. Disposition of the strain gauges, Moiré grids and acoustic emission sensors on the prototype and scheme for the cutting of specimens

was loaded in two stages. The strain gauges were zero adjusted at a pressure of 0.49 MPa for both loading cycles. During the first stage, the pressure reached the value of $p = 9.02$ MPa, that corresponds to the calculated circular stress $\sigma_\varphi = 399$ MPa (Table 2). The prototype was held under a pressure of 7.35 MPa for a period of two hours. At the end of the first stage, the prototype was unloaded. At the second loading, the pressure reached the value $p = 12.06$ MPa, that corresponds to the calculated circular stress $\sigma_\varphi = 533$ MPa. This stress is equal to the 75% of the nominal yield stress and corresponds approximately to the total design and water hammer load. After testing the prototype was unloaded, emptied and specimens for mechanical tests were cut out of the shell (numbers 5, 6, 7, 8, Fig 2). The same specimens were cut out of the trial samples (numbers 1, 2, 3, 4). The strains measured by strain gauges are given in Table 2.

Mechanical Tests

The mechanical properties were examined for both welding procedures (MAW and SAW) and both types of welded joints (longitudinal and circular).

Tensile properties were determined by testing two specimens of welded joints according to JIS Z 3121 and two specimens of weld metals according to JIS Z 3111. Instrumented impact tests were performed on three Charpy V specimens according to JIS Z 2202, in the temperature range between 20°C and – 80°C. V notches were positioned so that fracture could be propagated either in weld metal or in HAZ. NDT temperature according to ASTM E208/69 was determined by testing 6 specimen of the P3 type, using an energy of 540 J. The crack starter direction was perpendicular to the direction of the welded joint. Explosion bulge tests were performed on specimens of 500x500x47 mm, with a crack starter made according to ASTM E208. An explosive charge of 6 kg TNT was placed 200 mm above the specimens. The charge and the distance were determined on the basis of previous tests on the plates. Depth of the bulge, thickness on the top of the bulge and crack propagation were measured after each

TABLE 2 Prototype Strains measured by Strain Gauges

			FIRST LOADING STAGE						SECOND LOADING STAGE							
Pressure		p, MPa	2.94	7.35	7.35	8.33	9.02	0.49	2.94	7.35	9.02	9.81	10.80	11.47	12.06	**0.49**
Calculated Values		σ_φ, MPa	130	325	325	369	399		130	325	399	434	477	507	533	
		σ_z, MPa	65	163	163	184	199		65	163	199	217	239	254	267	
		$\epsilon_\varphi \cdot 10^{-4}$	524	1311	1311	1485	1608		524	1311	1608	1747	1930	2044	2149	
		$\epsilon_z \cdot 10^{-4}$	123	308	308	349	378		123	308	378	411	456	481	506	
		Strain gauge	$\epsilon \cdot 10^{-4}$ (%)						$\epsilon \cdot 10^{-4}$ (%)							
ϵ_φ	Plates	51	360	1131	1195	1360	1455	60	380	1200	1496	1655	1847	1999	2131	215
		53	485	1453	1527	1757	1895	188	489	1445	1790	1971	2212	2395	2560	273
		55	443	1364	1455	1655	1777	195	445	1365	1710	1872	2080	2238	2384	268
		56	368	1340	1438	1663	1812	280	431	1380	1764	1940	2201	2415	2605	485
		58	486	1497	1600	1845	2008	293	450	1490	1861	2072	2342	2550	2745	445
	Longitudinal Welds	1	191	692	815	927	1021	333	162	680	873	962	1059	1158	1236	174
		1°	572	1832	1955	2283	2554	445	581	1800	2369	2659	3130	3553	3950	1053
		2	585	1951	2095	2475	2781	785	547	1771	2234	2557	3093	3621	4066	1356
		4°	359	1150	1273	1476	1619	308	361	1270	1488	1743	1912	2066	2194	296
		5	271	987	1110	1274	1399	310	293	1071	1375	1480	1269	1775	1898	305
		6	410	1383	1518	1766	1955	373	434	1478	1863	2062	2326	2576	2792	560
		17	512	1644	1777	2082	2329	591	533	1657	2080	2349	2770	3180	3560	999
		18	487	1701	1837	2148	2401	682	497	1572	1973	2241	2678	3097	3476	1090
		20	394	1475	1608	1900	2138	637	438	1407	1778	2018	2366	2665	2923	755
		25	535	1642	1759	2020	2239	598	510	2126	2776	4266	4430	4599	4811	1097
		26	438	1435	1551	1778	1961	646	427	1871	2466	3928	4058	4197	4389	1091
		27	516	1689	1831	2130	2370	847	483	2044	2694	4200	4465	4736	5039	1528
		28	435	1409	1536	1768	1953	661	423	1856	2473	3910	4054	4211	4410	1169
		29							620	2849	3271	4833	5203	5575	5979	1673
		34							620	2492	3350	4821	5206	5597	6042	1728
	Circular Welds	52	513	1612	1690	1942	2097	188	543	1640	2037	2259	2525	2727	2910	350
		54	438	1402	1459	1663	1787	185	465	1412	1740	1921	2146	2320	2470	315
		57	458	1557	1630	1872	2030	195	492	1625	2020	2245	2531	2752	2954	485
		59	519	1684	1782	2045	2203	375	500	1605	2005	2210	2475	2680	2865	422
	Weld Crosses	2°	305	1084	1207	1403	1530	355	356	1240	1560	1708	1868	2008	2132	270
		5°	515	1736	1847	2130	2320	544	505	1600	2015	2230	2516	2764	3050	513
		6°	405	1323	1446	1694	2867	413	430	1380	1742	1936	2195	2405	2612	540
		35							522	2156	2887	4244	4348	4436	4586	829
		36							500	2148	2904	4373	4537	4702	4940	1171
ϵ_z	Longitudinal Welds	3	231	736	846		1039	234	215	789	993	1098	1206	1319	1412	235
		7	−91	−109	17	3	1	188	−57	52	81	64	15	2	−9	91
		9	156	567	681	795	905	399	139	557	703	819	933	1043	1142	358
		11	56	350	464	511	554	220	117	460	655	621	631	662	691	102
		16	−35	126	238	254	276	220	42	252	316	327	307	310	315	86
		19	91	559	681	778	857	333	168	639	816	871	932	992	1060	196
		21							175	1187	1588	3028	2913	2894	2947	747
		22		340	656	720	777		167	1191	1593	2978	2842	2788	2817	699
		24	134	350	454	484	523	306	116	991	1352	2676	2548	2479	2494	630
	Circular Welds	13	249	1009	1134	1313	1459	631	261	852	1067	1213	1402	1574	1734	504
		14	215	831	940	1079	1195	474	231	758	949	1065	1210	1362	1491	411
		15	−29	84	188	185	195	137	43	248	310	311	267	252	246	27
	Plate	23	279	921	1040	1179	1299	616	249	1416	1882	3409	3388	3413	3509	933

*

$$\sigma_\varphi = \frac{D\,p}{2s} \qquad \sigma_z = \frac{D\,p}{4s} \qquad \epsilon_\varphi = \frac{1}{E}[\sigma_\varphi - \nu\,\sigma_z] \qquad \epsilon_z = \frac{1}{E}[\sigma_z - \nu\,\sigma_\varphi]$$

s - Wall thickness

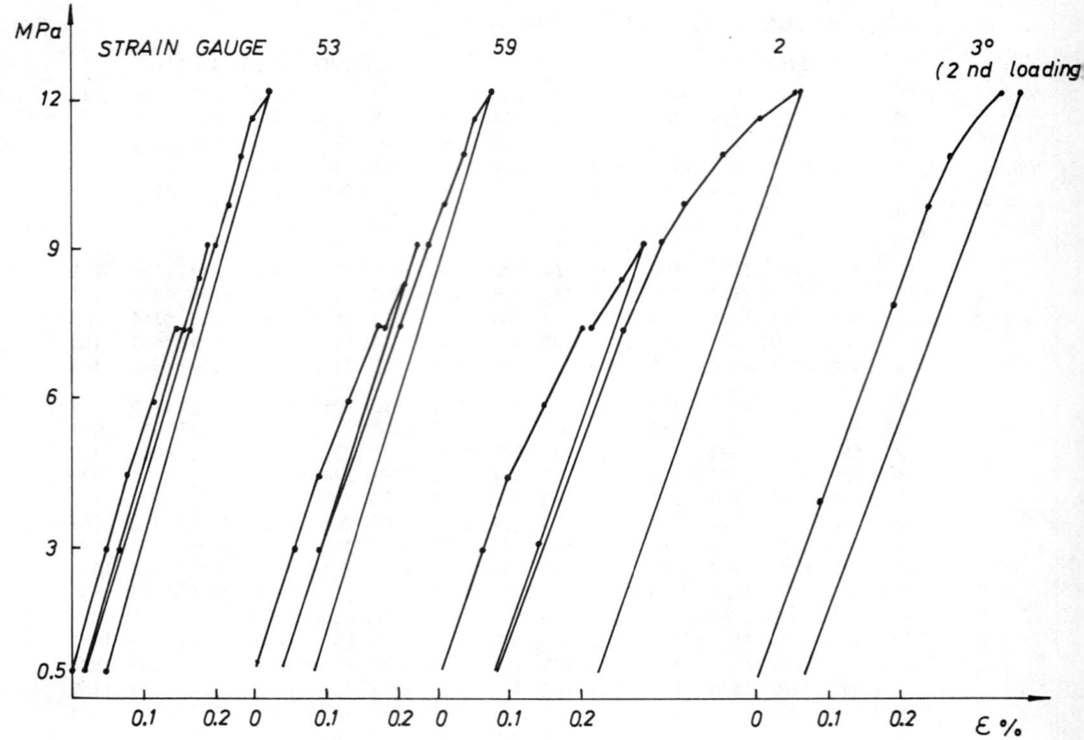

Fig. 3. Strains vs. pressure for some strain gauges on prototype

explosion. Hardness of welded joints was determined for all types of welds from the prototype and trial samples.

RESULTS

Deformations during Pressurizing of the Prototype

The behaviour of the prototype as a whole during pressurizing was satisfactory. The maximum local plastic deformation was recorded by strain gauges 2 and 27 and amounted to 0.214% and 0.24 respectively. The total plastic deformations were obtained by adding the residual deformations after the first and the second loading stages in Table 2 for a pressure of 0.49 MPa, upon unloading [e.g. for strain gauge 2 : $(785 + 1356) \cdot 10^{-4} = 0.214\%$]. Based on the high recorded residual deformation after the first loading stage by strain gauge 2, two additional strain gauges, 29 and 34, were placed on the cross weld LS1 – CM, near the knee, because it was felt that the highest plastic flow could occur at this part of the prototype. During the second loading cycle a maximum total (0.6%) and plastic (0.17%) deformations on the prototype were indeed found on strain gauges 29 and 34 (Table 2). After unloading, the reversed deformation at that place was about 0.43%. A straight and reverse component of elastic and plastic deformation, that occurred during the loading cycle, was contained in it. The recorded residual deformation of 0.17% is the minimum amount of plastic deformation that took place there. The real plastic deformation in the welded joint at this place is higher by an unspecified value, which is implicitly contained in the reverse elastic and plastic components. Reverse elastic deformation remained after unloading of the prototype in the welded joints in the form of redistributed internal stress. This was generally valid for all the welded joints. Diagrams in Fig. 3. show the variation of strains with pressure during prototype testing for strain gauges 53 (50 mm gauge length, plate, base metal), 59 (50 mm gauge length, circular weld), 2 (longitudinal weld) and 3° (large–strain strain gauge on longitudinal weld, diagram from x–y recorder, y connected with pressure transducer). Results obtained by the Moiré technique confirmed the general picture of prototype deformation. The amount of plastic deformation of the specimens cut from the prototype was estimated on the basis of the general distribution of plastic and total deformation, measured during prototype testing (Fig. 2, Tables 3 and 4).

TABLE 3 Results of Tensile Tests of Welded Joints and Weld Metals and Hardness of Welded Joints

Welded Joint	Specimen		Welded Joint Y.P. MPa	Welded Joint U.T.S. MPa	Weld Metal Tensile Properties Y.P. MPa	U.T.S. MPa	El. %	Reduction of Area ψ %	n · 10^{-2}	Hardness of Welded Joint HV HAZ max	HAZ min	Weld Metal (mean)
MAW–L	1		761	825	796	848	22,1	67.4	5.5	330	225	265
	5	ϵ_{pl} %	0.09		0.10–0.14							
			739	815	783	811	22.2	67.2	4.4	285	228	260
SAW–L	2		761	793	672	767	22	67.9	8.4	335	222	242
	6	ϵ_{pl} %	0.16		0.12–0.14							
			759	786	693	749	23.3	63.8	7.6	325	230	240
SAW–C	3		774	804	683	762	23.5	63.5	7.8	335	232	240
	7	ϵ_{pl} %	0.12		0.12–0.14							
			754	795	722	778	24	63.7	7.9	330	222	258
MAW–C	4		782	822	757	796	21.8	68.1	6.1	335	220	265
	8	ϵ_{pl} %	0.085		0.07–0.11							
			769	800	781	814	20	63.4	4.7	325	225	255

L – Longitudinal C – Circular ϵ_{pl} – Plastic strain (%)

TABLE 4 Results of Charpy V Impact Tests, DW Tests and Explosion Bulge Tests

Welded Joint	Specimen		Charpy V Impact Properties: Energy for Unit Area, Total ρ_v, MJ/m^2 +20°C	Total ρ_v, MJ/m^2 – 60°C	Crack Initiation	Crack Propagation	Transition Temp. [50%] °C	DW–NDT Temp. °C	Explosion Bulge Test: Number of explosions	Deformation after 3 explosions h, mm	Deformation after 3 explosions Δs, %	Total Deformation h, mm	Total Deformation Δs, %
MAW–L	1		1.46	0.36	0.38	1.07	–60	–113					
	5	ϵ_{pl} %			0.12			0.17					
			1.88	1.07	0.43	1.45	–80	–100					
SAW–L	2		1.27	0.50	0.23	1.04	–34	–103	6	34.5	4	90	9.0
	6	ϵ_{pl} %	0.05–0.11					0.12	0.08–0.09				
			1.15	0.56	0.15	1.0	–60	–92	5	43.5	5.5	73	9.0
SAW–C	3		1.22	0.725	0.25	0.98	–60	–110	5	36	3.5	58	7.0
	7	ϵ_{pl} %			0.14			0.135	0.11				
			0.82	0.43	0.24	0.58	–56	–84	4	32	4	61	8.0
MAW–C	4		167	0.97	0.42	1.25	–60	–113	7	39.5	5.5	96	18.0
	8	ϵ_{pl} %			0.07			0.075	0.10				
			1.63	0.79	0.44	1.2	–44	–110	5	54	9	95	21.0

Mechanical Tests Results

The most important results of the mechanical tests are given in Table 3 for static tests and in Table 4 for impact tests.

Figs. 4 – 7 illustrate typical results. In Charpy V instrumented impact tests it was possible to separate the energy for crack initiation and crack propagation (Fig. 4). The energy used for crack initiation is practically constant at all the test temperatures and significantly higher for MAW welds. The energy for crack propagation and the total absorbed energy are higher for MAW welds.

The NDT temperatures for all the specimens are very low. Fig. 5 shows some specimens, tested by the drop weight test. The characteristic view of specimens after explosion bulge testing is given in Fig. 6. It should be noticed that fracture propagates in the bond (a,b) and in the base metal (c). Fig. 7 compares the results obtained with base metal specimens having a crack starter in a rolling direction and transverse to it, with the results of welded joint specimens.

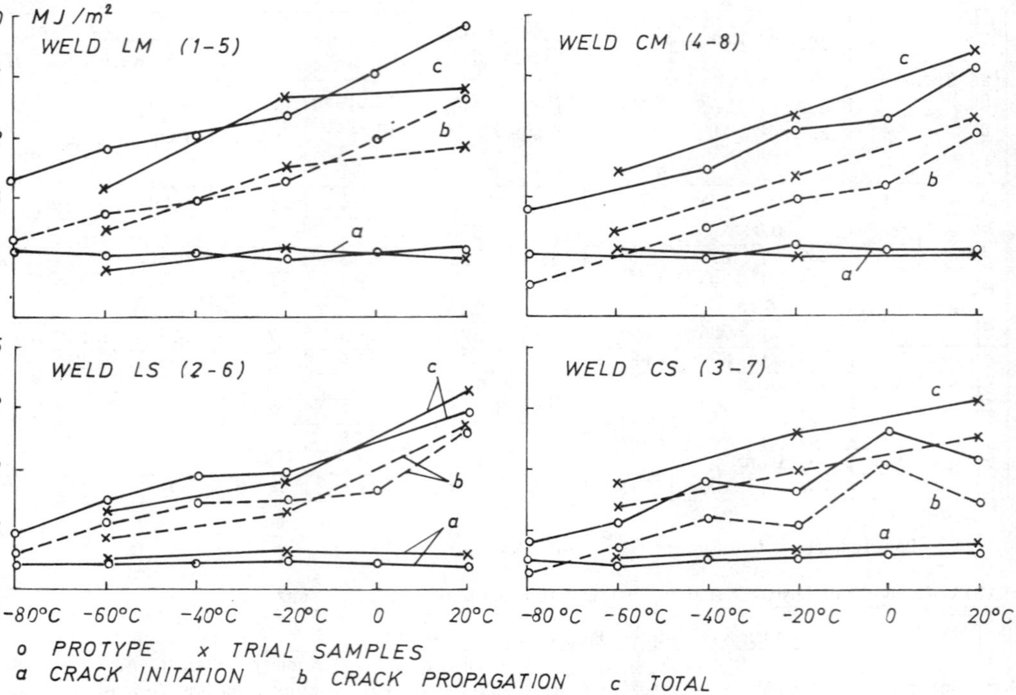

Fig. 4. Charpy V absorbed energy per unit area of weld metals

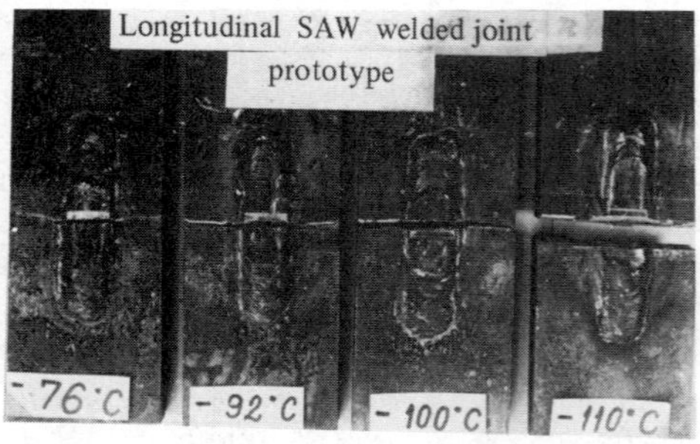

Fig. 5.. Determination of NDT temperature of weld metals by Drop Weight Test

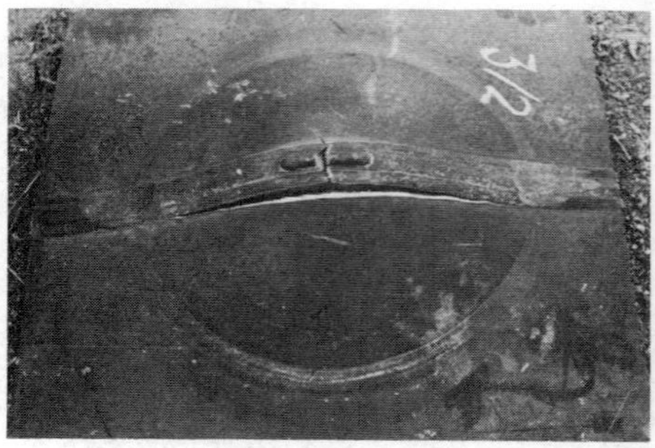

a.

c.

b.

Fig.. 6. Explosion Bulge Test

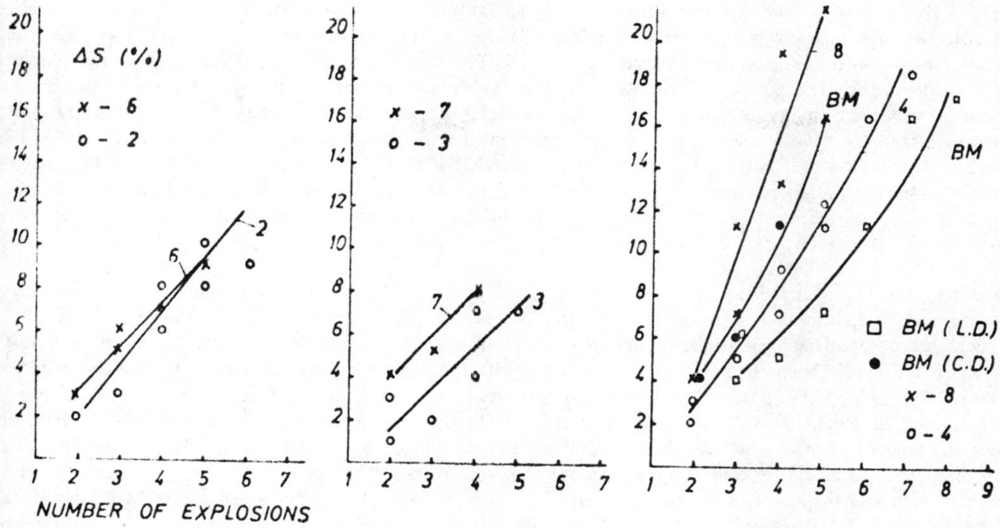

Fig. 7. Reduction of thickness Δ s (%) vs. number of explosions

INTERPRETATION AND DISCUSSION OF RESULTS

Recorded deformations that occurred during the pressurizing of the prototype up to design overloading of the Bajina Bašta penstock, which is 30% higher than the working pressure and amounts to 75% of the yield strength of SM80P steel, show that approximately calculated values in the direction of the main circumferential stresses are obtained only in steel plates. A significant plastic flow developed in the welded joints and particularly in the weld metal. The recorded residual deformations in the direction of the main circumferential stresses after prototype unloading, which are, as previously mentioned, the minimum amounts of real plastic deformations at these points are about 0.1%. The total residual deformations after both loading cycles at intersections of the circular and longitudinal welds, and in the vicinity of the knee, can exceed the yield strain of the weld metal (the above mentioned strain gauges 2, 29, 34 in Table 2). The recorded deformations in a direction parallel to the longitudinal axis of the prototype are not compatible with the calculated strains and, in this case, plastic flow also develops, particularly in the weld metal.

Plastic flow took place, as the stress – strain curves recorded by the strain gauges in Fig. 3 and acoustic emission measurements showed, only during the first loading of the prototype. The acoustic emission intensity was substantially reduced during repeated loading of the prototype up to the same pressure. The behaviour of the welded joints and weld metal was in accordance with the calculated values, i.e. approximately elastic, as illustrated by the second loading curves in Fig. 3 and by the recorded deformations up to a pressure of 9.02 MPa in the second loading cycle in Table 2. This is in agreement with the hydrostatic test of a similar welded prototype of HSLA steel (Gabrovšek, Rak, Šlibar, 1974).

The start of plastic flow at a stress level significantly below the yield stress can be accounted for by the presence of internal stresses in the welded joints, which are released during loading of the prototype and superimposed on hydrostatic pressure stresses. Internal stresses in the welded joints may be residual welding stresses, stresses retained in plates after cold forming, angular distortion stresses, misalignment stresses, restrained stresses, etc. As the measurements showed, residual stresses are active only during the first loading. As plastic deformation develops, residual stresses can be relieved and redistributed. The plastic flow is proof of partly relieved stresses and continued plastic flow after exceeding the maximum former pressure, as well as a large reverse deformation, recorded by several strain gaugesges during the second loading cycle (2, 27, 29, 34, 35, 36 in Table 2), proves stress redistribution.

Results in Tables 3 and 4 show the effect of initial plastic deformation brought about by prototype pressurizing on the fracture properties of welded joints.

The effect of residual plastic deformation on the tensile properties of welded joints is relatively small. The tensile strength of the specimens from the prototype (5, 6, 7, 8, Table 3) is somewhat lower than the tensile strength of the specimens from the trial samples (1, 2, 3, 4, Table 3) for all combinations of welded joints. The yield strength follows this trend, with the exception of specimens 1 – 5. The situation is different in the case of the tensile properties of weld metals. The position of the weld joint in the prototype, and not the plastic deformation, has an important influence on the tensile strength. The tensile strength of weld metal in longitudinal welds of the prototype is lower than the tensile strength of the weld metal of the trial samples, for SAW as well as for MAW (specimens 1 – 5 and 2 – 6, Table 3). In circular welds, weld metal of the prototype has a higher tensile strength than the weld metal in the trial samples (specimens 3 – 7 and 4 – 8). The same is valid for the yield point, with the exception of specimens 2 – 6. No similar regularities are noticed for elongation and reduction in area of the weld metal.

The values of the tensile properties of SAW weld metals (2 – 6 and 3 – 7) are considerably lower than those of MAW weld metals (specimens 1 – 5 and 4 – 8). But not even a reduction of 90 MPa in the yield point of SAW weld metal has a considerable influence on the yield point of welded joints. This illustrates the effect of a soft interlayer on the tensile strength of welded joints or the effect of an undermatching electrode, considered in recent literature (Bakshi, Shron, 1962). Satoh, 1974, emphasizes the importance of this effect for the welding of high strength steels. The index of work hardening, n, calculated for the weld metal from the tensile diagram, indirectly indicates the amount of plastic deformation of weld metals, induced during pressurizing of the prototype. It is higher for the weld metal of the trial samples than for the weld metal of the prototype, with the exception of specimens 3 – 7 in which no difference was observed. The index of work hardening is higher for SAW weld metals. The hardness of HAZ varies from 220 to 335 HV, while in the weld metal it is practically constant for the same welding material. The maximum hardness is somewhat higher for the trial samples and the minimum hardness is somewhat higher for the prototype, with the exception of specimens 3 – 7.

The impact values of the weld metal (Table 4, Fig. 4) are high and satisfactory for all the welds. The same regularities, observed in the tensile properties, are also noticed here. The transition temperature, calculated on the basis of 50% impact toughness at room temperature, is higher in the circular welds (3 – 7 and 4 – 8), and lower in the longitudinal welds (1 – 5 and 2 – 6) for the weld metals of the prototype.

The values of HAZ Charpy V impact toughness, not shown in the tables, fall within the scatter band for the prototype and for the trial samples, and confirm the satisfactory behaviour of the welded joints.

NDT temperatures are higher in all welds of the prototype compared with the trial samples. Because of the fact that the DW test is related to the behaviour of the welded joint, it can be concluded that the residual deformation of welded joints of the prototype reduces their resistance to brittle fracture. The welding procedure strongly influences the reduction in resistance to brittle fracture. For the same procedure, the amount of residual deformation determines the degree of reduction. NDT temperatures are considerably increased for SAW welds (for 2 – 6 and 3 – 7 specimens the increases are 11°C and 26°C respectively), above those for MAW welds (for 1 – 5 and 4 – 8 the increases are 13°C and 3°C respectively). For both the circular and longitudinal welds higher increases correspond to higher degrees of plastic deformation (0.17 and 0.135% for specimens 5 – 7 compared with 0.12 and 0.075% for specimens 6 – 8, Table 4).

The adverse effect of residual deformation on behaviour under fast fracture conditions, determined in the DW test, disappears when only weld metal is considered. Plastic deformation of weld metal can cause either an increase or a decrease in resistance to brittle fracture, measured by the variation of transition temperature. The variation of Charpy V transition temperature in the weld metal of the prototype is primarily dependent on the position of the welds in the prototype, and the relation between the main direction of elastic deformation and the longitudinal direction of the weld. The increase in transition temperature in the weld metals of the prototype (specimens 7 and 8) compared to the weld metals of the trial samples (specimens 3 and 4) in the case of circular welds is connected with a correspondence between the main direction of plastic deformation and the direction of the welds. In contrast, the decrease in transition temperature in the weld metals of the prototype (specimens 5 and 6) in relation to the weld metals of the trial samples (specimens 1 and 2) is due to the perpendicular direction of the main strains to the longitudinal weld axes. This unexpected improvement in brittle fracture resistance of the weld metal, expressed by the decrease in transition temperature, amounts to 20°C for MAW (specimens 1 – 5) and 26°C for SAW (specimens 2 – 6) welds.

The behaviour of all the types of welded joints tested by the explosion bulge test, was satisfactory (specimens 1 – 5 were not tested, because of a shortage of material). It should be noted that conditions for such a fast fracture do not occur in penstocks. In all the specimens tested, large plastic deformation precedes fracture propagation from the crack starter (Fig. 6). The crack initiated from the crack starter, propagated in the bond when the rolling direction and longitudinal weld direction were parallel and the crack starter normal to it (Fig. 6 a, b). With the crack starter ,positioned parallel to the rolling direction, in the case of a weld crosswise to the rolling direction, fracture propagated into a base metal. The crack propagated slowly and uniformly in successive explosions after initiation. The differences in behaviour of specimens, taken from the prototype and the trial samples, can be seen in Fig. 7. For the same number of explosions, specimens of the prototype (6, 7 and 8 in Table 4) have larger plastic deformation than specimens taken from the trial samples (2, 3 and 4 in Table 4), and sustain a lower number of explosions before fracturing. The number of explosions sustained before fracturing is the main result in the explosion bulge test. Therefore, the results obtained are in good agreement with the results of the DW tests, showing that the plasticity of welds is reduced by plastic deformation after pressurizing. The number of explosions before fracturing is higher for MAW welds than for SAW welds (Table 4).

Plastic deformation in the prototype welds decreases their resistance to brittle fracture, as one can conclude from the results of DW tests and explosion bulge tests. In the case considered this decrease in weld properties does not critically influence the safety of the Bajina Bašta penstock, due to the good plasticity of the selected steel and consumables and to the small effect of developed plastic deformation on the residual plasticity. This is well illustrated by the results of the explosion bulge test, that is used here in its technological modification with crack starter, as a substitute for the explosion burst test, currently used for pressure vessel safety control (Degenkolbe, 1968; Gabrovšek, Rak, Šlibar, 1974). The observed adverse effect of initial plastic flow on welded joints could be of critical importance in other situations and with other structural materials, and it might be taken into consideration in assessing similar structures.

The aim of the prototype investigation was to estimate the influence of initial plastic deformation on the resistance of welded joints and of the penstock as a whole to fast fracture. However, the results obtained showed non uniform and diverse influences of plastic deformation on weld properties. The amount of plastic deformation could be correlated with a decrease in fast fracture properties only by DW test results for each welding procedure; even then in some cases, the plastic deformation has no influence on the NDT temperature (specimens 4 – 8, Table 4). This disagreement suggests the existence of two opposite effects. The first one is negative and decreases the plasticity of welds which undergo plastic deformation. The second one is positive and reduces the effect of the first by a relief and redistribution of residual stresses in the welds. This positive effect can be used as a treatment for stress relief by pressurizing up to a determined level. The effect of plastic deformation on the weld metal is unexpected and a study of the mechanism operating would be interesting.

According to the test results, the weld metal will be work hardened when the directions of main plastic deformation and weld coincide, giving increasing tensile strength and the Charpy V transition temperature of the weld metal (circular welds, 3 – 7 and 4 – 8, Tables 3 and 4). When those directions are perpendicular (longitudinal welds 1 – 5 and 2 – 6) work softening takes place, tensile strength decreases and the transition temperature will be lower. The structural changes in the weld metal can explain such behaviour. The weld metal structure forms during intensive heat conduction, on the plate edges normal to the weld direction. Therefore the main dendritic axes and columnar crystals in the weld metal structure are oriented in the direction normal to the weld axis. Work hardening in circular welds takes place when plastic flow develops normal to the weld metal structure direction. Work softening takes place when plastic deformation develops in the direction of the weld metal.

The initiation of fast fracture, which, in penstocks, is more critical than its propagation (Hasebe, Koga, Okazawa, 1973). could not be analysed from the tests performed. The problem of fast fracture initiation, as well as the identification of residual stress components, requires further investigation of the influence of initial plastic deformation on the behaviour of the welded joints under fast fracture conditions.

CONCLUSION

A prototype for the Bajina Bašta Pumping–Up Water Power Plant penstock was constructed with the aim of testing its behaviour under hydrostatic pressurizing up to the occurrence of initial plastic deformation. The prototype was constructed of 47 mm steel plates of HT80 weldable steel.

Appreciable plastic deformation of about 0.1%, disproportionately large for the applied stress, occurred upon the attainment of penstock overloading pressure. The overloading pressure amounted to 12.06 MPa and produced stress equal to 75% yield stress of the steel. Plastic deformation was concentrated in the welded joints, predominantly in the weld metal, and locally, at the most critical points, exceeded the guaranteed yield strain of steel and welded joints. Plastic deformation occurred only upon the first loading. During subsequent loading up to the same pressure, the corresponding deformation was approximately elastic. The plastic flow was reactivated by a further increase in the load above the former maximum level.

The results of drop weight tests and explosion bulge tests showed that the initial plastic deformation slightly reduced the resistance of all types of welded joints to fast fracture. A change in the properties of welded joints after initial plastic deformation could be accounted for by two opposite effects. The first adversely affects the resistance, lowering the plasticity margin of the welded joints by the amount of plastic flow that had already occurred. The second beneficially affects the resistance by relief and redistribution of internal stress in the welded joints. The impact properties of the weld metal were also affected by plastic deformation. The plastic flow parallel to the longitudinal axis of the welded joints increased the Charpy V transition temperature of the weld metal. Surprisingly enough the plastic flow normal to the longitudinal axis of welded joints lowered it. The effect may probably be accounted for by structural anisotropy due to the directional solidification structure of the weld metal.

The results obtained pertain to the broader field of Post Yield Fracture Mechanics and show that initial plastic deformation definitely affects the resistance of welded joints to fast fracture. In the prototype examined the effect was not significantly harmful for the safety of the penstock, because of the high plasticity margin of the steel and the welding consumables selected, but in some critical cases it could be more important and should be considered in the future.

ACKNOWLEDGEMENTS

The authors are indebted to the GOŠA OMK Factory for permission to publish this paper. They also wish to thank Professor Dr H. Petershagen for his help in the interpretation of results.

REFERENCES

Bakshi, D.A. (1962). The static tensile strength of welded joints with a soft interlayer. Svar. Proiz. 5, 6 – 10.

Degenkolbe, J. (1968). Wasservergütete Schweissbare Baustahle – Herstellung und Eigenschaften. Schweizer Archiv. Januar 1968.

Deutscher Ausschuss für Stahlbau (1974). Anwendung der hochfesten schweissgeeigneten Feinkornbaustähle St E 47 und St E 70, Dast Richtlinie 011.

Gabrovšek, M., Rak, I., Šlibar, F. (1974). Development of production of fine–grain steels at Jesenice Iron and Steel Works – Problems of welding and welded structure stress annealing. International Symposium Weldability of heightened and high strength metals.

Hasebe, S., Koga, T., Okazawa, T. (1973). Low temperature burst test of large size welded steel. Sumitomo Search. 9

Horikawa, K. (1971). Fabrication of huge penstocks with 70, 80 kg/m² tensile strength steel, J. of Jap. Weld. S. 41. No 11.

Satoh, K. and Tyoda, M. (1973). Static tensile and brittle fracture strength of soft welded joints. Trans. JWRI. 2.

Technical standards for gates and penstocks. (1971). Water Gate and Penstock Association

Volume 3, ICM 3, Cambridge, England, August 1979

J INTEGRAL OF A MIXED MODE CRACK AND ITS APPLICATION

H. Ishikawa, H. Kitagawa* and H. Okamura**

University of Electro-Communications, 1-5-1 Chofugaoka, Chofu-shi, Tokyo 182, Japan
***Institute of Industrial Science, University of Tokyo, 7-22-1 Roppongi, Minato-ku, Tokyo 106, Japan**
****University of Tokyo, 7-3-1 Hongo, Bunkyo-ku, Tokyo 113, Japan**

INTRODUCTION

In recent work, the J integral proposed by Rice (1968a) has been successfully adopted as an elastic-plastic fracture criterion in fracture mechanics (Begley and Landes, 1972; Landes and Begley, 1974). Some numerical or experimental estimation procedures of the value of the J integral have been developed for its effective use (Bucci and co-workers, 1972; Rice, Paris and Merkle, 1973; Merkle and Corten, 1974; Hayes and Turner, 1974; Shir and Hutchinson, 1976). However, the J integral has in most cases applied only to the case of the first mode crack or crack-like notch. For instance, in the experimental determination of the fracture toughness of materials, cracked specimens of mode I are in current use. However, cracks or crack-like notches in actual structures are very often subjected to the conditions of combined mode I and mode II. Then, if one wants to apply the concept of the J integral ral to mixed mode cracks or notches in actual structures, it would be required to find a method to separate the J integral of a mixed mode crack or notch into two components of mode I and mode II.

However, most existing analytical or experimental estimating procedures cannot be effectively available for the separation. In this paper, an analytical method to estimate separately J_I related to the mode I deformation and J_{II} related to the mode II deformation from the J integral of a mixed mode crack or notch will be presented. Moreover, it is proved that both these J_I and J_{II} are path independent integrals in a linear elastic deformation field of an elastic or elastic-plastic material, as was the original J integral, and that $J = J_I + J_{II}$.

From the separation of the J integral into J_I and J_{II}, it can be demonstrated that there is a possibility that the concept of the J integral can also be applied to various classes of mixed mode fracture of ductile materials. For instance, by applying the separation method to a mixed mode crack in ductile materials, we may discuss how the fracture behaviour of the crack, such as an initiation condition of a stable crack or the direction of an initiated crack, is influenced by the history of plastic deformation at the crack tip. Moreover, by means of these J_I and J_{II}, in linear elastic material and also for small scale yielding stress intensity factors, K_I and K_{II} of a mixed mode crack can be obtained with fairly good accuracy.

ANALYTICAL PROCEDURE

The method for separating the J integral into J_I and J_{II} is as follows:
(1) First, this method proposes the analytical separation of stress σ_{ij}, strain ε_{ij}, traction vector $\vec{T}$, and displacement vector $\vec{u}$, near the tip of a crack into two components of mode I and mode II, respectively, that is, σ_{ij}^{I} and σ_{ij}^{II}, ε_{ij}^{I} and ε_{ij}^{II}, $\vec{T}^{I}$ and $\vec{T}^{II}$, and $\vec{u}^{I}$ and $\vec{u}^{II}$, where superscripts I and II show mode I and mode II, respectively.
(2) Next, by means of each component of stress, strain, traction vector and displacement vector, the J integral along an integration path in a linear elastic deformation field is also separated into two components, J_I and J_{II}.

Definition of the J integral

The J integral is defined for two-dimensional problems and is given by the equation

$$J = \int_{\Gamma}[W dx_2 - \vec{T}(\partial\vec{u}/\partial x_1)ds] \tag{1}$$

where W is the strain energy density as defined by

$$W = W(\varepsilon_{ij}) = \int_0^{\varepsilon}\sigma_{ij}d\varepsilon_{ij} \tag{2}$$

and, as shown in Fig. 1, Γ is any contour surrounding the crack tip, $\vec{T}$ is the traction vector defined by the outward normal $\vec{\ell}$ along Γ, $T_i = \sigma_{ij}\ell_j$, and ds is an increment of the arc length along Γ. In Eq. (1), x_1 and x_2 are rectangular coordinates normal to the crack front x_2 being perpendicular to the crack surface, as shown in Fig.1.

For any linear or nonlinear elastic material, Rice has proved that the J integral is path independent. That is, J has the same value for any contour surrounding the crack tip. The assumptions for having the path independent integral are:
(1) The strain energy density W defined by Eq. (2) is partially differentiable with respect to strain.
(2) The stress field is in equilibrium:

$$\sigma_{ij,j} = 0 \quad \text{(equilibrium of force)} \tag{3}$$

$$\sigma_{ij} = \sigma_{ji} \quad \text{(equilibrium of moment)} \tag{4}$$

(3) The strain is compatible, that is, it derives from a displacement field:

$$\varepsilon_{ij} = (u_{i,j} + u_{j,i})/2 \tag{5}$$

(4) The traction is zero on the surface of the crack.

The path independent nature of the J integral is generally true for any deformation mode of a crack, that is, mode I, mode II or mixed mode.

Separation of Stress, Strain and Displacement into Two Components of Mode I and Mode II

First, separate stress σ_{ij} at point P of coordinates (x_1,x_2) shown in Fig. 1 into two components, σ_{ij}^{I} and σ_{ij}^{II}, of mode I and mode II. Deformations of mode I and mode II are symmetric and skewsymmetric, respectively, with respect to x_1 axis. Then, the stress σ_{ij} at the point P can be separated in such a form that σ_{ij} is the sum of σ_{ij}^{I} and σ_{ij}^{II} that are expressed in terms of stress σ_{ij} at point P of coordinates (x_1,x_2) and stress σ'_{ij} at point P' of coordinates $(x_1,-x_2)$, as shown in Eq. (6) and Eq. (7).

$$\sigma_{ij} = \sigma_{ij}^{I} + \sigma_{ij}^{II} \tag{6}$$

$$\begin{Bmatrix} \sigma_{11}^{I} \\ \sigma_{22}^{I} \\ \sigma_{12}^{I} \end{Bmatrix} = \frac{1}{2}\begin{Bmatrix} \sigma_{11} + \sigma'_{11} \\ \sigma_{22} + \sigma'_{22} \\ \sigma_{12} - \sigma'_{12} \end{Bmatrix} , \quad \begin{Bmatrix} \sigma_{11}^{II} \\ \sigma_{22}^{II} \\ \sigma_{12}^{II} \end{Bmatrix} = \frac{1}{2}\begin{Bmatrix} \sigma_{11} - \sigma'_{11} \\ \sigma_{22} - \sigma'_{22} \\ \sigma_{12} + \sigma'_{12} \end{Bmatrix} \tag{7}$$

Here, ( )' denotes the value of stress, displacement, etc. at point P' that is a symmetrical point of point P with respect to the x_1 axis. For instance.

$$\sigma'_{ij} = \sigma'_{ij}(x_1,x_2) = \sigma_{ij}(x_1,-x_2) \tag{8}$$

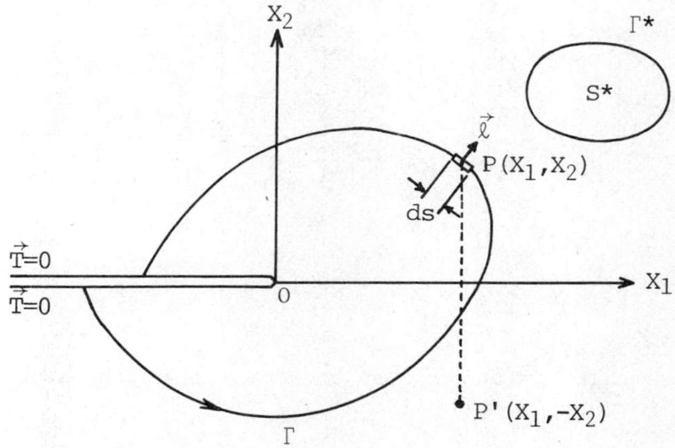

Fig. 1 Crack tip coordinate system and arbitrary line integral contour.

On the other hand, partially differentiating σ'_{ij} with respect to x_2, we have

$$[\partial\sigma'_{ij}/\partial x_2]_{(x_1,x_2)} = -[\partial\sigma_{ij}/\partial x_2]_{(x_1,-x_2)} \tag{9}$$

Therefore, substituting Eq. (7) into Eq. (4) and applying Eq. (9) to it,

$$\sigma_{ij,j}^{M} = 0 \qquad (M = I, II) \tag{3'}$$

and

$$\sigma_{ij}^{M} = \sigma_{ji}^{M} \qquad (M = I, II) \tag{4'}$$

are obtained, respectively. In other words, σ_{ij}^{M} (M = I, II) satisfy the equilibrium equation of force and that of moment.

Next, let us consider the separation of displacement $\vec{u}$ into two components, $\vec{u}^{I}$ and $\vec{u}^{II}$, of mode I and mode II. Because of the symmetrical deformation of mode I and skewsymmetrical deformation of mode II with respect to the x_1 axis, $\vec{u}^{I}$ and $\vec{u}^{II}$ can also be defined

$$u = \vec{u}^{I} + \vec{u}^{II} \tag{10}$$

$$\begin{Bmatrix} u_1^I \\ u_2^I \end{Bmatrix} = \frac{1}{2}\begin{Bmatrix} u_1 + u_1' \\ u_2 - u_2' \end{Bmatrix}, \qquad \begin{Bmatrix} u_1^{II} \\ u_2^{II} \end{Bmatrix} = \frac{1}{2}\begin{Bmatrix} u_1 - u_1' \\ u_2 + u_2' \end{Bmatrix} \tag{11}$$

On the other hand, using these displacement components, u_i^I and u_i^{II}, strain components, ε_{ij}^I related to u_i^I and ε_{ij}^{II} related to u_i^{II}, are defined by a strain-displacement relation such as

$$\varepsilon_{ij} = (u_{i,j}^M + u_{j,i}^M)/2 \qquad (M = I, II) \tag{5'}$$

Substituting Eq. (11) into Eq. (5'), we have

$$\begin{Bmatrix} \varepsilon_{11}^I \\ \varepsilon_{22}^I \\ \varepsilon_{12}^I \end{Bmatrix} = \frac{1}{2}\begin{Bmatrix} \varepsilon_{11} + \varepsilon_{11}' \\ \varepsilon_{22} + \varepsilon_{22}' \\ \varepsilon_{12} - \varepsilon_{12}' \end{Bmatrix}, \qquad \begin{Bmatrix} \varepsilon_{11}^{II} \\ \varepsilon_{22}^{II} \\ \varepsilon_{12}^{II} \end{Bmatrix} = \frac{1}{2}\begin{Bmatrix} \varepsilon_{11} - \varepsilon_{11}' \\ \varepsilon_{22} - \varepsilon_{22}' \\ \varepsilon_{12} + \varepsilon_{12}' \end{Bmatrix} \tag{12}$$

From Eq. (12), we obtain Eq. (13).

$$\varepsilon_{ij} = \varepsilon_{ij}^I + \varepsilon_{ij}^{II} \tag{13}$$

This equation shows that strain can also be separated in the same form as stress.

<u>Separation of the J Integral into J_I and J_{II} and Path Independence of J_I and J_{II}</u>

As an integration path of the J integral, let us consider here an integration path in a two-dimensional linear elastic deformation field. First, to separate th J integral into J_I and J_{II}, define the strain energy density, W_{MN}, by

$$W_{MN} = \int_0^{\varepsilon} \sigma_{ij}^M d\varepsilon_{ij}^N \qquad (M, N = I, II) \tag{14}$$

On the other hand, two components, T_i^I ($\vec{T}^I$) and T_i^{II} ($\vec{T}^{II}$), of mode I and mode II of the traction vector are related to σ_{ij}^I and σ_{ij}^{II}, respectively, and are given by

$$T_i = \sigma_{ij}\ell_j = (\sigma_{ij}^I + \sigma_{ij}^{II})\ell_j = T_i^I + T_i^{II} \tag{15}$$

or

$$\vec{T} = \vec{T}^I + \vec{T}^{II} \tag{15'}$$

where ℓ_j is the component of the unit normal vector, $\vec{\ell}$, drawn outwards on Γ. Substituting Eq. (6), Eq. (10), Eq. (13), Eq. (14) and Eq. (15') into Eq. (1), the J integral in Eq. (1) is transformed into

$$J = \int_\Gamma [\int_0^\varepsilon (\sigma_{ij}^{I}+\sigma_{ij}^{II}) d(\varepsilon_{ij}^{I}+\varepsilon_{ij}^{II}) dx_2 - (\vec{T}^{I}+\vec{T}^{II})\{\partial(\vec{u}^{I}+\vec{u}^{II})/\partial x_1\} ds] \qquad (16)$$

Then,

$$J = J_{II} + J_{I\,II} + J_{II\,I} + J_{II\,II} \qquad (17)$$

where

$$J_{MN} = \int_\Gamma [W_{MN} dx_2 - \vec{T}^{M}(\partial \vec{u}^{N}/\partial x_1) ds] \qquad (M,\ N = I,\ II) \qquad (18)$$

Next, let us prove that each of the four integrals J_{MN} is path independent. To prove the path independence, consider any closed integration path, Γ^*, enclosing a region, S^*, in the two-dimensional elastic deformation field, as shown in Fig. 1, the same as in the case of Rice's J integral. If the value, J_{MN}^*, of the integral J_{MN} along Γ^* is equal to zero, the integral J_{MN} is also path independent.

Gauss' divergence theorem for a vector, $\vec{A}$, is generally expressed as

$$\int_{\Gamma^*} A_i \ell_i ds = \iint_{S^*} A_{i,i} dS \qquad (19)$$

where A_i is the component of a vector $\vec{A}$. The right hand side of Eq. (19) is the area integral in S^*. Then, since

$$W_{MN} dx_2 = W_{MN} \ell_1 ds \quad \text{and} \quad \vec{T}^{M}(\partial \vec{u}^{N}/\partial x_1) ds = \sigma_{ij}^{M}(\partial u_i^{N}/\partial x_1)\ell_j ds\ ,$$

an application of Gauss' divergence theorem J_{MN}^* gives

$$J_{MN}^* = \iint_{S^*} [\partial W_{MN}/\partial x_1 - \{\sigma_{ij}^{M}(\partial u_i^{N}/\partial x_1)\}_{,j}] ds \qquad (20)$$

Substituting Eq. (14) into Eq. (20), we have

$$J_{MN}^* = \iint_{S^*} [\sigma_{ij}^{M}(\partial \varepsilon_{ij}^{N}/\partial x_1) - \sigma_{ij,j}^{M}(\partial u_i^{N}/\partial x_1) - \sigma_{ij}^{M}(\partial u_{i,j}^{N}/\partial x_1)] ds \qquad (21)$$

From Eq. (3'), the second term of the right hand side of Eq. (21) is found to be equal to zero. Applying Eq. (4') and Eq. (5') to the third term, we have

$$\sigma_{ij}^{M}(\partial u_{i,j}^{N}/\partial x_1) = \sigma_{ij}^{M}(\partial \varepsilon_{ij}/\partial x_1)$$

Therefore,

$$J_{MN}^* = 0 \qquad (M,\ N = I,\ II) \qquad (22)$$

for any closed integration path, Γ^*, in a linear elastic deformation field. This shows that each J_{MN} is path independent under the condition of $\vec{T}=0$ and $\ell_1=0$ on the flat crack surfaces.

Next, take any integration path, Γ, as a symmetrical contour, Γ_s, with respect to the x_1 axis, as shown in Fig. 2. In this case, there is the following relation between the unit normal vector, (ℓ_1, ℓ_2), drawn outwards at the point $P(x_1, x_2)$ on Γ_s and the unit normal vector, (ℓ_1', ℓ_2'), drawn outwards at the point $P'(x_1, -x_2)$ on Γ_s.

$$(\ell_1', \ell_2') = (\ell_1, -\ell_2) \qquad (23)$$

Then, using Eq. (7), Eq. (15) and Eq. (23), it is found that each component of the mode I or mode II traction vector, $\vec{T}(T_1,T_2)$, at the point P is related to that of the traction vector, $\vec{T}'(T_1',T_2')$, at the point P' by

$$\begin{Bmatrix} T_1^{I\,'} \\ T_2^{I\,'} \end{Bmatrix} = \begin{Bmatrix} T_1^{I} \\ -T_2^{I} \end{Bmatrix} \quad , \quad \begin{Bmatrix} T_1^{II\,'} \\ T_2^{II\,'} \end{Bmatrix} = \begin{Bmatrix} -T_1^{II} \\ T_2^{II} \end{Bmatrix} \tag{24}$$

In the same manner, by means of Eq. (7), Eq. (12) and Eq. (11) stress σ_{ij}, strain ε_{ij}, and displacement u_i at the point P can be related to stress σ_{ij}', strain ε_{ij}', and displacement u_i' at the point P', respectively. These relations are given, respectively, by

$$\begin{Bmatrix} \sigma_{11}^{I\,'} \\ \sigma_{22}^{I\,'} \\ \sigma_{12}^{I\,'} \end{Bmatrix} = \begin{Bmatrix} \sigma_{11}^{I} \\ \sigma_{22}^{I} \\ -\sigma_{12}^{I} \end{Bmatrix} \quad , \quad \begin{Bmatrix} \sigma_{11}^{II\,'} \\ \sigma_{22}^{II\,'} \\ \sigma_{12}^{II\,'} \end{Bmatrix} = \begin{Bmatrix} -\sigma_{11}^{II} \\ -\sigma_{22}^{II} \\ \sigma_{12}^{II} \end{Bmatrix} \tag{25}$$

$$\begin{Bmatrix} \varepsilon_{11}^{I\,'} \\ \varepsilon_{22}^{I\,'} \\ \varepsilon_{12}^{I\,'} \end{Bmatrix} = \begin{Bmatrix} \varepsilon_{11}^{I} \\ \varepsilon_{22}^{I} \\ -\varepsilon_{12}^{I} \end{Bmatrix} \quad , \quad \begin{Bmatrix} \varepsilon_{11}^{II\,'} \\ \varepsilon_{22}^{II\,'} \\ \varepsilon_{12}^{II\,'} \end{Bmatrix} = \begin{Bmatrix} -\varepsilon_{11}^{II} \\ -\varepsilon_{22}^{II} \\ \varepsilon_{12}^{II} \end{Bmatrix} \tag{26}$$

$$\begin{Bmatrix} u_1^{I\,'} \\ u_2^{I\,'} \end{Bmatrix} = \begin{Bmatrix} u_1^{I} \\ -u_2^{I} \end{Bmatrix} \quad , \quad \begin{Bmatrix} u_1^{II\,'} \\ u_2^{II\,'} \end{Bmatrix} = \begin{Bmatrix} -u_1^{II} \\ u_2^{II} \end{Bmatrix} \tag{27}$$

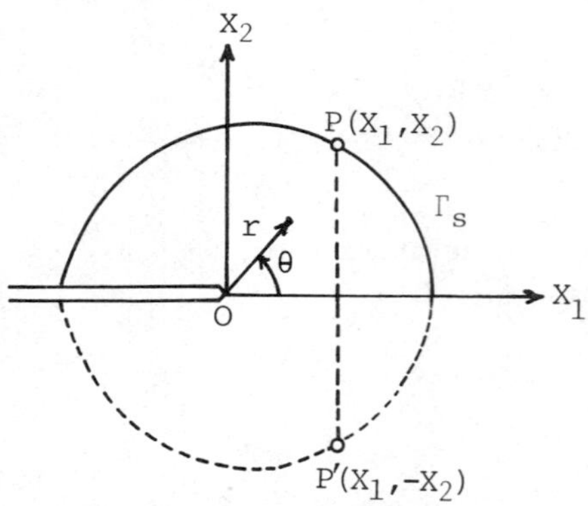

Fig. 2 Symmetrical line integral contour.

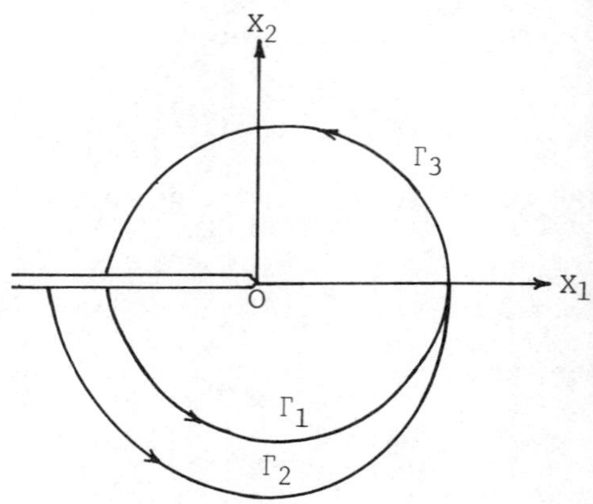

Fig. 3 Semi-circular line integral contours.

Using Eq. (24), Eq. (25), Eq. (26) and Eq. (27), we obtain the following equations:

$$W_{MN}' = \pm W_{MN} \tag{28}$$

$$\vec{T}^{M'}(\partial\vec{u}^{N'}/\partial x_1) = \pm\vec{T}^M(\partial\vec{u}^N/\partial x_1) \qquad (29)$$

where W_{MN}, $\vec{T}^M$, $\vec{u}^N$ are values at the point P and W'_{MN}, $\vec{T}^{M'}$, $\vec{u}^{N'}$ are values at the point P'. In Eq. (28) and Eq. (29), the positive sign denotes the case of M=N while the negative sign denotes the case of M≠N. Therefore, in the case of M≠N the integration values of Eq. (18) at symmetrical points with respect to the x_1 axis cancel each other out, and so we obtain

$$J_{MN} = 0 \qquad (M \neq N) \qquad (30)$$

Substituting Eq. (30) into Eq. (17),

$$J = J_{II} + J_{\text{II}\,\text{II}} \equiv J_I + J_{\text{II}} \qquad (31)$$

$$(J_{II} = J_I,\ J_{\text{II}\,\text{II}} = J_{\text{II}})$$

Since J_{MN} is path independent, Eq. (31) is generally true for any integration path in linear elastic deformation field of an elastic or elastic-plastic material.

Next, let us consider any three integration contours, Γ_1, Γ_2 and Γ_3, as shown in Fig. 3 (Γ_1 and Γ_3 are symmetrical with respect to the x_1 axis). As illustrated in Fig. 3, Γ_1 and Γ_2 are the contours (Γ_-) taken from the lower crack surface in a counterclockwise direction to the positive x_1 axis, and Γ_3 is the contour (Γ_+) taken from the positive x_1 axis in the same direction to the upper crack surface. Then, $\Gamma=\Gamma_1+\Gamma_3$ or $\Gamma=\Gamma_2+\Gamma_3$. In this case, from Eq. (29), it is found that the value of J_I for Γ_1 is equal to the value of J_I for Γ_3. Moreover, since J_I is path independent, the J_I value for $\Gamma_1+\Gamma_3$ is equal to that for $\Gamma_2+\Gamma_3$. Therefore, the value of J_I for Γ_2 is also equal to that for Γ_3. The same computation may be carried out for J_{II}.

Finally, the values of J_I and J_{II} for Γ are generally equal to double the values of J_I and J_{II} for any integration contour, Γ_- or Γ_+, respectively.

ENERGY INTERPRETATION OF J_{MN}

To clarify the meaning of the path independent integral, J_{MN}, let us consider a two-dimensional linear elastic material with a crack. Employing polar coordinates r,θ with origin at the crack tip, as shown in Fig. 2, stresses σ_{ij} and displacements u_i near the crack tip in linear elastic body are given as follows;

$$\begin{Bmatrix}\sigma_{11}\\ \sigma_{22}\\ \sigma_{12}\end{Bmatrix} = \frac{K_I}{\sqrt{2\pi r}}\begin{Bmatrix}\cos\frac{\theta}{2} - \cos\frac{\theta}{2}\sin\frac{\theta}{2}\sin\frac{3\theta}{2}\\ \cos\frac{\theta}{2} + \cos\frac{\theta}{2}\sin\frac{\theta}{2}\sin\frac{3\theta}{2}\\ \cos\frac{\theta}{2}\sin\frac{\theta}{2}\sin\frac{3\theta}{2}\end{Bmatrix} + \frac{K_{\text{II}}}{\sqrt{2\pi r}}\begin{Bmatrix}-2\sin\frac{\theta}{2} - \sin\frac{\theta}{2}\cos\frac{\theta}{2}\cos\frac{3\theta}{2}\\ \sin\frac{\theta}{2}\cos\frac{\theta}{2}\cos\frac{3\theta}{2}\\ \cos\frac{\theta}{2} - \cos\frac{\theta}{2}\sin\frac{\theta}{2}\sin\frac{3\theta}{2}\end{Bmatrix} + 0(r^0) \qquad (32)$$

$$\begin{Bmatrix}u_1\\ u_2\end{Bmatrix} = \frac{K_I}{2\mu}\sqrt{\frac{r}{2\pi}}\begin{Bmatrix}\cos\frac{\theta}{2}(\eta-1+2\sin^2\frac{\theta}{2})\\ \sin\frac{\theta}{2}(\eta+1-2\cos^2\frac{\theta}{2})\end{Bmatrix} + \frac{K_{\text{II}}}{2\mu}\sqrt{\frac{r}{2\pi}}\begin{Bmatrix}\sin\frac{\theta}{2}(\eta+1+2\cos^2\frac{\theta}{2})\\ -\cos\frac{\theta}{2}(\eta-1-2\sin^2\frac{\theta}{2})\end{Bmatrix} + 0(r^1) \qquad (33)$$

where K_I and K_{II} are stress intensity factors of mode I and mode II, respectively, μ is the modulus of shearing elasticity $\eta=(3-\nu)/(1+\nu)$ for plane stress and $\eta=3-4\nu$ for plain strain, ν is Poisson's ratio. We now evaluate the integral J_{MN} from these equations, Eq. (32) and Eq. (33), taking Γ to be a circle of radius a. Letting $a\to 0$, we obtain:

$$J_{I II} = J_{II I} = 0, \qquad J = J_I + J_{II} \tag{34}$$

$$J_I = (1+\eta)K_I^2/(8\mu), \qquad J_{II} = (1+\eta)K_{II}^2/(8\mu) \tag{35}$$

Not only for the circular integration path but also for any integration path Γ, Eq. (34) and Eq. (35) can also be obtained. From Eq. (35), it is found that in the linear elastic case (and also naturally for small scale yielding), J_I and J_{II} are equal to the energy release rate G_I of mode I and the energy release rate of G_{II} of mode II, respectively.

Next, let us consider nonlinear elastic material. In this case, the definition of W_{MN} in Eq. (14) is not sufficient for the case of mixed mode deformation of a crack. However, for the single mode deformation of a crack, that is, mode I or mode II deformation, the definition of W_{MN} (M=N) in Eq. (14) is sufficient, and it represents the strain energy density of a nonlinear elastic material. Then, in the case of single mode deformation, J_I or J_{II} is equal to Rice's J integral itself, and os it is path independent. Finally, for any nonlinear elastic body, J_I or J_{II} may be interpreted as the energy available for crack extention under pure mode I deformation or pure mode II deformation, respectively (Rice, 1968b; Budiansky and Rice, 1973).

CONCLUDING REMARKS

(1) For a mixed mode crack, Rice's J integral along an integration path in a linear elastic deformation field of an elastic or elastic-plastic material can be separated into J_I of mode I and J_{II} of mode II. A method of separation is proposed.
(2) It is proved that both these J_I and J_{II} are path independent integrals in a two-dimensional linear elastic deformation field, as is Rice's J integral, and that $J=J_I+J_{II}$. From this, it can be shown that there is a possibility that the concept of the J integral can be used as a failure criterion for ductile materials with a mixed mode crack, except for the case of general yielding.
(3) In linear elastic material and also for small scale yielding, J_I and J_{II} are equal to the energy release rate G_I of mode I and the energy release rate G_{II} of mode II, respectively, that is, $J_I=G_I=(1+\eta)K_I^2/(8\mu)$ and $J_{II}=G_{II}=(1+\eta)K_{II}^2/(8\mu)$. Then, combining these relations and the solution from numerical analysis of a mixed mode crack problem, such as finite element analysis, we can obtain stress intensity factors, K_I and K_{II}, of the mixed mode crack with fairly good accuracy.

ACKNOWLEDGEMENTS

The authors would like to express their hearty acknowledgements to the kind and valuable discussions given by Associate Professor K. Watanabe of the University of Tokyo, and Dr. S. Kubo of Osaka University.

REFERENCES

Begley, J. A. and J. D. Landes (1972). The J Integral as a Fracture Criterion. In Fracture Toughness, ASTM STP 514, American Society for Testing and Materials, pp.1-20.

Bucci, R. J. and co-workers (1972). J Integral Estimation Procedures. In Fracture Toughness, ASTM STP 514, American Society for Testing and Materials, pp.40-69.
Budiansky, B. and J. R. Rice (1973). Conservation Laws and Energy-Release Rate. Trans. ASME, Ser. E, 40, pp.201-203.
Hayes, D. J. and C. E. Turner (1974). An application of finite element techniques to post-yield analysis of proposed standard three-point bend fracture test pieces. Int. Journ. of Fracture, 10, pp.17-32.
Landes, J. D. and J. A. Begley (1974). Test Resuts from J-Integral Studies: An Attempt to Establish a J_{Ic} Testing Procedure. In Fracture Analysis, ASTM STP 560, American Society for Testing and Materials, pp.170-186.
Merkle, J. G. and H. T. Corten (1974). A J Integral Analysis for the Compact Specimen, Considering Axial Force as Well as Bending Effects. Trans. ASME, Ser. J, 96, pp.286-292.
Rice. J. R. (1968a). A Path Independent Integral and the Approximate Analysis of Strain Concentration by Notches and Cracks. J. Appl. Mech., 35, pp.379-386.
Rice, J. R. (1968b). Mathematical Analysis in the Mechanics of Fracture. In H. Liebowitz (Ed.), Fracture, Vol. 2, Academic Press, New York. pp.191-311.
Rice, J. R., P. C. Paris and J. G. Merkle (1973). Some Further Results of J-Integral Analysis and Estimates. In Progress in Flaw Growth and Fracture Toughness Testing, ASTM STP 536, American Society for Testing and Materials, pp.231-245.
Shih, C. F. and J. W. Hutchinson (1976). Fully Plastic Solutions and Large Scale Yielding Estimates for Plane Stress Crack Problems. Trans. ASME, Ser. H, 98, pp.289-295.

Volume 3, ICM 3, Cambridge, England, August 1979

ELASTIC-PLASTIC FRACTURE MECHANICS STUDY OF THERMAL SHOCK CRACKING

K. Hirano, H. Kobayashi and H. Nakazawa

Department of Physical Engineering, Faculty of Engineering, Tokyo Institute of Technology, Ōokayama 2-12-1, Meguro-Ku, Tokyo, Japan

INTRODUCTION

During the cold water injection through the emergency core cooling system (ECCS) following a postulated loss of coolant accident (LOCA), the inside wall of a reactor pressure vessel is subjected to high tensile thermal stresses, i.e., thermal shock. If the wall contains a crack-like defect at the inner surface, an unstable fracture may occur. Therefore, it is very important to investigate the initiation and propagation of thermal shock cracking from a structural design and a reactor safety viewpoint.

Many analytical and experimental studies concerning the thermal shock cracking have been carried out lately on the basis of linear-elastic fracture mechanics (Apblett, 1976; Blauel, 1974; Buchalt, 1975; Cheverton, 1978; Hasslman, 1971; Hirano, 1976, 1978; Kassir, 1971; Yagawa, 1978). However, thermal shock experiments on nuclear pressure vessel steels, and also a comparison of thermal shock fracture toughness with fracture toughness in the same high temperature range of the thermal shock experiment have barely been conducted. In addition to these facts, the linear-elastic fracture mechanics analysis of the fracture behavior in a high temperature range where the thermal shock cracking is of most interest is meaningless due to large plastic deformation at the crack tip.

This paper describes thermal shock experiments conducted on a nuclear pressure vessel steel (A533 Grade B Class 1), an AISI304 steel and a tool steel (JIS SKD62) using both a new thermal shock test facility and method. Analysis of their quasi-static thermal stress intensity factors is performed on the basis of linear-elastic fracture mechanics; and a thermal shock fracture toughness value, K_{tsc} is evaluated. Then, elastic-plastic fracture toughness tests are carried out in the same high temperature range of the thermal shock experiment, and a relation between the stretched zone width, SZW, formed as a result of the fatigue precrack tip plastic blunting and the J-integral is clarified. An elastic-plastic thermal shock fracture toughness value, J_{tsc}, is evaluated from a critical value of the stretched zone width, SZW_{tsc}, at the initiation of the thermal shock cracking by using the relation between SZW and J. The J_{tsc} value is compared with an elastic-plastic fracture toughness value, J_{Ic}, and the difference between these J_{tsc} and J_{Ic} values is discussed on the basis of fractography.

M.B.M.3—Q*

MATERIALS AND EXPERIMENTAL PROCEDURE

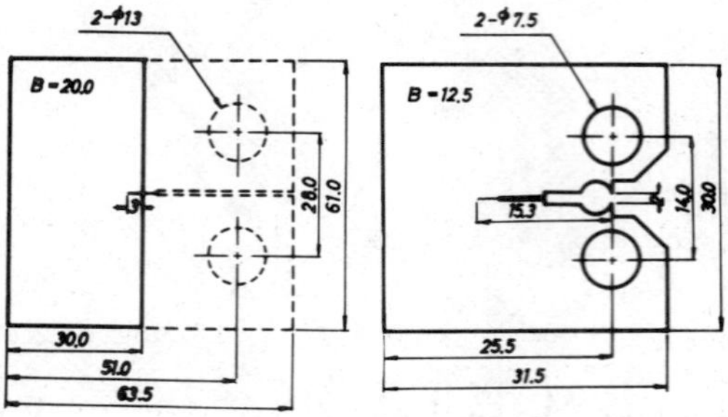

Fig. 1 Configuration and dimensions of test specimens

TABLE 1 Mechanical Properties

Material	σ_{ys} MPa	σ_B MPa	δ %	ψ %
A533B-1 steel	534	660	27.0	71.0
AISI304 steel	240	624	—	—
Tool steel	843	1255	9.0	22.0

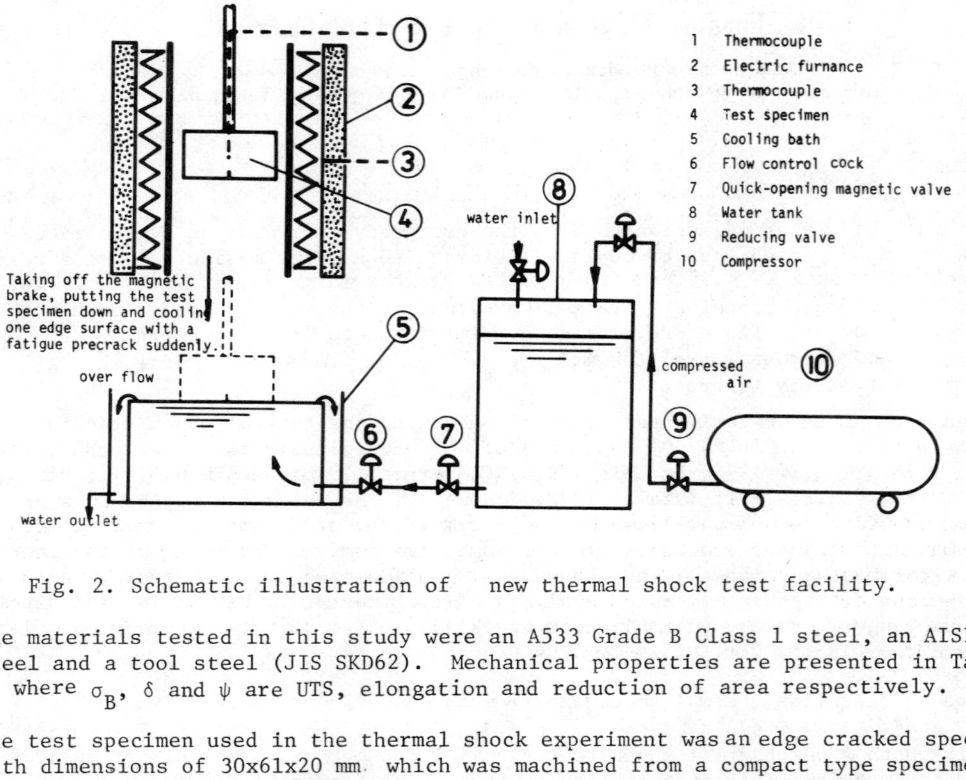

Fig. 2. Schematic illustration of new thermal shock test facility.

The materials tested in this study were an A533 Grade B Class 1 steel, an AISI304 steel and a tool steel (JIS SKD62). Mechanical properties are presented in Table 1, where σ_B, δ and ψ are UTS, elongation and reduction of area respectively.

The test specimen used in the thermal shock experiment was an edge cracked specimen with dimensions of 30x61x20 mm which was machined from a compact type specimen as shown in Fig. 1(a) after the fatigue precracking (for both the AISI 304 steel and tool steel, specimen thickness, B, is 5 mm.). Fatigue precrack length, a, was shorter than 5 mm. Thermal shock experiments were conducted by using a new thermal shock test facility (Hirano, 1976) as illustrated schematically in Fig. 2. Specimens were heated uniformly at different temperatures(in this study, maximum heating temperature, T_{max}, from 723 to 923K) in an electric furnace.
After one hour, one edge surface with a fatigue precrack was cooled suddenly in a cooling bath. After the thermal shock experiment, these specimens were fractured in liquid nitrogen, or were fractured by fatigue at room temperature. Then, a fractographic examination was conducted at fatigue precrack tip of the

midthickness of the specimen by using both a scanning and a transmission electron microscope in order to investigate whether the thermal shock cracking occurred or not.

The test specimen used in elastic-plastic fracture toughness tests was a 12.5 mm -J_{Ic} test specimen as shown in Fig. 1(b). All of the specimens were fatigue pre-cracked according to ASTM standards (E399-74, 1975) and for a distance of 3 mm to provide an overall crack-length-to-width ratio, a/W, of 0.60. Fracture toughness tests were conducted up to 673K by using an MTS closed-loop electrohydraulic materials test systems with a radiant heating furnace. An elastic-plastic fracture toughness was evaluated according to a J_{Ic} test procedure based on a fractographic method (Kobayashi, 1977). But in this study, these specimens were loaded at a rate such that the rate of increase of stress intensity factor on the straight part in the load-load line displacement curve (p-δ diagram) is nearly equal to that of the thermal stress intensity factor during the thermal shock condition ($\dot{K}_I$ = 15.5MNm$^{-3/2}$ /sec).

RESULTS AND DISCUSSION

Linear-Elastic Fracture Mechanics Studies

Analysis of stress intensity factor during the thermal shock condition. Figure 3 shows the analytical model of thermal stress intensity factor we have used to evaluate a thermal shock fracture toughness value, K_{tsc}, on the basis of linear elastic fracture mechanics. One edge surface with a crack, in a long rectangular plate, is subjected to uniform sudden cooling. The momentary temperature distribution across the plate measured relative to the initial temperature, $\Delta T(x,t)$, produces thermal stresses, $\sigma_y(x,t)$, parallel to the edge which are only dependent on the coordinate, x, and the time, t. They can be calculated by using linear elastic theory (Timoshenko, 1934).

$$\sigma_y(x,t) = -\alpha E\{\Delta T(x,t) - \frac{1}{W}\int_0^W \Delta T(x,t)dx - \frac{12(x-W/2)}{W^3}\int_0^W \Delta T(x,t)(x-W/2)dx\}$$

$$\sigma_x(x,t) = \tau_{xy}(x,t) = 0 \quad (1)$$

where, α is the linear coefficient of expansion of the material.

Thermal stress intensity factors have been calculated following the concept of superposition. These thermal stresses, Eq. (1), in the crack-free plate are used in conjunction with the known solution for a crack in an infinite plate loaded at its surfaces by an arbitrary pressure distribution (see for instance, Paris, 1965) to obtain the thermal stress intensity factor, $K_I(t)$. The analytical relation can be written in the following expression.

$$K_I(t) = 1.12\frac{2\sqrt{a}}{\sqrt{\pi}}\int_0^a \sigma_y(x,t)\frac{1}{\sqrt{a^2-x^2}}dx \quad (2)$$

Figure 4. shows one example of results calculated numerically from Eq. (2) by using both Eq.(1) and the temperature distribution measured in the thermal shock experiment. In this figure, K_{max} and ΔT_{max} are a maximum value of $K_I(t)$ with respect to time and a maximum temperature difference, $\Delta T_{max} = T_{max}-T_{water\ temp.}$, respectively. It can be seen from this figure that the normalized maximum thermal stress intensity factor, $K_{max}/\alpha E\Delta T_{max}\sqrt{W}$, increases and then decreases with increasing crack length.

Thermal shock fracture toughness. Figure 5 shows results of thermal shock experiment on a tool steel obtained on the basis of a linear-elastic fracture mechanics.

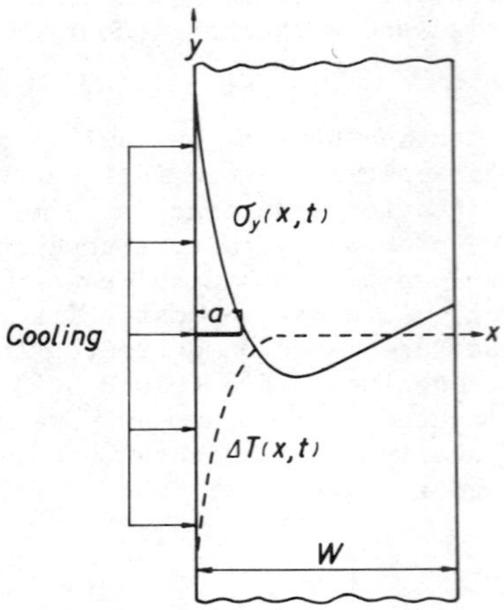

Fig. 3. Analytical model of thermal stress intensity factor subjected to a sudden cooling.

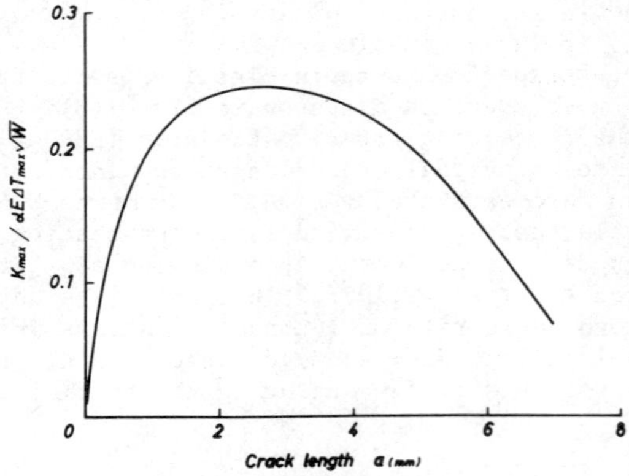

Fig. 4. Relation between normalized maximum thermal stress intensity factor $K_{max}/\alpha E\Delta T_{max}\sqrt{W}$, with respect to time and crack length, a.

In this figure, Δa_s and Δa_i are a crack extension measured at the surface of the specimen by using a photomicroscope and a crack extension measured at the midthickness of the specimen by using an electron microscope, respectively.

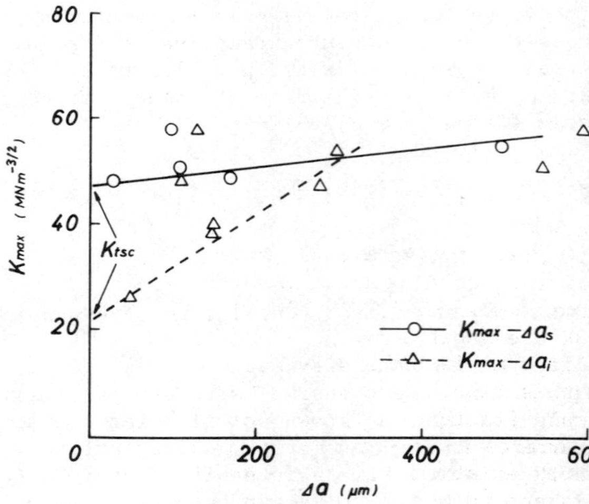

Fig. 5. Results of thermal shock experiments for a tool steel (JIS SKD62).

According to a fracture criterion on the basis of linear-elastic fracture mechanics, materials fracture when the stress intensity factor at a crack tip achieves a critical value, a plane strain fracture toughness value, K_{Ic}. As we have described in introduction, it is not reasonable to discuss the problems of thermal shock cracking at a high temperature from a linear-elastic fracture mechanics viewpoint. But there might be possibility of applying it to the high-strength low-toughness materials. We temporarily make the assumption that the above fracture criterion can be applied to the thermal shock cracking a tool steel. With this assumption, we define K_{tsc} as the stress intensity factor when the thermal shock cracking occurs at a fatigue precrack tip. It can be presumed from Fig. 5 that K_{tsc} for a tool steel is nearly $47 MNm^{-3/2}$. And it can be also seen from Fig. 5 that a crack extends at the midthickness of specimen even though there is no occurrence of crack extension at the surface of the specimen. Therefore, it is considered that K_{tsc} at the onset of a crack extension at the midthickness of the specimen is less than $22 MNm^{-3/2}$.

Elastic-Plastic Fracture Mechanics Studies

Elastic-plastic fracture toughness at a high temperature. As stated earlier, the above discussions concerning thermal shock fracture toughness from a linear-elastic fracture mechanics viewpoint is not valid for low strength and ductile materials owing to large plastic deformation at the crack tip. It is therefore necessary to discuss the thermal shock cracking on the basis of elastic-plastic fracture mechanics. We shall discuss results of the elastic-plastic fracture toughness tests obtained at a high temperature for an A533B-1 steel first.

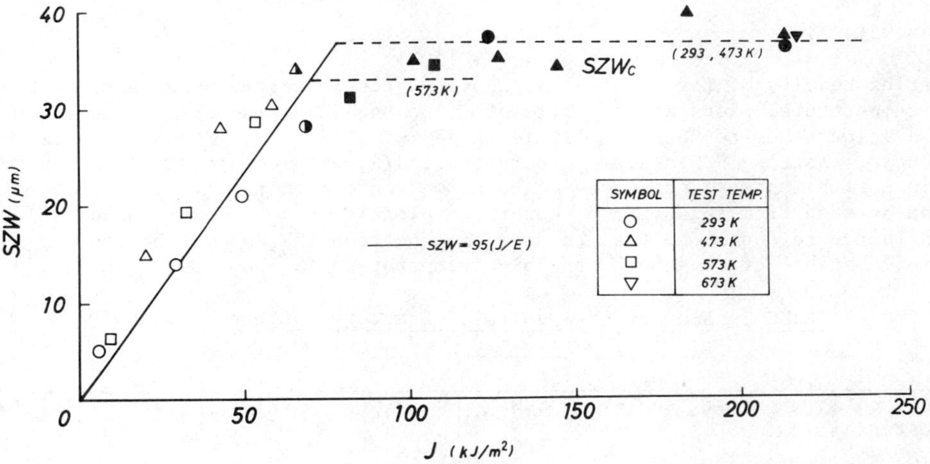

Fig. 6. Relation between SZW and J for an A533 Grade B Class 1 steel at room temperature (293K), 473, 573 and 673K.

Relations between the stretched zone width, SZW, and the J-integral (Rice, 1968, 1973) obtained in elastic-plastic fracture toughness tests at room temperature (293K), 473, 573 and 673K are shown in Fig. 6. The half-solid symbols and the solid symbols in this figure denote that the tear dimple fractured zone showed the occurrence of an actual stable crack growth existing locally or globally adjacent to the stretched zone formed as a result of fatigue precrack tip plastic blunting, respectively. Kobayashi (1977) has already reported results of elastic-plastic fracture toughness tests for a quenched and tempered martensitic steel (JIS SNCM8) at a low temperature. These results had shown that; (1) the relation between SZW and J before the initiation of stable crack growth was not dependent on test temperature, that is, yield stress, σ_{ys}; (2) a critical value of the stretched zone width, SZW_c, at the initiation of stable crack growth was not affected by the continued stable crack growth; (3) a J_{Ic} value could be evaluated from the SZW_c value; (4) an apparent plane-strain

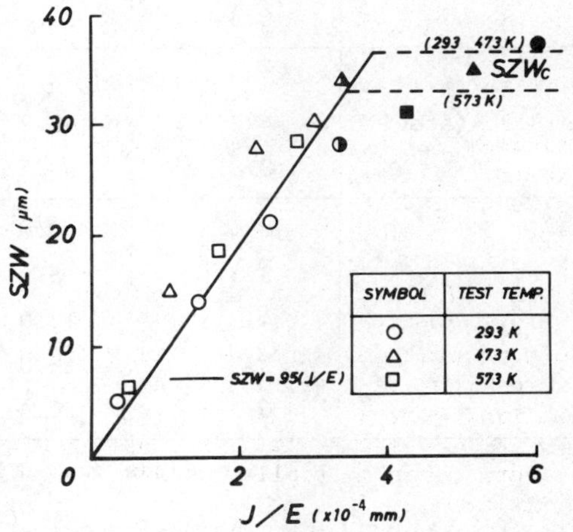

Fig. 7. Relation between SZW and J for A533 Grade B Class 1 steel at room temperature (293K), 473 and 673K.

fracture toughness value, $K_{Ic}(J)$, which was converted from the J_{Ic} value on the hypothesis that a small scale yielding condition might be retained at the midthickness of specimen, agreed with a K_{Ic} value. It can be seen from Fig. 6 that (1) and (2) summarized in the previous study are confirmed in this elastic-plastic fracture toughness test. Nakamura et al (1979) have recently shown that even in a large scale yielding condition beyond linear-elastic fracture mechanics, SZW formed before the initiation of a stable crack growth, corresponds to J/E for many metallic materials including an AISI304 steel and Al alloys, and that this relation may be written as the following expression.

$$SZW = 95(J/E) \tag{3}$$

Comparing results of Fig. 6 with the above expression drawn using a value of E at a room temperature, plots are slightly off, and the SZW value corresponding to the same J value tends to become a little larger with increasing test temperature. Thereupon, plotting SZW against a parameter, J/E, with consideration of the decrease in E with increasing test temperature, we can obtain results as shown in Fig. 7. It can be seen from this figure that these plots tend to agree with Eq. (3). It is therefore concluded that the relation between SZW and J/E before the initiation of stable crack growth at high temperature is also obeyed in Eq. (3).

TABLE 2 Summary of Results of Fracture Toughness Values at High Temperature

Msterial	Test temp. (K)	SZW_c (μm)	J_{Ic} (kJ/m²)	$K_{Ic}(J)$ ($MNm^{-3/2}$)	$K_Q(K_{Ic})$ ($MNm^{-3/2}$)
A533B-1 steel	293	36.5	79.1	135	—
	473	36.2	75.1	128	—
	573	33.6	67.2	119	—
	673	35.9	68.1	117	—
AISI304 steel	293	33.0	74.7	125	—
	473	41.9	91.1	136	—
	873	53.5	98.1	128	—
Tool steel	293	3.0	6.7	37	45*
	373	8.6	18.4	59	66
	473	12.0	25.1	68	92
	573	15.0	30.6	73	95
	673	16.4	32.3	75	94

*ASTM Standards valid K_{Ic}

Now, J_{Ic} values for an A533B-1 steel at high temperature can be obtained by reading off J values from Fig. 6 which correspond to these SZW_c values. These results are listed in Table 2. Results for both an AISI304 steel and a tool steel obtained by a similar J_{Ic} test procedure are also summarized in Table 2. These J_{Ic} values obtained here except for an AISI304 steel (B=5 mm), sufficiently satisfy the following equation which stands for valid specimen size requirements (Paris, 1972).

$$B,\ a,\ b \geqq 25(J_{Ic}/\sigma_{flow}) \tag{4}$$

where, b is an uncracked ligament and σ_{flow} is a stress taken half way between σ_{ys} and ultimate tensile stress, σ_B. It is therefore considered that these J_{Ic} values correspond to a plane-strain fracture toughness, G_{Ic}, when specimen size becomes large and a small scale yielding condition may be retained. So, assuming that a small scale yielding condition may be retained at the midthickness of the specimen until stable crack growth initiates, $K_{Ic}(J)$ values can be calculated from these J_{Ic} values according to the following expression.

$$K_{Ic}(J) = ((EJ_{Ic}/(1-\nu^2))^{1/2} \tag{5}$$

The results obtained are also listed in Table 2. And for reference, results of a plane-strain fracture toughness test only for a tool steel conducted according to ASTM standards (E399-74, 1975) are listed in Table 2. What is evident from the table is that J_{Ic} and $K_{Ic}(J)$ values for an A533B-1 and an AISI304 steels are nearly constant within the limits of this experiment, and that these values for a tool steel increase with increasing test temperature. And it can be seen also from the table that the agreement between K_{Ic} and $K_{Ic}(J)$ is fairly good at a room temperature and 373K where a small scale yielding condition may be retained at a fatigue precrack tip. That is to say, the validity of the evaluation of fracture toughness from SZW_c is confirmed.

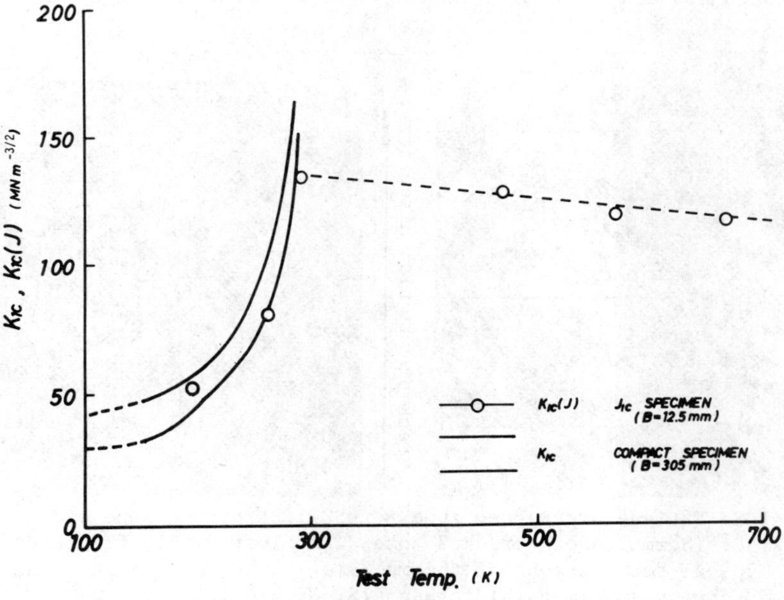

Fig. 8. Temperature dependence of fracture toughness values, K_{Ic} and $K_{Ic}(J)$, for an A533 Grade B Class 1 steel.

Plotting $K_{Ic}(J)$ values for an A533B-1 steel in Table 2 against test temperature results in Fig. 8. Both $K_{Ic}(J)$ values below room temperature obtained according to a similar J_{Ic} test procedure and K_{Ic} values obtained by Wessel (1969) using 12 in. (305 mm.) compact specimens are also plotted in Fig. 8. Figure 8 shows that the agreement between $K_{Ic}(J)$ values of a small specimen and K_{Ic} values of a large specimen is better in a lower temperature range than at room temperature where a small scale yielding condition may remain at a crack tip. And also it can be seen from Fig. 8 that $K_{Ic}(J)$ values are nearly constant in the higher temperature range than at room temperature. That is to say, it is concluded that upper shelf fracture toughness exists, and that fracture toughnesses within the same temperature range of thermal shock experiments are nearly constant. Logsdon (1977) has recently reported that $K_{Ic}(J)$ values increase slightly or decrease with increasing test temperature at a high temperature where upper shelf values exist in the Charpy impact absorbed energy. It is therefore necessary to discuss in detail the existence of upper shelf fracture toughness, and to explain the limitation of applying J_{Ic} concepts to the elevated temperature fracture toughness.

Elastic-plastic thermal shock fracture toughness. Hirano (1976) has already reported that a stretched zone formed as a result of fatigue precrack tip plastic blunting during thermal shock experiments. One example of many fractographs of thermal shock cracking is shown in Fig. 9(a). As seen from this fractograph, the stretched zone exists between the fatigue precracked zone and the thermal shock cracked zone. But the stretched zone formed during the thermal shock condition is not clear in comparison with that of fracture toughness specimens, Fig. 9(b), and exists locally adjacent to the fatigue precracked zone, because the fracture mode during the thermal shock cracking is intergranular cracking. It is therefore a further problem to understand the essential significance of such a stretched zone formed during the thermal shock condition and to study a quantitative analysis of it. But in order to evaluate fracture toughness during the thermal shock condition on the basis of elastic-plastic fracture mechanics, we hasten to calculate a J_{tsc} value from a SZW_{tsc} value examined in a thermal shock specimen by using the relation between SZW and J obtained in the J_{Ic} test at a high temperature.

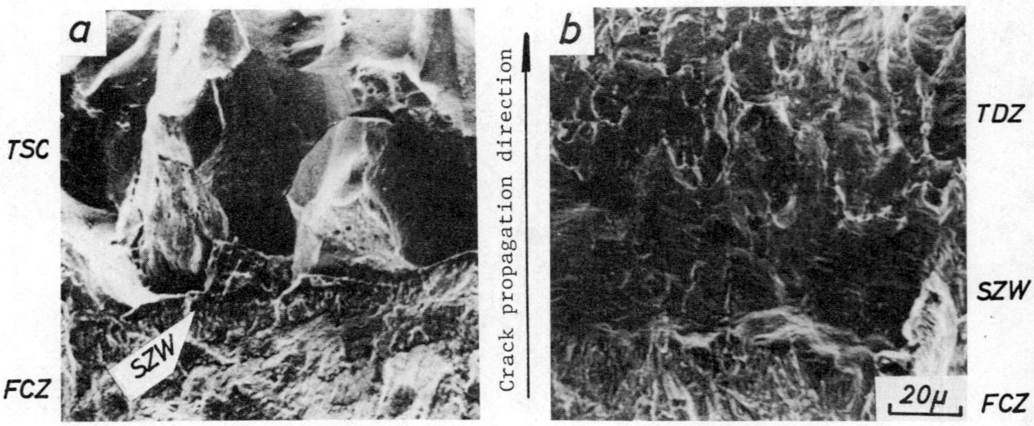

FCZ: Fatigue precracked zone SZW: Stretched zone width
TSC: Thermal shock cracked zone TDZ: Tear dimple fractured zone

Fig. 9. Typical fractographs of thermal shock specimen (a) and fracture toughness specimen (b) for an AISI304 steel.

Results of the thermal shock experiment on the basis of elastic-plastic fracture mechanics are summarized in Table 3. J_{tsc} values are evaluated from SZW_{tsc} values on the assumption that the relation between SZW and J obtained in a J_{Ic} test at a high temperature, Eq. (3), can be adopted in the thermal shock experiment. And $K_{tsc}(J)$ values are calculated from J_{tsc} values according to the following expression on the assumption that a small scale yielding condition may remain at a fatigue precrack tip at the initiation of the thermal shock cracking.

$$K_{tsc}(J) = ((EJ_{tsc}/(1-\nu^2))^{1/2} \tag{6}$$

For all materials tested in this study, these $K_{tsc}(J)$ values are rather smaller than K_{Ic}(J) values at a high temperature shown in Table 2. From this fact, it is concluded that both usual structural design codes and the evaluation method of safety in reactor pressure vessel construction (ASME Boiler and Pressure Vessel Code, Sec. XI, 1977) are not reasonable, because it is considered that a crack initiates when thermal stress intensity factor, K_I, during the thermal shock condition achieves a K_{Ic} value.

TABLE 3 Summary of Results of Thermal Shock Fracture Toughness Values

Material	SZW_{tsc} (μm)	J_{tsc} (kJ/m^2)	K_{tsc}(J) (MNm$^{-3/2}$)
A533B-1 steel	5.6	11.0	48
AISI304 steel	6.0	12.7	51
Tool steel	1.5	3.1	25

Why does the difference between K_{tsc}(J) and K_{Ic}(J) values arise? The reasons for the difference could be as follows;

(1) Difference of strain rate (or loading rate) between the thermal shock experiment and the fracture toughness test.

(2) Effects of the precipitation of a carbide on grain boundaries and the formation of brittle oxides on the crack tip surface during holding at high temperature in the thermal shock experiment.

(3) Difference of fracture mode.

For an A533B-1 steel, rate of increase of stress intensity factor in the fracture toughness test was nearly equal to that in the thermal shock experiment. Therefore reason (1) does not seem to be the cause of this difference. And in order to clarify the effect of reason (2), J_{Ic} tests were conducted at room temperature using specimens subjected to one or five thermal cycles from room temperature to 923K and examinations of the microstructure were done by using a photomicroscope. These results of the J_{Ic} test are shown in Fig. 10.

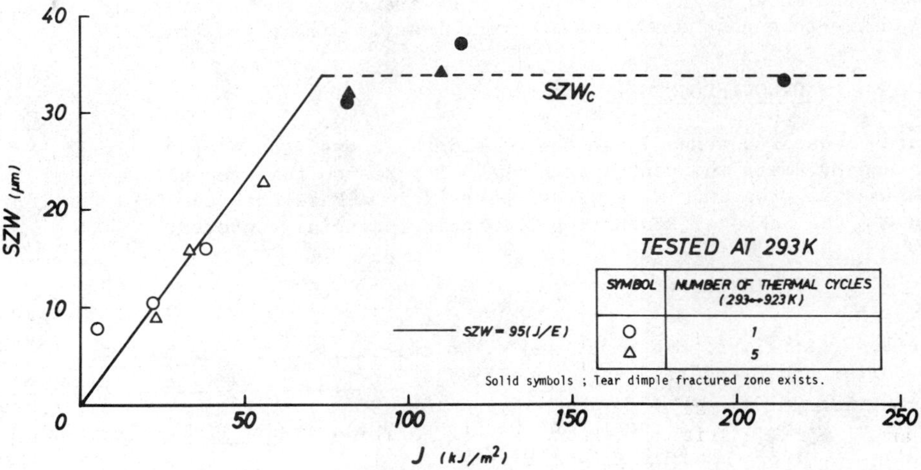

Fig. 10. Influence of thermal cycles on elastic-plastic fracture toughness.

There is little difference in J_{Ic} value between a reference specimen and a virgin material specimen, and the J_{Ic} value of a reference specimen is larger than the J_{tsc} value. No change in the microstructure was detected before and after the thermal shock experiment. And it is also known that an A533B-1 steel is not sensitive to hydrogen-embrittlement (Mager, 1977). Accordingly, reason (2) also does not seem to be a cause of this difference As shown in Fig. 9(a) and (b), the appearance of the fracture surface during the thermal shock cracking is intergranular, whereas that of fracture toughness specimen is predominantly dimple. However, we do not yet understand why the thermal shock cracking is intergranular, and how the difference of the fracture appearance corresponds to the difference of fracture toughness. Further detailed investigation of causes of the difference of fracture toughness is necessary. Finally, it should be emphasized here that an elastic-plastic thermal shock fracture toughness is smaller than an elastic-plastic fracture toughness in the same high temperature range, and that both usual structual design codes and standards for the evaluation method of safety in reactor pressure vessel construction are not reasonable, because it is considered that a crack initiates when the thermal stress intensity factor during the thermal shock condition achieves a plane-strain fracture toughness.

CONCLUSIONS

(1) Thermal shock fracture toughness for a tool steel (JIS SKD62) can be evaluated by means of the analysis of quasi-static thermal stress intensity factor on the basis of linear-elastic fracture mechanics.

(2) The relation between SZW and J before the initiation of stable crack growth in the elastic-plastic fracture toughness test at a high temperature can be expressed by the following expression regardless of test temperature.

$$SZW = 95(J/E).$$

(3) Elevated temperature fracture toughness values for an A533 Grade B Class 1 steel ranging from room temperature to 673K are nearly constant regardless of test temperature. It is confirmed that upper shelf fracture toughness exists.

(4) Thermal shock fracture toughness is smaller than elevated temperature fracture toughness in the same high temperature range of the thermal shock experiment.

(5) Fracture mode during the thermal shock cracking is intergranular, whereas that of fracture toughness specimens is predominantly dimple pattern.

ACKNOWLEDGEMENT

We are pleased to acknowledge the considerable assistance of Mr. Tsuda, Mr. Tachikawa and Mr. Nara who contributed their efforts to the accomplishment of the experimental program. We also are pleased to acknowledge the Department of Education and the Sakkokai Foundation for their financial supports.

REFERENCES

Apblett, W. R., E. D. Montrone, and W. Wolowldiuk(1976). Trans. ASME., 98, 291.
ASME Boiler and Pressure Vessel Code(1977). Section XI, American Society of Mechanical Engineers, New York.
Blauel, J. G., J. F. Kalthoff, and D. Stahn(1974). Trans. ASME., 96, Ser. J, 299.
Buchalt, C. B.(1974). ASTM STP, 560, 240.
Cheverton, R. D.(1978). Nuclear Safety, 19, 10.
E399-74(1975). Annual Book of ASTM Standards, Part 10.
Hasslman, D. P. H.(1971). Int. J. Fract. Mech., 9, 157.
Hirano, K., H. Kobayashi, and H. Nakazawa(1976). Preprint of the 54th Annual Meeting of the Japan Society of Mechanical Engineers, 760, 83.
Hirano, K., H. Kobayashi, and H. Nakazawa(1978). J. of the Society of Materials Science Japan, 27, 104.
Kassir, M. K.(1971). Trans. ASME., 93, Ser. E, 643.

Kobayashi, H., K. Hirano, H. Nakazawa, and co-workers(1977). Proceedings of the 4th International Conference on Fracture(edited by Taplin, D. M. R.), 3, Waterloo Univ., Canada, 583.
Logsdon, W. A., and J. A. Begley(1977). Engng. Fract. Mech., 9, 461.
Mager, T. R., D. M. Moon, and J. D. Landes(1977). Trans. ASME, 99, Ser. J, 238.
Nakamura, H., H. Kobayashi, and H. Nakazawa(1979). Trans. Japan Soc. Mech. Engrs., 45.
Paris, P. C., and G. C. Sih(1965). ASTM STP, 381, 30.
Paris, P. C.(1972). ASTM STP, 514, 21.
Rice, J. R.(1968). Trans. ASME, 35, Ser. D, 379.
Rice, J. R., P. C. Paris, and J. G. Merkle(1973). ASTM STP, 536, 231.
Timoshenko, S.(1934). Theory of Elasticity, 1st. ed. Mcgraw-Hill, New York. 209.
Yagawa, G., M. Ichimiya, and Y. Ando(1978). Preprint of the 56th Annual Meeting of the Japan Society of Mechanical Engineers, 780, 158.
Wessel, E.T., W. G. Clark, Jr., and W. H. Pryle(1969). Proceedings of the 2nd International Conference on Fracture, edited by P. L. Pratt et al., Chapman and Hall, 825.

Volume 3, ICM 3, Cambridge, England, August 1979

THE DIRECTIONAL STABILITY OF CRACK PROPAGATION

R. D. Streit and I. Finnie*

Lawrence Livermore Laboratory, Livermore, California 94550, U.S.A.
***University of California Department of Mechanical Engineering, Berkeley, California 94720, U.S.A.**

Since fracture toughness testing is time consuming and expensive, it is desirable to obtain as many data points as possible from a single specimen. This allows the variability of toughness values to be studied and also permits the effects of such variables as welds or heat treatment to be evaluated with a single specimen. For such tests to be feasible, two requirements must be satisfied. First, the specimen must exhibit propagational stability. That is, the crack must arrest after propagating a small distance. Within the framework of linear elastic fracture mechanics the propagational aspect of crack growth appears to be well understood. Various expressions for propagational stability have been presented in the literature and all may be shown to be equivalent to $(dK/da) < 0$ for stable crack growth. The second requirement, if many data points are to be obtained, is that the crack continue to propagate in its own plane. For example, in the double cantilever beam (DCB) specimen there is a strong tendency for the crack to deviate from the plane of symmetry. This is unfortunate because the DCB exhibits propagational stability when the specimen displacement is controlled. Therefore, "side grooves" are often used to control the crack's direction. This topic of directional stability has been examined by a number of researchers (for example, Benbow and Roesler, 1957; Berry, 1963; Gurney and Hunt, 1967; Kibler and Roberts, 1970; Mai, Atkins, and Caddell, 1975) over the past two decades. Despite this attention, the conditions which lead to directional instability do not appear to be well understood at the present time.

DIRECTION OF CRACK GROWTH UNDER MIXED MODE LOADING

Several theories have been proposed to predict the direction of crack growth for the case of combined Mode I and Mode II loading. Because this problem is related to the question of directional stability, it will be useful to review some of the literature.

The geometrical configuration often used for mixed mode studies is that of the infinite plate containing a crack of length 2a, shown in Fig. 1. The crack plane lies at an angle β to the direction of the loading axis whereas the direction of crack growth is defined by the angle θ. The stress field near the tip of the crack is given most conveniently by the Williams (1957) solution,

$$\sigma_\theta = \left[\cos(\theta/2)/(2\pi r)^{1/2}\right]\left[K_I \cos^2(\theta/2) - (3/2)K_{II}\sin\theta\right] + A_2\sin^2\theta + O(r^{1/2}) \qquad (1a)$$

$$\sigma_r = \left[\cos(\theta/2)/(2\pi r)^{1/2}\right]\left\{K_I[1 + \sin^2(\theta/2)] + K_{II}[3/2 \sin\theta - 2\tan(\theta/2)]\right\} + A_2 \cos^2\theta + 0(r^{1/2}) \quad (1b)$$

$$\tau_{r\theta} = \left[\cos(\theta/2)/(2\pi r)^{1/2}\right]\left[K_I \sin\theta + K_{II}(3\cos\theta - 1)\right] - A_2 \cos\theta \sin\theta + 0(r^{1/2}) \quad (1c)$$

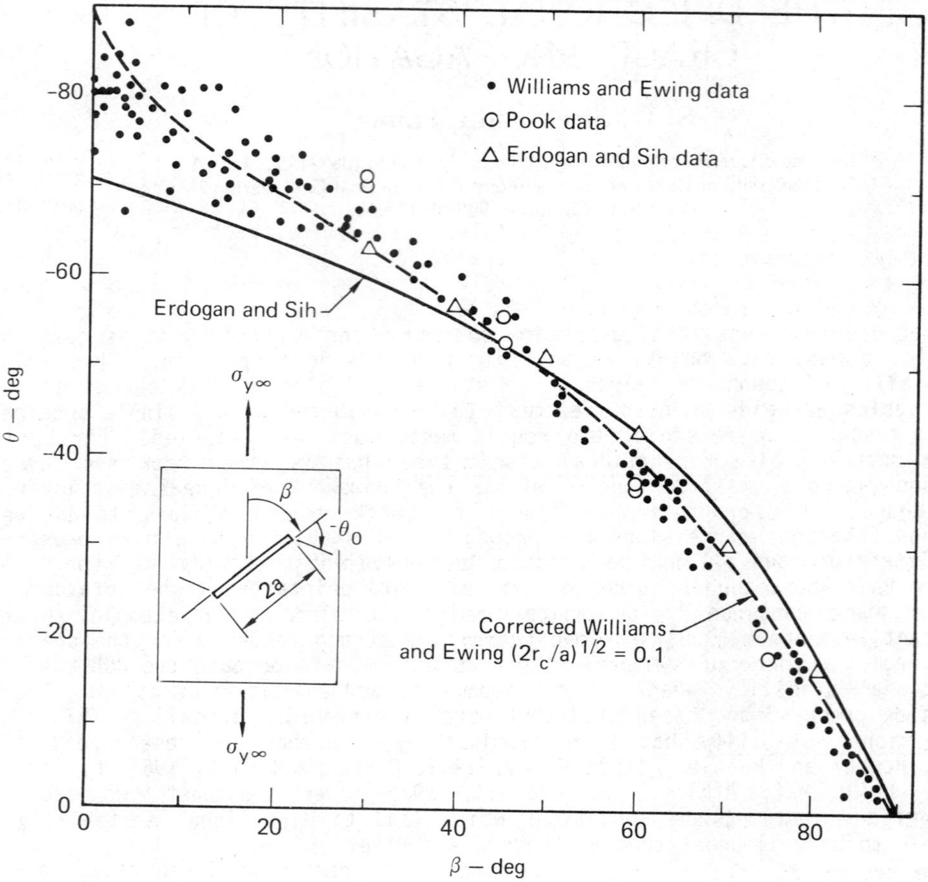

Fig. 1. Predicted angle of crack extension and available data.

The first attempt to predict the direction of crack extension is due to Erdogan and Sih (1963). They assume that the crack will extend in a radial direction from the parent crack tip in a plane perpendicular to the maximum value of the stress σ_θ. Considering only the singularity terms in Eqs. 1, they concluded that this direction coincided with the plane of zero shear stress and obtained the expression,

$$\sin\theta + (3\cos\theta - 1)\cot\beta = 0 \quad (2)$$

shown in Fig. 1. Later Williams and Ewing (1973) suggested that this prediction for the direction of crack growth could be improved by examining the values of σ_θ a small distance, r_c, ahead of the crack tip. By taking the first two terms in the series expansion for σ_θ and writing $\partial\sigma_\theta/\partial\theta = 0$ they were able to express the angle of crack growth as a function of the crack orientation and a dimensionless parameter $\alpha = (2r_c/a)^{1/2}$. After making a correction to their analysis as pointed out by Finnie and Saith (1973), the result is

$$\tan^2\beta + (3\cos\theta - 1)(\tan\beta)/\sin\theta - 16\alpha\sin(\theta/2)(1 - \tan^2\beta)/(3\tan\theta) = 0 \quad (3)$$

Equation 3 is also plotted in Fig. 1. For the arbitrary choice $\alpha = 0.1$, Eq. 3 is seen to fit the experimental data better than Eq. 2. The assumption that the crack growth direction can be predicted by examining the stresses at some distance from the crack tip appears reasonable. In ductile fracture voids initiate at second-phase particles ahead of the crack, while in cleavage fracture grains ahead of the crack may cleave if suitably oriented. Similar arguments may be made for polymeric or ceramic materials as well. In any event, it would be desirable to determine the distance r_c on physical grounds rather than by the currently employed curve fitting method.

As a final comment on the comparison of Eqs. 2 and 3, we note that the equivalence of $\partial\sigma_\theta/\partial\theta = 0$ and $\tau_{r\theta} = 0$, which is observed if only the singularity term is considered, no longer holds if the second term in the series is included. As a result, in general $\tau_{r\theta} \neq 0$ and σ_θ is not a principal stress for the value of θ given by Eq. 3.

Various other models have been offered to describe the direction of crack growth. Predictions have been made using energy criteria by several authors considering both a kinked and unkinked crack tip geometry. Further, Bilby and Cardew (1975) have proposed that a crack will kink to produce pure Mode I loading. A detailed comparison of the stress and energy approaches has been given by Streit (1978). For our present discussion, it is sufficient to note that all of these lead to predictions very similar to Eq. 3, particularly in the region near $\beta = 90$ degrees, i.e., predominantly Mode I loading.

DIRECTIONAL STABILITY MODEL

From the review of crack growth under mixed mode loading it seems reasonable to study crack stability by examining the behavior of σ_θ at some distance ahead of the crack. Since the vast majority of fracture toughness tests involve Mode I loading, attention will be restricted to this case, i.e, $K_{II} = 0$. Differentiating the expression for σ_θ, Eq. 1a, with respect to θ and equating to zero yields,

$$-[3K_I \sin\theta\cos(\theta/2)]/[4(2\pi r)^{1/2}] + 2A_2\sin\theta\cos\theta = 0 \quad (4)$$

One solution to Eq. 4 is $\theta = 0$. However, whether this corresponds to a maximum or minimum value of σ_θ has to be established.

Further examination of Eq. 4 shows that a second solution exists if $(9/128\pi)(1/r)(K_I/A_2)^2 \leq 1$. Defining the distance r_0 as the limit of the inequality, i.e.,

$$r_o = (9/128\pi)(K_I/A_2)^2 , \quad (5)$$

then we find that at radii less than r_o the maximum value of σ_θ is on $\theta = 0$. This implies that the crack will grow in the direction along $\theta = 0$ and therefore, be directionally stable. If, however, the radius of interest is greater than r_o, then the maximum value of σ_θ will deviate from the plane of symmetry and the specimen will be directionally unstable. This is true for all A_2 greater than or equal to zero. For negative A_2, σ_θ on $\theta = 0$ is a maximum value for all values of r. Using this definition of r_o, Eq. 1a may be rewritten in a dimensionless form (for the case $K_{II} = 0$) as,

$$\sigma_\theta/A_2 = (8/3)(r_o/r)^{1/2} \cos^3 (\theta/2) + \sin^2 \theta \quad (6)$$

which is shown graphically in Fig. 2. Thus, if we assume that the physical processes involved in fracture lead to crack initiation at a distance r_c ahead of the crack tip, the criterion for directional stability is

$$r_o \geq r_c . \quad (7)$$

From Fig. 2 it is seen that for r approximately r_o, the variation of σ_θ with respect to θ , in the vicinity of $\theta = 0$, is very slight. Thus, we might anticipate that there will be considerable scatter in observations on directional stability.

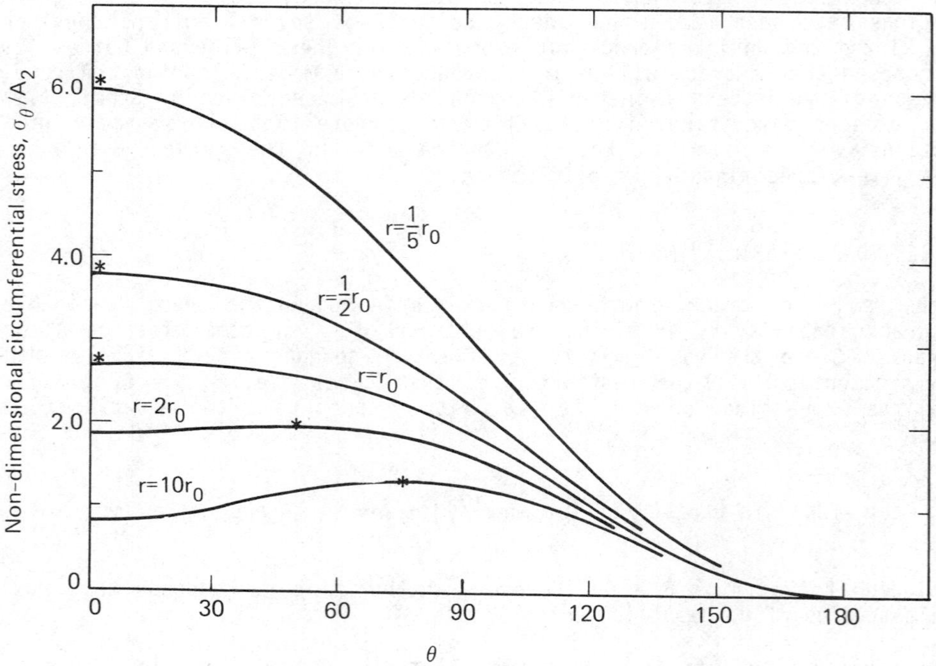

Fig. 2. Nondimensional circumferential stress σ_θ vs. the angular stress coordinates θ at varying radii in front of the crack tip. Location of the maximum σ_θ shown by*.

Before examining some of the other implications of Eq. 4, it will be interesting to discuss related work. Saith (1971) and Finnie and Saith (1973) used the condition that $\tau_{r\theta} = 0$ to obtain the value $r_o = (1/8\pi)(K_I/A_2)^2$. We note that this expression differs more than twofold from the present expression derived from $\partial\sigma_\theta/\partial\theta = 0$. Earlier, Cotterell (1966) attempted to predict directional stability by examining the direction of crack growth following a small angular deviation. On this basis he predicted directional stability for negative A_2 and instability for positive A_2. The present derivation shows that this requirement is over restrictive and that directional stability can be obtained for positive values of A_2 if r_o is greater than r_c. After noting that the compact tension specimen, which has positive A_2, exhibited directional stability, Cotterell (1970) adopted another approach. Using strength of materials solutions for beam bending to estimate the stresses at the crack tip, he proposed that a crack will be stable if the stress normal to the crack plane is greater than or equal to the stress in the direction of crack propagation. While this approach appears to provide a criterion for directional stability, it is also in many cases over restrictive.

Application of the Critical Distance Concept to Directional Stability

As a first example of the critical distance concept to directional stability let us consider the case of an infinite plate as in Fig. 1 with β = 90 degrees and a transverse stress $\sigma_{x\infty} = B\,\sigma_{y\infty}$ in addition to the opening stress of $\sigma_{y\infty}$. Here, $A_2 = \sigma_{x\infty} - \sigma_{y\infty} = \sigma_{y\infty}(B-1)$ and we obtain,

$$r_o = 9a/[128(B-1)^2] \qquad (8)$$

Eq. 8 is only valid for B greater than or equal to 1. Directional stability is ensured if B=1 since r_o goes to infinity and for B less than 1 it may be shown that there are no real solutions for r_o. In this case the specimen will always be stable. However, we see that for $B > 1$ directional stability can be obtained by either increasing the crack length or by decreasing B so that $r_o \geq r_c$. Kibler and Roberts (1970), in studying this configuration, found a transition from directional stability for $B < 1$ to directional instability as B increased beyond unity.

Although the infinite plate configuration possesses an analytical solution for the stress series coefficients, most commonly used specimen geometries must be handled by numerical means. Unfortunately, although many K_I values have been obtained and tabulated, the corresponding values of A_2 are rarely reported. However, Larsson and Carlsson (1973) have presented results for various specimens having a crack length of half the specimen width W. From their results we find that for the center-cracked and double-edge-cracked specimens no real solution exists for r_o, thus ensuring directional stability. For the bend and compact tension specimens r_o/W equals 10 and 0.13, respectively. Therefore, directional stability is also ensured for the bend specimen and would be expected in the compact tension. Earlier, Cotterell (1970) had given values of A_2 for various crack lengths in the compact tension specimen while Saith (1971) obtained photoelastic results on a DCB specimen which enabled us to calculate A_2. The values of r_o/W as a function of a/W for the two specimens are shown in Fig. 3. In the compact tension specimen, as already noted, r_o is a relatively large fraction of the specimen width and stable crack growth is expected, in agreement with experiment. For the DCB specimen r_o/W goes to zero for $a/W > 0.4$. Tests confirm that directional instability occurs in this region. This is unfortunate since the DCB specimen under displacement control shows a high degree of propagational stability and, in the absence of directional instability, would allow many data points to be obtained on a single specimen. A solution already mentioned is side grooving, but this is not always satisfactory and produces a complex three-dimensional stress distribution. A second approach to ensure directional stability, which has been known for some time, is to apply lengthwise compression to the specimen. It was pointed out by Finnie and Saith (1973) that this may be explained by noting that a lengthwise compressive stress σ_{ox} leads to an effective value of $A_2 = A_2 - \sigma_{ox}$ and hence, an increase in r_o which enhances directional stability.

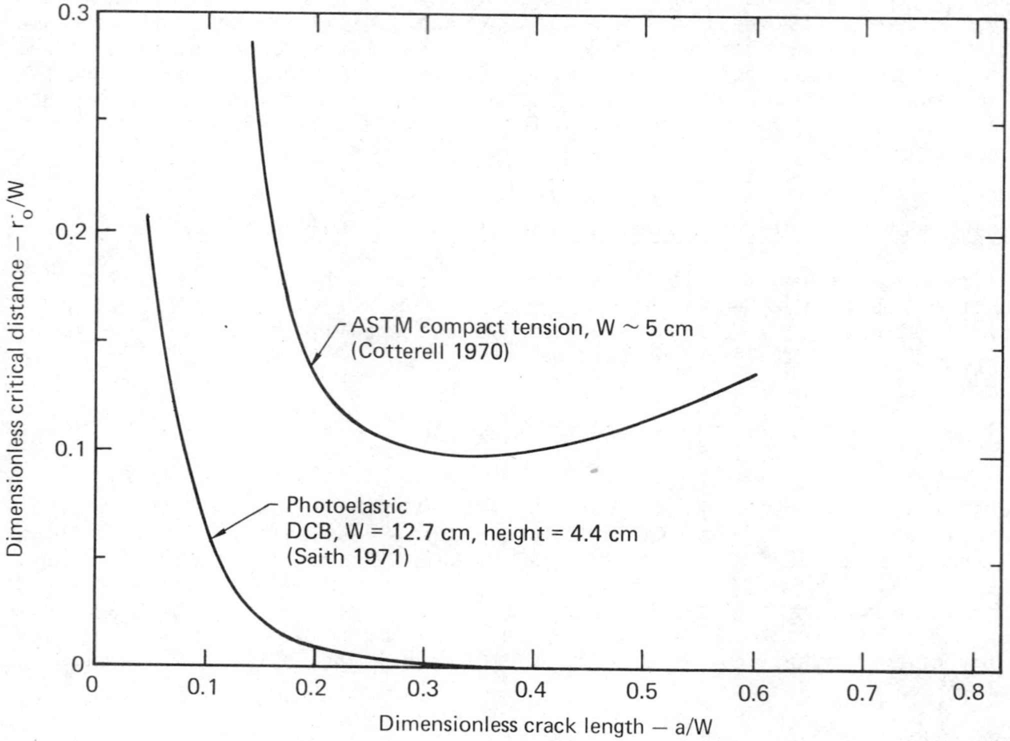

Fig. 3. Nondimensional critical distance vs crack length for the compact tension and double cantilever beam specimen.

A NEW SPECIMEN FOR PROPAGATIONAL AND DIRECTIONAL STABILITY

In practice it is difficult to apply lengthwise compression to a DCB type specimen without perturbing the opening mode load. For this reason we have developed a center-cracked specimen (CCS) which allows us to apply controlled lengthwise compression or tension independently of the crack opening load, Fig. 4, and therefore change the parameter r_o. Although this specimen exhibits directional instability when unconstrained, it can be made to be directionally stable with application of adequate lengthwise compression.

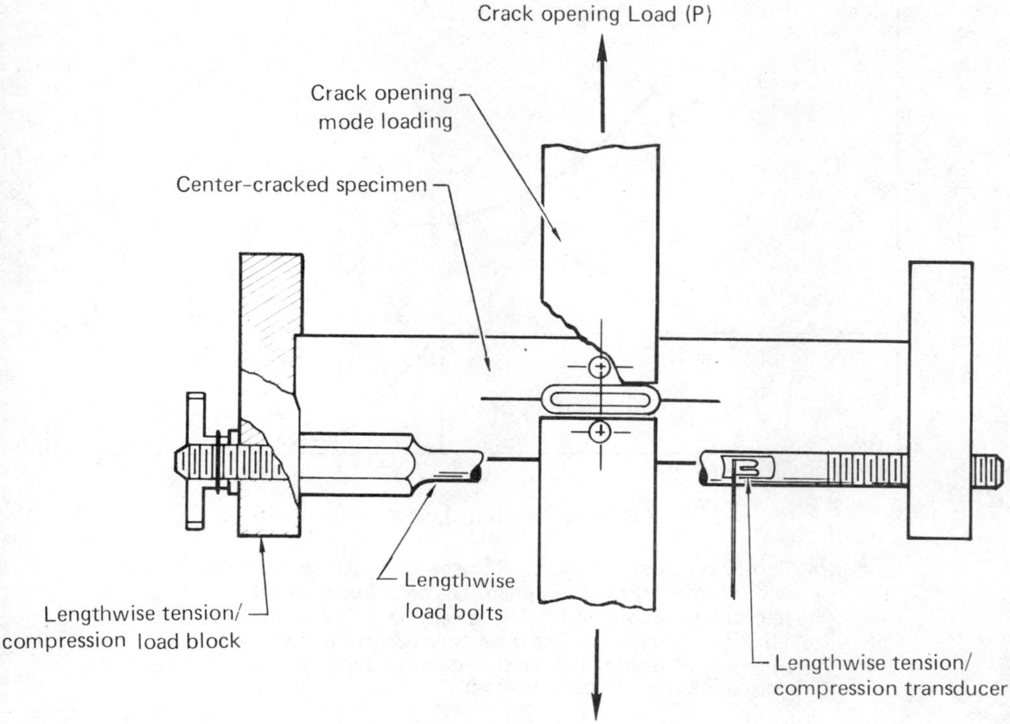

Fig. 4. The center-cracked specimen and load apparatus.

While the central focus of our current research is to investigate directional growth characteristics, the specimen and test setup must be designed so that propagational stability is achieved. As pointed out earlier, this can be accomplished by testing under specimen displacement control. That is, by controlling the rate of crack opening we can control the rate of crack propagation. In addition, actual measurement of the specimen crack opening enables us to determine its compliance and change of compliance as a function of crack length. From these measurements the crack length at any point in subsequent tests may be calculated.

Photoelastic analysis was employed to determine the state of stress in the vicinity of the crack tip. Using a bi-phase epoxy as the test material, and employing the stabilizing feature of the testing apparatus, natural cracks were grown to predetermined lengths. Subsequently, the specimens were loaded at elevated temperature and slowly cooled so as to freeze the stress state into the material. After sectioning, the region very close to the tip of the sharp crack was examined. By measuring the isochromatic fringe radius and angle at a location of maximum shear stress, the ratio K_I/A_2 was determined and the geometrical radius r_0 calculated for each crack length (Streit 1978). The curve of r_0 vs. crack length is shown in Fig. 5 with the limits on scatter indicated.

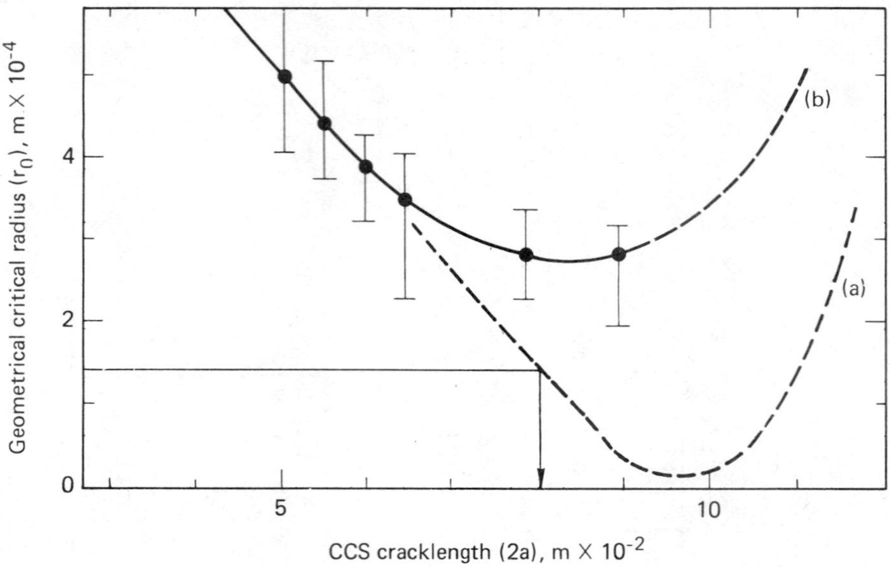

Fig. 5. Photoelastic determination of the geometrical radius r_0 for a 12.7 cm width center-cracked specimen. The solid line shows the results of the photoelastic epoxy tests. The lower line (a) shows the expected curve for stiffer specimens. The difference occurs due to "end effects" observed when the photoelastic specimen's elastic modulus is lowered during the stress freezing process.

Recalling that the crack will grow straight if $r_0 > r_c$, the location of initial instability should correspond to the condition $r_0 = r_c$. If, for example, the crack path begins to turn when 2a = 8 cm, we can determine from Fig. 5 that the critical radius r_c will be 140μm. In this manner we can determine the location of crack initiation and decide if the distances are reasonable on a microstructural scale. Because the fracture behavior of 7075-T651 aluminum is well documented, it was chosen as the test material to investigate this relationship. Although correlation of the measured distance r_c to a microstructural parameter -- inclusion spacing or grain size -- was not observed, this distance was found to be closely related to the size of the plastic zone, i.e., the location of the elastic-plastic interface. This suggests that the voids which lead to final fracture are being initiated throughout the plastic zone. If this is the case, the extent of the plastic zone determines the maximum distance from the crack tip where the fracture process can influence directional stability. Table I shows the relation between the critical distance r_c and the plastic zone along $\theta = 0$ for three different analyses. While all the elastic-plastic models show the proper trend, the detailed finite element analysis of Larsson and Carlsson (1973) correlates the best.

The difference between plastic zone size and critical radius can be explained by realizing that the values of r_p are the limit of the elastic-plastic interface. However, a finite amount of plastic deformation is required to initiate void nucleation from inclusions. Thus, we expect that micro-crack initiation will occur at distances somewhat smaller than the distance to the plastic interface, and indeed this is what we observe experimentally.

TABLE 1 Plastic Zone Along θ = 0 and Critical Radius at Incipient Directional Instability

Specimen	K_c (@ Instability) ($MPa\text{-}m^{1/2}$)	r_p Irwin (m) x 10^{-4}	r_p Larsson & Carlsson (m) x 10^{-4}	r_p Rice & Rosengren (m) x 10^{-4}	r_c Experimental (m) x 10^{-4}
Longitudinal T-L	45	3.8	2.2	0.09	1.27
Transverse Ungrooved (L-T)	52	5.1	2.8	0.12	2.54
Transverse Grooved (L-T)	32	2.0	1.1	0.05	0.000 (appprox)

CONCLUSIONS

Despite many alternative models, the original Erdogan and Sih (1963) hypothesis that a crack will grow in the direction perpendicular to the maximum circumferential stress σ_θ is seen to be adequate for predicting the angle of crack growth under the condition of mixed mode loading. Their predictions, which were based on the singularity terms in the series expansion for the Mode I and Mode II stress fields, can be improved if the second term in the series is also included. Although conceptually simple, their predictions of the crack growth direction fit very closely to the data obtained from many sources.

Incorporating the second term of the series expansion for the crack tip stresses into the prediction for crack extension, we make the assumption that fracture initiates at some distance r_c ahead of the parent crack tip. Using the maximum circumferential stress theory as our crack growth criterion, we consider the directional stability of a crack subjected to Mode I loading. Two solutions for the maximum value of σ_θ are found by setting $\partial\sigma_\theta/\partial\theta = 0$. The obvious solution is that σ_θ has a stationary value for $\theta = 0$, regardless of the distance in front of the crack. The other solution shows that for $r > r_o$, where $r_o = (9/128\pi)(K_I/A_2)^2$, the maximum value of σ_θ is located symmetrically about $\theta = 0$. In this region σ_θ on $\theta = 0$ is a local minimum. For $r \leq r_o$, however, σ_θ on $\theta = 0$ represents the only maximum. Thus, the criterion for directional stability can be simply stated by $r_o \geq r_c$.

It was found that r_c, determined experimentally, is slightly smaller than the plastic zone size along $\theta = 0$. This is reasonable because the highly strained region in the vicinity of the crack tip corresponds to the area where the void growth process is active.

In the future, we plan to present the photoelastic analysis of the specimen and discuss the influence of the lengthwise compression on fracture toughness.

ACKNOWLEDGEMENT

The authors wish to express their thanks to the Lawrence Livermore Laboratory for their help and funding throughout the project.

"Work performed under the auspices of the U.S. Department of Energy by the Lawrence Livermore Laboratory under contract number W-7405-ENG-48."

REFERENCES

Benbow, J. J., and F. C. Roesler (1957). Experiments on Controlled Fractures. Proc. Phys. Soc. (London), 70, Ser. B., 201-211.

Berry, J. P. (1963). Determination of Fracture Surface Energies by the Cleavage Technique. J. Appl. Phys. 34, 62-68.

Bilby, B. A., and G. E. Cardew (1975). The Crack with a Kinked Tip. Int. J. Fract., 11, 708-712.

Cotterell, B. (1966). Notes on the Paths and Stability of Cracks. Int. J. Frac. Mech., 2, 526-533.

Cotterell, B. (1970). On Fracture Path Stability in the Compact Tension Test. Int. J. Frac. Mech., 6 (2), 189-192.

Erdogan, F., and G. C. Sih. (1963). On the Crack Extension in Plates Under Plane Loading and Transverse Shear. J. Basic Eng., Trans. ASME, 85, Ser. D., 519-527.

Finnie, I., and A. Saith (1973). A Note on the Angled Crack Problem and the Directional Stability of Cracks. Int. J. of Fract., 9, 484-486.

Gurney, C., and J. Hunt (1967). Quasi-static Crack Propagation. Proc. Royal Soc., 299, Ser. A, Math. & Phys. Sciences, 508-523.

Irwin, G. R. (1964) Structural Aspects of Brittle Fracture. Appl. Mater. Res. 3 (2), 65-81.

Kibler, J. J., and R. Roberts (1970). The Effect of Biaxial Stresses of Fatigue and Fracture, ASME Trans. J. of Engin. Ind., 727-734.

Larsson, S. C., and A. J. Carlsson (1973). Influence of Non-Singular Stress Terms and Specimen Geometry on Small Scale Yielding at the Crack Tip in Elastic -Plastic Materials. J. Mech. Phys. Solids, 21, 263-277.

Mai, Y. W., A. G. Atkins, and R. M. Caddell (1975). On the Stability of Cracking in Tapered DCB Test Pieces. Int. J. Fract., 11 (6), 939-953.

Rice, J. R., and G. F. Rosengren (1968). Plane Strain Deformation Near a Crack Tip in a Power-Law Hardening Material. J. Mech. Phys. Solids, 16, 1-12.

Saith, A. (1971). On The Stability and Arrest of Cracks. Ph.D. Thesis, University of California, Berkeley, CA.

Streit, R. D. (1978) The Directional Stability of Crack Propagation. UCRL-52593, University of California, Berkeley, CA.

Williams, J. G., and P. D. Ewing (1972). Fracture Under Complex Stress - The Angled Crack Problem. Int. J. Fract. Mech., 8 (4), 441-446.

Williams, M. L. (1957). On the Stress Distribution at the Base of a Stationary Crack. J. Appl. Mech., 24, Trans. ASME, 109-114.

Volume 3, ICM 3, Cambridge, England, August 1979

THE TEARING AND CUTTING OF MILD STEEL PLATE WITH APPLICATION TO SHIP GROUNDING DAMAGE

H. Vaughan

Director of Naval Architecture, University of British Columbia, Canada

INTRODUCTION

It was about twenty years ago that nuclear power was first applied to the propulsion systems of ships, more precisely to sub-surface ships. The extension of nuclear power to surface ships has been contemporaneous with the change from transporting relatively safe cargoes such as grain and bulk oil, to include more dangerous cargoes, particularly liquid natural and petroleum gases, carried at extremely low temperatures in speciality ships in which a collision and possible breach of a container could be catastrophic. It is not implied that these dangerous cargoes increase in any way the risk of ship collision but that highly volatile and radio-active materials need to be handled with more care and protected or insulated more effectively. Ship designers have anticipated these needs and accordingly over the past twenty years a technology has emerged for the protection of dangerous cargoes against damage during transportation due to collision of the ship with another ship or due to grounding. A sufficiently severe collision will, of course, always prove disastrous.

The first and possibly the simplest safety feature for prevention of oil spillage has been the inclusion of a double bottom in modern tanker construction. The double bottom offers some increase in structural protection to the oil cargo but its main effectiveness is simply due to geometry: a significant number of incidents involving bottom damage to ships result in shallow penetration of the bottom plating, the double bottom then merely contains the liquid cargo safely above the penetrating object.

The main design feature for the protection of an inner cargo against ramming by another ship from the side has been the development and installation of side collision barriers, consisting of strong edge-plate grillages built into the ship sides, amounting in some instances to metre-thick structures. Complementing side strengthening, some ships are designed with soft bows so that should ramming occur energy is absorbed in significant proportions by both the striking and the struck ship. Figure 61 in Johnson and Mamalis (1978) shows the damage suffered by a relatively soft-bowed vessel, whereas, Fig. 60 shows how a relatively stiff-bowed vessel has sliced into the side of a large ferry, a vessel with no side protection.

Ship collisions may conveniently be classified into two categories. There are the more numerous minor collisions in which penetration is resisted by large in-plane forces in the plastically deformed but unruptured plates. Jones (1976) has

MS181

contributed significantly to this technology. Then there are the major collisions in which large volumes of structural elements are distorted and significant penetration occurs. Minorsky (1959) provided the first and standard method of estimating gross damage suffered by two colliding ships in oblique impact. Minorsky's method has been applied to the design of collision barriers for very large ships involving energy levels far in excess of those considered in his original study. It has also been verified using small and moderate scale tests by Akita and Kitamura (1972) and Woisin (1971).

An aspect of ship damage which has received little attention is that due to grounding. When a ship runs over ice or a sharp reef the damage is not volumetric so that Minorsky's method is inapplicable. It is more likely to consist of a long narrow slit in the bottom plating such as that shown in Fig. 1. This type of damage is no longer uncommon because of increasing traffic on routes which are partly ice-bound and also because of the increasing number of LNG tankers entering coastal regions. From the safety point of view, it is desirable to be able to predict a safe operating speed for a ship coming into harbour based on the requirement that should it run aground and be slit, damage is limited to the ship hull leaving the inner protected item undamaged. To design for such a requirement it is necessary to know the tearing strength of steel plate.

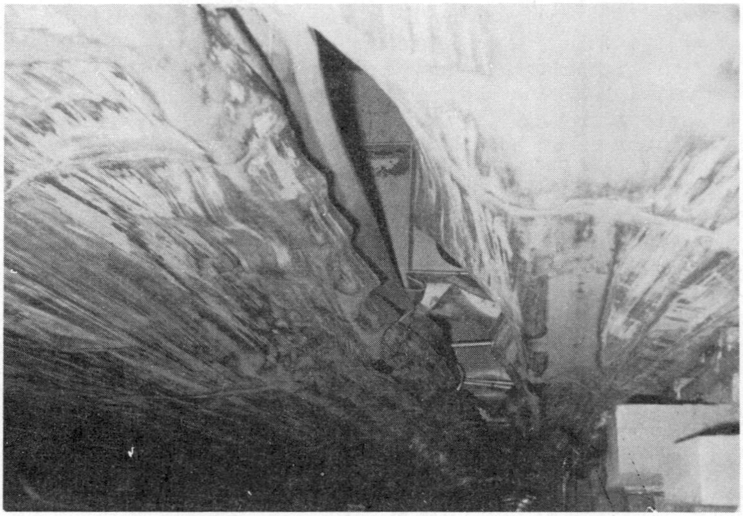

Fig. 1. Torn bottom-plating of ship due to grounding.

A simple method of estimating grounding damage was developed by Vaughan (1977) and used to predict safety speeds for a 125,000 m^3 LNG carrier running over and being slit by a number of idealised rigid projections. It was shown that the leading LNG tank, although set well aft of the bulbous bow, could possibly be damaged if grounding occurred at the relatively low speed of 1 m/sec. This investigation was reported in brief by Vaughan (1978).

In the above investigations Vaughan was able to make preliminary calculations for the tearing strength of steel plate without having to establish the dimensional effect of plate thickness on tearing strength. This investigation sets out specifically to determine this dimensional effect. The results of about seventy drop tests are presented in which a sharp rigid wedge enters the side of a plate.

Reproducible results were obtained and a simple formula for the tearing strength of mild steel plate is established.

THEORETICAL MODEL

Figure 2 shows a sharp rigid wedge penetrating a thin plate. The sides of the plate are attached to a strong frame so that the distance between the sides is fixed, thereby introducing a kinematic resistance to buckling. This situation models the bottom of a ship in which the bottom plating is welded over a relatively rigid rectangular grillage.

(a)

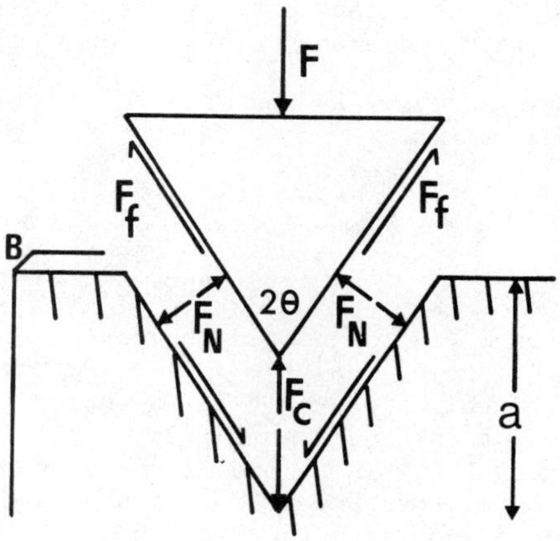

(b)

Fig. 2. Rigid wedge penetrating a thin plate. (The specimen has been reversed in 2(a) to show the curl of the plate.)

For extreme penetration ($a \gg B$) the elastic or recoverable part of the distortion is negligible. The damage to the plate then consists of a cut or tear of length 'a' and an area of plastically deformed plate A extending slightly wider than the superimposed plan of the wedge. It is required to determine the work done in penetrating the plate a given distance in terms of the free parameters of the system.

The forces acting on the wedge and plate are shown in Fig. 2(b). F is the external force required to push the wedge slowly into the plate, F_N is the normal force between the wedge and the plate which causes the plate to fold or curl away from the path, F_f is the associated frictional force and F_c is the force required to push a blade through the plate. A simple balance of forces gives

$$F = F_c + 2F_f \cos\theta + 2F_N \sin\theta \quad . \tag{1}$$

Integrating this equation with respect to 'a', gives

$$\int F\,da = a\,F_c + \int 2F_f \cos\theta\,da + \int 2F_N \sin\theta\,da \quad , \tag{2}$$

where it is assumed that F_c, the fundamental cutting force with which this article is concerned, is independent of 'a' but depends in some way on the plate thickness B. It is further assumed that F_N can be expressed as $F_N = F_0 \times a/\cos\theta$, and $F_f = \mu F_N$ where μ is a constant. F_0 is then the force associated with curling unit length of plate and $a/\cos\theta$ is the length of curled plate over which F_0 is acting.

Equation (2) then becomes

$$W = W_c + W_f + W_B$$

$$\equiv a\,F_c + \mu F_0 a^2 + F_0 a^2 \tan\theta \tag{3}$$

where W_c is the cutting energy, W_f the work done to overcome friction and W_B the work done to bend the plate. F_c and F_0 clearly depend in some way on B, the plate thickness. If it is assumed that $F_0 = \bar{F}B$, and note that $a^2 \tan\theta$ is the area of damaged plate, then $W_B = \bar{F}BA = \bar{F}V$ where V is the volume of damaged material. This is Minorsky's assumption. It is likely, however, that F_0 depends on some other power of B, probably B^2. Nevertheless, if as Minorsky did, attention is confined to plates which do not change greatly from a standard thickness, the assumption that W_B is proportional to the volume of damaged material is certainly acceptable.

Taking into account the dependence of F_c on B, the expression for W_c is written

$$W_c = a\,F_c = a\,\bar{F}_c B^{\alpha} \quad , \tag{4}$$

where $\bar{F}_c$ and α are unknown constants to be determined from experiment. The value of $\bar{F}_c$ will be appropriate only for the material used in the tests, namely mild steel, but the exponent α can be expected to hold for any material tested in plane stress since it is a measure of surface effect.

In the tests, only W the total work done can be measured so that W_c has to be isolated from the results. By lubricating the cutting wedge μ was made small and it is assumed from here on that W_f is negligibly small. Furthermore, it is seen that the cutting energy W_c scales linearly with 'a' but that the bending energy W_B scales as a^2, hence for small scale tests of the type reported here W_c will be dominant, but as 'a' increases W_B will become increasingly dominant. For extremely large penetrations of steel plate (e.g. a = 10 m, B = 10 mm) it is common to write $W = W_B$ + constant, this being Minorsky's equation. An intermediate case in which W_c and W_B are of equal magnitude has been considered by Vaughan (1977) based upon the published test records of Akita and Kitamura (1972).

As discussed in the introduction, there is a need to investigate W_c rather than W_B and the above remarks on scaling indicated to the author that if small scale tests were carried out (a = 0.1 m, B = 1 mm) then W, which is really the only quantity that can actually be measured, would in fact almost be equal to W_c. Consequently, this article is primarily concerned with the determination of $\bar{F}_c$ and α appearing in (4).

The experiments reported in the next section show that α = 1.5 and that equation (4) for mild steel plates reduces to

$$W_c = 5500\, a\, B^{1.5} \text{ Nm}, \quad \text{or equivalently}$$

$$F_c = 5500\, B^{1.5} \text{ N} \tag{5}$$

where B is measured in mm and a in metres.

EXPERIMENTAL RESULTS

The apparatus used is shown in Fig. 2(a). It consists primarily of a drop-table of known mass to which may be attached a rigid wedge. The plate specimen is held (bolted) in a rigid frame at the sides and the wedge is allowed to strike the upper edge of the plate. It should be noted that the frame is fixed at 10° to the vertical plane for several reasons. Firstly, this gives a bias to the direction in which the plate will curl as it is cut by the wedge. That is, it enables the bent plate to be pushed consistently away from the wedge and probably enables the tearing or cutting process to be examined more accurately. It also enables the wedge cutting-edge to move slightly across the edge of the plate (a sawing motion) as the wedge drops, thereby reducing wear on the cutting edge of the wedge. Finally, it is planned to use the apparatus for additional tests in which the plate has to be pierced before it can be cut. With the 10° slant to the frame such tests may be arranged merely by extending the plates vertically beyond the projection of the wedge.

In the tests three wedges were used of semi-angle 5°, 15°, and 30°. The wedges were machined from tool steel and were lubricated before each test. Four plate thicknesses were considered, 1.867 mm, 1.181 mm, 0.958 mm, and 0.782 mm, all plates being cold rolled mild steel sheet with the grain vertical.

Drop tests were carried out varying in order the drop height, the wedge angle, and the plate thickness. For each test the following quantities were recorded--total mass of drop-table, vertical drop, length of cut in plate specimen. Figure 3 shows the 1.181 mm specimens for the three wedge angles with progressive damage. The results of the experiments are shown in Fig. 4 where W, the work done to cut and bend the plate, is equated to the loss in potential energy of the drop table.

Figure 4 shows that the penetration 'a' is relatively independent of the wedge angle θ. This is consistent with equation (3), that the work done to penetrate the plate consists of a cutting part and a bending part, and for tests on this scale the cutting part is dominant. As discussed earlier, if penetrations of several metres were considered, with the wedge faces also extending that distance, the bending energy would be very prominent and the work done would then depend strongly on θ.

Figure 4 also shows that W versus 'a' is fairly linear except near the origin (from a discussion of scale effects given earlier it is to be expected that W varies as a^2 for large a). The reason for this non-linearity at the origin, a hardening in fact, is probably due to the establishing of a curl to the plate (see Fig. 2) which is established more quickly for larger θ as shown in Fig. 3. Rather than attempt to fit an analytic expression to the curve for small 'a' which for practical situations is not of importance, the curves have been extrapolated as shown. Equation (3) is now modified to

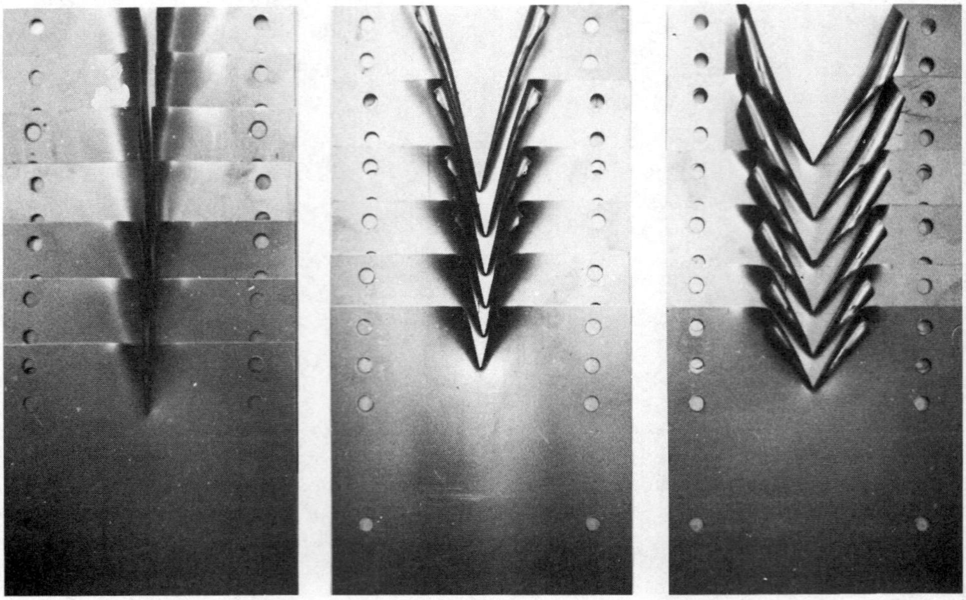

Fig. 3. Progressive damage on 1.181 mm plates using 5°, 15°, and 30° semi-angle wedges.

$$W = a F_c + F_0 a^2 \tan\theta - W^*(\theta, B) \qquad (6)$$

valid for (say) $a > 3$ cm and where W* is negligible for large 'a'. For interest the values taken from the figures are given in TABLE 1.

TABLE 1 Values of W* Versus θ and B

	1.867 mm	1.181 mm	0.958 mm	0.752 mm
$\theta = 5°$	220 Nm	140	115	90
$\theta = 15°$	170	107	87	68
$\theta = 30°$	120	75	60	50

Consider a series of tests in which the plate thickness is kept fixed. Differentiating (6) with respect to a but keeping A fixed ($A = a^2 \tan\theta$), gives

$$\left.\frac{\partial W}{\partial a}\right|_A = F_c \ . \qquad (7)$$

Now since $A = a^2 \tan\theta$, as 'a' changes in the above derivative θ must also be allowed to change so that $a^2 \tan\theta$ is constant. Since results are available only for three discrete values of θ, namely those corresponding to the three wedges, only three

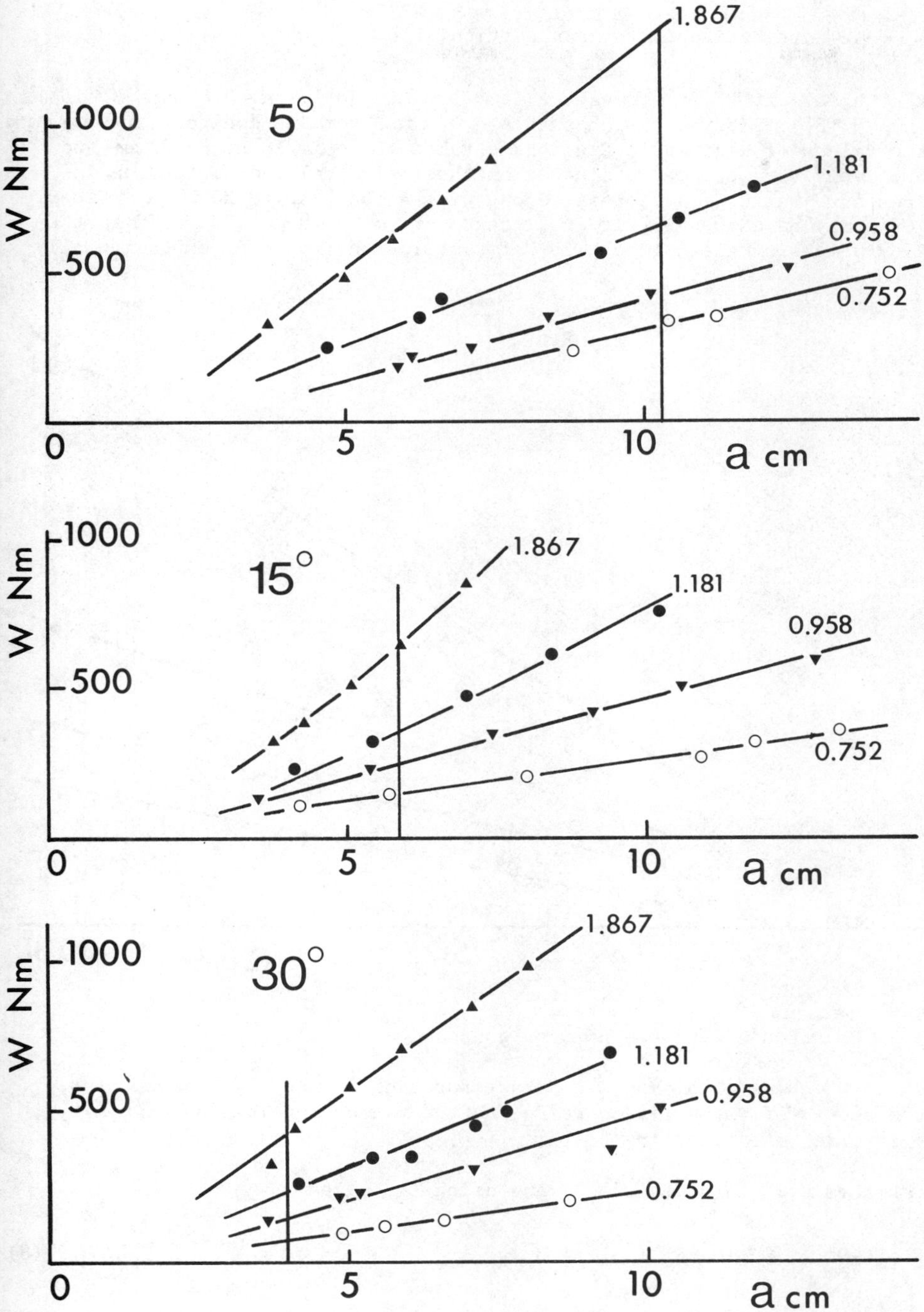

Fig. 4. External work versus penetration for 1.867, 1.181, 0.958, and 0.752 mm plates using 5°, 15°, and 30° wedges.

values of $\left.\frac{\partial W}{\partial a}\right|_A$ are available for any particular A .

Consider the three sets of (a, θ) values (a = 4 cm, θ = 30°), (a = 5.87 cm, θ = 15°), (a = 10.28 cm, θ = 5°). They are such that $A = a^2 \tan\theta$ remains constant at 9.24 cm^2. Hence if the values of W given in Fig. 4 are taken at a = 10.28 cm, 5.87 cm and 4 cm, respectively (see erected ordinates on these figures) and plotted as in Fig. 5, lines of W versus 'a' are obtained but with A constant (9.24 cm^2). Clearly any value of A may be chosen but if it differs very much from 9.24 cm^2 then the corresponding values of 'a' are likely to fall outside the range of experimentally available results.

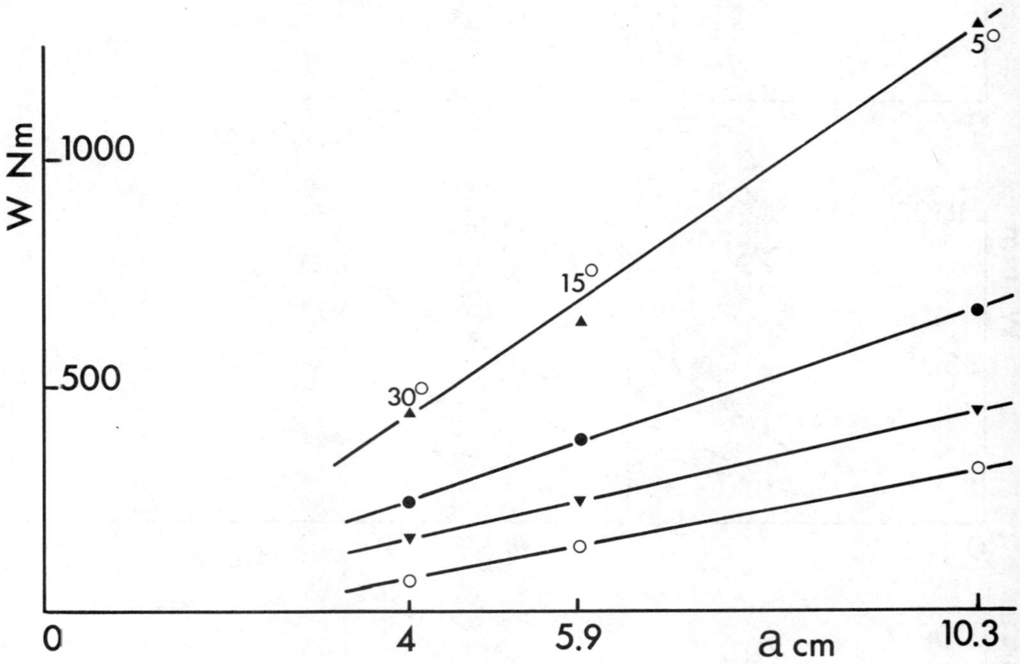

Fig. 5. W versus 'a' with $A = a^2 \tan\theta$ constant.

Figure 5 shows W versus 'a' holding A constant for each of the four values of B. From (7) the slopes of these lines are F_c and the values obtained are 140, 75, 46, and 39 in order of decreasing B.

Taking logarithms of both sides of (7) and using (4) gives

$$\log \left.\frac{\partial W}{\partial a}\right|_A = \log \bar{F}_c + \alpha \log B \quad . \tag{8}$$

The values of $\log \left.\frac{\partial W}{\partial a}\right|_A$ obtained from Fig. 5 are plotted against log B in Fig. 6.

Equation (8) predicts a straight line with slope α and intercept $\log \bar{F}_c$. Figure 6 gives $\alpha = 1.5$, $\log \bar{F}_c = 1.75$, that is, $\bar{F} = 55$. Hence $W_c = 55\, aB^{1.5}$ Nm where 'a' is in cm, B in mm.

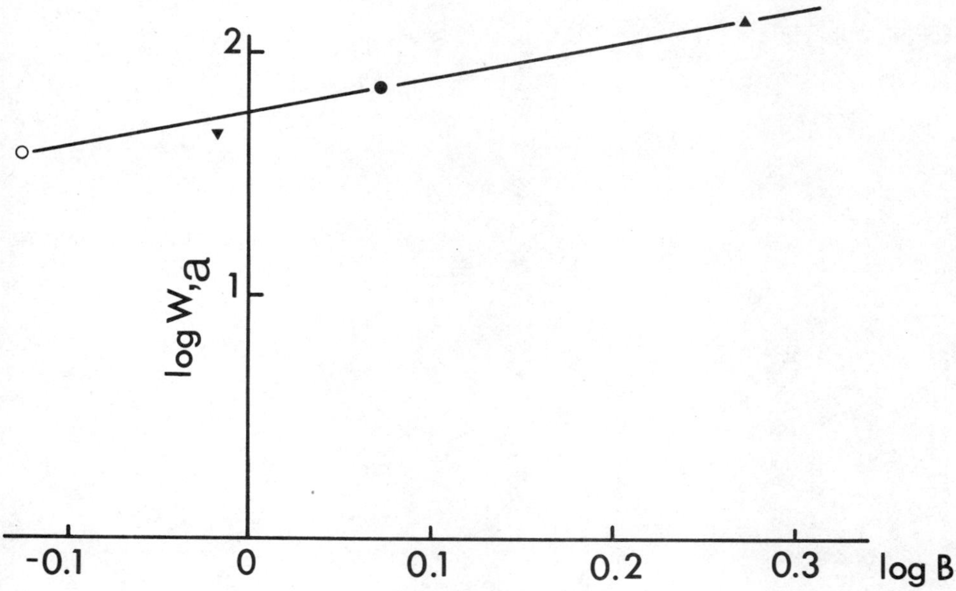

Fig. 6. Log $\left.\frac{\partial W}{\partial a}\right|_A$ versus log B. ($\left.\frac{\partial W}{\partial a}\right|_A$ is written $W,_a$ in the Fig.)

CONCLUSION

A series of tests have been conducted which enables the force required to cut a thin steel plate to be expressed in terms of the plate thickness. The force was deduced by measuring the work done to penetrate the plate a known distance. This external work was shown to consist of three parts, that necessary to cut the plate, that necessary to bend the plate and that necessary to overcome friction. In the experiments the friction contribution was neglected. A few isolated tests using more effective lubricants have noticeably reduced the total measured cutting force so that the value for F_c given by equation (5) is likely to be an upper bound for the true value.

REFERENCES

Akita, Y., and K. Kitamura (1972). A study on collision by an elastic stem to a side structure of ships. J. Soc. Naval. Arch. Japan, 131, 307-317.

Johnson, W., and A.G. Mamalis (1978). Crashworthiness of Vehicles. Mechanical Engineering Publications Ltd., London.

Jones, N. (1976). Plastic behaviour of ship structures. Trans. SNAME, 84, 115-145

Minorsky, V.U. (1959). An analysis of ship collision with reference to protection of nuclear power plants. J. Ship Research, 3, 1-4.

Vaughan, H. (1977). Damage to ships due to collision and grounding. Det norske Veritas, DnV Report 77-345.

Vaughan, H. (1978). Bending and tearing of plate with application to ship-bottom damage. J. Roy. Inst. Naval Arch., 97-99.

Woisin, G. (1971). Schiffbauliche Forschungsarbeiten fur die Sicherheit Kernenergiegetriebener Handelsschiffe. Jahrbuch der Schiffbautechnischen Gesellschaft, 65, 225-263.

Volume 3, ICM 3, Cambridge, England, August 1979

FRACTURE TOUGHNESS PREDICTIONS FOR NUCLEAR PRESSURE VESSEL STEELS

R. O. Ritchie, W. L. Server* and R. A. Wullaert*

Department of Mechanical Engineering, Massachusetts Institute of Technology, Cambridge, Massachusetts 02139, U.S.A.
***Fracture Control Corporation, Goleta, California 93017, U.S.A.**

INTRODUCTION

Stringent requirements on commercial light water reactor pressure boundary materials are imposed to provide adequate safety margins against the possibility of failure. Because of this concern, a large research effort over the last 10 to 15 years has been directed at fully characterizing the toughness of nuclear pressure vessel steels. These ferritic steels, such as SA533B-1, undergo a classical fracture mode transition from low energy brittle fractures at low temperatures (lower shelf) to high energy ductile fractures at higher temperatures (upper shelf). Fractographic observations (Van Stone, 1978) in such low strength steels have shown that, microscopically, low temperature (lower shelf) fractures occur by a transgranular cleavage mechanism along low energy cleavage planes, whereas at the upper shelf, fracture occurs by a mechanism of microvoid coalescence involving the initiation, growth and coalescence of small voids formed at inclusions and second phase particles.

Recently, attempts have been made to relate these microscopic failure modes to the mechanical criteria governing fracture, with the objective of predicting fracture toughness behavior from a knowledge of simple uniaxial mechanical properties. Failure at the lower shelf has been modeled as slip-initiated cleavage fracture using the critical stress criterion proposed by Ritchie, Knott and Rice (RKR) (1973). In a similar fashion, failure at the upper shelf has been modeled as fully ductile rupture (microvoid coalescence) using a critical strain criterion originally proposed by McClintock (1968) and adapted by MacKenzie and co-workers (1977). Successful application of such models requires a determination of critical stress and strain values from notched tensile and bend tests and a knowledge of precise analysis for the elastic-plastic stress and strain distributions ahead of sharp cracks, together with an evaluation of the microstructurally-significant size-scale controlling fracture, such as some multiple of the grain size or inter-particle spacing.

The present studies were instigated to determine the feasibility of applying such simple models to predict the variation of both lower and upper shelf static and dynamic fracture toughness with temperature and strain rate for two ferritic alloy steels utilized in the construction of nuclear pressure vessels, namely SA533B-1 and SA302B. The work constitutes an extension of studies by Parks (1976) who showed the applicability of the RKR model to predict lower shelf static toughness in SA533B-1.

FRACTURE MODELS

Critical Stress Model for Lower Shelf Toughness

The RKR model (Ritchie, Knott, Rice, 1973) proposes a local criterion for unstable slip-initiated cleavage ahead of sharp cracks which requires the local tensile stress (σ_{yy}) to exceed a critical fracture stress (σ_f^*) over a microstructurally-significant size-scale (ℓ_o^*). Accordingly, the low temperature fracture toughness can be related to the material's yield and fracture stresses. Ritchie, Knott and Rice (1973) derived quantitative predictions of the variation with temperature of the cleavage fracture toughness in mild steel with a characteristic distance of the order of two ferritic grain diameters. Other workers have since demonstrated the applicability of this model to failure in low alloy steels (Parks, 1976), higher carbon steels (Rawal, Gurland, 1977), mild steel over a range of grain sizes (Curry, Knott, 1976; Curry, 1978) and titanium alloys (Rack, 1976).

The model can be described in analytical form (Curry, 1978) by considering the HRR (Hutchinson, 1968; Rice and Rosengren, 1968) small scale yielding solution for the stress distribution ahead of a crack tip in a non-linear elastic material where the maximum principal stress (σ_{yy}) at distance x from a crack of length a is given by

$$\frac{\sigma_{yy}}{\sigma_o} = f.\left(\frac{J \cdot a}{\alpha I \cdot x}\right)^{\frac{1}{N+1}} , \qquad (1)$$

where σ_o is the flow stress (taken as the tensile yield stress σ_y), J a non-dimensionalized value of the J contour integral (Rice, 1968) to which the crack is loaded, α the yield strain, N the hardening exponent for a Ranberg-Osgood material, I a numerical constant weakly dependent on N, and f the angular dependence term. By expressing equation (1) in terms of stress intensity K, viz:

$$\frac{\sigma_{yy}}{\sigma_y} = f.\left[\frac{(1-\nu^2)}{\alpha I}\right]^{\frac{1}{N+1}} \left[\frac{x}{(K/\sigma_y)^2}\right]^{\frac{-1}{N+1}} \qquad (2)$$

and applying the RKR criterion that $K = K_{IC}$ when σ_{yy} exceeds σ_f^* over a characteristic distance $x = \ell_o^*$ from the crack tip, an expression for the fracture toughness is obtained, i.e.,

$$K_{IC} = \beta^{-\left(\frac{N+1}{2}\right)} . \ \ell_o^{*\,1/2} \left[\frac{\sigma_f^{*\,\frac{N+1}{2}}}{\sigma_y^{\frac{N-1}{2}}}\right] , \qquad (3)$$

where $\beta = f.\left[\frac{(1-\nu^2)}{\alpha I}\right]^{\frac{1}{N+1}}$ is the amplitude of the stress singularity available from the HRR solutions (Hutchinson, 1968).

Experimental determination of critical fracture stress (σ_f^*) values is achieved by fracturing V-notched bend bars at low temperatures (Knott, 1966; Wilshaw, Rau, Tetelman, 1968). Traditionally the value of σ_f^* has been calculated from slip-line field theory (which assumes non-hardening plane strain plasticity) to determine the maximum principle stress (σ_{yy}^{max}) at the notch root at failure (Hill, 1950). More recently, however, Griffiths and Owen (1971) and Owen and co-workers (1973) have derived accurate finite element solutions for the elastic-plastic stress distributions in V-notched 4-point bend specimens for hardening materials, and the determination of σ_f^* is now best achieved using such analyses.

Critical Strain Model for Upper Shelf Toughness

Ductile fracture can be modeled as being strain-induced, in that macrocrack growth via linking of voids occurs when some critical strain is exceeded ahead of the crack tip. However, as McClintock (1968) first pointed out, void growth during ductile fracture is also a strong function of stress state; in particular the ratio of mean to equivalent flow stress. Furthermore, in an analogous situation to cleavage fracture, the critical strain must be exceeded over a minimum amount of material which is characteristic of the scale of physical events involved. Accordingly, MacKenzie and co-workers (1977) have proposed a criterion for ductile failure where a critical fracture strain ($\bar{\varepsilon}_f^*$) is locally exceeded over some microstructurally-significant characteristic distance (ℓ_o^*) ahead of the crack tip, but which further recognizes that this critical strain is a strong function of stress state which itself varies with distance ahead of the crack tip.

The application of this model for the prediction of upper shelf fracture toughness involves experimental determination of the fracture strain ($\bar{\varepsilon}_f$) or ductility as a function of stress state, together with knowledge of the elastic-plastic strain distribution ahead of a sharp crack. Values of $\bar{\varepsilon}_f$ can be obtained using circumferentially-notched round tension specimens loaded uniaxially. With such test pieces failure initiates in the central region of the notch where the hydrostatic stress state is most severe, and different states of stress may be obtained by varying the notch acuity. Using the standard definitions of equivalent stress ($\bar{\sigma}$), mean or hydrostatic stress (σ_m) and equivalent plastic strain ($\bar{\varepsilon}_p = \int d\bar{\varepsilon}_p$), the Bridgeman analysis for the necked tensile specimen (Bridgeman, 1952) gives a constant strain distribution at the minimum cross section of

$$\bar{\varepsilon}_p = \int d\bar{\varepsilon}_p = 2 \ln \frac{r_o}{r_{min}} , \qquad (4)$$

for a stress-state at the center of the specimen of

$$\sigma_m/\bar{\sigma} = 1/3 + \ln\left(1 + \frac{r_{min}}{2\rho}\right) , \qquad (5)$$

where ρ is the notch root radius and r_o and r_{min} are the original and minimum radii of the notched cross-section respectively. By loading a series of such specimens with varying ρ under notch opening displacement control, values of $\bar{\varepsilon}_p$ at failure initiation ($\bar{\varepsilon}_f$) are obtained as a function of $\sigma_m/\bar{\sigma}$. The distribution of strain ($\bar{\varepsilon}_p$) and stress state ($\sigma_m/\bar{\sigma}$) ahead of a crack tip for small scale yielding are then taken from the blunting solution of Rice and Johnson (1970) as a function of distance ahead of the crack (x) normalized with respect to crack tip opening displacement δ (Fig. 1). Fracture toughness values are now predicted by determining the value of δ where the equivalent plastic strain ($\bar{\varepsilon}_p$) exceeds, over the characteristic distance (ℓ_o^*), the value of critical fracture strain ($\bar{\varepsilon}_f^*$) representative of the crack tip stress state at that position. Thus, if the Rice and Johnson (1970) strain distribution is written in the form

$$\bar{\varepsilon}_p = c_1 \frac{\delta}{x} \quad , \tag{6}$$

where c_1 is a constant determined from Fig. 1, then at initiation of ductile fracture, $\bar{\varepsilon}_p = \bar{\varepsilon}_f^*$ at distance $x = \ell_o^*$ when $\delta = \delta_c$. K_{IC} values may be related to the critical crack tip opening displacements (δ_c) using the small scale yielding formulation

$$\delta_c = 0.49 \frac{K_{IC}^2}{E\,\sigma_y} \quad , \tag{7}$$

where E is the elastic modulus and σ_y the yield strength. By combining equations (6) and (7), the critical strain model can be expressed in the form

$$K_{IC} = \text{constant}\ (\varepsilon_f^* \cdot \ell_o^* \cdot \sigma_y \cdot E)^{1/2} \tag{8}$$

Since the failure event for ductile rupture generally involves coalescence by shear localization between a number of voids nucleated at the largest particles, values of ℓ_o^* for such fracture are generally found to be some small multiple of the inter-particle spacing (Hancock, MacKenzie, 1976; McClintock, 1968).

MATERIALS AND EXPERIMENTAL PROCEDURES

Materials and Microstructure

The materials examined were two low strength alloy steels utilized in nuclear pressure vessel construction, namely the currently-used material SA533 Grade B Class 1 (HSST Plate 02) and the formerly used material SA302B (Surveillance Correlation Heat). The composition, heat treatment and standard mechanical properties of these steels are shown in Tables 1 and 2. It is apparent that aside from the lower residual element (P and S) and higher nickel content in SA533B-1, the chemistries of the two steels are not too dissimiliar, yet in terms of their respective toughness, SA533B-1 is clearly a significantly tougher material.

As with many low alloy steels quenched or normalized in thick plate section, the microstructure of both steels is primarily acicular tempered upper bainite with a small volume fraction of proeutectoid ferrite. Prior austenite grain sizes (d) were measured at 25 μm for SA533B-1 and 33 μm for SA302B. Although extensively investigated with respect to mechanical properties, there is a surprising lack of information in the literature with regard to the fine-scale microstructures of these steels. Particle morphology and distribution, however, has been studied by Van Stone (1978) in a similar plate of SA533B-1, indicating around 0.12 volume percent of manganese sulfide and aluminum oxide inclusions (5-10 μm in size) with equiaxed sub-micron cementite precipitates (1/2 μm in size) in bands along bainitic laths. No information could be found on the fine-scale microstructural details of SA302B.

Mechanical Properties

Most of the fracture toughness results were gathered from the EPRI Data Bank on nuclear pressure vessel steels compiled by Server and Oldfield (1978). ASTM valid static fracture toughness data (K_{IC}, $\dot{K} \sim 3$ MN-m$^{-3/2}$s^{-1}) for both steels were obtained using 1 in (25.4 mm) thick compact (termed 1-T) specimens and precracked

Charpy specimens in the transverse (T-L) orientation. Additional static K_{IC} data on HSST Plate 02 were obtained from longitudinally (L-T) oriented 1 in (25.4 mm) to 11 in (280 mm) thick (1T-11T) compact specimens tested as part of the Heavy Section Steel Technology (HSST) Program (Hunter, Williams, 1971); Mager, Thomas 1969; Shabbitts, 1970). Dynamic fracture toughness (K_{Id}, $\dot{K} \sim 10^5$ MN-m$^{-3/2}$s^{-1}) was also examined from the EPRI Data Bank (precracked Charpy, 1-T bend) and the HSST Program (1-T compact) (Shabbitts, 1970). The K_{IC} and K_{Id} data correspond to approximate crack tip strain rates of 10^{-2}s^{-1} and 3×10^2s^{-1}, respectively. Upper shelf fracture toughness data were obtained from the EPRI Data Bank. These values were computed from valid elastic-plastic J_{IC} tests on 1-T compacts using the multi-specimen R-curve procedure for defining crack initiation (Landes, Begley, 1976), where

$$K_{IC} \simeq K_{JC} = \left[\frac{J_{IC}\,E}{1 - \nu^2}\right]^{1/2} , \quad (11)$$

where E is the elastic modulus and ν is Poisson's ratio.

Tensile yield strength data for both steels were taken from the EPRI Data Bank, and the data of Steichen and Williams (1973) and Server (1978). Additional high strain rate data were derived from measurements of the general yield load on V-notch and precracked Charpy impact tests according to the analysis of Server (1978). Yield stress (σ_y) data as a function of temperature and strain rate were compiled, using the strain-rate temperature relationship of Bennett and Sinclair (1966), in terms of T ln $(A/\dot{\varepsilon})$, where T is the absolute temperature, $\dot{\varepsilon}$ the strain rate, and A a constant equal of 10^8s^{-1}. For the purposes of applying the fracture models to predict fracture toughness values from such data, all yield stress values used were corrected for approximate crack tip strain rates; that is $\dot{\varepsilon} \simeq 10^{-2}$s^{-1} for static K_{IC} tests, and $\dot{\varepsilon} \simeq 3 \times 10^2$s^{-1} for dynamic (impact) K_{Id} tests.

Experimental Determination of Critical Fracture Stress and Strain (σ_f^*, $\bar{\varepsilon}_f^*$)

Critical fracture stress (σ_f^*) values for lower shelf slip-initiated cleavage in both SA533B-1 and SA302B were determined from slow bend and impact four-point bend tests on 45° V-notched bars broken at temperatures of -73°C, -129°C and -193°C. These temperatures were chosen such that failure occurred catastrophically by unstable cleavage fracture well before general yielding. Slow bend tests were performed at an effective crack tip strain rate $\dot{\varepsilon}$ of approximately 10^{-4}s^{-1}, and impact tests were performed at $\dot{\varepsilon} \simeq 50$ s^{-1}. The nominal bending stress at the onset of catastrophic failure was measured in each test, and using the Griffith and Owen (1971) finite element stress distribution for this particular geometry of test piece, the critical values of the maximum principle stress at the crack tip ($\sigma_{yy}^{max} = \sigma_f^*$) were computed.[1] The values of the critical fracture stress (σ_f^*) obtained are listed in Table 3.

Fracture strain ($\bar{\varepsilon}_f$) values as a function of stress state ($\sigma_m/\bar{\sigma}$) for upper shelf ductile fracture in both steels were determined from a series of circumferentially-notched round tensile specimens, tested at 24°C, 71°C and 177°C. Test pieces unnotched and notched with root radii ρ of 1.27, 1.91, 2.54, 3.81 and 6.35 mm were loaded under longitudinal displacement control, monitored with an axial strain transducer mounted across the notch. In addition, a diametral strain transducer was mounted in the notch continuously to monitor changes in the minimum diameter of

[1]Values of yield stress used in the computation were corrected for the particular crack tip strain rates, as described above.

the notched cross-section. Failure initiation was defined as the point when the average axial stress (applied load divided by instantaneous cross-sectional area) dropped sharply prior to final failure. The fracture strain ($\bar{\varepsilon}_f$) was computed as the equivalent plastic strain ($\bar{\varepsilon}_p$) at this point, using equation 4. The corresponding stress state $\sigma_m/\bar{\sigma}$ at the center of the neck where failure initiated was determined using equation 5. A plot of fracture strain ($\bar{\varepsilon}_f$) versus stress state ($\sigma_m/\bar{\sigma}$) for both materials at room temperature is shown in Fig. 2 and clearly indicates the marked reduction in ductility as the hydrostatic stress state is increased. Further, it can be seen that whereas both steels show similar ductility values in uniaxial tension ($\sigma_m/\bar{\sigma}$ = 1/3), at highly triaxial stress states i.e., $\sigma_m/\bar{\sigma}$ = 1.2) SA302B shows a significantly smaller fracture strain. Similar behavior was also observed at 71°C and 177°C.

The critical fracture strain ($\bar{\varepsilon}_f^*$), used to predict fracture toughness and representing material ductility ahead of a crack tip, was taken as the value of $\bar{\varepsilon}_f$ corresponding to the particular stress state at the characteristic distance from the crack tip (from Fig. 1). For the cases studied, this corresponded to the fracture strain at values of $\sigma_m/\bar{\sigma}$ in excess of 1.2.

Estimation of Characteristic Distances (ℓ_o^*)

Precise determination of the critical microstructural size-scale for either cleavage or ductile fracture demands at least that the complete micro-mechanisms of failure are fully understood. Since the local mechanisms of the critical failure event in these steels are not known, it would be impossible from first principles to determine these characteristic distances. In their derivation of the critical stress model for cleavage fracture, Ritchie, Knott and Rice (1973) found best agreement with experimental data with a characteristic distance of two ferrite grain diameters ($\ell_o^* \simeq 2$ d = 120 μm) in mild steel, but there is no real fundamental physical significance for this. However, examination of lower shelf cleavage fractures in SA533B-1 and SA302B reveals that cleavage facets are of the order of the prior austenite grain size rather than, say, the bainite packet size. Thus it is not unreasonable to presume that the critical fracture initiation event must occur over the first few prior austenite grains from the crack tip. Accordingly for lower shelf toughness predictions characteristic distances were chosen between 2 d and 4 d with d, the prior austenite grain size, measured as 25 and 33 μm for SA533B-1 and SA302B respectively. These values are consistent with previous estimates by Parks (1976).

At upper shelf temperatures, the initiation event for ductile fracture is related to the displacements produced by internal necking between particles. Thus it does not seem possible that initiation would occur over distances smaller than the particle spacing. Since in the present steels it is the linkage by shear localization between large voids, nucleated at inclusions, which describes the initiation of actual ductile crack growth (Van Stone, 1978), it would be expected that the characteristic distance is some function of the inclusion spacing. Examination of upper shelf fractures in both steels shows that the spacing between major voids on fracture surfaces is roughly 50 μm in SA533B-1 (consistent with Van Stone (1978) inclusion distribution studies) and 100 - 150 μm in SA302B. However, it is not known in either case how many major voids coalesce with the main crack tip for the initiation of a critical crack. Estimates of ℓ_o^* for ductile fracture in somewhat similar steels range from 1 to 10 times the inclusion spacing (Pandey, Banerjee, 1978; Rawal, Gurland, 1977), indicating that in certain instances initiation of ductile crack growth involves the coalescence of a single void to the main crack whereas in other cases several voids are involved. Thus no a priori determination of ℓ_o^* for ductile fracture in the present steels could be made, and characteristics distances were chosen best to fit the experimental data.

PREDICTION OF TOUGHNESS

Lower Shelf Toughness

The dynamic fracture toughness (K_{Id}) for SA533B-1 steel, between -200 and 0°C, is illustrated in Fig. 3, together with calculated values, determined from the RKR critical cleavage stress model for characteristic distances of 2, 3, and 4 prior austenite grain diameters. Predictions are based on the known temperature dependence of yield strength calculated for approximate crack tip strain rates of $3 \times 10^{2} s^{-1}$, a critical cleavage stress σ_f^* of 2000 MPa (Table 3) assumed independent of temperature, a work hardening exponent n of 0.1 representing an average value for this steel over the temperature range -200 to 0°C, and plane strain crack tip stress analyses due to Rice and Tracey (1973), Tracey (1976) and Rice and Johnson (1970). Similar predictions for the temperature variation of lower shelf static fracture toughness (K_{IC}) of SA533B-1 and SA302B are shown in Figs. 4 and 5 for an approximate crack tip strain rate of $1 \times 10^{-2} s^{-1}$.

It is apparent from Figs. 3-5 that the RKR critical cleavage stress model provides a very good description of lower shelf toughness behavior in these steels for temperatures up to NDTT, although experimental data for SA302B are extremely sparse. Above the NDT temperature, however, RKR predictions tend to level off, due to the smaller temperature variation of yield strength, and markedly underestimate the data. As pointed out previously (Ritchie, Knott and Rice, 1973) failure in these steels above NDTT can no longer be considered as unstable cleavage, and involves ductile tearing between "stable" cleavage micro-cracks. Such failure is clearly not amenable to analysis using a purely critical stress model for failure.

Upper Shelf Toughness

The use of the critical strain model to predict upper shelf toughness required determination of the critical fracture strain ($\bar{\varepsilon}_f^*$) at each temperature in question. Values of ε_f^* were determined at three upper shelf temperatures (24, 71, 177°C) at a single strain rate ($\dot{\varepsilon} \simeq 10^{-2} s^{-1}$). Valid fracture toughness data, computed from initiation J_{IC} values, in this temperature range are extremely scarce. However, using the fracture strain as a function of stress state data (example given in Fig. 2) and the Rice and Johnson (1970) near-tip strain distribution (Fig. 1), predictions of the upper shelf static fracture toughness between room temperature and 200°C for both steels were made using the critical fracture strain model. Results for SA533B-1 and SA302B are illustrated in Fig. 4 and 5, respectively. For SA533B-1, the experimental data are best described by the attainment of a critical fracture strain ($\bar{\varepsilon}_f^*$) over a characteristic distance (ℓ_o^*) of 300 to 350 μm ahead of the crack tip (Fig. 4) whereas in SA302B, ℓ_o^* is of the order of 100 to 150 μm. The significantly lower upper shelf toughness of SA302B is thus consistent with a lower plane strain ductility and a smaller characteristic distance.

Several points are worthy of note from these critical strain fracture toughness predictions for the upper shelf. First, the critical strain model predicts an essentially constant fracture toughness at the upper shelf, consistent with most experimental data (Server, 1977, Williams, 1975). Second, the critical microstructural size scales (ℓ_o^*) for ductile fracture in SA302B and SA533B-1 are of the order of 1 and 6 to 8 times the planar inter-inclusion spacing respectively (i.e., 100-150 μm and 300-400 μm). This is consistent with the fractographic studies by Van Stone (1978) on SA533B-1, who found that the crack growth mechanism was not simply the nucleation of one void which coalesced with the main crack, but rather the nucleation and linkage of several isolated voids within the plastic zone. Similar conclusions have been reported for ductile fracture in other low strength steels (Pandey, Banerjee, 1978). Thus, for ductile fracture, the characteristic distance must be regarded as not merely reflecting the spacing between particles which nucleate major voids, but also the critical number of voids which coalesce with the main crack at the initiation of crack growth.

SUMMARY AND CONCLUSIONS

The present work has examined the application of simple fracture models to predict the temperature variation of plane strain fracture toughness in ferritic nuclear pressure vessel steels. These models attempt to combine accurate continuum mechanics solutions for the stress and strain distributions ahead of crack tips with necessarily simplified idealizations of the microscopic fracture mechanisms involved. For lower shelf cleavage fracture at temperatures below NDTT, fracture toughness values are consistent with the attainment of a critical fracture stress over a distance of 3 prior austenite grain diameters ahead of the crack tip, according to the RKR critical stress model for slip-initiated cleavage.

Similarly, for upper shelf ductile fracture at temperatures between ambient and 200°C, elastic-plastic initiation fracture toughness values under quasi-static loading are consistent with the attainment of a critical fracture strain (determined for highly triaxial stress states) over a characteristic distance related to the major inclusion spacing, according to the critical strain model for micro-void coalescence. Based on the magnitude of this distance determined from the present work, initiation of ductile fracture in SA302B appears to involve linkage of a single major void to the crack tip, whereas in SA533B-1 several voids within the plastic zone are involved.

Whilst it is felt that a priori predictions of macroscopic fracture behavior (i.e., fracture toughness values) using such models for local failure criteria are not at present feasible because of uncertainties in the magnitude of the microstructurally-significant dimension, the description of fracture toughness in terms of these models does provide some insight into the micro-mechanics and mechanisms of failure, and furthermore provides a vital connection between continuum plasticity and microstructural mechanisms of fracture.

ACKNOWLEDGEMENTS

The work was performed under contract No. RP886-1 from the Electric Power Research Institute, Palo Alto, California.

REFERENCES

Bennett, P. E. and G. M. Sinclair (1966). J. Basic Eng., Trans ASME Series D, 88, 518-524.

Bridgeman, P. W. (1952). Studies in Large Flow and Fracture, McGraw Hill, N.Y.

Curry, D. A. (1978). C.E.R.L. Technical Report No. RD/L/N 64/78, Central Electricity Research Laboratory, Leatherhead, England.

Curry, D. A. and J. F. Knott (1976). Metal Science, 10, 1-6.

Griffiths, J. R. and D. R. J. Owen (1971). J. Mech. Phys. Solids, 19, 419-531.

Hancock, J. W. and A. C. MacKenzie (1976). J. Mech. Phys. Solids, 24, 147-169.

Hill, R. (1950). The Mathematical Theory of Plasticity, Oxford University Press.

Hunter, C. W. and J. A. Williams (1971). Nuclear Engineering and Design, 17, 131-148.

Hutchinson, J. W. (1968). J. Mech. Phys. Solids, 16, 13-31.

Knott, J. F. (1966). J. Iron Steel Inst., 204, 104-111.

Landes, J. D. and J. A. Begley (1976). Westinghouse Scientific Paper 76-1E7-JTINTF-P3, Westinghouse Scientific Laboratories, Pittsburgh, PA.

MacKenzie, A. C., J. W. Hancock and D. K. Brown (1977). Eng. Fracture Mech., 9, 167-188.

Mager, J. R. and F. O. Thomas (1969). HSST Technical Report No. 5, Westinghouse Electric Corp., Pittsburgh, P.A.

McClintock, F. A. (1968). J. Applied Mech., Trans ASME Series E, 35, 363-371.

Owen, D. R. J., G. C. Nayak, A. P. Kfowi, and J. R. Griffiths (1973). Int. J. Num. Methods in Engng., 6, 63.

Pandey, R. K. and S. Banerjee (1978). Eng. Fract. Mechanics, 10, 817-829.

Parks, D. M. (1976). J. Eng. Matls. and Tech., Trans ASME Series H, 98, 30-35.

Rack, H. J. (1976). Matls. Sci. Eng., 24, 165-170.

Rawal, S. P. and J. Gurland (1977). Met. Trans. A., 8A, 691-698.

Rice, J. R. (1968). J. Applied Mech., Trans ASME Series E, 35, 379.

Rice, J. R. and M. A. Johnson (1970). In M.F. Kanninen (Ed.) Inelastic Behavior of Solids, McGraw Hill, N.Y., p. 641.

Rice, J. R. and G. F. Rosengren (1968). J. Mech. Phys. Solids, 16, 1-12.

Rice, J. R. and D. M. Tracey (1973). In S. J. Fenves (Ed.) Numerical and Computational Methods in Structural Mechanics, Academic Press, New York.

Ritchie, R. O., J. F. Knott and J. R. Rice (1973). J. Mech. Phys. Solids, 21, 395-410.

Server, W. L. (1977). Proc. of ASTM Symposium on Elastic-Plastic Fracture, Atlanta, GA, (ASTM STP in press).

Server, W. L. (1978). J. Eng. Matl. and Tech., Trans. ASME Series H, 100, 183-188.

Server, W. L. and W. Oldfield (1978). "Nuclear Pressure Vessel Steel Data Base", Fracture Control Corp. Topical Report to Electric Power Research Institute, EPRI NP-933.

Shabbits, W. O. (1970). HSST Technical Report No. 13, Westinghouse Electric Corp., Pittsburgh, PA.

Steichen, J. M. and J. A. Williams (1973). HSST Technical Report No. 32, Handford Engineering Development Laboratory, Richland, WA.

Tracey, D. M. (1976). J. Eng. Matls. & Tech., Trans ASME Series H, 98, 146-151.

Van Stone, R. H. (1978). General Electric Report No. SRD-78-116 for EPRI, General Electric Company, Schenectady, N.Y. pp. 5.1-5.13.

Williams, J. A. (1975). HSST Technical Report No. 36, Hanford Engineering Development Laboratory, Richland, WA.

Wilshaw, T. R., C. A. Rau, and A. S. Tetelman (1968). Eng. Fracture Mechanics, 1, 191-211.

TABLE 1 Composition and Heat Treatment of Nuclear Pressure Vessel Steels

Material	Heat Code	C	Mn	P	S	Si	Ni	Cr	Mo	V	Cu
		Chemistry in wt pct									
SA533B-1* (HSST Plate 02)	1bA	0.23	1.55	0.009	0.014	0.20	0.67	0.04	0.53	0.003	0.16
SA302B† (Surveillance Correlation Heat)	4bA	0.23	1.47	0.013	0.024	0.26	0.17	0.05	0.52	0.004	0.20

*Normalized 4 hrs at 914°C, air cooled; austenitized 4 hrs at 900°C, agitated water quenched; tempered 4 hrs at 664°C, water quenched; stress relieved 40 hrs at 612°C, furnace cooled.

†Austenitized 4 hrs at 900°C, water quenched; stress relieved 6 hrs at 650°C, air cooled.

TABLE 2 Mechanical Properties of Nuclear Pressure Vessel Steels (T-L Orientation)

Material	Tensile Properties at 24°C			NDTT	Charpy V-notch Energy, Upper Shelf
	Yield Stress (MPa)	U.T.S. (MPa)	Elongation* (pct)	(°C)	(J)
SA533B-1	481	642	25.4	-29	137
SA302B	465	625	25.2	-18	60

*for 50.8 mm gauge length

TABLE 3 Measured Values of Critical Cleavage Fracture Stress (σ_f^*) for Nuclear Pressure Vessel Steels

Material	σ_f^*, Static (MPa)	σ_f^*, Dynamic Impact (MPa)
SA533B-1	1830	2000
SA302B	1650	1900

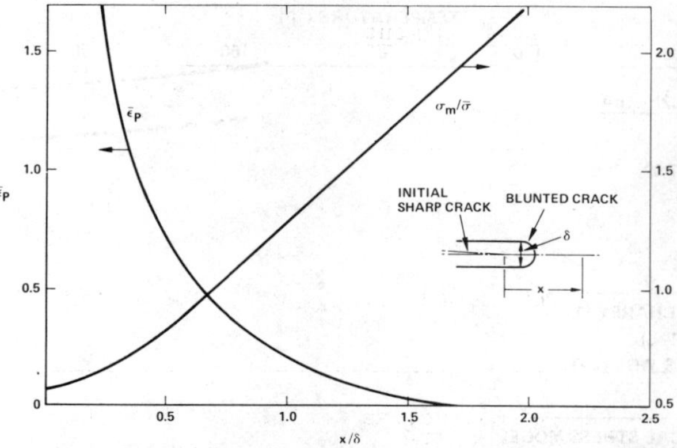

Fig. 1. Distribution of plastic strain ($\bar{\epsilon}_p$) and stress state ($\sigma_m/\bar{\sigma}$) near crack tip for Rice and Johnson small scale yielding solution.

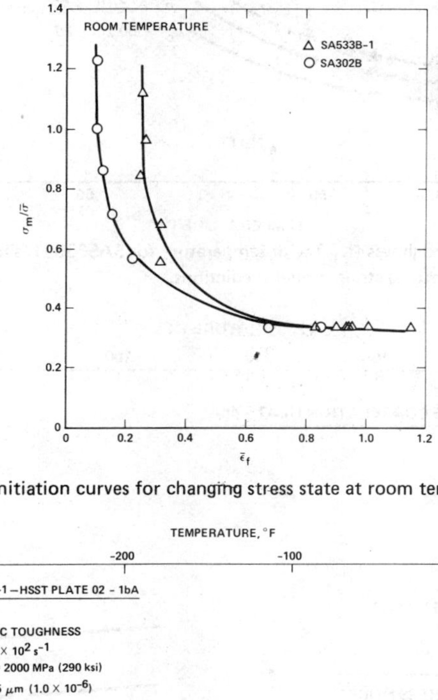

Fig. 2. Failure initiation curves for changing stress state at room temperature (24°C).

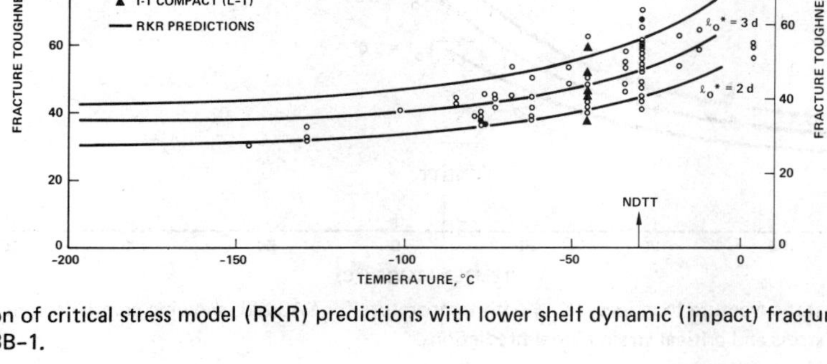

Fig. 3. Comparison of critical stress model (RKR) predictions with lower shelf dynamic (impact) fracture toughness values for SA533B-1.

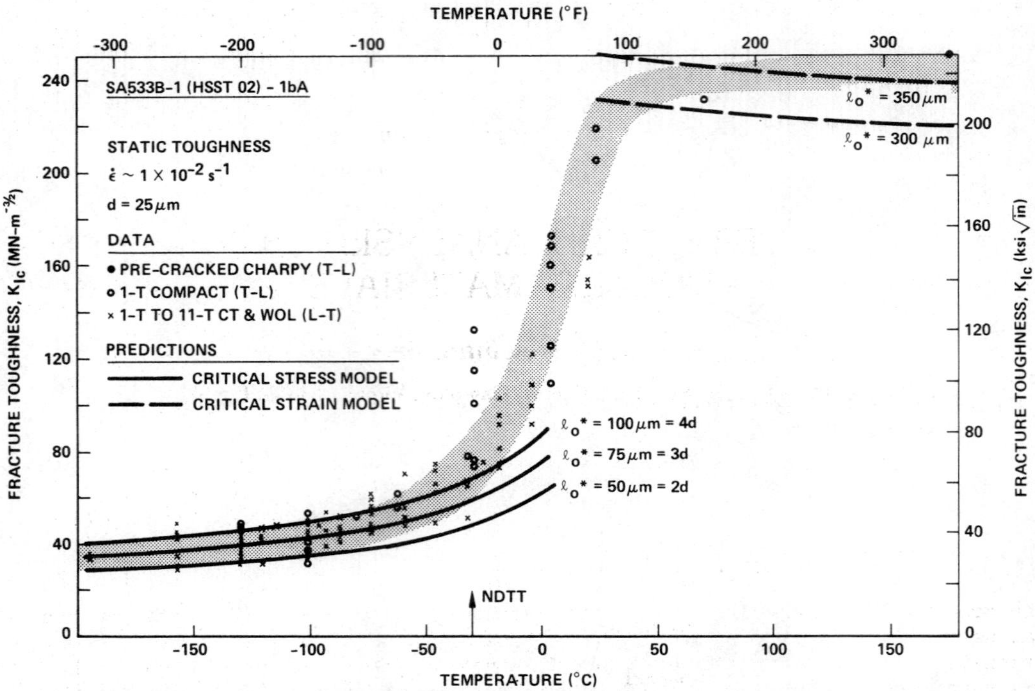

Fig. 4. Variation of static fracture toughness (K_{Ic}) with temperature for SA533B-1, showing comparison of experimental data with critical stress and critical strain model predictions.

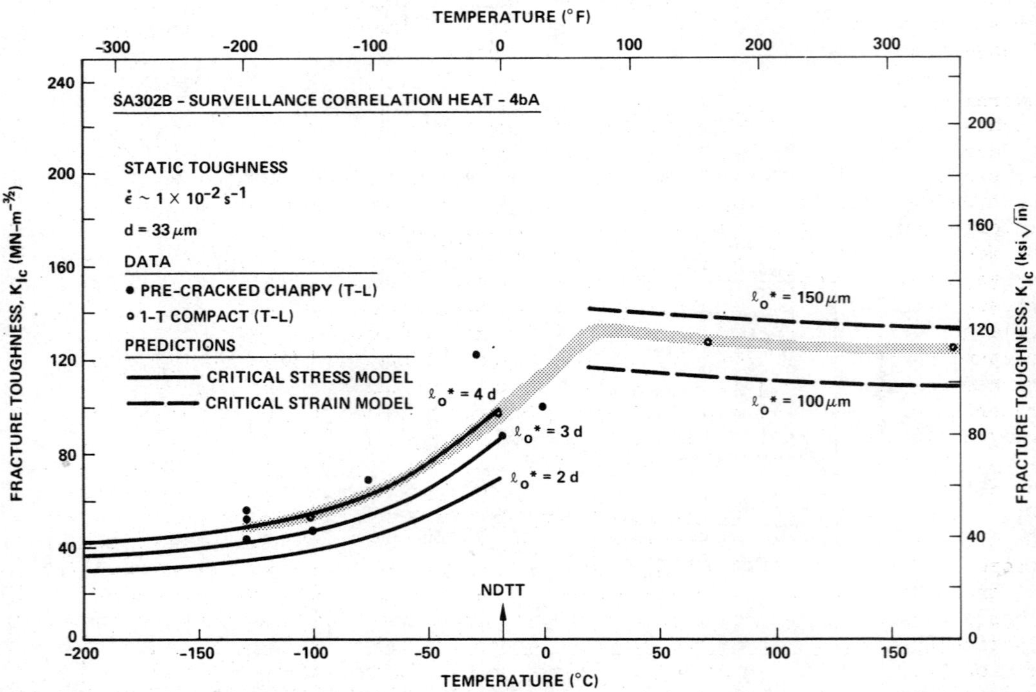

Fig. 5. Variation of static fracture toughness (K_{Ic}) with temperature for SA302B, showing comparison of experimental data with critical stress and critical strain model predictions.

Volume 3, ICM 3, Cambridge, England, August 1979

FRACTURE ANALYSIS OF DUCTILE MATERIALS

J. C. Newman, Jr.

NASA Langley Research Center, Hampton, Virginia 23665, U.S.A.

INTRODUCTION

The concepts of linear-elastic fracture mechanics have been very useful in correlating fracture data from cracked plates and structural components in which the crack-tip plastic deformations are constrained to small regions (plane-strain fracture, Brown (1966)). However, for high-toughness materials, large amounts of plastic deformation occur near the crack tip at fracture and the elastic stress-intensity factor at failure (K_{Ie}) varies with planar dimensions, such as crack length and specimen width (Srawley, 1964; Kuhn, 1970; Newman, 1973a). To account for variations in K_{Ie} with crack length and specimen width, the elastic-plastic stress-strain behavior near the crack tip must be considered.

Several equations for calculating the elastic-plastic stress-strain behavior at notches or cracks have been proposed. Among these are equations derived for notches by Hardrath and Ohman (1953), and by Neuber (1961). For cracks, equations have been derived by Hutchinson (1968) and by Rice and Rosengren (1968). The Hardrath-Ohman equation was later generalized for a cracked plate and was applied as a fracture criterion by Kuhn and Figge (1962). In a similar way, Newman (1973a, 1973b) used the Neuber relation and the elastic-stress distribution in the crack-tip region to derive a fracture criterion for a cracked plate. This criterion related the elastic stress-intensity factor at failure, the elastic nominal failure stress, the plastic-collapse (or plastic-hinge) stress, and two material parameters. The two-parameter fracture criterion (TPFC) was used to analyze failure of surface- and through-cracked sheet and plate specimens under tensile loading (Newman, 1973a). This criterion was rederived in a more general form by Newman (1973b) and was used to analyze failure of compact and notch-bend sheet specimens.

In the present paper, the TPFC was applied to single-edge-crack tension specimens (Fig. 1) made of a bolting steel and a rotor steel tested at room temperature (Chell, 1978). The thicknesses of the steel specimens were of the same order as the specimen widths. In previous analyses using the TPFC, the specimen width-to-thickness ratios (W/B) were usually much greater than 2 and the "plane-stress" plastic-collapse stress was used. However, for W/B ratios about unity, the plastic-collapse stress is strongly influenced by the state of stress on the net section and the collapse stress is closer to the "plane-strain" value. The present paper applies the TPFC to fracture of materials and of configurations where the use of the "plane-strain" plastic-collapse stress is necessary.

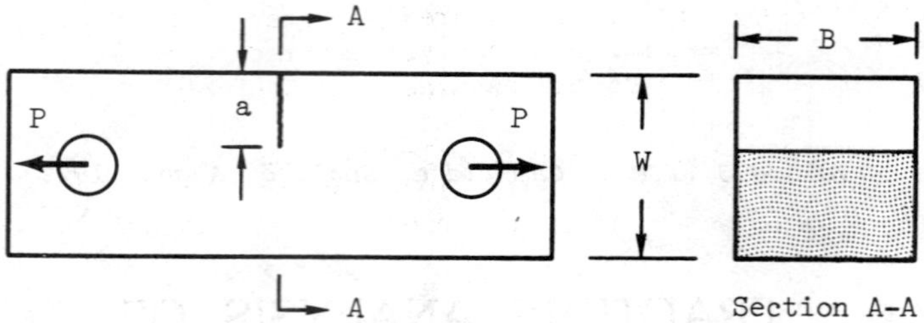

Fig. 1. Single-edge-crack tension specimen configuration.

NOMENCLATURE

F	stress-intensity boundary-correction factor
K_F	elastic-plastic fracture toughness, $MN/m^{3/2}$
K_{Ie}	elastic (Mode I) stress-intensity factor at failure, $MN/m^{3/2}$
m	fracture-toughness parameter
R	ratio of plane-strain-to-plane-stress collapse stress
S_n	nominal (net-section) failure stress, MN/m^2
S_u	plane-stress plastic-collapse stress, MN/m^2
S_u^*	plane-strain plastic-collapse stress, MN/m^2
W	specimen width, mm
λ	ratio of crack length to specimen width
σ_u	uniaxial tensile strength, MN/m^2
Φ	ratio of K_{Ie} to K_F

TWO-PARAMETER FRACTURE CRITERION

The equation for the elastic-stress distribution near a crack tip in an elastic material which contains the stress-intensity factor, K_I, and the square-root singularity is well known. The determination of K_I is the basis for linear-elastic fracture mechanics (LEFM). The stress-intensity factor for cracked components is a function of the load, the configuration, and the size and location of the crack. In general, the elastic stress-intensity factor at failure for any cracked (Mode I) configuration can be expressed as

$$K_{Ie} = S_n \sqrt{\pi a}\, F \tag{1}$$

where S_n is the nominal failure stress (computed from the maximum load at failure) and a is the initial crack length. The boundary-correction factor, F, accounts for the influence of various boundaries on stress intensity. (The Appendix gives the equations for S_n and F for the single-edge-crack tension specimen.) The use of the linear-elastic equation is restricted to conditions in which the plastic zone at the crack tip is very small compared to other dimensions of the body (brittle fracture). Consequently, to analyze ductile materials, the elastic-plastic behavior of the stresses and strains near the crack tip must be considered.

The TPFC, which accounts for the elastic-plastic material behavior, has been successfully applied (Newman, 1973a, 1973b, 1976) to various crack configurations and to a wide range of materials. This criterion is

$$K_F = \frac{K_{Ie}}{\Phi} \tag{2}$$

where

$$\Phi = 1 - m\frac{S_n}{S_u} \qquad \text{for} \quad S_n < \sigma_y \tag{3}$$

and

$$\Phi = \frac{\sigma_y}{S_n}\left(1 - m\frac{S_n}{S_u}\right) \qquad \text{for} \quad \sigma_y < S_n < S_u \tag{4}$$

K_F and m are the two material fracture-toughness parameters, K_{Ie} is the elastic stress-intensity factor at failure, σ_y is the tensile yield stress, and S_u is the "plane-stress" plastic-collapse stress. The plastic-collapse stress was computed from the load required to produce a plastic hinge (Timoshenko, 1956) on the net section. For the single-edge-crack tension specimen (Fig. 1), S_u is a function of the crack-length-to-specimen-width ratio. (See the Appendix.)

The fracture parameters K_F and m have been shown to be material properties that vary with material thickness, state of stress, temperature, and rate of loading. To obtain fracture constants that are representative for a given material and test temperature, the fracture data should be from material of the same thickness, and the tests should encompass a wide range of specimen width or crack length.

The function Φ in Eq. (2) reflects the influence of the nominal failure stress on K_F, the elastic-plastic fracture toughness. The variation of the denominator with nominal stress for a hypothetical material is shown in Fig. 2. When the nominal stress is less than the uniaxial yield stress, σ_y, the function Φ (ratio of K_{Ie} to K_F) is a linear function of nominal stress (solid line). The line has a negative slope, m. However, when the nominal failure stress is greater than the yield stress, the function Φ becomes nonlinear and depends upon the stress-strain curve of the material and the state of stress in the crack-tip region, as discussed by Newman (1973a). An estimate for this behavior is shown by the dash-dot curve and it is approximated by Eq. (4) (Newman, 1976). This equation is shown as the dashed curve. The vertical dashed line truncates the nominal stress at S_u. The results shown in Fig. 2 are for a given material of constant thickness. The three-dimensional fracture diagram in Fig. 3 illustrates the relationship between K_{Ie}, S_n, and specimen thickness (B) for a given material. The solid curve on the $S_n = 0$ plane shows the values of K_F plotted against specimen thickness. For a given thickness, the relationship between K_{Ie} and S_n is linear for $S_n < \sigma_y$. The three-dimensional surface is the locus of K_{Ie} values for various combinations of specimen dimensions. For large thicknesses, a constant value of plane-strain fracture toughness (K_{Ic}) is produced, in accordance with LEFM. Such behavior produces a plateau (shaded area) near the left extremity of the surface shown (with $m = 0$). For small thicknesses, fracture occurs under plane-stress conditions.

The fracture-toughness parameters, K_F and m, describe the crack sensitivity of the material. In tests of various materials and thicknesses, the parameter m has been found to range from zero to unity (Newman, 1973a, 1976). If m is equal to

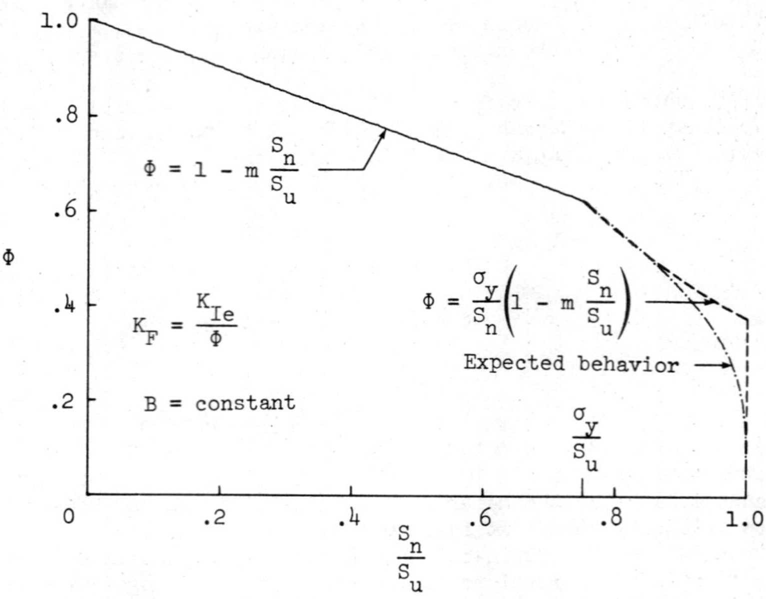

Fig. 2. Relationship between Φ and S_n/S_u.

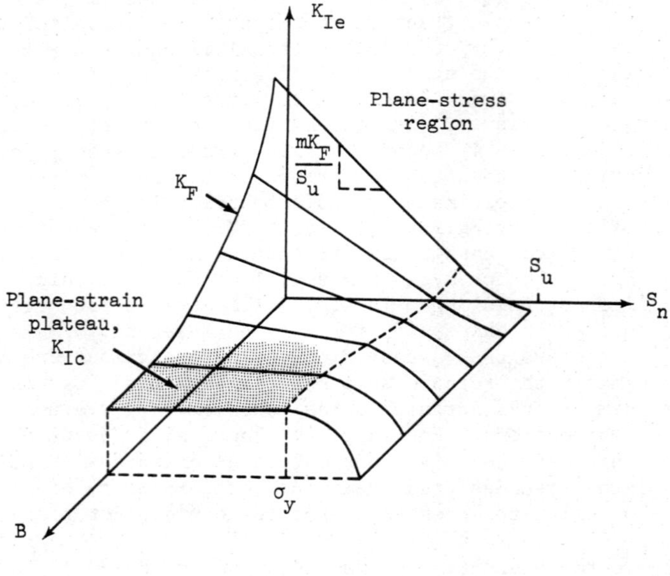

Fig. 3. Elastic stress-intensity factors at failure as a function of nominal failure stress and thickness.

zero in Eq. (2), K_F is equal to K_{Ie} (or K_{Ic}) and the equation represents behavior of low-toughness materials. For large m values, the equation represents behavior of high-toughness materials. For extremely tough materials, where m is equal to unity and K_F is much larger than K_{Ie}, the equation reduces to

$$S_n = \frac{S_u}{1 + \frac{K_{Ie} S_u}{K_F \sigma_y}} \approx S_u \qquad (5)$$

Thus, for specimens made of extremely tough materials, failure occurs when the nominal stress is near the plastic-collapse stress.

FAILURE PREDICTIONS

After the fracture-toughness parameters K_F and m had been determined from fracture tests on a given material, thickness, and test temperature, Eqs. (1) and (2) were used to predict nominal failure stresses for other cracked configurations. The failure stresses were calculated by substituting Eq. (1) into Eq. (2) and were given by

$$S_n = \frac{K_F}{\sqrt{\pi a}\, F + \frac{m K_F}{S_u}} \qquad \text{for } S_n < \sigma_y \qquad (6)$$

and

$$S_n = \sqrt{(m\gamma)^2 + 2\gamma S_u} - m\gamma \qquad \text{for } \sigma_y < S_n < S_u \qquad (7)$$

where

$$\gamma = \frac{K_F \sigma_y}{2 S_u \sqrt{\pi a}\, F} \qquad (8)$$

For single-edge-crack tension specimens (Fig. 1), the failure load, P_{max}, was calculated by

$$P_{max} = \frac{S_n B(W - a)}{1 + \frac{3\lambda}{1 - \lambda}} \qquad (9)$$

where S_n is given by Eqs. (6) or (7) and $\lambda = a/W$. Equation (9) was calculated from simple beam theory.

ANALYSIS OF FRACTURE DATA

Fracture data on single-edge-crack tension specimens made of a bolting steel and a rotor steel were taken from the literature (Chell, 1978) and were analyzed using the TPFC. The W/B ratios for these data ranged from 0.3 to 3. As previously mentioned, the TPFC has been successfully applied (Newman, 1973a, 1973b, 1976) to "plane-stress" fracture data (W/B ratios much greater than 2) and the "plane-stress" plastic-collapse stress S_u was used. However, for W/B ratios about unity, the plastic-collapse stress for high-toughness materials is closer to the "plane-strain" value. Thus, for W/B ratios near unity, the "plane-strain" plastic-collapse stress was used in the TPFC analysis.

An estimate for the ratio of plane-strain to plane-stress plastic-collapse stress (denoted as R) was attributed to Haigh and Richards by Chell (1978). The plane-strain plastic-collapse stress, S_u^*, was given by Chell (1978) as

$$S_u^* = RS_u \quad (10)$$

where
$$R = 1 + 1.585\lambda - 0.683\lambda^2 + 1.193\lambda^3 - 4.215\lambda^4 \quad (11)$$

and $\lambda = a/W$ for $\lambda < 0.7$. A plot of S_u and S_u^* normalized by the uniaxial tensile strength σ_u is shown in Fig. 4 as the lower and upper solid curves, respectively. The equation for S_u is given by Eq. (15). For small a/W ratios, S_u/σ_u and S_u^*/σ_u approach unity (uniaxial tension) and for a/W ratios near unity, S_u/σ_u approaches the pure bending case of 1.5. The ratio S_u^*/σ_u should also approach the pure bending case for a/W ratios near unity, as estimated herein by the dashed curve.

In the subsequent TPFC analysis of the single-edge-crack tension specimens (W/B ratios near unity), the plane-stress plastic-collapse stress, S_u, was replaced by S_u^*, the plane-strain value, in Eqs. (2) through (8).

Bolting Steel

Figure 5 shows the results of analyzing fracture data obtained by Chell (1978) on a bolting steel (Durehete 1055). The specimen width ranged from 8 to 25 mm and the material thickness was 25 mm. The tensile properties of the material are given in the figure. The figure shows the experimental (symbols) and calculated (solid curves) failure loads (P_{max}) plotted as a function of a/W for the three widths tested. The values of K_F and m used to calculate the solid curves were obtained from an analysis of these data using a best-fit procedure (Newman, 1973a). The failure loads were calculated from Eq. (9) using Eqs. (6) through (8). As previously mentioned, S_u in Eqs. (6) through (8) was replaced by S_u^*. The symbols on the vertical axis (a/W = 0) indicate the failure loads calculated using the ultimate

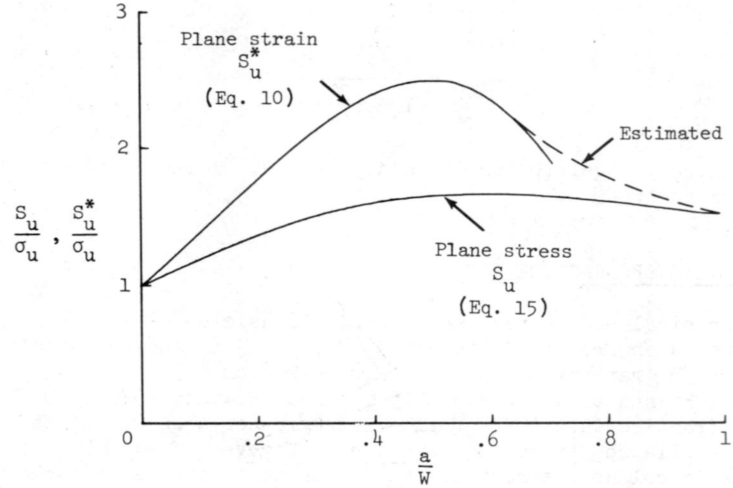

Fig. 4. Plane-stress and plane-strain plastic-collapse stress for the single-edge-crack tension specimen.

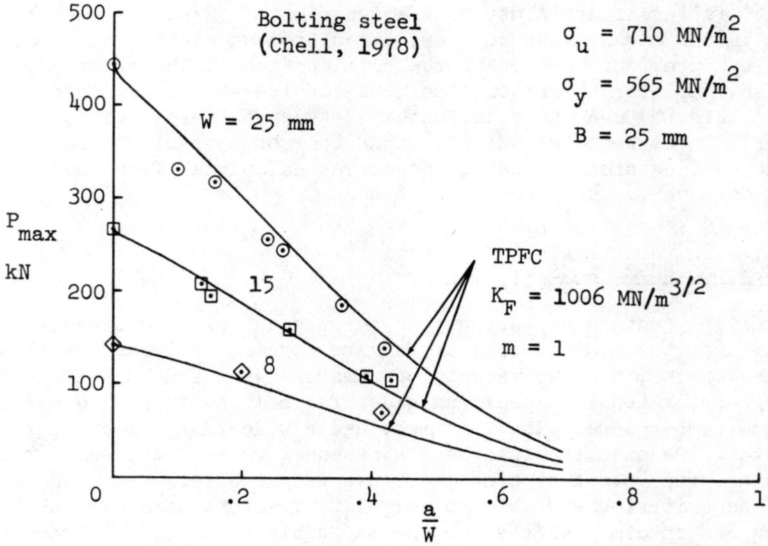

Fig. 5. Experimental and calculated failure loads as a function of a/W and W for a bolting steel.

tensile strength ($P_{max} = \sigma_u WB$). The curves calculated from the TPFC fit the experimental data well and indicate that the relationship between failure load and crack length for this material is nearly linear for a/W ratios less than about 0.4.

Rotor Steel

Figure 6 shows a similar analysis of fracture data (Chell, 1978) on a rotor steel (1 percent CrMoV). The material thickness was 10 mm and the specimen width ranged

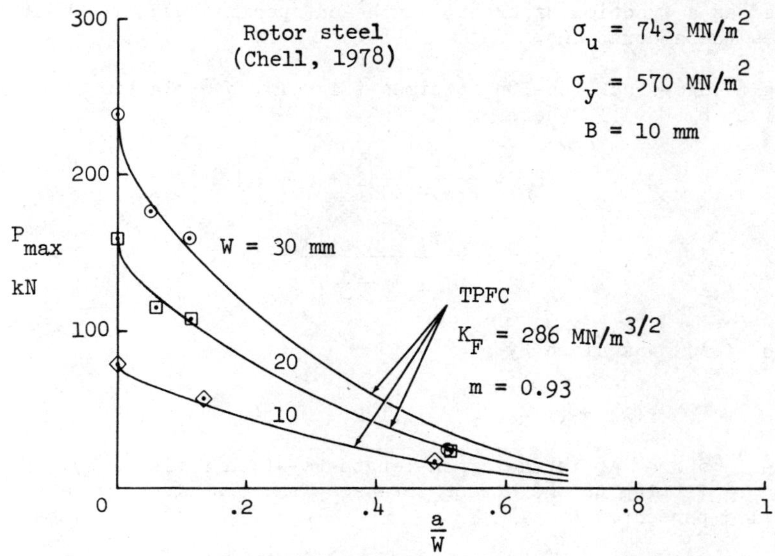

Fig. 6. Experimental and calculated failure loads as a function of a/W and W for a rotor steel.

from 10 to 30 mm. Again, the figure shows the experimental (symbols) and calculated (solid curves) failure loads plotted as a function of a/W for the three widths tested. The values of K_F and m were calculated using the same procedure as described in the previous section on the bolting steel. The elastic-plastic fracture toughness, K_F, for the rotor steel was considerably lower than that for the bolting steel (Fig. 5) and the relationship between failure load and crack length was nonlinear. These results indicated that the rotor steel was more crack sensitive than the bolting steel. Again, the curves calculated from the TPFC fit the experimental data well.

CONCLUDING REMARKS

The two-parameter fracture criterion that relates the elastic stress-intensity factor at failure, the elastic nominal failure stress, the plastic-collapse (or plastic-hinge) stress, and two material parameters was used to analyze fracture data on single-edge-crack tension specimens made of a bolting steel and a rotor steel tested at room temperature. The specimens had a wide range of crack lengths and specimen widths. Because the specimen thicknesses were of the same order as the specimen widths, the "plane-strain" plastic-collapse stress was used in the fracture criterion. The criterion was able to correlate fracture data over a wide range in crack lengths and specimen widths for the materials considered. Previously, the criterion had been successfully applied to many crack configurations and to a wide range of materials and thicknesses. Therefore, the criterion should be useful to designers who must select materials for application and who must calculate failure stresses in flawed components.

APPENDIX: ELASTIC STRESS-INTENSITY FACTOR, NOMINAL STRESS DEFINITION, AND PLASTIC-COLLAPSE STRESS FOR THE SINGLE-EDGE-CRACK TENSION SPECIMEN

In the application of Eqs. (1) and (2) to single-edge-crack tension specimens, the stress-intensity factor, the nominal stress, and the plastic-collapse stress must be determined as a function of crack length and specimen width. The following section gives these equations.

For the single-edge-crack tension specimen (Fig. 1), the elastic stress-intensity factor is given by Eq. (1) where

$$S_n = \frac{P}{B(W - a)}\left[1 + \frac{3\lambda}{1 - \lambda}\right] \tag{12}$$

and

$$F = \frac{(1 - \lambda)f(\lambda)}{1 + \frac{3\lambda}{1 - \lambda}} \tag{13}$$

The function $f(\lambda)$ was given by

$$f(\lambda) = (1 - \lambda)^{-3/2}[1.122 - 0.36(1 - \lambda)\sqrt{\lambda}] \tag{14}$$

for $0 \leq \lambda < 1$ where λ is the crack-length-to-width ratio (a/W). Equation (14) was obtained by fitting to the stress-intensity factors given by Tada (1973) and was within ±1.5 percent.

The "plane-stress" plastic-collapse stress, S_u, for the single-edge-crack tension specimen is a function of λ and is given by

$$S_u = \left\{\left[\sqrt{1 + \left(\frac{\lambda}{1 - \lambda}\right)^2} - \frac{\lambda}{1 - \lambda}\right]\left[1 + \frac{3\lambda}{1 - \lambda}\right]\right\}\sigma_u \quad (15)$$

This was computed from the load required to produce a fully plastic hinge (Timoshenko, 1956) on the net section using the ultimate tensile strength. A plot of Eq. (15) is shown in Fig. 4.

REFERENCES

Brown, W. F., Jr., and J. E. Srawley (1966). Plane strain crack toughness testing of high strength metallic materials. Am. Soc. for Testing and Materials, ASTM STP-410, 1-65.

Chell, G. G., and R. S. Gates (1978). A study of failure in the post yield regime using single-edge notched tension specimens. Int. J. of Fracture, 14, 233-247.

Hardrath, H. F., and L. Ohman (1953). A study of elastic and plastic stress concentration factors due to notches and fillets in flat plates. National Advisory Committee for Aeronautics, NACA TN 2566.

Hutchinson, J. W. (1968). Singular behavior at the end of a tensile crack in a hardening material. J. of Mech. Phys. Solids, 16, 13-31.

Kuhn, P., and I. E. Figge (1962). Unified notch-strength analysis for wrought aluminum alloys. National Aeronautics and Space Administration, NASA TN D-1259.

Kuhn, P. (1970). Residual tensile strength in the presence of through cracks or surface cracks. National Aeronautics and Space Administration, NASA TN D-5432.

Neuber, H. (1961). Theory of stress concentration for shear-strained prismatical bodies with arbitrary non-linear stress-strain law. J. Appl. Mechs., 28, 544-550.

Newman, J. C., Jr. (1973). Fracture analysis of surface- and through-cracked sheets and plates. Engr. Fracture Mechs. J., 5, 667-689.

Newman, J. C., Jr. (1973). Plane stress fracture of compact- and notch-bend specimens. National Aeronautics and Space Administration, NASA TM X-71926.

Newman, J. C., Jr. (1976). Fracture analysis of various cracked configurations in sheet and plate materials. Am. Soc. for Testing and Materials, ASTM STP-605, 104-123.

Rice, J. R., and G. F. Rosengren (1968). Plane strain deformation near a crack tip in a power-law hardening material. J. of Mech. Phys. Solids, 16, 1-12.

Srawley, J. E., and W. F. Brown, Jr. (1964). Fracture toughness testing methods. Am. Soc. for Testing and Materials, ASTM STP-381, 133-198.

Tada, H., P. C. Paris, and G. R. Irwin (1973). The stress analysis of cracks handbook. Del Research Corporation.

Timoshenko, S. (1956). Strength of Materials. D. VanNostrand Co., New York.

Volume 3, ICM 3, Cambridge, England, August 1979

DETERMINATION OF PLASTIC STRAINS AND THE RATE OF STRAIN ENERGY DISSIPATION IN THIN CRACKED PLATES

N. V. Novikov and A. L. Maystrenko

The Institute for Superhard Materials of the Ukrainian Academy of Sciences, Kiev, U.S.S.R.

INTRODUCTION

Crack propagation in thin sheet metals is accompanied by a crack tip plastic zone which is considerable in size but substantially smaller than the linear dimensions of the crack. Cherepanov (1967), Rice (1968a, 1968b) and Hutchinson (1968) have shown that the Cherepanov-Rice J-integral can be used as a single parameter to describe the stress-strain state at the crack tip when plastic flow occurs.

There are several experimental methods for J-integral determination (Landes, 1972, 1974). These methods are based on finding the potential energy difference on testing two identical specimens differing in the length of initial cracks, a_o, with the same displacement of the loading points. These methods make it possible indirectly to estimate the damage accumulation preceding macroscopic failure and to relate it to the values of the path independent J-integral.

Suppose that, in practice, the change in the total potential energy, U, put in from the outside is dissipated by plastic strains at the crack tip (Olesiak, 1966). Then the rate of strain energy dissipation per unit area of newly formed surface will be physically equal to the J-integral:

$$J = -\frac{1}{B}\frac{\partial U_p}{\partial a} \qquad (1)$$

A method of dividing grids (Novikov, 1977) has been used to investigate plastic strain distribution at the tip of equilibrium and developing cracks under static tension in cracked specimens cut from thin sheets (B=0.72 mm) of D16T aluminium alloy.

EXPERIMENTAL PROCEDURE

To determine the plastic strain distribution at a crack tip the surface of a sheet specimen with an effective area of 160 x 300 mm was covered with a dividing grid produced by photolithography from operating copies of precision originals made with a space error not more than 0.25 µm. The original grids with filling area of 40 x 40 mm contained 6400 squares of side 0.5 mm and a line thickness of 5µm.

Tensile tests on single edge notch (SEN) specimens with a fatigue crack were conducted in an Instron (model 1126) testing machine of rigid type with a cross-head travel grid of 1 mm/min. The use of low strain rate made it possible repeatedly to photograph a region adjacent to the crack tip with different values of loading. Consecutive measurements of photographs have been performed to determine coordinates for the points of the distorted dividing grid and to calculate the values of the principal strains, ε_1 and ε_2

CORRELATION PROCEDURE

Investigations were carried out on thin sheets of plastic alloy producing a plane stress state at the crack tip. It is assumed that the body material is incompressible (γ = 0.5), having Young's modulus $E = 6.86 \times 10^4$MPa for D16T at 293K and yield stress for simple shear θ = 314.8 MPa (with the A.Y. Ishlinsky hypothesis $\tau_y = 0.667\ \theta_y$). With allowance made for the fact that in the case under consideration the crack edges are free of loading and that during crack extension tangents of stress and displacement along the Y-axis are equal to zero, we find that the real displacements in the rectangular area S($-W/2 \leq x,y \leq W/2$, where "W" is the width of specimen)attain a functional minimum (Parton, 1974).

$$R = \frac{1}{6}E \iint_{S_I} F^2\, dx\, dy + \tau_y \iint_{S_2} \left(F - \frac{3}{2}\frac{\tau}{E} y\right) dx\, dy\ , \qquad (2)$$

where

$$F = 2\left(\varepsilon_x^2 + \varepsilon_y^2 + \varepsilon_x \cdot \varepsilon_y + \frac{1}{4}\gamma_{xy}^2\right)^{\frac{1}{2}} =$$

$$= 2\left(\varepsilon_1^2 + \varepsilon_2^2 + \varepsilon_1 \cdot \varepsilon_2\right)^{\frac{1}{2}}\ ,$$

S_1 is the region of elastic strains and S_2 the region of plastic strains. The boundary between regions S_1 and S_2 will be determined from the condition:

$$F = F_o = \frac{3\tau}{E} y = 3\tau' \qquad (3)$$

The considerations above provide a basis for determining the external boundaries of plastic zones at the crack tips, which may be outlined from a knowledge of the fields of principal strains, ε_1, and, ε_2, found experimentally at different stages of fracture. Figures 1 and 2 show a distribution of lines for equal values of F/τ' within a plastic region for cracks of different initial length - a_o/W = 0.415 and 0.512. The figures illustrate plastic zone formation at the crack tip under conditions of crack equilibrium and in the case of its subcritical growth. The graphs have dimensionless coordinates X* and Y*:

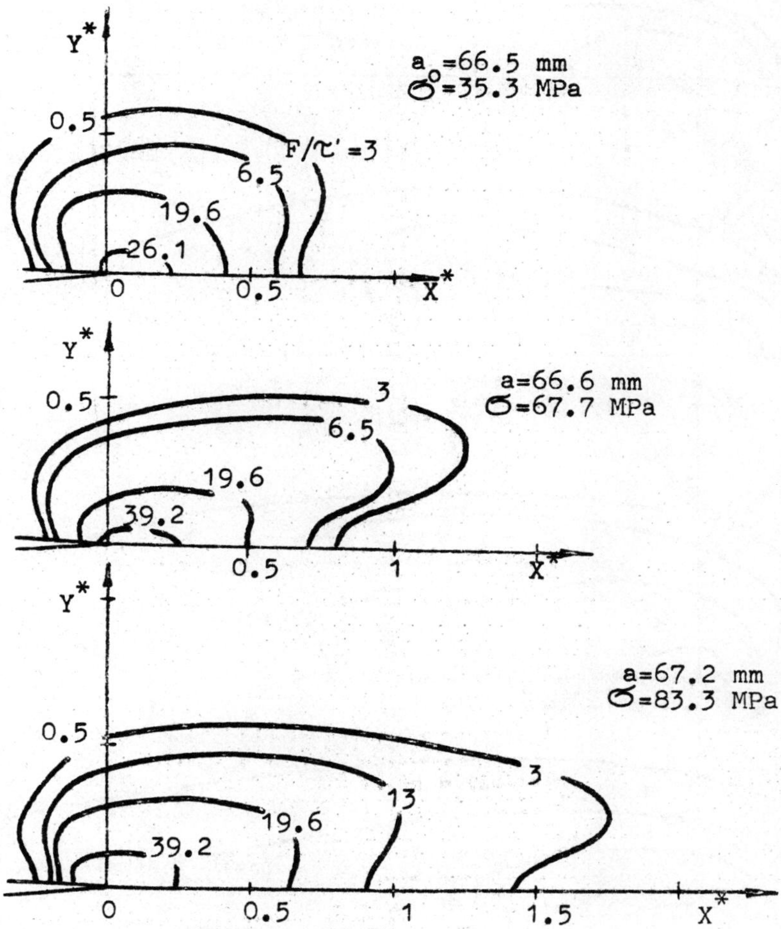

Fig. 1. Lines of equal values of F/τ' at a crack tip ($a_o/W = 0.415$) at different stages of its growth.

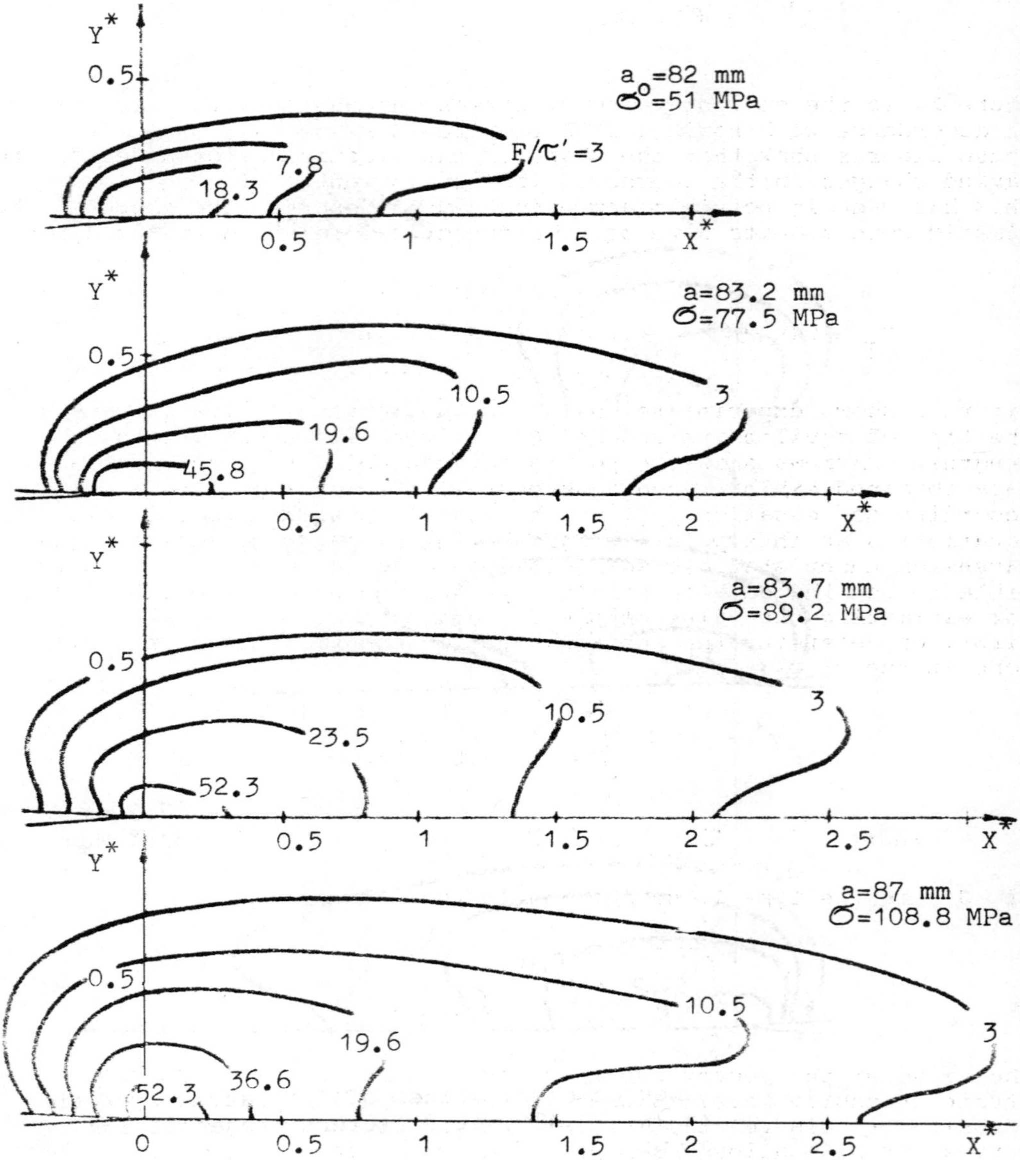

Fig. 2. Lines of equal values of F/τ' at a crack tip ($a_o/W = 0.512$) at different stages of its growth.

$$X^{*} = 8x\sigma^{2}_{y} / \pi K_{I}^{2} ,$$

$$Y^{*} = 8y\sigma^{2}_{y} / \pi K_{I}^{2} , \qquad (4)$$

where K_I is the current value of stress intensity factor, calculated in accordance with Bowie (1965) requirements.
These figures show that the shape of the plastic region undergoes no marked changes in the course of its development.
This has made it possible approximately to describe the shape of the plastic zone and its size using a functional relation of the type:

$$r_y = \left\{ 0.25(\pi - \theta) + 0.95\, e^{-2\theta} \sin 3\theta \right. \frac{\pi K_I^2}{8\sigma_y^2} . \qquad (5)$$

Figure 3 shows experimental data for the length of plastic zone at the tips of equilibrium cracks ($\theta = 0$) with different initial lengths. Arrows show the points corresponding to crack initiation. Data obtained exhibit good agreement with calculated curves plotted according to equation (5), which makes it possible to use this equation as an interpolation formula for calculating plastic zone extension under static loading in thin cracked sheets. It is possible to use the data on strain distribution within the plastic zone for estimating the value of the Cherepanov-Rice J-integral.
First, we determine the specific work of plastic strain in a plastic zone at the crack tip:

$$W_p = \int_0^{\gamma_{oct}} \tau_{oct}\, d\gamma_{oct} , \qquad (6)$$

where $\tau_{oct} = G' \gamma_{oct}$;

If G' is the type II secant module then

$$W_p = \int_0^{\gamma_{oct}} G' \gamma_{oct}\, d\gamma_{oct} . \qquad (7)$$

The value of the secant shear modulus $G' = G(\varepsilon_i)$ as a function of strain intensity level is used in equation (7). A functional expression of this relation has been found from experimental strain curves for D16T alloy under practical conditions. The curves have been plotted in coordinates of stress and strain intensities, from which the following polynomial approximation has been found:

$$G' = G\,(2716\,\varepsilon_i^4 - 1360\,\varepsilon_i^3 + 242.8\,\varepsilon_i^2 - 18.5\varepsilon_i + 0.58) . \qquad (8)$$

Since $\gamma_{oct} = \sqrt{2}\, \varepsilon_i$, (9)

Substituting (8) and (9) into equation (7) and changing the variable of integration we obtain:

$$Wp = 2 \int_0^{\sqrt{2}\varepsilon_i} G'(\varepsilon_i)\,\varepsilon_i\, d\varepsilon_i =$$

$$= G(7242\varepsilon_i^6 - 3077.3\varepsilon_i^5 + 485.6\varepsilon_i^4 - 34.9\varepsilon_i^3 + 1.2\varepsilon_i^2) \quad (10)$$

Because of our assumption of material incompressibility ε_i is found from ε_I, and ε_2 as:

$$\varepsilon_i = \frac{2\sqrt{3}}{3}\sqrt{\varepsilon_1^2 + \varepsilon_2^2 + \varepsilon_1\varepsilon_2} \quad (11)$$

We may now compute the strain energy within the plastic region at the crack tip from:

$$U_p = \int_0^{2\pi}\int_0^{r_y} W_p(r,\theta)\, r\, dr\, d\theta\,. \quad (12)$$

An algorithm for a "MIR-2" computer has been written using equations (5), (10) and (12). A program has allowed us to use experimental values of the principal strains for the calculation of Up independently in each distorted element of the dividing grid where the combination of ε_I and ε_2 strains satisfies the condition $F/\tau' \geqslant 3$. Then, summing "discrete" values we find the value of work done, Up, for the total plastic zone. Having found the values of Up for the two lengths of initial crack considered with the same displacement of loading points (δ = 0.32 mm) and differentiating along crack length we obtain the value of J-integral:

$$J = -\frac{1}{B}\frac{U_p}{a} = 0.0239 \text{ MN m}^{-1}\,, \quad (13)$$

which corresponds to a value of the stress intensity factor of $K_I \approx 40.6$ MN $m^{-3/2}$ which is close to the critical value of stress intensity factor K_{Ic} for this material found by a conventional method using compact tension specimens of 50 mm thick.

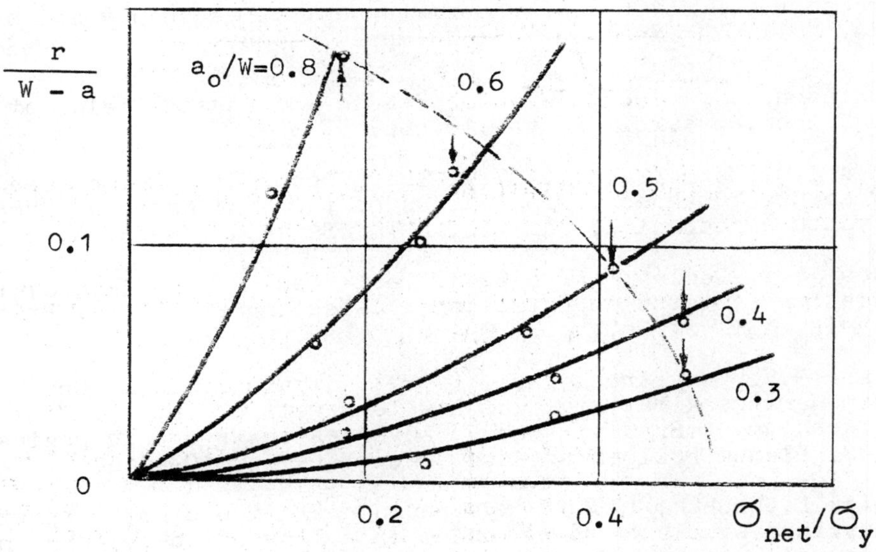

Fig. 3. Relation of plastic zone extension to the level of applied stress, obtained by experiment (points) and calculation (lines).

CONCLUSIONS

The results presented here provide evidence that the method described above may be used to estimate the value of J-integral under conditions of developed plastic flow at a crack tip. Moreover, the method makes it possible to determine the J-integral indirectly in structures of complex configuration by analysing only the region of plastic strains at the crack tip.

REFERENCES

Bowie O.L., Neal D.M., (1965). J.Appl.Mech., Trans.ASME, Ser.E,N3, pp.268-270.

Cherepanov G.P. (1967). O rasprostranenii treshchin v sploshnoi srede, Prikl.Mat.Mekh.,vol.31,N3, pp.476-488.

Hutchinson J.W. (1968). Plastic Stress and Strain Fields at Crack Tip. J.Mech.Phys.Solids, vol.16, pp.337-347.

Landes J.D.,Begley J.A. (1972). The Effects of Specimen Geometry on J_{Ic} Fracture Toughness, Proceedings of the 1971 National Symposium on Fracture Mechanics, part II, ASTM STP514,pp.24-39.

Landes J.D., Begley J.A. (1974). Test Results from J-integral Studies: An Attempt to Establish a J-Testing Procedure. Fracture Analysis, ASTM STP560, pp.170-186.

Novikov N.V., Maystrenko A.L. (1977). Strain Field at Crack Tip for Strain-Hardening Materials at Low Temperature. In K.D. Timmerhaus, R.P. Reed and A.F. Clark (Ed.), Advances in Cryogenic Engineering, vol.22, Plenum Publishing Corp., New York, pp.109-112.

Olesiak L.,Wnuk M. (1966). Dysypacia energii w osrodku sprezysto-plastycznym wywolana wystepwaniem czczeliny osiwo-symetryczney. Rozprawy Inzyniekie, 3, 14, pp.441-447.

Parton V.L., Morozov E.M. (1974). Mekhanika uprugo-plasticheskogo razrusheniya. Nauka, Moscow.

Rice J.R., (1968a).A Path Independent Integral and the Approximate Analysis of Strain Concentration by Notches and Cracks, Trans.ASME, J.Appl.Mech.,pp. 379-386.

Rice J.R., (1968b). Mathematical Analysis in the Mechanics of Fracture, Treatise on Fracture, vol.2, Academic Press Inc., New York, pp.191-311.

Volume 3, ICM 3, Cambridge, England, August 1979

SOME ASPECTS OF THE TIME DEPENDENT DUCTILE FRACTURE OF LINE PIPE STEELS

S. Tsuru and S. J. Garwood*

Nippon Steel Corporation, 6-3 Otemachi 2 Chome, Chiyoda-ku, Tokyo, 100 Japan
***The Welding Institute, Abington, Cambridge, U.K.**

INTRODUCTION

In many industries, particularly for nuclear applications and transmission pipelines, there is a growing interest in the description of ductile tearing. Many ductile materials, such as stainless steel and steels operating above the ductile-brittle transitional regions have relatively low values of toughness at the initiation of ductile crack extension. With further tearing dramatic increases in toughness are often exhibited, indicating that the use of initiation values for allowable flaw size determinations may be extremely conservative, particularly in displacement controlled applications. This philosophy has led to the recent attention to the crack growth resistance concept or R curve (Garwood, 1978; Garwood and Turner, 1977; Novak, 1974; Tanaka and Harrison, 1978). The R curve describes the apparent increase in resistance to further crack extension which is exhibited as a ductile tear extends. The rise in resistance often allows substantial amounts of stable tearing to be experienced before a ductile instability occurs. In these situations it is conceivable that an R curve approach can be used to describe the instability condition. However, if a stable ductile tear is allowed to exist in the real structure, it is necessary to consider the time dependent characteristics of ductile fracture. Although this phenomenon is well documented in the literature (Duffy, Eiber and Maxey, 1969; Elliot, Walker and May, 1971; Fearnehough and co-workers, 1971; Green and Knott, 1975; Schulze, 1977) it does not seem that its mechanism is clearly understood.

The purpose of this programme was to assess the significance of load control type situations on ductile tears in line pipe steels. This is highly relevant to transmission pipelines where tears may have initiated from existing defects during hydrostatic testing and in operation a load control mode is experienced due to the pressurisation of the pipeline.

EXPERIMENTAL PROCEDURE

Materials and Specimens

Two types of line pipe steel to API-5L-X65 specifications were used. The first type (13mm thick) had a relatively low toughness and a high inclusion content, whereas the second type (20mm thick) had a relatively high toughness and less inclusions. Hereafter, they are referred to as steel (1) for the dirtier steel and steel (2) for the cleaner steel.

Mechanical properties and chemical composition of the steels are shown in Tables 1 and 2.

TABLE 1 Mechanical Properties of Steels used

Steel	Specimen direction	σ_y (N/mm^2)	σ_u (N/mm^2)	El. (%)	σ_y/σ_u (%)	RA (%)
X65 (1)	T-L	425	588	29.5	72.3	45
	L-T	386	603	33.0	64.0	63
X65 (2)	T-L	584	607	27.3	96.2	72

TABLE 2 Chemical Compositions of Steels used

Steel	C	Si	Mn	P	S	Al	Nb	V	Others
X65 (1)	0.19	0.23	1.49	0.010	0.015	0.041	0.06	0.06	
X65 (2)	0.06	0.34	1.55	0.016	<0.005	0.036	0.03	0.09	Cr: 0.28

Full thickness single edged notched bend (SENB) specimens were employed. Specimens of steel (1) were extracted in the T and L orientations (Fig. 1) (e.g. T means that the axis of specimen is transverse to the rolling direction). Specimens of steel (2) were extracted in the T orientation only.

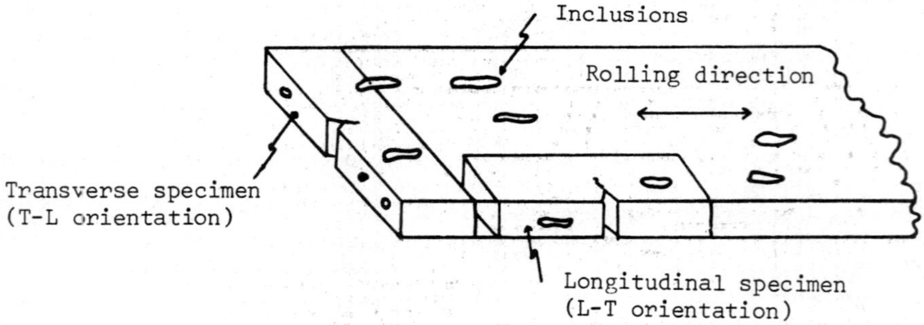

Fig. 1. Specimen extraction.

All specimens were of a geometry with W = 2B, S = 4W and were notched and fatigue pre-cracked to an initial crack depth, a, of 0.3W according to the British Standard Draft for Development DD19 (1972).

Testing Procedure

All tests were conducted at room temperature in three-point bending using a servo-hydraulic universal testing machine.

Specimens in which a constant load was to be sustained were loaded in two ways. Some specimens were loaded up to the chosen combination of load and displacement using a constant loading rate of 20kN/min. (This is roughly equivalent to a displacement rate of 1.5mm/min.) The load was then held constant. The other specimens were loaded up to the selected point at a constant displacement rate of

1.5mm/min and then, after decreasing the load by about 15%, load control was selected and the load increased again to the required value with a constant loading rate of 20kN/min and then held constant. A stepless change from displacement to load control was impossible on the test facility employed.

These two loading procedures were used in order to achieve different combinations of holding loads and displacements. Most of the load controlled tests were conducted using the latter loading pattern.

Specimens which were to sustain a constant displacement were subjected to a constant displacement rate of about 1.5mm/min until the attainment of the chosen cross-head displacement position, which was then held constant.

Experimental Measurements

Load, clip-gauge displacement, load-point displacement and crack extension were monitored while holding a specimen under constant load or displacement.

The load-point displacement was measured by the method described by Dawes (1976). Crack extension was monitored using a DC Potential Difference Method. The initial potential and the current were about 0.5mV and 50A (= ~20A/cm^2), respectively.

RESULTS

Constant Load Behaviour

General characteristics. Typical displacement-time behaviour exhibited in this programme is shown for steel (1) in the T-L orientation at different load levels (Fig. 2). Figure 3 shows the corresponding crack growth-time curves. In these figures, symbols A-G represent approximate stages on the load-displacement curve where specimens were sustained under constant load. Hence the specimen referred to as C-3 is the third specimen tested in region C of the load-displacement curve. When a specimen was sustained below the load required to initiate a tear, no change in V_g was experienced and no crack growth observed. Specimens held at a load level above the tear initiation load showed increases in both V_g and crack length. At loads just above the level required to initiate tearing, both V_g and crack length gradually stop increasing. However, as the holding load level approaches maximum load the crack growth rate increases with continued extension of the crack and final fracture occurs.

When specimens were held beyond the maximum load point (stages E, F and G), they were deformed under displacement control to the required value of clip gauge displacement V_g and then held under load control at a slightly lower load than the final load attained under displacement control. This was necessary since specimens held under load control at the final load achieved under displacement control were found to fracture instantaneously.

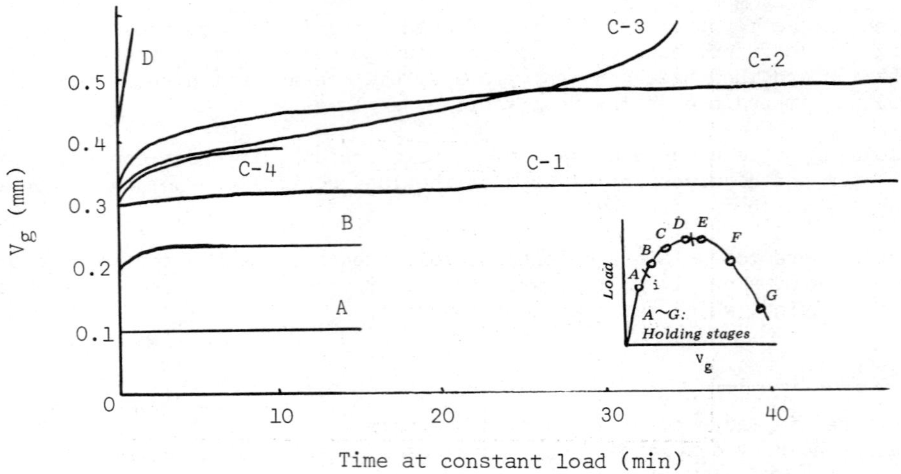

Fig. 2. Increment of clip-gauge displacement at constant load (steel (1), T-L orientation)

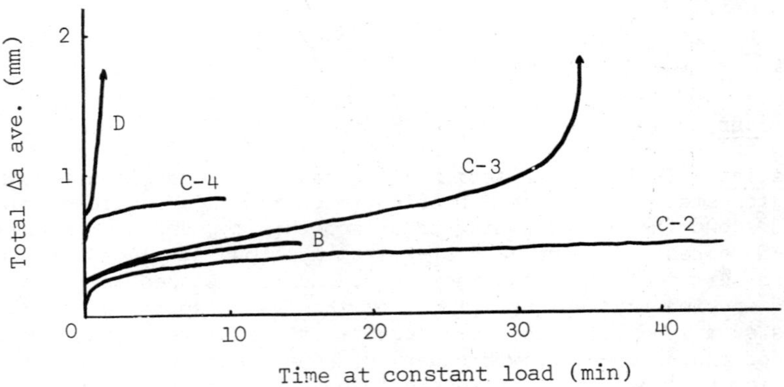

Fig. 3. Crack growth at constant load (steel (1), T-L orientation)

The curves from specimen C-3 (Figs 2 and 3) show typical examples of behaviour during load hold. This specimen was held at a clip gauge displacement approximately mid-way between the value required for the initiation of tearing and maximum load. These curves have a similar form to a typical creep curve.

Two points of some significance can be noted from these time dependency curves:

i) The increasing displacement rates and crack growth rates at the steady crack growth stage are slow, in the region of 10^{-3} to 10^{-2}mm/min, whereas in a conventional displacement control test the normal loading rate and the crack growth rate are around 1.0mm/min.

ii) Under load hold conditions, displacements and crack lengths achieved before the loading rate reaches the normal rate in a conventional test (say 1.5mm/min) are much larger than those obtained at the maximum load point in the conventional test.

These two characteristics are shown in Figs 4 and 5.

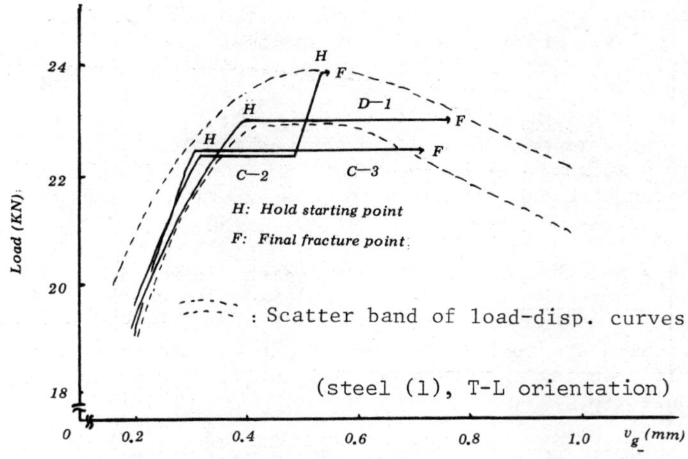

Fig. 4. Load displacement curve during load holding

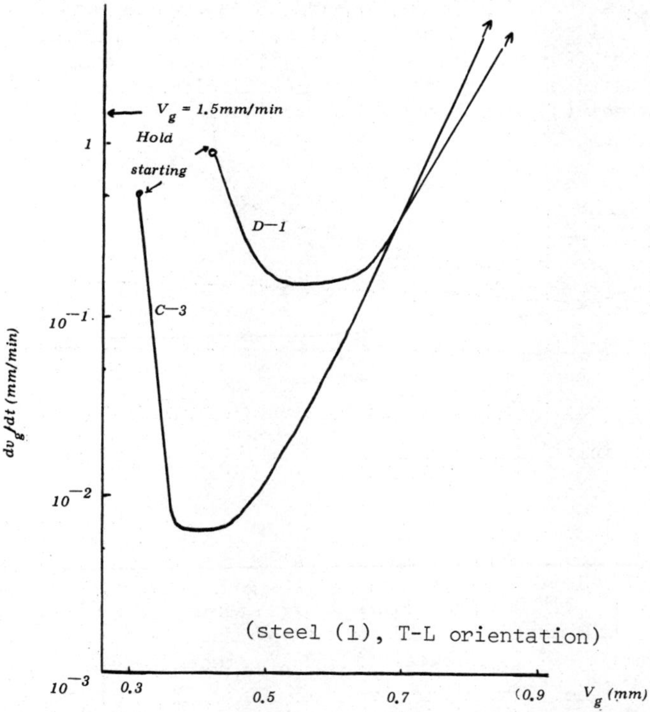

Fig. 5. Variation of dV_g/dt during load holding

Figure 4 indicates how the clip gauge displacement increases under constant load. Here, the final fracture point is defined as the point where dV_g/dt has accelerated to 1.5mm/min. (This represents the normal loading rate in the conventional test.) Before specimens reach the final fracture point, however, V_g increases beyond the maximum load value as depicted in Fig. 4.

Figure 5 indicates the variation of dV_g/dt during load hold. At first dV_g/dt decreases due to the change from the initial displacement loading rate of 1.5mm/min to load hold conditions. Following the initial decrease, dV_g/dt then reaches a constant value. The level of constant dV_g/dt seems to be dependent upon the load hold level and in the case of specimen C-3, dV_g/dt at the steady rate was 6.2×10^{-3}mm/min. With continued clip gauge displacement dV_g/dt eventually begins to increase.

Specimens which were held far beyond the maximum load point (stages F and G) seemed to show slightly different behaviour. Only relatively small increases in the displacement and the crack length were experienced before the final fracture took place.

The clip gauge and the load-point displacement behave in an ostensibly similar fashion.

The above results are for the T-direction in the steel (1). Steel (1) in the longitudinal direction and the steel (2) in the transverse direction showed very similar results.

Threshold factors. In order to get some idea about the factors affecting the threshold load value, i.e. the load level below which failure does not occur under load hold, the time to failure under constant load was analysed. Figure 6 shows the effect of dividing the load hold level by the maximum load. As observed by other investigators (Duffy, Eiber and Maxey, 1969; Fearnehough and co-workers, 1971), there seems to be a threshold load level below which no failure occurs. In this material, steel (1), it is about 95% of the maximum load for both T and L orientation.

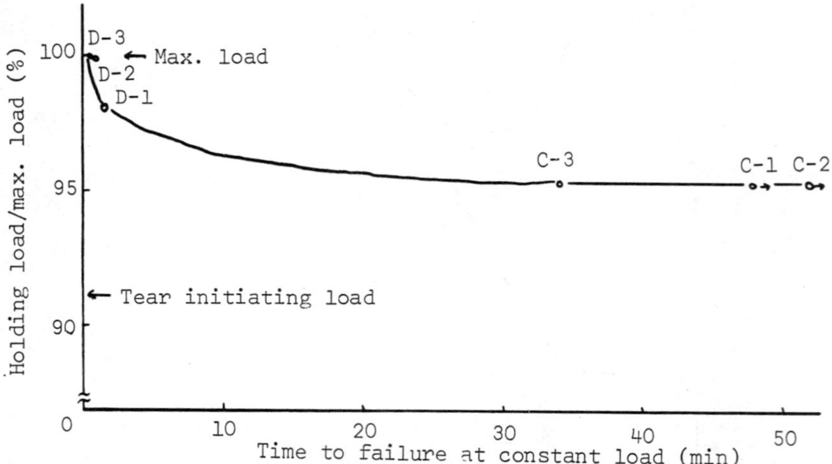

Fig. 6. Holding load level vs time to failure at constant load (steel (1), T-L orientation)

There also seems to be a threshold level of COD which is consistent with the observations of Green and Knott (1975) and Schulze (1977). In these materials, it is about 60% of the COD at maximum load (COD max) for the steel (1) and about 70% for the cleaner steel, steel (2). The initiation values of COD (δ_i) at the initiation of tearing and at maximum load (δ_m) are given in Table 3.

TABLE 3 COD Measurements for the Steels used

Steel	Orientation	δi (Initiation of tearing value), mm	δm, mm
1	T-L	0.10	0.22
	L-T	0.28	0.78
2	T-L	0.27	0.38

Constant Displacement Behaviour

General characteristics. When specimens were held under constant displacement, a drop in the load was observed (Fig. 7). The total amount of load drop was 6-12% of the initial load. Most of that drop occurred within the first few minutes of holding at constant displacement and a constant load was obtained after 10 or 20 minutes. During this period no crack growth was evident from the DC Potential Difference (DCPD) monitoring device. This behaviour is shown in Fig. 8. Here, a specimen was loaded up to the maximum load point and held at constant displacement. During this holding period, a load drop was observed, but no crack growth was experienced by DCPD.

After the load had stopped decreasing, the control mode of the machine was changed from displacement to load control. As soon as the mode was changed, both displacement and crack length began to increase. Although there was a small difference in load (about 0.2%) due to the change of control mode, the level of load and displacement before, and after, the change of control can be considered to be virtually identical. Therefore, it seems reasonable to conclude that in this specimen no ductile crack growth occurs under a constant displacement condition (i.e. fixed grips).

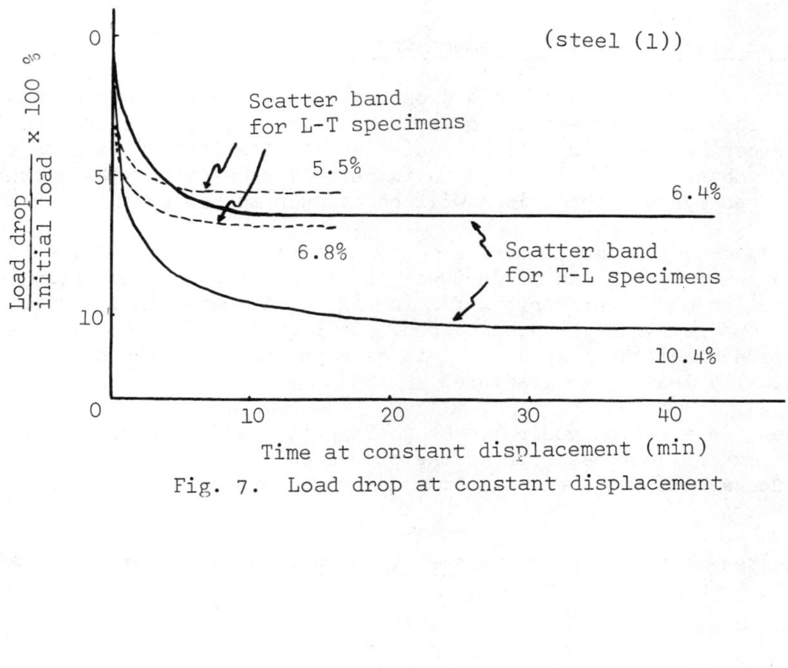

Fig. 7. Load drop at constant displacement

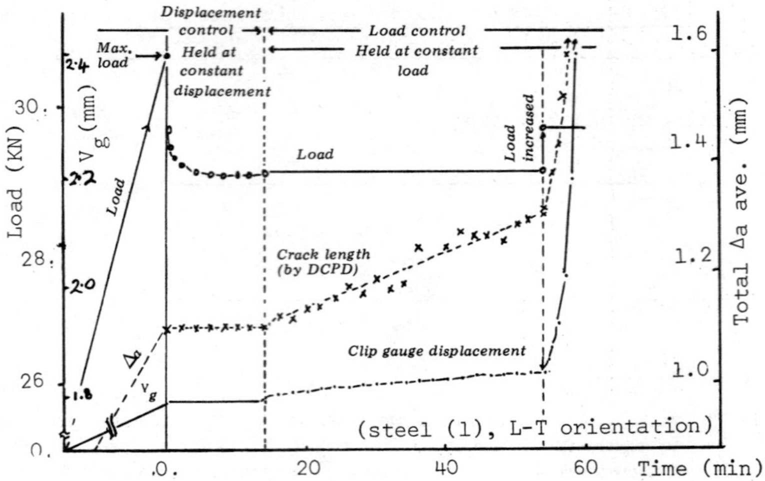

Fig. 8 . Specimen behaviour for both constant load and constant displacement

Possible explanations for the load drop. As there would appear to be insignificant crack growth under constant displacement, two possible causes for the load drop are:

i) the effect of material dependent properties such as a stress relaxation, or a reduction of strain rate at the crack tip. As soon as constant displacement conditions are imposed, a macroscopic strain rate of zero is enforced; therefore some kind of stress relaxation may well occur.

ii) the effect of external factors such as the characteristics of the machine or the loading rig.

DISCUSSION

The behaviour exhibited in this programme can be explained almost entirely by strain rate dependence of material and is discussed below in terms of an assumed static load-displacement curve.

The Use of an Assumed Static Load-Displacement Curve

From a consideration of Fig. 4 and the load drop during holding at constant displacement, a static load-displacement curve may be derived. This is expressed in Fig. 9. This curve can be considered to be equivalent to the load-displacement curve which would be obtained by loading at an infinitely slow rate. The difference between static and dynamic conditions will be as much as several per cent.

Using this assumed load-displacement curve a possible explanation of all the time dependent phenomena noted herein can be derived. If a specimen is held under constant load in the region above the curve X-Y-Z in Fig. 9, it means that a specimen is overloaded when compared with the static load. Consequently the specimen begins to deform and, provided that the displacement is free to increase, the crack will begin to grow, eventually leading to fracture.

If a specimen is held in a region encircled by points i, X and Y (Fig. 9), the crack will begin to extend, but, as soon as the displacement reaches the static load-displacement curve, the specimen will stop deforming and crack growth will cease.

If load hold is started at the same displacement, but with a different load, this being achieved by loading at different loading rates (points B1 and B2 in Fig. 9), the specimen at point B1 will fracture, but the other will not.

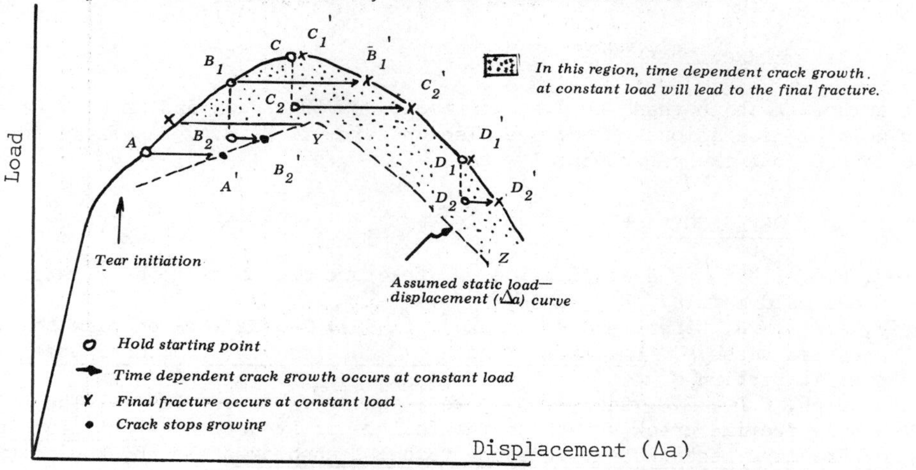

Fig. 9. The assumed static load-displacement curve.

Furthermore, if this assumed static load-displacement curve concept is correct, the load required for the initiation of tearing can be different from that experienced under normal test conditions. (The displacement may not be different, however.) This implies the possibility of time dependent fracture occurring in a specimen under load controlled conditions at a load below the tear initiation load obtained from the load displacement curve determined with the normal loading rate.

CONCLUSIONS

Constant Load Behaviour

1. The behaviour of a specimen is controlled by both the load and displacement imposed upon it. However, in time dependent fracture, the load appears to play a more important role, since for a given displacement in load control it is the magnitude of the load which governs the time to failure.
2. There appears to be a threshold load level above which time dependent crack growth leads to a final fracture.
3. If a static load-displacement curve is assumed, almost all time dependent phenomena observed here seem to be explained in terms of the relation between the load and displacement of a specimen at a given loading rate and the static equivalent.
4. The difference in the shape and density of the inclusions present does not seem to affect the time dependent behaviour to any large extent for the steels used in this test programme.

Constant Displacement Behaviour

5. Under constant displacement conditions no obvious time dependence was observed. This indicates that, for the material and testing conditions employed herein, time dependent ductile fracture does not occur for fixed grip conditions.
6. When a specimen was loaded to, and then held at, a particular displacement, some amount of load drop was observed (6-12% of the initial load). Part of this (about one half) may be attributed to the effect of the load relaxations at the

crack tip, the remainder being due to plasticity effects under the loading rollers.

ACKNOWLEDGEMENTS

The authors wish to thank Dr J.D. Harrison, Dr J.F. Knott and Mr G.D. Fearnehough for helpful discussions. They are also grateful to Mr B.A. Wakefield and his staff who carried out the experimental work.

REFERENCES

Dawes, M.G. (1976). The application of fracture mechanics to brittle fracture in steel weld metals. PhD, CNAA.

Duffy, A.R., R.J. Eiber, and W.A. Maxey. (1969). Recent work on flaw behaviour in pressure vessels. Practical Fracture Mechanics for Structural Steel. Chapman & Hall. Section M.

Fearnehough, G.D., G.M. Lees, J.M. Lowes, and R.T. Weiner. (1971). The role of stable ductile crack growth in the failure of structure. Practical Application of Fracture Mechanics to Pressure Vessel Technology. IMechE, London. p. 119.

Garwood, S.J. (1978). The effect of specimen geometry on crack growth resistance. Presented at 11th Nat. ASTM Symp., Blacksburg, Va.

Garwood, S.J., and C.E. Turner. (1977). The use of the J integral to measure the resistance of mild steel to slow stable crack growth. Fracture, 1977, Vol. 3, pp. 279-84.

Green, G., and J.F. Knott. (1975). On effects of thickness on ductile crack growth in mild steel. J. Mech. Phys. Solids, Vol. 23. p. 169.

Novak, S.R. (1974). Resistance to plane stress fracture (R-curve behaviour) of A572 structural steel.

Schulze, H.D. (1977). Stable crack growth under sustained load. Structural Mechanics in Reactor Technology, Brussels. Paper G2/5.

Tanaka, K., and J.D. Harrison. (1978). An R-curve approach to COD and J for an austenitic steel. Welding Research Abroad, Vol. 24, No. 3.

Elliott, D., E.F. Walker and M.J. May. (1971). The determination and applicability of COD test data. Practical Application of Fracture Mechanics to Pressure Vessel Technology. I.Mech.E., London, p. 217.

Volume 3, ICM 3, Cambridge, England, August 1979

A RELATION BETWEEN CRACK TIP PLASTIC BLUNTING AND THE J-INTEGRAL

H. Kobayashi, H. Nakamura and H. Nakazawa

Department of Physical Engineering, Faculty of Engineering, Tokyo Institute of Technology, Tokyo, Japan

INTRODUCTION

In recent years, several attempts have been made to extend fracture mechanics into the elastic-plastic regime. The first attempt was the crack opening displacement (COD) method (Wells, 1961). It was somewhat lacking in a rational physical and analytical basis. Measurements and applications were thus left unclear. More recently, a quantity called the J-integral has been introduced by Rice (1968) and its attainment of a critical value has been advanced by Begley and Landes (1972) as a fracture criterion. J can be regarded as a measure of the intensity of the stress and strain fields surrounding the crack tip within the elastic-plastic region, but outside the intensely deformed non-linear zone indicated in Fig. 1. It plays a similar role to that of the stress intensity factor K_I, a measure of the intensity of the surrounding elastic field outside the plastic zone in linear elastic fracture mechanics (LEFM) (Rice, 1976; Paris, 1977).

For the LEFM case, a relationship between the real crack tip opening displacement (CTOD) and K_I of the form

$$\delta = \frac{(1 - \nu^2)}{\lambda} \frac{K_I^2}{E\sigma_{ys}}, \tag{1}$$

under plane-strain conditions, has been known for a long time; ν is Poisson's ratio, E is Young's modulus, σ_{ys} is the yield stress for ideally plastic materials and λ is about 2. Since

$$J = (1 - \nu^2) \frac{K_I^2}{E} \tag{2}$$

for the LEFM case, it is suggested that the following equation can stand for the elastic-plastic fracture mechanics (EPFM) case,

$$\delta = \frac{1}{\lambda} \frac{J}{\sigma_{ys}}. \tag{3}$$

Recently, a finite-element analysis of large crack tip geometry changes has been carried out by McMeeking (1977). The result shows the relationship between δ and J to be of the form

$$\delta = \frac{1}{\lambda} \frac{J}{\sigma_{flow}}, \tag{4}$$

where σ_{flow} is the flow stress for hardening materials and $1/\lambda = 0.55 - 0.65$. The result also shows a significant path dependence of J, very close to the crack tip, on contours of radius less than the size of the intensely deformed non-linear zone,

$$R_n = 2\delta. \tag{5}$$

A direct measurement of δ is difficult generally, and virtually impossible in a routine test. When a fatigue pre-cracked specimen is subjected to a monotonically increasing load, the crack tip blunts and fracture initiates at the critical value of δ. Fractographic examination of this fracture surface reveals that there exists a featureless transition region between the fatigue pre-crack and overload fracture regions. This region, referred to as a stretched zone, has been interpreted as the result of the crack tip plastic blunting, and the critical value of the stretched zone width (SZW_c) obtained from a specimen broken by overload has been correlated with the critical value of the crack opening stretch δ_c as well as a plane-strain fracture toughness value (K_{Ic}) for the LEFM case (Brothers, 1971; Kobayashi, 1976).

Recently, an elastic-plastic fracture toughness (J_{Ic}) criterion and its test method have been developed by Begley and Landes (1972, 1974) and ASTM task Group E24.01.09 (Clarke, 1979). In their test method attention has been directed mainly to processes of stable-fracture and an assumed J versus Δa blunting line

$$a = \frac{1}{2}\delta = \frac{1}{2}\frac{J}{\sigma_{flow}} \tag{6}$$

has been proposed. Furthermore, an experimentally determined J versus Δa resistance curve has been utilized, as shown in Fig. 2, where Δa is the crack extension and σ_{flow} is the average of σ_{ys} in tension (offset = 0.2%) and the tensile strength (σ_B). On the other hand, as the present authors (1977a) have indicated in previous work, a stretched zone appears as the loading progresses and its width (SZW) reaches a critical value (SZW_c) when fracture initiates. Moreover, SZW_c is not affected by any following stable crack growth so that the J_{Ic} value can also be obtained by examination of a relation between SZW and J before and after the fracture initiation. In previous work, the authors (1977a, 1979) proposed a new J_{Ic} test method based on this concept and confirmed its validity for a wide range of metallic materials from high- to low-strength. Also, the fractographically derived J versus SZW blunting line before fracture initiation was examined quantitatively. The results showed that the blunting line depends not on σ_{ys} but on Young's modulus (E) and is given by the following expression;

$$SZW = 95\,\frac{J}{E}. \tag{7}$$

In the last two years, the J_{Ic} tests proposed by the authors have been carried out by ourselves (1977a, 1978a, 1978b, 1979), Iwadate (1977, 1978), Miyamoto and Hijikata (1978a, 1978b), Hashimoto (1978), Ohtsuka (1978), Kishi (1978), Ohji (1978) and

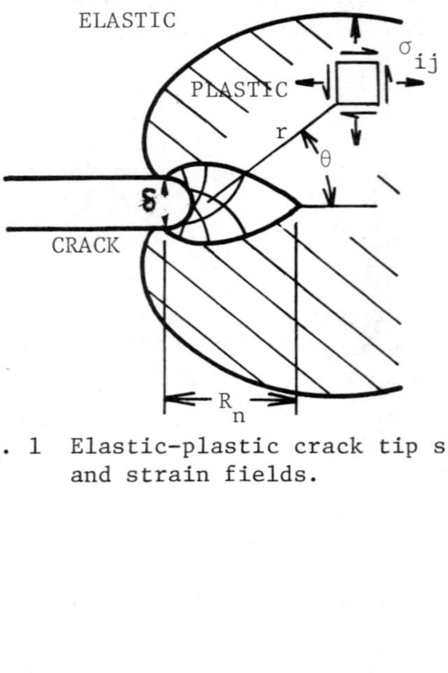

Fig. 1 Elastic-plastic crack tip stress and strain fields.

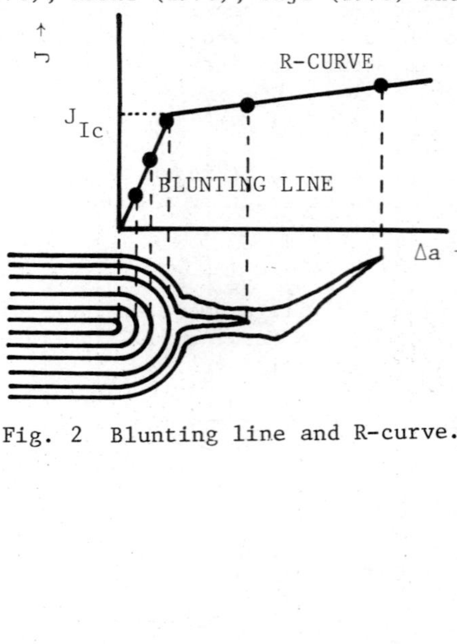

Fig. 2 Blunting line and R-curve.

Yokomaku (1978) in Japan. The objective of this paper is to summarize Japanese J_{Ic} test results and to present the broad applicability and usefulness of Eq. (7). A brief description of the significance and limitation of Eq. (7) is also included.

EXPERIMENTAL PROCEDURE

The procedure of the proposed J_{Ic} test method is summarized as follows.

1) A bend type specimen is needed. This includes 3- or 4-point bend bars, compact specimens, or WOL specimens.
2) The specimen should be deeply notched with a/W = 0.6 and pre-cracked in fatigue (part I in Fig. 3(a)). The cyclic load for the fatigue pre-crack is set at 30 - 40% of the load corresponding to the anticipated J_{Ic} value.
3) Load statically several specimens to different displacement values where fracture does not initiate (part II in Fig. 3(a)).
4) Unload partially each specimen and mark the front of the stretched zone formed at part II by subsequent fatigue cracking (part III in Fig. 3(a)). Then, pull the specimen apart by overload (part IV in Fig. 3(a)).
5) Measure the SZW formed in part II. Calculate a J value from the load versus load line displacement record (Fig. 3(b)) using the estimation procedure of Rice (1973) or Merkle (1974). Plot SZW versus J and draw a best fit blunting line to the points before the fracture initiation (Fig. 3(c)).
6) Pull a virgin specimen apart by overload and measure SZW_c.
7) Mark J_{Ic} as a J value at the intersection of the line SZW = SZW_c and the extrapolation of the blunting line developed in 5) (Fig. 3(c)).

Fractographic examinations were conducted mainly with the help of a transmission electron microscope (TEM), using a two-stage plastic-carbon replica with chromium shadowing, but a scanning electron microscope (SEM) was sometimes used, as it was by some other researchers. The observation was performed at the midthickness of specimens where plane-strain conditions exist. A representative fractograph show-

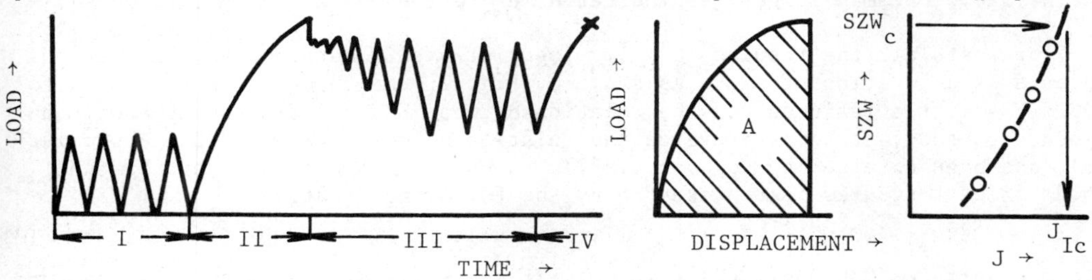

(a) (b) (c)

Fig. 3 Procedure of proposed J_{Ic} test method.

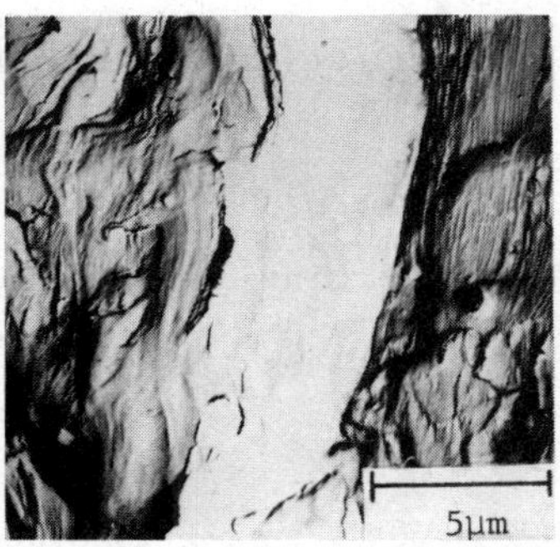

Fig. 4 Fractograph showing stretched zone (Ti-6Al-4V).

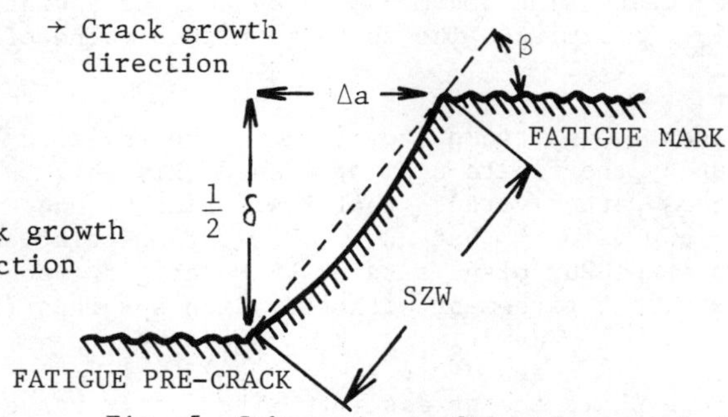

Fig. 5 Schematic profile of stretched zone.

ing the general features of the stretched zone between the fatigue pre-crack region and the subsequent fatigue marked region is shown in Fig. 4. It appeared that SZW measured by SEM generally takes a maximum value when the macroscopic fracture surface makes some oblique angle β (= 35-45°), with the electron beam, rather than being normal to the beam, and this maximum value is in excellent agreement with SZW measured by TEM. Therefore, the crack tip blunting angle is assumed to be 2β = 70-90°, as shown in Fig. 5. The same results have been obtained by using stereo fractographs (Broek, 1974) and sectioning (Kobayashi, 1974; Maruyama, 1976; Kodaira, 1978). Then

$$\delta = 2 \sin \ SZW \approx (1.1\text{-}1.4)\ SZW \tag{8}$$

$$a = \cos\beta \ SZW \approx (0.8\text{-}0.7)\ SZW \tag{9}$$

assuming that β = (35-45°). The result of the finite-element analysis by McMeeking (1977) shows somewhat lower values of $\Delta a/\delta$, 0.45-0.54.

RESULTS AND DISCUSSION

Table 1 shows materials, specimens and methods used in the Japanese J_{Ic} tests, and it also shows the results of SZW_c, where the abbreviations used are as follows; R.T.: room temperature, CT: compact type specimen, 3B: 3-point bend bar, 4B: 4-point bend bar, W: specimen width, B: specimen thickness, R: J-integral estimation procedure by Rice (1973), M: J-integral estimation procedure by Merkle (1974), T: transmission electron microscope, and S: scanning electron microscope.

Fig. 6 presents all the results on a log-log plot of SZW obtained for the various materials as a function of J/σ_{flow}. As shown by Fig. 6, the relation between SZW and J/σ_{flow} shows an apparent dependent on σ_{flow}/E: there is a tendency for SZW to become smaller for the same value of J/σ_{flow} as σ_{flow}/E becomes smaller. Therefore it is evident that the relation between δ and J does not obey Eq. (4) or Eq. (6). It should be noted that if J/σ_{ys} instead of J/σ_{flow} is taken as a parameter, the scatter band in Fig. 6 becomes larger, as indicated in previous work (Kobayashi, 1979).

Fig. 7 presents all the results on a log-log plot of SZW obtained for the various materials as a function of J/E. As shown by Fig. 7, the change of E from 70.6 to 206 GPa has little influence on the relation between SZW and J/E, and there is no influence of the test temperature on the relation between SZW and J/E. A similar result has been obtained by Logsdon (1977). A curve fitted to the data by the method of least squares is represented by the following equation,

$$SZW = 400 \left(\frac{J}{E}\right)^{1.18}, \tag{10}$$

where the units of SZW and J/E are in mm. If we assume a relationship of the form

$$SZW = C_1 \frac{J}{E}, \tag{11}$$

the mean value becomes $\overline{C}_1$ = 89 and the deviation is $54.7 \leq C_1 \leq 143$ for 90% confidence limits. The above-mentioned value of C_1 is in excellent agreement with 95 in Eq. (7).

Recently an attempt to simulate the relation between SZW and J/E has been carried out by the finite-element method (Miyoshi, 1977) with qualitative success. It seems rather strange that SZW formed by crack tip plastic blunting does not depend on σ_{ys} or σ_{flow} but on E. However, a relation like this is well known in fatigue. In stage 2b, plane strain, LEFM fatigue crack growth, growth rates (da/dN) are in excellent agreement with striation spacings (S) and obey

$$\frac{da}{dN} = S = C_2 \ (1 - \nu^2) \left(\frac{\Delta K}{E}\right)^2, \tag{12}$$

where ΔK is the stress intensity factor range and C_2 is about 9 on the assumption that ν = 1/3 (Kobayashi, 1977b). This relation is obeyed regardless of σ_{ys} (Bates, 1969) and of the test temperature (Tobler, 1977). Since the relation between J and

TABLE 1 Materials, Specimens, Methods and Results

Materials	σ_{ys} (MPa)	σ_B (MPa)	σ_{flow} (MPa)	E (GPa)	Temp. (K)	Spec. type	W (mm)	B (mm)	J	EM	SZW_c (μm)
Kobayashi, 1977a											
SNCM8	1220	1300	1260	206	133	3B	20-40	10-20	R	T	2.3
(4340)	1110	1200	1150	206	203	3B	20-40	10-20	R	T	4.3
	1070	1140	1100	206	R.T.	CT,3B	25.5-51	5.5-25.5	R	T	6.0
Kobayashi, 1978a, 1978b, 1979											
10B35	1640	1850	1740	206	R.T.	CT	25.5	12.5	R	T	6.0
	1330	1380	1360	206	R.T.	CT	25.5	12.5	R	T	7.0
	706	794	750	206	R.T.	CT	25.5	12.5	R	T	9.8
A533B-1	533	657	595	206	R.T.	CT	25.5	12.5	R	T,S	37
				206	373	CT	25.5	12.5	R	T,S	33
				197	473	CT	25.5	12.5	R	T,S	36
				190	573	CT	25.5	12.5	R	S	34
A533B-1	522	647	585	206	R.T.	CT	51	1.5-19	R	S	45
304 S.S.	245	627	436	191	R.T.	CT	51	5.0	R	T	
	299	696	498	191	R.T.	CT	25.5-153	12.5	R	S	
	284	588	436	191	R.T.	CT	51	19	R	S	
Ti-6Al-4V	941	1000	971	113	R.T.	CT	25.5-38	1.5-20	R	T	23
5083-O Al	168	338	253	70.6	R.T.	CT	51	3.2-25.5	R	T	18.2
Iwadate, 1977											
3.5Ni-Mo-V	620	758	689	206	273	CT	51	25.5	R	S	29
	620	758	689	206	R.T.	CT	51	25.5	R	S	50
				206	373	CT	51	25.5	R	S	118
Iwadate, 1978											
3.5Ni-Cr-	774	931	853	206	233	CT	51	25.5	M	S	
Mo-V	735	872	804	206	296	CT	51	25.5	M	S	90
	676	794	735	206	423	CT	51	25.5	M	S	89
	657	764	711	206	453	CT	51	25.5	M	S	82
Miyamoto and Hijikata, 1978a, 1978b											
3.5Ni-Mo-V	686	813	750	206	373	CT	50	25	M	T,S	43
Cu-2.5Co-Be	745	872	809	127	R.T.	CT	50	10	M	T	22
2024-T6 Al	426	588	507	73.5	R.T.	CT	30.5	18	M	T	24
Hashimoto, 1978											
A508-3	500	657	579	206	223	3B	50	25	R	T,S	25
	450	621	536	206	273	3B	50	25	R	T,S	29
	442	590	516	206	R.T.	3B	50	25	R	T,S	31
Ohtsuka, 1978											
A387	624	738	681	206	R.T.	4B	50	15	R	S	70
SB42	348	459	404	206	R.T.	4B	50	15	R	S	37
Kishi, 1978											
Maraging	1890	1960	1930	188	R.T.	CT	28	14	R	S	23
Ohji, 1978											
HT80				206	123	3B	15	7.5	R	S	41
				206	173	3B	15	7.5	R	S	55
				206	201	3B	15	7.5	R	S	54
	764	813	789	206	293	3B	15	7.5	R	S	59
HT60				206	123	3B	15	7.5-15	R	S	51
				206	143	3B	15	7.5-15	R	S	75
				206	168	3B	15	15	R	S	127
				206	201	3B	15	15	R	S	205
				206	243	3B	15	15	R	S	180
	529	637	583	206	293	3B	15	7.5-15	R	S	180
Yokomaku, 1978											
Maraging	1880	1920	1900	188	R.T.	CT	25.5	2.0-7.5	R	S	14.5
12Cr-1Mo	730	872	801	206	R.T.	CT	40	16	R	S	21
304 S.S.	316	664	490	201	R.T.	CT	25.5-50	12.7-25.4	R	S	160

K_1 is given by Eq. (2) for the LEFM case, the following equation can also stand for the LEFM case,

$$\frac{da}{dN} = S = C_2 \frac{\Delta J}{E}, \tag{13}$$

where ΔJ is the apparent cyclic J-integral and $\Delta J = J$ in the case that the stress ratio (R) is 0 and C_2 is the material constant for a given value of R.

For comparison, the striation spacings (S) when R is about 0 for various materials (Kobayashi, 1977b, 1977c, 1977d, 1978c, 1978d, 1978e, 1978f) have been plotted in Fig. 6 and Fig. 7 in relation to J/σ_{flow} and J/E, respectively, where all the data

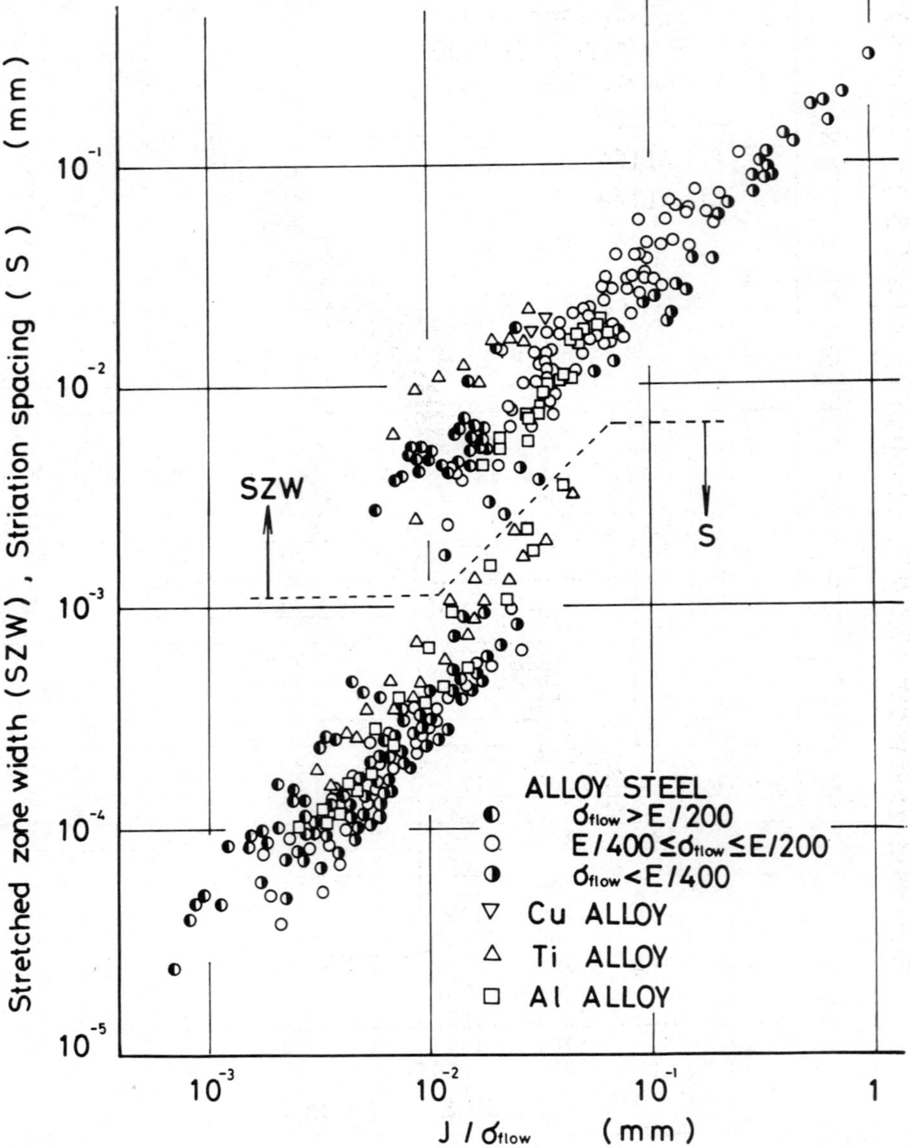

Fig. 6 Comparison between stretched zone width, SZW, and striation spacing, S, as a function of J/σ_{flow}.

on S satisfy small scale yielding conditions. As shown by Fig. 6, the relation between S and J/σ_{flow} also shows an apparent dependence on σ_{flow}/E: the same result as the relation between SZW and J/σ_{flow}. On the other hand, as shown in Fig. 7, the change of E from 70.6 to 206 GPa has also little influence on the relation between S and J/E. A curve fitted to the data by the method of least squares is represented by the following equation,

$$S = 18.5 \left(\frac{J}{E}\right)^{1.06}, \tag{14}$$

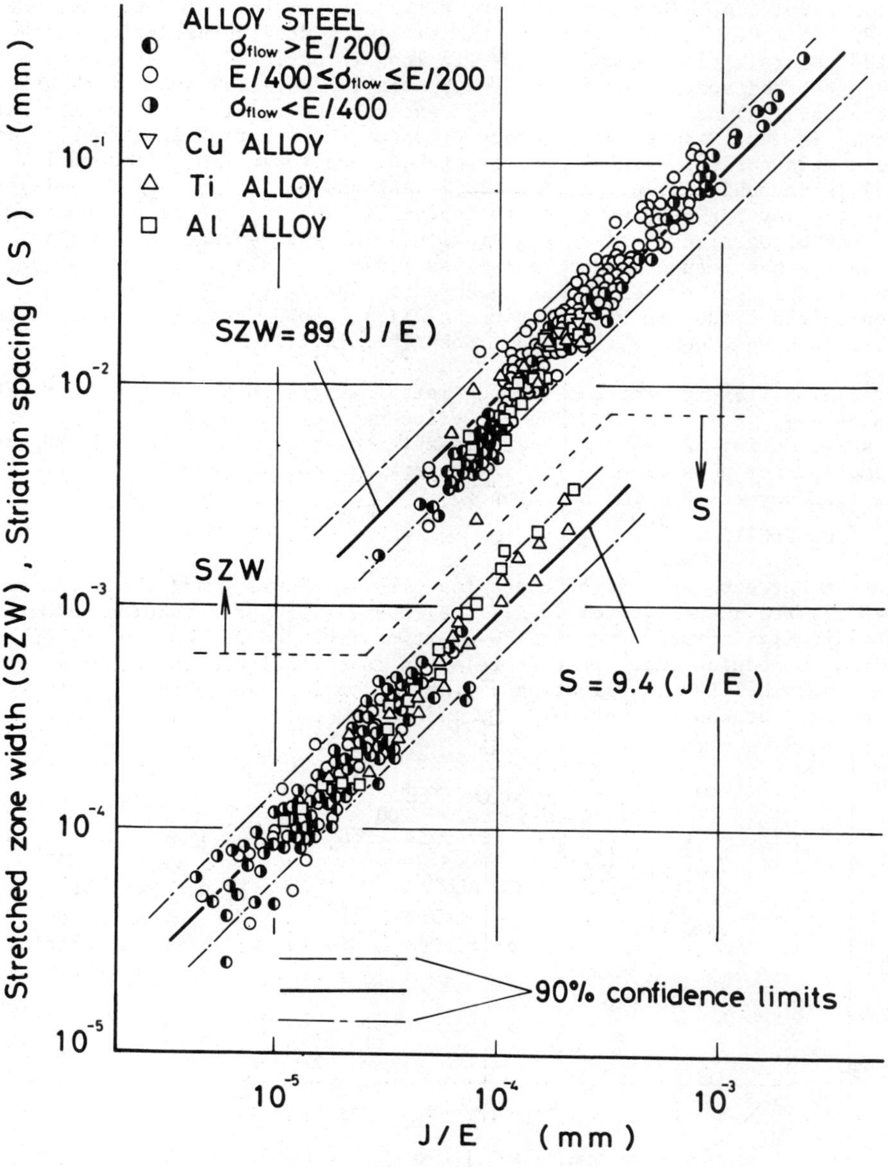

Fig. 7 Comparison between stretched zone width, SZW, and striation spacing, S, as a function of J/E.

where the units of S and J/E are in mm. If we assume the equation of the form

$$S = C_3 \frac{J}{E}^{1.06}, \tag{15}$$

the mean value becomes $\overline{C}_3 = 9.4$ and the deviation is $6.0 \leq C_3 \leq 14.8$ for 90% confidence limits.

That is to say, despite the difference between the monotonic load and the cyclic load, the widths of the stretched zone and of the striation formed by the same micromechanism, glide plane decohesion, are characterized by the same fracture mechanics parameter. However, as is evident from Fig. 7, the widths of these two for the same value of J/E are as different as approximately one order of magnitude, $C_1/C_3 = 89/9.4 = 9.5$. The reason for the smaller width of the striation should be attributed to crack closure under a cyclic load (Elber, 1971). It can be supposed that the stretched zone formation begins by obeying Eq. (15) and shifts gradually to obeying Eq. (11) as the monotonic load exceeds the cyclic load for the fatigue pre-crack. In fact, there is an intersection of the J versus SZW blunting line with the J-axis corresponding to the stretched zone formation, as shown in Fig. 3(c) (Kobayashi, 1977a; Keller, 1977) and the deviation from Eq. (11) becomes larger as the J value becomes smaller (Kobayashi, 1979). The high exponent in Eq. (10) may be attributed to this deviation. The influence of the cyclic loading history on SZW has been argued by ourselves (1978f, 1979), Keller (1979) and Hertzberg (1979). Likewise, it has been found that, for the EPFM case, striation formation shifts gradually from obeying Eq. (15) to obeying Eq. (11) under constant cyclic loading (Kobayashi, 1978d; Schwalbe, 1977).

Finally, all the results for the various materials are shown as a plot of SZW_c as a function of σ_{flow}/E in Fig. 8. The J_{Ic} value can be evaluated roughly from Eq. (11) by substituting $SZW = SZW_c$. There is a tendency for the value of SZW_c to become smaller for a particular type of material as σ_{flow} becomes larger. This is the same tendency as that for K_{Ic}, δ and J_{Ic}.

CONCLUSION

The relation between measures of crack tip plastic blunting and the J-integral have been investigated quantitatively. All the Japanese J_{Ic} test results show that the fractographically derived J versus SZW blunting line depends not on σ_{ys} or σ_{flow} but on Young's modulus (E). If this relation is established for other materials, there is a possibility of evaluating J_{Ic} more simply by measuring SZW_c from a single specimen broken by overload.

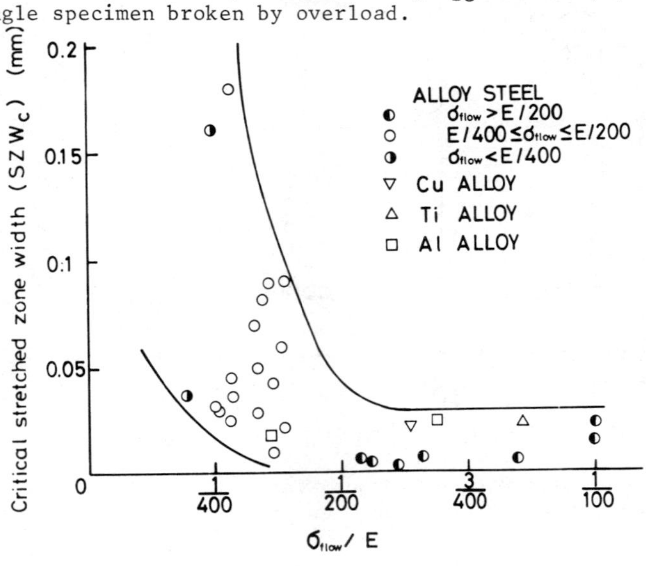

Fig. 8 Critical stretched zone width, SZW_c, as a function of σ_{flow}/E.

ACKNOWLEDGEMENTS

The task Group under JSME (the Japan Society for Mechanical Engineers) Committee 1P-SC21 on "Non-Linear Fracture Mechanics (chairman: Professor Miyamoto)" has been making an effort towards standardization of the J_{Ic} test method for two years (Kobayashi, 1978g, 1978h). Recently, JSME Committee S-781 on "Standard Method of Test for Elastic-Plastic Fracture Toughness (chairman: Professor Miyamoto)" was organized and the task is now being continued. Methods that have been used to obtain crack extension measurements include the fractographic method and the δ_c concept (Ohtsuka, 1975, 1977), the heat tinting method (Iwadate, 1977, 1978; Kodaira, 1978; Sakata, 1978; Miyamoto and Hijikata, 1978a, 1978b; Hashimoto, 1978; Yokomaku, 1978), the electrical potential method (Miyoshi, 1975; Kodaira, 1978; Nakano, 1978; Miyamoto and Hijikata, 1978a, 1978b; Matsushita, 1978; Hashimoto, 1978; Ohji, 1978; Yokomaku, 1978) and the acoustic emission method (Kishi, 1978; Ohtsuka, 1978). The authors wish to thank Professor Miyamoto and other members of JSME Committee S-781 for providing advanced copies of their papers and their constructive comments.

REFERENCES

Bates, R. C., W. G. Clark, Jr., and D. M. Moon (1969). ASTM STP 453, pp. 192-214.
Begley, J. A., and J. D. Landes (1972). ASTM STP 514, pp. 1-23.
Broek, D. (1974). Engineering Fracture Mechanics, 6, 173-181.
Brothers, A. J., M. Hill, M. T. Parker, W. A. Spitzig, W. Wiebe, and U. E. Wolff (1971). ASTM STP 493, pp. 3-19.
Clarke, G. A., W. R. Andrews, J. A. Begley, K. Donald, G. T. Embley, J. D. Landes, D. E. McCabe, and J. H. Underwood (1979). A procedure for the determination of ductile fracture toughness values using J integral techniques. Submitted to J. of Testing and Evaluation, ASTM.
Elber, W. (1971). ASTM STP 486, pp. 230-242.
Hashimoto, M., T. Tatara, and K. Kusaba (1978). In Elastic-Plastic Fracture Mechanics, Pre-print of 15th JSME Symp., No. 780-9, 23-28.
Hertzberg, R. W. (1979). On the relationship between fatigue striation spacings and stretch zone width. Submitted to Inter. J. of Fracture.
Hijikata, A., S. Yoshioka, and M. Demizu (1978b). Report of JSME Committee S-781, No. 13.
Iwadate, T., T. Karaushi, and J. Watanabe (1977). ASTM STP 631, pp. 493-506.
Iwadate, T., T. Karaushi, J. Watanabe, M. Horiuchi, and Y. Terajima (1978). In Elastic-Plastic Fracture Mechanics, Pre-print of 15th JSME Symp., No. 780-9, 35-40.
Keller, H. P., and D. Munz (1977). ASTM STP 631, pp. 217-231.
Kishi, T., S. Saito, U. Mishima, and R. Horiuchi (1978). In Elastic-Plastic Fracture Mechanics, Pre-print of 15th JSME Symp., No. 780-9, 81-86.
Kobayashi, H., H. Nakazawa, and H. Kawamura (1974). Pre-print of the annual meeting of JSME, No. 740-1, 103-106.
Kobayashi, H., H. Nakazawa, and A. Nakajima (1976). Correlation of plane-strain and plane-stress fracture toughness with fractographically derived stretched zone width in high strength steel. In H. Miyamoto and others (Ed.), Strength and Structure of Solid Materials. Noordhoff. pp. 115-128.
Kobayashi, H., K. Hirano, H. Nakamura, and H. Nakazawa (1977a). A fractographic study on evaluation of fracture toughness. In D. M. R. Taplin (Ed.), Proc. 4th Inter. Conf. on Fracture, Vol. 3. Waterloo, Canada. pp. 583-592.
Kobayashi, H. (1977b). J. of JSME, 80, 492-497.
Kobayashi, H., M. Sugiura, R. Murakami, H. Nakazawa, T. Katagiri, and Y. Iwasa (1977c). Trans. of JSME, 43, 416-425.
Kobayashi, H., R. Murakami, and H. Nakazawa (1977d). In G. C. Sih and C. L.Chow (Ed.), Proc. Inter. Conf. on Fracture Mechanics and Technology, Vol. I. Sijthoff and Noordhoff. pp. 205-219.
Kobayashi, H., H. Nakamura, and H. Nakazawa (1978a). The J-integral approach to elastic-plastic fracture toughness of base metals and weldments of A533B and

304 steels. In K. Iida (Ed.), Criteria for Preventing Service Failure in Welded Structures, Proc. 3rd Inter. Symp. of the Japan Welding Society. pp. 191-196.
Kobayashi, H., K. Hirano, H. Nakazawa, A. Nara, and H. Nakamura (1978b). Pre-print of the annual meeting of JSME, No. 780-12, 205-207 and unpublished work.
Kobayashi, H., K. Fujita, A. Komine, and H. Nakazawa (1978c). Pre-print of the annual meeting of JSME, No. 780-4, 113-115.
Kobayashi, H., E. Shibuya, and H. Nakazawa (1978d). In Elastic-Plastic Fracture Mechanics, Pre-print of 15th JSME Symp., No. 780-9, 105-110.
Kobayashi, H., H. Nakamura, T. Yamashita, and H. Nakazawa (1978e). Pre-print of the annual meeting of JSME, No. 780-13, 206-208.
Kobayashi, H., A. Hirano, H. Nakamura, and H. Nakazawa (1978f). Pre-print of the annual meeting of JSME, No. 780-13, 232-234.
Kobayashi, H. (1978g). In Elastic-Plastic Fracture Mechanics, Pre-print of 15th JSME Symp., No. 780-9, 1-10.
Kobayashi, H. (1978h). J. of the Society of Materials Science, Japan, 27, 495-505 and 711-720.
Kobayashi, H., H. Nakamura, and H. Nakazawa (1979). The J-integral evaluation of stretched zone width and its application to elastic-plastic fracture toughness test. In H. Miyamoto (Ed.), Recent Researches on Mechanical Behavior of Solids. University of Tokyo Press. (in press).
Kodaira, T., N. Nakajima, and M. Matsumoto (1978). J. of the Iron and Steel Institute of Japan, 64, 877-889.
Landes, J. D., and J. A. Begley (1974). ASTM STP 560, pp. 170-186.
Logsdon, W. A., and J. A. Begley (1977). Engineering Fracture Mechanics, 9, 461-470.
Maruyama, T., S. Tominaga, and H. Togawa (1976). Pre-print of the annual meeting of JSME, No. 760-2, 33-36.
Matsushita, H., and T. Miyoshi (1978). In Elastic-Plastic Fracture Mechanics, Pre-print of 15th JSME Symp., No. 780-9, 17-22.
McMeeking, R. M. (1977). J. Mech. & Phys. of Solids, 25, 357-381.
Merkle, J. G., and H. T. Corten (1974). Trans. ASME, Ser. J, 96, 286-292.
Miyamoto, H., A. Hijikata, S. Yoshioka, M. Kumazawa, and H. Tohda (1978a). In Elastic-Plastic Fracture Mechanics, Pre-print of 15th JSME Symp., No. 780-9, 11-16.
Miyoshi, T., T. Uno, and M. Shiratori (1975). Pre-print of the annual meeting of JSME, No. 750-1, 191-194.
Miyoshi, T. (1977). Pre-print of the annual meeting of JSME, No. 770-2, 30-32.
Nakano, Y. (1978). J. of the Iron and Steel Institute of Japan, 64, 891-898.
Ohji, K., K. Ogura, H. Uchida, and T. Fujimoto (1978). Pre-print of the annual meeting of JSME, No. 780-12, 196-198.
Ohtsuka, A., T. Miyata, S. Nishimura, and Y. Kashiwagi (1975). Engineering Fracture Mechanics, 7, 419-428.
Ohtsuka, A., T. Miyata, and S. Nishimura (1977). In G. C. Sih and C. L. Chow (Ed.), Proc. Inter. Conf. on Fracture Mechanics and Technology, Vol. I. Sijthoff and Noordhoff. pp. 539-552.
Ohtsuka, N., H. Kamiyama, and M. Nakano (1978). In Elastic-Plastic Fracture Mechanics, Pre-print of 15th JSME Symp., No. 780-9, 29-34.
Paris, P. C. (1977). ASTM STP 631, pp. 3-27.
Rice, J. R. (1968). Trans. ASME, Ser. E, 35, 379-386.
Rice, J. R., P. C. Paris, and J. G. Merkle (1973). ASTM STP 536, pp. 231-245.
Rice, J. R. (1976). In F. Eedogan (Ed.), The Mechanics of Fracture, AMD-Vol. 19. ASME. pp. 23-53.
Sakata, M., S. Aoki, M. Kanzawa, N. Ogure, and K. Tateishi (1978). Trans. ASME, Ser. H, 100, 128-133.
Schwalbe, K.-H. (1977). Engineering Fracture Mechanics, 9, 547-556.
Tobler, R. L., and R. P. Reed (1977). In K. D. Timmerhaus and others (Ed.), Advances in Cryogenic Engineering, Vol. 22. Plenum. pp. 35-46.
Yokomaku, T., H. Toyoda, and S. Ohta (1978). Report of JSME Committee S-781, No. 11.
Wells, A. A. (1961). In Proc. of the Crack Propagation Symposium, Vol. 1. Cranfield College of Aeronautics. pp. 210-230.

Volume 3, ICM 3, Cambridge, England, August 1979

EFFECT OF SPECIMEN SIZE AND GEOMETRY ON DUCTILE FRACTURE CHARACTERIZATION

J. D. Landes and D. E. McCabe

Westinghouse R & D Center, Pittsburgh, Pennsylvania 15235, U.S.A.

INTRODUCTION

Ductile fracture characterization is important for the evaluation of many large structures. Methods recently developed have allowed fracture mechanics concepts to be extended into the post yield regime. Of several methods proposed, the J integral approach to ductile fracture characterization (Rice, 1968; Begley, 1972; and Landes, 1977) has become popular because it can be easily determined experimentally and can be used to evaluate the fracture potential of cracked structures.

J can be determined from an experimental load versus load point displacement record by two methods. The most widely used method involves the determination of the work done on the specimen from the area under the load-displacement record. This work is used in formulas developed by Rice, et al. (1973) to determine values of J which closely approximate the line integral value (Landes, 1979a). A second method of approximating J has been developed from the extension of linear elastic concepts (McCabe, 1979 and Chell, 1976). The nonlinear part of the load displacement record is taken as a measure of effective crack advance. The effective crack length is determined from a secant to the load-displacement curve. This crack length along with the value of load is used to determine a plasticity corrected stress intensity factor, K_R, value from which J is calculated. This method of determining J has been demonstrated to yield values which are nearly identical to those determined by energy methods for effective crack extensions up to about one half of the uncracked ligament (McCabe, 1979).

Fracture characterization by J integral is usually made from a crack growth resistance curve where J is plotted as a function of physical crack advance (Landes, 1977). Crack advance is generally taken from an averaged value of crack length determined either from a measurement on the fracture surface or by an unloading compliance method. R curves are sometimes plotted in terms of effective crack advance when the objective is to use the R curves to predict maximum load instabilities.

An R curve plotted in terms of physical crack advance is shown schematically in Fig. 1. From this R curve four steps in the ductile fracture process can be identified. They are: (1) blunting of the pre-existing crack tip, (2) initiation of stable growth from the blunted crack tip (labelled J_{Ic}), (3) stable crack growth and (4) ductile instability. Each of these steps is an integral part of the ductile fracture process. Steps 1 through 3 are used to determine J_{Ic} which is a single valued fracture criterion. However, J_{Ic} may give an overly conservative measure of fracture potential and instability (step 4) could be a more realistic fracture criterion.

Methods have been proposed to handle each of the four steps in the fracture process. In this paper these four steps will be discussed, methods of determination presented, and problems in evaluating each step will be illustrated.

CRACK BLUNTING

Crack tip blunting is used primarily in the determination of J_{Ic}. Blunting is related to crack tip opening displacement which in turn is related to J. An approximate representation of crack blunting on the R curve is given by the expression

$$J = 2\sigma_Y \Delta a \tag{1}$$

where σ_Y is an effective yield strength and Δa is the apparent crack advance due to blunting (Landes, 1977). Crack opening displacement is not always related to J by a single expression (Robinson, 1976). Therefore, experimental measurement of apparent blunting sometimes follows the line given by Eq. 1 and sometimes differs significantly. Figure 2 shows an example of an R curve where Eq. 1 accurately predicted blunting. Other studies have shown examples where the coefficient in Eq. 1 could be as large as 3.7 (Kodaira, 1977).

In the J_{Ic} method, the crack blunting line is used as a reference line for determining the point of crack initiation. For materials with low to moderate

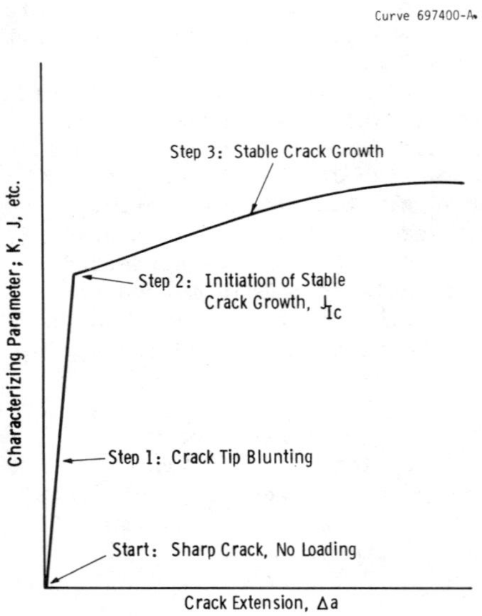

Fig. 1 Four steps of ductile fracture process on an R-curve.

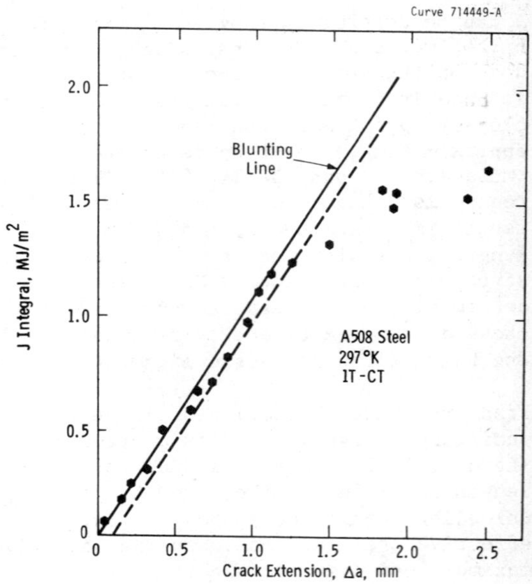

Fig. 2 J vs. Δa R-curve for A508 Cl 2 steel from 1T compact specimens (297 K).

toughness, the value of J_{Ic} is not strongly dependent on the slope of the blunting line and an accurate representation of blunting is usually not necessary. For some very high toughness materials, the J_{Ic} value will be more influenced by the blunting line slope. However, for these materials, the toughness characterization is more properly represented by an instability criterion and the slope of the R curve is more important than J_{Ic}.

CRACK INITIATION

J_{Ic} is often used as the single valued ductile fracture criterion. It is overly conservative for the very high toughness materials, but in many cases it is completely adequate for a fracture characterization. A comprehensive J_{Ic} test procedure has been reported; (Clarke, 1979), and the method is presently being developed into an ASTM standard.

The procedure for determining J_{Ic} from an R curve is shown in Fig. 3. A minimum of four points on a plot of J versus crack extension are experimentally determined, all of which must fall between two limits. A linear curve is fit to these points and extrapolated to the point of intersection with the blunting line, J_{Ic}. A lower limit of crack extension is chosen to assure that actual crack extension is distinguishable from blunting. An upper limit is chosen to keep the total amount of crack extension small enough so that the data is confined to the initial linear part of the R-curve. The slope of the R curve often decreases with large crack extensions. Points are subsequently rejected if they fall outside the range given by the intersection of the fitted line with the dashed lines, Fig. 3. The points on the R curve and the J_{Ic} point are further screened by size criteria applied to the thickness and remaining ligament of the test specimen given by

$$B, b \geq 25 \frac{J_{Ic}}{\sigma_Y} \tag{2}$$

where B is specimen thickness, b is uncracked ligament length and σ_Y is an effective yield strength. A complete description of the procedure is given by Clarke (1979).

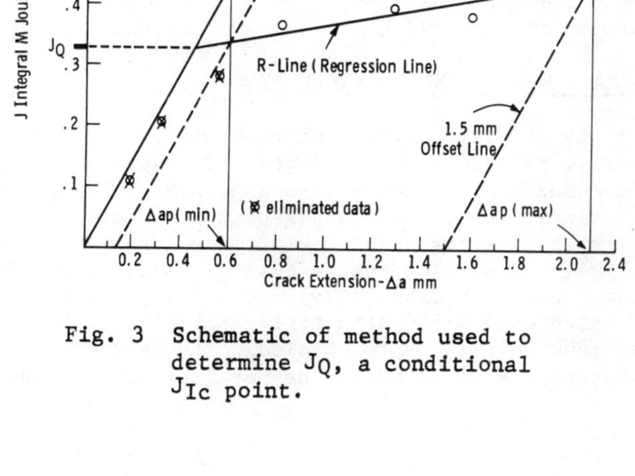

Fig. 3 Schematic of method used to determine J_Q, a conditional J_{Ic} point.

The J_{Ic} fracture criterion is presently taken to be a material property independent of specimen size and geometry within the constraints of the size criterion. The criterion, however, is based on a limited amount of test data and can be subject to revision.

Some evidence can be presented to support claims of size and geometry independence of J_{Ic}. Figure 4 shows the independence of J_{Ic} value and R curve slope on specimen size for an A508 Class 2 steel (Landes, 1978). Figure 5 compares J_{Ic} values determined from a bend loaded geometry, CT, and a pure tension loaded geometry, CCP, (Begley, 1976). The J_{Ic} value for the two specimen types is coincident but the subsequent R curve slopes are different. Other examples are available in the literature to illustrate the size and geometry independence of J_{Ic} however a much larger sampling of different materials and sizes is needed before size limitations such as those in Eq. 2 can be confidently set (Landes, 1979b).

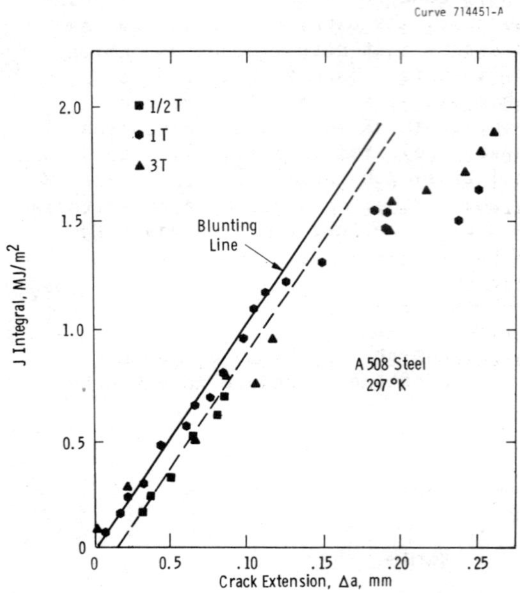

Fig. 4 J vs. Δa R-curve for A508 Cl 2 steel from three sizes of compact specimens tested (297 K).

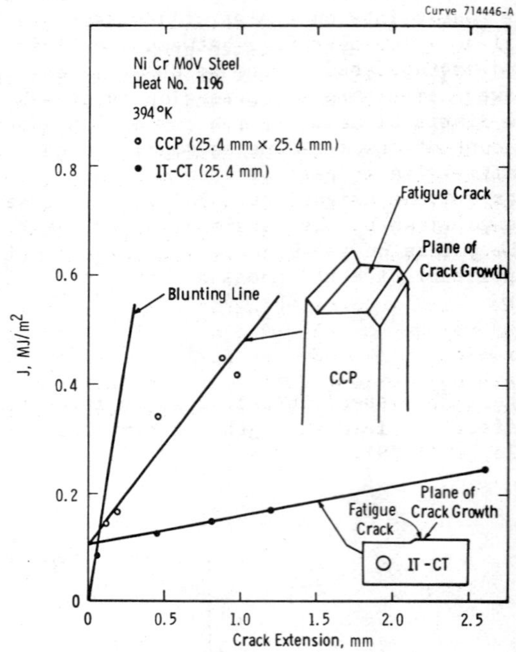

Fig. 5 J vs. crack extension for a CCP specimen and a 1T-CT specimen - NiCrMoV at 394 K.

STABLE CRACK GROWTH AND INSTABILITY

Ductile instability is predicted from R-curve behavior of materials by comparing calculated crack drive rate to R-curve slopes. The approach to calculation of crack drive curves depends upon whether crack growth of the R-curve is expressed in terms of the physically observed crack growth or effective crack growth where crack tip plasticity effects are added on. In the latter case, crack drive can be calculated for fully elastic bodies because all plasticity effects are incorporated in the R-curve (McCabe, 1979). The fully plastic R-curve is developed using either compliance determined effective crack size or by adding on a calculated plastic zone adjustment to physical crack size. The alternative form of R-curve representation where J versus Δa physical is plotted, requires that crack drive must reflect the contributions of crack tip plasticity since this is absent in the R-curve characterization.

The two methods of R-curve representation are shown for HY130 steel in Fig. 6. The data development was from an experiment to determine if elastic-plastic to fully plastic R-curves are independent of specimen size and/or initial crack size (McCabe, 1978). Compact specimens were made from 25 mm thick plate and planar dimensions were varied from 1T to 12T size. The solid lines are the 12T generated R-curves and data points are from a 1T size specimen. All of the data development obtained on various specimens was not as ideal as shown here, but the evidence was clear that there was no effect of specimen size or initial crack size on either type of R-curve representation. Of special interest is the fact that much of the 1T specimen data was developed beyond maximum load. The extent of the R-curve data was not limited by the validity of the analysis so much as by the inability of the specimen to further sustain a reasonable load level.

At present, it is not entirely clear that elastic-plastic R-curves are applicable across different specimen geometries. The data shown in Fig. 5 suggest that only J_{Ic} is independent of geometry and that R-curve development under appreciable stable growth is different. A mitigating consideration, however, is that the remaining ligament in the CCP specimen was quite small and 45 degree shear bands had developed from the crack tip to the free surfaces. As a result, the Δa_p was measured at 45 degrees to the original crack plane. This type of deformation is not representative of wide panel tests in general and the crack growth resistance development is expected to be appreciably different with larger ligaments. Several test programs are presently in progress to explore this consideration.

The crack geometry that is most relevant to heavy section instability analysis is the partial thickness crack, PTC. The application of R-curve analysis and prediction procedures for this configuration represents a most challenging task. Perhaps the most difficult problem is to define the severity of stress intensification where the crack front is under a K gradient from the free surface to the

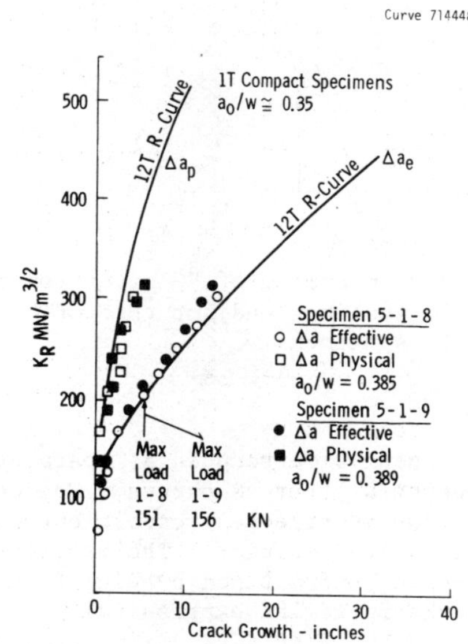

Fig. 6 R curve for a 1T size compact specimen of HY130.

most advanced point through the thickness. In addition, a plastic flow field, unique to this crack geometry, extends from the crack tip to the back surface, and especially for deep cracks this ligament yielding may develop at relatively low nominal stress levels. A preliminary study of a strictly empirical nature has been tried with encouraging results. If one is willing to accept the proposition that attention can be drawn away from local crack tip processes, a compliance argument could be developed to justify the use of equivalent through-thickness (TTC) crack size determinations from surface crack-opening measurements. For example, Fig. 7 shows R-curves developed in wide panel tests, three of which were generated from PTC cracks of the sizes shown. The test records were obtained from published work (Finger, 1976). The records for both TTC and PTC panels were treated as if they all were TTC. The fact that R-curve development was identical in magnitude and shape independent of initial flaw size or shape encourages further exploration. The computation process of course can be reversible, and when the R-curve is known to be geometry and crack direction independent, then TTC generated R-curves could be used to predict instability stress for vessels containing surface flaws.

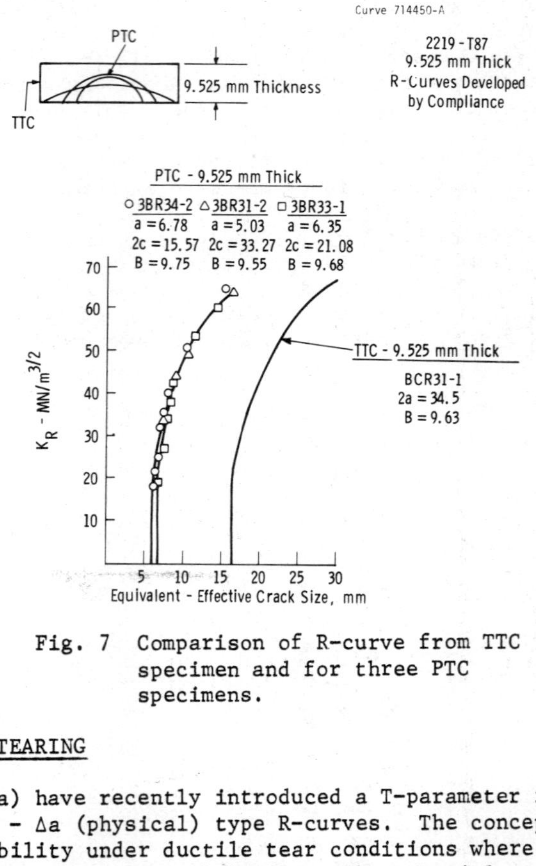

Fig. 7 Comparison of R-curve from TTC specimen and for three PTC specimens.

DUCTILE TEARING

Paris, et al. (1979a) have recently introduced a T-parameter instability criterion which is based on J - Δa (physical) type R-curves. The concept is designed to predict crack instability under ductile tear conditions where the material follows upper shelf transition temperature behavior. The material toughness characterization is the slope of the R-curve, normalized by elastic modulus, E, and flow strength, σ_o, according to the following expression:

$$T\ (\text{material}) = (dJ/da)E/\sigma_o^2 \qquad (3)$$

Crack drive is determined in terms of T (applied) from calculated specimen elastic displacements and available plastic deformation formulas. Initial T (applied) developments were coupled with limit load analysis and the controlling variables were specimen dimensions as indicated in Table 1. Initial crack size is not a dominating consideration as it is in conventional R-curve analysis. More recently, T (applied) has been determined for confined plasticity cases and initial crack size enters the analysis as a variable of secondary importance.

TABLE 1 T(Applied) Expressions for Common Geometries (Paris, 1977a)

Geometry	T (applied)	Symbols Definitions
Center Cracked Panel	$T = 2L/W$	W - Specimen Width
		L - Total Length
Double Edge Cracked Panel	$T = 12L/W$	B - Thickness
		b - Uncracked Ligament
Deep Surface Crack	$T = 4C/B$	C - Surface Crack Half Length
Three Point Bend Specimens	$T = 2b^2L/W^3 - \theta_c E/\sigma_o$	θ_c - Angle change due to Crack Growth
		E - Elastic Modulus
		σ_o - Material Flow Strength

To illustrate how T controlled instability works in practice, Paris, et al. had developed a test jig for testing three point bend bars that can be modified with different spring bars to provide variable T (applied), (Paris, 1979b). This test apparatus is shown in Fig. 8. A NiCrMoV rotor steel having a T (material) of 36 was tested at several levels of T (applied). Examples of test records of stable and unstable cracking created by the stored elastic energy in the test system are shown in Fig. 9. Stored elastic strain energy did not force instability before maximum load in this experiment since all specimens developed about the same maximum load within normal experimental variability.

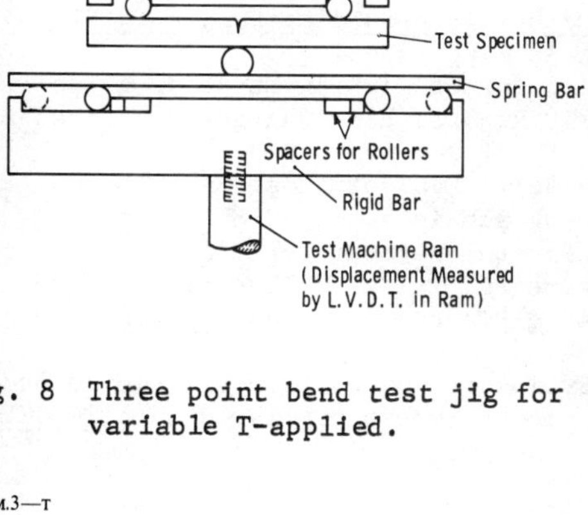

Fig. 8 Three point bend test jig for variable T-applied.

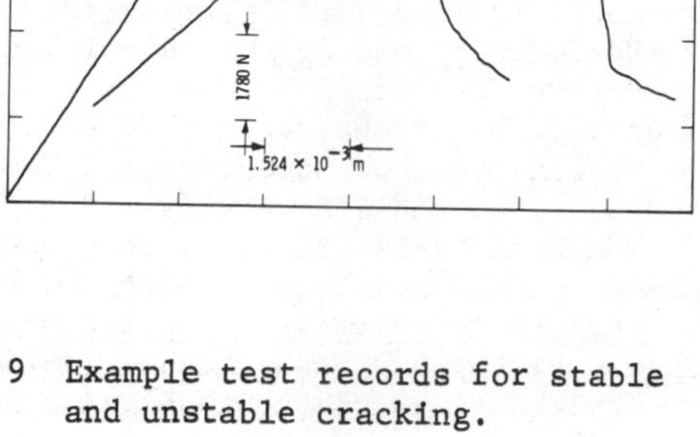

Fig. 9 Example test records for stable and unstable cracking.

The T-parameter concept is related to conventional R-curve analysis for displacement controlled conditions. For conventional R-curve predictions, the K_I for crack drive can be altered from negative to positive K_I gradient, dK_I/da, under displacement control depending upon specimen length dimensions and hence stored elastic strain as in T-parameter analysis. The limiting condition for such analysis is for infinitely long specimens which will simulate load controlled behavior of shorter specimens having the same cross-section. The T-parameter relates to the critical specimen length or size parameters that will develop ductile crack instability beyond maximum load for systems under displacement controlled loading.

CONCLUSIONS

1. Four steps in the ductile fracture process were identified; crack blunting; initiation of stable cracking, J_{Ic}; stable crack growth; ductile instability.
2. J_{Ic} which corresponds to initiation of stable tearing is an acceptable fracture criterion for many applications. However, for high toughness materials it may be over conservative. J_{Ic} is shown to be a material constant independent of specimen geometry and size if the proper minimum size limitations are exceeded.
3. R curves for ductile materials developed beyond the J_{Ic} point can be used to predict ductile instability. These R curves appear to be independent of specimen planar dimensions but may not be independent of specimen geometry.
4. Instability predictions can be made by putting plasticity effects into the driving force, such as in the T parameter, approach or into the R curve for a more conventional approach.

REFERENCES

Begley, J. A. and J. D. Landes (1972). The J Integral as a Fracture Criterion, in Fracture Toughness, ASTM STP 514, ASTM, pp 1-20.

Begley, J. A. and J. D. Landes (1976). Serendipity and the J Integral, International Journal of Fracture, Vol. 12, pp. 764-766.

Chell, G. G. (1976). An Assessment of Some Methods of Obtaining Fracture Toughness Values from Invalid Test Data, Materials Science and Engineering, Vol. 24, pp. 233-238.

Clarke, G. A., et al. (1979). A Procedure for the Determination of Ductile Fracture Toughness Values Using J Integral Techniques, to be published in Journal of Testing and Evaluation, ASTM.

Finger, R. W. (1976). Proof Test Criteria for Thin-Walled 2219 Aluminum Pressure Vessels. NASA CR 135037, Vol. II.

Kodaira, T., N. Makagimer, and M. Matsumoto (1977), Interrelation Among J Values, COD and Lateral Contraction, Japan Atomic Energy Research Institute, JAERI - Memo 7161.

Landes, J. D. and J. A. Begley (1977). Recent Developments in J_{Ic} Testing, Developments in Fracture Mechanics Test Methods Standardization, ASTM STP 632, W. F. Brown, Jr. and J. G. Kaufman, Eds., ASTM pp. 57-81.

Landes, J. D. and G. A. Clarke (1978). Development of Testing Methodology to Measure J_{Ic} and J_R Curves - Third Quarterly Report, HSST Quarterly Progress Report for Period Ending May 31, 1978.

Landes, J. D., H. Walker and G. A. Clarke (1979a). Evaluation of Estimation Procedures Used in J-Integral Testing, to be published in ASTM STP 668.

Landes, J. D. and J.A. Begley (1979b). Experimental Methods for Elastic-Plastic and Post Yield Fracture Toughness Measurements, to be published.

McCabe, D. E., and J. D. Landes (1978). Elastic-Plastic R-Curves, Journal of Engineering Materials and Technology, ASME.

McCabe, D. E., and J. D. Landes (1979). An Evaluation of Elastic-Plastic Methods Applied to Crack Growth Resistance Measurements, to be published in ASTM STP 668.

Paris, P. C., H. Tada, A. Zahoor, and H. Ernst (1979a). The Theory of Instability of the Tearing Mode of Elastic-Plastic Crack Growth, Elastic Plastic Fracture, ASTM STP 668, J. D. Landes, J. A. Begley and G. A. Clarke, Eds., American Society for Testing and Materials, pp. 5-30.

Paris, P. C., H. Tada, A. Zahoor, and H. Ernst (1979b). Initial Experimental Investigation of Tearing Instability Theory, Elastic Plastic Fracture, ASTM STP 668, J. D. Landes, J. A. Begley and G. A. Clarke, Eds., American Society for Testing and Materials, pp. 251-265.

Rice, J. R. (1968). A Path Independent Integral and the Approximate Analysis of Strain Concentration by Notches and Cracks, Journal of Applied Mechanics, Transactions, ASME, Vol. 35, pp. 379-386.

Rice, J. R., P. C. Paris and J. G. Merkle (1973). Some Further Results on J-Integral Analysis and Estimates, Progress in Flaw Growth and Fracture Toughness Testing, ASTM STP 536, ASTM, pp. 231-245.

Robinson, J. N. (1976). An Experimental Investigation of the Effect of Specimen Type on the Crack Opening Tip Displacement and J Integral Fracture Criterion, International Journal of Fracture, Vol. 12, pp. 723-737.

Volume 3, ICM 3, Cambridge, England, August 1979

TRANSITION FROM SLOW TO FAST FRACTURE UNDER POST-YIELD CONDITION

M. P. Wnuk

Mechanical Engineering, South Dakota State University, Brookings, SD, U.S.A.

INTRODUCTION

Conditions under which fast fracture may occur in ductile solids are described in terms of a new quantity, named stability index, λ. This index, defined as $(dR/da - \partial R_A/\partial a)(\partial R_A/\partial a)^{-1}$, encompasses the tearing modulus of Paris $(E/\sigma_o^2)(dJ/da)_i$ determined at the point of <u>incipient</u> crack extension, and other geometry dependent quantities. Stability index serves as a measure of the "distance" of any given state of the system considered (specimen containing an extending crack) from the critical state. Instability occurs at $\lambda = 0$. Symbols R and R_A denote the extent of the plastic zone developed during a quasi-static crack growth and considered either as a measure of material toughness (R) or as an intensity of the applied field (R_A). Symbol "a" denotes the current crack half-length.

The DBCS line plasticity model is extended to a moving crack situation. According to the present theory the crack advances in a sequence of small but finite steps; the single "jump" being on the order of Neuber's particle size, Δ. The work done against the restraining forces during the final act of decohesion is assumed to be a material constant, $\hat{w}_\Delta$. Thus, we distinguish between "true" material resistance to fracture, $\hat{w}_\Delta$, and the "apparent" resistance measured by the total energy release rate, $\hat{w}_R$, which includes all of the work done (reversible and irreversible) associated with the deformation processes accompanying an incremental crack extension. While the former quantity ($\hat{w}_\Delta$) remains invariant to crack growth, the latter ($\hat{w}_R$) increases in the course of crack extension; hence the notion of a rising resistance to fracture in ductile solids. The ratio of the two quantities $\hat{w}_\Delta/\hat{w}_R$, has been determined quantitatively within the framework of the present model, and then employed to estimate the screening action of the plastic zone. It is shown that most of the energy available for fracture is absorbed in this "buffer" zone during the preliminary crack advance. Such effects are particularly prominent in large scale yielding.

There are no restrictions imposed on the size of the plastic zone except that it should be much greater than the process zone size, Δ. Therefore, the solutions suggested here cover both the limiting cases, i.e. (a) the contained yielding case, when $R/a \ll 1$, and (b) large scale (or unlimited) yielding case, when $R/a \gg 1$. Interestingly, both these asymptotic cases have the closed-form solutions, as discussed in the text. For the intermediate range of the ratio R/a the governing non-linear differential equation of the first order (valid for a center-cracked infinite width plate under Mode I loading):

$$\frac{dR}{da} = \frac{a + R}{a} \{\frac{\hat{R}}{\Delta} - \frac{1}{2} \log [\frac{4eRa}{\Delta(2R + a)}]\}, \ R = R(a)$$

can be integrated numerically for a given set of the initial data R_i and a_o, thus generating an R-curve valid for any R/a ratio which may be observed experimentally.

It is noteworthy that the initial slope of an R-curve, $(dR/da)_i$, is a material constant for the case of small scale yielding (R<<a), while for the other extreme case (R>>a) such slope becomes geometry dependent.

Subcritical Crack in Small Scale Yielding Range

Remarkable agreement has been noted between equations governing the ductile fracture process, derived from the Wnuk (1972-1978) model of a quasi-static crack, and the most recent results of Rice and Sorensen (1978), derived from considerations of Prandtl slip line field associated with a quasi-static plane strain crack. The mathematics of Wnuk's model is based on the "final stretch" concept proposed in 1972, according to which an incremental displacement $\hat{\delta}$ is continually generated at the crack tip during the process of subcritical crack growth. Such a criterion for crack extension is equivalent to the condition of a constant crack opening angle, as suggested by Green and Knott (1976), Rice and Sorensen (1978) and, most recently by Shih, deLorenzi and Andrews (1978). To enable computational success wihtout resorting to numerical studies, the concept of final stretch is linked with the Dugdale-Bilby-Cottrell-Swinden model of yielding associated with a moving crack (only a quasi-static problem is considered). It may be shown that the crack opening near the crack tip is in fact of the form r log(const/r) at a distance r from the tip, whether one uses Rice's or Wnuk's model. In particular, both models predict an R-curve which, within small scale yielding range and when the crack tip process zone is fully enveloped by the yielded region, obeys the following differential equation

$$dR/da = (1/2)\log (R_{ss}/R), \ R<<a \tag{1}$$

Here, R denotes the extent of the plastic zone in front of the crack, while R_{ss} is its steady-state limit. This limit, a material constant pertinent to the slow stable cracking phenomenon, is related to Wnuk's opening stretch $\hat{\delta}$ and the size of the process zone Δ (Neuber's constant) as follows

$$R_{ss} = (\Delta/4e) \exp (\pi E\hat{\delta}/4\sigma_Y\Delta) \tag{2}$$

While the constant Δ is determined by the micro structure of the material, the "effective" yield strength σ_Y, encountered along the crack front, depends on the stress state prevailing in the regions adjacent to the crack. Thus, within a certain range of specimen thickness, σ_Y and R_{ss} may exhibit thickness dependence; obviously, both constants are not insensitive to work-hardening. Such effects could be considered in a more precise manner through the use of a modified DBCS model, which accounts for variations of yield strength due to variable thickness and strain hardening. Here, we shall confine the attention to a non-hardening case and a uniform σ_Y across the specimen thickness; allowing however for σ_Y to be a certain multiple of the uniaxial yield stress σ_o, measured in a conventional way. Denoting the ratio σ_Y/σ_o by η and admitting the relation between the J-integral and the crack tip opening to be of the form

$$J = \alpha \ \delta_t \ \sigma_Y \tag{3}$$

(where α is a numerical constant), equations (1) and (2) may be re-written as follows

$$dJ_R/da = \alpha\eta^2(4\sigma_o^2/\pi E)\log(J_{ss}/J_R)$$

$$J_{ss} = (2\alpha\sigma_Y^2\Delta/\pi eE)\exp(\pi E/4\sigma_Y)\hat{\delta}/\Delta] \tag{4}$$

Here, index "R" is added to emphasize that we consider the J-integral as a measure of material resistance to cracking in the subcritical range. Such material property, of course, is a certain function of crack advance $J_R = J_R(a)$.

To conclude this section we note that the governing equation (1) is a special case of a more general equation given for a tensile quasi-static crack by Wnuk in 1973, i.e.,

$$R - \sqrt{R(R-\Delta)} + \Delta\frac{dR}{da} + \frac{\Delta}{2}\log\frac{\sqrt{R} + \sqrt{R-\Delta}}{\sqrt{R} - \sqrt{R-\Delta}} = \hat{R} \tag{5}$$

Here, the constant $\hat{R}$ is directly proportional to the opening stretch $\hat{\delta}$, namely $\hat{R} = (\pi E/8\sigma_Y)\hat{\delta}$. Unlike eq. (1), the equation above is not subjected to the restriction $\Delta << R$, and, therefore, it applies to both the limits:

I. A quasi-brittle solid, in which case R is much less than a, but sufficiently large so that the crack tip process zone is small compared to R. With $\Delta<<R<<a$ equation (5) becomes

$$dR/da = \frac{\hat{R}}{\Delta} - \frac{1}{2} - \frac{1}{2}\log(4R/\Delta) \tag{5a}$$

which is equivalent to equations (1) and (2).

II. Representation of a brittle solid is obtained when the process zone and the plastic zone are allowed to shrink to a common "end zone", better known under more appropriate name for the case, i.e., "cohesive zone". Such cohesive zone is autonomous during the crack growth and independent of the imposed loadings. It is readily seen that when R reduces to Δ, and when R does not vary with a, i.e., when the derivative dR/da vanishes, equation (5) predicts

$$R = \hat{R} \text{ or } \delta = \hat{\delta} \tag{5b}$$

In this way, with δ_{IC} substituted for $\hat{\delta}$, the classical LEFM criterion of brittle fracture $\delta = \delta_{IC}$ or $K = K_{IC}$ is recovered. We may, therefore, conclude that the Griffith-Irwin theory of fracture is contained as a limiting possibility within the framework of non-linear fracture mechanics considered here. Note also that the Neuber constant Δ should not be allowed to vanish completely, even in the limit of brittle behavior, since then the left-hand side of equation (5) reduces to zero, rendering crack extension impossible. Recognition of necessity of finiteness of the cohesive zone is not new. It has been noted for both brittle and quasi-brittle behaviors by numerous authors, beginning from Barenblatt (1959), Rice (1965), and ending with more recent observations of Swinden (1970), Kfouri and Miller (1976) and Wnuk (1977). The last two works listed suggest that an admissible fracture criterion can be obtained by an energy balance approach if the smallest increment of crack growth is considered a material property, and if the quantity $\mathcal{G}^\Delta$ (or $2\hat{w}_\Delta$ in Wnuk's notation) is equated to a certain critical value. This may be thought of as a "quantization" of the fracture process, consisting essentially in the inclusion of a finite-sized zone of decohesion into the fracture model.

A more extensive discussion of crack growth in the near-yield or post-yield conditions is given in the following section.

Subcritical Crack in Large Scale Yielding Range

Considering the substructure of the plastic zone extending ahead of the slowly growing crack and applying the "final stretch" concept to such a crack, Wnuk (1978a, 1978b) extended the range of validity of his model of a quasi-static crack into the domain of arbitrarily large plasticity. The discriminating quantity, of course, is the ratio of the plastic zone size to crack length, R/a. While no a priori assumptions about the R/a ratio were made, it was necessary to consider the crack tip process size Δ small against the extent of the plastic region. Such a requirement coincides with that made previously by Sumpter and Turner (1976), Hutchinson and Paris (1977) and Rice and Sorensen (1978) in their studies of slow cracking in fully plastic range. Generalization of equation (5a) for an arbitrary R/a ratio leads to the following non-linear differential equation

$$\frac{dR}{da} = \frac{a + R}{a} \{\frac{\hat{R}}{\Delta} - \frac{1}{2} \log [\frac{4eRa}{\Delta(2R + a)}]\} \tag{6}$$

It may be readily shown that for R<<a the eq. (6) reduces to eq. (5a), which is valid within the small scale yielding range. It should be emphasized, however, that unlike eq. (5a), which is invariant to geometry alterations, eq. (6) holds only for the configuration considered in the deriviation process, i.e., that of a center cracked panel of infinite width. Since the loading parameter Q $(=\pi\sigma/2\sigma_Y)$ and the extent of the plastic zone R are related for this configuration through the Dugdale formula

$$Q = \text{arc cos } (\frac{a}{a + R}) \tag{7}$$

(in which both Q and R are treated as functions of a), the differential relation (6) can be transformed to the (Q,a) plane as follows, cf. Wnuk (1977):

$$\frac{dQ}{da} = \frac{a(dR/da) - R}{(a + R) \sqrt{2aR + R^2}}, \quad R = R(a), \quad Q = Q(a) \tag{8}$$

Here, dR/da should be replaced by the right-hand side of equation (6). Both equations (6) and (8) were integrated simulteneously with help of a computer and some of the resulting integral curves, i.e. the R-curves and the Q-curves, are shown in Fig. 1. The lower and upper toughness bounds shown in Fig. 1 are

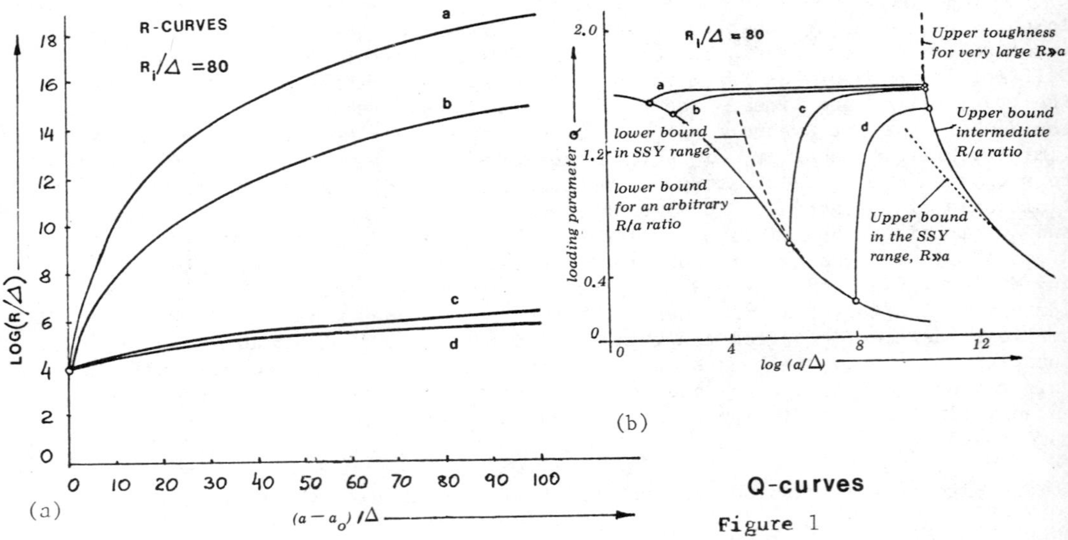

Figure 1

described by these formulae

(a) initiation locus at an arbitrary R/a ratio

$$Q_i = \cos^{-1}(a_o/(a_o + R_i)) \tag{9a}$$

(b) initiation locus in the small scale yielding range

$$Q_i = (2R_i/a_o)^{1/2} \tag{9b}$$

(c) terminal instability locus at arbitrary R/a ratio

$$0 = \{(a_f + R_f)\{\frac{\hat{R}}{\Delta} - \frac{1}{2}\log[\frac{4eR_f a_f}{\Delta(2R_f + a_f)}]\} - R_f\}_{R_f = a_f[\sec Q_f - 1]} \tag{9c}$$

At the limit of R>>a the latter expression reduces to

$$\frac{\hat{R}}{\Delta} - \frac{1}{2}\log(2ea_f/\Delta) - 1 = 0 \tag{9d}$$

or

$$a_f/\Delta = \exp[2(n-1)] \tag{9e}$$

(vertical line in Fig. 1). At the other limiting case of R<<a, one obtains from equation (c)

$$\{a_f[\frac{\hat{R}}{\Delta} - \frac{1}{2}\log(4eR_f/\Delta)] - R_f\}_{R_f=(1/2)a_f Q_f^2} = 0 \tag{9f}$$

Although the two asymptotic results are distinctly different, the terminal instability locus, valid for an arbitrary R/a ratio, is seen to converge smoothly and without any substantial deviations into the corresponding asymptotic solutions applicable at both extreme behaviors considered. This is shown by the graphs in Fig. 1. Table 1 supplements these graphs with numbers which summarize the essential findings pertinent to ductile fracture and transition to unstable crack propagation. It is seen that a stable crack, initiated from a fairly large initial defect contained in a solid of high ductility (i.e., runs c and d in Fig. 1), will continue to grow under a monotonically increasing load until it reaches a critical size substantially greater than the initial defect dimension. This leads to a considerable increase both in toughness parameter and in the loading parameter (for example, for run d the former increases by a factor 4.11 x 10^3 while the latter has increased 536%). Such phenomenon appears to be not uncommon for high ductility/low toughness materials, and while the problem is not easily susceptible to a mathematical analysis, it offers a desirable margin of safety.

For comparison, five other "runs" were carried out for the large scale yield situation. The first three of these ductile fracture runs are in the range of large R/a ratios, and therefore, could be treated as asymptotic cases resulting from eq. (6) when R>>a. As expected, the governing equation (6) simplifies then

$$\frac{dR}{da} = \frac{R}{a}\{\frac{\hat{R}}{\Delta} - \frac{1}{2}\log(\frac{2ea}{\Delta})\} \tag{10}$$

An equivalent form involving the J-integral rather the resistance parameter R was given by Wnuk (1978a), and it reads

$$\frac{dJ_R}{da} = \frac{\hat{J}}{\Delta} + \frac{J}{a} - \frac{4\sigma_Y^2}{\pi E}\alpha\log(\frac{2e^3 a}{\Delta}) \tag{11}$$

Table 1. Transition to unstable fracture in a very ductile solid ($R_i = 80\Delta$, $n = 6.15$). Data shown are computed from the eqs. (6)&(8) valid for an arbitrary ratio R/a.

Run #	R_i/Δ / R_f/Δ	a_o/Δ / a_f/Δ	R_i/a_o / R_f/a_f	λ_i	Q_i	Q_f	$\frac{Q_f - Q_i}{Q_i}$ %
a	80	4.51	17.74	4.7172	1.5174	1.5980	5.31
	2.95×10^{13}	2.95×10^{4}	1.00×10^{9}				
b	80	10	8.00	4.6591	1.4595	1.5857	8.65
	1.24×10^{12}	2.95×10^{4}	4.20×10^{7}				
c	80	400	0.20	21.692	0.5857	1.5639	167
	5.96×10^{6}	2.99×10^{4}	199				
d	80	3000	0.0267	139.13	0.2284	1.4519	535.7
	3.29×10^{5}	3.95×10^{4}	8.33				

$$n = \frac{1}{2}\{\log 2 - 1 + \sqrt{2(R_i/\Delta) - 1}\}$$

Equations applicable to LSY case

$\lambda_i = n - 1 - (1/2)\log(a_o/\Delta)$

$a_f = a_o \exp(2\lambda_i)$

$R_f = R_i \exp(2\lambda_i + \lambda_i^2)$

$Q_i = \cos^{-1}[1 + (R_i/a_o)]^{-1}$

$Q_f = \cos^{-1}[1 + (R_i/a_o)\exp\lambda_i^2]^{-1}$

Equations applicable to SSY case

$R_{ss}/\Delta = (1/2)\exp(2n)$, $Q_i = \sqrt{2R_i/a_o}$, $Q_f = \sqrt{2R_f/a_f}$

$\lambda_i = (a_o/2R_i)[\log(R_{ss}/\Delta) - \log(R_i/\Delta)] - 1$

a_f and R_f follow from numerical integration of eqs.

$dR/da = (1/2)\log(R_{ss}/R)$

$dQ/da = [\log(2R_{ss}/Q^2 a) - Q^2]/(2Qa)$

Here, the opening constant $\hat{J}$ equals $\alpha\hat{\delta}\sigma_Y$. Both the equations above have closed form solutions, which were found as follows

$$R(a) = R_i\left(\frac{a}{a_o}\right)^{\lambda_i + 1} \exp\left\{-\frac{1}{4}\log^2\left(\frac{a}{a_o}\right)\right\}$$

$$J_R(a) = J_i\left(\frac{a}{a_o}\right) + \alpha\frac{8\sigma_Y^2}{\pi E}\, a\left\{\lambda_i \log\left(\frac{a}{a_o}\right) - \frac{1}{4}\log^2\left(\frac{a}{a_o}\right)\right\} \tag{12}$$

Here, the quantity $\lambda_i + 1$ $(=(dR/da)_i(a/R)_i)$ equals the contents of the curly bracket of equation (10) evaluated at $a = a_o$ (see the next section for details), i.e.,

$$\lambda_i = \frac{\hat{R}}{\Delta} - 1 - \frac{1}{2}\log\left(\frac{2ea_o}{\Delta}\right) \tag{13}$$

or

$$\lambda_i = n - 1 - \frac{1}{2}\log(a_o/\Delta) \tag{14}$$

Here, the constant

$$n = (\hat{R}/\Delta) - (\tfrac{1}{2})\log(2e) \tag{15}$$

plays the role of a dimensionless tearing modulus.

The upper toughness limit (R_{ss}) predicted for the small scale yielding situation may never materialize in an experiment involving specimen sizes too small to ensure a containment of plasticity effects (indeed, very few experimentalists have ever seen the upper plateau for R_{ss} or J_{ss} predicted from the small scale yielding solution). The theory presented here allows one to extend the limits of applicability of the earlier solutions (of Wnuk (1972), Rice and Sorensen (1978)) valid within the contained plasticity range.

For large structures, however, such as nuclear reactor pressure vessels, there exists an interesting possibility of tapping the reserve strength that the lower strength/higher toughness materials possess in their inelastic region. This becomes apparent upon an examination of graphs shown in Fig. 2. Three fracture

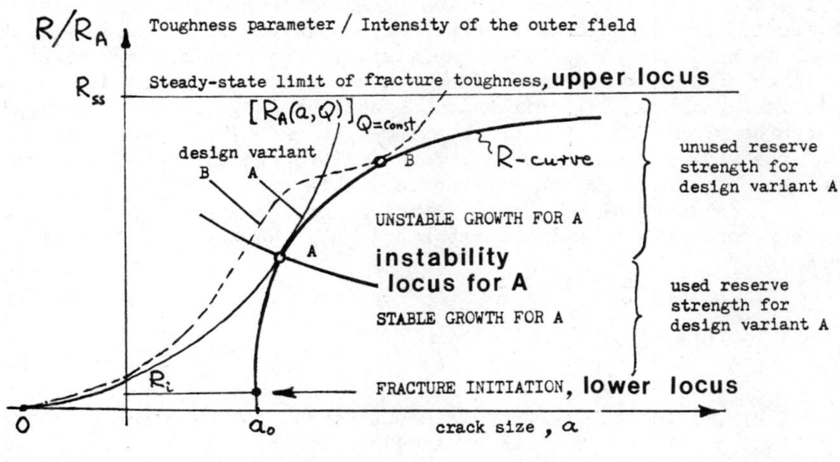

Fig. 2. Two possible theoretical strength limits are shown for a component made of the same material but with different designs, A and B.

loci, i.e., the initiation, terminal and the upper plateau steady-state limit of fracture toughness are shown. The domain contained between the middle and upper lines depicted in Fig. 2. may serve as a measure of the unused portion of the theoretical strength of a ductile material. It is noteworthy that while the two extreme lines Fig. 2 represent material properties pertinent to fracture initiation and the maximum strength, respectively, the middle line can be shifted up and down depending on geometry and loading configuration of a structural component. Therefore, it is here where design and materials science are expected to interact.

Stability Analysis

Symbol λ_i, employed in the equations (12), denotes the initial stability index. It is a quantity essential for stability analysis, since it measures the "distance" of the crack initiation point from that corresponding to an unstable crack

$$\lambda_i = \begin{cases} \frac{\pi E}{8\sigma_Y^2}\left[\left(\frac{dJ_R}{da}\right)_i - \left(\frac{\partial J_A}{\partial a}\right)_i\right], & \text{or} \\ \left[(dR/da)_i/(\partial R_A/\partial a)_i\right] - 1 \end{cases} \tag{16}$$

It is seen that the quantity λ_i depends on material properties (E, σ_Y, $(dJ_R/da)_i$) and on the geometrical and loading configurations (symbols J_A and R_A are used to denote the intensity of the applied field). It may be shown that when the quantity λ (index "i" is dropped whenever a current value of stability index is considered) is greater than zero, fracture extends in a stable manner, as then it is also $dQ/da>0$. A negative or zero λ is tantamount to unstable fracture. A more detailed discussion of ductile instabilities in tensile fracture is given by Wnuk (1978b).

Energy Considerations

Although Rice (1965, 1978) pointed out certain paradoxes which result from the energy balance considerations in connection with elastic-ideally plastic solids and structureless sharp-tip cracks, it is nevertheless an interesting problem to look at. One feasible approach to the problem consists in modification of the crack model by an inclusion of a finite process zone adjacent to the crack front, and embedded within the larger plastic zone ("fine structure" of the end zone is thus introduced into the model). When the energy balance concept is then re-applied to such modified crack model, all phases of fracture development in inelastic solids, initiation, stable crack growth and terminal instability, can be predicted. The basic assumption underlying these calculations concerns the work of decohesion $\hat{w}_\Delta$, done against the restraining forces $S(x,t)$ within the process zone, and considered within the framework of the present model to be a material property (invariant to the extent of subcritical crack growth, but sensitive to thickness variations). Wnuk (1972) suggested the following formula

$$\hat{w}_\Delta = \left\{ \int_{t-\delta t}^{t} S[x,\tau]\dot{u}_y[x,\tau]d\tau \right\}_{x = x_P} \tag{17}$$

Here, the symbol $u(x,\tau)$ denotes the displacement normal to the crack face at the point x and instant τ, while $x = x_P$ is a fixed coordinate of the control point P. A material element located at this point enters the decohesion zone at the instant $\tau = t - \delta t$, and undergoes fracture an instant later, at $\tau = t$. The increment δt represents time interval required for the advancing crack tip to traverse its own process zone, i.e., $\delta t = \Delta/\dot{a}$. With a constant restraining force $S = \sigma_Y$, the work of decohesion becomes simply a product of the effective yield stress σ_Y and the opening constant $\hat{u}$ ($=\hat{\delta}/2$), i.e.

$$\hat{w}_\Delta = \sigma_Y \int_{t-\delta t}^{t} \dot{u}_y(x,\tau)d\tau = \sigma_Y\{u_y(x_P,t) - u_y(x_P,t-\delta t)\} = \frac{1}{2}\sigma_Y\hat{\delta} \tag{18}$$

An equivalent form of $\hat{w}_\Delta$, involving the steady-state toughness level R_{ss}, was suggested by Wnuk (1977)

$$\hat{w}_\Delta = \frac{2\sigma_Y^2}{\pi E_1}\Delta\{\log(4R_{ss}/\Delta) + 1\} \tag{19}$$

$$E_1 = \begin{cases} E & \text{plane stress} \\ E(1-\nu^2)^{-1} & \text{plane strain} \end{cases}$$

If other than S=const. assumptions are made, and, in particular, when a certain release mechanism is allowed to relax the restraining force from its maximum value to zero, the results given by eqs. (17) and (18) are merely multiplied by a numerical factor. For example, this factor is 1/2, if Kfouri and Rice's (1977) assumption of linear release mechanism is followed. It should also be noted that the quantity $2\hat{w}_\Delta$, as defined by eq. (17), can be shown to be identical with the energy separation rate of a "quantized" crack, $\mathcal{G}^\Delta$. In this way, in lieu of Griffith-Irwin criterion of brittle fracture, $\mathcal{G} = \mathcal{G}_{IC}$, we arrive at the criterion applicable to inelastic solids, i.e., $\mathcal{G}^\Delta = 2\hat{w}_\Delta$. This criterion covers all stages of development of fracture in ductile materials, including the initiation and the subcritical crack growth up to the point of terminal instability. When ductility is reduced, and only brittle behavior considered, the classical Griffith-Irwin equations are recovered. For this limit $2\hat{w}_\Delta$ is identified with $\mathcal{G}_{IC}$, while $\Delta \to R$, and $dR/da \to 0$, as discussed in the preceding section.

The Wnuk model of a quasi-static tensile crack suggests

$$\mathcal{G}^\Delta = \frac{8\sigma_Y^2}{\pi E}\{R - \sqrt{R(R-\Delta)} + \Delta\frac{dR}{da} + \frac{\Delta}{2}\log\frac{\sqrt{R} + \sqrt{R-\Delta}}{\sqrt{R} - \sqrt{R-\Delta}}\} \tag{20}$$

In derivation of this formula the extent of the plastic zone R was assumed small vs. the crack length, but no restrictions were placed on R vs. Δ relation. If the ratio R/Δ is restricted to large numbers, i.e., when the process zone is considered to be enveloped by a larger plastic zone (as would be expected in a ductile material), then the present model suggests

$$\mathcal{G}^\Delta = \frac{8\sigma_Y^2}{\pi E} \times \begin{cases} \{\frac{a}{a+R}\,\Delta\,\frac{dR}{da} + \frac{\Delta}{2}\,\log[\frac{4eRa}{\Delta(2R+a)}]\}, & \Delta \ll R \text{ arbitrary } R/a \quad (20a) \\ \{\Delta\frac{dR}{da} + \frac{\Delta}{2}\log(\frac{4eR}{\Delta})\}, & \Delta \ll R \ll a \quad (20b) \\ \{\frac{a}{R}\,\Delta\,\frac{dR}{da} + \frac{\Delta}{2}\log(\frac{2ea}{\Delta})\}, & \Delta \ll R \gg a \quad (20c) \end{cases}$$

It is readily seen that the expressions given above when equated to the constant $2\hat{w}_\Delta$ reduce to the equations of motion for the three cases of subcritical crack growth considered previously, i.e., small scale, intermediate, and large scale yielding situations.

Somewhat similar concepts, involving "generalized energy release rate based on computational process zone concept" (equivalent to $\mathcal{G}^\Delta$) and its critical value pertinent to stable crack growth (equivalent to $2\hat{w}_\Delta$ of this paper) were advanced by Kanninen et al. (1977) in their numerical studies of stable crack growth in ductile solids. They also noticed, in accord with an earlier observation of Broberg (1975), that the portion of the energy flow reaching the crack tip region diminishes with crack extension. This effect is thought to result from the screening action of the plastic zone accompanying the growing crack. Let us describe such energy screening action in quantitative terms.

It has been postulated that the quantity $\hat{w}_\Delta$, defined by eq. (17) is a material constant. On the other hand, we could compute the amount of energy $\hat{w}_R$ dissipated within the plastic zone during an incremental crack advance. This latter quantity, unlike $\hat{w}_\Delta$, is strongly geometry dependent and it could in fact be used as a measure of the outer field intensity (just as the J-integral is). The work $\hat{w}_R$ within the framework of our model can be computed as follows

$$\hat{w}_R = \int_a^{a+R} S(x,a) \frac{\partial u_y(x,a)}{\partial a} dx \tag{21}$$

which, for $S = \sigma_Y$ = const., and the geometrical configuration considered here becomes, cf. Wnuk (1971)

$$\hat{w}_R = 2\sigma_Y u_y(a,a) + 2\sigma_Y \int_a^{a+R} u_y(x,a)dx =$$

$$= \frac{4\sigma_Y^2}{\pi E} a \, \{Q\tan Q + \log\cos Q\}_{Q = Q(a)} \tag{22}$$

Function $Q = Q(a)$ in the expression above (the Q-curve) is found by numerical integration of equation (9), if an arbitrary ratio R/a is involved. Thus the total rate of energy release associated with stable crack extension

$$\hat{w}_R = dW/da \tag{23}$$

becomes a measure of the apparent material resistance while the true toughness, as pointed out by Kanninen et al. (1977), for which the quantity $\hat{w}_\Delta$ appears to be a proper measure, remains constant during the subcritical growth stage. The $\hat{w}_R$ vs. a curve shown in Fig. 3 is indeed of an R-curve shape. It is not a mere coincidence; as we proceed to show the total energy release rate, $\hat{w}_R$, associated with an incremental crack extension, is indeed directly proportional to the extent of the plastic zone, $R = R(a)$, at least it is so in both limiting situations of small and large scale yielding. Therefore, the early interpretations of an R-curve as given by McClintock and Irwin (1965) and that given by Rice (1968) are equivalent within a certain range of R/a ratio. (The former would in essence be identical with plotting $\hat{w}_R$ vs. Δa while the latter corresponds to a plot of R vs. Δa). Terminal instability point on the $\hat{w}_R$ vs. Δa curve may be established through a standard procedure of finding a tangency point between the curves representing the material resistance and the applied field parameter , respectively, both plotted against the extent of crack growth.

Simultaneous plot of the energy screening ratio (ESR) is given in Fig. 3 . The energy screening ratio is defined as the ratio of the work of decohesion to the total energy dissipated within the plastic zone while the crack length increases by da, i.e.,

$$ESR = \frac{\hat{w}_\Delta}{\hat{w}_R} \tag{24}$$

This ratio serves as a measure of that portion of energy flow which "penetrates" the plastic zone and becomes available for the final act of decohesion. Substituting eq. (19) for $\hat{w}_\Delta$, and expression (22) for $\hat{w}_R$, we obtain

$$\mathrm{ESR} = \frac{\frac{1}{2}\sigma_Y\hat{\delta}}{\frac{4\sigma_Y^2}{\pi E}\, a\{Q\tan Q + \log\cos Q\}_{Q\,=\,Q(a)}} \tag{25}$$

or (26)

$$\mathrm{ESR} = \frac{\hat{R}}{a}\,\{Q\tan Q + \log\cos Q\}^{-1}_{Q\,=\,Q(a)}$$

The loading parameter Q can be readily eliminated from this expression. Substituting the known Dugdale relations

$$\begin{aligned} Q &= \cos^{-1}\,[1/(1 + R/a)] \\ \tan Q &= \sqrt{2(\tfrac{R}{a}) + (\tfrac{R}{a})^2} \\ \sec Q &= 1 + \frac{R}{a} \end{aligned} \tag{27}$$

we obtain for an arbitrary R/a ratio

$$\mathrm{ESR} = \frac{\hat{w}_\Delta}{\hat{w}_R} = \frac{\hat{R}}{a}\,\{\sqrt{2(\tfrac{R}{a}) + (\tfrac{R}{a})^2}\,\cos^{-1}(\frac{1}{1+R/a}) + \log\,(1 + \frac{R}{a})\}^{-1}_{R\,=\,R(a)} \tag{28}$$

Simple approximate formulae follow from eq. (28) for the two easy asymptotic situations

$$\frac{\hat{w}_\Delta}{\hat{w}_R} = \begin{cases} \dfrac{1}{3}\,\dfrac{\hat{R}}{R(a)} & \text{for } R \ll a \\[2ex] \dfrac{2}{\pi}\,\dfrac{\hat{R}}{R(a)} & \text{for } R \gg a \end{cases} \tag{29}$$

Finally, it should be noted that the EŚR at fracture is not a geometry independent quantity (as some researchers have predicted), but it turns out to be a function of geometrical and loading configurations in <u>both</u> ranges of behavior, i.e., in large as well as small scale yielding case. Assuming $\hat{R} \simeq R_i$ and substituting

$$R_f = R_i\,\exp(2\lambda_i + \lambda_i^2) \tag{30}$$

for the limiting case of very large yielding (R>>a) into the second equation in (29) we obtain the following estimate for the critical energy screening ratio

$$\left(\frac{\hat{w}_\Delta}{\hat{w}_R}\right)_f \simeq \frac{2}{\pi}\,\exp\,(-2\lambda_i - \lambda_i^2) \tag{31}$$

This formula suggest an indirect way of determining λ_i for a specified material and geometry of the test piece. Even though, it may appear a difficult experimental task to measure in a laboratory test the ratio $\hat{w}_\Delta/\hat{w}_R$, and its critical values, Cotterell (1977, 1978) has shown that such an approach to fracture of very ductile materials is indeed feasible.

If the logarithmic term, log (R/a), is also retained in the second equation of (29) then the asymptotic form for ESR in the large scale yielding range becomes

$$\left(\frac{\hat{w}_\Delta}{\hat{w}_R}\right)_{R>>a} \simeq \frac{\hat{R}}{a}\left[\frac{\pi}{2}\frac{R}{a} + \log(R/a)\right]^{-1} \tag{32}$$

while eq. (31) is re-written as follows

$$\left(\frac{\hat{w}_\Delta}{\hat{w}_R}\right)_f \simeq \left\{\frac{\pi}{2}\exp\left(\lambda_i^2 + 2\lambda_i\right) + \frac{a_o}{R_i}\left[\log\left(\frac{R_i}{a_o}\right) + \lambda_i^2\right]\right\}^{-1} \tag{33}$$

REFERENCES

Wnuk, M. P., (1972). Accelerating Crack in a Viscoelastic Solid Subject to Subcritical Stress Intensity, in Proceedings of Int. Conf. on Dynamic Crack Propagation, Lehigh University, pp. 273-280.

Wnuk, M. P., (1973). Prior to Failure Extension of Flaws in a Rate Sensitive Tresca Solid, in ASTM STP 536, pp. 64-75.

Wnuk, M. P., (1974). Quasi-Static Extension of a Tensile Crack Contained in a Viscoelastic-Plastic Solid, Trans. ASME, J. Appl. Mechanics. Vol. 41, pp. 234-242.

Wnuk, M. P., (1978). Stable and Unstable Cleavage Fracture in Fully Yielded Components, Mechanics Research Communications, Vol. 5, No. 5, pp. 169-275.

Rice, J. R. and E. P. Sorensen, (1978). Continuing Crack Tip Deformation and Fracture for Plane-Strain Crack Growth in Elastic-Plastic Solids, J. Mech. Physics of Solids, in print.

Green, G and J. F. Knott, (1976). The Initiation and Propagation of Ductile Fracture in Low Strength Steels, Trans. ASME., J. Eng. Mat. & Technology pp. 37-46.

Shih, C. F., H. G. deLorenzi and W. R. Andrews, (1978). Studies on Crack Initiation and Stable Crack Grwoth, Report No. 78CRD129, Power Generation and Propulsion Laboratory, General Electric Co., Schenactady, N.Y.

Barenblatt, G. I. (1959). Stable Cracks Formed in the Brittle Fracture Process (in Russian), Prikladnaya Matematika i Mekhanika, Nos. 3, 4, and 5. English Versions: Advan. Appl. Mech. Vol. 7, 1962.

Rice, J. R. (1965), in Proceedings of the Ist Int. Conference on Fracture, Sendai 1965, Vol. I, pp. 283-308, publ. by Japanese Society for Strength and Fract. Mechanics. Tokyo 1966.

Swinden, K. H., (1970). The Energy of Fracture, Int. J. Fract. Mechanics, Vol. 6, pp. 445-446.

Kfouri, A. P. and K. J. Miller, (1976). Crack Separation Energy Rates in Elastic-Plastic Fracture Mechanics, Proc. Inst. Mech. Engrs., Vol. 190, 48/87, pp. 571-584.

Wnuk, M.P. (1977). On Assessment of Fracture Toughness Near or After General Yield, Final Progress Report (Part A) prepared for NSF, Grants GH 43605 and DMR 74-0231602, South Dakota State University.

Sumpter J.D.G. and C.E. Turner, (1976). Use of the J-contour Integral in Elastic-Plastic Fracture Studies by Finite Element Methods, J. Mech. Eng. Science, Vol. 18, No. 3, pp. 97-112.

Hutchinson, J. W. and P. C. Paris, (1977). Stability Analysis of J-Controlled Crack Growth, presented at the ASTM Symposium on Elastic-Plastic Fracture (Atlanta, No. 1977), submitted for publication in Proceedings as ASTM Special Technical Publication.

Rice, J. R., (1978). The Mechanics of Quasi-static Crack Growth, Brown Univ. Division of Engineering, TR No. 63, October 1978, to be published in Proceedings of 8th U. S. Natl. Congress of Applied Mechanics.

Kanninen, M. F., E. F. Rybicki, R. B. Stonesifer, D. Broek, A. R. Rosenfield, C. W. Marshall and G. T. Hahn, (1977). Elastic-Plastic Fracture Mechanics

for Two-dimensional Stable Crack Growth and Inst Problems, Presented at ASTM Symposium in Atlanta, November 1977, ASTM-STP, in press.
Kfouri, A. P. and J. R. Rice, (1977), Elastic-Plastic Separation Energy for Crack Advance in Finite Growth Steps, Proc. ICF4, U. of Waterloo, Vol. 1, pp. 43-60.
Broberg, K. B. (1975). On Stable Crack Growth, J. Mech. Phys. Solids, Vol. 23, pp. 215-237.
Wnuk, M. P., (1971). Subcritical Crack Growth, Int. J. Fract. Mechanics, Vol. 7, No. 4, pp. 383-407.
McClintock F. A. and G. R. Irwin, (1965), in ASTM STP 381, pp. 84-113, ASTM, Philadelphia.
Rice, J. R. (1968). Mathematical Analysis in the Mechanics of Fracture, in "Fracture, an Advanced Treatise", ed. Liebowitz, Vol. 2, pp. 191-311.
Cotterell, B. (1977). Plane Stress Ductile Fracture, in Proceedings of Int. Conf. on Fract. Mechanics and Technology, U. of Hong Kong, Vol. 2, pp. 785-795.
Cotterell, B. (1978), private communication.

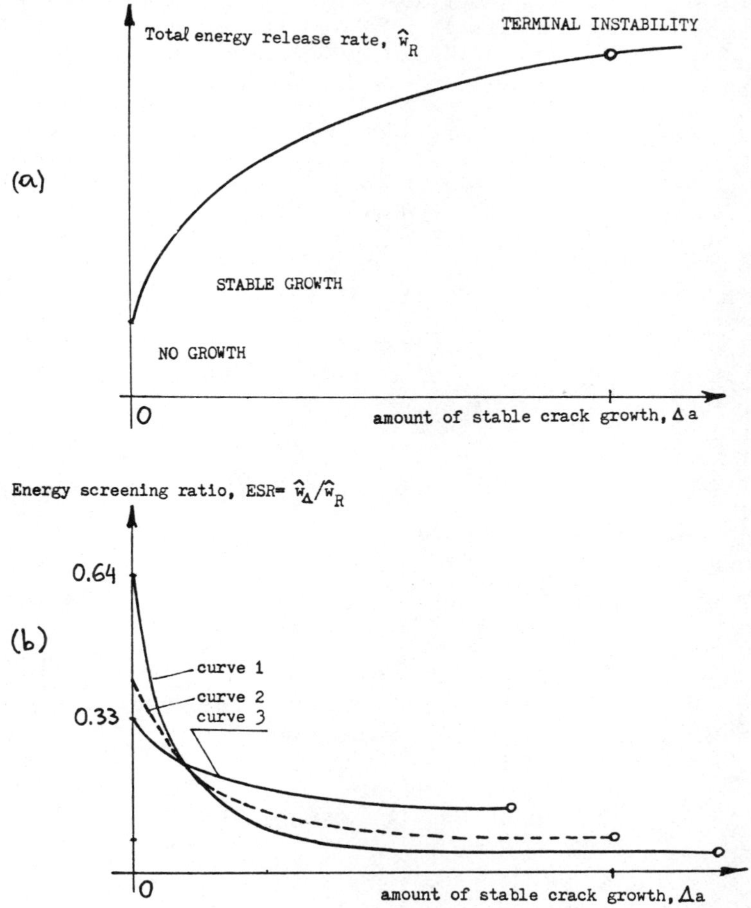

Fig. 3 (a) Total energy release rate in inelastic solid, $\hat{w}_R$, shown as a function of stable crack advance. This quantity serves as a measure of the apparent fracture toughness encountered in a ductile solid.

(b) Energy screening ratio, ESR= $\hat{w}_\Delta/\hat{w}_R$, shown as a function of stable crack advance. Curves 1 and 3 correspond to large and small scale yielding solutions, respectively, while curve 2 represents an exact solution valid for an arbitrary R/a ratio. Points of terminal instability are marked by circles.

Volume 3, ICM 3, Cambridge, England, August 1979

PREDICTION OF FAST FRACTURE IN Zr-2.5% Nb PRESSURE TUBES USING ELASTIC-PLASTIC FRACTURE MECHANICS

L. A. Simpson and B. J. S. Wilkins

Atomic Energy of Canada Limited, Whiteshell Nuclear Research Establishment, Pinawa, Manitoba R0E 1L0, Canada

INTRODUCTION

The CANDU[1] nuclear reactor system uses cold-worked Zr-2.5% Nb pressure tubes as primary coolant containment. At present, critical crack length data are obtained by burst-testing full size tube sections (Langford and Mooder, 1978; Wilkins, Barrie and Zink, 1978). Burst-testing programs are time consuming, costly and experimentally awkward, and the development of a technique for predicting critical crack length from a small-specimen test would allow the collection of many more data, and improve the statistical basis of failure predictions. It would also provide a quick and easy method of assessing unusual material or environmental conditions, which may arise in the future, and provide a framework for predicting critical conditions in new tube sizes, which might become relevant through changes in reactor design.

Because of the ductile condition of the pressure tube alloy under most circumstances, linear elastic fracture mechanics does not provide a suitable fracture criterion. Furthermore, Zr-2.5% Nb can accommodate considerable stable crack extension under rising load, hence, a fracture criterion based on initiation of crack extension (Simpson, 1977) is extremely conservative. Therefore, an R-curve approach has been selected, for relating small specimen behavior to the instability of pressure tube cracks.

In most reactors, the pressure tubes are 107 mm in diameter with a wall thickness of 4.1 mm. Because it is necessary to test actual tube material to obtain the relevant mechanical properties and microstructures, the size of the test specimen is limited because the criterion for in-plane dimensions, required for conventional R-curve testing[2], cannot be satisfied. Thus, in calculating the crack growth resistance, K_R, the linear elastic expresssion for stress intensity factor, cannot be used. In a previous study, Simpson and Clarke (1979) used elastic-plastic crack tip parameters to characterize the crack growth resistance. Most of the paper dealt with the crack opening displacement, δ, at the actual crack tip, which was converted to units of K through

$$K_R = m\,(\sigma_y \cdot E \cdot \delta)^{\frac{1}{2}} \qquad (1)$$

1 CANada Deuterium Uranium

2 ASTM Recommended Practice for R-Curve Determination (E561-75T)

where m is a constant shown to be close to unity. This approach followed, in principle, work by McCabe (1977) and Tanaka and Harrison (1978), although the experimental methods for measuring crack length and δ were different. The results, obtained on compact tension specimens, indicated that the R-curve shape was independent of crack length and specimen size. Predictions of the critical crack length, in a pressure tube at 300°C, were made by matching the R-curves to a crack-driving-force curve for an axially cracked pressure tube. The results were in reasonable agreement with some data from an earlier burst-testing program. However, some concern was expressed because the burst-testing data were obtained from irradiated material, and were also from a different shipment of tubing. Thus the material properties may have been different from those of the small, unirradiated specimens.

In this work, R-curves are determined on specimens cut from actual burst-tested tube sections. These R-curves are then used to predict the crack length of the tube at the experimentally determined burst pressure. This approach eliminates uncertainties due to material variations and provides a sound test of the R-curve method. The R-curves are based on measurements of δ, as in the earlier work, and also on J-integral determinations during stable crack extension. For the latter case, the method of Garwood, Robinson and Turner (1975) is employed, which provides an expression for calculating J at successive stages of stable crack extension from a single load, P, load point deflection, δ_p, curve

$$J_n = J_{n-1} \frac{W-a_n}{W-a_{n-1}} + \frac{2\,(U_n - U_{n-1})}{B\,(W-a_{n-1})} \tag{2}$$

where J_n = J at the nth point along the P-δ_p curve
W = specimen width
a_n = crack length corresponding to point, n
U_n = area under the P-δ_p curve up to point, n.

The initial value of $J = J_o$ is taken at or prior to the initiation of stable crack growth, and is calculated from the Rice, Paris and Merkle (1973) equation

$$J_o = \frac{2\,U_o}{B\,(W-a_o)} \,. \tag{3}$$

Equation 2 is based on the assumption that the crack-tip deformation field of a specimen, which has undergone an increment of stable crack extension to a_n, is the same as for a non-linear elastic specimen, of initial crack length a_n loaded to identical values of P and δ_p (with no crack extension). Since this is not proven, the true meaning of J_n is not clear although earlier measurements (Simpson and Clarke, 1979; Simpson, 1979) show a promising consistency between J_n and δ after as much as 4 mm of stable crack extension. Therefore, this approach will also be used here to develop J-integral-based R-curves ($J_R = J_n$) and to test their ability to predict critical crack length.

EXPERIMENTAL

Burst Testing

Tests were made on 460 mm long sections of tubing cut from two standard, unirradiated pressure tubes (Nos. 1866 and 1960) and sealed at each end. A through-wall axial slit was cut at the mid-length of each test section. The slits were cut under oil, using an electric discharge machine, and were typically 0.38 to 0.45 mm wide. Hydraulic pressure (oil) was maintained in the test section by a thin aluminum-alloy liner tube. Sandwiched between the liner tube and the pressure tube, underneath the slit, was a cold-rolled, stainless steel shim of 0.25 mm thickness. The slit was sharpened by pressure cycling and the tube was then pressurized until it burst. The measured burst pressure was corrected for the strengthening effect of the liner by assuming that the liner had attained general yield. Three sections each were tested at 20°C and 240°C, the latter temperature being the inlet temperature of the reactor coolant. Crack lengths of approximately 25, 50 and 75 mm were tested at each temperature.

R-curve Determination

Compact tension specimens (W = 34 mm, B = 3.8 mm, a/W = 0.45 or 0.60 mm) were machined from undeformed portions of the burst-test specimens after careful flattening. These specimens were then pre-cracked by fatigue.

Tensile specimens were also machined from each tube section and the yield stress, σ_y, and ultimate tensile stress, σ_u, were determined at the relevant temperature.

K_R was determined from measurements of δ and Eqn. 1, using the procedures described by Simpson and Clarke (1979). Briefly, the specimen was photographed at frequent stages during the test, while crack extension was monitored using the electrical-potential-drop (P.D.) method. At each loading stage, δ was determined by extrapolating a plot of crack face displacement (measured on the photographs) against distance from the original crack tip, back to the current position of the crack front. K_R was then calculated for each stage and plotted against Δa, the total change in crack length from the initial value. At 240°C a test furnace containing a window was used to facilitate the photography.

J_R was also calculated for the deeply cracked specimens. δ_p was determined from the photographs of crack face displacement and a P-δ_p curve was constructed. Initially U_n was measured using a planimeter but it was later found that these areas could be computed just as accurately by integrating the data numerically.

RESULTS

Burst Testing

The results of burst testing are summarized in Table 1, where the burst pressure is expressed in terms of the nominal hoop stress, σ_B, which that pressure would exert on an uncracked tube. Also tabulated are the measured values of σ_y and σ_u which are averaged to give a flow stress $\bar{\sigma}$. The crack, being double ended, has its length expressed as 2a to be consistent with the small specimen terminology. The remainder of the table is explained later.

The fracture surfaces of tubing burst at 20°C are shown in Fig. 1a. Fracture was unstable in all cases with very little bulging around the crack, except for the short crack, which showed modest bulging before failure. At 240°C, considerable bulging occurred for all crack lengths. Fig. 1b shows the 44 mm crack tested at 240°C (specimen 1960/2). This crack, and the shorter one, both became unstable at maximum pressure, but the liner failed almost instantly through the large crack opening, and the crack arrested. For the 74 mm crack, the opening was so large that the liner failed prior to instability.

TABLE 1

Material Properties and Burst-Test Results

SPEC. No.	TEST TEMP. °C	σ_B MPa	σ_y MPa	σ_u MPa	$\bar{\sigma}$ MPa	2a mm	M	σ_{COLL} MPa
1866/5	20	317	735	846	791	44.8	1.96	418
1866/15	20	242	811	917	864	74.4	2.89	298
1866/27	20	481	787	881	834	24.5	1.36	613
1960/2	240	289	521	595	558	43.6	1.91	423
1960/5	240	466	532	612	572	24.2	1.35	292
1960/7	240	206	485	607	546	73.5	2.87	190

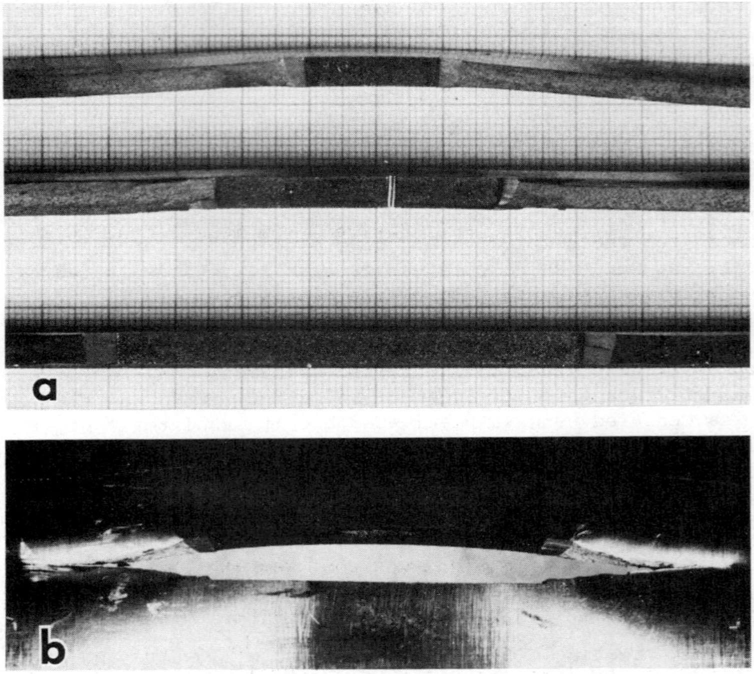

Fig. 1. Typical burst-test fractures. a: 20°C, b: 240°C

K_R Determinations

Typical R-curves for 20°C and 240°C are shown in Fig. 2. As noted by Simpson and Clarke (1979), the shape is independent of crack length, in spite of the small specimen size, and consequent proximity of the back surface to the crack tip. The curves for the other tube sections were very similar, and are not shown here. The only difference was a tendency to steeper R-curves and higher plateaus with decreasing $\bar{\sigma}$ at 20°C. This trend was not distinguishable above the scatter at 240°C, however, the range in $\bar{\sigma}$ was only half (percentage basis) that at 20°C.

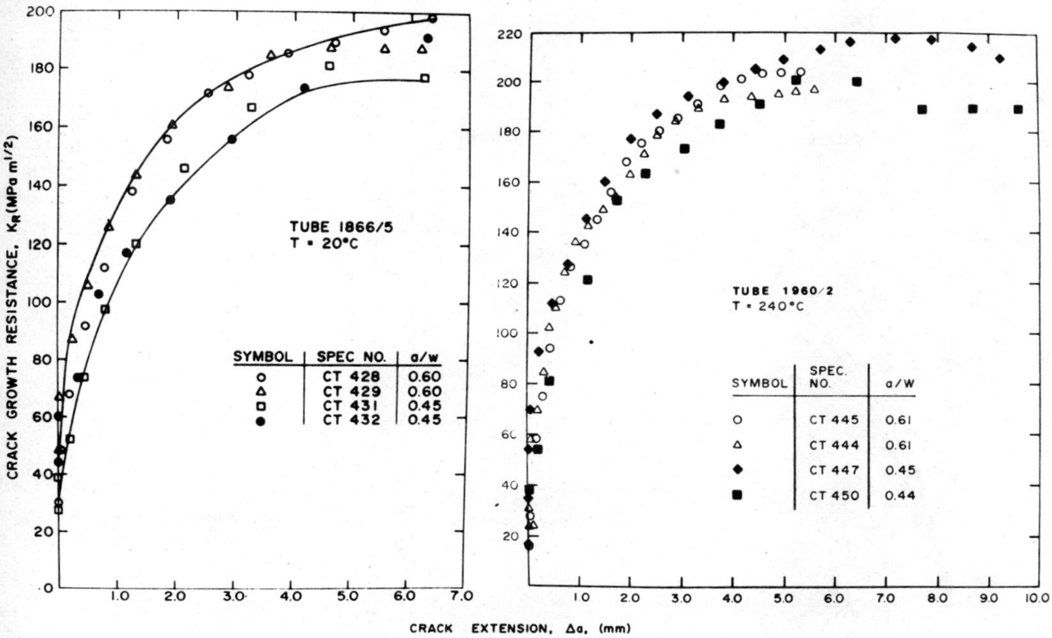

Fig. 2. Typical R-curves derived from measurements of δ

Initiation of crack extension, in all of the tests, occurred in the range K_R = 45 to 55 MPa $m^{1/2}$, independent of temperature to within the scatter of the measurements. These convert to values of δ_i (initiation C.O.D.) of 40 to 60 μm through Equation 1, which is consistent with Simpson's (1977) direct measurements of $\delta_i \sim 50$ μm at both 20°C and 300°C.

J_R Determination

The J_R curves are plotted in Fig. 3 for all the tubes. A significant effect of $\bar{\sigma}$ is evident at 20°C, much larger than was observed in the K_R determinations although, much of the difference is related to the fact that, dimensionally, $J_R \propto K_R^2$. At 240°C, the data for all these tubes fall on a single curve for up to 3 mm of crack extension. The initiation values of J_o = 25 kJ/m^2 at 20°C and 30 kJ/m^2 at 300°C convert to initiation stress intensity factors of ≃ 50 MPa $m^{1/2}$ through the relationship (Simpson, 1979),

$$K = (EJ)^{1/2} , \qquad (4)$$

which is consistent with the initiation values for K_R above.

Critical Crack Length Prediction

Determinations of the critical crack lengths, corresponding to the observed burst stresses in the tubes, were made by comparing the R-curves to crack-driving-force (C.D.F.) curves for the cracked tube. C.D.F. was calculated as a function of crack length using the expression (Kiefner and others, 1973)

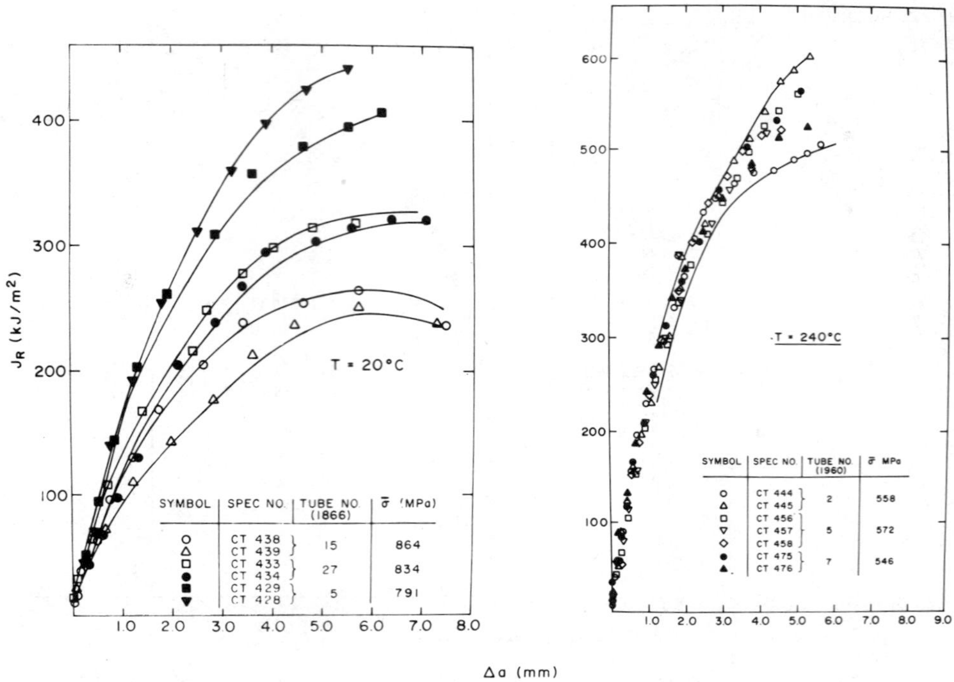

Fig. 3. Typical R-curves derived from measurements of J

$$K = \left[\frac{8a\bar{\sigma}^2}{\pi} \cdot \ln \left(\sec \frac{\pi M \sigma_H}{2\bar{\sigma}}\right) \right]^{\frac{1}{2}} \tag{5}$$

where σ_H = applied hoop stress in the tube

$$M = 1 + 1.255 \frac{a^2}{Rt} - 0.0135 \frac{a^4}{(Rt)^2} \tag{6}$$

where R = tube radius
t = tube wall thickness.

This is essentially a linear elastic expression with a plastic zone correction to the crack length and a factor M which reflects the additional bending forces at the crack tip due to the curvature of the tube. The C.D.F. curves for the J_R approach were obtained by substituting Eqn. 5 into Eqn. 4 and solving for J.

In Fig. 4, the C.D.F. curves for σ_H = actual burst value, are compared with the upper and lower bounds of the K_R data, to determine the critical crack lengths for the 20°C series. The crack length is read off the abscissa when the appropriate R-curve is just tangent to the C.D.F. curve. Hence, at 20°C the predicted crack lengths are 16-20 mm, 33-36 mm and 37-46 mm, which can be compared with the actual values (Table 1) of 25, 48 and 74 mm. Thus, the actual crack lengths are somewhat underestimated. These data, plus those at 240°C, and similar data using the J_R method, are summarized in Figs. 5 and 6 for 20°C and 240°C respectively. In addition, predictions of critical crack length at reactor operating pressure (120 MPa) are also included. In all cases, the J_R and K_R approaches

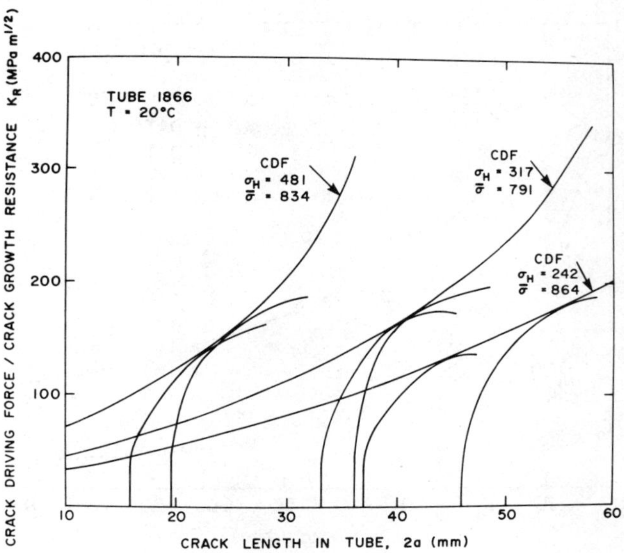

Fig. 4. Determination of critical crack lengths using K_R data for 20°C.

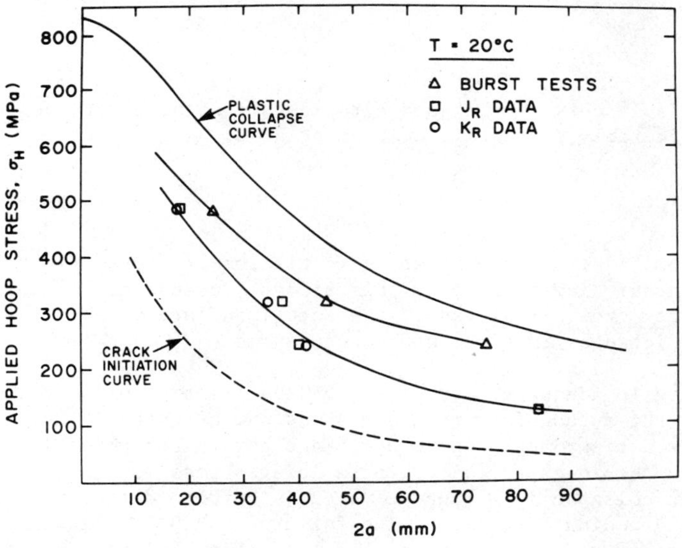

Fig. 5. Critical crack length predictions at 20°C using J_R and K_R.

give nearly identical predictions but consistently underestimate the critical crack lengths, especially for the 75 mm cracks.

The dashed lines in Figs. 5 and 6 are plots of failure stress against crack length if an initiation criterion for failure is assumed. They were calculated using Eqn. 5, and a K_i = 50 MPa m$^{1/2}$ to obtain σ_H. The uppermost solid line in each figure represents a plastic collapse criterion, or the situation when the secant term in Eqn. 5 becomes infinite. Solving for σ_H under these circumstances gives

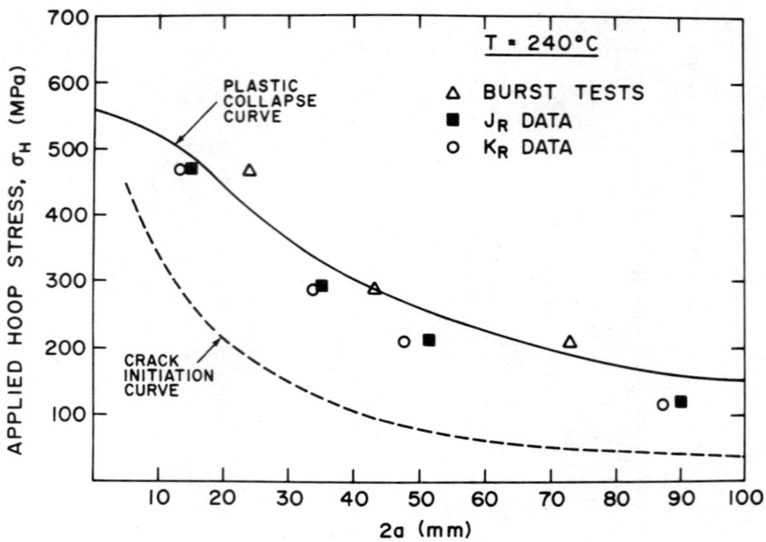

Fig. 6. Critical crack length predictions at 240°C using J_R and K_R.

$$\sigma_H = \sigma_{COLL} = \frac{\bar{\sigma}}{M} \quad . \tag{7}$$

The plastic collapse curve is in good agreement with the burst-test data at 240°C, but considerably overestimates them at 20°C.

DISCUSSION

It is clear, from Figs. 5 and 6, that an initiation-fracture criterion is wholly unsuitable for predicting failure of this alloy. Because of its ability to increase its crack growth resistance, the actual failure stresses are more than twice those predicted by initiation based analysis.

At 240°C, failure is adequately predicted by the plastic collapse curve (Eqn. 7), even though the actual failure process was one of unstable crack extension. However, this is a risky approach to use since any factor which lowers the crack growth resistance could cause failure at a lower stress (or crack length). The onset of plastic collapse is a concept which is hard to define precisely, especially in work-hardening materials. The failures at 20°C fall well below the plastic collapse curve, even though one of them (Fig. 1a) was accompanied by considerable bulging. While the R-curve approach has not predicted the failure condition precisely, it errs on the side of conservatism, yet, not to the degree that it becomes impractical or wasteful of the load-carrying capability of the structure.

The reasons for the discrepancies between R-curve predictions and the actual crack lengths, could involve the R-curve methods used here, the C.D.F. curves or both. The complete geometry independence of the R-curves determined here has yet to be proven. While they have been shown to be independent of crack length and size for the compact tension specimen in this and previous work (Simpson and Clarke, 1979), it will still be necessary to test other specimen configurations as well. A recent paper by Garwood (1978) reported a geometry dependence in a comparison of the three-point-bend and centre-cracked-tension (C.C.T.) specimens using J_R and δ-based approaches on pipeline steel. The C.C.T. specimen showed the steeper R-curves, which if it was also used for Zr 2.5% Nb, would result in higher crack length predictions. The C.C.T. specimen may provide a better simulation of the crack-tip deformation state in the tube because, unlike the compact tension specimen (with its large bending component), the far-field stress is uniform tension.

The precise significance of the K_R and J_R parameters measured here can also be questioned. While crack-opening displacement is conceptually an ideal way of defining a critical crack-tip strain for fracture, in practice, difficulties arise in obtaining accurate measurements and, in fact, deciding exactly where to take the measurement. The method used here yields an average through-thickness value of displacement at the point of the "average" crack front (the latter is usually curved for the first few mm of extension). This may well be the best approach since a precise measurement at any point in the specimen would neglect the effect of other areas behaving differently, and it is the total balance of forces and displacements that ultimately governs failure.

The meaning of J, after crack extension has occurred, is a widely debated subject. Even strong advocates of the J_R method (Garwood, Pratt and Turner, 1978) refer to it as a "J-like" term and report that it includes energy dissipated remote from the crack tip. This would preclude J_R from having the path-independent property of the usual J-integral.

In defense of the R-curves, it is significant that for all eight determinations of critical crack length, the K_R and J_R approaches yielded almost identical predictions, even though the methods used were quite different. This would suggest that the same quantity was being measured in both cases, and is a reasonable measure of crack growth resistance, at least for the compact tension geometry.

The other possible source of the discrepancy in crack length predictions is the C.D.F. curves. The expression used in Eqn. 5 is essentially a linear elastic one, containing a plastic zone correction. While such an approach can significantly extend the applicability of linear elastic equations (Irwin and Paris, 1977), obviously as the secant term becomes very large ($M\sigma_H = \bar{\sigma}$), the suitability of the expression is questionable. An alternative method of expressing C.D.F., by calculating J from the contour integral definition using finite element analysis, is under development.

A final concern about Eqn. 5 involves the curvature correction factor M. Bernard and Henry (1974) have shown that patching a flaw with a steel shim, in a burst test, has the effect of lowering the curvature-induced stress enhancement at the crack. They measured M experimentally by comparing fatigue crack growth rates in a tube with standard da/dN - ΔK data taken from flat specimens. M measured this way was significantly lower (up to 50%) than the theoretical value given by Eqn. 6. Similar attempts in this laboratory to confirm this behavior for Zr-2.5% Nb have been inconclusive in quantitative terms although, indications are that some reduction in M is occurring. Reduction of the C.D.F. curves by reducing M would result in larger critical crack length predictions. The fact that the largest errors in crack length prediction occur for the largest M values (Table 1) supports the above concern.

SUMMARY

Most of the recent studies of elastic-plastic R-curve methods fall short of applying the proposed method to an actual engineering structure. This paper provides a stiff test of the ability of both δ-based and J-integral-based R-curve methods to predict the failure condition for a pressure tube. Both methods provide critical crack length predictions which err conservatively but are much more generous than initiation-based criteria. A number of sources for these errors have been suggested and their further investigation is both warranted and recommended. These include geometry dependence and the validity of the crack driving force expressions.

REFERENCES

Bernard, J. and B. Henry (1974). Experimental study on the effect of curvature on the stress intensity factor in cylindrical shells. Euratom Report No. EUR 5137f, Ispra, Italy.

Garwood, S.J. (1975). The effect of specimen geometry on crack growth resistance. Proceedings of the 11th National Symposium on Fracture Mechanics. Blacksbury, Va. to be published in an ASTM special technical publication.

Garwood, S.J., P.L. Pratt and C.E. Turner (1978). Measurements of slow stable crack growth in structural steels and prediction of unstable fracture. OECD/NEA Specialists Meeting on Elasto-Plastic Fracture Mechanics. UKAEA, Daresbury, U.K.

Garwood, S.J., J.N. Robinson and C.E. Turner (1975). The measurement of crack growth resistance curves (R-curves) using the J-integral. Int. Journ. of Fracture, 11, 528-530.

Irwin, G.R. and P.C. Paris (1977). Elastic-plastic crack tip characterization in relation to R-curves. In D.M.R. Taplin (Ed.), Fracture 1977, Vol. 1, University of Waterloo Press, Waterloo, Canada. pp. 93-100.

Kiefner, J.F., W.A. Maxey, R.J. Eiber and A.R. Duffy (1973). Failure stress levels of flows in pressurized cylinders. In J.G. Kaufman (Ed.), Progress in Flow Growth and Fracture Toughness Testing, ASTM STP 536, American Society for Testing and Materials, Philadelphia. pp. 461-481.

Langford, W.J. and L.E.J. Mooder (1978). Fracture behavior of zirconium alloy pressure tubes for Canadian nuclear power reactors. Int. J. Pres. Ves. and Piping, 6, 275-310.

McCabe, D.E. (1978). Determination of R-curves for structural materials using non-linear mechanics methods. In J.M. Barson (Ed.), Flaw Growth and Fracture, ASTM STP 631, American Society for Testing and Materials, Philadelphia. pp. 245-266.

Rice, J.R., P.C. Paris and J.G. Merkle (1973). Some further results of J-integral analysis and estimates. In J.G. Kaufman (Ed.), Flaw Growth and Fracture Toughness Testing, ASTM STP 536, American Society for Testing and Materials, Philadelphia. pp. 231-245.

Simpson, L.A. (1977). Initiation COD as a fracture criterion for Zr-2.5% Nb pressure tube alloy. In D.M.R. Taplin (Ed.), Fracture 1977, Vol. 3, University of Waterloo Press, Waterloo, Canada. pp. 705-711.

Simpson, L.A. (1979). The relationships between stress intensity factor, crack opening displacement and J-integral in Zr-2.5% Nb. Submitted to Trans. A.S.M.E., J. Eng. Mater. Tech.

Simpson, L.A. and C.F. Clarke (1979). An elastic-plastic R-curve description of fracture in Zr-2.5% Nb pressure tube alloy. Symposium on Elastic-Plastic Fracture, Atlanta 1977, appearing early 1979 in ASTM STP 668 American Society for testing Materials, Philadelphia.

Tanaka, K. and J.D. Harrison (1978). An R-curve approach to COD and J for an austenitic steel. Int. J. Pres. Ves. and Piping, 6, 177-207.

Wilkins, B.J.S., J.N. Barrie and R. Zink (1978). Slit burst testing of cold-worked Zr-2.5% Nb pressure tubing for CANDU-PHW reactors. Atomic Energy of Canada Limited Report, AECL-6195.

Volume 3, ICM 3, Cambridge, England, August 1979

DETERMINATION OF FRACTURE TOUGHNESS OF PLASTIC MATERIALS UNDER CONDITIONS OF SIMILARITY OF PLASTIC INSTABILITY

V. Ivanova, V. Kobzov and L. Maslov

Laboratory of the Strength of Metallic Materials, Baikov Institute of Metallurgy Academy of Science, Moscow, U.S.S.R.

INTRODUCTION

When determining the fracture toughness of constructional steels of medium strength certain difficulties arise, since specimens of large size are needed to assure the condition of plane strain at the crack tip and special instruments are required to test such specimens. However, to extend the range of constructional steels of medium strength, which are most widely used in industry, one has to know with certainty their fracture toughness (K_{IC}) determined on small specimens. It is accepted that a reliable estimate of the fracture toughness for small specimens is possible, in accordance with linear fracture mechanics, only for specimens of high strength. The difficulties associated with the determination of fracture toughness for medium-strength steels stimulated the development of methods of estimation for specimens which do not meet the requirements of linear fracture mechanics. The basis of most of these methods also lies in concepts of linear fracture mechanics (for example, the J-integral method). Thus, in the literature there are many estimates of fracture toughness data for plastic materials by different methods which frequently are not reliable enough because of their incompatibility. This is due to the fact that in the case of plastic metals the fracture toughness can be determined from parameters corresponding to either the conditions of plastic instability or fracture instability. It is necessary to determine fracture toughness under locally similar fracture conditions. The following criteria could be used as similarity criteria of subcritical crack growth:

1. The μ_L - criterion of local stress similarity ahead of a critical crack length, that is, $\mu_L = K_{I*}^{N-1}/K_{I*}^{N}$. Here K_{I*}^{N-1} is the threshold stress intensity factor above which subcritical crack growth occurs; K_{I*}^{N} is the plain strain K_I threshold above which plastic instability occurs.
2. The M - criterion of the geometric similarity of critical crack lengths, that is $M = \ell_{N-1}/\ell_N$. Here ℓ_{N-1} and ℓ_N are critical crack lengths corresponding to K_{I*}^{N-1} and K_{I*}^{N} respectively.
3. The R - criterion of plastic instability, that is $R = \sqrt{\ell_c/2r_c}$. Here ℓ_c is the plastic instability zone size; $2r_c$ is the critical plastic zone size above which plastic instability occurs. The plastic instability process ahead of a critical crack is an automodel process. Therefore only one criterion of similarity is required to calculate the K_{IC} value if subcritical crack growth occurs. The fracture instability, as is known (Irwin, 1972) is controlled by the critical value of the stress intensity factor K. Recent work (Ivanova, 1977) indicates that for plastic materials characterised by a plastic-brittle transition, it is necessary to estimate a plane strain K_I threshold above which a macroplastic zone starts

(denote this value as K_{IC}^{min}) and the threshold above which crack initiation occurs at the plastic instability zone (denote this value as K_{IC}^{max}).

The difference $A = K_{IC}^{max} - K_{IC}^{min}$ characterises the reserve of static fracture toughness at the subcritical crack growth condition. Irwin (1972) discovered that the size of the plastic deformation zone at the crack tip depends on the stress intensity factor K according to the following equation:

$$2r_y = \frac{1}{\pi}\left[\frac{K}{\sigma_r}\right]^2 \qquad (1)$$

In the case of plane stress σ_T is assumed to be equal to the yield stress under uniaxial tension (σ_y). The plastic zone in this case will be a circle with the front edge of the crack at its centre, so $2r_y$ is the diameter of this circle. As a result of the elastic constraint, in the case of plain strain, the actual yield stress in the plastic deformation zone is higher than σ_y and so when determining the critical size of the plastic deformation zone for given K one has to take into account the yield stress $\sigma_T.= \sqrt{3}.\sigma_y$. For the cases when subcritical crack growth is observed before final fracture, $\sigma_T = \sigma_y.\beta$. Here the coefficient β is a function of the plastic constraint factor and is expressed as (Ivanova, 1972):

$$\beta = K_{IC}/K_{I*}^{R} \qquad (2)$$

K_{I*}^{R} is the threshold intensity factor for which nucleation of a plastic deformation zone of critical size $2r_c$ controls crack initiation. Ivanova (1977) found that the critical size of the plastic zone under elastic-plastic stress conditions is a function of the plastic constraint factor and is expressed as

$$2r_c = \left[\frac{K_{I*}^{R}}{\sigma_y}\right]^2 \frac{1}{\pi} \qquad (3)$$

For iron alloy $K_{I*}^{R} = 40.5\ MNm^{-3/2}$ (Ivanova, Maslov, 1976) if $\ell_N/\ell_{N-1} = 2$. Recent results (Ivanova, 1977) indicate that at constant plastic constraint the factor

$$2r_c/\ell_c = C^2/(1-C^2) \qquad (4)$$

where $C = \tau_0/\sigma_I$ and τ_0 is the octahedral stress ahead of a crack. If we write $\sqrt{2r_c/\ell_c} = \frac{1}{R}$, then

$$R = \sqrt{\ell_c/2r_c} = \frac{1}{C}\sqrt{1-C^2} \qquad (5)$$

The R - ratio is a criterion of similarity of plastic instability ahead of a crack during subcritical crack growth. When the macroplastic instability arises ahead of a crack the stress intensity factor K increases from K_Q^{N-1} to K_Q^N up to start of cracking. Here K_Q^{N-1} is the threshold value of K corresponding to the moment of macroplastic zone starting and K_Q^N is the threshold value of K_Q corresponding to the crack start. As was shown previously (Ivanova, 1977), the maximum size of the macroplastic instability zone $\ell_c = \ell_c^{max}$, above which the transition from constrained plastic flow to unconstrained occurs, corresponds to the stress state characterised as $\tau_0/\sigma_I = \tau_s/\sigma_y = 0.575$ (for materials obeying the Huber-Mises theory). Here τ_s is the shear yield stress. Thus the attainment of unconstrained plastic flow is characterised by the value $R = R_{max} = 1.4$ and $K_Q = K_{IC}^{max}$. On the other hand, the transition to plane strain is governed by the condition $R = 1$; so the value of K_Q corresponds to K_{IC}. The values of K_{IC}^{min} are determined from the relation (Ivanova, 1977):

$$K_{IC}^{min} = K_{IC}^{max}.\ \Delta^{1/4} \qquad (6)$$

where Δ = 0.11 for iron alloys, Δ = 0.22 for titanium alloys, and Δ = 0.22 for aluminium alloys (Ivanova, Terentyev, 1975). The purpose of these experiments is to show that the R-criterion is a necessary and sufficient criterion for similarity of plastic instability.

EXPERIMENTAL AND DISCUSSION

The materials studied were: A - austenitic Cr-Mn-Steel (σ_y = 618 MPa); B - 0.4% C-Steel (σ_y = 383 MPa); C - austenitic 18Cr-10Ni-Steel (σ_y = 422 MPa); D - austenitic Cr-Mn-N-Steel (σ_y = 491 MPa); E - welded joint of steel D with a notch at the middle of the weld zone; F - welded joint of steel D with notch near the weld zone. The fracture toughness was investigated on 0.015 m wide WOL-specimens. The method of investigation was as follows. First, a fatigue precrack 0.003 m long was grown in the WOL-specimens, then by static preloading with an "Instron" TT-CM-L machine up to $P = P_k$ a plastic deformation zone ℓ_c was grown. For the first specimen the plastic zone was equal to $(\ell_c)_1$, for the second to $(\ell_c)_2$ and so on. From 7 to 10 specimens of each material were tested by this technique. After that, another fatigue crack was grown in each of the specimens to establish the position of the plastic deformation zone, ℓ_c, occurring during static preloading, and then the specimens were fractured by static tension. As a result of such a loading the fracture surface of the tested specimen is of the form schematically shown in fig. 1.

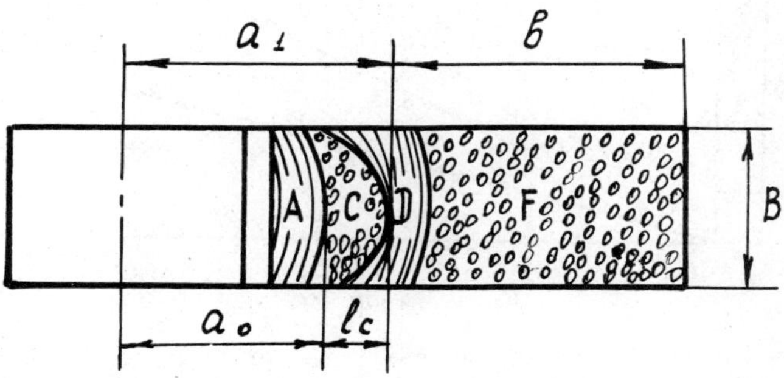

Fig. 1 Sketch of specimen fracture surface.

In Fig. 1 A is the position of the fatigue precrack; C is the macroplastic unstable zone; D is the secondary fatigue crack; F is the final zone of fracture. From the value of the load P_k corresponding to the end of the preloading the value of K may be calculated using the following relation of linear fracture mechanics:

$$K_Q = \frac{P_k\sqrt{a_1}}{Bb}\left[29.6 - 185.5\frac{a_1}{b} + 655.7\left(\frac{a_1}{b}\right)^2 - 1017\left(\frac{a_1}{b}\right)^3 + 638.9\left(\frac{a_1}{b}\right)^4\right] \quad (7)$$

Here a is the length of the crack (in accordance with Fig. 1). Then the value of $2r_c$ is calculated from relation (3). The path-independent integral characterising the elastic-plastic deformation field intensity at the crack tip was determined also. The experimental method of determining the parameter J_{IC}, in accordance with Landes and Begley (1974) involves plotting the dependence of the J-integral on the increment of crack length when testing the specimens up to different values of a. The critical value $J = J_{IC}$ is determined by extrapolation of this dependence to a value corresponding to the moment when the crack starts to move in the stretching zone. In this case, the value of the J-integral is calculated from the relation

$$J = 2E/(B.b) \tag{8}$$

where E is the plastic work for the nucleation of the crack and its growth up to Δa; B is the specimen thickness, and b is the width of the working section as realized under the fracture instability conditions. For the elastic-plastic fracture conditions this method allows one to estimate J_C. The method of determining J_C can be significantly simplified if the acoustic emission method is used for establishing the crack start. In this case J_C is calculated by using the following equation:

$$J_C = \frac{2E_*}{B(W-a_1)} \tag{9}$$

Here E_* is the crack start energy; B, W are the thickness and width of the specimen respectively, and a_1 is the crack length (Fig. 1). The experimental data show that there is a linear dependency between K_Q values and the R-criterion (Fig. 2). The value of K_{IC} equal to K_Q is obtained for R = 1, and $K_Q = K_{IC}^{max}$ for R = 1.4.

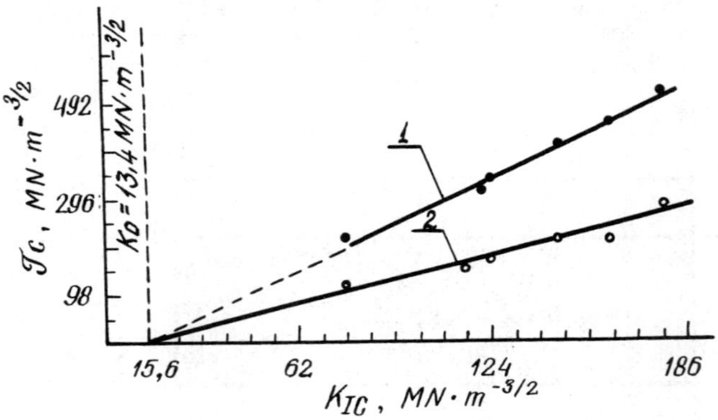

Fig. 2 Threshold stress intensity factor (K_Q) versus R-criterion (see Table 1).

Table 1 presents data for the fracture toughness corresponding to K_{IC}^{min}, K_{IC}^{max} (R = 1.4), K_{IC} (R = 1), J_C and $\overline{J_C}$. Here $\overline{J_C}$ is the average J_C value determined by using the acoustic emission method. Comparison of the parameters K_{IC}, J_C and J_C exhibited a good correlation (Fig. 3), the tangent of the angle of slope (θ) of the straight line $K_{IC} - J_C$ being dependent on the constraint of the plastic deformation at the crack tip. In Fig. 3 J_C on line 1 was determined by Landes and Begley's (1974) method, and on line 2 by using the acoustic emission method. From Fig. 3 it follows that the analytical relation between K_{IC} and J_C can be expressed as follows:

$$K_{IC} = K_0 + \theta\ J_C \tag{10}$$

where K = 13.5 $MNm^{-3/2}$ (for iron alloys). Here θ is a coefficient dependent on the stress state at the crack tip

Recent results (Shur, Bishutin, Gelmiza, 1978) indicate that critical values of the J-integral strongly depend on the bulk extension of the stresses at the crack tip which changes when the sample thickness is varied. This result showed that $K_{IC}^{min} \leq K_{IC} \leq K_{IC}^{max}$.

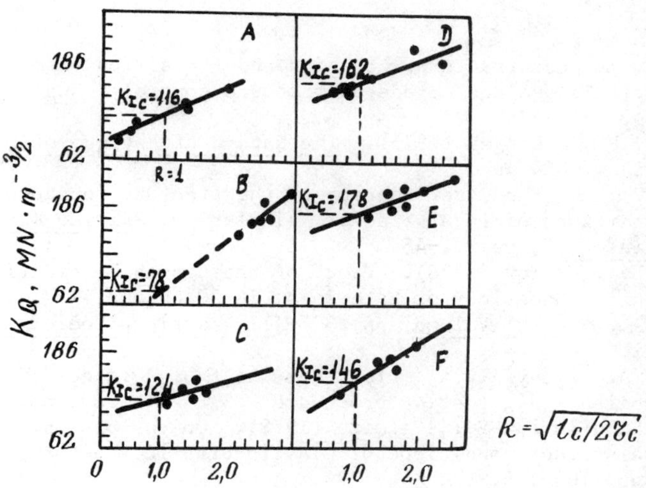

Fig. 3 K_{IC} versus J_C for different plastic constaint factors

Table 1. The Fracture Toughness Values (K_{IC} and J_C) obtained under the conditions of plastic instability similarity

Material	MPa	$2r_c$ 10^{-3} m	K_{IC}^{min}	K_{IC}^{max} $MNm^{-3/2}$	K_{IC}	J_C KNm^{-1}	$\overline{J_C}$
A Cr-Mn-St	618	1.35	116	132	76.5	314	156
B 0.4% C	383	3.54	78	107	58.5	216	117
C 18Cr-10Ni-St	422	2.93	124	133	77	334	145
D Cr-Mn-N-St	491	2.17	162	175	102	441	213
E welded joint	638	1.28	178	189	109	500	277
F welded joint	491	2.17	146	166	96	392	220

CONCLUSION

The analysis presented confirms the conclusion drawn previous (Ivanova, 1977) that, in the case of subcritical crack growth, to determine fracture toughness the similarity condition of local stress state is necessary. This result showed that the R-criterion is a necessary and sufficient criterion of similarity of plastic instability. It allows one to estimate K_{IC}^{min}, K_{IC}^{max} (R = 1.4) and K_{IC} (R = 1). The difference $A = E_{IC}^{max} - K_{IC}^{min}$ characterises the reserve of the static fracture toughness at the subcritical crack growth condition. The critical value of K_{IC} lies in the range $K_{IC}^{min} < K_{IC} < K_{IC}^{max}$. If subcritical crack growth is absent then $K_{IC}^{min} = K_{IC} = K_{IC}^{max}$. A correlation exists between the criterion of K_{IC} and the J_C-integral.

REFERENCES

Irwin, G.R. (1968). Linear fracture mechanics and transition from a toughness fracture to a brittle one and methds of fracture control. Engineering Fracture Mechanics, 1, pp. 241-257.

Ivanova, V.S. and V.P. Terentjev (1975). The Nature of Fatigue of Metals. Metallurgia, Moscow, 455 pp.

Ivanova, V.S. (1977). On the determination of the fracture toughness of metals and alloys under conditions of local fracture similarity. Fisiko-Khimicheskava mekhanika materialov, S, pp. 31-45.

Ivanova, V.S. and L.I. Maslov (1978). Proc. of the Second International Conference on Mechanical Behaviour of Materials, 16-20 August, 1975. Boston, Massachusetts, U.S.A., Special Volume, pp. 377-411, American Society for Metals, Metals Park, Ohio.

Landes, J.D. and J.A. Begley (1974). Test results from J-integral studies: an attempt to establish a J_{IC} testing procedure. STP ASTM 560, pp. 170-186.

Shur, D.M., G.A. Rishutin and V.I. Gelmiza (1978). On the possibility of determining J-integrals under conditions of elastic-plastic fracture. Problemi prochnosti, 8, pp. 104-108.

Volume 3, ICM 3, Cambridge, England, August 1979

J-INTEGRAL ELASTIC-PLASTIC FRACTURE MECHANICS TECHNOLOGY IN THE U.S. NAVY

J. P. Gudas, J. A. Joyce* and H. H. Vanderveldt**

David W. Taylor Naval Ship and Research and Development Center, Annapolis, Maryland, U.S.A.
***United States Naval Academy, Annapolis, Maryland, U.S.A.**
****Naval Sea Systems Command, Washington, D.C., U.S.A.**

INTRODUCTION

The United States Navy has historically been interested in the development of fracture safe materials for ship construction, and in developing fracture mechanics criteria for design considerations. Recently, a substantial research effort has been directed to the development of J-Integral technology. The purpose of this paper is to review recent Navy advances in this area. The main points will include a discussion of the applicability of the J-Integral to fracture in ductile materials, the development of test procedures for J_{Ic}, recent results in the areas of J-controlled crack growth, dynamic J-Integral properties, and instability criteria based on the J_I versus crack growth resistance curve.

J-INTEGRAL ELASTIC-PLASTIC FRACTURE TOUGHNESS

The application of the J-Integral as an elastic-plastic fracture toughness criterion is based on the non-linear elastic Hutchinson (1968) and Rice and Rosengren (1968) singularity solution which gives the stress and strain components near a sharp crack in a material which behaves in accordance with incompressible deformation theory plasticity where the unaxial stress-strain relation is given by:

$$\sigma = \bar{\sigma}_1 \, (\varepsilon_p)^N \tag{1}$$

such that:

$$\frac{\sigma_{ij}}{\bar{\sigma}_1} = \left(\frac{J_I}{\bar{\sigma}_1 \, I_N r}\right)^{\frac{N}{N+1}} \tilde{\sigma}_{ij}(\theta)$$

$$\varepsilon_{ij} = \left(\frac{J_I}{\bar{\sigma}_1 \, I_N r}\right)^{\frac{1}{N+1}} \tilde{\varepsilon}_{ij}(\theta) \tag{2}$$

where $\tilde{\sigma}_{ij}(\theta)$, $\tilde{\varepsilon}_{ij}(\theta)$ and I_N are functions of θ and N, and r and θ define a polar coordinate system about the crack tip. Only J_I in Equation 2 depends on the applied boundary conditions or specimen geometry and thus J_I sets the intensity of the stress and strain singularity in a manner completely analogous to the stress intensity factor K_I in the Williams (1957), Irwin (1957) elastic solution. Begley and Landes (1972) proposed that a critical value of J_I exists (for each material) such that crack extension occurs when $J_I > J_{Ic}$. For the special case that N = 1 in Equation 1, Equation 2 reduces to the Irwin-Williams elastic solution, and for this case Rice (1968) has demonstrated that:

$$J_I = \frac{(1-\nu^2)\ K_I^2}{E} \tag{3}$$

Rice (1968) and Merkle and Corten (1974) have shown that for compact specimens, J_I can be evaluated approximately by:

$$J_I = \frac{\beta A}{Bb} \tag{4}$$

where A is the area under the load-point displacement curve, B is the specimen thickness, b is the uncracked ligament, and β is a multiplicative factor, approximately equal to 2, which includes effects of axial force and bending.

U.S. Navy interest in the J-Integral was first directed to the development of test procedures allowing the routine evaluation of J_{Ic}. The direction of research was then changed to consider J-controlled crack growth and the use of J_I versus crack growth data as a possible instability criterion.

J_{Ic} TEST METHODS FOR MATERIALS CHARACTERIZATION

The first evaluation of Navy structural alloys (Gudas and Joyce (1978)) utilized the Landes-Begley (1973) multi-specimen procedure. This method employs similarly cracked compact or bend specimens which are loaded to various crack opening displacement levels. Single J_I and crack extension measurements derived from each specimen are used to construct a plot to extrapolate back to the point of crack initiation for J_{Ic}. Although this procedure has been and continues to be applied successfully, the development of a single specimen, routine test procedure for generating J_{Ic} was seen as necessary for general use of the J-Integral criterion.

The U.S. Navy development of the computer-interactive unloading compliance test procedure was a modification of the unloading compliance technique of Clarke and coworkers (1976). A schematic of the test arrangement is shown in figure 1 (Joyce and Gudas (1979)).

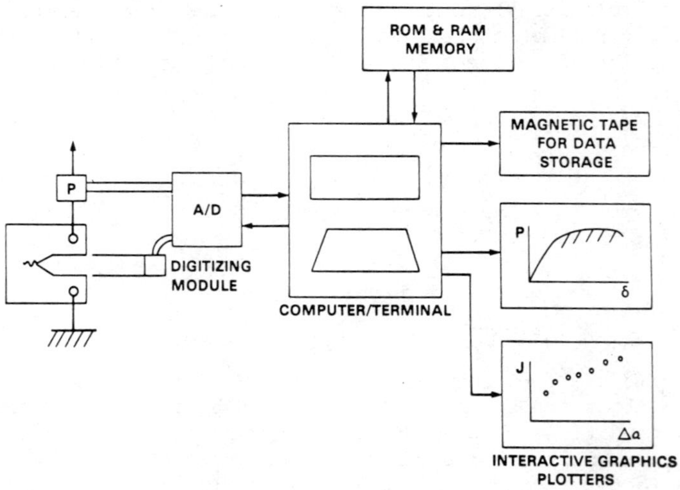

Fig. 1. Schematic of computer interactive J_{Ic} test arrangement.

In this setup, load cell and clip gauge signals are conventionally amplified, digitized and made available to a microprocessor with magnetic tape cartridge and CRT graphics capability. A single specimen J_{Ic} test is accomplished by carrying out a series of approximately 10% unloadings during a normal fracture mechanics type test. A typical load versus displacement plot is presented in figure 2. At each unloading, estimated crack length, change in crack length, J_I and a statistical analysis of the crack length estimate are determined and presented on the CRT screen. This method develops equivalent J_{Ic} values when compared to the multiple specimen method (Gudas and Joyce (1978)). Beside producing J_{Ic} from a single specimen, this technique provides for storage of digitized load-displacement data and lends itself readily to reanalysis of test results. Further, the interactive nature of the testing allows for accurate placement of data points, particularly after crack growth initiates.

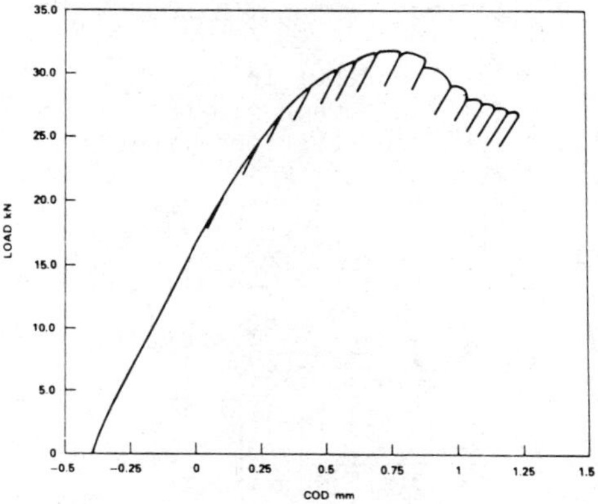

Fig. 2. Load versus displacement plot for unloading compliance J_{Ic} test.

This procedure has been employed in evaluating over fifty Navy structural alloys, weldments and product forms. It has also been employed in evaluating graphite/aluminum composites and in addressing other mechanics related problems. Figure 3 presents results of an investigation of notch tip geometry effects in HY-130 steel (Joyce and Gudas (1979)). This figure shows typical J_I versus crack growth data obtained with computer interactive unloading compliance test procedure. It also shows that the apparent J_{Ic} of HY-130 steel was elevated 300% by machined notches while dJ_I/da was insensitive to notch tip geometry.

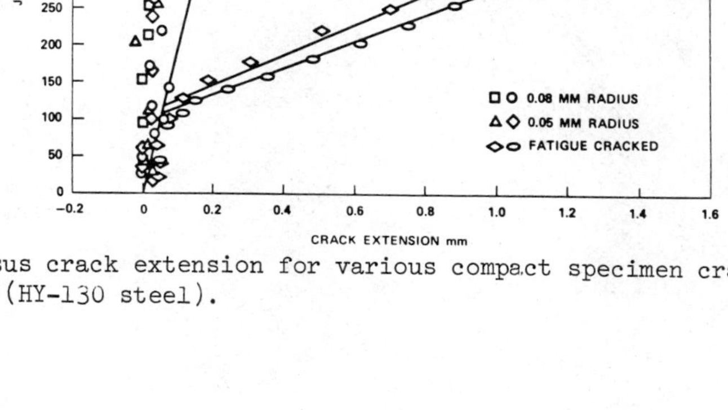

Fig. 3. J_I versus crack extension for various compact specimen crack tip geomitries (HY-130 steel).

J CONTROLLED CRACK GROWTH

Recent work by Hutchinson and Paris (1979) has shown that the J singularity is applicable to the analysis of crack growth if the following conditions are satisfied:

$$\omega = \frac{b}{J}\frac{dJ}{da} >> 1; \quad \frac{b\sigma_o}{J} >> 1 \tag{5}$$

The first condition is a requirement for applicability of deformation theory of plasticity, the second is a specimen size requirement. One objective of this work is to utilize rationally the large amount of toughness available in present engineering materials beyond the point of crack initiation. The J_I-R curve then, becomes more than a tool to define J_I, and takes on value of its own as a method to index materials resistance to crack extension beyond crack initiation.

The J_I-R curve, however, to be truly useful in the regime of crack growth, must be corrected to include the effects of crack growth. A new analysis by Ernst and coworkers (1979) including these corrections gives J and Δa as the integral of the following differential expressions:

$$dJ = \left[\frac{2b}{W}F_1 - \frac{b^2}{W^2}\frac{\partial F_1}{\partial(a/W)}\right] d\Delta + \left[-\int_o^{\Delta}\frac{2}{W}F_1\,d\Delta + \int_o^{\Delta}\frac{4b}{W^2}\frac{\partial F_1}{\partial(a/W)}\,d\Delta - \int_o^{\Delta}\frac{b^2}{W^3}\frac{\partial^2 F_1}{\partial(a/W)^2}\,d\Delta\right] da \tag{6}$$

and

$$da = \frac{\frac{b^2}{W^2}\frac{\partial F_1}{\partial(\Delta/W)}\,d\Delta - dP}{\frac{2b}{W}F_1 - \frac{b^2}{W^2}\frac{\partial F_1}{\partial(\Delta/W)}} \tag{7}$$

where:

$$\frac{PW}{Bb^2} = F_1\left(\frac{\Delta}{W}, \frac{a}{W}, \frac{H}{W}, \frac{B}{W}, \text{Mat'l Properties}\right) \tag{8}$$

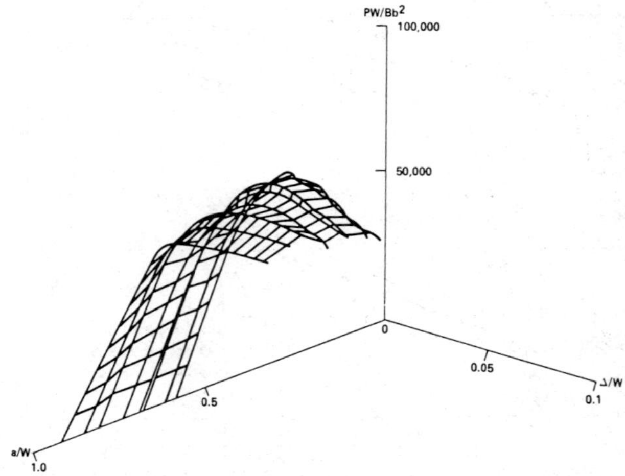

Fig. 4. Key curve function for HY-130 steel

The key curve function, F_1, is a non-dimensional load displacement relationship for the material and specimen geometry being tested. Such an experimentally obtained calibration function for an HY-130 steel is shown in figure 4. This "function" is an assemblage of load displacement records from eight 1/2T compact specimens taken up to the point where crack initiation begins. The key curve function of figure 4 is then used in combination with the load displacement records of geometrically similar but larger compact specimens of the same material to determine the J_I-R resistance curve without the need for specimen unloadings or other crack length measurements. All corrections for crack extension, tensile loading components (Merkle-Corten correction (1974)), and rotational effects are included in equations 6 and 7. Typical results for 1TCT compact specimens are shown in figures 5 and 6. The data points shown are results of unloading compliance testing on these same specimens.

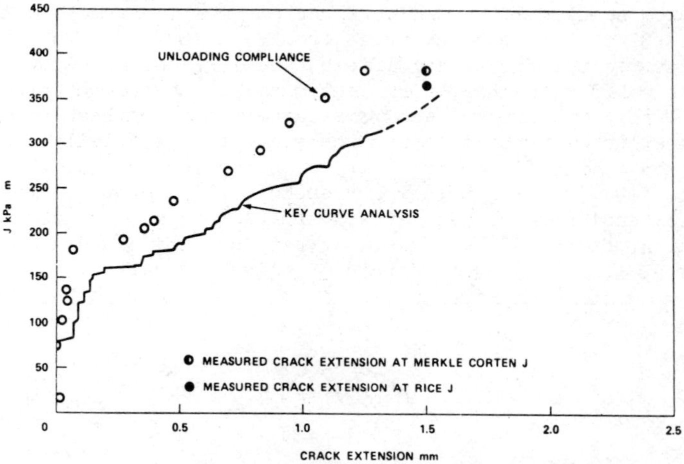

Fig. 5. Comparison of J_I-R curves from unloading compliance and key curve analyses for non-face grooved HY-130 compact specimens.

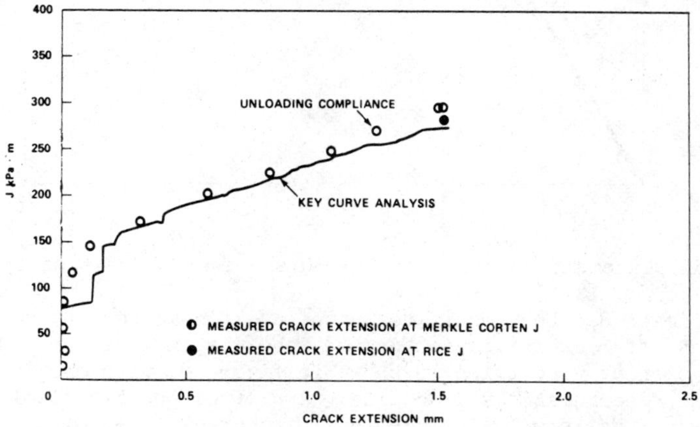

Fig. 6. Comparison of J_I-R curves from unloading compliance and key curve analyses for 12% face grooved HY-130 compact specimens.

The major conclusions resulting from this analysis are that first, the J_{Ic} values found by both methods are similar. Second, the key curve analysis accurately predicts the final measured crack lengths corresponding to the solid points in figures 5 and 6. Third, the tensile J component in these specimens is considerably smaller than that which is predicted by Merkle and Corten (1974). Fourth, the slopes of the J_I-R curves beyond J_{Ic} depend on the crack length in compact specimens, and finally, the specimens face grooved to eliminate crack tunnelling, give the most accurate results when a single specimen unloading compliance method is used.

DYNAMIC J_I TESTING

The key curve J_I-R curve evaluation technique described in the previous section was developed specifically to allow evaluation of J_I-R curves from dynamic tests. This new method eliminates the need for unloadings or other techniques to obtain crack length estimates at various values of COD and load. A feasibility study has been conducted to determine if it is possible to obtain J_I-R curves at higher rates from compact specimens. Tests were run on non-face grooved 1T compact specimens of HY-130 at test machine speds from 2. x 10^{-5}m/sec to 8.5 X 10^{-4}m/sec. Load displacement records for three different test speeds are shown in figure 7. The slowest test was done using the unloading compliance technique of Joyce and Gudas (1979). When the key curve analysis was applied to these load displacement records, good results were obtained for the slowest, essentially static, test with a close prediction of the final crack extension and J_{RICE} value where $\beta = 2$ in equation 4 (Rice (1968)). For the higher speed tests, however, gross underestimates of crack extension resulted. It was apparent from these results that the key curve function, for HY-130 at least, is rate sensitive and a function corrected for test rate would be required to obtain dynamic J_I-R curves.

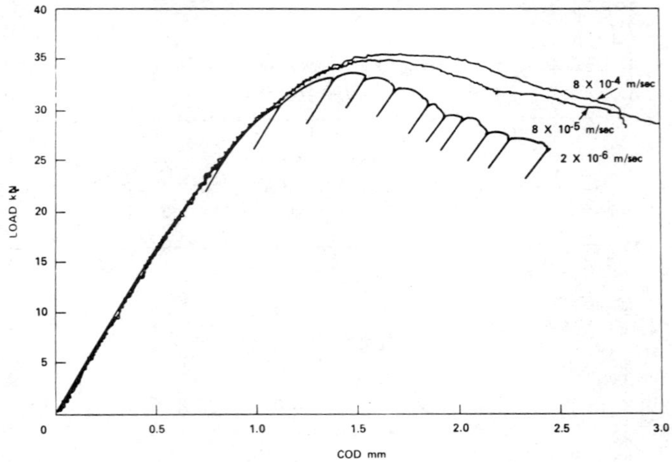

Fig. 7. Load-displacement records for HY-130 steel at three test speeds.

These same specimens had been modified in one additional way in an attempt to locate the point of crack initiation on the load displacement record. Two measurements of crack opening were taken as shown in figure 8, with one just behind the crack tip and the second at the load line. A plot of the two displacements is shown in figure 8. For small displacements, the crack faces of the compact specimen are rotating about an elastic center of rotation, giving a straight line relationship between the two displacements. When crack extension occurs, a tearing component is added to each displacement which changes the slope of the plot in figure 8. To determine the point of crack initiation, linear regression best fit lines were fit to the data well below and well above the transition region and extrapolated to an intersection as shown on figure 8. This value of load line displacement is then taken to the load displacement plot to determine J_{Ic} from Equation 4.

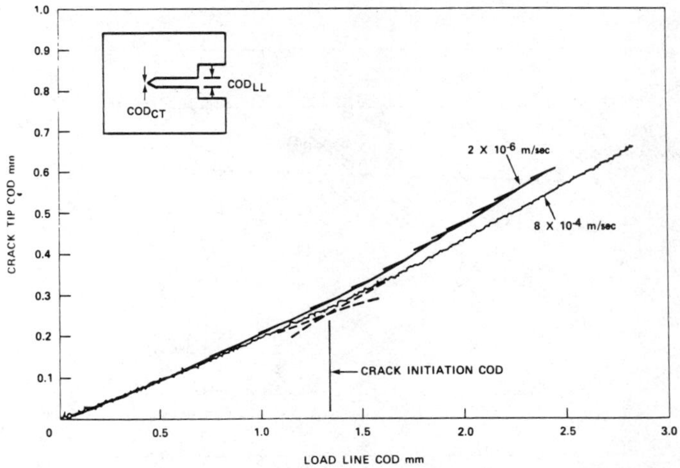

Fig. 8. Crack tip versus load line COD for HY-130 steel compact specimens.

Application of this method for the above high rate specimens up to 8.5×10^{-4}m/sec showed no effect of rate on J_{Ic}. This method shows promise, however, for the determination of the effect of high rate on the ductile to brittle transition temperature.

J_I-R CURVE TEARING MODULUS CONCEPTS

In several important U.S. Navy applications, considerable slow stable crack extension can be tolerated under extreme conditions, but rapid instability must be avoided. Recent developments by Paris and coworkers (1978a) promise to allow application of the J_I-R curve to predict the onset of such instabilities. Paris and coworkers (1978b) define a material property called a tearing modulus by:

$$T_{mat} = \frac{dJ}{da} \cdot \frac{E}{\sigma o^2} \tag{9}$$

where dJ/da is the slope of the J_I-R curve beyond crack initiation. For instability, this criterion then requires $T_{mat} \leq T_{applied}$ where $T_{applied}$ is a function of the geometry of the structure and the deformation state applied which must be evaluated for each structural detail. The same authors also present limited data to demonstrate the applicability of tearing instability to laboratory bend bar tests. Results obtained duri-g the development of the computer interactive unloading compliance J_I test method indicated that the slope of the J_I versus crack growth resistance curves were apparently elevated by crack tunnelling in many alloys (Gudas and Joyce (1978)). With the advent of the tearing instability concept, experiments were performed to determine the effects of face grooving and crack length on J_{Ic} and the tearing modulus of HY-130 steel using the compact specimen (Gudas, Joyce and Davis (1978)). The objective of these tests were to evaluate the sensitivity of the tearing modulus to compact specimen geometry and produce a conservative measurement of T_{mat}. Figures 9 and 10 are a summary of J_{Ic} and tearing modulus measurements of HY-130 steel for specimens with crack lengths ranging from 0.55 to 0.8 a/W, and face grooves of 12.5 to 25%. It can be seen that with scatter, J_{Ic} is independent of the variables, but face grooving was seen to decrease the tearing modulus at all crack length ratios. These results were later duplicated with ASTM A533 B steel. In the case of both steels, face grooving of sufficient depth caused straightening of the crack front during slow stable crack growth. These results suggest that tearing instability analyses should consider specimen geometry effects as well as the nature of crack growth in a structure in developing structural integrity assessments.

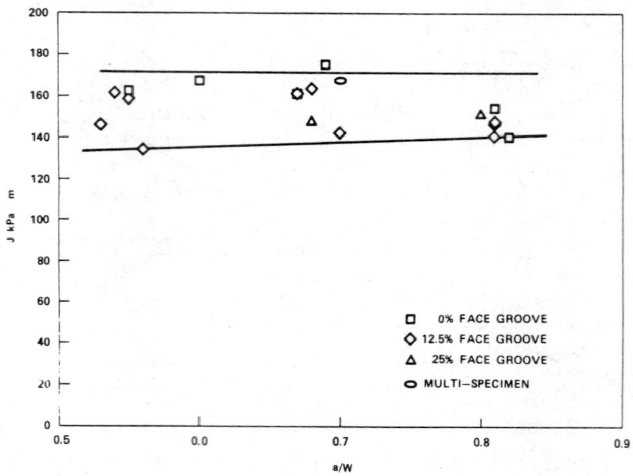

Fig. 9. J_{Ic} for HY-130 steel as related to compact specimen crack length and face groove depth.

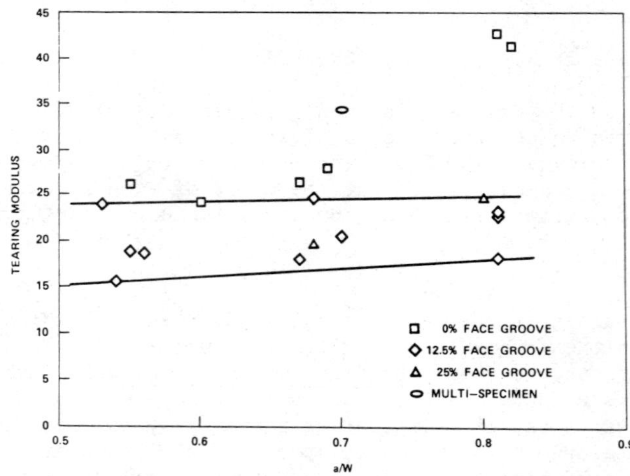

Fig. 10. Tearing modulus for HY-130 steel as related to compact specimen crack length and face groove depth.

SUMMARY

The J-Integral is seen as a valid criterion to index the elastic-plastic fracture toughness of U.S. Navy structural alloys. The computer interactive unloading compliance test procedures has resulted in the routine development of J_{Ic} data in the manner of other fracture mechanics data. The J-Integral is also seen as a promising concept to treat crack growth beyond initiation in elastic-plastic materials. Results of Navy research have shown that J_I versus crack growth resistance curves can be constructed accounting for crack growth. Such J_I-R curves are seen as potentially useful in selecting materials and setting defect criteria. Similarly, the tearing modulus concept is seen as one use of the crack growth resistance data to assess the stability of a structure. U.S. Navy emphasis is now being placed in the area of application of the J-Integral and its application to actual structural integrity considerations.

REFERENCES

Begley, J.A. and J.D. Landes (1972). The J-Integral as a failure criterion. In Fracture Toughness, ASTM STP 514. Philadelphia, Pa. pp. 1-20.

Clarke, G.A., W.K. Andrews, P.C. Paris and D.W. Schmidt (1976). Single specimen tests for J_{Ic} determination. In Mechanisms of Crack Growth, ASTM STP 590. Philadelphia, Pa. pp. 27-42.

Ernst, H., P.C. Paris, M. Rossow and J.W. Hutchinson (1979). The analysis of load-displacement relationships to determine J-R curve and tearing instability material properties. In C.W. Smith (Ed.), Fracture Mechanics, ASTM STP 677, Philadelphia, Pa.

Gudas, J.P. and J.A. Joyce (1978). Elastic-plastic fracture toughness (J_{Ic}) of high strength steels and titanium alloys. Report 78/054. David W. Taylor Naval Ship Research and Development Center, Annapolis, Md.

Gudas, J.P., J.A. Joyce and D.A. Davis (1979). Investigation of specimen geometry modifications to determine the conservative J_I-R curve tearing modulus using the HY-130 steel system. In C.W. Smith (Ed.), Fracture Mechanics, ASTM STP 677, Philadelphia, Pa.

Hutchinson, J.W. (1968). Singular behavior at the end of a tensile crack in a hardening material. J. Mechanics and Physics of Solids, 16, 13-31

Hutchinson, J.W. and P.C. Paris (1979). Stability analysis of J controlled crack growth. In J.D. Landes, J.A. Begley and G.A. Clarke (Eds.), Elastic Plastic Fracture, ASTM STP 668. Philadelphia, Pa. pp. 37-64

Irwin, G.R. (1957). Analysis of stresses and strains near the end of a crack traversing a plate, J. Appl. Mech., 24, 361-364.

Joyce, J.A. and J.P. Gudas (1979). Computer Interactive J_{Ic} Testing of Navy Alloys. In J.D. Landes, J.A. Begley and G.A. Clarke (Eds.), Elastic-Plastic Fracture, ASTM STP 668, Philadelphia, Pa. pp. 451-468.

Landes, J.D. and J.A. Begley (1973). Test results from J-Integral studies: an attempt to establish a J_{Ic} testing procedure. In Fracture Analysis, ASTM STP 560. Philadelphia, Pa. pp. 170-186.

Merkle, J.G. and H.T. Corten (1974). A J-Integral analysis for the compact specimen, considering axial force as well as bend effects. J. Pressure Vessel Tech., 96, 286-292.

Paris, P.C., H. Tada, A. Zahoor and H. Ernst (1978a). The theory of instability of the tearing mode of elastic-plastic crack growth. In J.S. Landes, J.A. Begley and G.A. Clarke (Eds.), Elastic-Plastic Fracture, ASTM STP 668. Philadelphia, Pa. 5-36.

Paris, P.C., H. Tada, A Zahoor and H. Ernst (1978b). An initial experimental investigation of tearing instability theory. In J.D. Landes, J.A. Begley and G.A. Clarke (Eds.), Elastic Plastic Fracture, ASTM STP 668. Philadelphia, Pa. 251-266.

Rice, J.R. (1968). A path independent integral and the approximate analysis of strain concentration by cracks and notches, J. Appl. Mech., 35, 379-386.

Rice, J.R. and G.E. Rosengren (1968). Plane and strain deformation near a crack tip in a power low hardening material. J. Mechanics and Physics of Solids, 16, 1-12.

Williams, M.L. (1957). On the stress distribution at the base of a stationary crack. J. Appl. Mech., 24, 109-114.

Volume 3, ICM 3, Cambridge, England, August 1979

CHARACTERIZATION OF CRACK INITIATION AND GROWTH WITH APPLICATIONS TO PRESSURE VESSEL STEELS

C. F. Shih, W. R. Andrews* and J. P. D. Wilkinson

Corporate Research and Development and *Materials and Processes Laboratory, General Electric Company, Schenectady, New York 12345, U.S.A.

INTRODUCTION

This paper summarizes progress made in a three year research program, sponsored by the Electric Power Research Institute (EPRI), to develop a methodology for plastic fracture. The ultimate goal in the development of such a methodology is to establish an improved basis for analyzing the effects of flaws (postulated or detected) on the safety margins of pressure boundary components of light water nuclear steam supply systems. The primary application emphasis is for reactor pressure vessels. Application to thinner section components such as piping can be foreseen.

The program undertaken consists of two distinct phases of study:

1. Selection, through analysis and laboratory experiments, of viable criteria for crack initiation and growth under large scale yielding conditions.

2. Application of the methodology in predicting the behavior of flawed pressure vessels and piping.

Ferritic pressure vessel steels like A533B and A508 have a ductile-to-brittle fracture transition temperature which is close to room temperature. Below the transition temperature the brittle fracture mode is usually cleavage. In this regime, life predictions can be made using linear elastic fracture mechanics and plane strain fracture toughness K_{IC}. Above the transition temperature, the fracture mechanism is usually dimpled rupture, and large plastic strains are required for crack initiation and propagation. Raising the temperature above the transition temperature first increases the toughness which is then followed by a region of relatively constant toughness - the upper shelf. The purpose of this research is to select and verify fracture criteria which can be used to predict the initiation and stable crack growth behavior of pressure vessels operating in the upper shelf temperature range where crack propagation will occur under conditions of large scale plasticity.

The experimental results from the Heavy Section Steel Technology program, and tests on large plates at the Southwest Research Institute, showed that the growth of part-through cracks in A533B steels remained flat for a considerable amount of crack extension; shear fracture developed only after the attainment of maximum load. In this investigation, attention has therefore been focused on flat fracture under essentially plane strain and large scale plasticity conditions.

The progress of ductile fracture from an existing sharp-tipped flaw may be separated into four regimes: the blunting of the initially sharp crack tip, crack initiation, stable crack growth, and unstable crack propagation. For high toughness materials on the upper shelf, there is significant crack tip blunting and substantial stable crack growth. More importantly, the 'resistance' of the material to further crack extension increases with increasing crack growth. In this case, the assessment of the margin of safety of flawed structures based on the onset of crack extension is conservative, since the 'toughness' of the material will increase significantly with relatively small crack extension.

In this investigation, a close coupling was maintained between testing and analysis. Tests were conducted on a variety of fracture specimens of A533B steel at 93°C with varying geometry and crack lengths, in order to obtain fracture data for widely different deformation fields in the fully plastic state. Crack growth analyses using finite element models of the test configurations were carried out in order to isolate viable fracture criteria from the test results. A detailed discussion of the potential fracture criteria evaluated in the study, the strategy employed in the evaluation, procedures for measuring J, COD and crack extension, detailed examination of the analytical and experimental results, and finally an assessment of these fracture parameters that were found viable are given in an EPRI Special Report, and published papers (Shih and co-workers, 1977, 1978).

The following parameters were examined during the study: the J-integral, crack opening displacement (COD), and crack opening angle (COA), work density in a process zone, energy separation rate, and critical strain over a characteristic length. The J and COD parameters were also evaluated in the context of a resistance curve approach.

The following main conclusions emerged from these investigations:

1. Macroscopically flat fracture surfaces with a straight leading edge can be produced by employing side-grooves on test specimens. Side-groove depths of 25 percent of the specimen thickness are recommended, since they promote an essentially uniform plane strain constraint along the crack front while producing minimal effect on specimen compliance and stress intensity factor (Shih, deLorenzi, and Andrews, 1977a).

2. The experimentally determined J and COD (measured at the original crack tip) resistance curves appear to be independent of specimen size and initial crack length when plane strain flat fracture occurs and if certain minimal size requirements are met (Shih, deLorenzi, and Andrews, 1977b; Andrews and Shih, 1977).

3. Analytical investigations also reveal that J and COD can be employed to characterize crack initiation and growth. The slope of the J-resistance curve (dJ/da) appears to be constant for a relatively short interval of crack extension, while the slope of the COD-resistance curve ($d\delta/da$, also called the average or nominal crack opening angle) and the local crack opening angle (COA) remain practically constant over the entire range of crack extension explored in our experimental and analytical investigations. The J-based criteria appear to be valid for limited range of crack growth; the range of validity depends on strain hardening properties and crack configuration. For A533B steels at the upper shelf, the J criterion is valid for crack growth up to 6% of the original uncracked ligament for test specimens subjected primarily to bending, e.g. compact specimen. The COD-based criteria appear to be valid for larger amounts of crack growth (Shih, deLorenzi, and Andrews, 1977b; Andrews and Shih, 1977).

4. The tearing modulus ($T_J = (E/\sigma_o^2)dJ/da$) proposed by Paris and co-workers (1977) as a measure of material resistance to stable growth is constant over relatively

short intervals of growth. Our investigations suggest that a tearing modulus based on the COA ($T_\delta = (E/\sigma_o^2)d\delta/da$) is an attractive alternative. The latter modulus is measurable directly and appears to be constant over the entire range of stable growth. Fracture toughness associated with crack initiation is measured by J_{IC} or δ_{IC} while fracture resistance associated with crack growth is measured by T_J or T_δ. The two parameter characterization of fracture properties by J_{IC} and T_J or δ_{IC} and T_δ are analogous to the characterization of material deformation properties by the yield stress and the strain hardening exponent (Shih, deLorenzi, and Andrews, 1977b).

5. A simple explicit expression relating the crack tip opening displacement δ_t to J has been derived based on the assumption that the Hutchinson-Rice-Rosengren (HRR) singularity dominates over finite intervals at the crack tip. The coefficient d_n that relates δ_t to J is a function of the hardening exponent and σ_o/E. Values of d_n for plane strain and plane stress conditions are given in Figs. 9 and 10 (Shih, 1978).

6. Small scale yielding finite element calculations for several hardening exponents were carried out. The values of d_n determined from the finite element results are in good agreement with the values derived from the HRR singularity. Further support for the J and COD relationship was found through fully plastic finite element analyses for the cracked bend bar (CBB) employing three different hardening exponents ($n = 3$, 10 and ∞). The latter results also suggest that the HRR singularity does indeed characterize the stress and strain state of the crack tip region in the CBB for the complete range of deformation (Shih, 1978).

7. By exploiting a deformation theory of plasticity, a relationship between the slope of the J resistance curve dJ/da and the crack opening angle $d\delta/da$ has been derived. Available finite element results for the growing crack seem to support this relationship (Shih, 1978).

8. Extensive experimental data on crack initiation and growth have been summarized. The predicted crack tip opening displacement at initiation and the crack opening angle associated with stable crack growth using the simple relationship between J and δ_t and between dJ/da and $d\delta/da$ (derived from deformation theory of plasticity) appear to be in reasonably good agreement with the actual experimental measurements. These results are summarized in Table 2.

9. Detailed finite element solutions showed that the size of the region dominated by the HRR singularity is substantially larger in the cracked bend bar (or compact specimen) than in the center-cracked panel. The study suggests that the proposed minimum specimen size for J toughness testing which requires all relevant specimen dimensions to exceed 25 to 50 times J/σ_o is adequate to ensure a valid J characterization of the crack tip region in the cracked bend bar and compact specimen. However the specimen dimensions will have to be considerably larger to ensure a J characterization of the crack tip region in a center-cracked panel. McMeeking and Parks (1977) suggested a size requirement of 200 times J/σ_o. While the loss of dominance of the HRR singularity in the center-cracked panel occurs rather gradually, it appears that a size requirement of 200 J/σ_o, though arbitrary, is consistent with the detailed solutions we have generated (Shih and German, 1978).

10. Analytical and numerical studies reveal that the requirements for J as a characterizing parameter for crack growth are rather stringent. In addition to the size requirement described in the preceding paragraph, the slope of the J-resistance must be sufficiently large. The dimensionless parameter $(1/\sigma_o)(dJ/da)$ must be significantly larger than 0.1 for J-controlled crack growth (Shih and Dean, 1978).

11. Available experimental data suggest a relationship between the material resistance to crack initiation and the resistance to stable tearing. The tearing

modulus T_J (based on J) increases with increasing fracture toughness J_{IC} and the relationship is essentially linear in the high toughness regime. A similar relationship is observed between the tearing modulus T_δ (based on COD) and the COD at crack initiation δ_{IC}. These general trends are also supported by a simple crack growth model (Shih and Wilkinson, 1979).

STRATEGY

In order to ascertain which parameters are viable fracture criteria, the results of tests were analyzed with detailed two dimensional plane strain finite element computations. The tests were done on a variety of specimen geometries including compact specimens, and center-cracked panels with varying remaining ligaments. The tests were designed to provide a variety of plastic deformation fields. The finite element analysis was carried out using the ADINA code (Bathe, 1975) suitably modified to allow crack tip blunting and growth, and so that the appropriate fracture parameters could be computed (deLorenzi and Shih, 1977). The two phases of analysis initial filter, and evaluation - are discussed below.

Requirements for Viable Fracture Parameters

In the two phases of the investigation, the fracture parameters are assessed on the basis of several requirements:

1. The parameter must be a measure of the stress and deformation state in the vicinity of the crack tip.

2. The critical value of the parameter must be independent of initial crack length and specimen geometry.

3. The parameter must be employable in instability analyses.

4. The parameter should be applicable to three-dimensional crack geometries, for example an elliptic surface flaw.

5. The parameter should be generalizable to mixed mode fracture, though the critical value will depend on the relative ratio of shear to flat fracture.

6. The parameter should remain constant during crack extension. This is an attractive but not an essential feature. In an R-curve approach, the resistance parameter will increase with crack extension.

7. The parameter must be relatively insensitive to finite element modelling, to mesh size, process zone size, and load/displacement increment size.

8. The parameter should be computable within reasonable computational cost.

9. The parameter should be obtainable from global measurements remote from the crack, but if this is not possible, it must be obtainable from local measurements near the crack tip.

These requirements are the basis of the subsequent discussions on experimental and analytical results.

Phase I - Initial Filter

The objective of this first analytical phase is to observe the behavior of potential fracture criteria when crack growth in the test specimen is forced to follow the measured relationship between load line displacement (LLD) and crack extension. From experimental measurements, it is possible to determine the crack extension, $a - a_o$, as a function of the measured load line displacement (Fig. 1). In the finite element model, the rate of crack growth is prescribed to follow the experimentally measured LLD - $(a - a_o)$ relationship. The process is repeated for the entire finite element crack growth calculations. The various potential fracture parameters, J and dJ/da at near-field and remote contours, crack opening displacement and angle, energy separation rates, work density over a process zone, etc., are computed during the crack growth simulation.

As an illustration of the procedure, the results of calculations and experiments on a 4T compact specimen with 25 percent side grooves are given here. The specimen (specimen T52, A533B steel tested at 93°C) has a ratio $a/W = 0.58$ and $W - a_o$ = 86.0 mm. Complete details of these and other calculations are given by Shih and co-workers (1978). Calculated results are compared to experimental observations in Figs. 3 through 6, for the J-integral, dJ/da, crack opening displacement, and crack opening angle. The resulting load versus LLD relationship compares well with the test measurements as shown in Fig. 2.

The Evaluation Phase

In the next series of calculations, the COD based criterion was employed to govern the crack extension for the specimen. The crack initiates at a critical value of the COD, δ_i, and its propagation is determined by the constant average angle α_a. Likewise, in another series of calculations, the critical value of J for initiation was used, and propagation was governed by the J-resistance curve. A comparison of the load-displacement relationship obtained from analysis and test is given in Fig. 7. Agreement is remarkably good, illustrating that both the J-based and the COD-based criteria are appropriate for predicting crack initiation and growth.

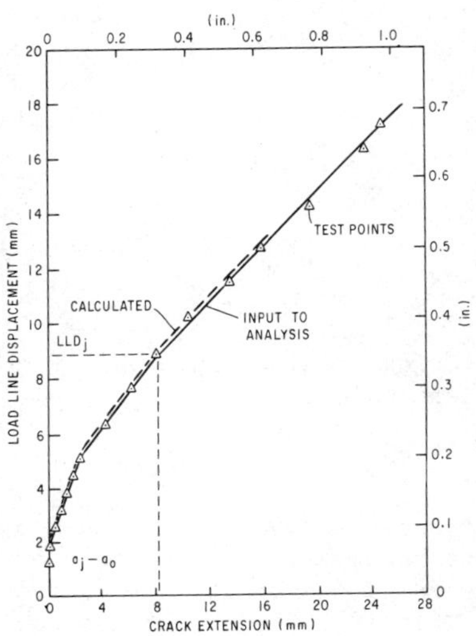

Fig. 1. Load-line displacement vs. crack extension.

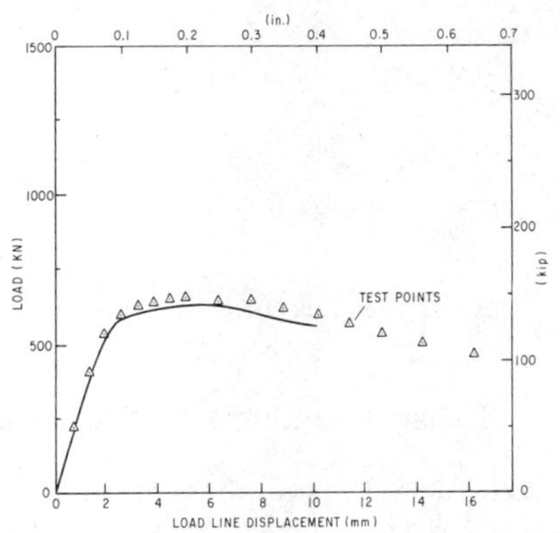

Fig. 2. Applied load vs. load-line displacement.

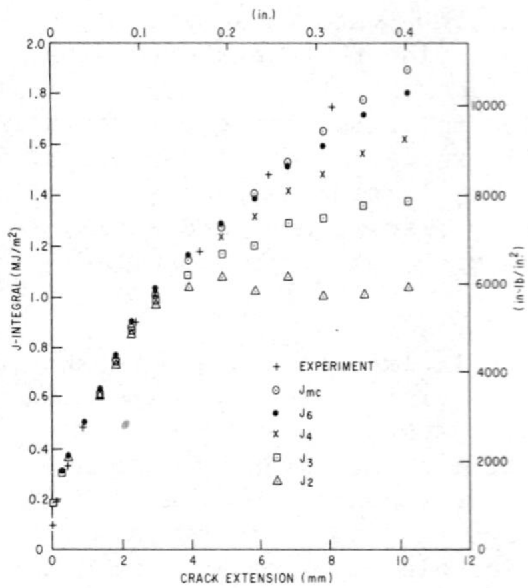

Fig. 3. J-integral evaluated at different contours.

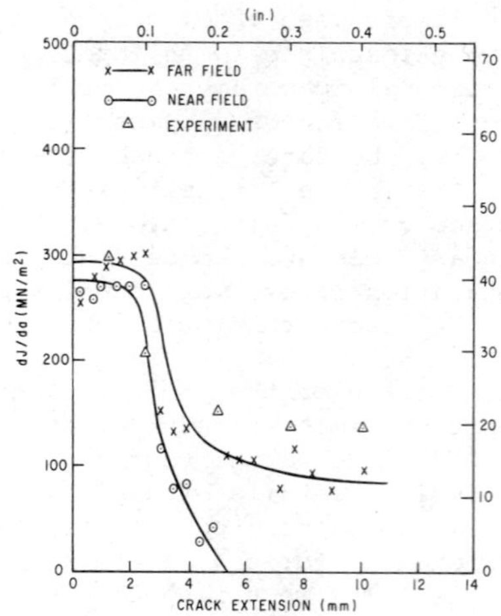

Fig. 4. dJ/da versus crack extension.

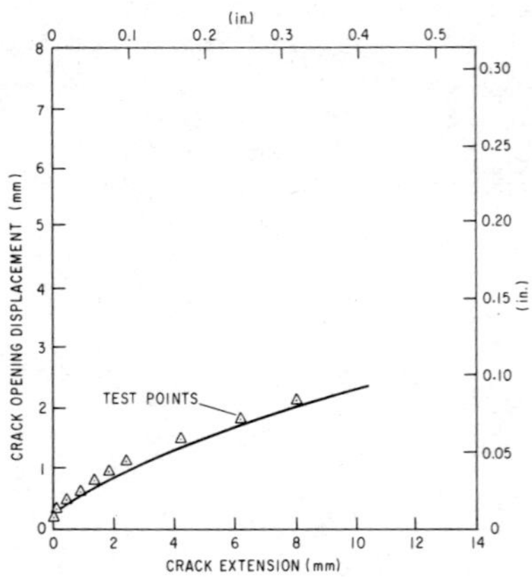

Fig. 5. COD vs. crack extension.

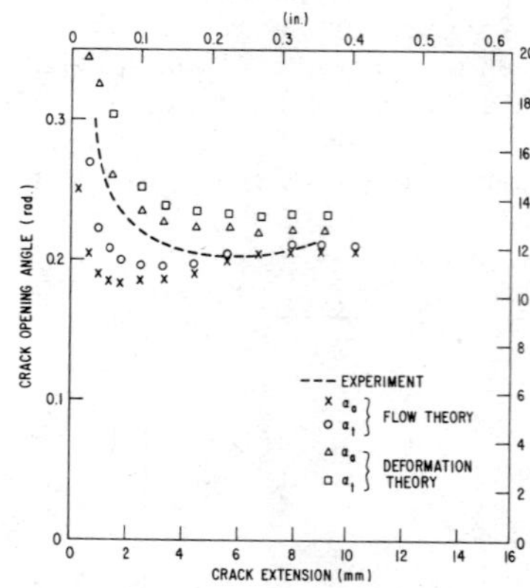

Fig. 6. Crack opening angle vs. crack extension.

ASSESSMENT OF FRACTURE PARAMETERS

The requirements that viable fracture criteria must satisfy have been discussed previously. A comparison of the contending fracture parameters on the basis of these requirements is summarized in Table 1. In terms of the first six requirements, all the parameters are about equally good. This is not unexpected since all these parameters, when appropriately defined and employed, characterize the crack near-field. However, in terms of the last four requirements, the trend is clear. From the computation and measurement point of view, the viable candidates are the J-integral, and the crack opening displacement and angle.

The work of Begley and Landes (1972) led to the characterization of crack

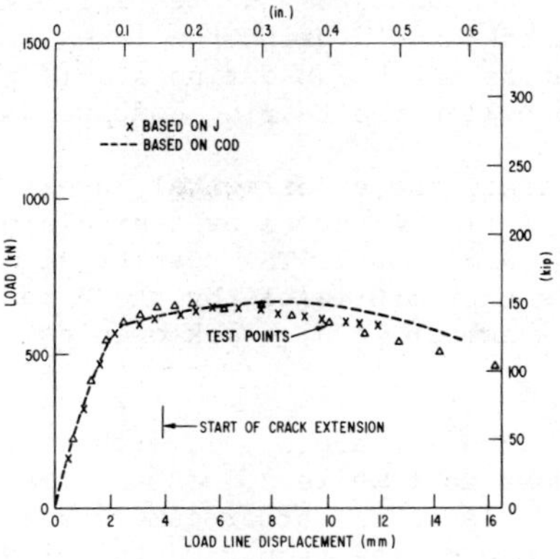

Fig. 7. Comparison of LLD relationships calculated with a J-resistance curve.

TABLE 1 A Comparison of Some Fracture Parameters

	J-Integral	Crack Opening Displacement	Local Crack Opening Angle	Work Done in a Process Zone	Generalized Energy Release Rate
Measure of Near-Tip Field	Yes	Yes	Yes	Yes	Yes
Constant During Crack Extension	No*	No*	Yes	Yes	Yes
Independent of Specimen Geometry	Mildly Dependent	Mildly Dependent	Mildly Dependent	Insufficient Data	Mildly Dependent
Applicability to Instability Analyses	Yes†	Yes	Yes	Yes	Yes
Generalizable to Mixed Mode Fracture	Probable	Uncertain	Uncertain	Probable	Probable
3D Crack Geometries	Uncertain	Probable	Probable	Probable	Uncertain
Sensitivity to Mesh, Process Zone and Increment Size	No	Mildly	Moderately	Moderately	Yes
Computation Cost (Estimated Ratio)	1	1	1	1	3
Direct Local Measurement	No	Yes	Yes (Difficult)	No	No
Direct Global Measurements	Yes	No	No	No	No

*J and COD are employed as resistance parameters. †For limited crack extension.

initiation in terms of J_{IC} under both small scale and large scale yielding conditions. Subsequent experimental investigations by Clarke and co-workers (1976) and by Griffis and Yoder (1976) revealed the presence of stable crack growth under large scale yielding conditions; they characterized the onset of growth using J_{IC}.

Paris and co-workers (1977) demonstrated the feasibility of characterizing the onset of growth and stable crack growth using the J-resistance curve. They also proposed to characterize fracture toughness during stable growth by the dimensionless parameter $(E/\sigma_o^2)dJ/da$ also called the tearing modulus T_J.

The results from our analytical and experimental investigations support the characterization of the onset of flat fracture by a single parameter, namely J_{IC} or δ_{IC} under large scale yielding conditions. The results also strongly suggest that stable flat crack growth is characterizable by the J-resistance or the COD-resistance curve. Furthermore the average or local crack opening angle is fairly constant during stable growth.

A typical variation of J, dJ/da, δ_t and α_a, α_t with crack extension is illustrated in Fig. 8. This figure shows that while J_{ff} (J far field) continues to increase, J_{nf} (J near field) begins to level off after some crack extension. Thus $(dJ/da)_{nf}$ falls to zero and violates one of the requirements for J-controlled growth expressed by Eq. (7). On the other hand, both the crack opening angles, α_a and α_t after an initial transient remain constant for a considerable range of crack growth.

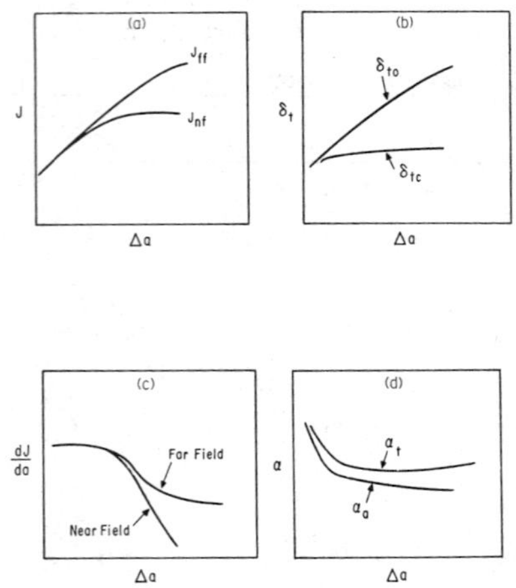

Fig. 8. Typical behavior of proposed fracture parameters during crack growth: a) J-resistance curves for near and far fields, b) COD-resistance curves based on opening at original crack tip (δ_{to}) and at a fixed distance behind current tip (δ_{tc}), c) dJ/da derived from near and far field J, d) COA's α_a and α_t, derived from δ_{to} and δ_{tc}.

This satisfies the requirement for COA-controlled growth. It also appears that beyond the initial phase of crack growth, a constant critical value of the average and local COA may be employed to characterize stable growth. This is not unexpected since the COA is derived from more fundamental considerations.

The J-based criteria are valid for limited amounts of crack growth. In the case of A533B steel in the upper shelf region, this amount appears to be about 6% of the original remaining ligament for compact specimens. The range of validity will differ from material to material depending on the strain hardening exponent and specimen geometry. The COD-based criteria are valid for larger amounts of crack growth.

The tearing modulus ($T_J = (E/\sigma_o^2)dJ/da$) proposed by Paris and co-workers (1977) as a measure of material toughness during stable growth is constant over relatively short intervals of growth. Based on our investigations a tearing modulus based on the COA ($T_\delta = (E/\sigma_o)d\delta_t/da$) is proposed. The latter modulus is physically more appealing and appears to be constant over the entire range of stable growth. Large values of T_δ correspond to higher material toughness associated with crack growth. The importance of the tearing moduli T_J and T_δ is that they can characterize the material toughness associated with crack growth, whereas crack initiation is characterized by J_{IC} or δ_{IC}. This two-parameter characterization of fracture toughness properties, namely J_{IC} and T_J, or δ_{IC} and T_δ, is analogous to the characterization of deformation properties of a material by the yield stress and the strain hardening exponent.

RELATIONSHIP BETWEEN J-INTEGRAL AND THE CRACK OPENING DISPLACEMENT

In the previous sections, the experimental and analytical evidence was presented which supports the characterization of crack initiation and stable crack growth by parameters based on the J-integral and the crack opening displacement. In this section, the relationship between these parameters is presented.

By exploiting the HRR singularity Shih (1978) derived the following relationship between the crack tip opening displacement and J, namely,

$$\delta_t = d_n J/\sigma_o \tag{1}$$

where $d_n = (\alpha\ \sigma_o/E)^{1/n}\ (\tilde{u}_x + \tilde{u}_y)^{1/n}\ \tilde{\delta}/I_n$. The dimensionless quantities $\tilde{u}_x$, $\tilde{u}_y$, $\tilde{\delta}$ and I_n are determined from the HRR singularity. Values of d_n for a range of hardening exponents n and σ_o/E are shown in Figs. 9 and 10 for plane strain and plane stress.

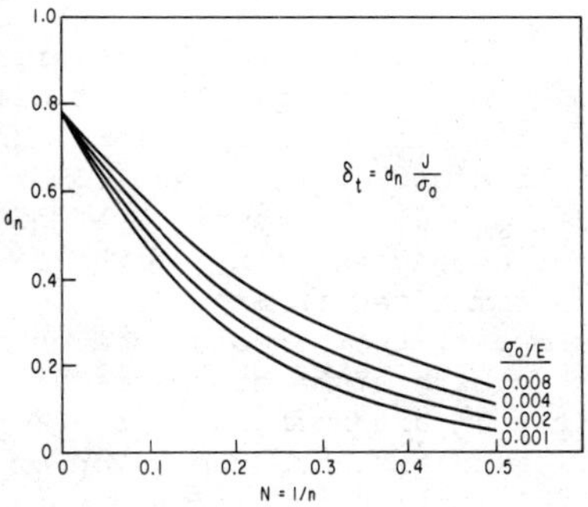

Fig. 9. Variation of d_n with n and σ_o/E for plane strain with $\alpha = 1$.

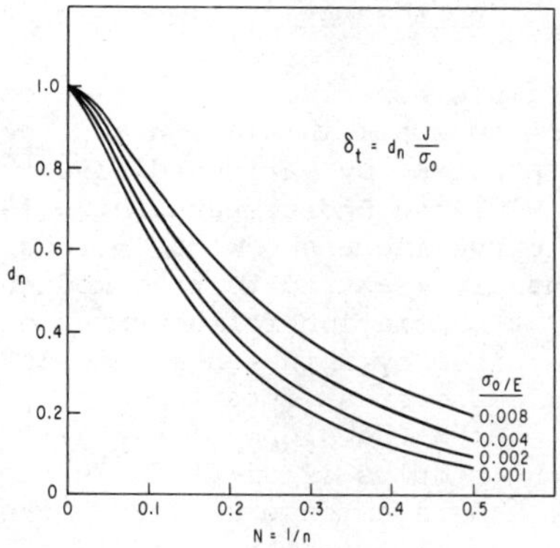

Fig. 10. Variation of d_n with n and σ_o/E for plane stress with $\alpha = 1$.

Under somewhat more restrictive conditions, the slope of the COD-resistance curve or the crack opening angle $d\delta/da$ is related to the slope of the J-resistance curve by

$$d\delta/da = [n/(n + 1)](d_n/\sigma_o)dJ/da \tag{2}$$

Expressions (1) and (2) are obtained on the assumptions that the HRR singularity dominates in the crack tip region and that deformation plasticity theory is applicable to small amounts of crack growth. Experimental data giving the values of J at initiation, the J resistance curve and COD measurements are provided in investigations by Andrews and Shih (1977), Shih and co-workers (1977), Robinson (1976), Garwood (1977), Garwood and Turner (1977), Griffis (1975), Kass and co-workers (1974). Based on these experimental data, namely J_c and dJ/da, the critical value of δ and $d\delta/da$ are predicted using (1) and (2). The predicted values appear to be consistent with the actual experimental measurements. A summary of the results is given in Table 2. The plane strain values of d_n were used in most of the cases examined; in situations where the specimens dimensions do not appear to meet plane strain conditions, the plane stress values of d_n were employed.

TABLE 2 Comparison of Predicted and Measured Values of COD and COA

						d_n		Predicted COD $\delta_c = \frac{d_n}{\sigma_o} J_c$				Predicted COA		
Material	E GN/m^2	σ_y MN/m^2	n	α	J_c MJ/m^2	Plane Strain	Plane Stress	Plane Strain	Plane Stress	Measured COD mm	$\frac{dJ}{da}$ MJ/m^3	$\frac{d\delta}{da} = \frac{d_n}{\sigma_o}\left(\frac{n}{n+1}\right)\frac{dJ}{da}$	Measured COA	Reference
A533B, 93°C	200	419	13	.75	.16-.17	.553	-	.21-.35	-	.22-.34	213-305	.26-.37	.27-.36	Andrews and Shih (1977)
A533B, 260°C	200	419	13	.75	.19-.25	.553	-	.25-.33	-	.3	180-270	.22-.33	.3	Andrews and Shih (1977)
En24 Low Alloy Steel (Q+T)	200	1025	12.5	1.	.058	.61	.84	.034	.046	.057	-	-	-	Robinson (1976)
					.09			.053		-				Garwood (1977)
En32 Mild Steel	200	275	5	1.	.16-.18	.29	.41	.17-.19	.24-.27	.27-.32	-	-	-	Robinson (1976)
	200	284	5	1.	.12	.30	.42	.125	.175	.2	160	.14 (plane strain)	.25	Garwood and Turner (1977)
												.195 (plane stress)		
HY80	200	584	9	1.	.16	.507	-	.138	-	.140*	209	.163	.205	Griffis (1975) (*our estimate)
AP15L x 65 Pipeline Steel	200	509	13	1.	.35	.57	.8	.39	.546	.42	650	.67	.64	Garwood (1977)
Zircaloy	96	330	13	1.	.087	.59	-	.156	-	} .147			.246	Kass (1974)
			9	1.		.514	-	.136						

RELATIONSHIP BETWEEN CRACK INITIATION AND GROWTH PARAMETERS

There is ample experimental data which suggest a relationship between the material resistance to crack initiation and the resistance to stable tearing. In particular the data provided by Paris and co-workers (1977), Knott and co-workers (Clark and others, 1978) and Andrews and Shih (1977) show that the slope of the J and COD resistance curve increases with increasing values of J_{IC} and δ_{IC} respectively. Experimentally measured tearing modulus T_J and the corresponding value of the J-integral at initiation for a number of materials are summarized in Fig. 11. Details are given by Shih and Wilkinson (1979). The slope of the J-resistance curve is determined over an interval corresponding to crack growth of 6% of the remaining ligament. Both the slope of the J-resistance curve and J_c are normalized by E/σ_o^2. The tearing modulus $(E/\sigma_o^2)dJ/da$ is dimensionless while $(E/\sigma_o^2)J_c$ has dimension of distance, and is a measure of the extent of plastic deformation. The trend suggested by the experimental data is clear - the tearing modulus rises rapidly at lower values of J_c and then an almost linear relationship exists between the tearing modulus and J_c. Experimentally measured tearing modulus T_δ (based on the COD-resistance curve) and δ_c are summarized in Fig. 12. The trend of data is similar to Fig. 11.

A crack growth model suggested by Rice and Sorensen (1978) may be modified to suggest a relationship between J_c and dJ/da. If the critical opening displacement δ, at some fixed distance $\Delta\ell$ behind the advancing crack is proportional to J_c/σ_o, then this expression emerges

$$T_J = (E/\sigma_o^2)dJ/da = \bar{J}_c - (\beta/\alpha)\ell n(\lambda e\, \bar{J}_c) \qquad (3)$$

where $\bar{J}_c = (E/\sigma_o^2)J_c/\Delta\ell$ and e is the natural logarithm base. From available solutions, $\alpha \simeq 0.5$, $\beta \simeq 4.0$; the loose parameters are λ and $\Delta\ell$. The solid line in

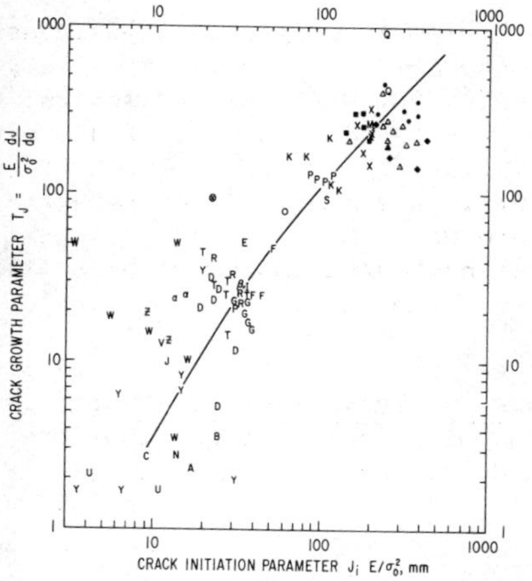

Fig. 11. Relationship between J at initiation and the crack growth parameter T_J.

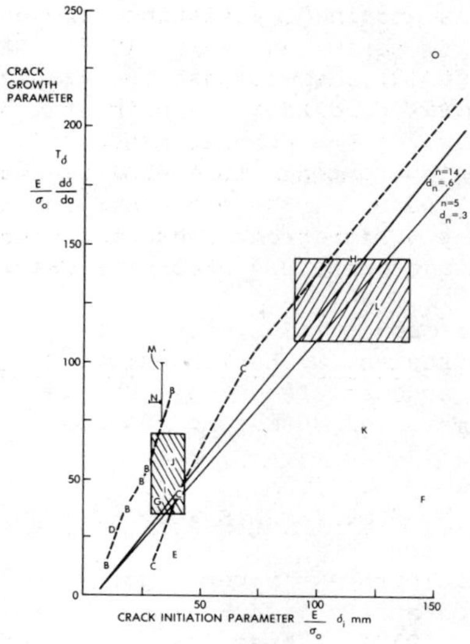

Fig. 12. Relationship between crack opening displacement at initiation and crack opening angle during growth.

TABLE 3 Guide to Fig. 11

A	2024 Aℓ
B	7075 Aℓ
C	6 Aℓ 4V Ti
D	A471
E	Pyromet 538
F	A470
G	AISI 403
J	En24
K	A217
M	En32
N	6061-T651 Aℓ
O	HY80
P	A469
Q	Pipeline Steel
R	A453
S	A453 Weld
T	A540
U	Inco LEA
V	3105 SS Welds
W	Inco X750
⊗	AISI 3105
Y	Kromarc 58 SS
Z	Inco 706
■	A533B Submerged Arc Weld
●	A533B Manual Arc Weld
×	A533B
△	A533B, 93°C
◆	A533B Weld
▲	A533B, 200°C
α	A286, 427°C, 538°C
β	A286, 25°C

TABLE 4 Guide to Fig. 12

B	HY80 with Prestrain
C	316 SS with Cold Work
D	HY130
E	ph Steel
F	Mild Steel
G	HY130
H	En32B
I	Zircaloy-4
J	HY80
K	q and t Steel
L	A533B, 200°F
M	HY100
N	HY130
O	Pipeline Steel
////	A533B, 93°C
\\\\	HY100

Fig. 11 is obtained by setting λ to 0.1 and $\Delta\ell$ to 0.7 mm in expression (3).

Expression (3) also suggests a relationship between δ_c and $d\delta/da$ if the latter is proportional to dJ/da:

$$T_\delta = (E/\sigma_o)d\delta/da = \bar{\delta}_c - (\beta/\alpha)\ell n(e\lambda\,\bar{\delta}_c) \qquad (4)$$

where $\bar{\delta}_c = (E/\sigma_o)\delta_c/\Delta\ell$. The solid line in Fig. 12 is obtained from (4) by setting λ to 0.1 and $\Delta\ell$ to 0.7 (the same values were employed in Fig. 11).

The relationship is between initiation and growth parameters expressed by (3) and (4) in agreement with the general trend of the experimental data, and as such is meant to suggest trends only but is not for the purpose of precise correlations between material resistance to crack initiation and growth.

REQUIREMENTS FOR J-CONTROLLED CRACK GROWTH

The sufficient requirements for J-controlled crack growth have been dealt with by Hutchinson and Paris (1977) and by Shih and co-workers (Shih and German, 1978; Shih and Dean, 1978). These are summarized below:

$$B,\ b \geq MJ/\sigma_o \qquad (5)$$

$$\Delta a << R \qquad (6)$$

$$(R/J)dJ/da >> 1 \qquad (7)$$

Requirement (5) is basically a size requirement and (6) requires that the amount of crack extension be small compared to the size R of region dominated by the HRR singularity. In addition the slope of the J-resistance curve must be large (7). Now the size R of the region dominated by the HRR singularity is typically proportional to the uncracked ligament b when fully plastic conditions prevail. Thus (7) may be modified to give

$$\omega \equiv (b/J)dJ/da >> 1 \qquad (7a)$$

In general, R is several times the crack tip opening displacement and this leads to

$$\psi \equiv (1/\sigma_o)dJ/da >> 0.1 \qquad (7b)$$

Analytical and numerical studies also suggest that if (7b) and (5) are satisfied then condition (6) is automatically satisfied, i.e. (6) is redundant. The evidence and requirements for J-controlled growth and the applications of the J approach to cracked structures are discussed in detail by Shih and Dean (1978).

REFERENCES

Andrews, W. R. and C. F. Shih (1977). Thickness and side-groove effects in J-resistance curves, ASTM Symposium on Elastic-Plastic Fracture, Atlanta, Ga., to appear in ASTM STP.

Begley, J. A. and J. D. Landes (1972). In Fracture Toughness, ASTM STP 514, 1-23, 24-39.

Clark, G., S. M. El Soudani, W. G. Ferguson, R. F. Smith, and J. F. Knott (1978). Ductile crack initiation in pressure vessel steels, Proc. Inst. Mech. E., Paper C90/78, 121-132.

Clarke, G. A., W. R. Andrews, P. C. Paris, and D. W. Schmidt (1976). In Mechanics of Crack Growth, ASTM STP 590.

deLorenzi, H. G. and C. F. Shih (1977). Application of ADINA to elastic-plastic fracture problems, Proc. ADINA User's Conf., MIT, Cambridge, Massachusetts.

Garwood, S. J. (1977). The measurement of crack growth resistance using yielding fracture mechanics, Ph. D. Thesis, London.

Garwood, S. J. and C. E. Turner (1977). The use of the J-integral to measure the resistance of mild steel to slow stable crack growth, Fracture 1977, 3, ICF4, Waterloo, Canada, 279-284.

Griffis, C. A. (1975). Elastic-plastic fracture toughness: a comparison of J-integral and crack opening displacement characterization, J. Press. Vessel and Tech., 278-283.

Griffis, C. A. and G. R. Yoder (1976). Initial crack extension in two intermediate strength aluminum alloys, J. Eng. Mat. Tech., 152-158.

Hutchinson, J. W. and P. C. Paris (1977). Stability analysis of J-controlled crack growth, ASTM Symposium on Elastic-Plastic Fracture, Atlanta, Ga., to appear in ASTM STP.

Kass, J. N., J. A. Begley, and H. Andrejasik (1974). Crack initiation and growth in plane strain, fully plastic zircaloy, J. Test. Eval., 2, 304-316.

McMeeking, R. M. and D. M. Parks (1977). On criteria for J-dominance of crack tip fields in large scale yielding, ASTM Symposium on Elastic-Plastic Fracture, Atlanta, Ga., to appear in ASTM STP.

Paris, P. C., H. Tada, A. Zahoor, and H. Ernst (1977). The theory of instability of the tearing mode of elastic-plastic crack growth, U. S. NRC Rept. NUREG-0311.

Rice, J. R. and E. P. Sorensen (1978). J. Mech. Phys. Solids, 26, 163-186.

Robinson, J. N. (1976). An experimental investigation of the effect of specimen type on the crack tip opening displacement and J-integral fracture criteria, Int. J. Fracture, 12, 723-737.

Shih, C. F. (1978). Relationship between the J-integral and crack opening displacement for stationary and extending cracks, General Electric Rept., to appear.

Shih, C. F., W. R. Andrews, H. G. deLorenzi, R. H. VanStone, S. Yukawa, and J. P. D. Wilkinson (1978). Crack initiation and growth under fully plastic conditions: a methodology for plastic fracture, T. U. Marston (Ed.), in EPRI Ductile Fracture Research Review Document, EPRI NP-701-SR.

Shih, C. F. and R. H. Dean (1978). On J-controlled crack growth: evidence, requirements and applications, General Electric Rept., to appear.

Shih, C. F., H. G. deLorenzi, and W. R. Andrews (1977a). Elastic compliances and stress-intensity factors for side-grooved compact specimens, Int. J. Fracture, 13, 544-548.

Shih, C. F., H. G. deLorenzi, and W. R. Andrews (1977b). Studies on crack initiation and stable crack growth, ASTM Symposium on Elastic-Plastic Fracture, Atlanta, Ga., to appear in ASTM STP.

Shih, C. F. and M. D. German (1978). Requirements for a one-parameter characterization of the crack tip fields by the HRR singularity, General Electric Rept., to appear.

Shih, C. F. and J. P. D. Wilkinson (1979). Relationship between crack initiation and growth parameters, General Electric Rept., to appear.

AUTHOR INDEX

The following is an index to all the authors of plenary and workshop papers included in the full proceedings. The first number indicates the volume in which the paper is found and the second the page in that volume.

SUBJECT INDEX

The following is an index to all plenary and workshop papers included in the full proceedings. The first number indicates the volume in which the paper is found and the second the page in that volume.